Plant Breeding: The Arnel R. Hallauer International Symposium

Plant Breeding: The Arnel R. Hallauer International Symposium

Editors

Kendall R. Lamkey, Michael Lee

Blackwell
Publishing

Kendall R. Lamkey, Ph.D., is the Pioneer Distinguished Chair in Maize Breeding and Director of the Raymond F. Baker Center for Plant Breeding, Agronomy Department, Iowa State University. His research is focused on the origin, maintenance, and utilization of genetic variation for important agronomic and grain quality traits in maize.

Michael Lee, Ph.D., is Professor and Chair of the Plant Breeding and Genetics Panel, Agronomy Department, Iowa State University. Lee's research focuses on developing and utilizing genetic techniques and principles to complement the programs in maize breeding and genetics with the most recent advances in applied plant molecular genetics.

Blackwell Publishing Professional
2121 State Avenue, Ames, Iowa 50014 USA

Orders: 1-800-862-6657
Office: 1-515-292-0140
Fax: 1-515-292-3348
Web site: www.blackwellprofessional.com

Blackwell Publishing Ltd
9600 Garsington Road, Oxford OX4 2DQ, UK
Tel.: +44 (0)1865 776868

Blackwell Publishing Asia
550 Swanston Street, Carlton, Victoria 3053, Australia
Tel.: +61 (0)3 8359 1011

First edition, 2006

Library of Congress Cataloging-in-Publication Data

Plant Breeding: The Arnel R. Hallauer International
Symposium (2003 : Mexico City, Mexico)
Plant breeding: the Arnel R. Hallauer International
Symposium/editors Kendall R. Lamkey, Michael Lee.—1st ed.
p. cm.
Includes bibliographical references.
ISBN-13: 978-0-8138-2824-4 (alk. paper)
ISBN-10: 0-8138-2824-4
1. Plant breeding—Congresses. I. Lamkey, Kendall R.
II. Lee, Michael. III. Title.

SB123.A75 2003
631.5′2—dc22

2005025635

The last digit is the print number: 9 8 7 6 5 4 3 2

Contents

Preface

The *Arnel R. Hallauer International Symposium on Plant Breeding* was held in Mexico City on 17–22 August 2003. The chapters in this book resulted from the papers presented at that symposium. The chapters are organized in the book in the same order that they were presented at the symposium. Many people were responsible for organizing this symposium and to list them all would mean that some would be left out. We would all agree, however, that this symposium would not have happened without the vision, dedication, and hard work of Dr. Shivaji Pandey. Dr. Julien de Meyer was responsible for organizing all aspects of the conference. The success of the conference was due to Dr. de Meyer's organizational skills and attention to detail, and for that we owe him a debt of gratitude.

The world of plant breeding has experienced dramatic changes during the span of Arnel Hallauer's career. At the institutional level, international centers of crop improvement have emerged and declined, legal and ethical issues have become routine considerations, the private sector has developed and consolidated, and the public sector (national programs, federal governments, universities) has diversified and placed greater emphasis on basic research as opposed to varietal development. Changes in infrastructure (e.g., off-season nurseries, service laboratories) and technology (e.g., computers, machinery, analytical methods, transgenic methods) enable the declining number of plant breeders to evaluate more germplasm in more ways in more environments and to identify genotypes that exhibit optimal adaptation to the needs of society, the demands of nature, and the desires of the market. Nascent developments in basic biological and informational sciences, as exemplified by the gradual annotation of entire genomes and their gene products, have provided additional tools and options for various aspects of plant breeding.

Yet, the essential activity of plant breeding remains constant: the development of germplasm with a superior aggregate phenotype for a given target environment. As the mediation of many important phenotypes will likely remain unknown to contemporary science, direct selection on a continuous basis using well-established methods by well-supported and integrated plant-breeding programs may be the best choice of approaches to crop improvement.

The contents of this book reflect the status of major challenges, approaches, and accomplishments of plant-breeding programs from around the world as told by several hundred scientists of plant breeding who gathered to honor a great teacher, practitioner, and researcher of that discipline, Arnel R. Hallauer.

Kendall R. Lamkey
Michael Lee

Plant Breeding:
The Arnel R. Hallauer
International Symposium

1

Plant Breeding: Past, Present, and Future

Theodore M. Crosbie, Vice President, Global Plant Breeding, Monsanto
Sam R. Eathington, Director of Breeding Applications, Monsanto
G. Richard Johnson, Sr., Science Fellow, Monsanto
Marlin Edwards, Global Lead, Breeding Technology, Monsanto
Robert Reiter, Director of High Throughput Genotyping, Monsanto
S. Stark, Lead, Seed Breeding and Biotech Statistical Services, Monsanto
Radha G. Mohanty, Senior Statistician, Seed Breeding and Biotech Statistical Services, Monsanto
Manuel Oyervides, R&D Director, Latin American Corn/Global Sorghum, Monsanto
Robert E. Buehler, Program Director, Trait Development Pipeline, Monsanto
Alan K. Walker, Global Soybean Breeding Director, Monsanto
Raymond Dobert, Regulatory Affairs Manager-Oilseeds, Monsanto
Xavier Delannay, Director, Ag Technology, Monsanto Protein Technologies
Jay C. Pershing, Corn Rootworm Project Lead, Monsanto
Michael A. Hall, Line Development Western Lead, North America Corn Breeding, Monsanto
Kendall R. Lamkey, Professor, Iowa State University

Introduction

As part of the Hallauer Symposium we have been asked to address three questions on plant breeding: what is it? what has it done? and what can it do? We have approached these questions with graduate education in mind and with the view that understanding the context of any particular concept is essential to understanding the details of science. Our objective is to lead the reader to the details of plant breeding science but not to get lost in them. In this chapter, we have described and contrasted the past, present, and future of plant breeding of important field crops from a commercial point of view. Dozens of authors have offered definitions of plant breeding in the published literature, and we have resisted the temptation to add yet another personal nuance to the stack. Bernardo (2002) offers the most universal description in our view: "Plant breeding is the science, art, and business of improving plants for human benefit."

Plant breeding has played a seminal role in the advancement of human civilization. The domestication and continuous improvement of plants and animals meant an ever-increasing segment of the human population could focus their inventive creativity on improving other aspects of civilization. The benefits of this phenomenon are completely obvious in well-developed countries, and the attending social unrest and anarchy in countries with severe food shortages are unquestionable. The smaller the percentage of people involved in food production the more rapidly a civilization has advanced and, inversely, the less social strife its people have endured. As the world moves from six billion people to a much larger number, the importance of plant breeding can only increase. We have divided plant breeding into three technical eras based on the methodology used to achieve genetic gain. Throughout history, breeders have improved the harvested crop in the field primarily by phenotypic mass selection and replicated progeny selection, and today direct genotypic selection is finally emerging as a reality. The breeding mission has been based on the concept that any given phenotype is the summation of several factors. As plant breeders, we write $P = G + E + G^*E + e$, where P is the phenotypic performance, G is contribution of the genotype, E is the environmental effect, G^*E is the interaction of the genotype with its environment, and e represents accumulated measurement errors. All breeders, regardless of their century, have devised various methods to cope with the frustratingly elusive nature of the components of phenotypic performance in an effort to estimate the genotypic or breeding value of individuals. Since the arrival of flowering plants, these components have remained timeless pieces of the puzzle. The quest for genetic improvement has not changed but the methodological choice to estimate breeding value and to achieve genetic gain through selection has changed and continues to change even today.

Era 1: Domestication and phenotypic mass selection

In the first era, early humans domesticated our current crops by mass selection of the female phenotype. The domesticators essentially invented agriculture and transformed human civilization from one of nomadic hunting to a more sedentary lifestyle based on gardening. Farming and farming tools were invented during the *Neolithic* or New Stone Age, and by 10,000 years ago, our hunter–gatherer ancestors had reached all but the most remote areas of the globe (Sykes, 2001). In the blink of a geological eye, human life had been changed beyond recognition for all time. Small bands of people across the globe originally survived on whatever they could gather, and then, according to Sykes (2001), the domestication of wild crops and animals began independently in several different parts of the world.

As the glaciers retreated for the last time, cereal grains were domesticated about 11,000 years ago from wild grasses in the Near East in what is now known as the Fertile Crescent. Early people also inevitably used mass selection with without control of the male (Hallauer and Miranda Fo, 1981) to domesticate beans (*Phaseolus vulgaris* L.) in India, rice (*Oryza sativa* L.) in China, sorghum (*Sorghum bicolor* L.) in west Africa, millet (*Pennisetum americanum* L.) in Ethiopia, sugar cane (*Saccharum* sp.) and taro (*Colocasia esculenta*) in New Guinea, maize (*Zea mays* L.) in Central America, and squash (*Cucurbita* sp.) and sunflowers (*Helianthus annuus*) in the eastern United States (Sykes, 2001). Given social roles in these early societies, it is likely that many of the first plant breeders were women and that their children served as their field technicians while the most able men were off hunting. Within a few thousand years, without any understanding of genetics and with the power of visual selection, our ancestral mothers created the germplasm base for modern food production.

So profound was the importance of maize to early South American civilizations that it was given religious meaning and significance. It is nearly impossible to imagine our modern world without domesticated maize, which is used directly or indirectly to produce much of the food on our tables as well as for fuel to deliver it to us. Once domesticated, maize spread from its center of origin to the agricultural corners of the globe to feed and be improved by all of the world's farmers.

Upon his retirement as president of the University of Chicago, Dr. George W. Beadle resumed his early interest in the ancestry of maize. As a graduate student with R.A. Emerson, he began studying the cytogenetics of maize–teosinte crosses in 1928, and they concurred with A. Vinson's 1877 hypothesis that wild teosinte was the direct ancestor of cultivated maize (Beadle, 1980). In the mid-1970s, after 40 years of debate in the literature about the origin of maize, he reconstructed their 1930s hypothesis using a primitive Mexican maize variety, Chapalote, and Chalco, the most cornlike variety of Mexican teosinte (Figure 1.1). We are indebted to Dr. Linda Pollak at Iowa State University, Walter Goeppinger, a Boone, Iowa, farmer, and Mrs. George Beadle for preserving Figures 1.1 and 1.2. In the mid-1980s, Mr. Goeppinger introduced Dr. Pollak to Mrs. Beadle who gave her all of Dr. Beadle's seed, breeding records, and photographs. Dr. Beadle personally took these photographs as part of the breeding experiments for his 1980 paper in which he showed that Chalco and Chapalote differed by only about five major genes.

Seventy years after Beadle and Emerson found normal chromosome pairing during meiosis in maize–teosinte crosses, Matsuoka et al. (2002)

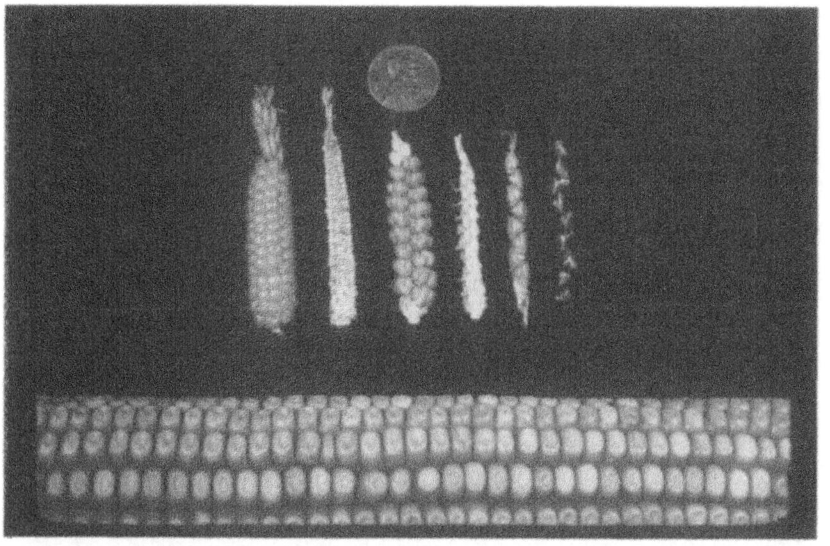

Figure 1.1 Dr. George W. Beadle's genetic reconstruction of the evolution of modern maize from teosinte. Photo taken by Dr. Beadle circa 1978.

used microsatellite-based phylogenetic analyses to confirm the Vinson–Emerson–Beadle hypothesis and showed that a single domestication event of *Zea mays* ssp. *parviglumis* resulted in cob maize in contrast to the multiple and independent domestications of most crops and animals. Matsuoka et al. (2002) also presented evidence that this most likely happened around 9188 BP in the highlands between the states of Oaxaca and Jalisco in Mexico and that the early diversification of maize occurred in the highlands before spreading to the lowlands at a later date. Interestingly, today ssp. *parviglumis* is not found in the highlands where its nearest maize relatives are found today, but it is found in the Balsas River drainage below 1800 m altitude, leaving an unwritten chapter in the history of maize (Matsuoka et al., 2002).

Mutational changes in as few as five restricted genomic regions account for most of the inflorescence differences between teosinte and maize and likely facilitated the transformation to cob maize (Beadle, 1939; Doebley and Stec, 1991). A locus on the long arm of chromosome 1, purportedly Tb1, ensures that the primary lateral inflorescence develops into a female rather than a male flower. A change from a two-ranked to a four-ranked inflorescence was necessary for cob formation to occur, and Doebley (1994) attributes genetic control to a locus on the short arm of chromosome 2. Suppression of teosinte cupulate fruitcase formation was necessary for cob formation in the origin of maize, and Doebley (1994) speculated that a mutation in a regulatory locus on the short arm of chromosome 4 would have been required to affect the complicated genetic array involved in this transformational event.

Less clear are studies on changes in spikelet pairing and ear disarticulation. According to Doebley (1994), most studies have been complicated by the concurrent distichous-polystichous segregation across teosinte × maize crosses made with maize lines of variable kernel row numbers. Most of the evidence, however, suggests that the principal genetic factors reside on the long arm of chromosome 3 and the short arm of chromosome 5, respectively (Doebley and Stec, 1991).

Figure 1.2 is another reconstruction by Dr. Beadle and depicts his theory on the effect of selection for a second set of mutations underpinning the evolution of primitive cob maize into modern maize. The oldest, most primitive cobs recovered from several caves in the Tehuacán Valley are quite uniform, less than 2 inches in length, and have eight rows of six to nine kernels each. Comparative morphologic studies indicate that primitive maize farmers using phenotypic mass selection were responsible for 2×, 5×, and 2× increases in the number of kernel rows, ear length, and kernel weight, respectively.

Both waves of genetic improvement by primitive farmers formed the foundation for a significant portion of the world's current food and feed supply. However it occurred, the selection and breeding of cob maize turned out to be a corner-

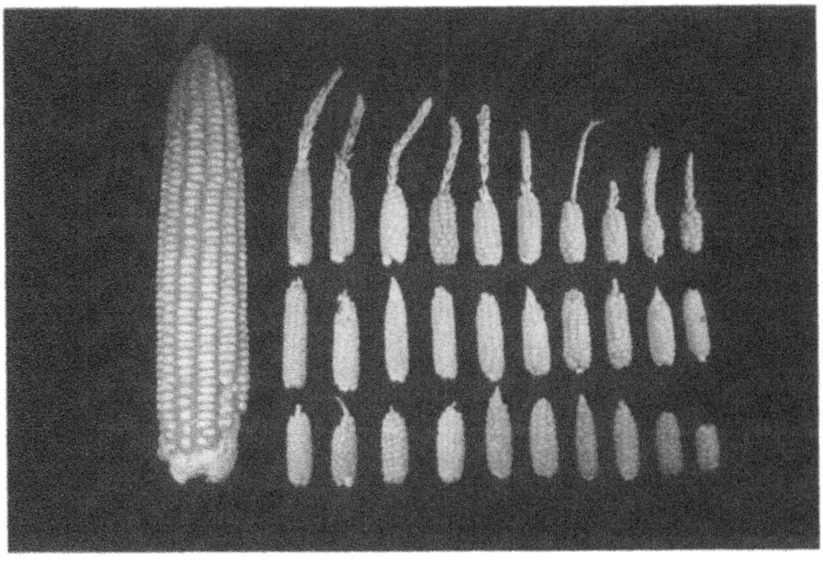

Figure 1.2 Comparison of cobs from primitive and modern maize varieties depicting the increase in yield potential of maize. Photo taken by Dr. Beadle circa 1978.

stone of immense value to recorded civilization and was extraordinary by any scientific standard. The efforts of these early breeders resulted in a dramatic 20-fold increase in yield potential, albeit over many millennia, which dwarfs even the most amazing accomplishments of modern science. Using this germplasm base, modern maize breeders quadrupled national maize yields in the United States since breeding became an organized science a century ago, based on replicated progeny selection systems. Without the feats of these early artisans, we would have a very different global economy today.

The cost and value of these two epic events in early plant breeding is difficult, if not impossible, to quantify but are easily seen and appreciated. The cost was simply the labor of multitudes of people, within a rather short period of geologic time, trying to survive the very elements of nature that propel natural selection. Whereas, small grains were the main plant food staple for other economies, maize was central to the physical and religious nourishment of the peoples of South America. Without it, it is not clear how or how well these cultures would have survived and flourished.

As is well known, an Austrian monk, Gregor Mendel, conducted studies and made observations on the inheritance of certain characteristics of sweet peas. His work, finished in 1866, lay dormant and unnoticed for 40 years until its rediscovery in 1900. Early geneticists recognized the value of his single-gene inheritance model in explaining observed variations in plants and animals. Consequently, Mendelian genetics formed the foundation for modern plant breeding.

Throughout the time Mendel's work gathered dust in a monastery, farmers in the United States were practicing a form of mass selection by saving the best seed ears from their own fields. U.S. government estimates of national corn yields (http://www.usda.gov/nass/pubs/histdata.htm) show that national yields from 1866 to 1910 changed little as a result of farmer selection, as evidenced by a regression coefficient near zero (b = 0.065; Figure 1.3). Yields may actually have trended down (b = -0.165; Figure 1.3) for the next 25 years, despite the improvements in husbandry usually associated with the change from horses to mechanically powered farming methods. Perhaps the expansion of corn production into western Corn Belt dry land areas, such as Kansas and Nebraska, contributed to this slight decrease.

While selection for simply inherited traits had been very successful in domesticating corn and in selection of modern maize types, farmer selection in the U.S. Corn Belt was not successful in producing noticeable genetic gains. Apparently, mass selection of individual plants was not successful in improving quantitative traits such as yield.

One positive outcome of the farmer selection era in the Corn Belt, however, was the fortuitous formation of distinct heterotic groups in maize. Farmers' saving their "best ears" as their own seed and for corn contests at the local county fairs seg-

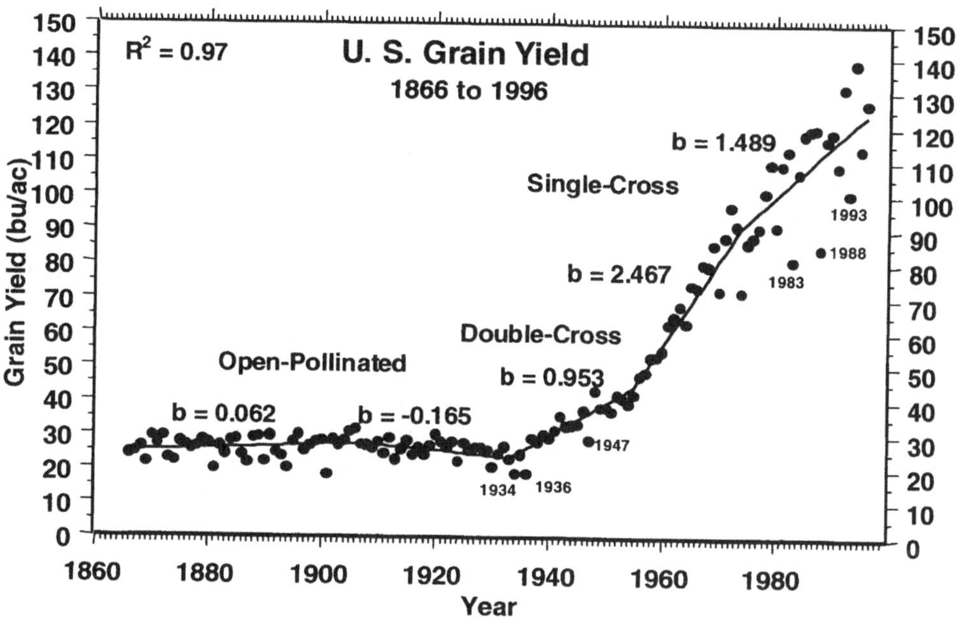

Figure 1.3 USDA-estimated average corn yield per acre for corn harvested for grain from 1866 to 1996.

mented and isolated gene pools across the Corn Belt. The resulting differences in gene frequency and types of gene action among the hundreds of populations formed a foundation for early breeding studies and progress. It also led to the formation and identification of the many heterotic groups, such as Reid and Lancaster, that are exploited today by corn breeders.

Era 2: Replicated progeny testing

Crabb (1993) chronicled the early years of corn breeding in the United States. The leaders in hybrid corn are well known to all breeders. George Shull, Edward East, Donald F. Jones, George N. Hoffer, Merle Jenkins, James R. Holbert, and Henry A. Wallace are often mentioned. Their efforts and accomplishments were extraordinary and their lifetime spending on breeding would get lost as rounding error in any modern day corporate research budget. The concept of replicated progeny testing became a common tool in breeding programs as these breeders and their counterparts studied the effects of inbreeding and pedigree selection.

Popular textbooks such as Hallauer and Miranda Fo (1981) and Bernardo (2002) outline the dozens of methodological variations of the replicated progeny test that often surface on Ph.D. preliminary exams. Breeders innovated these approaches in an effort to (1) exploit various types of gene action and genetic variance, (2) reduce cycle times, (3) optimize crossing and testing schemes for self- and cross-pollinated crops, and (4) find the most economical ways to maximize genetic gain. At the heart of it, breeders attempt to cross "good-by-good" and select the best as rapidly as possible. The phenotypic simplicity and the statistical complexity of this elegant concatenation have fueled late-night discussions in graduate student offices and in corporate offices alike for nearly a century.

Early corn breeders used Mendel's model to analyze phenotypic variation, and early statistical geneticists expanded the single gene model to explain continuous variation in quantitative traits such as grain yield and other agronomic traits. By necessity, breeders used a replicated progeny test to study continuous variation and the practice became a common tool in virtually every breeding program replacing phenotypic mass selection as the primary breeding method. Inbreeding studies and their implications prompted the use of double crosses, and national maize yields immediately began to improve steadily at a rate of nearly 1 bushel/acre/year (b = 0.953) from the mid-1930s to about 1960, when single-cross use overtook double crosses (Figure 1.3). During the so-called double-cross era, corn breeders also were making the improvements in inbred performance per se necessary for their use in profitable single-cross productions systems.

In the early 1960s an entirely new farming system swept across the Corn Belt, boosting annual improvements in maize yields to a rate of nearly 2.5 bushel/acre/year ($b = 2.467$, Figure 1.3). Major improvements in weed control, increased plant density, earlier and more reliable planting dates, increased rates of nitrogen fertilizer and balanced fertility programs coupled with modern single-cross hybrids changed national corn yields at a breathtaking rate. The increases in corn production were also paralleled by a huge increase in breeding effort by the private sector. Companies attracted by the profit potential of single-cross seed corn rapidly built breeding, production, and sales capacities unseen in previous decades and began a shift from public to private enterprise in nearly all aspects of breeding and agronomic research and development that has continued to the present day.

Sometime in the late 1970s, the rate of yield improvement apparently slowed to 1.5 bushels/acre/year ($b = 1.489$; Figure 1.3). We speculate that this slowdown was associated with little if any further improvements in agronomic practices other than a steady increase in plant density of 260 plants per acre per year (Figure 1.4) because other factors such as nitrogen fertilizer were declining (Figure 1.5) and very effective weed control had been achieved.

Figure 1.6 shows that the amount of yield variability across years has remained relatively constant from 1866 to 1995, contrary to populist views that monoculture and the use of single genotypes puts our total production more at risk than did the use of more heterogeneous varieties in crop rotations. Droughts and freezes account for most of the significantly lower years that visually appear in clusters of three in each half century. The residual analysis does not show any difference in yield variability among open-pollinated, double-cross, and single-cross eras, suggesting that all of the science in the world cannot completely counteract weather as the dominant force on corn yields.

Return on corn-breeding investment

Russell (1993) summarized 10 papers covering 13 studies showing genetic gain from breeding accounted for 56–94% of the yield improvement from the 1930s to the 1970s. On average, 75% of yield improvement in these studies was attributed

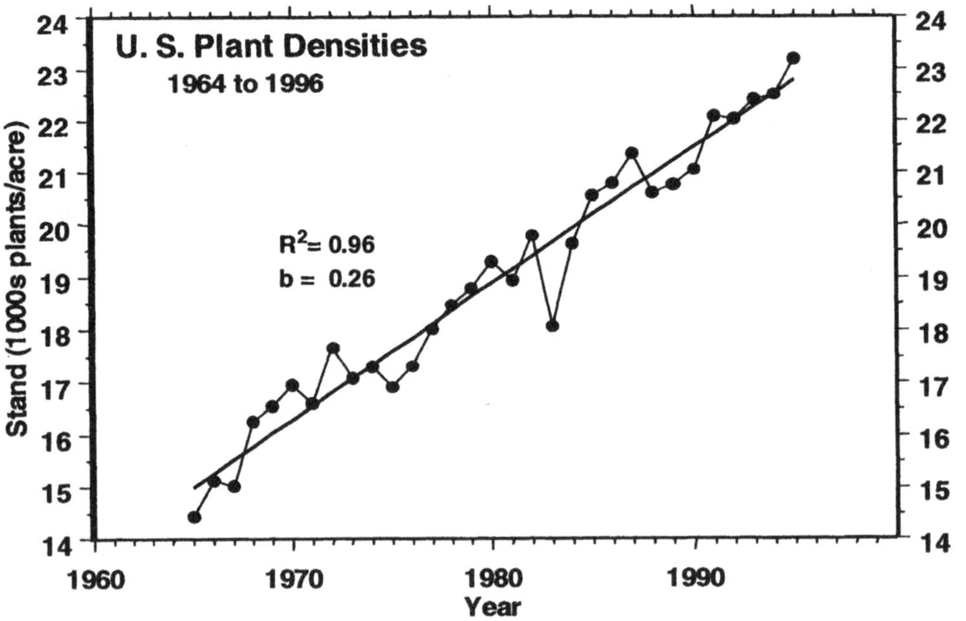

Figure 1.4 Average plant densities for U.S. corn production from 1964 to 1996.

to genetic gain from breeding, and the remainder was generally attributed to improved farming practices.

In a 1994 survey of plant breeding in the United States, Frey (1996) reported that 91 companies funded 510 science person years (SYs) in corn breeding, which approximated the number of corn-breeding programs in these companies. Public institutions accounted for an additional 35 SYs for corn, although nearly all public institutions con-

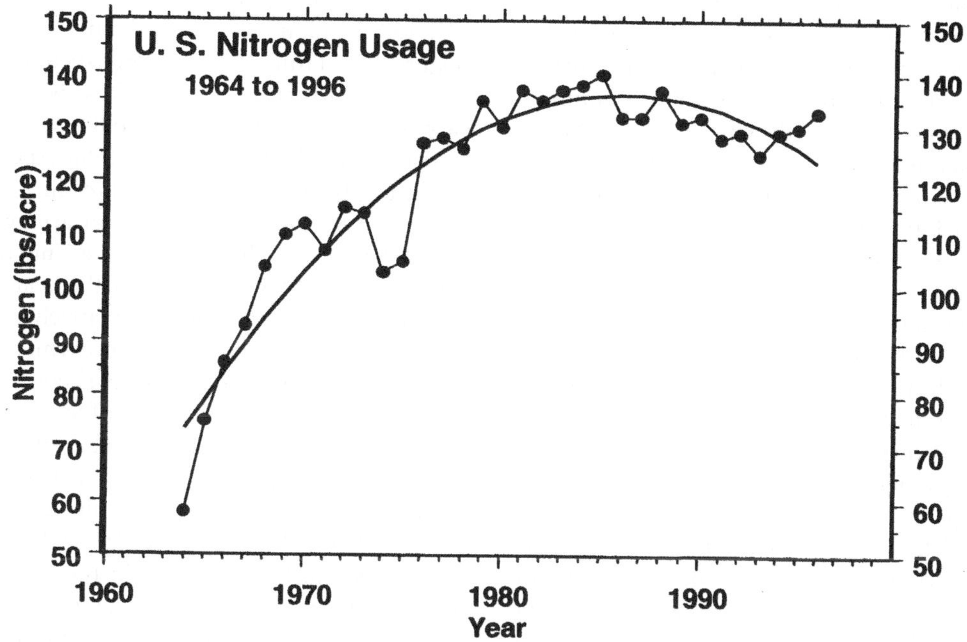

Figure 1.5 Nitrogen application rates per acre for U.S. corn production from 1964 to 1996.

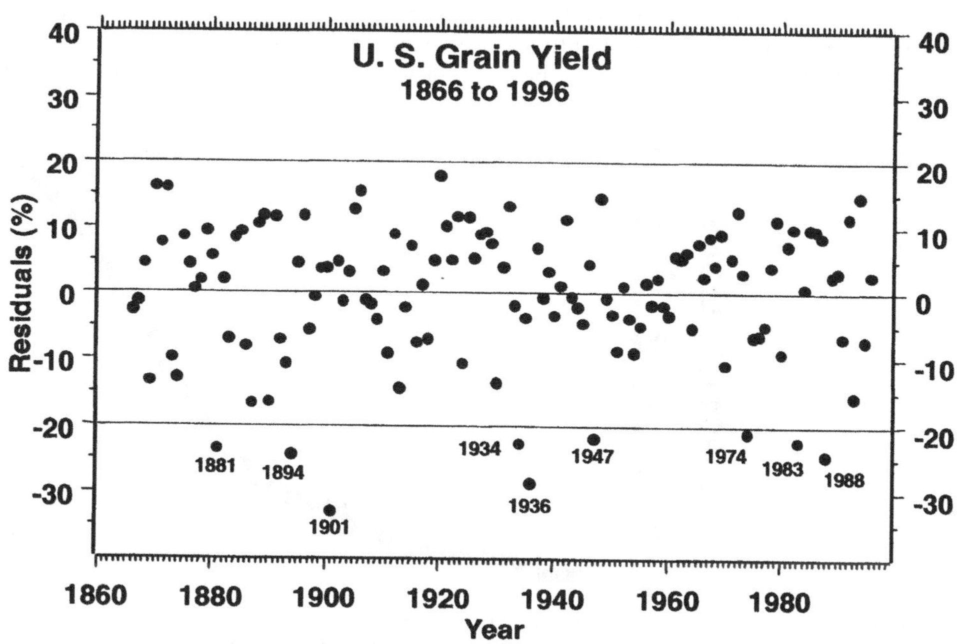

Figure 1.6 Residual deviations from regression for average U.S. corn yield from 1866 to 1996 expressed as a percentage of predicted yields by year.

duct research on breeding rather than on hybrid development per se. Frey (1996) did not supply estimates of plant-breeding spending by crop but did estimate that total plant-breeding spending by the private sector was $338 million for all crops. An extrapolation of Frey's numbers would predict a total industry expenditure of $156 million in 1994 for corn breeding, which could account for only direct costs and may not include normal overhead and infrastructure costs. Frey's numbers probably do not include the cost of capital. We estimate that in 1995, prior to major integration of breeding and biotech, the private and public sectors spent $200 million annually on corn breeding per se. Our estimates are based on the published amounts in company annual reports and our own knowledge of the fully loaded costs associated with private sector breeding programs. Applying a 30% overhead factor to Frey's survey results gives a spending estimate nearly identical to our estimate of $200 million for 545 programs, with larger companies spending between $400,000 and $500,000 per year per breeding program in 1995.

Using data from a number of sources, we estimated annual corn-breeding spending from 1930 to 1995 and expressed it in constant 1984 dollars based on an agricultural price index from Huffman and Evenson (1993). Data sources were company annual reports for larger publicly traded companies, estimates from Kalton and Richardson (1983), and U.S. Department of Agriculture (USDA) data (http://www.usda.gov/nass/pubs/histdata.htm). This period was chosen because it was possible to separate breeding and biotechnology program costs and because yield improvements during this period were relatively unaffected by biotechnology.

We estimate that total spending on public and private corn breeding was approximately $3.0 billion (1984 dollars) from 1930 to 1996 and that 87% ($2.6 billion) was spent from 1960 to 1996 (Figure 1.7). From 1930 to 1960, spending averaged $12.8M/year in 1984 dollars, whereas the pace averaged $74.3M/year in 1984 dollars from 1961 to 1996. Virtually all of the total expenditure can be attributed to hybrid corn breeding.

Figure 1.8 shows the market year average price per bushel for #2 yellow corn in the United States from 1930 to 1995 as dollars of the year and in constant 1984 dollars. We conservatively assumed that two-thirds of the yield improvements were due to genetic gain, and Figure 1.9 shows accumulated genetic gain as a percentage of total harvested grain production. The annual value of accumulated genetic gain (1984 dollars) increased from 1930 to 1950, slowly declined in value for the next 20 years despite huge increases in yields annually, and jumped to historic highs in the early 1970s, only to slowly decline in constant dollar value (Figure 1.10). Even though both spending in real terms (Figure 1.7) and accumulated genetic gain (Figure 1.9) continued to increase over time, the value of accumulated gain did not show the same increase in value (Figure 1.10).

The cumulative value of genetic gain from corn

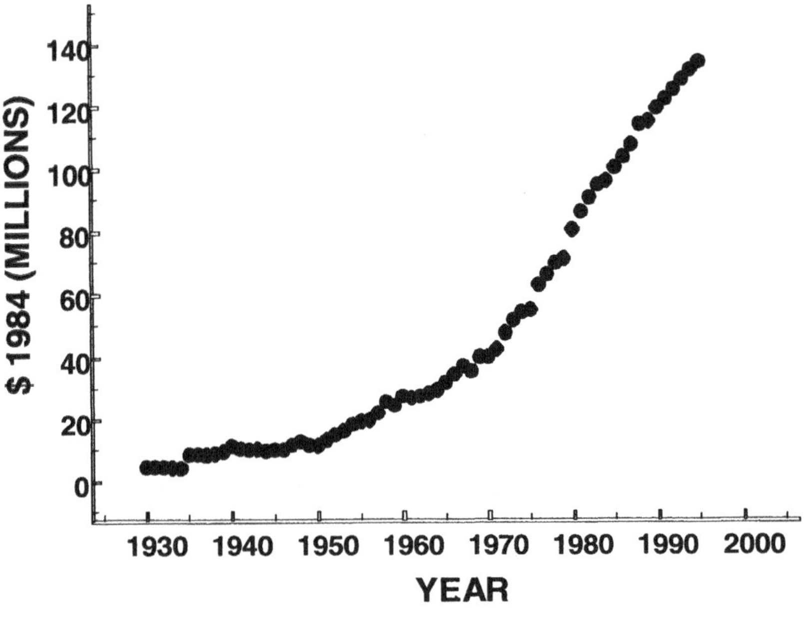

Figure 1.7 Estimated annual spending for public and private corn breeding in the United States from 1930 to 1996 expressed in constant 1984 dollars.

breeding is estimated at $260 billion in dollars of the year, and 89% of the total value was realized between 1960 to 1995 (Figure 1.11), which was nearly identical to the breeding spend ratio for the two halves of the overall period. The total value of genetic gain in 1984 dollars was estimated to be $460 billion (Table 1.1), but 36% of the gain in constant dollars was realized in the first 30 years versus 11% when expressed in dollars of the year.

Our estimates, which are intended only to be directional, are that industry corn-breeding spending of $3 billion produced an improvement in U.S.

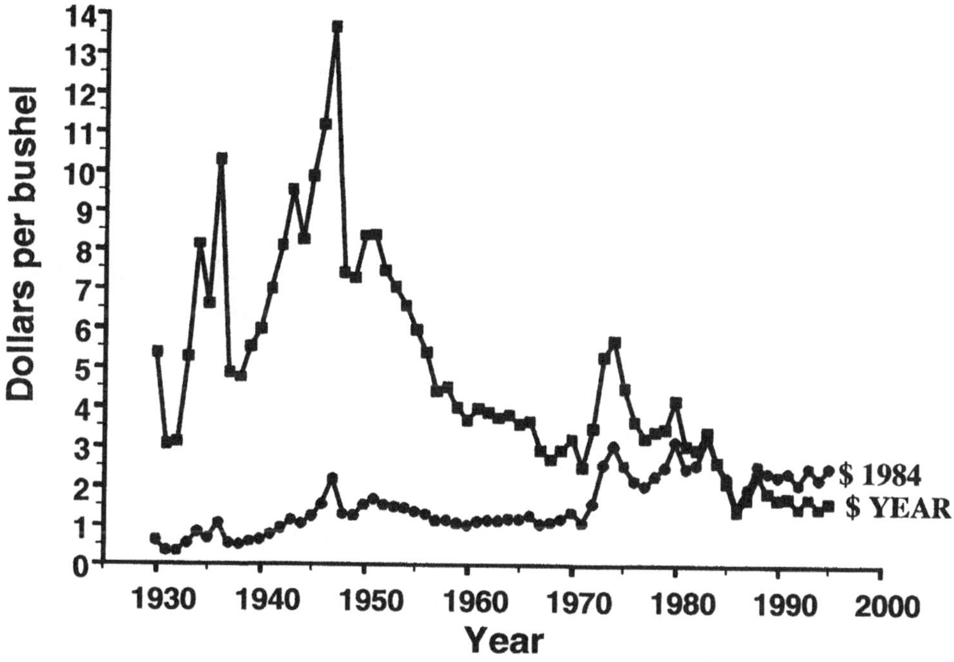

Figure 1.8 Market-year average price for corn in the United States from 1930 to 1996 expressed in dollars of the year and in constant 1984 dollars.

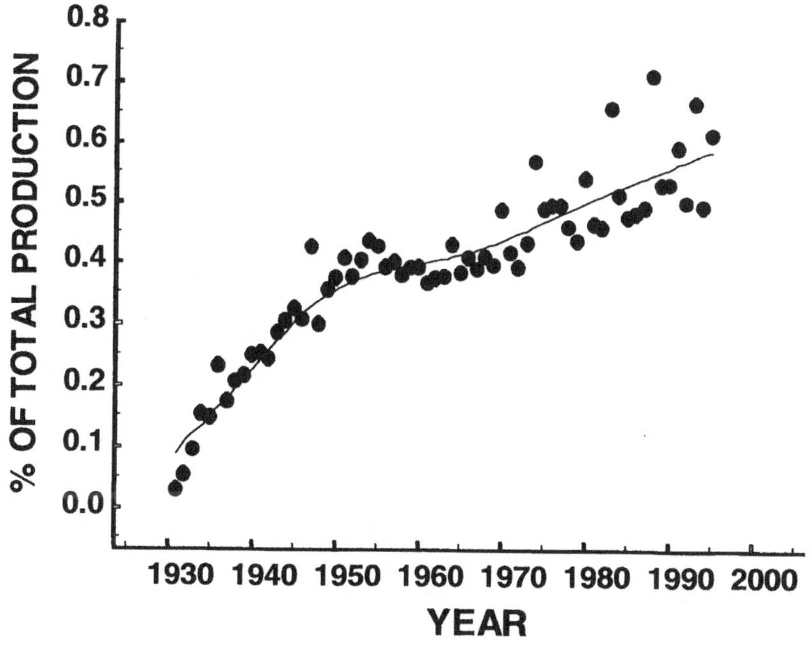

Figure 1.9 Estimated genetic gain expressed as a percentage of total U.S. corn production by year from 1930 to 1996.

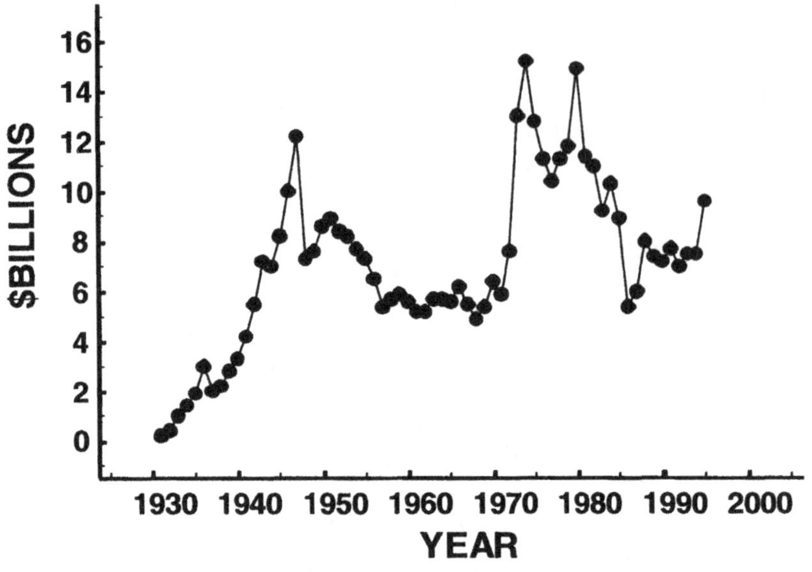

Figure 1.10 Estimated value of accumulated genetic gain by year from 1930 to 1996 expressed as constant 1984 dollars.

Table 1.1 Estimated cost and value of public and private corn breeding from 1866 to 1996

	DC Era	SC Era	Total
Cost	\$0.4B	\$2.6B	\$3.0B
(\$1984)	(13%)	(87%)	
Value	\$166B	\$294B	\$460B
(\$1984)	(36%)	(64%)	
Value	\$29B	\$232B	\$261B
(\$ Year)	(11%)	(89%)	

farm gate value of nearly \$0.5 trillion when expressed in 1984 constant dollars. This return on investment does not quantify the enormous financial impact the germplasm has had outside the U.S. Corn Belt, its role in reducing overall consumer food costs, or its value in avoiding environmental costs that inevitably would have been incurred by bringing marginal land into production had corn yields not quadrupled.

Impact of genetic improvement on market share

Even though enormous genetic gains have been made in corn in most countries, we are unaware of any studies in the literature directly relating improvements in product performance to commercial success in a competitive market like the United States. The enigmatic association between product performance and a company's market share has intrigued industry observers in most countries for many years. Company A will gain market share even though their newest products do not appear to be that much better, and Company B's market share will continue to slide even though they have recently released significantly improved products. In any given year, there has always appeared to be a confusing relationship between product performance and market share, and it has not been clear how the market-valued product performance versus other things such as sales and marketing programs, relative pricing strategies, and a myriad of factors that make up brand loyalty. The commercial proposition for most businesses, however, is that the company will fund each of these activities relative to their ongoing impact on sales and profitability.

We investigated the relationship between relative product performance and share of market for two large, successful seed corn companies in the United States. We obtained North America hybrid corn market share and yield performance information for Pioneer Hi-Bred from a 1998 Prudential Securities report. Historically, Pioneer was well-known for disclosing an estimate of their performance advantage, which was based on thousands of side-by-side, on-farm yield comparisons summarized nationwide into a simple, aggregate yield advantage claim for their product portfolio over all competitors.

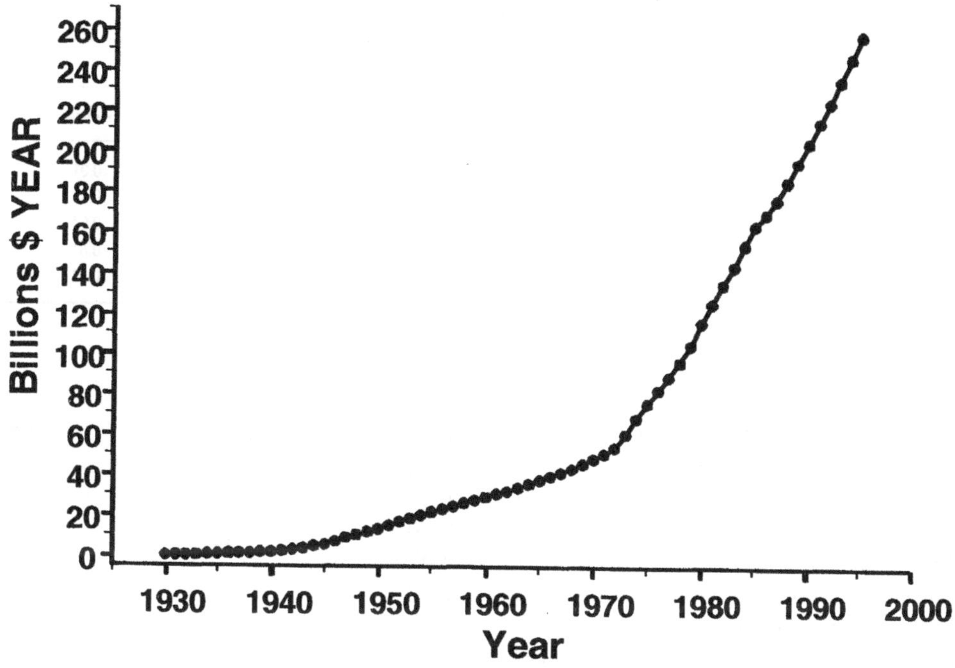

Figure 1.11 Estimated cumulative value of genetic gain in U.S. corn from 1930 to 1996 expressed in the dollars of the year.

Historical DEKALB® data were used to model market share as a function of yield performance. Yield performance in company strip trials was characterized for each of a set of relative maturity (RM) groups (≤90, 95–100, 105, 110, ≥115 days) based on the four, top-selling DEKALB® hybrids within the RM group in any given year. The performance of the four DEKALB® hybrids was measured relative to the four popular Pioneer hybrids that provided a large number of head-to-head observations in the DEKALB® Genetics corn yield trials database. Head-to-head averages within a given year and RM were combined to the DEKALB® hybrid level (across the Pioneer comparison hybrids), using a weighted average where the weights were the inverse of the variance. Unweighted averages were used to combine across DEKALB® hybrids within an RM group and across RM groups. The resulting by-year, across-hybrid, and across-RM average yield differences were used with by-year DEKALB® corn hybrid market share data in the regression modeling.

We used information for each company to construct a linear regression model of market share. Pioneer market share was modeled for the years 1990–1998, and DEKALB® market share was modeled for the years 1990–2002. However, before interpreting this model, we hypothesized that a temporal lag might exist between the performance advantage and the resulting market share impact. The rationale was that growers were typical consumers and would prefer products with proven performance records and that most farmers would base purchase decisions on one or more years of experience with a hybrid. Farmers might increase their purchase or decrease their purchase depending on the change in relative performance of the products. In other words, a lag would exist between the commercial release of products and their impact on market share, particularly when comparing the aggregate portfolio performance of Pioneer to a plethora of hybrids representing the rest of the market.

We modeled lag phases from one to five years in an effort to determine if there was a temporal lag between relative product performance and share of market. The *R*-square values for the lags from one to five years suggested that the third year provided the strongest correlation (Table 1.2) for both companies, although the trend was much more obvious for Pioneer than for DEKALB®. That is, if Pioneer's relative portfolio advantage over the competition increased or decreased by 1 bushel per acre, the effect on market share was most obvi-

Table 1.2 Regression of market share on yield performance for two seed corn brands in the United States

	Regression Equation		R^2	
Year Lag	Monsanto	Pioneer	Monsanto	Pioneer
1	share = 10.56 + 0.54 * yld diff	share = 39.59 + 0.26*yld diff	0.42	0.01
2	share = 10.61 + 0.47 * yld diff	share = 32.09 + 1.55*yld diff	0.37	0.58
3	share = 11.10 + 0.63 * yld diff	share = 32.05 + 1.66*yld diff	0.78	0.94
4	share = 11.33 + 0.54 * yld diff	share = 34.03 + 1.31*yld diff	0.66	0.61
5	share = 11.00 + 0.26 * yld diff	share = 39.98 + 0.22*yld diff	0.17	0.02

ous three years later. The regression of Pioneer market share on yield difference, using the three-year lag, is presented in Figure 1.12. The R^2 of 0.94 indicates a very strong association between overall portfolio performance changes and market share changes three years later for this period of time for Pioneer. The slope of the regression line estimated for the three-year lag model suggests that Pioneer's share of market changed significantly with changes in portfolio performance advantage relative to competitors. There are many company-specific factors affecting the nature and strength of this association as evidenced by the differential results for the two analyses.

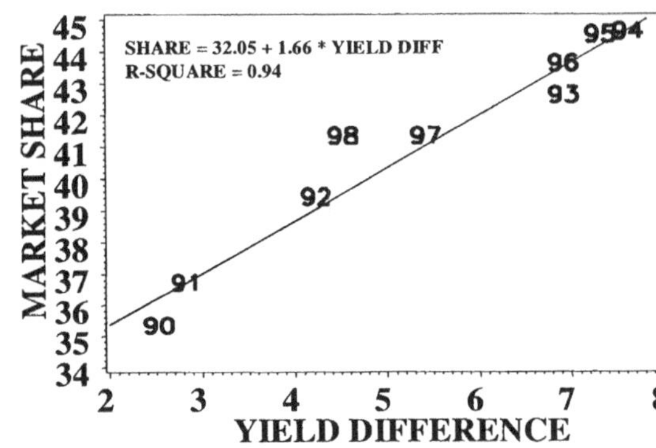

Figure 1.12 Pioneer share vs. 3 years before yield difference.

The DEKALB® analysis (Table 1.2) showed some similarity to the Pioneer trends in that the strongest association between performance changes and market share was found for the three-year lag model. The two analyses, however, are not directly comparable because of significant model differences. Whereas the Pioneer analysis presumably includes comparisons to hybrids from a large sampling of competitive companies, the DEKALB® analysis included only head-to-head comparisons to an equal number of competitive Pioneer hybrids. A lesser difference is the number of observations included in the regression models. In the Pioneer analysis, 9 observations were used in the models for lag 1 to lag 5. In the DEKALB® analysis, the number of observations decreased from 12 for lag 1 to 8 for lag 5.

Studies of two major hybrid seed corn brands in the United States suggest that overall performance changes may be related to company market share, but clearly other factors also play important roles. Performance is necessary but not the sole driver of market penetration. The influence on market share of a hybrid or a group of hybrids released at the same time appears to be greatest three years after their introduction to the market place. We view the implications of these results as more of a confidence factor than as a literal predictor of market dynamics because of the many factors that can influence share of market. By monitoring relative performance changes it is possible to anticipate or at least not be surprised by some change in your own company's market share potential as well as for your competitors who show a significant improvement or decline in relative performance. A company might have more confidence in producing more seed for sale and in providing additional sales and marketing resources if performance of their products is improving relative to a major competitor's. A more conservative approach to production and inventory control might be appropriate if breeding gains were to fall behind key competitors.

*Understanding G*E in plant breeding*

The use of replicated progeny trials powered genetic gain in crops such as corn and also enabled breeders to better understand the role and importance of $G*E$ in selection and breeding. Breeders and farmers have long known that the best hybrid in one year in a sample of ten similar locations might not be the best in another year or when averaged across several years at the same locations. The best variety of soybeans (*Glycine* max L.) across several years in a set of western locations might not be the best one in a similar set of eastern locations within the same maturity group band. The vexation of G^*E has humbled every practicing breeder at some point, regardless of their intellect, fame, and fortune and greatly complicates the process of selecting superior genotypes in breeding programs. Unless G^*E is dealt with effectively, the potential genetic gains of plant-breeding programs will not be realized and delivered to the marketplace.

In the absence of significant G^*E and with experimental error at reasonably low levels, average phenotypic performance across environments provides a good representation of genotypic performance. Consequently, relative performance of genotypes can be determined from differences in these phenotypic performances. However, in the presence of significant G^*E interaction, relative genotypic performance can only be characterized for specific environments. In either case, it is essential that a representative sample of environments be taken to characterize the genotypic response to environmental variation adequately. With an inadequate sample of environments, absence of G^*E is not conclusive evidence that relative performance of genotypes is not environment dependent. The intensity and scope of environmental sampling should be guided by prior information on the expected level of G^*E for genotypes and environments similar to those under current study.

While experimental error also complicates characterization of genotypic performance, it can be reduced by experimental design and/or analytic methodologies. Within-location replication is also useful, since it allows separation of G^*E from experimental error, thereby enabling better characterization of G^*E. Unfortunately, G^*E interaction cannot be reduced or mitigated by design or analysis methods because G^*E is an inherent attribute of the given genotypes in the given environments. As such, plant breeders have no choice but to deal with the ubiquitous presence of G^*E in their trials and nurseries. Typically, breeders do this by (1) conducting well-designed trials at uniform and representative sites, (2) sampling as many environments as is financially prudent for the target market area for the new products under test, (3) using efficient analytical methodologies to quantify G^*E accurately in each trial set; and (4) incorporating the resulting information on G^*E effectively into their decision process.

The term environment in G^*E interaction can be used to represent many things and is frequently used in a very broad sense. For example, one might use G^*E to describe differential genotypic response to various geographic locations in a given year. There is also a temporal component to G^*E, since the same geographic location will have a different environment in different years and at different times of the year (e.g., effect of planting date). To deal explicitly with both the spatial and the temporal aspects of environment, some researchers separate these so that $G^*E = G^*L + G^*Y + G^*L^*Y$, where G^*L is genotype by location interaction, G^*Y is genotype by year interaction, and G^*L^*Y is genotype by location by year interaction. Without loss of generality, we will simply use G^*E in this discussion.

To complicate things further, an environment might also be defined as an entire collection of geographic locations. These collections of locations might simply be locations in close proximity within a geographic market area. They might also be identified by their similarity in edaphic, abiotic, or other climatic conditions.

The potential impact of G^*E on plant-breeding operations was studied in DEKALB Genetics advanced corn hybrid-breeding trials data from 1991 to 1993. An indirect measure of the impact of G^*E on breeding locations was determined by calculating the magnitude of G^*E interaction variance relative to total genotypic variance (i.e., the sum of pure genotypic variance and G^*E interaction variance). Balanced experiments were used in these analyses, where all genotypes in the experiment were included in testing at all locations. For each experiment in each year, variance components were estimated and the percentage of G^*E variance in total genotypic variance was determined. Experiments were grouped by RM, the groupings running from 80 RM to 120 RM in increments of 5 RM. The resulting information allowed investigation of effect of RM and year on relative magnitude of G^*E variance to total genotypic variance. Quadratic polynomial regressions were fit to the data for each year to illustrate general G^*E trends.

In 1991, there was a very clear increase in the percentage of G^*E variance relative to total genotypic variance as RM increased from early to late zones (Figure 1.13). This would imply that a greater sampling effort would be required as RM increases to achieve the same level of predictive accuracy. However, in 1992 (Figure 1.14) and 1993 (Figure 1.15) the pattern was quite different. In those years, there was a tendency toward a decreased percentage of G^*E variance in the medium relative maturities. This indicates that, while a fixed sampling plan cannot be optimal for all years, such a plan needs to take into consideration the historical variability in G^*E.

Note that any patterns in the percentage of G^*E variance across relative maturities or years are driven not only by differences in environments, but also by differences in the genotypes in the various experiments. Also note that there is tremendous variability in the percentage G^*E across experiments within any given RM grouping in any given year.

Perhaps the simplest approach to characterize G^*E accurately is to determine average performance of each of the genotypes in each of the geographic locations and then attempt to determine visually if any patterns exist in the relative performances of hybrids at the various locations. Meaningful patterns are often extremely difficult to detect, especially with large numbers of genotypes and locations. One might also attempt this by using hybrid averages for predefined subsets of locations such as different geographic regions or different levels of edaphic factors. Although this reduces the dimension of the outputs that must be reviewed, it is often equally difficult to apply if the collection of environmental subsets is large. More seriously, these predefined subsets might not represent the actual environments responsible for producing G^*E interactions and the underlying pattern in G^*E might not be detected at all.

A number of other more sophisticated statistical methodologies have been developed to provide more efficient characterization of G^*E. One of the more popular is the additive main effects and multiplicative interaction (AMMI) methodology (Gauch, 1992), which has been shown often to provide a comprehensive description of the underlying structure in G^*E. There are also a number of related methodologies, for example, the shifted multiplicative model (Gauch, 1992), that further refine this general modeling approach.

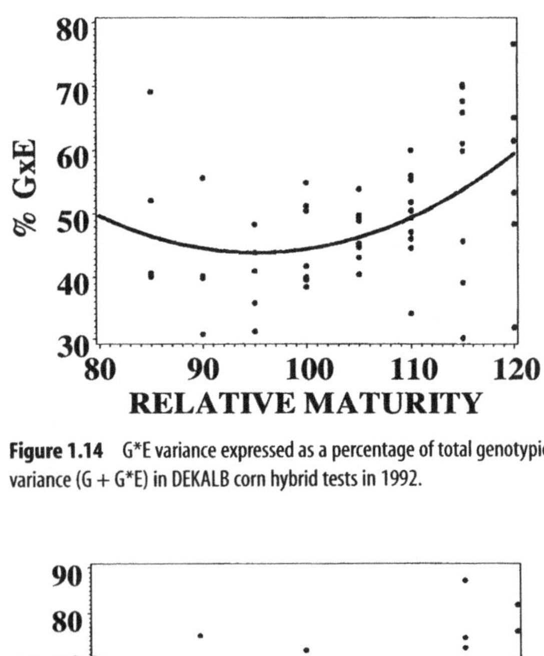

Figure 1.14 G*E variance expressed as a percentage of total genotypic variance (G + G*E) in DEKALB corn hybrid tests in 1992.

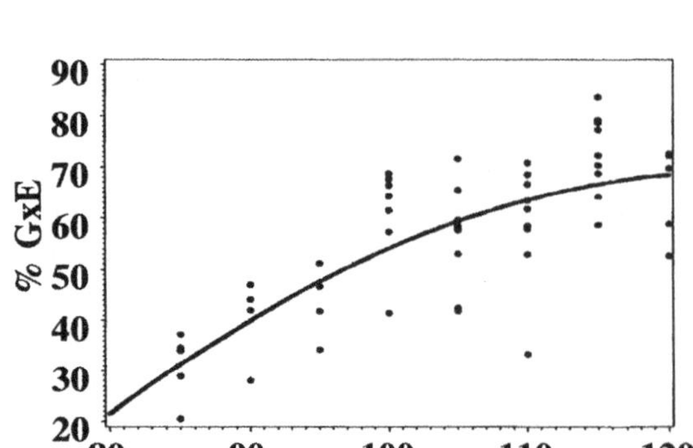

Figure 1.13 G*E variance expressed as a percentage of total genotypic variance (G + G*E) in DEKALB corn hybrid tests in 1991.

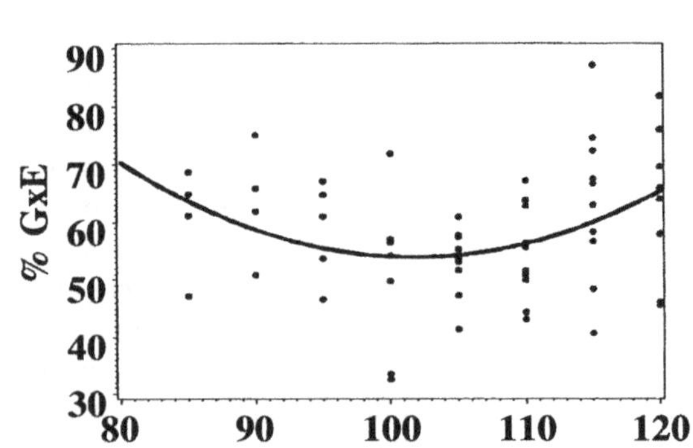

Figure 1.15 G*E variance expressed as a percentage of total genotypic variance (G + G*E) in DEKALB corn hybrid tests in 1993.

Geostatistics in corn hybrid performance assessment

For a variety of reasons that are obvious to a practicing plant breeder, it is not practical to plant and harvest only perfect experiments. Seed supply, timing of seed deliveries from winter nurseries, and less than perfect field and weather conditions often mean that the breeder is selecting among entries that have not always been in the same locations with equal replication. Reducing the data set to only those comparisons for which there are an equal number of replications of good data in the same experiments at the same locations would mean throwing away a lot of expensive data in a commercial breeding program. Neither finance nor breeding directors are inclined to toss data of this nature because, inevitably, our commercial colleagues also want comparisons of varieties that were not planted in the same tests or at the same number of locations. Classical statistical field analyses are of limited help in these situations. Consequently, breeders and statisticians have devised new ways to make valid comparisons with the somewhat woolly data structures that occur even with the best of planning in the real world. One approach focuses on the spatial structure in the environment and uses geostatistical analysis to characterize wide-area performance and performance differences of genotypes.

Hybrids grown at testing locations in close proximity may exhibit a spatial autocorrelation in hybrid yields. The existence of spatial autocorrelation in hybrid yields can be demonstrated by a study of the variogram (Cressie, 1993). The variogram for wide-area yield trials data from a given hybrid shows the squared difference in yields from pairs of locations at various separation distances. In the presence of spatial autocorrelation, these squared differences tend to be smaller at shorter distances, indicating that yields at small separation distances are more highly correlated than those at greater separation distances, contradicting a very basic assumption of independence in the analysis of variance.

A study of variograms of yields of major DEKALB® brand commercial hybrids showed very clear evidence of between-location spatial autocorrelation in approximately 80% of the hybrids. The spatial autocorrelation in yield suggests that hybrid yield trials data might be amenable to geostatistical modeling. This indicates that modeling the spatial autocorrelation in yield data by geostatistical methodologies might lead to improved inferences for hybrid performance assessment.

In the traditional statistical analysis methodologies used in hybrid performance assessment (e.g., Bradley, et al., 1988), inferences are made on the yield performance of an experimental hybrid relative to various comparison hybrids on a pairwise basis. Data from a given yield testing location are used if both hybrids are tested at that location (i.e., location-matched data). For inferences on a geographic region, data from a given testing location are used if they fall within the region. For single location inference and for inference on an entire geographic region, the simple mean yield difference is calculated, followed by a paired t-test of statistical significance.

This traditional methodology has several shortcomings. First, it does not effectively model the spatial trend in hybrid yield data, and it ignores large-scale (across-location) spatial autocorrelation. Potentially informative yield data from testing locations where just one of the hybrids appears are not used. Traditional analyses omit informative yield data from testing locations outside of the region when making inferences on performance within a geographic region. Finally, it does not differentiate between two general types of inferences, that is, estimation and prediction.

In making pairwise hybrid comparisons with geostatistical methodologies, all yield observations for the pair of hybrids are used regardless of whether the observations are location-paired or not. A linear mixed model is constructed from the following components: (1) large-scale performance is modeled by fixed-effect trend surfaces by cubic polynomial surfaces for each of the two hybrids; (2) the spatial autocovariance among the locations is modeled for each of the two hybrids, independently; and (3) the spatial cross-covariance is modeled between the locations for the first hybrid and the locations for the second hybrid. Covariance parameters are estimated by the method of restricted maximum likelihood (Patterson and Thompson, 1971; Patterson and Thompson, 1974), and fixed effects parameters are estimated by the method of generalized least squares (Searle, 1971).

As mentioned above, two general types of inferences can be made from this model. First, one can estimate the long-term average performance difference in the hybrids by estimating the difference in the fixed-effect trend surfaces. This difference

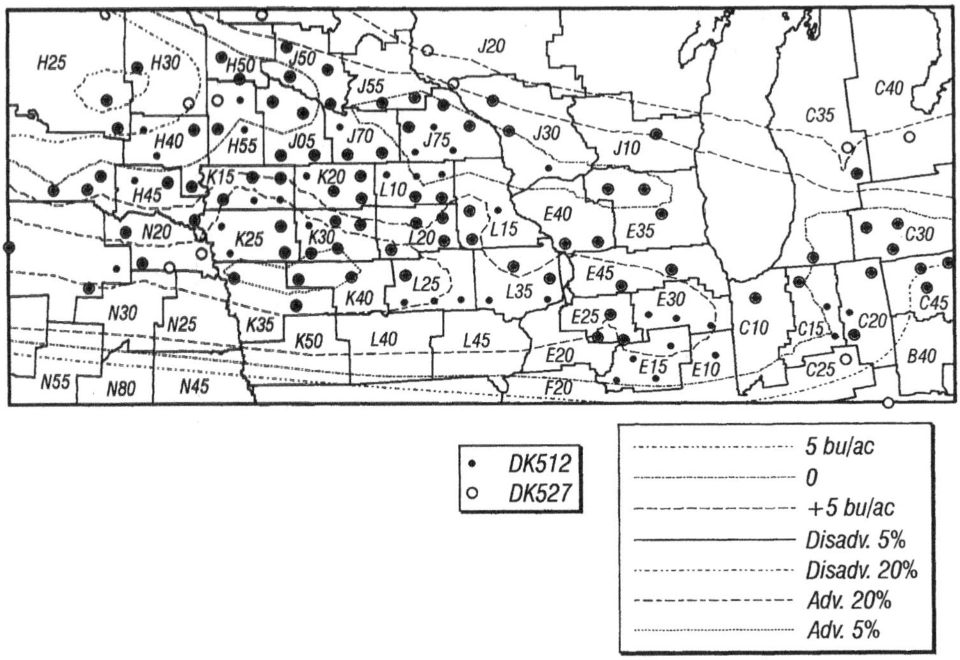

Figure 1.16 Contour plot of expected yield.

can be estimated either for a specific location or for an entire geographic region by integrating the difference in the trend surfaces over the region. Alternatively, one can predict the average performance difference in the hybrids for a given year and in a specific location or geographic region. Prediction for a location is done by universal cokriging (Cressie, 1993), and prediction for an entire region is done by universal block cokriging (Cressie, 1993).

Estimation and prediction surfaces are presented graphically by a contour plot of the yield difference surface, overlaid by another contour plot showing various levels of statistical significance. An example contour plot is presented in Figure 1.16. For inferences on a geographic region, the mean yield difference and its statistical significance are presented in a choropleth map (Figure 1.17).

Based on extensive validation studies, the geostatistical methods have been shown to produce inferences that are approximately 37% more precise than those from traditional methods based on a simple mean and *t*-test on location-matched data. The gain in precision translates to a 37% increase in the effective number of testing locations by replacement of the traditional analysis method by the geostatistical method. Given the expense of hybrid-yield testing, the geostatistical methodology provides a very large and cost-effective gain in the efficiency of a commercial corn-breeding operation.

Era 3: Direct genotypic selection

Transgenic breeding

The breeding of transgenic plants ushered in a new era in plant breeding based on direct genotypic selection. First sold in the United States in 1996, transgenic varieties offered new solutions to old problems, such as resistance to nonselective herbicides conveying superior weed control, European corn borer (*Ostrinia nubilalis*), and the corn rootworm (*Diabrotica* spp). Transgenic breeding changed the ways in which breeders and breeding organizations work and has had a major impact on production agriculture in countries growing these crops. For nearly 70 years, universities and companies in the United States bred and released cultivars of all of the major crops with little regulatory oversight. Even in countries with official trial and registration systems, varieties were not regulated in any significant way by those countries' equivalents to agencies such as the Environmental Protection

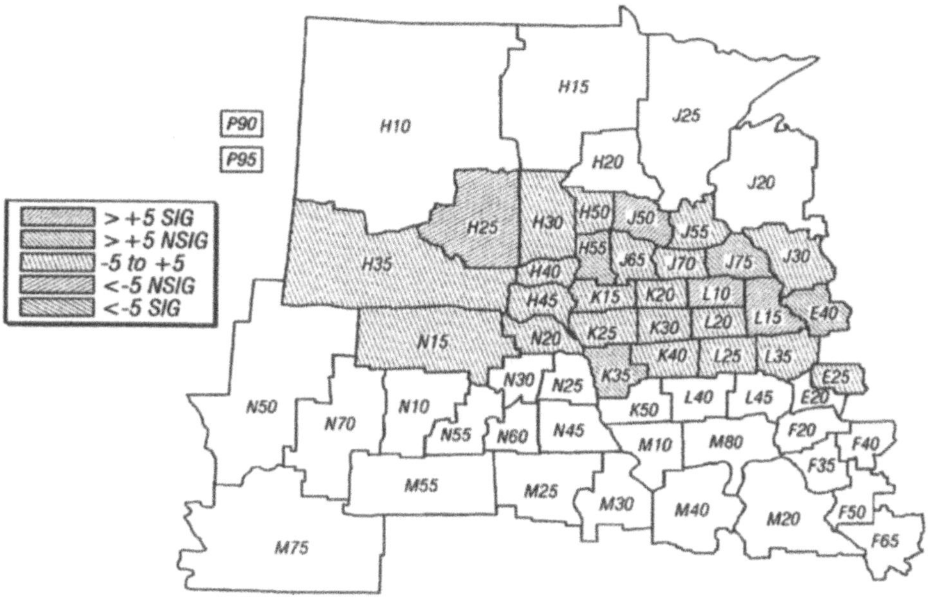

Figure 1.17 Chloropleth map of expected yield differences between two hybrids.

Agency (EPA) or the Food and Drug Administration (FDA) in the United States. A company did not need permission from the USDA or the Ministry of Agriculture in another country to plant experimental varieties for observational purposes, and seed for testing was easily shipped among many countries around the world with only standard phytosanitary quarantine restrictions.

With the advent of transgenic crops, breeders suddenly needed permission from appropriate regulatory agencies to conduct field tests of certain cultivars. Another key difference was that grain from nonregulated tests continued to be sold into commerce, while at the same time grain from regulated tests had to be destroyed. Regulatory oversight applied throughout the entire breeding process with a new transgenic trait (Figure 1.18). For example, over 1,700 proximate analysis tests were completed to determine that Roundup Ready® soybeans were substantially equivalent to conventional varieties for all nutritional and compositional traits (Table 1.3). Before this new era, breeders had taken for granted the natural variation for protein, oil, and starch and thousands of varieties were planted on millions of acres each year without regulatory oversight on these traits or their individual components. Despite the stringent standards for substantial equivalence, certain countries no longer regarded these new varieties as commodity grains, and import approvals were necessary for grain carrying specific genetic traits such as Roundup Ready® soybeans and Yieldgard® corn. To date, regulatory approvals have been granted in 19 countries for various types of uses ranging from food and feed to production. Whereas it had been and continues to be a consumer's responsibility to know if they are allergic to a specific commodity food product, transgenic varieties now had to be tested in a myriad of expensive ways for known and potential allergenic characteristics.

Breeders and breeding, however, crossed an agricultural Rubicon in 1996 with the commercial introduction of Roundup Ready® soybeans, and farmers were not going to let them turn back. The acreages planted to biotech crops increased dramatically each year due to market demand (Figure 1.19), despite complicated and expensive regulations and import restrictions on some of the most popular products. Companies such as Monsanto employed hundreds of scientists and other professionals to manage the regulatory requirements at a cost approximating the annual spending on conventional breeding for these crops. The global message, however, was that crops improved through the use of biotechnology were here to stay and breeding companies simply had to reinvent their business and breeding organizations to enable the new commercial paradigm.

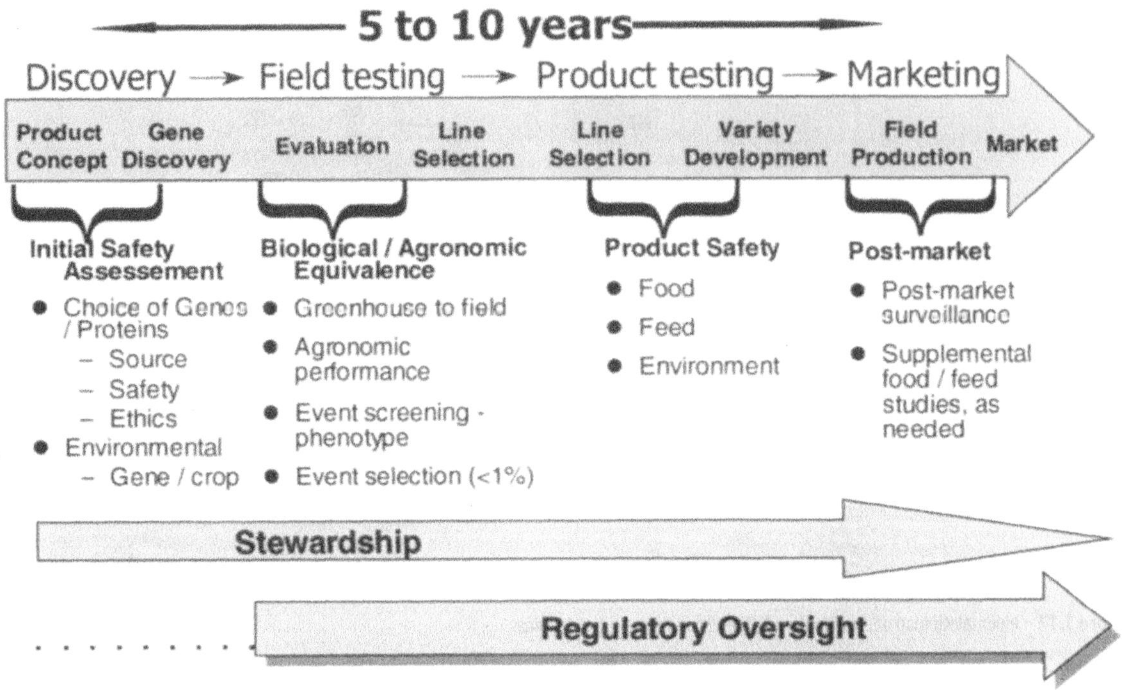

Figure 1.18 Impact and timing of regulatory oversight on the plant-breeding process with a new transgenic trait.

With the commercial sale of transgenic crops, biotechnology moved from an interesting complex of sciences to a classic disruptive technology as described by Christensen (2000). For many years, chemical treatments were used to control insect pests in cotton, and several companies enjoyed a lucrative insecticide business due to the large number of chemical treatments required to control cotton insects. Numerous researchers in multiple organizations worked for many years on expression of genes from *Bacillus thuringensis* (*Bt*) in plants without much success until Perlak et al. (1991) showed that synthetic gene technology could provide expression levels high enough to provide excellent insect control. Even then, the chemically oriented management teams of many companies regarded the new technology as too expensive, less efficacious, less convenient, and generally more work. By 2000, however, Bollgard® cotton, which expresses the Cry1Ac protein, was grown on more than one-

Table 1.3 Over 1700 independent analyses were performed to demonstrate that Roundup Ready® soybeans are compositionally equivalent to commercial soybeans

				Protein		
Component	Beans	T Meal	Defat Flour	Isolate	Conc	Refined Oil
Proximate analysis	SE	SE	SE	SE	SE	
Amino acid comp		SE				
Fatty acid comp	SE					SE
Trypsin inhibitors	SE	SE	SE			
Lectins	SE	SE				
Phytoestrogens	SE	SE				
Urease	SE	SE	SE			
Stachyose, raffinose		SE				
Phytate		SE				
Nitrogen solubility		SE				

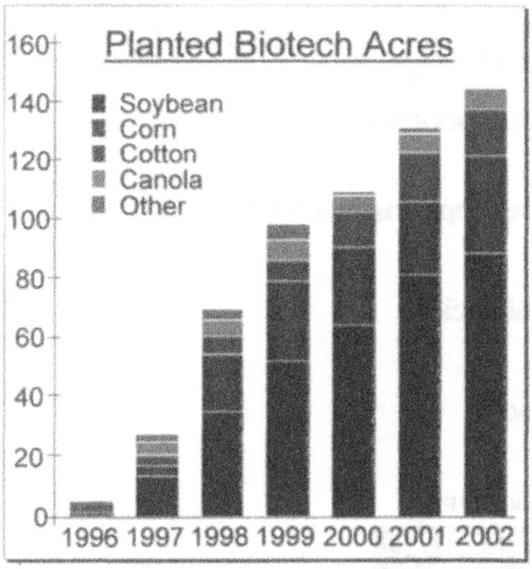

■ Crop acre share

Biotech Crops	ISAAA 2002 WW	USDA 2002 U.S.	USDA 2003 (P) U.S.
Soybean	51%	75%	81%
Cotton	20%	71%	73%
Corn	9%	34%	40%

Figure 1.19 Worldwide adoption of transgenic cultivars for major crops. Source: ISAAA, USDA Actual 2002 and Projected 2003, Historical Center for Food and Agriculture Policy.

third of the cotton acreage in the United States and had displaced a significant portion of the cotton insecticide market (Perlak et al., 2001). Similarly, the Roundup Ready® system had also displaced over 75% of the traditional weed-control systems for soybeans within six years of its introduction (Walker, 2002).

To further illustrate the impact of transgenes on breeding and agriculture, we will (1) review some aspects of the invention and development of the Roundup Ready® soybean system, (2) compare conventional and transgenic breeding approaches to achieve resistance to the corn rootworm (CRW) complex (*Coleopteran*, *Diabrotica* spp.), and (3) discuss an integrated breeding model based on the commercial development of Yieldgard® Rootworm.

Roundup Ready® soybeans

Roundup Ready® soybeans represented the first time that an understanding of a biochemical pathway was truly essential to the breeding and commercialization of major new crop cultivars. Roundup Ready® soybeans also paved the regulatory process road for all future transgenic crop cultivars.

In 1970, Monsanto researcher John Franz discovered that phosphonomethylglycine was a nonselective herbicide. The compound was known in the literature and had been patented for other uses. At the time, he was looking in his research studies at a variety of aminomethyl phosphonic acids that exhibited minimal herbicidal activity, with little success, and it was a surprise eventually to find a compound such as glyphosate. Monsanto discovered herbicidal activity in this class of chemistry because similar compounds degraded to glyphosate *in planta* and exhibited bioactivity in new bioassays performed for seven days versus the industry standard three-day assay. Researchers at the time were not quite sure what to do with a compound, renamed glyphosate, that killed monocots and dicots alike but Roundup® herbicide was launched in 1974, and for many years was mostly viewed as an expensive specialty agrochemical. In the mid-1980s, Monsanto sales and marketing personnel began to promote Roundup® in new ways, such as the burn-down market in western fallow areas. Within 10 years, Roundup® herbicides became the largest single agricultural product in the United States, with annual sales exceeding $2.5 billion on a global basis.

It was only logical that researchers would try to find a source of resistance to Roundup® herbicide so that the herbicide could be used to control weeds in crops. It had all of the characteristics of a desirable weed-control solution but no crop plant was resistant. Attempts to find natural sources of plant resistance failed, and researchers turned their attention to sources outside the plant kingdom and focused on a transgenic solution.

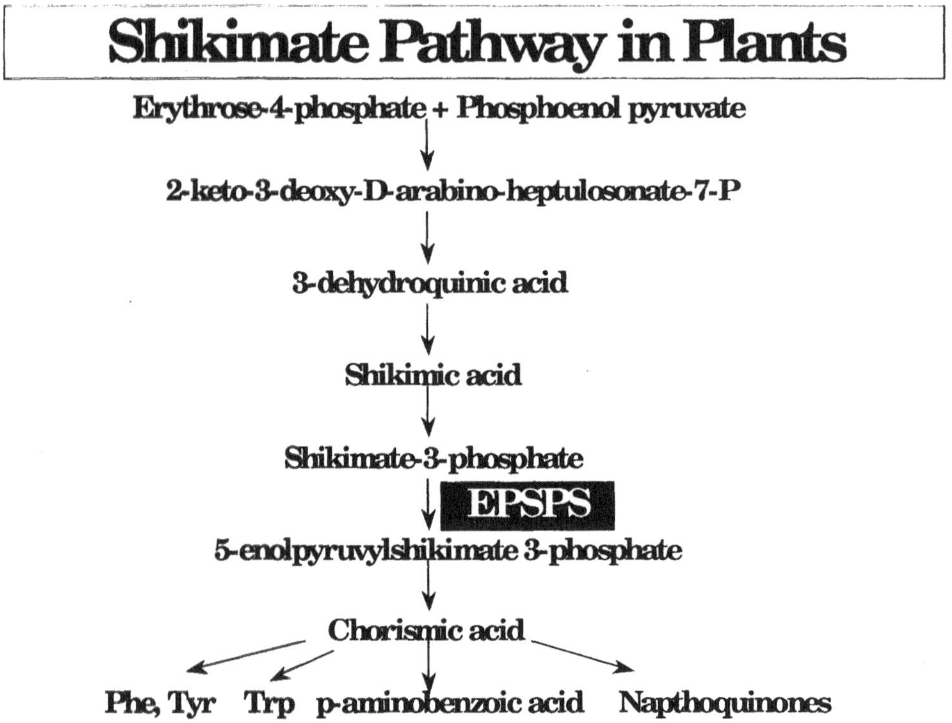

Figure 1.20 Schematic of the shikimate synthesis pathway in plants.

Glyphosate, the active ingredient in Roundup® agricultural herbicide, kills plants by inhibiting the activity of the enolpyruvylshikimate-3-phosphate synthase (EPSPS) enzyme (Figure 1.20). Glyphosate is proposed to be a transition-state inhibitor of EPSPS (Schönbrunn et al., 2001) and is the only known practical inhibitor of the enzyme. This enzyme is part of the shikimic acid pathway present in plants, bacteria, and fungi but not in animal life forms, and so glyphosate was predicted to be innocuous to animals, which has proven to be the case. The shikimate pathway in plants is critical because it is responsible for a wide array of essential aromatic compounds including amino acids, hormones, coenzymes, various quinones and lignin, and a host of secondary metabolites (Haslam, 1993).

Researchers at Monsanto investigated two approaches to achieving resistance to glyphosate (Figure 1.21), and bacterial cultures from diverse sources were screened for tolerance to glyphosate. A specific strain of *Agrobacterium* sp. (CP4) was found to contain an EPSPS enzyme with a different amino acid sequence than the plant EPSPS enzymes (*CP4-EPSPS*), where glyphosate was a very poor inhibitor for this new class of EPSPS enzymes (Padgette et al., 1996).

Horsch et al. (1984) produced the first genetically transformed plant, and by 1986 Monsanto researchers had produced glyphosate-tolerant petunia, tobacco, and tomato plants (Figure 1.22). Micro-particle bombardment transformation was performed in 1990 on Asgrow soybean cultivar A5403 with a vector containing the *CP4* gene. A transformation event known as 40-3, along with several others, was first selected in the greenhouse during the winter of 1990–1991 at the Monsanto Research Center in Chesterfield, Missouri. The line, which became known as 40-3-2, was derived from a specific R1 selection from event 40-3 that was homozygous for the *CP4-EPSPS* gene (Padgette et al., 1995). The insert containing *CP4* behaves as a single dominant gene and later was localized on linkage group D1b (U19) of the USDA genetic map of the soybean (Cregan et al., 1999a).

During the summer of 1991, soybean line 40-3-2 was planted in the field where it was confirmed to be homozygous for the gene and tolerant to Roundup® agricultural herbicides. Roundup Ready® soybeans are unaffected by Roundup® for-

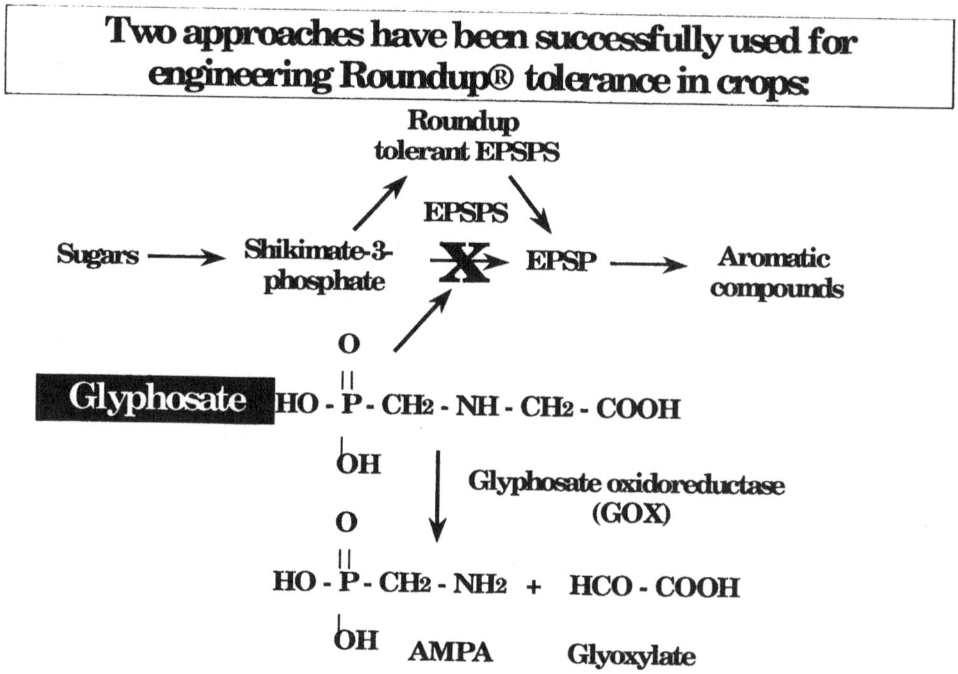

Figure 1.21 Two approaches to achieving Roundup® tolerance in plants.

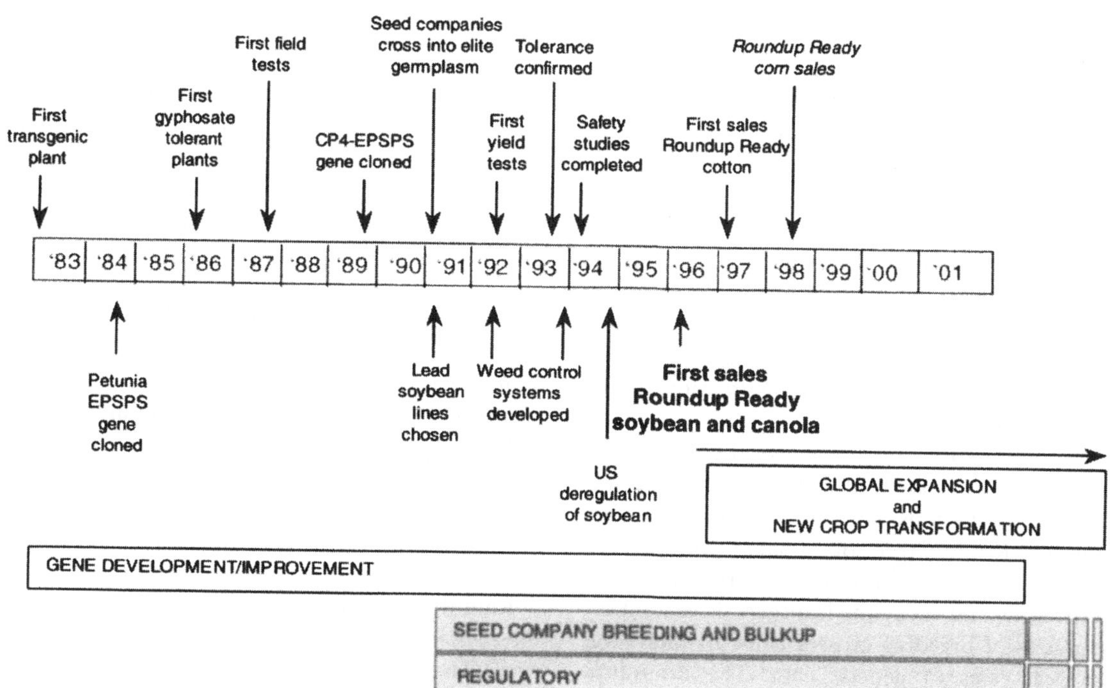

Figure 1.22 Development timeline for Roundup Ready® soybeans.

mulations at labeled rates because the CP4-EPSPS enzyme allows for continued flux through the shikimate pathway in the presence of glyphosate, which is the active ingredient. A soybean plant with the added *CP4-EPSPS* gene synthesizes two different EPSPS enzymes. The native plant EPSPS enzyme is inhibited by glyphosate, but the second bacterial CP4-EPSPS enzyme is not inhibited at labeled rates of glyphosate (Figure 1.21).

An extensive breeding and backcrossing program was initiated in 1991 between Monsanto and Asgrow researchers. Other soybean-breeding companies were also included in this effort in order to ensure that the trait was broadly available to farmers. Crosses between susceptible and tolerant genotypes were made on a large scale. Line 40-3-2 was backcrossed three times or forward crossed to a wide range of genetic backgrounds over all maturity groups to ensure that the Roundup Ready® trait would be available in a diverse set of genetic backgrounds. Sneller (2003) studied the impact of Roundup Ready® on the genetic diversity of soybeans and used coefficient of parentage to confirm that introduction of the Roundup Ready® trait had no negative effect on the overall diversity of soybean cultivars (Sneller, 2003).

At least six breeding companies initially sold Roundup Ready® soybeans in 1996, with the majority of companies introducing Roundup Ready® soybeans in 1997. The agronomic performance of this trait was thoroughly evaluated prior to the initial release. For example, data from 58 environments showed no yield penalty with rates of Roundup® herbicides twice the level needed to control weeds (Delanney et al., 1995). Roundup Ready® soybeans were evaluated under USDA-authorized field trials in 14 states in 1995, and following the completion of U.S. regulatory reviews and approvals in key export markets, the first commercial sales were made in 1996 to more than 10,000 farmers.

By 2000, most breeders used the Roundup Ready® gene as a base trait in a high percentage of breeding populations. Forward breeding with the Roundup Ready® trait on a large commercial breeding scale was relatively straightforward and inexpensive, and today the transgene is present in thousands of breeding crosses while maintaining historical rates of genetic gain. The Roundup Ready® system also enables breeders to evaluate traits such as yield more effectively and efficiently because this system minimizes potentially confounding variables such as weed competition and crop injury from other herbicides. Currently, most soybean breeders treat all generations with formulations of Roundup® herbicide, and a commercially released variety typically would have been screened for approximately 10 successive generations of breeding and seed increase for herbicide tolerance.

Since 2000, companies have been replacing varieties released in 1996–1999 with higher-yielding varieties with traits such as improved tolerance to soybean cyst nematode (*Heterodera gylcines* Ichinohe) (SCN) and disease resistance. Currently, there are over 250 seed companies in the United States and Canada licensed to sell Roundup Ready® soybeans. Today there are over 1,000 Roundup Ready® soybean varieties commercially available in the United States and Canada. Walker (2002) estimated that since 1996 approximately 800 Roundup Ready® varieties have been used commercially and replaced with new Roundup Ready® varieties.

Since its introduction in the United States in 1996, the Roundup Ready® trait and herbicide system has been adopted widely due to the simplicity and effectiveness that it offers in managing weeds. In the 2003 growing season, 81% of the soybeans—approximately 60 million acres of the 73.7 million acres of the soybeans grown in the United States—were Roundup Ready® soybeans (USDA, 2003). In Argentina, where the adoption rate is estimated to be nearly 99%, Roundup Ready® soybeans were grown on nearly 30 million acres in 2002 (James, 2002). In addition to the United States and Argentina, regulatory approvals for the commercial production of Roundup Ready® soybeans were obtained in five additional countries in 2002—Canada, Mexico, Romania, Uruguay, and South Africa. Globally, Roundup Ready® soybean occupied 90.1 million acres (36.5 million hectares) in 2002, representing 62% of the global transgenic crop area of 58.7 million hectares for all crops (James, 2002). Since 1996, Roundup Ready® soybeans have been produced on over 335 millions acres (135 million hectares) globally.

Roundup Ready® soybeans may be one of the most intensively studied and reviewed food crops ever placed on the market. An extensive set of compositional and nutritional analyses were conducted and established that Roundup Ready® soybeans are comparable to other soybean cultivars (Padgette et al., 1994; Padgette et al., 1996; Taylor et al., 1999; Burks and Fuchs, 1995; Hammond et al., 1996; Nair et al., 2002). Environmental effects of Roundup Ready® soybeans and the Roundup Ready® soybean system have also been extensively studied (reviewed in Carpenter, 2001). These studies have found that Roundup Ready® soybeans pose no greater environmental impact than conventional soybean varieties, and, as described below, certain beneficial environment effects have been observed. Over the past decade the data on the characteristics and safety of Roundup Ready® soybeans have been reviewed by over 35 regulatory agencies in over 20 countries. These reviews have generally concluded that Roundup Ready® soybeans are the same ("substantially equivalent") as other soybeans in nutrition, composition, safety, and how they function in food and feed products.

The large-scale adoption of Roundup Ready® soybeans has had a beneficial effect on the environment for the following main reasons.

- Glyphosate is ranked by EPA in the lowest toxicity category for pesticides due to its favorable characteristics, which include:
  - glyphosate is specific to plants, fungi, and bacteria, and it does not have any negative effect on animals or humans, which do not depend on the EPSPS enzyme in their metabolism;
  - glyphosate binds quickly to soil particles after application and thus does not leach to the ground water; and
  - glyphosate gets degraded rapidly by soil bacteria to basic and harmless compounds.
- As a result of these characteristics, the use of Roundup® herbicide replaces other herbicides with higher toxicity, longer residual activity, and the potential to contaminate ground water. A recent study (Nelson and Bullock, 2003) modeled environmental effects based on the relative mammalian toxicity of various herbicides and their commercial application volumes and concluded that the Roundup Ready® soybean system was more environmentally benign than conventional weed control systems.

In addition to the benefits provided by Roundup® herbicide per se, the Roundup Ready® soybean system provides the following additional environmental and economic benefits:

- a greater than 20% reduction in the levels of foreign matter in harvested Roundup Ready® soybeans relative to soybeans grown with conventional weed control (Shaw and Bray, 2003). Reduced foreign matter means improved farm economics and benefits to the agricultural grain trade and processing industry.
- improved efficacy in weed control compared with herbicide programs used in conventional soybeans, as specific pre-emergent herbicides that are used for prevention are replaced by a broad-spectrum postemergent herbicide that can be used on an as-needed basis (Nelson and Renner, 1999; Roberts et al., 1999). The introduction of Roundup Ready® soybeans in the United States has eliminated 19 million herbicide applications per year—a decrease of 12%, even though the total soybean acres increased by 18% from 1996–1999 (Carpenter, 2001). This decrease in herbicide applications means that growers make fewer trips over their fields to apply herbicides, which translates into ease of management and reduced fossil fuel use.
- a reduction in weed-control costs for the farmer. It's been estimated that U.S. soybean growers saved $216 million in 1999 compared with 1995, the year before Roundup Ready® soybeans were introduced, including the technology fee (Carpenter, 2001).
- high compatibility with integrated pest management (IPM) and soil conservation techniques such as no-till cropping systems (Stark et al., 2001; Fawcett and Towery, 2002), resulting in a number of important environmental benefits, including reduced soil erosion and improved water quality (Baker and Laflen, 1979; Hebblethwaite, 1995; CTIC, 1998); an improvement in the ability of fields to serve as wildlife habitat (Fawcett and Towery, 2002); improved soil structure with higher organic matter (Kay, 1995; CTIC, 2000); improved carbon sequestration (Reicosky, 1995; Reicosky and Lindstrom, 1995); and reduced CO2 emissions (Kern and Johnson, 1993; CTIC, 2000). Implementation of this system has corresponded with a decrease in the number of perennial weeds in soybean cropping systems (Sprague 2002).

The development of Roundup Ready® soybeans represented a major breakthrough in plant breeding and in agricultural systems in the last part of the twentieth century. The new combination of sciences required to invent and commercialize Roundup Ready® soybean cultivars broadened and redefined plant breeding in very significant ways. Biochemistry, organic and physical chemistry, cell biology, and molecular genetics and biology, for example, joined forces with the traditional plant-breeding sciences as requisite tools for new millennium plant breeders. The development of Roundup Ready® soybeans also enabled a step-change in weed control in soybean production and in environmental stewardship, which has had a dramatic effect on the agricultural input supply business. The retail distribution channel, the choice of tillage implements, and the labor hours per acre, for example, have been influenced by the development and adoption of Roundup Ready® soybean cultivars. Taken in total, the impact of Roundup Ready® soybeans rivals the impact of hybrid corn.

Insect-protected corn

Environmental and economic impact of the corn rootworm complex

Corn is the largest U.S. crop in terms of acreage planted and net crop value. In 2002, for example, the U.S. corn crop covered 79 million acres and had a net value of $21 billion (National Corn Growers Association, 2003). CRW is one of the most destructive insect pests in the U.S. Corn Belt and is comprised primarily of the western corn rootworm (WCR), *D. virgifera virgifera* LeConte, and the northern corn rootworm (NCR), *D. barberi* Smith and Lawrence. CRW adults oviposit on the soil surface of cornfields, and the eggs overwinter in a state of diapause. Larval eclosion occurs in the spring following oviposition. Larvae damage corn by feeding on the roots, which reduces the ability of the plant to take up water and nutrients from the soil (Riedell, 1990) and causes harvest difficulties due to root lodging (Spike and Tollefson, 1991).

From the standpoint of insecticide use, the CRW is the most significant insect pest problem of corn in the U.S. Midwest (Office of Pest Management Policy, 1999). Metcalf (1986) described CRW as a billion dollar pest complex based on costs associated with the purchase of soil insecticides and crop losses due to CRW damage. Historically, farmers have primarily used soil insecticides or crop rotation to mitigate CRW damage. The most common insecticide application regime for controlling CRW is at the time of planting, and the most widely used insecticides have been organophosphates such as chlorpyrifos, terbufos and tebupirimphos, and the synthetic pyrethroids, such as tefluthrin and cyfluthrin (Hartzler, 1997; Office of Pest Management Policy, 1999). The National Agricultural Statistics Service and the Economic Research Service of the USDA have compiled statistics on corn insecticide use across 18 states comprising 73.8 million acres of corn during the 2000 crop season (National Agricultural Statistics Service, 2001). These figures indicate that chemical insecticides registered for CRW control were applied on over 31% of this corn acreage in 2000. CRW control accounted for the largest insecticide usage in any one crop, totaling approximately 9.8 million pounds of active ingredient.

Historically, crop rotation has provided highly effective protection from rootworm damage; however, two primary factors increasingly limit the usefulness of this management strategy. First, researchers have confirmed that populations of both NCR and WCR can exhibit extended diapause (Krysan et al., 1984; Levine et al., 2002), in which a portion of the eggs are able to survive through the non-corn years of crop rotation to yield larvae that feed on the roots of first-year corn. Second, and of critical importance, crop rotation is no longer effective in east central Illinois and northern Indiana due to the rapid spread of a new race of WCR that, unlike previous populations, preferentially oviposits in existing stands of soybeans (Levine et al., 2002). Eggs that are oviposited in soybeans in one crop year hatch the following crop year to damage the rotational corn crop. Based on the rapid expansion of this CRW variant population since its initial discovery in 1993, it is expected to continue to spread throughout the Corn Belt. Both of these factors have increased growers' reliance on chemical insecticides for CRW control in cases in which crop rotation previously provided effective control.

Breeding for corn rootworm resistance

For many years, the widespread use of insecticides and crop rotations meant that little selection pressure for CRW resistance likely was applied in standard breeding nurseries and yield trials. The sophisticated biology of the CRW complex and the challenge of evaluating root damage on large numbers of progeny also meant that selection was tedious, and genetic progress for CRW resistance was limited to specialized programs.

Conventional breeding approaches for resistance to CRW resulted in germplasm with only moderate levels of resistance to rootworm feeding (Knutson et al., 1999). The predominant mechanism for resistance is tolerance rather than antibiosis, with the more resistant genotypes having larger root systems with improved ability to regenerate from CRW damage (Branson et al., 1982).

Rogers et al. (1975) reported only low levels of tolerance among 25 commercial hybrids. In a two-year study of 11 commercial hybrids representing three decades of breeding (the 60s, 70s, and 80s), Riedell and Evenson (1993) reported that a larger root system and a decrease in root lodging indicated better rootworm tolerance in the 1980-era hybrids. However, even with this level of tolerance, substantial yield loss occurred in the presence of moderate to heavy rootworm incidence when corn was grown in low to moderate plant densities. Gray and Steffey (1998) evaluated 12 popular hybrids over a four-year period. Their results supported the conclusion that a larger root system as measured in July and August allowed hybrids to tolerate rootworm feeding better. They reported that compensatory root regrowth was positively correlated with yield when moisture was inadequate. When moisture was adequate, this compensation was potentially at the expense of yield.

Conventional breeding developed products with slightly better tolerance to CRW damage, but 75 years of selection for improved roots did not develop a level of tolerance sufficient to reduce the need for insecticides or crop rotations. The development of methods for rearing CRWs provided a means for testing corn crops exposed to various levels of artificial CRW infestation, which allowed the development of a root damage-rating system (Sutter and Branson, 1980). Sufficient marker technology to search global germplasm sources for quantitative trait loci (QTL) conferring host plant resistance is also available, but there is little evidence to date that the corn genome possesses genes conferring useful levels of resistance to the CRW. If useful QTLs for CRW could be found, however, they would offer breeders additional opportunities to develop conventional resistance that could be used in conjunction with transgenic control in refuge acres or in geographies that do not permit the growth of transgenic products (Bohn, 2003).

Given the limited success in conventional breeding programs, a transgenic approach to CRW control was pursued to find a single dominant gene for antibiosis. In the mid- to late 1980s, many groups began the search for genes from *Bacillus thuringiensis* (Bt) that would control CRW. The experimental path to find and confirm the desired expression *in planta* proved to be a daunting research challenge for anyone who picked up the genetic gauntlet. Monsanto and its collaborators built on the findings from numerous research projects over more than two decades in an effort to develop CRW control in maize. Corporate stamina, support, and vision proved to be essential in solving a very important problem for corn growers. Increasing public concerns relative to insecticide usage and the emergence of insect variants were key factors of encouragement for researchers throughout the research process.

Conventional breeding failed to develop CRW control comparable to soil-applied insecticides, but a molecular biology approach achieved CRW control in elite corn germplasm. As will be discussed later, a transgene also allows breeders to continue making progress on the myriad of other traits under selection and can be added to any product as desired, based on adaptation and market demand. A variety of Cry 3Bb1 amino acid sequence variants were identified by English et al. (2000) that exhibited improved CRW control when compared with the native Cry1Bb1 protein. One variant has been expressed in plants, and one event (MON863) has recently been approved for commercialization. This event delivered excellent control of CRW when expressed in a broad panel of elite inbred lines and hybrids. The MON863 Cry3Bb1 protein variant is highly specific for control of CRW, and no adverse affects have been observed after testing the variant protein in a wide array of nontarget organisms. The Cry3Bb amino acid sequence variant protein produced in event MON863 has also been shown to break down rapidly in soil (time to 90% dissipation = 2.48 days), and the event has received food and feed approval from regulatory agencies in the United States, Canada, and Japan.

A generalized transgenic breeding timeline

Commercializing a new transgenic event requires a multidisciplinary teamwork of plant breeders, molecular biologists, biochemists, plant physiologists,

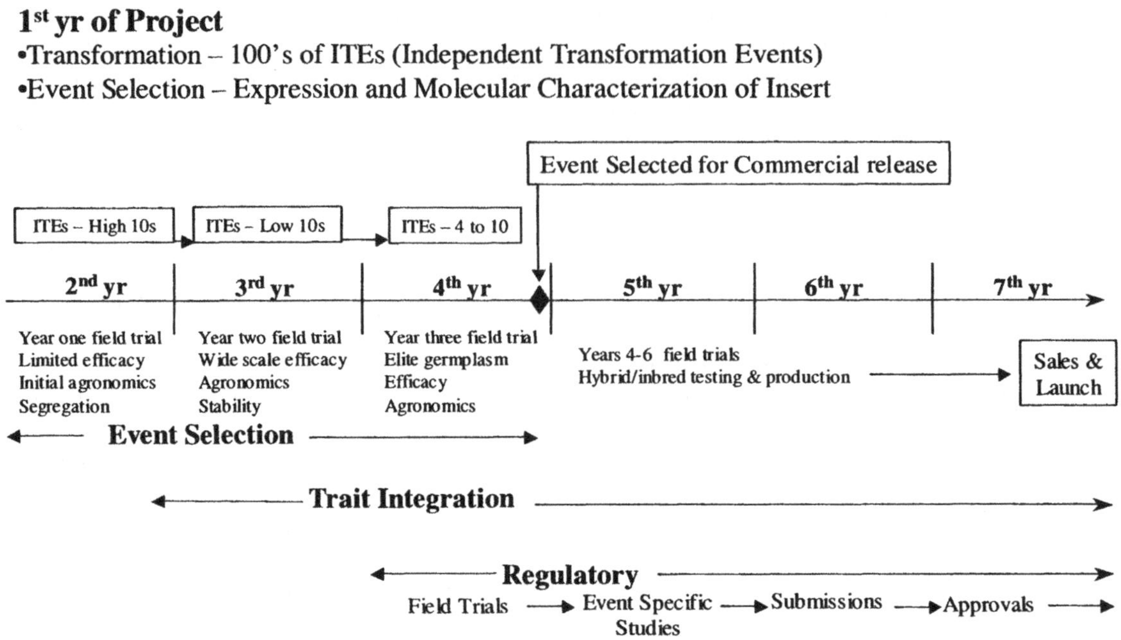

Figure 1.23 A generalized breeding timeline for developing and commercializing a typical transgenic insect-protected trait in corn.

agronomists and many other scientists from a range of disciplines working on several concurrent activities, including event selection, trait integration, and regulatory studies. These efforts ultimately lead to the identification of a single independent transformation event (ITE) with the required functionality and regulatory approvals for commercial sales and release. The following is a brief description of this process illustrating the activities and timelines for developing a typical transgenic corn trait from gene discovery to commercialization (Figure 1.23).

The key disciplines involved in developing a transgenic product include molecular biology for designing and constructing plant expression vectors used in transformation, transformation capabilities for producing ITEs, breeding for trait evaluation and for converting elite germplasm to contain the trait, and concurrent regulatory science based studies designed to meet data requirements for regulatory oversight and approval.

The product development process starts after the proof of concept phase, where the transgene has been demonstrated to express at levels expected to be efficacious and, when expressed, does not result in obvious deleterious effects to the plant. The transgene is then advanced to commercial scale transformation where hundreds or thousands of independent transformation events will be produced.

Typically, the time required from transformation through regeneration of the required ITEs (R0 plants, or the first generation of plants developed from tissue culture after transformation) is approximately 4–6 months, depending on the efficiency of the transformation system. The objective of this phase is to produce a sufficient number of ITEs to have a reasonable probability of finding a few events that meet all technical and regulatory criteria. Tissue from each event of the R0 plants is analyzed for key criteria, including expression of the protein of interest and molecular characterization of the transgene insert(s). While these analyses are being conducted, the R0 plants typically are outcrossed to a tester and backcrossed to the recurrent parent to evaluate both trait efficacy and conformation to expected segregation ratios. The time frame for transformation through analysis of the R0 progeny is about 12 months, during which the numbers of ITEs are reduced significantly based on advancement criteria.

Often as many as 100 events are advanced to the field for further evaluation of trait efficacy, gross inbred and hybrid agronomics and yield equivalency. The first year field trials can be characterized by having a large number of ITEs in limited germplasm backgrounds across a few environments. Events that meet the trial criteria and also meet stringent molecular criteria are advanced into a trait integration program for incorporation into a wider range of elite lines. Typically, 10-30 ITEs are selected from the first year field trials to complete the second year of the development process.

The objective of the second-year field season is similar to the first year, but with fewer events in a limited but expanded number of genotypes and across a wider array of environments. Typically, the second year of field selection reduces the number of ITEs to a manageable number of events that can be incorporated into a broad array of elite lines. Often, the regulatory field trials are initiated after the second year of field trials, that is, by the end of the third year of the development process.

In the third year of field-testing and selection, the events are evaluated for efficacy and agronomic equivalency in a much wider range of elite germplasm across a wide range of corn-growing environments. The objective of these wide-area trials is to identify a single event for commercial release. Additionally, the candidate events are also placed in regulatory field trials where a number of plant tissue types are collected for each event. The tissues collected from these trials are subjected to a wide range of tests to satisfy specific global regulatory data requirements necessary for regulatory approval. These include spatial and temporal expression of the protein of interest in the plant, molecular characterization of the DNA insert, biochemical composition, nontarget organism toxicity, and environmental fate.

In the case of YieldGard® corn rootworm, additional testing was conducted with purified Cry3Bb1 protein produced through bacterial fermentation. Once the bacterial-produced protein was determined to be biologically and physicochemically equivalent to the plant-produced protein, it was also subjected to a battery of regulatory tests to establish its safety in food and feed and toward nontarget organisms. Lastly, Monsanto conducted a set of large-animal feeding studies to establish the nutritional equivalency of corn plants transformed with the selected event compared with their conventional counterpart for market acceptance purposes.

At the end of the third-year field season, or the fourth year in the development process, the event for commercial release had been identified. Once the event had been identified, the regulatory studies mentioned above were initiated with the tissues collected from the regulatory field trials held in the third-year field season.

The process to develop a typical transgenic corn trait takes four years from transformation to selection of the event for commercial release. However, an additional two to three years is required to obtain the necessary regulatory approvals to commercialize corn hybrids with this trait.

Unlike traits such as the Roundup Ready® trait in soybeans, transgenic corn traits are not currently handled as a base trait in most breeding programs. As discussed earlier, the Roundup Ready® gene has been incorporated into a high percentage of soybean-breeding populations, and backcrossing programs have largely been eliminated. With most corn traits, however, it is not presently obvious which traits might be used in forward breeding, and breeders have largely opted to backcross transgenic traits into finished or nearly finished elite lines. Ironically, progress in plant breeding has been advanced significantly by this old-fashioned methodology that was previously dismissed by many plant breeders as a viable method to increase genetic gain. As is often the case, one should not assess genes or methodologies in isolation. Many organizations have developed trait integration (TI) functions to incorporate the desired combinations of events into the final commercial product rather than incorporating this work into forward-breeding programs. There are many advantages to this broader breeding strategy. From a germplasm perspective, inbred development breeders can stay focused on germplasm development, and elite lines are readily transferable to other world areas irrespective of transgenic approval. From a transgenic perspective, this strategy also allows more flexibility in stacking combinations of transgenic events to meet market needs. TI programs typically encompass teams for inbred conversion, molecular marker genotyping, hybrid performance and efficacy testing, quality assurance/quality control (QA/QC), information management, and regulatory compliance.

Inbred conversion is typically conducted in a continuous nursery environment, and the objective is to return as close to the recurrent parent as possible in the shortest period of time. Rapid nursery cycling in combination with marker-assisted backcrossing is commonly used to shorten timelines. The delay in time to market versus the conventional counterpart is a function of resources, but a one-year lag is not uncommon.

The objective of hybrid equivalency testing is to verify agronomic performance versus the recurrent parent and to confirm transgenic trait efficacy. Ideally, events that are commercialized should perform in all germplasm and event combinations. This step is recommended particularly as new events are brought to market.

The QA/QC focus of a TI program is more than simply having positive or negative assays. In a broader sense, it is the process of managing transgenic events in a breeding program. QA consists of the overall processes or best practices employed to ensure the overall quality of the final product from a transgenic perspective. QC consists of the actual testing conducted to verify that the desired quality is actually achieved. The end goal is 100% purity of the desired transgenic event(s) and the absence of any other events. Hall et al. (2001) detailed industry QA/QC practices used in transgenic breeding programs.

Information management is critical in any breeding program, however, there are special considerations in working with transgenic germplasm. A robust information management system is needed to track breeding material. This is largely because of the rapid cycling of continuous nurseries, but also due to the need to predict accurately when converted inbreds will be available. Other information essentials include consistent nomenclatures and electronic safeguards in trial-management software to assist in regulatory compliance. It is critical that transgenic breeders have secured the appropriate movement and environmental release notifications when working with regulated material. Isolation and gene-containment measures must also meet regulatory guidelines.

Marker-based genotypic selection

For as long as breeders have been estimating genotypic effects from replicated progeny tests, they have also desired to practice direct genotypic selection for favorable alleles found in normal breeding populations. Marker-aided breeding can be categorized as follows: (1) marker-aided selection (MAS), where the marker is linked to a gene of interest (GOI); (2) marker-aided backcrossing (MABC) to recover the recurrent parent with a GOI; (3) marker-aided recurrent selection (MARS) for QTLs, where a panel of polymorphic markers are linked to QTLs; and (4) selection involving marker(s) in the gene (MGOI) for simple or complex traits. The totality of these uses means that a large, robust marker platform is needed to handle millions of data points in a very timely fashion, particularly when selecting for quantitatively inherited traits such as yield, which have low to moderate heritability and significant G^*E variances in large breeding programs. Table 1.4 shows the typical ranges for key parameters in large, global corn-breeding programs. Hundreds of thousands of genotypes are grown in millions of plots each year. Consequently, there has been considerable interest in developing marker platforms for plants to handle economically tens of millions of data points annually.

Table 1.4 Ranges for key parameters of a global corn breeding program

Pipeline Step	Number of Items
Breeding populations	2,000–5,000
Segregating lines	200,000–500,000
Finished lines	10,000–30,000
Test crosses	200,000–500,000
Finished hybrids	20,000–50,000
Nursery rows	1,000,000–2,000,000
Yield trial plots	2,000,000–4,000,000

Genotyping Platforms

Amazing progress has been made on marker platforms that began with classical genetic maps. The evolution of genotyping methods parallels those in the computer and electronics industry. Both have been characterized by periods of steady improvement punctuated by technological step changes. Plant genotyping platforms (Figure 1.24) have followed the disruptive technology theory of Christensen (2000). In each case the newer, more expensive, less effective technology was refined and improved and eventually surpassed the established technology in ease of use, cost effectiveness, and functionality only to be disrupted by another newer technology.

The earliest examples of using genetic markers were those based upon phenotype (Sax, 1923; Emerson et al., 1935). These genetic markers or mutations produced phenotypes that were simple to score such as seed coat and flower color, leaf striping, trichome formation, plant stature, and a

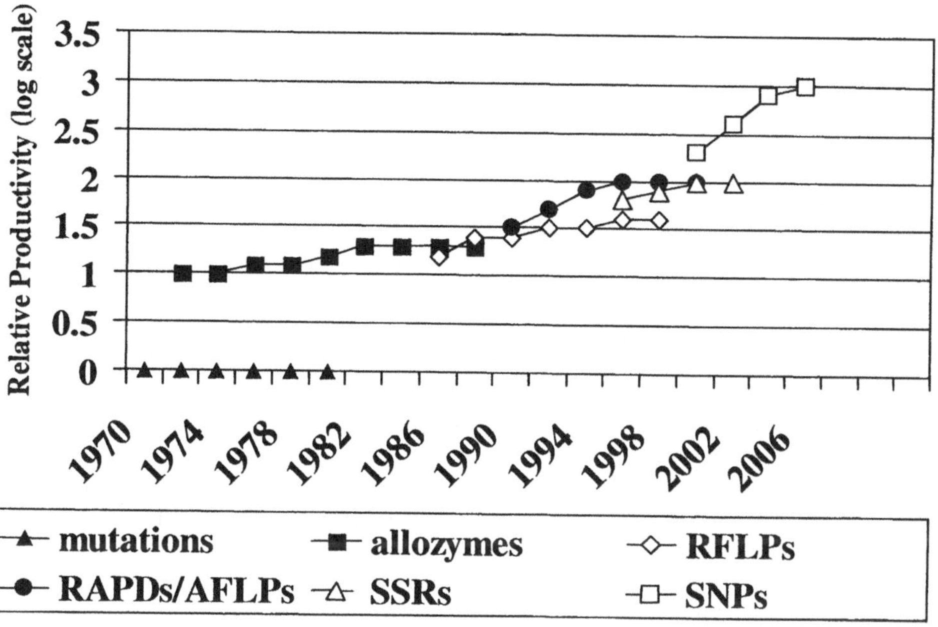

Figure 1.24 Development periods and relative productivity of six methods of genotyping in plants.

host of other attributes. Many of these attributes are not neutral phenotypes, but instead have negative impact on plant performance. Genome-wide genetic studies were thus severely handicapped because plants carrying more than a few markers could barely survive and some genetic abnormalities were nearly lethal in a homozygous state. Despite these handicaps, large numbers of mutations have been described and used as genetic markers in a number of genetic studies (Coe et al. 1988). Using these phenotypic markers for selection was based more upon serendipity than experimental planning, primarily because the markers first had to be introgressed into the germplasm under selection if they were not already present in the germplasm. More severe variants of single gene mutations are chromosomal aberrations (Burnham, 1966). These aneuploid and translocation stocks have proven invaluable at helping to assign individual markers to chromosome arms (Patterson, 1982). Without the classical genetic maps and associated knowledge, however, it would have been more difficult to map and deploy the marker systems that followed.

Following the use of phenotypic markers, the first breakthrough technology was the discovery and use of allozymes (Stuber et al., 1988; Tanksley and Rick, 1980). Allozymes were the first example of the exploitation of neutral variation. They were present in all germplasm, so the need for specific introgression was obviated. This step change in abundance and neutrality was particularly important because it enabled meaningful, genome-wide analyses. They did, however, require some laboratory equipment and their analysis was more complex than scoring kernel color. Crude protein extracts were run using starch gel electrophoresis to separate allelic variants, followed by *in situ* enzymatic detection of the allozyme of interest. The method was initially slow and somewhat expensive. Later with refinement, productivity increased significantly. Allozymes have been used extensively to study genetic variability (Stuber and Goodman, 1983). Today, allozymes still remain a standard for the determination of both genetic variability and genetic purity in the seed industry.

The first genome-wide genetic studies of quantitative traits were performed using allozymes (Edwards et al., 1987; Tanksley et al., 1982). It was, however, impractical to use allozymes for genetic analysis and selection in breeding because they were not available in sufficient numbers and were not very informative. The number of available enzyme assays and the frequency of monomorphism, particularly in narrow breeding crosses, limited the utility of allozymes.

Phenotypes and protein variants are indirect indicators of DNA variation. The first practical technology that directly assayed DNA variability was restriction fragment length polymorphisms (RFLPs) (Botstein et al., 1980; Helentjaris et al., 1985; Patterson et al., 1988). RFLPs afforded the opportunity to conduct both detailed, genome-wide genetic analysis, and more importantly, genotypic selection for quantitative traits (Patterson et al., 1988). Hundreds of RFLP markers could be developed rapidly, by screening additional clones containing low copy DNA. RFLPs were somewhat expensive to process, typically performed using radioisotopes, labor-intensive, and required large amounts of high quality DNA. Despite these limitations, the ability to dissect genetically and to select for traits of interest with such new found precision was irresistible. A number of commercial and public sector programs built significant competencies in RFLP markers and used these capabilities to conduct key experiments in the mapping and selection of traits and in performing marker assisted backcross conversions.

The introduction of DNA amplification-based procedures was the next significant breakthrough in genotyping techniques. The first DNA amplification procedures that provided genome-wide marker distribution were random amplified polymorphic DNAs (RAPDs) (Williams et al., 1990; Welsh and McClelland, 1990) and AFLPs (Vos et al., 1995). These methods were fast and more cost effective than RFLPs and required very limited capital set up. Another advantage was that they required no *a priori* sequence information of the amplification target sites. Again a surge in activity resulted with the creation of many new genetic maps in most crop species and hundreds of QTL studies. Challenges remained however. Because of the non-targeted nature of the genomic sequences amplified by these methods, there was no assurance that amplicons of similar fragment size observed in different samples were allelic. This was of little concern in bi-parental populations, but made the extension of results across germplasm more difficult because some amplicons were likely identical only in state and lacked identity by descent. Secondly, in order for these techniques to be reproducible, considerable effort was necessary to achieve stringent control of all aspects in the laboratory process.

Reports of the presence and use of simple tandem repeats (STRs) in humans (Edwards et al., 1992) quickly drove similar efforts in crop species and the third step change in genotyping took place. The use of STRs, more commonly referred to as simple sequence repeats (SSRs) in plants (Akkaya et al., 1992) required larger initial investments in marker discovery (Figure 1.24) through sequencing of cloned DNA containing selected tandem repeat motifs. The timing was consistent with the introduction and proliferation of second generation, robust, automated sequencing equipment like the ABI377 (Applied Biosystems, Foster City, CA). After significant efforts in marker discovery, a number of species now have hundreds, or more commonly thousands of SSR markers available for use in genetic mapping and trait selection. SSRs were sequence target specific, highly informative and relatively robust and generally required only polymerase chain reaction (PCR) amplification and fragment separation using some type of gel or capillary separation technology. These features resulted in rapid adoption, in particular for large-scale selection in plant breeding. Today a number of large commercial breeding programs continue to use SSRs as their main genotyping method. Despite this, the drive to identify cheaper, more robust technology continues.

Again, following the lead in mammalian species, interest is now shifting to the detection and use of single nucleotide polymorphisms (SNPs) (Figure 1.24). SNPs and insertion/deletions (indels) by far represent the vast repository of sequence variation that exists in DNA. In humans there are millions of candidate SNPs available for detection (Kwok, 2001). Crop species also are expected to harbor large numbers of polymorphisms, although the frequency of SNPs is likely dependent upon the reproductive nature, domestication, and breeding history of the crop in question. Interest in SNP detection in humans has resulted in a number of competing detection technologies becoming available for use in plants (Kwok, 2001; Jenkins and Gibson, 2002). These range from fluorescence resonance energy transfer-based detection procedures using Taq polymerase 5′ nuclease activity (TaqMan®, Applied Biosystems, Foster City, CA), or structural recognition cleavage (Invader®, Third Wave Technologies, Madison, WI), single base or multibase extension (SNPStream®, Beckman-Coulter, Fullerton, CA; Pyrosequencing®, Uppsala, Sweden); MALDI-TOF MS, (Sequenom, San Diego, CA); ligase-mediated amplification and hybridization to beads (Illumina, La Jolla, CA); and direct amplification and allele-specific hybridization to very-high-density oligonucleotide arrays (Affymetrix, Santa Clara, CA). Other methods abound, including those using some type of fragment separation technique, but those listed are among the more widely used.

Kwok (2001) and Jenkins and Gibson (2002) have written concise reviews of these more common SNP genotyping methods. The challenge remains to sort through the myriad of detection technologies and identify those best suited for activities as diverse as single gene genotypic selection and large-scale fingerprinting. What SNPs lack in relative information per marker, they provide in abundance and diagnostic simplicity, enabling researchers to reduce costs through automation. Our own efforts in the discovery and development of SNP markers has allowed us to assemble a first SNP map in corn (Figure 1.25). As anticipated, the distribution of SNPs is essentially random. The level of information from such a SNP platform is sufficient to allow two elite lines from the same heterotic group to be differentiated, on average, by about 20% of markers from a random set of SNPs. A set of 1000 SNPs is sufficient to enable full genome mapping in breeding populations with significant linkage disequilibrium (LD). The abundance of SNP polymorphisms in most crops makes possible the analysis of populations approaching linkage equilibrium, which require higher marker densities.

The relative cost of developing individual markers varies significantly among the three most popular marker platforms today (Figure 1.26). Significant progress has been made in reducing the cost of developing SNP markers, but they remain much more expensive to develop than RFLP and SSR markers. Despite the cost differences, SNP platforms are preferred in plant-breeding programs because of concomitant productivity gains (Figure 1.24). The future is likely to see routine whole-genome screening of all material variation in and around genes and, ultimately, routine whole-genome sequencing. Adoption of these technological shifts likely will occur whenever the perceived or realized benefits exceed the cost of creating and using the genotypic information. In plant breeding, this benefit can ultimately be measured through selection gain. The future, therefore, holds much promise in marker-assisted selection as the cost, speed, and accuracy of genotyping methods continues to rapidly evolve.

Marker-based selection

One of the first routine uses of markers in breeding programs was MAS. An example of MAS involves soybean cyst nematode (SCN), one of the most economically destructive pathogens of soybean. Production losses from SCN-attributed infestations exceed 3 million metric tons from the 10 countries with the greatest soybean production. In the United States alone, an estimated 5600 metric tons of production was lost due to SCN in 1998, and 3630 metric tons in 2002. The decrease in yield loss may be due to the effect of deploying more cultivars with the *rhg1* gene (Pratt and Wrather, 1998; Wrather et al. 2001).

Greenhouse and field-based screening procedures have been developed but are often costly, low throughput, labor intensive, and have poor reliability. Molecular marker loci associated with major SCN resistance loci provide an alternative first-pass screening procedure that improves the efficiency and capacity of the breeding process. Several RFLP markers were found near the SCN resistance locus (*rhg1*) on LG-G and have been shown to be useful in marker-assisted selection (Concibido et al., 2003; Webb et al. 1995). Cregan et al. (1999a, 1999b) identified the SSR markers, Satt309 and Sat168, as being closely linked to the *rhg1* locus. Monsanto confirmed the association of Satt309 with SCN resistance in mapping populations and concluded that the four alleles at Satt309 were 95% predictive of resistance. MAS technology has enabled a companywide early generation marker-assisted selection program for *rhg1* using linked molecular markers.

Marker-assisted backcrossing

Backcrossing is a routine procedure used in plant breeding to incorporate genetic factors targeted to specific traits (Harlan and Pope, 1922; Allard, 1960; Fehr, 1987). Backcross breeding methodologies have been used to integrate disease resistance into a number of crops such as maize, for example, Northern leaf blight caused by the fungus *Exserohilum turcicum* using *Ht* genes (Hooker, 1963, 1975, 1977a, 1977b); common corn rust caused by *Puccinia sorghi* (Russell 1965); phytopthora rot caused by *Phytophthora megasperma* in soybean (Bernard and Cremeens, 1988a, 1988b); rust and

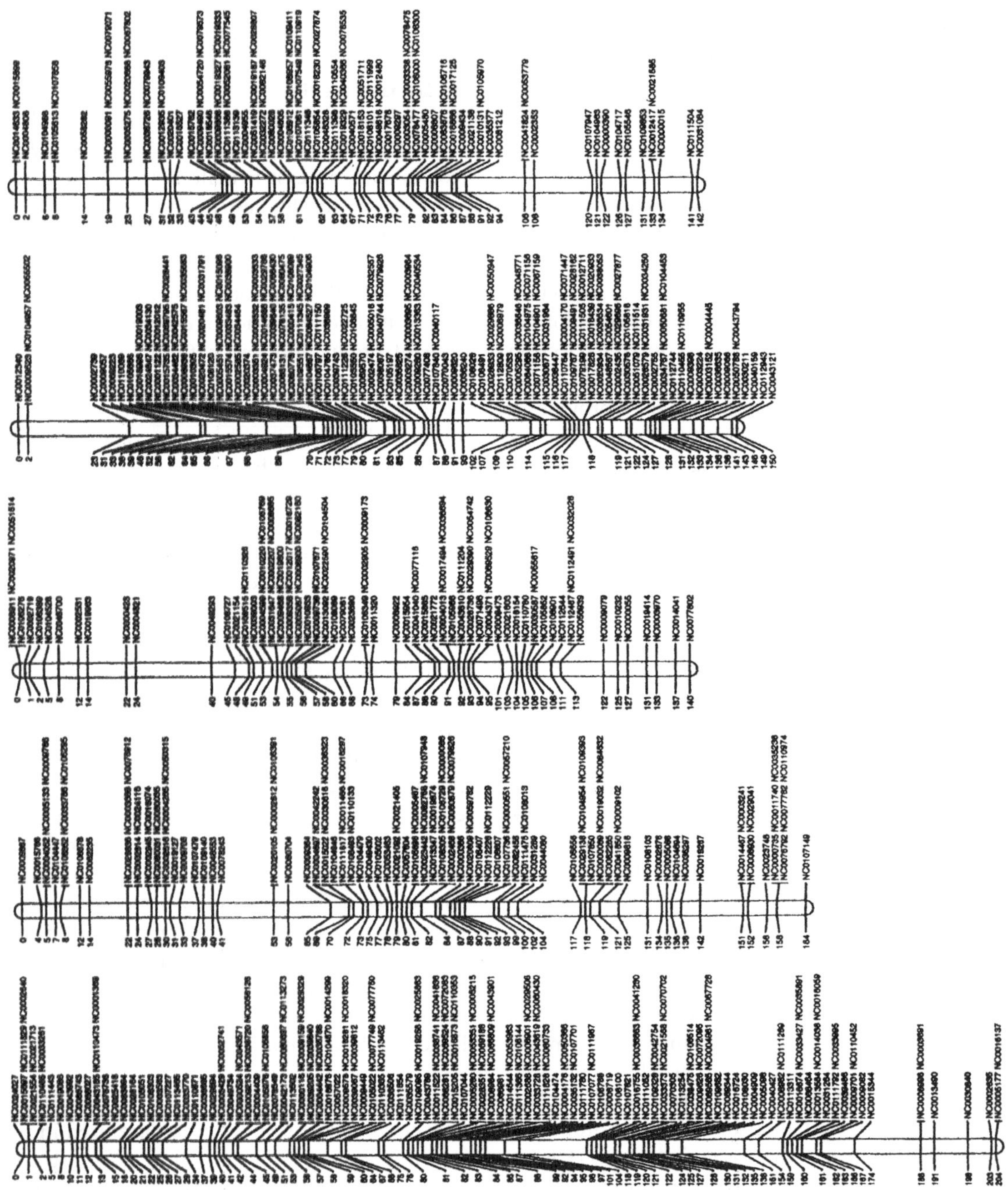

Figure 1.25 Monsanto's genetic map for maize with 2300 SNP markers and an estimated map size of 1604 cM.

powdery mildew resistance in wheat (*Triticum aestivum* L. em Thell.) (Sharma and Gill, 1983); and tobacco mosaic virus in tomato (*Lycopersicon peruvianum*) (Young and Tanksley, 1989). In addition, quality traits such as modified starch (waxy) in corn and physiological characteristics such photoperiod response in sorghum (communication from Tropical Agricultural Research Station in Mayaguez, Puerto Rico) and cotton (*Gossypium hirsutum* L.) (Liu et al., 2000) have been modified through backcross breeding methodologies.

Prior to molecular marker technology, pheno-

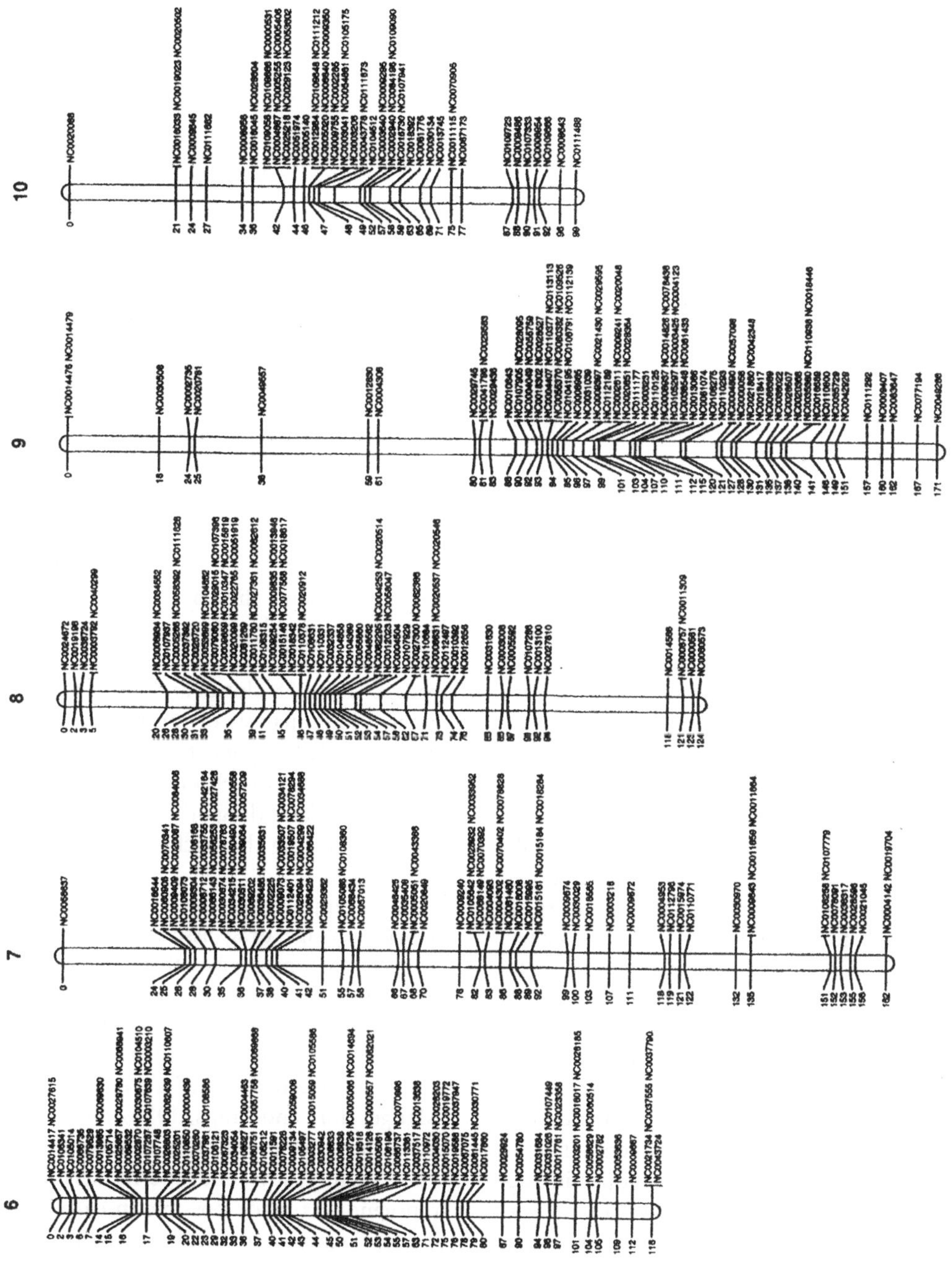

Figure 1.25 *(continued)*

typic selection was done at each stage of the process to identify plants possessing the target trait and a high level of resemblance to the recurrent parent. With recessive traits, slow and complex procedures using a progeny test were required to track the target trait at each generation (Fehr, 1987).

With the discovery and implementation of molecular marker tools, MABC became a reality. Molecular marker information is used to (1) track target allele(s) from the donor parent that are difficult or impossible to select for during the backcrossing process, (2) identify plants that have fa-

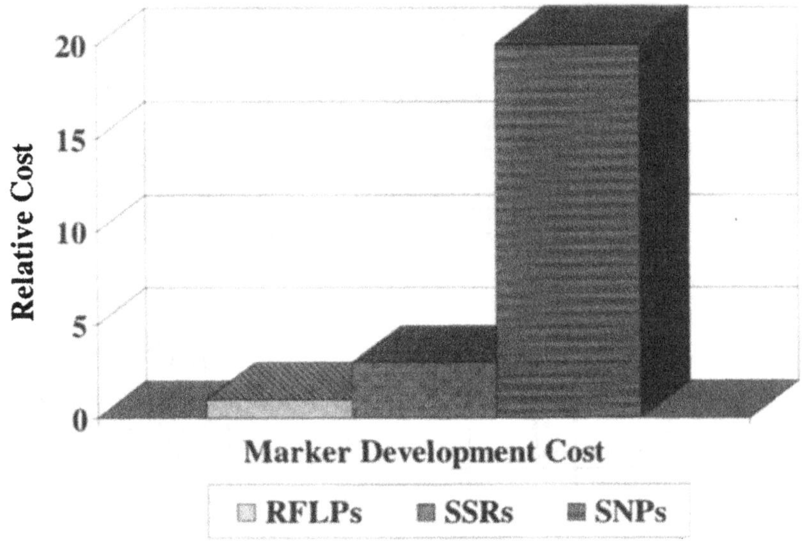

Figure 1.26 The relative cost of developing individual markers for RFLP, SSR, and SNP markers in 2003.

vorable recombinant gametes between the donor and recurrent alleles flanking the genomic region being introgressed, and (3) identify plants that have a high proportion of desirable genome from the recurrent parent. Soller and Plotkin-Hazan (1977) outlined the general concept of using linked marker loci to follow the introgression of favorable exotic quantitative alleles into elite germplasm. Early research using isozymes (Tanksley and Rick, 1980; Tanksley et al., 1981) outlined the utility of using genomewide polymorphic markers to reduce the number of backcross generations needed to recover the recurrent parent alleles. Additional theoretical research and computer simulations (Hillel et al., 1990; Hospital et al., 1992; Visscher et al., 1996; Hospital and Charcosset, 1997; Frisch et al., 1999a; and Frisch et al., 1999b) outlined optimal breeding schemes that allow rapid recovery of recurrent parent alleles in a few generations of backcrossing, reduce linkage drag (Brinkman and Frey, 1977) of donor alleles linked to the target genomic region, and stage genotyping to minimize cost.

DNA fingerprinting of nonmarker-assisted conversions have demonstrated the difficulty of effectively selecting against linked genomic regions and completely recovering the recurrent parent alleles. Young and Tanksley (1989) genotyped a series of tomato (*Lycopersicon peruvianum*) cultivars with RFLPs and observed that the linkage drag associated with the Tm-2 gene ranged from 4 to over 51 cM. Using marker data from intermediate backcross generations, they observed that plants with desirable recombinants around the *TM-2* gene were rarely selected during the backcrossing process. Using 58 clones and three restriction enzymes, RFLP analysis of the maize lines B14 and B14A revealed higher than expected levels of donor parent for the HindIII RFLP patterns possibly due to linkage drag or selection for other characteristics during the backcrossing program (Lee et al., 1990).

MABC approaches enable researchers to better manage linkage drag associated with transgenic conversions compared with conventional phenotypic selection. In one such example, we observed an association between a transgene and white endosperm kernels. Genetic linkage analysis revealed that the transgene inserted 2 cM from the y1 locus (Mangelsdorf and Fraps, 1931) on the short-arm of chromosome 6. Selection of recombinant genotypes with yellow endosperm kernels and the transgene was made possible in early generations due to diagnostic molecular marker tools and enabled complete recovery of the recurrent parent genotype. The use of markers also enabled rapid and efficient cycling of backcross generations because selections could be made based on the genotype before pollination. Since white endosperm is typically recessive to yellow endosperm, it is doubtful in our view that conventional visual selection during backcrossing would have been as rapid or successful as was the MABC approach.

Based on over 500 MABC projects, the average

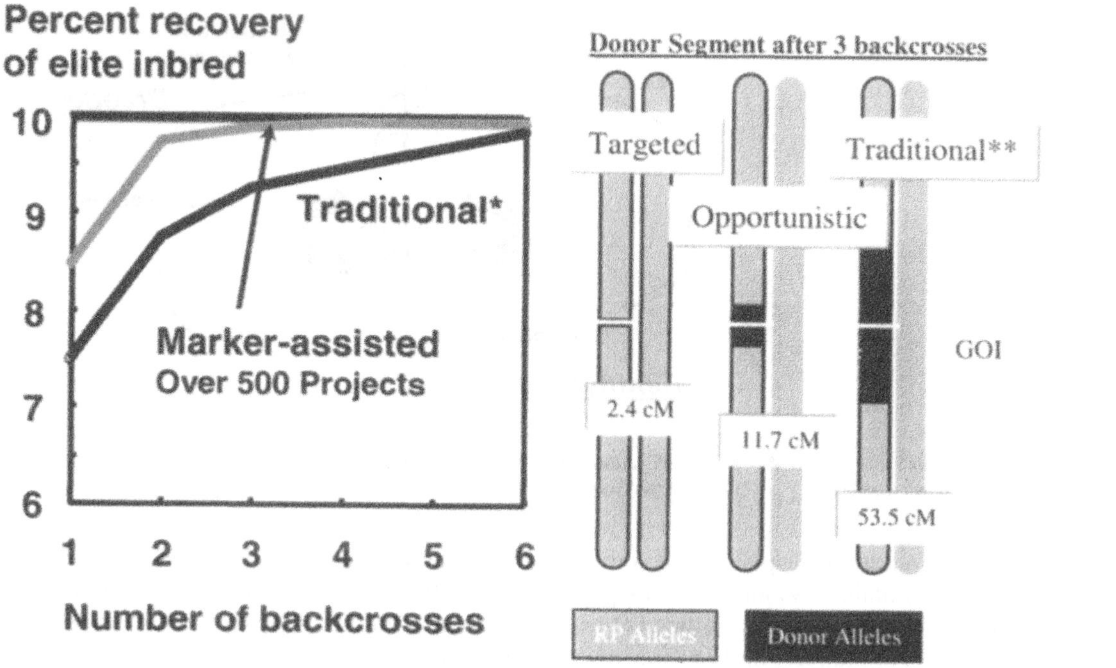

* = assuming no selection

** = assuming no selection, BC3 with random gene location and 1.66 M length (Stam & Zeven, 1981)

Figure 1.27 Comparison of MABC and phenotypic selection for backcrossing a single gene into a recurrent parent.

recurrent parent recovery is 90.0% at the BC1 (backcross 1) generation, 98.0% at the BC2 generation, and 99.5% at the BC3 generation (Figure 1.27). In these studies, marker selection targeted to reduce linkage drag around the GOI averaged 2.4 cM of donor genome in the resulting conversions compared with 11.7 cM for serendipitous selection for recombinant gametes versus an expected value of 53.5 cM for unselected backcross conversions (Stam and Zeven, 1981) (Figure 1.27). Utilizing optimized breeding schemes, marker-assisted conversions have a 95% probability of being agronomically equivalent to the recurrent parent. This rapid introgression of transgenic traits combined with a high probability of success has made MABC a standard practice in many plant-breeding programs.

The recent fusion of quantitative genetics theory, plant-breeding methodologies, and molecular biology has rekindled interest in indirect selection, an old tool in plant breeding. As molecular marker strategies have evolved from a "bandwagon" (Simmonds, 1991) to a reality, researchers also worked to develop new statistical genetic theory and conducted experiments to evaluate the utility of directed genotypic selection tools. Almost all applications have focused on the potential value of these markers to serve as indirect selection tools in MARS (Figure 1.28).

MARS for the improvement of a primary phenotypic trait is done either through indirect selection on markers linked to the primary trait, or via selection indices where direct selection on the primary trait is combined with indirect selection on the markers (Lande and Thompson, 1990).

The basic equation for genetic gain from index selection is

$$\Delta G = k^{*}\text{sqrt}\,(B'^{*}G^{*}w)$$

where k = selection intensity;

G = genotypic covariance matrix;

P = phenotypic covariance matrix;

w = trait weight vector;

B = vector for the solution of $G^{*}w = P^{*}B$, i.e., $B = \text{inv}(P)^{*}G^{*}w$.

For index or indirect selection, the element of the trait weight vector corresponding to the primary trait is set equal to one. Since, if considered as traits, markers have no economic value, all weights corresponding to the markers are set equal

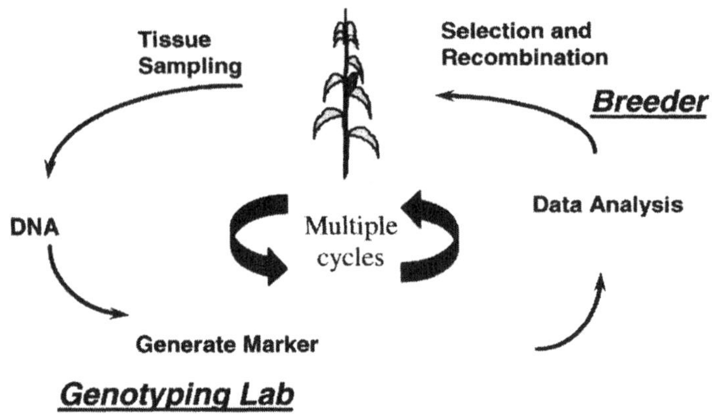

Figure 1.28 A generalized approach to marker-assisted recurrent selection. ❖ Multiple cycles of marker assisted selection followed by directed recombination to increase the frequency of favorable marker alleles associated with agronomic traits

to zero. For index selection, the phenotypic covariance matrix will contain rows and columns pertaining to both the primary trait and the markers (secondary traits). For indirect selection, only rows and columns pertaining to the molecular markers are retained. By some obvious algebraic manipulations, the above equation can be expressed in terms of heritabilities and genetic and phenotypic correlations (Falconer, 1960), allowing discussion in a more traditional quantitative genetic lexicon.

Excluding the possibility of misclassification, marker genotypes are 100% heritable. Consequently, if the genetic correlations of the markers with the primary phenotypic trait are sufficiently high, indirect selection on markers alone, or in combination with the primary trait in a selection index, will produce greater gain than selection on the primary trait itself. The genetic correlation of the primary trait with the markers is dependent on the physical proximity of the marker loci to the loci of genes controlling expression of the primary trait, the degree of LD in the population undergoing selection, and the numbers of individuals in each marker genotypic class. Larger values of each factor will act to enhance the strength of the genetic correlations between the primary trait and the markers. The remaining factor determining the efficacy of markers as aids to selection is the heritability of the primary trait. All other factors constant, phenotypic traits with low heritability will receive relatively more benefit from index selection with markers than will traits with high heritability. Traits with low heritability are often quite susceptible to environmental perturbation during development and are usually governed by many genes, each with relatively small effect. Generally speaking then, complex traits are those most appropriate for index selection that includes markers. Both simple and complex traits may be suitable for indirect selection.

Marker-aided selection is particularly applicable to populations of F2 individuals or F2-derived lines, because LD between the marker loci and the primary trait loci will be at a maximum for a segregating population. In these populations, two alleles will be segregating at each marker locus to produce the two parental and the heterozygote marker genotypes. The effect on the primary trait attributable to segregation at the marker loci will then be just a function of the differences between the marker class means and can be estimated simply as additive and dominance effects at each marker locus. If the F2 individuals or F2-derived lines have been crossed to a tester, only additive effects are relevant. If not, both additive and dominance effects will be applicable. Epistasis may be considered, but is usually ignored. Generally, the estimates of genetic effects at the marker loci will be combined in a linear model to produce a prediction of the breeding or genotypic value of each individual or line evaluated in the population. Usually, marker loci will be selected for inclusion into the linear model on the basis of a statistical test for significance. Once the linear model for estimation of breeding value based on marker effects has been derived, the equation can be used to predict the breeding value of any marker-genotyped member of the population or any marker-genotyped offspring resulting from inter-mating selected members of the population. The prediction of breeding value of the offspring will be biased due to the decay of LD occasioned by intermating, but with tight physical linkage of the marker loci and trait loci, the decay of LD will be slight and the bias small. If so, the equation can reasonably be used in several cycles of indirect recurrent selection based solely on marker genotypes. The practical utility of this scheme is that selection and recombination can proceed unencumbered by the need to evaluate the primary trait anew after each cycle of recombination. If off-season nurseries or greenhouse facilities are employed, two or three cycles of recurrent selection can be achieved in a single year, thus greatly reducing cycle time and accelerating selection gain per year (Figure 1.27).

Many of the quantitative traits that constitute the primary focus of plant breeding are very complex in inheritance, with variation believed to be attributable to dozens if not hundreds of underlying genes. It is not unusual to identify 20 chromosome regions affecting yield or other key agronomic traits in a bi-parental, marker-based mapping project in maize. If only 20 key genes segregate independently in a breeding project, the favorable gene combination for all 20 loci occurs in an F2 at such a low frequency that growing the F2 population over the entire U.S. corn acreage would be insufficient to provide a 95% chance that the most favorable genotype would occur. Even if the F2 population were randomly inbred to fixation, several million inbred lines would be required to have reasonable chance of recovering the favorable genotype. Clearly, breeders rarely, if ever, recover the optimum genotype from their breeding crosses. With low heritabilities, small sample sizes, and breeding approaches involving rapid inbreeding, the simple goal of achieving a gene combination significantly better than the parental genotypes is an ambitious undertaking with relatively low odds of success. By employing genetic markers in a recurrent selection scheme as discussed above, our aim is to improve the fixation rate of favorable QTLs by using recurrent cycles of marker-based selection. In a practical sense, we would like to accomplish this within reasonable experiment sizes and within and among modestly sized populations and to use three or more generations per year in multiseason nurseries or greenhouses.

Estimates of variance and covariance were used to predict genetic gain for several selection methods in 43 elite corn-breeding populations (Figure 1.29). The addition of SSR marker information

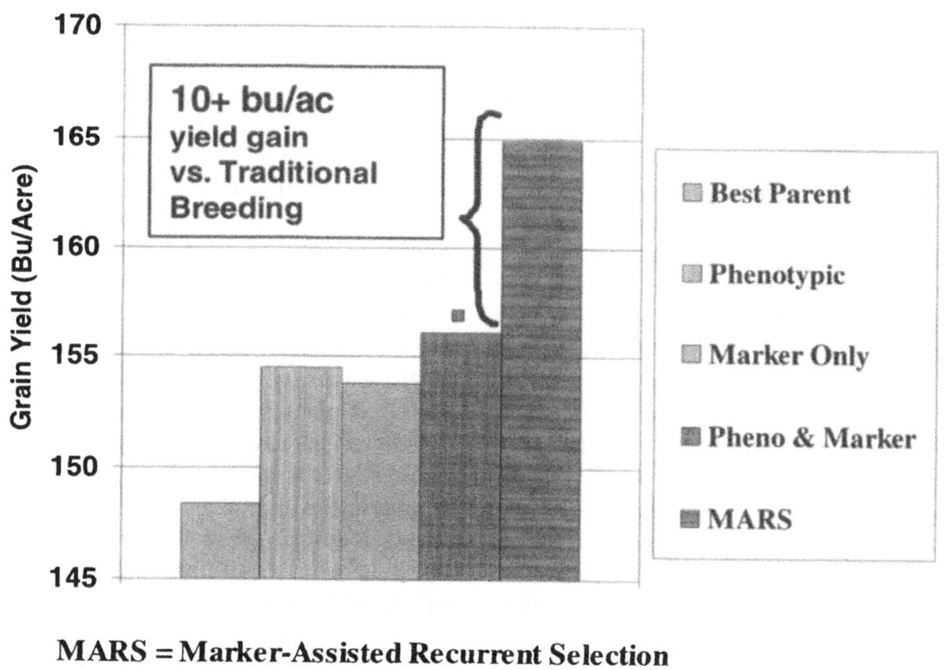

Figure 1.29 Predicted gain for several methods of selection averaged across 43 corn-breeding populations.

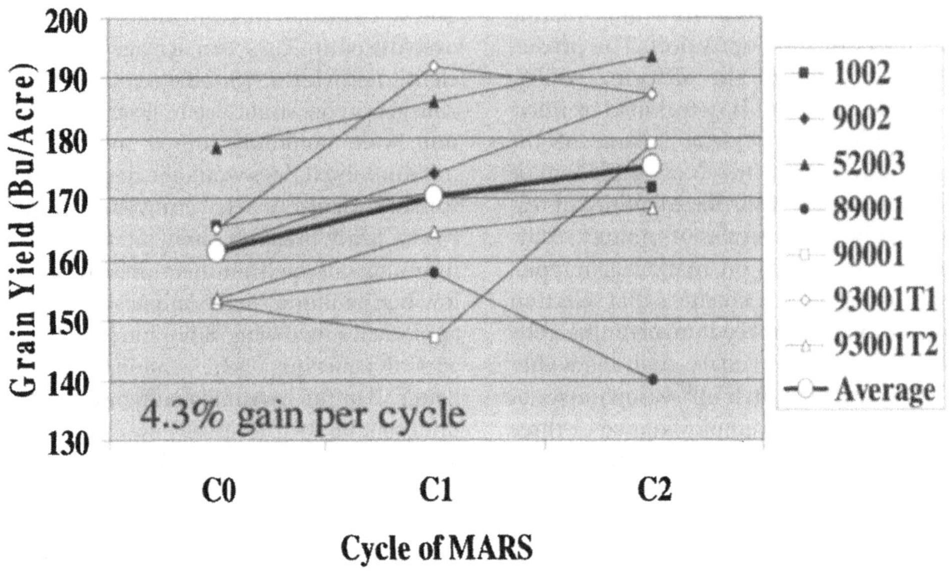

Figure 1.30 Two cycles of MARS increased topcross grain yield of selected lines from seven breeding populations by 13.8 Bu/acre vs. the topcross performance of the C0 of the original populations

was predicted to provide small, incremental improvements over conventional phenotypic selection in one selection cycle, but the use of MARS was predicted to double the rate of yield improvement (Figure 1.29). Data collected from top crosses of selected lines from seven breeding populations demonstrated that actual gain from two cycles of MARS was 13.8 bushels/acre higher than the C0 (Cycle 0) of the original populations (Figure 1.30). These results also were reported by Johnson (2002) and demonstrate that MARS is effective in increasing the frequency of favorable marker alleles, which in turn provides significant improvement over conventional selection for the trait or traits of interest.

Utilization of global germplasm sources

A key component to success for any breeding program is having useful genetic variation upon which selection can be applied. Commercial plant breeders have continued to identify, evaluate, and integrate useful sources of germplasm into breeding programs. In numerous situations, disease resistance has been identified in exotic (related species) or unadapted germplasm and integrated into elite germplasm pools. In many cases, this was done though backcross-breeding strategies that are now being enhanced with the use of molecular markers to accelerate recovery of the recurrent parent and reduce linkage drag associated with the integration of the targeted genomic regions.

In addition to disease resistance, plant breeders routinely try to extract useful favorable alleles for grain yield from unadapted sources of germplasm. Successful commercial products have been developed, but in general it is difficult to recover improved varieties. Recovery of improved varieties from a breeding cross is directly related to the distribution of the contribution of favorable alleles from the parental lines and the breeding scheme (Bailey and Comstock, 1976; Dudley, 1982). Evaluation of a large number of germplasm sources to determine which sources carry favorable alleles is the first challenge plant breeders face.

Most unadapted sources of maize have numerous unfavorable characteristics, such as short-day-length sensitivity, increased root and stalk lodging, increased plant and ear height, and unfavorable dry-down characteristics. In addition, every germplasm source is unique for its favorable and unfavorable alleles and the linkage phase between these alleles, requiring plant breeders to use modified breeding schemes compared with those used for elite-by-elite breeding populations. Given the opportunity germplasm sources (both exotic and unadapted) provide to long-term genetic gain in plant breeding (Tanksley and McCouch, 1997), researchers continue to evaluate how molecular markers can be used to extract favorable alleles from these sources.

Recent research has demonstrated that exotic sources are capable of contributing favorable alleles to the improvement of tomatoes, rice, and soybeans. QTL mapping and introgression studies have demonstrated that *Lycopersicon esculentum* can be improved by favorable alleles from *L. pennellii* (Vicente and Tanksley, 1993), *L. pimpinellifolium* (Tanksley et al., 1996), *L. hirsutum* (Bernacchi et al., 1998), and *L. peruvianum* (Fulton et al., 1997). QTLs from wild rice were identified to improve cultivated rice (Xiao et al., 1998), while favorable alleles for grain yield (Concibido et al., 2003) and seed protein (Sebolt et al., 2000) in soybeans (*Glycine max*) were identified in *Glycine soja*, with genetic background effects observed for both QTLs. These studies demonstrate both the utility of exotic sources to contribute favorable alleles and the value of molecular marker tools to identify and integrate exotic QTLs into elite germplasm.

Breeders typically have used phenotypic observations for key morphological and physiological traits along with test-cross schemes to evaluate general combining ability. Dudley (1984a) developed a formal method to determine the ability of germplasm sources to contribute favorable alleles to a target hybrid. This methodology has been modified and adjusted for different breeding schemes (Dudley, 1984b, 1987; Bernardo, 1990; Metz, 1994). QTL mapping studies in corn at Monsanto have confirmed that most exotic or unadapted germplasm sources contribute a low frequency of favorable QTLs relative to elite U.S. germplasm. In one study involving 6 elite-by-exotic breeding crosses, 87% of the markers linked to favorable QTLs came from the elite parent versus 13% from the exotic parent. In another study involving 140 elite-by-elite breeding crosses the favorable marker linkages were evenly distributed among the parents. This supports the practice and importance of backcrossing to the elite line prior to selection in breeding populations composed of exotic or unadapted germplasm sources. Molecular marker tools allow breeders to understand the parental contribution of favorable QTLs in each breeding population, enabling a breeder to tailor the selection process to improve the probability of recovering a superior variety.

Most global plant-breeding programs actively use elite germplasm from other world regions as a source of new QTLs (Figure 1.31). At a macro

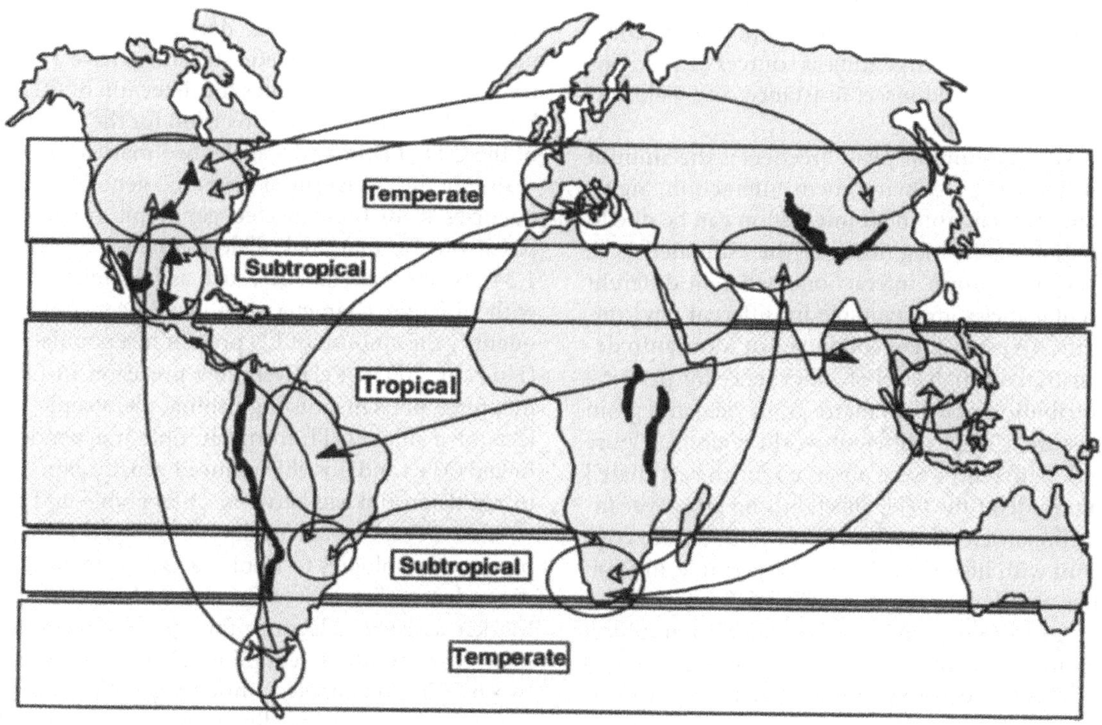

Figure 1.31 Use of germplasm in a typical global maize-breeding program.

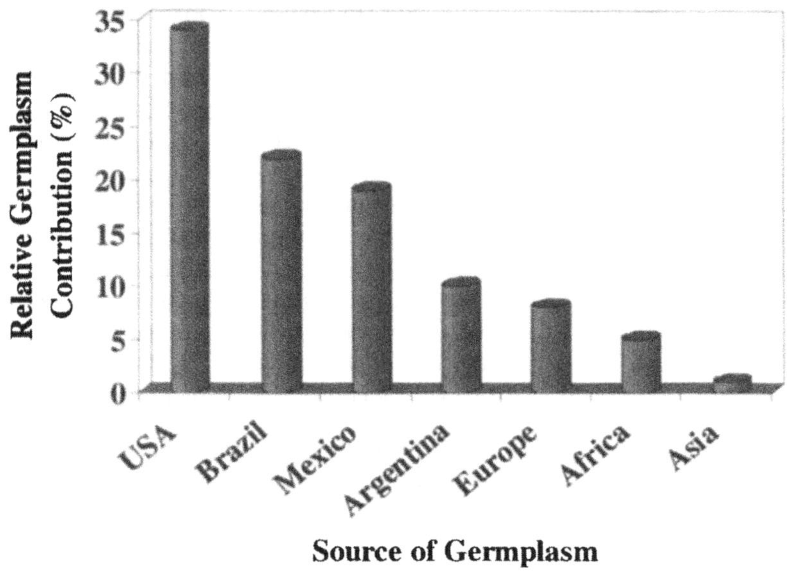

Figure 1.32 Relative contribution of several different germplasm sources to Monsanto's commercial portfolio in Argentina.

level, a geographic region can contribute commercial parental lines or commercial products to other geographic regions within similar latitude. Introgression of germplasm from different geographic regions has produced tremendous commercial success in countries such as Argentina (Figure 1.32). Germplasm from the United States, Brazil, and Mexico has been extremely important to commercial success in Argentina as sources of yield potential, disease and insect resistance, and yield stability, respectively.

A key question for plant breeders is the amount and type of QTL/environment interaction. Significant QTL/environment interaction can be due to variation in the magnitude of the estimated QTL effect or crossover interactions, wherein different parental alleles are favorable in different environments. An experiment conducted at Monsanto demonstrated a high level of congruence for parental contribution of hybrid maize grain yield and grain moisture QTLs in different world regions (Figure 1.33). Differences were observed for the estimated magnitude of the QTL effect, but no crossover interactions were identified. These findings are consistent with historical use of elite germplasm from other world areas (Figure 1.31). QTL information derived from experiments that span geographical regions will enable identification and use of global QTLs. Once identified, these global QTLs can be transferred across geographical regions to enhance the global rate of genetic gain in plant breeding.

Trait associations in molecular-breeding strategies

Mapping of quantitative trait loci to specific genomic regions enables marker-assisted selection in plant-breeding programs or serves as a starting point for map-based cloning of causal genetic factors. Typical QTL mapping projects use standard breeding population structures of F2, backcross, or recombinant inbred lines derived from inbred parental lines. These genetic structures have a high level of LD, which facilitates the detection of QTLs, but results in low level of precision for the location of the QTL. LD in a two-allele, two marker system is simply the deviation of observed genotypic frequencies from the expected genotypic frequency based on the marginal allele frequencies (Figure 1.34). Numerous statistics, such as D' and R^2, each with different characteristics, have been used to quantify the amount of LD present in a population (Hudson, 2001). Relatively low precision in QTL mapping prevents distinguishing pleiotropic effects of a single QTL from multiple independent linked QTLs and possible reduced genetic gain due to repulsion linkage between unfavorable and favorable QTLs. Graham et al. (1997) used fine mapping methodologies to resolve a QTL into two independent QTLs that were in repulsion linkage. Marker-assisted selection for specific recombinants that result in coupling phase linkage between QTLs can enable additional genetic gain opportunities in plant-breeding programs. Map-based cloning projects are facilitated by subcentimorgan

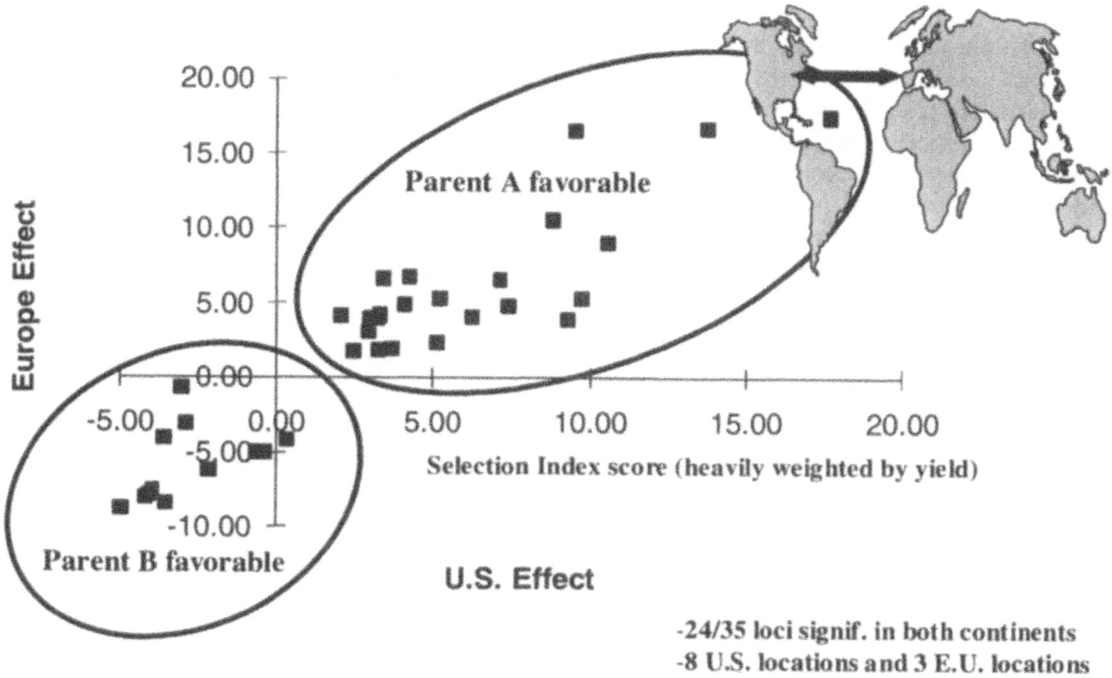

Figure 1.33 Directional consistency of QTLs between the United States and Europe.

Locus 1 (columns) × **Locus 2** (rows)

	A(.5)	a (.5)
B(.5)	.5	0
b(.5)	0	.5

$D'=1$
complete association

$$D' = \frac{p_{11}p_{22} - p_{12}p_{21}}{D_{max}}$$

$$D_{max} = \begin{cases} Min(p_1q_2, q_1p_2), & D > 0 \\ Min(p_1q_1, q_2p_2), & D < 0 \end{cases}$$

	A(.5)	a (.5)
B(.5)	.375	.125
b(.5)	.125	.375

$D'=.5$
partial association

	A(.5)	a (.5)
B(.5)	.25	.25
b(.5)	.25	.25

$D'=0$
no association (Equilibrium)

Figure 1.34 Calculation of linkage disequilibrium between two loci.

resolution (Falconer and Mackay, 1996). Therefore, methodologies to improve the precision of QTL location are necessary to enhance both marker-assisted plant-breeding methodologies and map-based cloning projects.

Fine mapping of QTL position can be categorized into mathematical, recombinational, and substitution mapping approaches (Paterson, 1998). Recombinational approaches reduce the amount of LD in the breeding population. This can be accomplished by random mating the breeding population prior to QTL mapping. This approach has been used in mapping of quality traits in Illinois chemical strains (Dudley personal communication, 2003); however, this approach requires a substantial number of generations to reduce the LD for tightly linked loci. An alternative approach is to use historical recombinations present in the breeding germplasm.

To understand the potential of association studies in plant germplasm requires the understanding of the LD patterns in the genetic material. Tenaillon et al. (2001) observed very rapid intralocus decay of LD and very little evidence of interlocus LD for 21 loci on chromosome 1 for a diverse set of 25 maize samples. A variable rate of LD decay was observed in six genes in a broad set of 102 maize inbreds (Remington et al., 2001). Ching et al. (2002) did not observe a rapid decay in LD in 18 genes in a set of 36 elite maize inbreds from a U.S. breeding program. Germplasm in elite breeding programs has been derived by selection and gone through numerous bottlenecks, both of which create LD.

Spurious associations due to structure in the population of germplasm have restricted the utility of association studies (Risch and Merikangas, 1996; Pritchard, 2001). Experimental designs using family-based tests such as the transmission disequilibrium test (Spielman et al., 1993) along with statistical methods to account for population structure (Pritchard et al., 2000; Thornsberry et al., 2001) are methods to reduce spurious associations. Thornsberry et al. (2001) used 141 SSRs to account for population structure to reduce spurious associations for flowering time in maize.

A study of the intralocus LD in 140 U.S. diverse, non-BSSS maize inbreds using 98 SSRs marker loci revealed 13% of the marker to marker R^2 values were greater than 0.2 (Figure 1.35). The R^2 values indicated significant LD across and within chromosomes. Therefore, it is likely that association studies in elite germplasm will require use of experimental designs and statistical methods to account for population structure due to selection and bottlenecks. Association studies will enable better precision of QTL location, but will require substantial marker density to enable genomewide

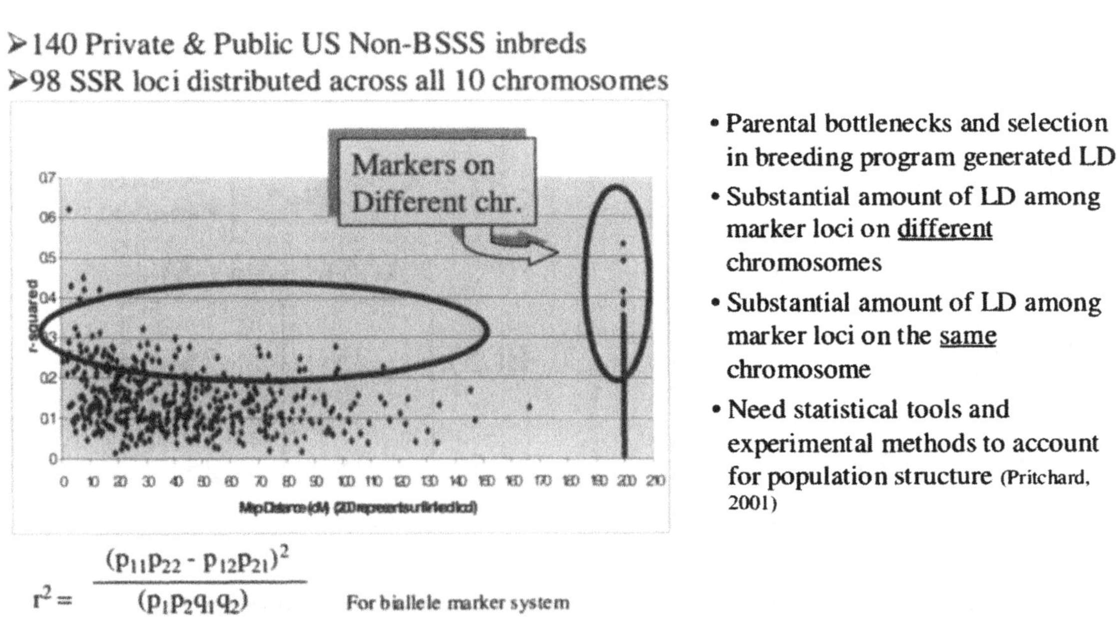

Figure 1.35 Linkage disequilibrium in 140 US public and private non-BSSS inbred lines of maize.

Table 1.5 Estimated selection gain in net photosynthetic rate (NPR) for NPR selection alone and in combination with glutamate dehydrogenase (GDH) expression (Johnson, 2002)

	Repeatability				
Selection Regimen	−2 Log Likelihood	NPR μmol $CO_2m^{-2}s^{-1}$	GDH Rel. MRNA Concn.	Genetic Correlation NPR*GDH	Relative Selection Gain
NPR	6057	0.10	—	—	1.00
NPR+GDH	4949	0.15	0.28	1.00	2.49

QTL scans. The required density will be dependent on the LD patterns within the study population. Using a population structure with very rapid decay will require markers every 100 to 200 base pairs (Tenaillon et al., 2001). This is in great contrast to F2 population structures that have a maximum level of LD and typically need a few hundred markers for QTL mapping in maize. Studies to identify causal genetic factors for QTLs will probably start with population structures that maximize LD followed by population structures with better genetic resolution, such as elite germplasm populations, and finally use population structures with very limited intralocus LD to pinpoint causal factors.

New approaches in genotypic selection and breeding

As described in the central dogma, genetic information passes from DNA to RNA by transcription, followed by translation of RNA to proteins, the proteins being the basis of a complex metabolism in which the proteins interact among themselves, with various organic and inorganic compounds, and with the RNA and DNA itself to ultimately produce trait phenotypes. Schork (1997) presented a schematic portrayal, roughly in the form of an equilateral triangle. DNA and the genes formed the base of the triangle. Located at the apex of the triangle were clinical symptoms of a (complex) disease, for which we need only to substitute a complex agronomic trait such as yield to make the schema relevant to plant breeding. Proceeding from the base to the apex a hierarchy of integrated interacting biochemical and physiological factors connect the genes at the base to the primary phenotype at the apex. Schork (1997) referred to these biochemical and physiological factors connecting the genes and the primary phenotype as intermediate phenotypes. Schork also surmised that the effect of environment on the intermediate phenotypes diminished with depth in the integrated hierarchy. The effect of the environment may not actually diminish, but, rather, the specificity of environmental effects on particular intermediate phenotypes may just become more sharply focused. Nonetheless, these intermediate phenotypes comprise another entire body of secondary traits, in addition to molecular markers, that could aid in the selection for the ultimate, primary trait. These traits can be used in indirect or index selection in the same way as markers. Heritabilities will likely be less than 100%, and the distribution of the trait values will generally be continuous, rather than categorical. The very important difference, though, is that the genetic correlations will have a biochemical and physiological foundation, rather than a basis due just to physical linkage on the chromosome.

Being positioned at a level just above sequence and the genes, messenger RNA (mRNA) transcript profiling is an obvious candidate for functional genomic application to plant breeding. Results from an experiment conducted on inbred corn lines evaluated under drought and irrigated conditions indicated that selection on an index that included transcript level from an NAD-specific glutamate dehydrogenase gene could significantly enhance selection for favorable response of net photosynthetic rate under drought (Table 1.5).

Though direct selection on gene transcript level may be a long-term eventuality through use, for example, of microarrays or real-time PCR, additional genomic tools enable opportunities for shorter term and perhaps more practical applications. Of paramount importance is the coding sequence imbedded in the full-length mRNA of a transcript phenotype. SNPs can obviously be sought within the exons of the expressed genes, but availability of a gene-specific sequence provides scope for further development. Coding sequence enables the construction of primers for both conventional and long-range PCR. Amplification of DNA employing these primers could conceivably provide sequence across introns, and the additional employment of universal primers might provide sequence in the upstream, promoter regions of the gene. Thus, the entire genomic sequence in and around specific genes may be attainable and available for SNP discovery, even before a crop has been completely sequenced. Not all SNPs will have functional significance, but any functional differences in alleles of a gene will be associated with SNP polymorphisms. Transcript profiling, then, may lead to the development of molecular markers that have real functional significance, not simply an inferred probabilistic significance due to correlation with expression of a phenotype under conditions of LD in a population. The polymorphisms within the gene would be intrinsically linked to the gene and would mark real functional variation due to allelic variation in the gene.

Proteins and metabolites constitute higher orders of intermediate phenotypes in closer proximity to the primary phenotype and should be fully amenable for use in direct or indirect selection. The important task is to assay the genetic correlations and heritabilities experimentally to determine adequacy of the intermediate phenotypes as selection aids.

SNPs within the coding and regulatory regions provide the means to quantify phenotypic variation ascribable to specific genes across the population, enabling a detailed description of the development of the phenotype. If a phenotypic ideal can be defined within a population in terms of imbedded SNP patterns within and among the genes controlling development, selection based on the genotypic ideotype could be more powerful than selection based on replicated phenotypes or on markers merely linked to anonymous QTLs.

Acknowledgments

The authors would like to thank the following people for contributions to the chapter: Joe Raab, Mike Graham, Steve Johnson, Dale Wickersham, Mark Messmer, Mike Roth, Mike Stern, Doug Sammons, and T.K. Ball.

References

Akkaya, M.S., A.A. Bhagwat, P.B. Cregan. 1992. Length polymorphisms of simple sequence repeat DNA in soybean. Genetics 132:1131–1139.

Allard, R.W. 1960. Principles of plant breeding. John Wiley and Sons, Inc., New York.

Bailey, T.B., Jr., and R.E. Comstock. 1976. Linkage and the synthesis of better genotypes in self-fertilizing species. Crop Sci. 16:363–370.

Baker, J.L., and J.M. Laflen. 1979. Runoff losses of surface-applied herbicides as affected by wheel tracks and incorporation. J. Environ. Qual. 8:602–607.

Beadle, G.W. 1939. Teosinte and the origin of maize. J. Hered. 30:245–247.

Beadle, G.W. 1980. The ancestry of corn. Sci. Am. 242(1): 112–119.

Bernacchi, D., T. Beck-Bunn, Y. Eshed, J. Lopes, V. Petiard, J. Uhlig, D. Zamir, S.D. Tanksley. 1998. Advanced backcross QTL analysis in tomato. I. Identification of QTLs for traits of agronomic importance from *Lycopersicon hirsutum*. Theor. Appl. Genet. 97:381–397.

Bernard, R.L., and C.R. Cremeens. 1988a. Registration of "Williams 82" soybean. Crop Sci. 28:1027.

Bernard, R.L., and C.R. Cremeens. 1988b. Registration of "Corsoy 79" soybean. Crop Sci. 28:1027.

Bernardo, R. 1990. An alternative statistic for identifying lines useful for improving parents of an elite single cross. Theor. Appl. Genet. 80:105–109.

Bernardo, R. 2002. Breeding for Quantitative Traits. Stemma Press. Minneapolis, Minnesota.

Bohn, M. 2003. Towards maize resistance to stem borers and rootworms. Annual Corn Breeders School, University of Illinois, Champaign, IL.

Botstein, D., R.L. White, M. Skolnick, and R.W. Davis. 1980. Construction of a genetic linkage map in man using restriction fragment length polymorphisms. Am. J. Hum. Genet. 32:314–331.

Bradley, J.P., K.H. Knittle, and A.F. Troyer. 1988. Statistical methods in seed corn product selection. J. Prod. Agric. 1:34–38.

Branson, T.F., G.R. Sutter, and J.R. Fisher. 1982. Comparison of a tolerant and susceptible maize inbred under artificial infestations of *Diabrotica virgifera virgifera*: Yield and adult emergence. Environ. Entomol. 11: 371–372.

Brinkman, M.A., and K.J. Frey. 1977. Yield component analysis of oat isolines that produce different grain yields. Crop Sci. 17:165–168.

Burks, A.W., and R.L. Fuchs. 1995. Assessment of the endogenous allergens in glyphosate-tolerant and commercial soybean varieties. J. Allergy Clin. Immunol. 96:1008–1010.

Burnham, C.R. 1966. Cytogenetics in plant improvement. Pp. 139–187. *In* K.J. Frey, ed.: Plant Breeding. Iowa State Univ. Press, Ames, IA.

Carpenter, J.E. 2001. Case studies in benefits and risks of agricultural biotechnology: Roundup Ready soybeans and Bt field corn. National Center for Food and Agricultural Policy, Washington, DC.

Ching, A., K.S. Caldwell, M. Jung, M. Dolan, O.S. Smith, S. Tingey, M. Morgante, A.J. Rafalski. 2002. SNP frequency, haplotype structure and linkage disequilibrium in elite maize inbred lines. BMC Genetics 3:19.

Christensen, C.M. 2000. The Innovators Dilemma. Harper Collins, New York.

Coe, E.H., M.G. Neuffer, and D.A. Hoisington. 1988. The genetics of corn. Pp.83–258. *In* G.F. Sprague and J.W. Dudley, eds.: Corn and Corn Improvement. Third Edition. Agronomy Mono. 18. Amer. Soc. Agron. Madison, WI.

Concibido, V.C., B. La Valle, P. Mclarid, N. Pineda, J. Meyer, L. Hummer, J. Yang, K. Wu, and X. Delannay. 2003. Introgression of a quantitative trait locus for yield from *Glycine soja* into commercial soybean cultivars. Theor. Appl. Genetics 106:575–582.

Crabb, R. 1993. The Hybrid Corn-Makers. Rutgers Univ. Press.

Cregan, P.B., T. Jarvik, A.L. Bush, R.C. Shoemaker, K.G. Lark, A.L. Kahler, Kaya N. Van Toi. 1999a. An integrated genetic linkage map of the soybean. Crop Sci. 39:1464–1490.

Cregan, P.B., J. Mudge, E.W. Fickus, D. Danesh, R. Denny, and N.D. Young. 1999b. Two simple sequence repeat markers to select for soybean cyst nematode resistance conditioned by the *rhg1* locus. Theor. Appl. Genet. 99:811–818.

Cressie, N. 1993. Statistics for Spatial Data. Revised Edition, John Wiley and Sons, New York.

CTIC. 1998. Crop Residue Management Survey. Conservation Technology Information Center. West Lafayette, Indiana.

CTIC. 2000. Top Ten Benefits. Conservation Technology Information Center. West Lafayette, Indiana.

de Vicente, M.C., and S.D. Tanksley. 1993. QTL analysis of transgressive segregation in an interspecific tomato cross. Genetics 134:585–596.

Delanney, X., T.T. Bauman, D.H. Beighly, M.J. Buettner, H.D. Coble, M.S. DeFelice, C.W. Derting, T.J. Diedrick, J.L. Griffin, E.S. Hagood, F.G. Hancock, S.E. Hart, B.J. LaVallee, M.M. Loux, W.E. Lueschen, K.W. Matson, C.K. Moots, E. Murdock, A.D. Nickell. M.D.K. Owen, E.H. Paschal II, L.M. Prochasta, P.J. Raymond, D.B. Reynolds, W.K. Rhodes, F.W. Roeth, P.L. Sprankle, L.J. Tarochione, C.N. Tinius, R.H. Walker, L.M. Wax, H.D. Weigelt, and S.R. Padgette. 1995. Yield evaluation of a glyphosate-tolerant soybean line after treatment with glyphosate. Crop Sci. 35:1461–1467.

Doebley, J. 1994. Genetics and the morphological evolution of maize. The Maize Handbook. Springer-Verlag.

Doebley, J., and A. Stec. 1991. Genetic analysis of the morphological differences between maize and teosinte. Genetics 129:285–295.

Dudley, J.W. 1982. Theory for transfer of alleles. Crop Sci. 22:631–637.

Dudley, J.W. 1984a. A method for identifying lines for use in improving parents of a single cross. Crop Sci. 24:355–357.

Dudley, J.W. 1984b. A method for identifying populations containing favorable alleles not present in elite germplasm. Crop Sci. 24:1053–1054.

Dudley, J.W. 1987. Modification of methods for identifying inbred lines useful for improving parents of elite single crosses. Crop Sci. 27:944–947.

Edwards, A., H.A. Hammond, L. Jin, C.T. Caskey, and R. Chakraborty. 1992. Genetic variation at five trimeric and tetrameric tandem repeat loci in four human population groups. Genomics 12:241–253.

Edwards, M.D., C.W. Stuber, and J.F. Wendel. 1987. Molecular marker facilitated investigations of quantitative trait loci in maize. I. Numbers, genomic distribution and types of gene action. Genetics. 116:113–125.

Emerson, R.A., G.W. Beadle, and A.C. Fraser. 1935. A summary of linkage studies in maize. Cornell Univ. Agric. Exp. Stn. Mem. 180 pp.

English, L.H., S.M. Brussock, T.M. Malvar, J.W. Bryson, C.A. Kulesza, F.S. Walters, S.L. Slatin, M.A. Von Tersch, C. Romano. 2000. Insect-resistant transgenic plants. United States Patent 6,023,013 B1.

Falconer, D.S. 1960. Introduction to Quantitative Genetics. The Ronald Press Co., New York.

Falconer, D.S., and T.F.C. Mackay. 1996. Introduction to Quantitative Genetics. Longman, London.

Fawcett, R. and D. Towery. 2002. Conservation Tillage and Plant Biotechnology: How New Technologies Can Improve the Environment by Reducing the Need to Plow. Conservation Technology Information Center. West Lafayette, IN.

Fehr, W.R. 1987. Principles of Cultivar Development. McGraw-Hill, Inc.

Frey, K.J. 1996. National Plant Breeding Study. ISU Sp. Rpt 98.

Frisch, M., M. Bohn, and A. Melchinger. 1999a. Comparison of selection strategies for marker-assisted backcrossing of a gene. Crop Sci. 39:1295–1301.

Frisch, M., M. Bohn, and A. Melchinger. 1999b. Minimum sample size and optimal positioning of flanking markers in marker-assisted backcrossing for transfer of a target gene. Crop Sci. 39:967–975.

Fulton, T.M., T. Beck-Bunn, D. Emmatty, Y. Eshed, J. Lopes, V. Petiard, J. Uhlig, D. Zamir, and S.D. Tanksley. 1997. QTL analysis in an advanced backcross of *Lycopersicon peruvianum* to the cultivated tomato and comparisons with QTLs found in other wild species. Theor. Appl. Genet. 95:881–894.

Gauch, H. G., Jr. 1992. Statistical Analysis of Regional Yield Trials: AMMI Analysis of Factorial Design. Elsevier, Amsterdam.

Graham, G.I, D.W. Wolff, and C.W. Stuber. 1997. Characterization of a yield quantitative trait locus on chromosome five of maize by fine mapping. Crop Sci. 37:1601–1610.

Gray, M.E. and K. L. Steffey. 1998. Corn rootworm (Coleoptera: Chrysomelidae) larval injury and root compensation of 12 maize hybrids: An assessment of the economic injury index. J. Econ. Entomol. 91:723–740.

Hall, M., M. Meghji, G. Parker, C. Peterman, and C. Yates. 2001. Strategies to maximize quality assurance (QA) when integrating transgenes in breeding programs. Annual Corn Breeders School, University of Illinois, Champaign, IL.

Hallauer, A.R., and J.B. Miranda Fo. 1981. Quantitative genetics in maize breeding. Iowa State Univ. Press. Ames, IA.

Hammond, B.G., J.L. Vicini, G.F. Hartnell, M.W. Naylor, C.D. Knight, E.H. Robinson, R.L. Fuchs, and S.R. Padgette. 1996. The feeding value of soybeans fed to rats, chickens, catfish and dairy cattle is not altered by genetic incorporation of glyphosate tolerance. J. Nutri. 126:717–727.

Harlan, H.V., and M.N. Pope. 1922. The use and value of backcrosses in small grain breeding. J. Hered. 13:319–322.

Hartzler, R. 1997. Pesticides used in Iowa Crop Production, 1995. Pest Management Bulletin PM-1718 of the Iowa State University Cooperative Extension Service. Downloaded through Web access at www.pme.iastate.edu/piap.

Haslam, E. 1993. Shikimic Acid: Metabolism and Metabolites. John Wiley & Sons Ltd., Chichester.

Hebblethwaite, J.F. 1995. The Contribution of No-Till to Sustainable and Environmentally Beneficial Crop Production: A Global Perspective. Conservation Technology Information Center. West Lafayette, Indiana.

Helentjaris, T., G. King, M. Slocum, C. Siedenstrang, and S. Wegman. 1985. Restriction fragment length polymorphisms as probes for plant diversity and as tools for applied plant breeding. Plant. Mol. Biol. 5:109–118.

Hillel, J., T. Schaap, A. Haberfeld, A.J. Jeffreys, Y. Plotzky, A. Cahaner, and U. Lavi. 1990. DNA fingerprints applied to gene introgression in breeding programs. Genetics 24:783–789.

Hooker, A.L. 1963. Inheritance of chlorotic-lesion resistance to *Helminthosporium turcicum* in seedling corn. Crop Sci. 660–662.

Hooker, A.L. 1975. *Helminthosporium turcicum* as a pathogen of corn. Rep. Tottori Mycol. Inst. (Jpn.) 12:115–125.

Hooker, A.L. 1977a. A plant pathologist's view of germplasm evaluation and utilization. Crop Sci. 17:689–694.

Hooker, A.L. 1977b. A second major gene locus in corn for chlorotic-lesion resistance to *Helminthosporium turcicum*. Crop Sci. 17:132–135.

Horsch, R.B., R.T. Fraley, S.G. Rogers, P.R. Sanders, A. Lloyd, and N. Hoffmann. 1984. Inheritance of functional foreign genes in plants. Science 223: 496–498.

Hospital, F., C. Chevalet, and P. Mulsant. 1992. Using markers in gene introgression breeding programs. Genetics 132:1199–1210.

Hospital, F.C., and A. Charcosset. 1997. Marker-assisted introgression of quantitative trait loci. Genetics 147:1469–1485.

Hudson, R.R. 2001. Linkage disequilibrium and recombination. *In* Handbook of Statistical Genetics. John Wiley and Sons, Inc.

Huffman, W.E., and R.E. Evenson. 1993. Science for Agriculture. ISU Press.

James, C. 2002. Global Status of Commercialized Transgenic Crops. International Service for the Acquisition of Agribiotech Applications (ISAAA). ISAAA Briefs No. 27: Preview. ISAAA, Ithaca, NY.

Jenkins, S., and N. Gibson. 2002. High-throughput SNP genotyping. Comp. Funct. Genom. 3: 57–66.

Johnson, G.R. 2002. Marker assisted selection. Long Term Selection Symposium, University of Illinois, Urbana, Illinois.

Kalton, R.R., and P. Richardson. 1983. Private sector plant breeding programs: A major thrust in U.S. agriculture. Diversity 5:16–18.

Kay, B.D. 1995. Soil quality: Impact of tillage on the structure of tilth of soil. *In* Farming for a Better Environment. Pp. 7–9. Soil and Water Conservation Society. Ankeny, Iowa.

Kern, J.S., and M.G. Johnson. 1993. Conservation tillage impacts on national soil and atmospheric carbon levels. Soil Sci. Soc. Am. J. 57:200–210.

Knutson, R.J., B.E. Hibbard, B.D. Barry, V.A. Smith, and L. L. Darrah. 1999. Comparison of screening techniques for western corn rootworm (Coleoptera:Chrysomelidae) host-plant resistance. J. Econ. Entomol. 92:714–722.

Krysan, J.L., J.J. Jackson, and A.C. Lew. 1984. Field termination of egg diapause in *Diabrotica* with new evidence of extended diapause in *D. barberi* (Coleoptera: Chrysomelidae). Environ. Entomol. 13:1237–1240.

Kwok, P.-Y. 2001. Methods for genotyping single nucleotide polymorphisms. Annu. Rev. Genomics Hum. Genet. 2:235–258.

Lande, R., and R. Thompson. 1990. Efficiency of marker-assisted selection in the improvement of quantitative traits. Genetics 124:743–756.

Lee, E.A., M. Lee, and K.R. Lamkey. 1990. RFLP analysis of isogenic lines B14 and B14A. Maize Genetics Newsletter 64:20–21.

Levine, E., J.L. Spencer, S.A. Isard, D.W. Onstad, and M.E. Gray. 2002. Adaptation of the western corn rootworm to crop rotation: Evolution of a new strain in response to a management practice. Am. Entomol. 48:94–107.

Liu, S., R.G. Cantrell, J.C. McCarty, Jr., and J.M. Stewart. 2000. Simple sequence repeat-based assessment of genetic diversity in cotton race stock accessions. Crop Sci. 40:1459–1469.

Mangelsdorf, P.C., and G.S. Fraps. 1931. A direct quantitative relationship between vitamin A in corn and the number of genes for yellow pigmentation. Science 73:241–242.

Matsuoka, Y., Y. Vigouroux, M.M. Goodman, J. Sanchez G., E. Buckler, and J. Doebley. 2002. A single domestication for maize shown by multilocus microsatellite genotyping. PNAS 99(9):6080–6084.

Metcalf, R.L. 1986. Foreword. pp. vii–xv. *In* Methods for the study of pest *Diabrotica*. J.L. Krysan and T. A. Miller (eds.) New York, Springer-Verlag. 260 pp.

Metz, G. 1994. Probability of net gain of favorable alleles for improving an elite single cross. Crop Sci. 34:668–672.

Nair, R.S., R.L. Fuchs, and S.A. Schuette. 2002. Current methods for assessing safety of genetically modified crops as exemplified by data on Roundup Ready Soybeans. Toxicol. Pathol. 30(1):117–125.

National Agricultural Statistics Service, Economic Research Service, United States Department of Agriculture. 2001. Agricultural Chemical Usage. 2000 Field Crops Summary. http://usda.mannlib.cornell.edu/reports/nassr/other/pcu-bb/agch0599.pdf.

National Corn Growers Association. 2003. The World of Corn. www.ncga.org.

Nelson, G.C., and D.S. Bullock. 2003. Simulating a relative environmental effect of glyphosate-resistant soybeans. Ecol. Economics 45(2):189–202.

Nelson, K.A., and K.A. Renner. 1999. Cost-effective weed management in wide- and narrow-row glyphosate resistant soybean. J. Prod. Agric 12:361–465.

Office of Pest Management Policy and the Pesticide Impact Assessment Program. 1999. United States Department of Agriculture. http://cipm.ncsu.edu/cropprofiles/.

Padgette, S.R., D.B. Re, B.G. Hammond, R.L. Fuchs, S.G. Rogers, L.A. Harrison, D.L. Nida, M.W. Naylor, K.H. Kolacz, N.B. Taylor, and J.E. Ream. 1994. Safety, compositional and nutritional aspects of glyphosate-tolerant soybeans: Conclusion based on studies and information evaluated according to FDA's consultation process. Pp. 1–107. Monsanto Company, St. Louis, Missouri.

Padgette, S.R., K.H. Kolacz, X. Delanney, D.B. Re, B.J. LaVallee, C.N. Tinius, W.K. Rhodes, Y.I. Otero, G.F. Barry, D.A. Eichhotz, V.M. Peschke, D.L. Nida, N.B. Taylor, and G.M. Kishore. 1995. Development, identification, and characterization of a glyphosate-tolerant soybean line. Crop Sci. 35:1451–1461.

Padgette, S.R., N.B. Taylor, D.L. Nida, M.R. Bailey, J. MacDonald, L.R. Holden, and R.L. Fuchs. 1996. The composition of glyphosate-tolerant soybean seeds is equivalent to that of conventional soybeans. Journal of Nutrition 126:702–716.

Paterson, A.H. 1998. High-resolution mapping of QTLs. *In* Molecular Dissection of Complex Traits. CRC Press, New York.

Patterson, A.H., E.S. Lander, J.D. Hewitt, S. Peterson, S. Lincoln and S.E. Tanksley. 1988. Resolution of quantitative traits into Mendelian factors by using a complete linkage map of Restriction Fragment Length Polymorphisms. Nature 335: 721–726.

Patterson, E.B. 1982. The mapping of genes by the use of chromosome aberrations and multiple gene marker stocks. Pp. 85–88. *In* W.F. Sheridan ed.: Maize for Biological Research. Plant Molecular Biol. Assoc., Charlottesville, VA.

Patterson, H.D., and R. Thompson. 1971. Recovery of inter-block information when block sizes are unequal. Biometrika 58:545–554.

Patterson, H.D., and R. Thompson. 1974. Maximum likelihood estimation of components of variance. Proceedings of the 8th International Biometric Conference: 197–207. Biometric Society, Washington, DC.

Perlak, F.J., M. Oppenhuizen, K. Gustafson, R. Voth, S. Sivasupramaniam, D. Heering, B. Carey, R.A. Ihrig, and J.K. Roberts. 2001. Development and commercial use of Bollgard(r) cotton in the USA-early promises versus today's reality. Plant J. 27(6):489–501.

Perlak, F.J., R.L. Fuchs, D.A. Dean, S.L. McPherson, and D.A. Fischhoff. 1991. Modification of the coding sequence enhances plant expression of insect control protein genes. PNAS 88:3324–3328.

Pratt, P.W., and A.J. Wrather. 1998. Soybean disease loss estimates for the Southern United States during 1994–1996. Plant Dis. 82:114–116.

Pritchard, J.K. 2001. Deconstructing maize population structure. Nature Genet. 28:203–204.

Pritchard, J.K., M. Stephens, N.A. Rosenberg, and P. Donnelly. 2000. Association mapping in structured populations. Am. J. Hum. Genet. 67:170–181.

Reicosky, D.C. 1995. Impact of tillage on soil as a carbon sink. *In* Farming for a Better Environment. Soil and Water Conservation Society. Ankeny, Iowa.

Reicosky, D.C., and M.J. Lindstrom. 1995. Impact of fall tillage on short-term carbon dioxide flux. Pp 177–187. *In* Soils and Global Change. Lal, R., J. Kimble, E. Levine, and B.A. Stewart (eds.). Lewis Publishers, Chelsea, MI.

Remington, D.L, J.M. Thornsberry, Y. Matsuoka, L.M. Wilson, S.R. Whitt, J. Doebley, S. Kresovich, M.M.Goodman, and E.S. Buckler IV. 2001. Structure of linkage disequilibrium and phenotypic associations in the maize genome. PNAS 98:11479–11484.

Riedell, W.E. 1990. Rootworm and mechanical damage effects on root morphology and water relations in maize. Crop Sci. 30:628–631.

Riedell, W.E., and P.D. Evenson. 1993. Rootworm feeding tolerance in single-cross maize hybrids from different eras. Crop Sci. 33:951–955.

Risch, N., and K. Merikangas. 1996. The future of genetic studies of complex human diseases. Science 273:1516–1517.

Roberts, R.K., R. Pendergrass, and R.M. Hayes. 1999. Economic analysis of alternative herbicide regimes on Roundup Ready soybeans. J. Production Agriculture 12:449–454.

Rogers, R.R., J.C. Owens, J.J. Tollefson, and J.F. Witkowski. 1975. Evaluation of commercial corn hybrids for tolerance to corn rootworms. Environ. Entomol. 4:92–922.

Russell, W.A. 1965. Effect of corn leaf rust on grain yield and moisture in corn. Crop Sci. 5:95–96.

Russell, W.A. 1993. Achievements of maize breeders in North America. International Crop Sci. CSSA Publications.

Sax, K. 1923. The association of size differences with seed-coat pattern and pigmentation in *Phaseolus vulgaris*. Genetics 8:552–560.

Schönbrunn E., S. Eschenburg, W.A. Shuttleworth, J.V. Schloss, N. Amrhein, J.N.S. Evans, and W. Kabsch (2001). Interaction of the herbicide glyphosate with its target enzyme EPSP synthase in atomic detail. *Proc. Natl. Acad. Sci. U.S.A.* 98:1376–1380.

Schork, N.J. 1997. Genetics of a complex disease. Am. J. Respir. Crit. Care Med. 156:S103–S109.

Searle, S.R. 1971. Linear Models. Wiley, New York.

Sebolt, A.M., R.C. Shoemaker, and B.W. Diers. 2000. Analysis of quantitative trait locus allele from wild soybean that increases seed protein concentration in soybean. Crop Sci. 40:1438–1444.

Sharma, H.C., and B.S. Gill. 1983. Current status of wide hybridization in wheat. Euphytica 32:17–31.

Shaw, D.R., and C.S. Bray. 2003. Foreign material and seed moisture in glyphosate-resistant and conventional soybean systems. Weed Tech. 17(2):389–393.

Simmonds, N.W. 1991. Bandwagons I have known. TAA Newsletter. December: 7–10.

Sneller, C.H. 2003. Impact of transgenic genotypes and subdivision on diversity within elite North American soybean germplasm. Crop Sci. 43:409–414.

Soller, M., and J. Plotkin-Hazan. 1977. The use of marker alleles for the introgression of linked quantitative alleles. Theor. Appl. Genet. 51:133–137.

Spielman, R.S, R.E. McGinnis, and W.J. Ewens. 1993. Transmission test for linkage disequilibrium: the insulin gene region and insulin-dependent diabetes mellitus. Am. J. Hum. Genet. 52:506–513.

Spike, B.P., and J.J. Tollefson. 1991. Yield response of corn subjected to western corn rootworm (Coleoptera: Chrysomelidae) infestation and lodging. J. Econ. Entomol. 84:1585–1590.

Sprague, C.L. 2002. Roundup Ready Crops: Are We Limiting Our Options. 32nd Annual Soybean Research Conference of the American Seed Trade Association, Chicago, IL.

Stam, P., and A.C. Zeven. 1981. The theoretical proportion of the donor genome in near-isogenic lines of self-fertilizers bred by backcrossing. Euphytica 30:227–238.

Stark, S.B., R.G. Mohanty, and J.M. Ver Hoef. 2001. ASA Conservation Tillage Study. American Soybean Association, St. Louis, MO.

Stuber, C.W., J.F. Wendel, M.M. Goodman, and J.S.C. Smith. 1988. Techniques and scoring procedures for starch gel electrophoresis of enzymes from maize *(Zea mays* L.). N.C. Agric. Res. Serv., N.C. State Univ., Tech. Bull. 286:1–87.

Stuber, C.W., and M.M. Goodman. 1983. Allozyme genotypes for popular and historically important inbred lines of corn, *Zea mays* L. USDA Agric. Res. Results, Southern Ser., No. 16.

Stuber, C.W., J.F. Wendel, M.M. Goodman, and J.S.C. Smith 1988. Techniques and scoring procedures for starch gel electrophoresis of enzymes from maize *Zea mays* L. North Carolina State Agric. Exp. Stn. Res. Bull. 286.

Sutter, G.R., and T.F. Branson. 1980. A procedure for artificially infesting field plots with corn rootworm eggs. J. Econ. Entomol. 73:135–137.

Sykes, B. 2001. The Seven Daughters of Eve. W.W. Norton and Company. London.

Tanksley, S.D., and S.R. McCouch. 1997. Seed banks and molecular maps: Unlocking genetic potential from the wild. Science 277:1063–1066.

Tanksley, S.D., S. Grandillo, T.M. Fulton, D. Zamir, Y. Eshed, V. Petiard, J. Lopes, T. Beck-Bunn. 1996. Advanced backcross QTL analysis in a cross between an elite processing line of tomato and its wild relative *L. pimpinellifolium*. Theor. Appl. Genet. 92:213–224.

Tanksley, S.D., H. Medina-Filho, and C.M. Rick. 1981. The effect of isozyme selection on metric characters of an interspecific backcross of tomato—basis of an early screening procedure. Theor. Appl. Genet. 60:291–296.

Tanksley, S.D., and C.M. Rick. 1980. Isozymic gene linkage map of the tomato: Applications in genetic and breeding. Theor. Appl. Genet. 57:161–170.

Tanksley, S.D., and C.M. Rick. 1980. Isozyme gene linkage map of the tomato: Applications in genetics and breeding. Theor. Appl. Genet. 57:161–170.

Tanksley, S.D., J. Medina-Filho, and C.M. Rick. 1982. Use of naturally-occurring enzyme variation to detect and map genes controlling quantitative traits in an interspecific backcross of tomato. Heredity 49:11–25.

Taylor, N.B., R.L. Fuchs, J. MacDonald, A.R. Shariff, and S.R. Padgette. 1999. Compositional analysis of glyphosate-tolerant soybeans treated with glyphosate. J. Agric. and Food Chem. 47:4469–4473.

Tenaillon, M.I., M.C. Sawkins, A.D. Long, R.L. Gaut, J.F. Doebly, and B.S. Gaut. 2001. Patterns of DNA sequence polymorphism along chromosome 1 of maize (*Zea mays* ssp. *mays* L.). PNAS 98:9161–9166.

Thornsberry, J.M., M.M. Goodman, J. Doebly, S. Kresovich, D. Nielsen, and E.S. Buckler IV. 2001. *Dwarf8* polymorphisms associatwith variation in flowering time. Nat. Genet. 28:286–289.

USDA. 2003. Crop Production Acreage—Supplement June 2003. National Agricultural Statistics Service, Agricultural Statistics Board, U.S. Department of Agriculture. Washington, DC. (http://usda.mannlib.cornell.edu/reports/nassr/field/pcp-bba/acrg0603.txt).

Visscher, P.M., C.S. Haley, and R. Thompson. 1996. Marker-assisted introgression in backcross breeding programs. Genetics 144:1923–1932.

Vos, P., Hogers, R., Bleeker, M., Reijans, M., van de Lee, T., Fornes, M., Frijters, A., Pot, J., Peleman, J., Kuiper, M., and Zabeau, M. 1995. AFLP: A new technique for DNA fingerprinting. Nucleic Acids Res. 23:4407–4414.

Walker, A.R. 2002. Impact of the Roundup Ready Gene on Soybean Breeding Roundup Ready Report Card: Genetic Gain. 32nd Annual Soybean Research Conference of the American Seed Trade Association, Chicago, Illinois.

Welsh, J. and McClelland, M. 1990. Fingerprinting genomes using PCR with arbitrary primers. Nucleic Acids Res. 18:7213–7218.

Williams, J.G.K., A.R. Kubelik, K.J. Livak, J.A. Rafalski, and S.V. Tingey. 1990. DNA polymorphisms amplified by arbitrary primers are useful as genetic markers. Nucleic Acids Res. 18:6531–6535.

Wrather, J.A., W.C. Stienstra, and S.R. Koenning. 2001. Soybean disease loss estimates for the United States from 1996 to 1998. Can. J. Plant Pathol. 23:122–131.

Xiao, J., J. Li, S. Grandillo, S. Nag Ahn, L. Yuan, S.D. Tanksley, and S.R. McCouch. 1998. Identification of trait-improving quantitative trait loci alleles from a wild rice relative, *Oryza rufipogon*. Genetics 150:899–909.

Young, N.D., and S.D. Tanksley. 1989. RFLP analysis of the size of chromosomal segments retained around the *Tm-2* locus of tomato during backcross breeding. Theor. Appl. Genet. 77:353–359.

2

Who Are Plant Breeders, What Do They Do, and Why?

James G. Coors, Department of Agronomy, University of Wisconsin-Madison

Even though plant breeders have an intuitive sense of what they do and what function they perform, the general scientific community and the public at large have little understanding of the essential nature of plant breeding. The first section of this chapter reviews historical trends relating to numbers of plant breeders. However, a great deal is left unsaid by merely reviewing survey results or historical patterns. So, while the first section provides a necessary starting point for a discussion about the plant-breeding profession, the more interesting issues—why do plant breeders do what they do, and should they even try to do it—are left hanging. Therefore, the last two sections address the more fundamental issues of what plant breeding actually accomplishes and how it fits in with the modern era of genomic science.

Who are the plant breeders, and how many are there?

There has been a recent increase in private plant-breeding expenditures in industrialized countries to the extent that private investment may now surpass public expenditures by a considerable margin (Heisey et al., 2001). This trend is particularly acute in the United States Based on Frey's National Plant Breeding Study — I (Frey, 1996); in 1994 there were a total of 2,241 science person years (SYs) devoted to plant-breeding research and development in the U.S. Of these, 1,499 were in the private sector, and 742 were in the public sector. From 1990 to 1994, the net loss from state agricultural experiment stations (71% of the total public sector involvement) was estimated to be 2.5 SY/year, while private industry increased at 32 SY/year. Over this period, private industry spent approximately $338 million on plant-breeding research annually (61%), while the public sector spent approximately $213 million/year (39%). There are many reasons for these trends, and among them are the following: (1) there is an increasing emphasis on basic (versus applied) research in the public sector because of the need to attract funds from federal granting agencies; (2) new organizations with single-interest focus (environment, consumer, etc.) are diluting the public-funding base; (3) funding for public agricultural research has not kept pace with increasing research and development costs; and (4) intellectual property restrictions have lessened public access to elite germplasm.

The consequences of the decrease in public sector plant breeding may be particularly severe for minor crops. As the public sector shrinks, many of the minor agronomic and horticultural crops risk becoming plant-breeding orphans. The private sector has embraced biotechnology to the extent that its near-term focus must be on relatively simply inherited traits and on major crops grown in the developed world as a necessary strategy to recoup the substantial research investments made in recent years. Given the negative public sentiment toward biotechnological innovations such as genetic transformation, those crops directly consumed by humans, many of which are classified as minor crops, will probably not receive much attention in the near future. Unfortunately, despite the moniker "minor," most minor crops are important components of the agricultural system, for example, perennial grasses and forage legumes, and

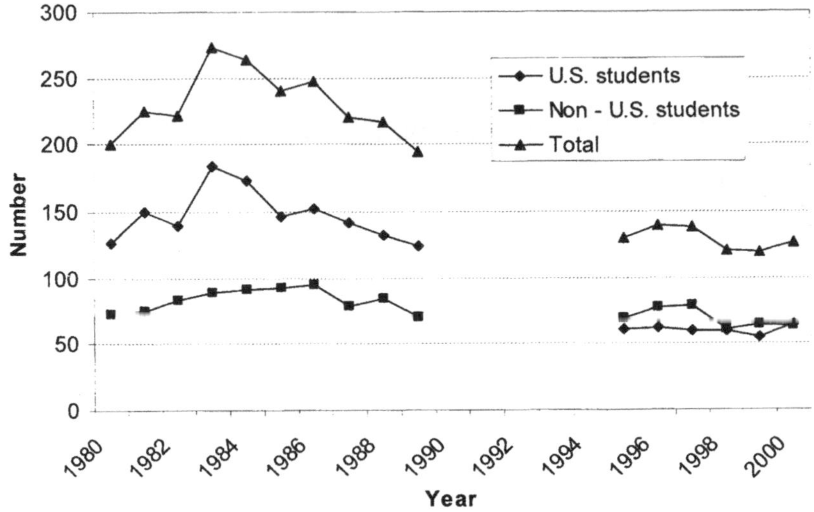

Figure 2.1 Plant-breeding graduate degrees awarded by U.S. universities from 1980 to 1989 (Collins and Phillips, 1991) and from 1995 to 2000 (Guner and Wehner, 2003).

many so-called minor crops can become major crops in a relatively short time, for example, alfalfa and soybeans.

One of the more ominous features of the Heisey et al. study (2001) and the Frey (1996) survey is that the public infrastructure supporting the education of plant breeders destined for either public or private service appears to be eroding. There have been a number of surveys of graduate training over the past several decades addressing this issue, but two of the most recent best depict the current situation—the Collins and Phillips (1991) survey performed over the 1980–1989 time period and the Guner and Wehner (2003) survey, which focused on 1995–2000.

The Collins and Phillips (1991) survey was sent to all public land grant and 1890 colleges. Responses were received from 84 departments from 46 institutions in 42 states. Institutions in 2 states did not respond, and 6 indicated no plant-breeding activity. The Guner and Wehner (2003) survey was sent to 71 land grant universities, and 52 indicated that they had capacity for plant-breeding training. In contrast to the Collins and Phillips (1991) survey, the Guner and Wehner (2003) survey had a specific statement requesting that students working mostly in molecular genetics not be counted as involved in plant-breeding research. Responses were received from 82 departments from 47 institutions in 47 states. Institutions in 3 states did not respond, and 7 reported that they had no degree programs involving plant breeding.

Based on the coverage and response rates, the two surveys seem roughly comparable, and they are graphed together in Figure 2.1. Collins and Phillips (1991) reported that there was no real change in numbers of graduate students from 1980 to 1989, but there was a trend upward in early 1980s followed by a downward trend toward the end of the decade. Collins and Phillips (1991) were not certain that the latter trend was real. The Guner and Wehner (2003) survey appears to support the downward trend starting in the mid-1980s, but from 1995 on there was little change. Some caution is needed comparing trend lines, however. In particular, it is difficult to determine what effect the molecular genetics disqualifier had in the Guner and Wehner (2003) survey and whether a similar statement would have affected the earlier survey.

One trend does seem obvious. The number of non-U.S. graduate students in plant breeding in the period from 1995 to 2000 equals or exceeds U.S. students, whereas in the 1980s there were nearly twice as many U.S. students as non-U.S. students. There may be several reasons for this. Non-U.S. graduate students in plant breeding are often funded by their home institutions, making them very attractive to cash-strapped U.S. plant-breeding programs. But also, many U.S. plant breeders may find non-U.S. students better acquainted with agricultural issues and better motivated to perform the public service of plant breeding.

Thirty-seven institutions were common to the two surveys, and the figures were broken down by institution for the two time periods (Figure 2.2).

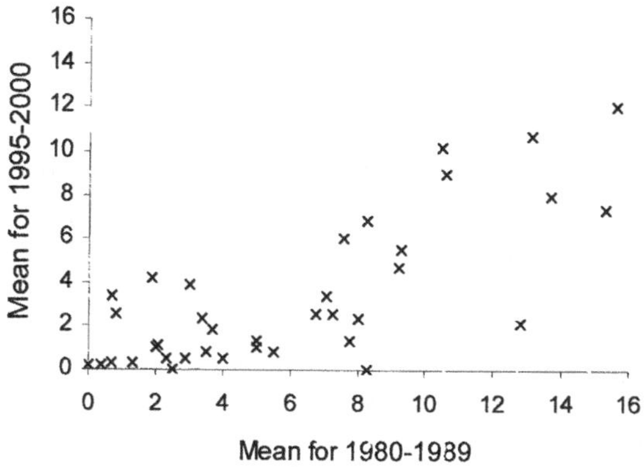

Figure 2.2 Number of plant-breeding graduate degrees awarded per year at 37 U.S. universities. Data for 1980–1989 are from the survey of Collins and Phillips (1991), and data from 1995 to 2000 are from the survey of Guner and Wehner (2003). The 37 land-grant universities are those in common for the two surveys. The institutions in the circle are the 10 that produced the highest number of plant-breeding graduate degrees (master's and doctorate degrees for both U.S. and non-U.S. students) based on the 1995–2000 survey.

Most of the major plant-breeding training institutions in the 1980s remained strong in the late 1990s, although most experienced a decrease in the number of graduate degrees awarded. Fortunately, the top 10 institutions in 1995–2000 represented a diversity of regions in the United States, with the possible exception of the far western United States. It is also seems that several institutions may have downsized their plant-breeding programs to a considerable extent.

One final way to quantify trends in plant-breeding activity is to review registration articles in *Crop Science* as tallied by the Germplasm Retrieval Information Network (GRIN, 2003). Since 1926 there have been over 10,000 such registration articles. Many plant breeders in the United States and elsewhere publish brief registration manuscripts in *Crop Science* and then deposit the referenced germplasm in the U.S. National Plant Germplasm System (NPGS). From an academic standpoint, registration manuscripts count in the tally of a public sector scientist's publications, which encourages registration, especially among young scientists seeking promotion. On the other hand, individuals also receive professional credit for registering intellectual property with their university's intellectual property office, which may preclude registering germplasm (along with the required seed deposit in the NPGS). In other words, some caution is needed when interpreting these data.

For all classes of registrations and for all crops combined, registration activity leveled off sometime in the early to mid-1990s (Figure 2.3). If it were not for the addition of the genetic stock category, the trend would be downward. Surprisingly, this is not due to a lessening of cultivar or parental line releases, which are the most adapted and immediately useful germplasm. These classes have remained relatively stable over a long period beginning in the 1970s. Instead, germplasm registrations leveled off beginning in the mid-1990s.

If the so-called major and minor crops (as defined by the National Agricultural Statistics Service,

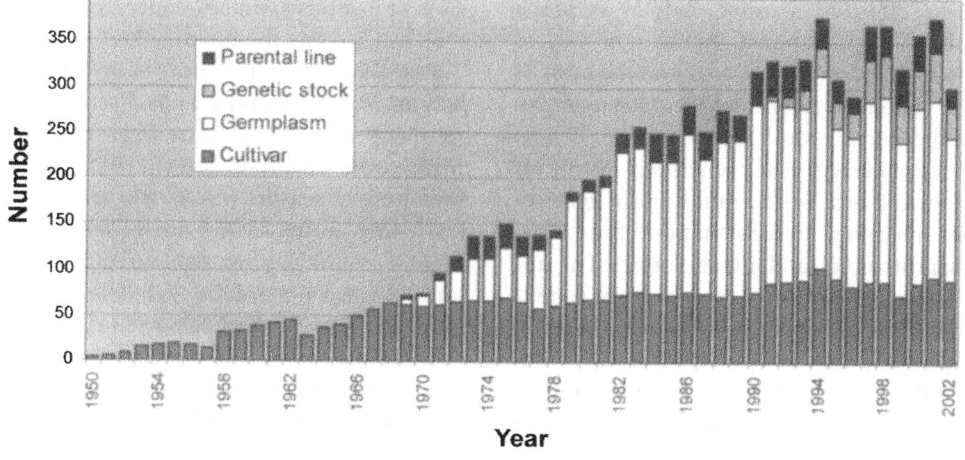

Figure 2.3 *Crop Science* registration manuscripts published from 1950 to 2002. Data presented are five-year trailing means for all crops (GRIN, 2003).

USDA) are separated out, the trends are somewhat similar for the two categories (data not provided). The germplasm class peaked sometime in the early to mid-1990s for the minor crops, perhaps a little earlier than that for major crops. However, the number of minor crop cultivar registrations trends slightly upward up to the present time, which is encouraging relative to the recommendations for increased public effort for minor crops coming from the National Plant Breeding Studies of Frey (2000).

Six crops, alfalfa, cotton, maize, soybean, sorghum, and wheat, make up about two-thirds of the total of 4,739 germplasm registrations from 1967 (the year the germplasm category came into use) to 2002. Germplasm registrations for alfalfa, cotton, and maize have trended downward from the mid-1990s, while sorghum, soybeans, and wheat remain level or have increased slightly. These trends reflect somewhat the relative importance of public and private sector plant-breeding involvement. Molecular genetic approaches may also be supplanting germplasm enhancement activities for those crops experiencing a decrease in germplasm registrations.

Why do plant breeders do what they do?

Operational model

There has been a substantial transformation in how genetics relates to plant breeding. Until recently, the focus was on plants and phenotypes, and phenotypic selection was the *raison d'etre* for plant breeders. Plant breeders relied on disciplines such as statistical genetics that, in some vague but nonetheless effective manner, helped improve germplasm. The operational model was that of form follows function; that is, select on the basis of phenotype (function), and changes in the underlying genotype (form) would follow.

The focus of current plant genetics is mostly on genes and genotypes. We are in the era of gene sequencing, mapping, transformation, functional genomics, proteomics, and metabolomics. The underlying assumption of most current plant geneticists is that if the genotype is well enough understood, improved plants and phenotypes will follow without undue exertion. The new vision of a plant breeder is that of a true engineer who assembles the appropriate set of nucleotide sequences in the construction of an ideal genotype. The engineering approach to plant improvement most closely follows the function follows form model. The ultimate goal is to engineer plants from the sequence up, locus by locus, rather than, as some would claim, work backward by using plants and their phenotypes to modify the underlying genotype.

What is not often recognized is that the change from the form follows function model to that of function follows form is a profound philosophical transformation in how scientists view the biological world. Form follows function clearly has been the Darwinian operational model underlying evolutionary advance starting with the first replicating molecule over 3 billion years ago. Only in the last several decades has it become conceivable to work from the genetic sequence back up to the whole organism. This transformation seems to be happening by default, without any discussion or challenge.

What is the real relationship between the genotype and phenotype? This is a particularly acute issue for students intrigued with the promise of the plant breeding and plant genetics disciplines. Recent generations of students have been generally highly disciplined academically, and most new students have broad and thorough understanding of genetic technology, far surpassing that of any past generation. Current students now also typically come from biological rather than agricultural disciplines. They have grown up mostly in air-conditioned urban settings; they tend to have little understanding of agriculture in general; and their notion of the "environment" is relatively unsophisticated. Most new students have the optimistic sense that the genotype is now directly controllable or it shortly will be. Understanding the genotype has become the essential and ultimate target.

Past plant-breeding students, on the other hand, had an ingrained and practical sense of the environment since many came from rural areas and many were involved directly with farming. To them, every season was a new season, and they knew that in any season no sequence of environmental events is ever repeated again. They also knew what a phenotype was. They helped plant, cultivate, harvest, and sell phenotypes. However, they faced a vexing limitation in that the genotype was only a concept. They knew it existed and that quantitative genetic models could be used to help breeders select more efficiently, but that was the end of it.

This transition is not particular to students. It has also occurred for their faculty mentors. What are the implications of this transition, and more to the point, what has plant breeding now become?

What is plant breeding in the modern era?

Typically, most discussions of how plant-breeding works start with the idea of a breeding population from which adapted cultivars are derived. A breeding population might be a broad-based population or a narrow-based population, such as an F2 generation of a cross between two lines. Plants derived from the breeding population might then be improved by pedigree selection to create the adapted cultivar. Some breeding approaches more tightly focus on trait introgression, and the breeding population used to start the process may actually be single inbred lines or even single plants. Backcrossing or some form of molecular genetic transformation can be used to insert one or few genes of value. The essential feature common to all approaches is that there is a starting germplasm source and that adapted varieties are to be derived from it in some fashion.

Trait introgression has become very important in the current era. It is now mostly a gene-oriented, mechanistic approach, and, as such, it is intellectually attractive and regarded as a more rational approach than merely relying on chance events inherent to the sexual cycle (segregation and recombination). Trait introgression uses *a priori* structural knowledge of genes and proteins and provides predictable outcomes. It also works well, for the most part.

But both pedigree selection and trait introgression are one-dimensional approaches, and plant breeders must work in at least two dimensions. Not only is it important to develop adapted cultivars from current breeding populations, plant breeders must also provide future generations a continuous supply of ever-improved breeding populations. Breeding populations, in whatever form, serve as the base platform for plant improvement, either by means of selection or trait introgression, and they will remain in this role for the foreseeable future.

We don't know all we need to know about the genetic control of even the most well-defined and simple metabolic pathways, so the notion that merely adjusting the genetic architecture of a common, stagnant germplasm base will suffice is simply foolish, although there is tremendous commercial reward for operating in exactly this fashion. Recycling selected materials to form new breeding populations has been a major long-term responsibility of plant breeders, but since reliance on the sexual cycle is now regarded as somewhat suspect, and perhaps even irrational, at least relative to modern genetic approaches, fewer and fewer plant breeders seem to want to do it.

As Knight (2003) points out, other forces are also at work that undermine the plant-breeding profession. In both the public and private sectors, reward structures are strongly skewed toward short-term objectives, for example, gene discovery, papers, patents, and promotion, rather than addressing more substantive and long-term problems. Knight glibly suggests redefining plant breeders as "open-source molecular agronomists," as a means of providing some sort of professional cachet, but much more is needed.

What is required is that there be a thorough re-examination and reinvigoration of the intellectual foundation of the plant-breeding discipline. The basic problem is that conventional plant breeding is not usually considered an overly scientific pursuit. To a large extent, success relies on factors of chance such as mutation, recombination, genetic drift, and the environment. Chance events cause the most problems for the current scientific generation. Under the engineering operational model, how can random events serve any purpose? The sexual cycle is inexact and, therefore, outmoded. Why rely on random recombination and mutation, if we can ultimately assemble the precise base sequences we need?

Plant breeders must recognize that their strength lies in what are now two unique attributes: (1) respect for the phenotype, and (2) an understanding of the creative power of selection. The challenge is to bring new intellectual rigor to the understanding of the phenotype and selection in an appealing and fruitful way.

What is the scientific rationale for plant breeding?

Ironically, the most intriguing justification for the plant-breeding approach to problem solving comes from disciplines closely tied to engineering and computational programming (e.g., artificial intelligence, evolutionary computation, and computational ecology). These disciplines are attempting to use the current understanding of molecular genetics, developmental biology, and evolutionary biology to address some of the most complex engineering/computational issues of the day. They do so by evoking the concept of evolvability as a way of embracing mechanisms promoting productive change.

The concept of evolvability

To a plant breeder, evolvability is an organism's capacity to generate heritable phenotypic variation. More generally, evolvability can be thought of as the process by which complex systems acquire the capacity to discover and perpetuate beneficial adaptations (Stewart, 1997). Living organisms are exquisitely evolvable, and many researchers in nonbiological disciplines are intrigued by the possibility of harnessing evolvability on a broader scale.

Computer programmers dealing with highly complex tasks such as prediction of climatic change or those in artificial intelligence who want to imbue computer code with the ability to learn are designing systems of computer algorithms in such a manner that one can use genetic operations to more efficiently arrive at optimal code than would be possible by a standard programming approach (Wagner and Altenberg, 1996). To greatly simplify, the goal of evolutionary computation is to design code such that it can handle random coding variants. One can then choose among the variants based on how efficiently programs accomplish some computational task. Selected variants are then recombined in some fashion to create the next round of possible solutions for continued improvement. Repeated iterations of this procedure can provide increasingly efficient solutions for highly complex tasks. The analogies to selective breeding are obvious. The computer code is the genotype, the function performed by the code is the phenotype, random coding mistakes or variants represent mutations, and random replacement of algorithms among selected variant systems of code represents recombination. Most intriguing is the fact that the operational model has shifted from function follows form to form follows function. These disciplines have the power plant geneticists so desperately want—the ability to create the underlying code specifying precisely any outcome—yet they are looking at evolutionary paradigms to more efficiently achieve their goals.

Obviously, the situation has been oversimplified. Computer code will not respond to an evolutionary approach unless programs are suitably designed (Marrow, 1999). Random coding mistakes and scrambling of algorithms are not, in and of themselves, creative forces and will quickly disable most computer programs. Conditions must be appropriate for such random forces to be creative rather than destructive. Those involved with evolutionary computation have recognized that a thorough understanding of evolutionary biology is needed to provide some perspective on what these conditions might be.

What enhances evolvability?

There are many core biological processes that have been highly conserved across eukaryotes and even all life forms. For example, based on extensive evaluations of genomic synteny across plant taxa, it is becoming clear that perhaps more than 90% of plant genes in any given species have close homologs within most other plant genomes (Bennetzen, 2000). But what does this really mean? Darwin would be pleased to know that we now have ample genetic evidence that all organisms trace back to a common source. The more important question, though, is what is it about genome organization that starts with such homology yet provides such immense diversity in plant morphology and adaptation. The conventional view is that conserved features have been selected for efficient function and optimal design. However, as we learn more about metabolic systems, it is beginning to look like a significant number of "highly conserved developmental mechanisms are characterized by not being programmed for a particular specialized job and in some cases by profligate inefficiency" (West-Eberhard, 1998).

Just as with complex computer code, genes provide the instructions for carrying out specific functions in a complex living system. If molecular requirements for gene function are numerous and extremely precise, the system becomes highly constrained. Changes in amino acid or base sequence are likely to be catastrophic. Something must be acting to deconstrain systems of core biological processes such that organisms can evolve.

Deconstraining mechanisms

There are a number of likely deconstraining mechanisms that ultimately shape the genotype–phenotype map in such a way as to preserve a great deal of phenotypic plasticity even though the underlying genetic systems may be highly conserved. Those interested in these issues use concepts such as "exploratory behavior," "hierarchical redundancy," "modularity," and "weak linkage" to explain how evolvable systems come about.

Exploratory behavior is well covered by the excellent review of evolvability by Kirschner and Gerhart (1998). Of course, the sexual cycle is inherently exploratory. It is a fundamentally stochastic process of creating variants and allowing selection to pick among the most successful. But the sort of exploratory behavior that Kirschner and Gerhart (1998) refer to occurs across all developmental stages and levels of organization. One example involves the kinetics of mitotic microtubule formation during the process of cell division, a highly conserved process throughout eukaryotes. Spindle microtubules connect to the kinetichores of chromosomes and mediate chromosomal segregation to the spindle poles. However, the process is far from straightforward. Spindle microtubules are dynamic and turn over with a half-life of 60–90 s. There is a rapid transition of microtubule ends from polymerizing to depolymerizing states. Since chromosomes are located somewhat randomly throughout the cell, random microtubule searches are required, and far more microtubules must be initiated than there are chromosomes. If a polymerizing microtubule contacts a kinetichore, fine; otherwise the microtubule depolimerizes, and the search goes on.

The dynamic structure of microtubule searches provides a very robust system because it reaches a functional state regardless of initial arrangement of chromosomes. It is a highly flexible system because it tolerates different cellular arrangements, and it allows an unlimited range of alternative cellular conformations. The process is fundamentally stochastic rather than mechanistic, and this is typical of exploratory behavior. There is an overproduction of random variants followed by selective use of only a few. In a more general sense, exploratory behavior is characterized by a system of random events that promote epigenetic variation that can become fixed by somatic selection (Kirschner and Gerhart, 1998).

Hierarchical redundancy seems to be a universal property of living organisms. Gene duplication is a well-known mechanism allowing divergence of gene function, but less well appreciated is the multiplicity of redundant systems at all levels of organization that serve essentially the same purpose—allowing divergence of function in response to varying internal and external conditions. Redundancy is particularly effective in concert with modularization. For example, repetition of morphological modules allows populations of cells to become independent. The evolution of multicellular organisms (Metazoa) is a case in point. For the first 3 billion years all life was unicellular. At some point, though, a number of independent single-celled organisms came into closer and closer contact, and some cells diverged slightly and took on specialized functions in response to particular microenvironmental stimuli. Once this happened and there was some benefit to the larger group, the race toward cellular specialization and new multicellular morphologies began. It was probably no mere coincidence that the Cambrian explosion closely followed the appearance of multicellular organisms (Gould, 1990).

Plants are really nothing more than repeating morphological modules termed phytomers. Repetition of morphological modules provides a degree of compartmental independence. Compartmentation allows weakly linked components to change function slightly (through mutation, epigenetic variation, and transcriptional regulation) and begin exploring alternate roles. Repetition of morphological modules allows populations of cells to become independent, reducing the deleterious effects of mutations, and increasing the potency of selection within modules. Phytomers represent a higher-order redundancy that provides a means of phenotypic accommodation that is very robust, yet also highly evolvable because any given change in extracellular or intracellular signal is not likely to cause a catastrophic failure in overall enzymatic, cellular, or morphological organization (West-Eberhard, 1998).

The nature of interactions, either among genes, molecules, pathways, or higher-order modules such as phytomers, strongly influences the evolvable potential of an organism. In general, as a biological system becomes more and more complex, interactions among components must become weaker (Conrad, 1990). Multiple weak interactions are complementary to redundancy in that if any one connection is broken, the system can remain functioning. Weak interactions allow for gradual transformation of function rather than complete dysfunction in the presence of mutation or some other genetic or environmental challenge.

Kirschner and Gerhart (1998) use the comparison of transcriptional regulation between prokaryotes and eukaryotes to highlight the more weakly linked (i.e., less-constrained and more evolvable) nature of the latter. In order to initiate gene expression, RNA polymerase is activated and bound to the transcription initiation site, but this process depends upon the binding of other components. In prokaryotes, there is a high degree of binding specificity for these components, the binding sites must be near the transcriptional initiation site, and the overall regulatory system is relatively simple and the control quite stringent. The eukaryotic system has a great many more transcriptional inputs involving proteins binding at multiple enhancer sequences located both near and far from the initiation site. The binding specificity can be relatively low. Multiple inputs are essential to regulate genes in response to the variable conditions eukaryotic organisms face during development. But individual inputs are less well linked to the regulatory network than that which is typical with prokaryotes.

Evolvable Features of the Lignin Pathway

The lignin pathway provides several examples of how an evolvable system of organization operates for a single metabolic process in plants that is important for both breeders and geneticists. Lignin is a core component of plant cell walls, and it is important for a number of reasons including water transport, structural integrity, rigidity, and pest resistance. High levels of lignin typically reduce the nutritional quality of forages and increase the difficulty in pulping of forest products. Lignin is under intense scrutiny by plant breeders and geneticists interested in altering lignin composition (Baucher et al., 1998).

Lignin is a highly complex molecule typically formed from three monolignols, sinapyl, conferyl, and *p*-coumaryl alchohols. Lignification occurs in three discrete steps. First is the biosynthesis of monolignols. The enzyme peroxidase then converts monolignols to free radicals, which are transported to the cell wall. Finally the monolignols in the cell wall are polymerized by an oxidative coupling process (Hatfield and Vermerris, 2001).

Monolignol precursors of lignin can be formed by any of several interconnected metabolic routes. In the past several years, many of the enzymes involved in the lignin biosynthesis have been sequenced and cloned and their function well characterized (Chabbert et al., 1994b; Halpin et al., 1998; Lapierre, 1993; Li et al., 2000; Marita et al., 2003; Vermerris and Boon, 2001; Vignols et al., 1995). Several researchers have attempted to limit monolignol production by down-regulating certain enzymes such as cinnamoyl-CoA reductase, caffeic acid *O*-methyl transferase, or cinnamyl alcohol dehydrogenase. However, it has been difficult to predict with certainty the result of any given enzymatic perturbation in the monolignol pathway. In some instances even novel phenolic components can be recruited as substitute monolignols, and the resulting lignin polymer may well have nearly the same properties as the original form (Marita et al., 2001; Ralph et al., 1998, Ralph et al., 2001). In more general terms, the system is weakly linked, and the genotype-to-phenotype map is imprecise. It appears from recent lignin research that weak linkage between gene function and metabolic outcome may actually be advantageous, since it may enhance the tolerance, flexibility, and robustness of metabolic regulation.

Peroxidase activity underlies the second step in lignin formation, the conversion of monolignols to free radicals. Peroxidase is highly conserved across bacteria, fungi, plants, and animals. In plants, peroxidase is a flexible enzyme used for many functions apart from the lignin pathway. In corn (*Zea mays* L.), for example, there are at least 13 different peroxidase genes having many distinct roles and different tissue specificities (Maize GDB, 2003). Peroxidase is typical of many redundant, flexible, and versatile proteins that have broad target specificity and can impose varying levels of inhibition/activation, depending on external conditions. These sorts of flexible and versatile proteins contribute to evolvability because they make it easier to develop new targets and regulatory roles than it would be to change highly specific and constrained proteins.

Once peroxidase converts monolignol precursors to free radicals, and these precursors are transported to the cell wall, the complex cross-linking in the plant cell wall to form the final lignin polymer may be controlled by little more than chemical conditions at the time the free radicals are formed. There may be few regulating enzymes of any sort (Hatfield and Vermerris, 2001; Ralph et al., 2001).

This is highly contentious research that has led to the so-called "lignin war" (Rouhi, 2001). Some researchers are very skeptical. How can nature be so haphazard in the assembly of the second most-abundant biopolymer in plants (Davin and Lewis, 2000)? The response is that haphazard processes may actually be essential for such critical functions as those involved in the structural integrity of many different tissues, as well as defense against a large array of plant pests. Exploratory mechanisms that have low systematic requirements for achieving highly complex functional outcomes contribute greatly to the overall evolvability of living organisms. And, of course, the most evolvable metabolic systems are those that now exist.

Should plant breeders continue breeding plants?

Plant breeders should take heart that those in fields such as artificial intelligence or evolutionary computation, who have the sort of knowledge and tools geneticists most covet, the complete understanding of the underlying controlling code, and the ability to modify it at will, have become intrigued with the power plant breeders already possess, the use of the sexual cycle and selection, to address some of the most complex technological issues of the day.

Exploratory behavior, hierarchical redundancy, modularity, and weak linkage have provided clues to those in evolutionary computation on how to imbue coding systems with the capacity to discover and perpetuate beneficial adaptations. What are the implications for plant breeders and geneticists? There are at least five:

1. The function performed by evolvable systems of complex code must be only imperfectly and, in some cases, even haphazardly related to the underlying coding sequence itself.
2. The genotypic–phenotypic map is necessarily inexact or evolvability is not possible.
3. Phenotypic plasticity and loosely drawn genotypic–phenotypic maps will not make functional genomics any easier.
4. Highly evolvable traits will probably not be the initial focus of functional genomics simply because these sorts of traits will be the most difficult to handle.
5. Highly evolvable traits would probably be those most directly affecting reproductive fitness, and these are usually the traits of most interest to plant breeders(e.g., plant vigor and seed yield).

We are dealing with a biological world in which stochastic processes have reigned supreme for more that three billion years. The Darwinian revolution showed us how, even in the face of such forces, or perhaps more accurately stated, precisely because of them, biological life has achieved the remarkable ability of self-organization. Furthermore, this self-organization is fundamentally based on flexibility and plasticity at all levels. The acknowledgment of this is what truly distinguishes plant breeders from genetic engineers. It is a deeply profound distinction that few appreciate or comprehend. As Conrad (1990), a computer scientist, so aptly comments:

> The organizations that are best suited to evolution are precisely those that are the most ill suited to the classical standards of scientific description.

Plant breeders already know that multiple phenotypes can be conditioned by a single genotype, and multiple genotypes can give rise to the same phenotype. There is not a one-to-one correspondence between genotype and phenotype, nor should there be. Plant breeders know that the phenotype is what matters in the end and that selection based on the phenotype is precisely the process that has given rise to the evolvable nature of the plants they work with. Plant breeders know that sex is an admittedly disruptive process, but one that, when coupled with selection, is extremely creative.

The challenges confronting public plant breeders are not due to any deficiencies in their application of genetics or defects in their traditional approaches, but rather to economic, sociological, and philosophical factors that are diverting them from the task of creating novel plant germplasm. For the foreseeable future the biological justification for continuing conventional selection remains intact, and the practical consequences of shifting course are disturbing.

All of humankind has benefited greatly from one of the most cost-effective technologies ever devised, plant breeding. The benefits have been widely distributed to both the developed and the developing world. Recent biotechnological approaches to plant improvement have come at great expense, and the benefits appear to have a more limited distribution. Many would argue that we are only in the initial phase of developing exciting new technologies with tremendous future potential. Perhaps, but it seems that we should more closely evaluate the nature of this argument and better examine its underlying premise.

Acknowledgments

The author is very grateful to Judy Grotenhuis and Steve Eberhart, National Center for Genetic Resources Preservation formerly the National Seed Storage Laboratory, for assisting with the GRIN database.

This chapter evolved from a similar presentation by the author to the American Seed Trade Association in December, 2001, entitled "Changing Role of Plant Breeding in the Public Sector" (Coors, 2002).

References

Baucher, M., B. Monties, M. Van Montagu, and W. Boerjan. 1998. Biosynthesis and genetic engineering of lignin. Crit. Rev. Plant Science. 17:125–197.

Bennetzen, J.L. 2000. Comparative sequence analysis of plant nuclear genomes: microcolinearity and its many exceptions. Plant Cell 12:1021–1029.

Chabbert B., M.T. Tollier, B. Monties. 1994b. Biological variability in lignification of maize: expression of the brown midrib *bm2* mutation. J. Sci. Food Agri. 64:455–460.

Collins, W.W., and R.L. Phillips. 1991. Plant breeding training in the United States. USDA Rep. 591 (Rev.) National Plant Genetic Resources Board. Office of Under Secretary of Science and Education, U.S. Govt. Print. Office, Washington, DC.

Conrad, M. The geometry of evolution. 1990. BioSystems 24: 61–81.

Coors, J.G. 2002. Changing Role of Plant Breeding in the Public Sector. p. 48–66. In Proc. 56th Annu. Corn Sorghum Res. Conf., 5–7 December, 2001. Chicago, IL.

Davin, L.B., and N.G. Lewis. 2000. Dirigent proteins and dirigent sites explain the mystery of specificity of radical precursor coupling in lignan and lignin biosynthesis. Plant Physiology. 123:453–461.

Frey, K.J. 1996. National Plant Breeding Study-I: Human and financial resources devoted to plant breeding research and development in the United States in 1994. Special Report 98. Iowa Agriculture and Home Economics Experiment Station. Ames, IA.

Frey, K.J. 2000. National Plant Breeding Study-IV: Future priorities for plant breeding. Special Report 102. Iowa Agriculture and Home Economics Experiment Station. Ames, IA.

Gould, S.J. 1990. Wonderful Life: The Burgess Shale and the Nature of History. Norton, New York, NY.

Guner N., and T.C. Wehner. 2003. Survey of U.S. land-grant universities for training of plant breeding students. Crop Sci. 2003 43: 1938–1944.

GRIN, 2003. Germplasm Retrieval and Information Network. http://www.ars-grin.gov/npgs/searchgrin.html.

Halpin, C., K. Holt, J. Chojecki, D. Oliver, B. Chabbert, B. Monties, K. Edwards, and G. A. Foxon. 1998. Brown-midrib maize (bm1)—a mutation affecting the cinnamyl alcohol dehydrogenase gene. Plant J. 14:545–553.

Hatfield, R., and W. Vermerris. 2001. Lignin formation in plants. The dilemma of linkage specificity. Plant Physiology. 126:1351–1357.

Heisey, P.W., C.S. Srinivasan, and C. Thirtle. 2001. Public sector plant breeding in a privatizing world. ERS Agriculture Information Bulletin No. 772. Economic Research Service/ USDA. Beltsville, MD.

Kirschner, M., and J. Gerhart. 1998. Evolvability. Proc. Natl. Acad. Sci. 95:8420–8427.

Knight, J. 2003. Crop improvement: A dying breed. Nature. 421:568–570.

Lapierre, C. 1993. Application of new methods for the investigation of lignin structure. p. 133–166. In Jung, H.G., D.R. Buxton, R.D. Hatfield, and J. Ralph (eds.). Forage cell wall structure and digestibility. ASA. Madison, WI.

Li, L., J.L. Popko, U. Toshiaki, and V.L. Chiang. 2000. 5-Hydroxconiferyl aldehyde modulates enzymatic methylation for syringyl monolignol formation, a new view of monolignol biosynthesis in angiosperms. J. Biol. Chem. 275:6537–6545.

Maize GDB. 2003. Maize Genetics/Genomics Database project. http://www.maizegdb.org/

Marita, J.M., W. Vermerris, J. Ralph, and R.D. Hatfield. 2003. Variations in cell wall composition of maize brown midrib mutants. J. Agric. Food Chem. 51:1313–1321.

Marita, J.M., J. Ralph, C. Lapierre, L. Jouanin, and W. Boerjan. 2001. NMR characterization of lignins from transgenic poplars with suppressed caffeic acid O-methyltransferase activity. J. Chem. Soc., Perkins. I. 1:2939–2945.

Marrow, P. 1999. Evolvability: Evolution, computation, biology. p. 30–33. In Wu, A.S. (ed) Proc. 1999 Genetic and Evolutionary Computation Conf.

Ralph, J., R.D. Hatfield, J. Piquemal, N. Yahiaoui, M. Pean, C. Lapierre, and A.M. Boudet. 1998. NMR characterization of altered lignins extracted from tobacco plants down-regulated for lignification enzymes cinnamylalcohol dehydrogenase and cinnamoyl-CoA reductase. Proc. Nat. Agron. Soc. 95:12803–12808.

Ralph, J., C. Lapierre, F.C. Lu, J.M. Marita, G. Pilate, J. Van Doorsselaere, W. Boerjan, and L. Jouanin. 2001. NMR evidence for benzodioxane structures resulting from incorporation of 5-hydroxyconiferyl alcohol into lignins of O-methyltransferase-deficient poplars. J. Agric. Food Chem. 49:86–91.

Rouhi, A.M. 2001. Only facts will end the lignin war. Chem. Engin. News 79:52–56.

Stewart, J. 1997. The evolution of genetic cognition. J. Soc. Evol. Systems 20:53–73.

Vermerris, W., and J.J. Boon. 2001. Tissue-specific patterns of lignification are disrupted in the brown midrib2 mutant of maize (*Zea mays* L.). J. Agric. Food Chem. 49:721–728.

Vignols, F., J. Rigau, M.A. Torres, M. Capellades, and P. Puigdomenech. 1995. The brown midrib3 (bm3) mutation in maize occurs in the gene encoding caffeic acid O-methyltransferease. The Plant Cell 7:407–416.

Wagner, G.P., and L. Altenberg. 1996. Perspective: Complex adaptations and the evolution of evolvability. Evolution 50:967–976.

West-Eberhard, M.J. 1998. Evolution in the light of developmental and cell biology, and vice versa. Proc. Nat. Agron. Soc. 95:8417–8419.

3

Social and Environmental Benefits of Plant Breeding

Donald N. Duvick, Affiliate Professor of Agronomy, Iowa State University

The first plant breeding: Domestication of wild species

Social Benefits

Hunter–gatherer societies in all parts of the globe created domesticated crops from some of the plant species that furnished their food. For example, they domesticated wheat (*Triticum* spp. L.) in southwestern Asia, rice (*Oryza sativa* L.) in eastern Asia, sorghum (*Sorghum bicolor* [L.] Moench) in northeastern Africa, bananas (*Musa* spp.) in Melanesia, and maize (*Zea mays* L.) in Mesoamerica (Denham et al., 2003; Harlan, 1992). We have no written record of why or how this was accomplished for any crop but we can speculate that the domesticators had definite social benefits in mind, for example, more food with less labor, increased convenience of harvest, or increased reliability of food supply. Specific goals no doubt varied with time, place and people, and of course with the crop (Harlan, 1992).

Environmental benefits

The first domesticators probably believed that their successful plant breeding also had provided environmental benefits. Thus, to make lush gardens and productive fields where forest or grassland or swamp had ruled supreme would be, in their eyes, a decided improvement in their surroundings, an environmental benefit.

Unintended consequences

As with any deliberate change, unexpected and sometimes undesirable consequences no doubt accompanied and sometimes nullified the intended benefits of these domestications. Archaeological studies identify a period of accelerated soil erosion on the steep slopes of a lake in central Mexico (O'Hara et al., 1993). The onset of this event coincided with the first appearance of maize pollen at that site, circa 3500 years BP, and investigators infer that improper cultivation of maize brought on the erosional episode. And in the Mesopotamian Valley of southwestern Asia, salinity presumably caused by improper irrigation put a stop to the production of wheat at some time prior to about 2300 BP; cereal production shifted to salt-tolerant barley (*Hordeum vulgare* L.) (p. 172, Harlan, 1992).

Continued plant breeding: Diversification of cultivars, from domestication to the Industrial Age

Social benefits

During the millennia that followed each of the domestications, farmers and gardeners on all continents continually altered the nature of their domesticated crops to provide adaptation to new places and new environments (p. 172, Harlan, 1992), to improve quality and/or appearance of the product, and certainly to improve yield and stability of yield.

For example, wheat production expanded westward from its origins in southwestern Asia to southern Europe, as well as to other regions. The first cultivars[1] of bread wheat (*Triticum aestivum*

[1] I use the term "cultivar" to indicate both "primitive cultivars" (also called "landraces") and "improved cultivars." The first term usually refers to products of farmer breeders, the second to products of today's professional plant breeders.

L.) were not adapted to the cooler temperatures and longer summer days of central and northern Europe. But in the Middle Ages, responding to desires of the northern French nobility and the rising class of bourgeoisie for this "rich food," farmer–selectors developed cultivars adapted to central and even some parts of northern Europe (pp. 419–422, Bertrand et al., 1975).

Farmer–breeders in all parts of the world presumably made their new cultivars by straightforward mass selection, without benefit of what today is called scientific plant breeding. Residual genetic diversity, chance hybridizations, and random mutations worked together to provide sources of genetic diversity for the perceptive selectors, our cultural ancestors.

It seems likely that our ancestors were acutely aware of the social benefits (even if they did not use the term) that ensued when plant breeding helped them to grow favored crops in new lands; entrepreneurial members of growing populations could establish themselves sustainably in new locations, and the ethnic group—clan, tribe or nation—was thereby increased in numbers and power.

Environmental benefits

As with the first domesticators, our ancestors apparently believed during the ensuing millennia of expanding crop cultivation that a positive environmental benefit of plant breeding was its contribution of productive new crop cultivars to fit new lands, thereby enabling transformation of wilderness into productive—and convenient—gardens, orchards, and croplands.

For example, during the Middle Ages, the northern expanse of western Europe was transformed from a land of mostly primeval forest into a land of primarily cultivated fields with forest remnants confined to lands unsuited for farming: rocky hills, mountains, or swamps. Nobles in central France engaged entrepreneurial peasants—*pionniers*—to move to the forest edge from whence they would establish villages and commence to carve out new fields and pastures (pp. 431–439, Bertrand et al., 1975), more or less like the American pioneers created cropland from the trans-Appalachian forest wilderness of present-day Ohio and Indiana in the early decades of the nineteenth century. As noted previously, plant breeding often played an essential role in the pioneering settlement of new lands; the settlers—the pioneers—usually needed to develop new cultivars suited to these new growing conditions, to marginally or drastically different day-lengths, weather patterns, and/or soils. Without adapted cultivars there would be no point in striving to transform wilderness into farmland.

Unintended consequences

Although transformation of wilderness into productive farmland was a desired end, it also had the unfortunate consequence of reducing and/or eliminating supplies of important products that came from the forests, grasslands, and swamps. In northern Europe, for example, berries and nuts typically were harvested from forests in medieval times. Also, the forests were the only source of fuel for cooking and heating; no city could exist without a nearby forest (pp. 362–367, Braudel, 1979). Thus, the desired environmental change, transforming forest to cropland, also had the undesired consequence of depleting sources of nuts and berries and, especially, of fuel.

Plant breeding, from the dawn of the Industrial Age until 1900, the beginning of "globalization"

Social benefits

The Industrial Age, beginning in Western Europe in the latter part of the eighteenth century and then spreading globally throughout the nineteenth century, markedly increased the size and importance of urban conglomerates and extensively increased the amount and importance of global communication and transportation systems.

The sharp increase in urban populations meant that increasing numbers of farmers had to produce crops in amounts well beyond their immediate needs in order to feed the city dwellers as well as themselves. (Of course, one might argue that without the potential to create surplus food, the increases in urban populations could not have occurred. Whichever the case, urbanites and farmers strongly affected each others' well-being.) Surpluses of wheat, rice, potatoes (*Solanum tuberosum* L.), and maize, as well as of other crops amenable to storage and transportation, were produced in much greater amounts than in earlier times to satisfy demands of a constantly enlarging urban market. Commercial agriculture markedly increased in importance at the expense of subsistence agriculture (pp. 72–89, Evans, 1998). Fewer and fewer peasants produced food for family only, plus (sometimes) a relatively small number of nobles and/or urban dwellers.

Not only population increase, but also the global transportation/communication revolution affected the activities of crop producers. The farmers' products now could be shipped worldwide to satisfy needs and tastes of urban consumers in far-off lands. As a consequence, it became profitable to grow bread wheat, for example, in the Great Plains of the United States and Canada to be shipped by rail and by steamship to burgeoning urban centers in Europe and eastern North America. Eager farmers plowed up the native short-grass prairies of the Great Plains and replaced them with wheat fields. Here again, plant breeding played an essential role; cultivars had to be bred and/or imported that were specifically suited to the weather and soils of the newly plowed semi-arid wheat lands of North America (e.g., Cox, 1991).

Similar stories could be told for other major crops in other parts of the world as farmers turned from subsistence agriculture to commercial production of major food crops.

However, it also is true that subsistence agriculture still was the essential way of life for many people in those parts of the world that were least touched by the Industrial Revolution, in particular in the tropics and subtropics of all continents. But even here the need for cultivars adapted to new growing conditions sometimes arose, as smallholders were pushed by larger enterprises from their original holdings onto less desirable lands. New cultivars or variations of the favored originals were needed, with the ability to cope with lower soil fertility, greater likelihood of drought, or new kinds of disease or insect pests.

But for much of the world, commercial crop production became predominant, and as commercial production expanded globally, it often mandated—and was dependent upon—development of new crop cultivars suited to new production areas. For better or for worse, plant breeding was a key player in the Industrial Revolution and the accompanying growth of cities and invention of all kinds.

One point should be noted: food supplies were increased primarily by increase of cropland area. Although higher yields in some locales did contribute to increased food supply in the Industrial Age, the primary answer to the global call for more food, especially in the earlier years of this era, was the same as in earlier times: expand the area of croplands by transforming more and more wilderness into farm fields (p. 114, Evans, 1998) and then develop new cultivars suited to those new territories.

Thus the primary gift of plant breeding, its primary social benefit during the Industrial Age—the nineteenth century—was a diverse assemblage of new cultivars with adaptation to new growing conditions in new lands.

(Not only breeding of new cultivars, but also global diffusion and exchange of cultivars per se were important factors for increases in food production in new lands and old. Sometimes the cultivars were moved and used with no change, as in the case of clonal crops such as sugar cane, but very often the move, after a pause, was followed by a cycle of productive genetic change, as with Turkey wheat and its daughter cultivars in Kansas [Cox, 1991].)

Environmental benefits

Environmental goals remained the same as ever throughout the nineteenth century: to transform wild, "unproductive," land into fields and gardens (and cities). As in earlier times, to transform a trackless wilderness into a "land of milk and honey" was a decided environmental improvement. Plant breeding played a critical role in enabling such an environmental benefit.

Unintended consequences

Extensive land conversion, especially strong in the nineteenth century, gave rise to far-reaching problems, but they were not appreciated until a century later.

In the later decades of the twentieth century, warnings that we were running out of potentially productive wilderness areas began to appear (Evans, 1998). Most of the remaining wilderness, it was said, was of a kind that could not be transformed into productive farmland even with extensive remediation such as drainage or changing the soil's nutrient balance; in fact, it probably should not be converted under any circumstances, because its soils would only deteriorate if plowed and planted.

And naturalists and others with appreciation of native ecosystems began to call for permanent conservation of some of the wilderness areas—forests, prairies, savannahs, wetlands—before they were lost forever. They stated that for many reasons wilderness is valuable in its own right, more so in some instances than as a source of future croplands, pastures, or forestry products (Leopold, 1949).

So to the extent that plant breeding during the Industrial Age had enabled extensive expansion of commercial farming and food crop production, it also had enabled the undesirable consequence of excessive destruction of wilderness, of native ecosystems and their accompanying benefits.

Not everyone believed that we could no longer increase arable land area in sustainable fashion; researchers pointed out that considerable potential for adding good arable land exists in parts of South America and central Africa, albeit natural areas would be lost if such conversions were made (Alexandratos, 1999).

But all discussants agreed that more than ever before, now, at the dawn of the twenty-first century, we must increase food supplies primarily by increasing yields, not by expanding arable land area. Analysts' projections showed that increases in food supply for burgeoning (and often undernourished) populations for the next several decades should come primarily from increases in crop productivity—from higher yields—especially in the developing countries (Crosson and Anderson, 2002; Rosegrant et al., 2001).

This was a sea change from earlier days; for the first time, increase in food supplies was to come primarily from higher yields, not from more farmland.

And plant breeding, in conjunction with yield-promoting management inputs, would be a major contributor to these higher yields. A general rule of thumb is that plant breeding contributes about 50% to yield gains, and management contributes the other 50% (Coffman and Bates, 1993), although the ratio varies significantly from crop to crop, with farming type, and with geographic location.

Plant breeding in the present era: Globalization

I define *globalization*, the present era, as beginning in the early years of the twentieth century, coincident with the rediscovery of Mendel's laws and the development of "scientific plant breeding" or, as I often call it, "professional plant breeding." Professional plant breeders consider plant breeding as their primary (or only) occupation. They use science, art, and intuition to develop new cultivars. They produce essentially the same results—new cultivars—as those created by our ancestors, but can do so more swiftly and (usually) more precisely.

The first half of the twentieth century saw significant increases in global interchange of information, ideas and goods, with major interruptions for two worldwide wars. This acceleration of the long-term trend to globalization was intensified even more during the latter half of the century: nations, individuals, science, technology, and commerce were strongly (although not always gladly) interconnected globally via air transport, electronic communications, international corporate structures, and numerous intergovernmental organizations.

The first half of the twentieth century also saw significant progress in development of the basic principles of science-based, professional plant breeding. Ensuing years have been devoted to the utilization and elaboration of these principles, as well as to the exploitation of the breakthrough possibilities presented by the introduction (starting in about the 1950s and 1960s) of affordable and plentiful supplies of synthetic nitrogen fertilizers, pesticides, and herbicides.

These latter changes gave plant breeders the opportunity to produce cultivars with much greater yielding ability than ever before, cultivars that could respond to new and higher levels of soil fertility, pest control, and weed control (Cassman, 1999; Rosegrant et al., 2001).

As in the Industrial era, these changes were not uniformly distributed among the different countries and/or regions of the globe (Crosson and Anderson, 2002). By and large, the industrialized countries have been much more likely than the developing countries to use synthetic agricultural chemicals as well as the new products of professional plant breeding. But islands of change exist in nearly all of the developing countries; one cannot categorize a country as a homogeneous entity in regard to its use of these new production tools.

Development of the ability to sharply increase yield per unit of land area coincided with the realization (noted earlier) that we no longer can or should break out new lands for agricultural production, that the curve of global increase in farmland area has leveled out or must soon do so; yield per unit area must be increased more or less in proportion to increases in population if new mouths are to be fed (Crosson and Anderson, 2002; Evans, 1998).

Note that, in actuality, global food supplies had been growing by increases in yield rather than by increases in area of arable land since about 1960 (p. 205, Evans, 1998), so we already had a basis for believing that we could discontinue or at least slow down conversion of wilderness to farmland.

Of course other alternatives to ensure adequate food supply per capita also exist at least in concept, for example, birth control and/or a more just and equitable distribution of food supplies. But it has seemed obvious (at least to some observers) that until more of these methods are advanced enough to solve the problem, greater yields per unit area will have to carry the burden.

This new paradigm for increase of food supplies brought new meaning to thoughts about the social and environmental benefits of plant breeding. Plant breeding's assistance in the movement of settlers to newly cleared lands was no longer considered an indisputable social and environmental benefit. And heightened concerns about social justice led to new ways of thinking about plant breeding's utility to rich versus poor agricultural producers.

As a consequence of these new points of view, a host of social and environmental benefits have been credited to plant breeding in recent years, as well as numerous attributions of unintended and often undesirable outcomes. At times one person will describe a particular consequence of plant breeding as a "benefit," but another person will call that same consequence an "undesirable outcome."

Such contradictions should not be surprising, for statements about benefit or harm are often normative statements, arbitrary judgments about what is desirable or undesirable, and as with any such judgment, a given consequence—actual or potential—of plant breeding can be looked upon as good or bad, depending on which ideological door one uses for entry.

I will list and comment upon some of the social and environmental benefits that currently are attributed to plant breeding and also will discuss a few of the unexpected and/or undesired consequences that are attributed to plant breeding.[2]

Social benefits

Plant breeding can provide numerous social benefits such as improvement in economic well-being and social justice, better food and/or feed quality and safety, cultivars that respond well to new methods of agronomic management, flexibility in response to demands for new kinds of cultivars, and new methods for economical and energy-efficient production of nonfood products.

Social benefits: Higher yields

Yields of major cereal crops (and also soybean) have increased significantly in most parts of the world during the past half-century, and plant breeding has accounted for a substantial portion of those gains. More profit for commercial farmers, more food for subsistence farmers, and lower food prices for everyone are some of the positive results attributed to the genetic yield gains that have been attained during the past several decades (e.g., Byerlee and Moya, 1993; Evans, 1998; Specht et al., 1999).

As noted earlier, increases in crop yield per unit area will be essential in years to come, to feed a growing global population. Cereal demand, for example, will increase globally by 1.3% per year for the next two decades, and most of that increase will need to come from higher yields, not more arable land (Rosegrant et al., 2001). The increase in yield will depend upon improvements in both management and genetics, but as noted earlier, improvements in genetic yielding ability probably will account for 50% or more of the increase for most crops. Higher-yielding cultivars have been and will be an important social benefit of plant breeding.

Social benefits: Social justice

The products of professional plant breeding—improved cultivars—have given major assistance to small, poor farmers in developing countries. They are aided when they can grow cultivars with higher

[2]For many of the items in this list I am indebted to a number of friends who were kind enough to answer my request for comments about social and environmental benefits of plant breeding as well as about any unintended/undesired results. Others helped as well, by informal review of the document. I hereby acknowledge and give thanks for these collective contributions.

yield, more nutritious content, and with less labor per unit of food that is produced (Byerlee and Moya, 1993; Crosson and Anderson, 2002; Frisvold et al., 2003; Hazell and Haddad, 2001; Khush, 1995).

Social benefits: Food quality and safety

Cultivars with increased resistance to insect and disease attack will make sounder grain, tubers, or whatever organ is used for food and so will be less likely to have invasion of fungi with associated mycotoxins (CAST, 2003). Plant breeding has provided and will continue to provide cultivars with improved pest resistance (e.g., Rudd et al., 2001; Walker, 1966).

Plant breeding has done much and can do much more to improve the nutrient content of food crops. An outstanding example of such a beneficial change is modification of rapeseed (*Brassica campestris* L. and *B. napus* L.) to remove certain toxic compounds (erucic acid and glucosinolates) while maintaining its content of "healthy" oils with low level of saturation (Busch et al., 1994). This undertaking, carried out in Canada in the 1960s and 1970s, used conventional plant breeding. The product is now known universally as canola.

In a second example, recombinant DNA technology potentially can improve nutritional value of the globally important food crop, rice. A combination of transgenes has enabled biosynthesis of provitamin A (beta-carotene) in the rice endosperm, which is normally carotenoid free (Ye et al., 2000). The next task will be to incorporate this change into farm-ready cultivars.

Recent examination of irradiated peanut (*Arachis hypogaea* L.) cultivars has identified strains with lower levels of major allergens; these strains potentially can be used to develop peanut cultivars with significantly reduced allergenic potential (Schmitt et al., 2003).

Social benefits: Interactions with agronomic practices

One of the outstanding achievements of modern plant breeding has been its use to develop cultivars able to take major advantage of yield-enhancing cultural practices. Thus, the Green Revolution wheat and rice cultivars (first introduced in the mid-1960s) were bred for superior stem strength (aided by short stature) and so were able to cope with the lush growth resulting from higher levels of fertilizer nitrogen; the short plants with strong stems did not lodge, and so the added grain yield could be harvested from erect plants (Dalrymple, 1985). Maize was bred to withstand the pressures of close spacing (which can induce lodging and barrenness in susceptible genotypes) and thus make better use of increased application rates of fertilizer, nitrogen in particular (Duvick, 1984b). Cultivars that are best able to take advantage of high-yield management practices are highly popular with farmers and are widely planted.

Social benefits: Flexibility in response to diverse and changing needs

Professional plant breeders have greatly increased the number of useful phenotypic and genotypic variants within each crop, often doing so in response to consumer demand as well as farmer demand. (Of course, farmer demand is often created by consumer demand.) Crops can be bred to withstand higher levels of salinity, to have more drought tolerance, to resist new pest problems (see later section), to have new and more desirable ratios of saturated and unsaturated oils, or to produce a familiar winter squash (*Cucurbita pepo* L.) in a smaller and more convenient size (e.g., Duvick et al., 1981; Edmeades et al., 1997; Jahn, 2003; Rajaram et al., 1997).

The speed at which these changes were made, as well as the potential number and diversity of new products, is greatly increased by the new breeding aids now available, including the varied technologies that have been spawned by molecular biology. And because of rapid turnover—rapid replacement—of cultivars the temporal genetic diversity (genetic diversity in time) is great as well (Duvick, 1984a).

The rapid turnover—a demonstration of the flexibility of plant breeding—is forced in part by competition among commercial seed firms, each desirous of having the best new cultivar on the market, but it also results from speed and efficiency of public programs; zealous public sector breeders (when properly funded) continue to develop replacement cultivars with improved yield, quality, pest tolerance, or whatever traits are most desired by farmers and consumers.

Social benefits: Nonfood products

Pharmaceuticals, edible vaccines, or biodegradable plastics potentially can be produced by transgenic cultivars of productive crop species; the cultivars can produce the compounds with sun power rather than with power from fossil fuel (e.g., Lee et al., 2001; Saruul et al., 2002). Lowered expense of production and greater purity of product seem possible. A major requirement (and caution) for production and distribution of these nonfood products is that they should not contaminate food or feed versions of the same crop.

Environmental benefits

Plant breeding in the present era can provide environmental benefits in many fields, such as conservation of natural systems, increase of useful biodiversity, reduction of global climate warming, energy conservation, water conservation, soil conservation, reduction of pesticide use, reduction of fertilizer use, and coping with adverse soil conditions.

Environmental benefits: Conservation of natural systems

High-yield cultivars can "spare land for nature" (Waggoner, 1994). World population growth is projected to level off eventually, but not immediately. "The world population is expected to grow from 5.8 billion people in 1997 to 7.5 billion people in 2020" (Rosegrant et al., 2001). Most of the increase will be in the developing countries. If yields can be increased on present cropland area sufficiently to feed this constantly increasing global population, it will not be necessary to convert present wilderness to cropland. Breeding for increased yielding ability, in combination with improved management techniques, will play an important role in raising yield per unit area in ecologically acceptable ways (Cassman et al., 2003). Higher yields will be especially beneficial in developing countries that otherwise would need to grow crops in marginal environments unsuited for crop production.

Breeding of trees for timber and pulp production in managed plantations (e.g., Sedjo, 1999) will reduce pressures for deforestation of ecologically important native forestlands. As well, higher crop yields per se have allowed forest expansion in Europe and the United States in recent years, because less cropland is needed for food production (Waggoner and Ausubel, 2001).

Environmental Benefits: Increase of Useful Biodiversity

Plant breeders increase the supply of useful crop genetic diversity as they use diverse and often exotic sources of germplasm to build up primary and secondary germplasm-breeding pools to serve as sources of new cultivars (Duvick, 1984a). As noted earlier, the breeders also provide useful temporal genetic diversity as they continually replace old cultivars with improved new cultivars.

Interspecific crosses can increase the number and complexity of genetic choices for improvement of crops, thereby increasing the genetic diversity among cultivars.

In the same manner, transgenesis can increase the number and complexity of genetic choices for improvement of crops, with attendant benefits to biodiversity.

The existence of both public and private sector plant breeding increases the opportunities for producing cultivars for a wide variety of crops and adaptation regions. The concentration of commercial plant breeding on a relatively small number of major crops frees up scarce public sector funds that otherwise would need to be spent on those crops, and those funds can be used to augment breeding of minor crops, new crops, and crops for small ecological niches, thus increasing the total amount of useful biodiversity of crop plants.

Environmental benefits: Abatement of global climate warming

Plant breeding can be used to create cultivars to be used as biofuels, renewable sources of energy. The use of biofuels on a large scale also could reduce the increase of emission of greenhouse gases because of increased soil carbon storage.

An example of such breeding for North America could be development of productive cultivars of switchgrass (*Panicum virgatum* L.), a prairie grass native to the midwestern prairies of North America (McLaughlin et al., 2002). The Asian grass genus Miscanthus may be a good source of cultivars for use as biofuels in various regions in Europe (Clifton-Brown et al., 2001).

One should keep in mind that because of many interactions, biofuel production and/or use can produce negative as well as positive benefits; for example, nitrogen fertilization of switchgrass fields would increase soil carbon sequestration but increase nitrous oxide emissions (McCarl and Schneider, 2001). Attention to the whole rather than to individual parts will be needed to obtain net benefits from biofuels breeding, just as for any other breeding program.

Environmental benefits: Energy conservation

Plant breeding substitutes genetics for synthetic chemicals and fossil fuel energy. As noted in other sections, breeding for insect and disease resistance reduces or eliminates the need to apply synthetic pesticides, all of which require supplies of fossil fuel for their production. Likewise, to the extent that drought-tolerant cultivars reduce the need for irrigation, they conserve the fossil fuel energy that would have been used to apply the irrigation water.

Environmental benefits: Conservation of water resources

Plant breeders cannot breed cultivars that thrive with no water, but they can and have bred cultivars with greatly improved ability to cope with drought, either episodic or season long (Edmeades et al., 1997; Rajaram et al., 1997; Toorchi et al., 2003). Because high-yielding crops transpire and evaporate little more water than low-yielding crops, breeding that increases yield increases water use efficiency (p. 43, Waggoner, 1994). Although expanded irrigation has been responsible for much of the recent increase in food production, forecasters say that if cereal production (for example) goes on with present methods, "Water scarcity for irrigation will intensify, with actual consumption of irrigation water worldwide projected to grow more slowly than potential consumption. . . ." They conclude that "crop breeding for rain-fed environments is crucial to future cereal yield growth" (Rosegrant et al., 2002). Thus, the need for drought-tolerant cultivars (at various levels of tolerance) will increase, and the ability of plant breeding to make such cultivars will be much appreciated, a benefit to the environment (and a social benefit as well).

Environmental benefits: Soil conservation

In temperate latitudes, plant breeders can aid no-till agriculture by developing crops able to germinate at lower soil temperatures. The cover of dead vegetation in the spring insulates the soil from heating by the sun, and some crops may germinate poorly or grow too slowly in resultant cooler soil temperatures. Therefore, cultivars with ability to germinate at lower soil temperatures will aid use of no-till agriculture and indirectly provide a benefit to soil conservation.

Plant breeding potentially could develop cultivars with increased utility as cover crops or increased ability to out-compete weeds, thereby reducing the need for mechanical cultivation. Such cultivars would be useful for farmers who abjure use of synthetic herbicides and depend upon mechanical cultivation for weed control. Soil erosion on their land could be reduced because of better ground cover and less tillage.

Environmental benefits: Reduction of pesticide use

Plant breeding has continually created disease- and insect-resistant cultivars (Clements et al., 2003; Rudd et al., 2001; Walker, 1966), although it always seems that more are needed to keep up with the constantly changing biotypes of pest organisms (Ghislain et al., 1997). Clearly, use of such cultivars reduces the need for and amount of application of synthetic or organic pesticides and benefits the environment. As well, cultivars with bred-in resistance to a more environmentally benign herbicide can reduce need for application of less-desirable ones (Waggoner, 2004).

Environmental benefits: Reduction of fertilizer use

Recent studies have shown that newer cultivars are more efficient than older cultivars in use of fertilizer, at least of nitrogen fertilizer (Castleberry et al., 1984; Edmeades et al., 1997; Hasegawa, 2003; R. Ortiz-Monasteriao et al., 1997). All cultivars yield less when fertilizer rates are below optimum, but the new cultivars yield more than the old ones. This means that if or when fertilizer application amounts are lowered from current levels, new cultivars will provide higher yields than would the old ones. I would predict, also, that if selection for yield at low fertilizer rates were intensified, one could develop cultivars with even higher yield potential at suboptimal rates of application.

Environmental benefits: Coping with adverse soil conditions

Plant breeders can develop cultivars that perform satisfactorily in adverse soils with problems such as salinity, micronutrient deficiency, or toxic metals (Duvick et al., 1981; Edmeades and Deutsch, 1994; Lasat, 2002; Villagarcia et al., 2001; Wang et al., 2003). It also seems likely that by means of genetic engineering and/or conventional breeding one could develop cultivars that take up and sequester toxic elements (phytoremediation), enabling their use in land cleanup operations (Bennett et al., 2003).

Unintended or undesired consequences

As stated earlier, opinions are not undivided about the effects of modern plant breeding. Criticisms abound, at least in some circles. Space limitations will not allow me to speak to all or even most of the criticisms, but I will comment on two of them.

Criticism: High-yield cultivars for developing countries

Critics say that social justice and environmental well-being are blighted when high-yield cultivars are introduced, particularly in developing countries. This criticism began in the early 1970s, at the time of the first successes of high-yield rice and wheat cultivars (the Green Revolution cultivars).

Critics with concern for social justice say that larger operators (especially in developing countries) are the chief beneficiaries of the new high-yield cultivars because they can more likely afford additional yield-enhancing technologies that complement the new cultivars; their profits from the new cultivars and technologies are used to buy or control even more land, thus forcing the smaller farmers off their land (see summaries in Crosson and Anderson, 2002; Evans, 1998; Hayami and Ruttan, 1985; Ruttan, 2004).

Furthermore, critics say that seed of the high-yield cultivars is more likely to be owned (through intellectual property rights) and sold by for-profit corporations, again favoring larger, wealthier farmers with sufficient money to invest in such improved seeds (RAFI, 1994; Shiva, 1996).

I disagree with these criticisms, at least in their simplistic forms, and instead concur with those researchers whose more holistic analyses show that (a) there is no single consequence, good or bad, to introduction of high-yield cultivars and accompanying management practices to developing countries, and (b) on the whole, rural communities, as well as urban dwellers, have benefited[3] (e.g., Byerlee, 1994; Byerlee and Moya, 1993; Crosson and Anderson, 2002). Everybody has gained an environmental benefit as well, in so far as high yields have spared conversion of unsuited wild areas into cultivated fields.

An example of the complexity of outcomes is the situation in the Punjab of India, where the economic improvement brought on by the Green Revolution has resulted in a shortage of farm laborers for work such as transplanting rice. As a consequence, temporary workers migrate from less-prosperous states such as Bihar, with consequent enhancement of their incomes and the income of their home state. But the influx of migratory workers also has brought on problems that often accompany such movement, for example, difficult social interactions between people with differing language and customs (Kaur et al., 1999).

Critics with concern for the environment say that cultivars that take maximum advantage of yield-enhancing agronomic practices will encourage farmers to add too much fertilizer, to irrigate wastefully and improperly, and in general to try to make maximum yields at the expense of ecologically sound management. Some of the critics say that high-yield cultivars will fail in the absence of such extravagant support; they cannot cope with drought, low fertility, or disease or insect attack. Critics have blamed the high-yield cultivars and the breeding (and breeders) that made them for the problems that followed the introduction of Green Revolution technology and cultivars: problems such as excessive soil salinity, or overuse, or dangerous use of fertilizer and insecticides.

I disagree with such conclusions. Plant breeding and its products, like any tool, can be used wisely or unwisely. The newly made products of maize and wheat breeding were grown in environmentally unsound ways in Mesoamerica and in Mesopotamia thousands of years before hybrid maize and Green Revolution wheat cultivars were produced. The pioneering farmers in Mesoamerica and Mesopotamia, as well as some of today's pioneers, had to learn ecologically sound production practices the hard way.

Although successful modern cultivars yield more than the old ones when well fertilized, well watered, and protected from pest damage, they also (as noted earlier) can significantly outperform the older ones when biotic or abiotic stress is present. Despite claims to the contrary, modern cultivars do not demand coddling. This is not surprising if one considers that popular (i.e., consistently successful) cultivars are those that have outperformed all others in good times and bad (e.g., Duvick et al., 2004).

[3]As stated by Ruttan (2003), "... the less securely grounded early impressions of green revolution impacts have remained pervasive in the popular literature and in public consciousness, even though the private and social rates of return to the investment in research and development that led to the green revolution have been high by any standard."

Criticism: Genetic engineering for beneficial traits

Some people argue that one should not use transgenic breeding to make a crop more nutritious, or safer to eat, or more resistant to disease and insects, or more tolerant of biotic stress, or higher yielding in general.

Thus, potential transgenic rice cultivars with high levels of the precursor to vitamin A (Gura, 1999), or transgenic maize hybrids with proven lower levels of mycotoxins as a consequence of lower levels of insect damage (Clements et al., 2003), are looked upon with disfavor by significant numbers of people because they believe that as yet undiscovered dangers from transgenic cultivars could be worse than the perceived benefits.

Concerned individuals fear that transgenic cultivars—as a class—could be harmful to human health or to the environment or to both. Suggested courses of action range from (a) banning transgenic cultivars outright to (b) delaying approval of them until intensive long-range tests have proven beyond any doubt that they will not harm people or the environment.

Aside from concerns about health and environmental problems, some people object to transgenic cultivars because of deep-seated convictions that genetic engineering is intrinsically wrong, perhaps even immoral, or at the least inherently uncontrollable and unpredictable in human hands.

Thus, conflicting opinions about use of genetic engineering as a tool of plant breeding often are based on differing ideological, ethical, and religious convictions as much as on science. They will not be resolved without rigorous attention to, and widespread discussion of, these convictions. Although scientific evidence has been and should be further provided to test validity of claims of safety or danger (e.g., Kaeppler, 2000), this alone will not change the minds of individuals with strong convictions based on normative ethics, on personal judgments about what ought to be. (For review and commentary on this topic, see Comstock, 2000; Comstock, 2002; Persley, 2003.)

Conclusion

Plant breeding can provide numerous social and environmental benefits through its contributions to objectives such as economic well-being and social justice, improved quality and safety of food and feed products, new methods for efficient production of nonfood products, conservation of natural systems, increase of useful biodiversity, water conservation, soil conservation, energy conservation, reduction of global climate warming, reduction of pesticide use, reduction of fertilizer use, and coping with adverse soil conditions.

The contribution of plant breeding to increased crop yields may be plant breeding's most important benefit. Science-based professional plant breeding, developed and practiced during the past century, has enabled accelerated rates of increase in yield of major food crops. This potential has been exploited in the industrialized nations of the world and in some developing countries. The yield gains have facilitated various social changes, with details and direction depending much on interactions with local social and economic conditions. On the whole the changes have been for the better. Crop yields alone do not necessarily have a straightforward effect (good or bad) on social and economic well-being, but higher yields can increase the odds that socioeconomic problems can be solved or alleviated.

Increased crop yields also can reduce the need for conversion of wilderness to farmland in developing countries with growing populations and growing food needs, thus "sparing land for nature." On the other hand, improper use of yield-enhancing inputs that complement the high-yield varieties can degrade the land and harm people.

Social and environmental effects of plant breeding can be regarded as beneficial or harmful, depending upon one's point of view. "It is all in the eye of the beholder."

The inescapable fact, however, is that modern plant breeding is an effective and economical tool for making changes in quantity and quality of food and other renewable products, and as a tool it can be employed to whatever social and environmental ends are desired by society, or by those who control society.

References

Alexandratos, N. 1999. World food and agriculture: Outlook for the medium and longer term. Proc. Natl. Acad. Sci. U S A 96:5908–5914.

Bennett, L.E., J.L. Burkhead, K.L. Hale, N. Terry, M. Pilon, and E.A.H. Pilon-Smits. 2003. Analysis of transgenic Indian mustard plants for phytoremediation of metal-contaminated mine tailings. J. Environ. Qual. 32:432–440.

Bertrand, G., C. Bertrand, G. Bailloud, M.L. Glay, and G. Fourquin. 1975. Histoire de la France Rurale: La Formation des Campagnes Françaises des origines au XIVe siècle Seuil, Tours.

Braudel, F. 1979. The Structures of Everyday Life: The Limits of the Possible. 1985 ed. Harper & Row, New York.

Busch, L., V. Gunter, T. Mentele, M. Tachikawa, and K. Tanaka. 1994. Socializing nature: Technoscience and the transformation of rapeseed into canola. Crop Science 3:607–614.

Byerlee, D. 1994. Modern Varieties, Productivity, and Sustainability: Recent Experience and Emerging Challenges. ISBN: 968-6923-32-2. CIMMYT, 06600 México, D.F. México.

Byerlee, D., and P. Moya. 1993. Impacts of International Wheat Breeding Research in the Developing World, 1966–1990. CIMMYT, México, D. F.

Cassman, K.G. 1999. Ecological intensification of cereal production systems: Yield potential, soil quality, and precision agriculture. Proc. Nat. Acad. Sci. USA 96:5952–5959.

Cassman, K.G., A. Dobermann, D.T. Walters, and H. Yang. 2003. Meeting cereal demand while protecting natural resources and improving environmental quality. Annu. Rev. Environ. Resour. 28:315–358.

CAST. 2003. Mycotoxins: Risks in Plant, Animal and Human Systems. Task Force Report 139. Council for Agricultural Science and Technology, Ames, Iowa.

Castleberry, R.M., C.W. Crum, and C.F. Krull. 1984. Genetic yield improvement of U.S. maize cultivars under varying fertility and climatic environments. Crop Science 24:33–36.

Clements, M.J., K.W. Campbell, C.M. Maragos, C. Pilcher, J.M. Headrick, J.K. Pataky, and D.G. White. 2003. Influence of Cry1Ab protein and hybrid genotype on fumonisin contamination and fusarium ear rot of corn. Crop Science 43:1283–1293.

Clifton-Brown, J.C., I. Lewandowski, B. Andersson, G. Basch, D.G. Christian, J.B. Kjeldsen, U. Jorgensen, J.V. Mortensen, A.B. Riche, K.-U. Schwarz, K. Tayebi, and F. Teixeira. 2001. Performance of 15 *Miscanthus* genotypes at five sites in Europe. Agronomy Journal 93:1013–1019.

Coffman, W.R., and D.M. Bates. 1993. History of Crop Improvement in Sustainable Agriculture, p. 19–32. *In* M. B. Callaway and C. A. Francis, eds. Crop Improvement for Sustainable Agriculture. University of Nebraska Press, Lincoln, NE.

Comstock, G.L. 2000. Vexing Nature? On the Ethical Case Against Agricultural Biotechnology. Kluwer Academic Publishers, Boston.

Comstock, G.L., (ed.) 2002. Life Science Ethics, pp. 1–380. Iowa State Press, Ames, Iowa.

Cox, T.S. 1991. The contribution of introduced germplasm to the development of U.S. wheat cultivars, p. 25–47. *In* H. L. Shands and L. E. Wiesner, eds. Use of Plant Introductions in Cultivar Development. Crop Science Society of America, Inc., Madison, WI (USA).

Crosson, P., and J.R. Anderson. 2002. Technologies for meeting future global demands for food. Discussion Paper 02-02. Resources for the Future, Washington, DC.

Dalrymple, D.G. 1985. The development and adoption of high-yielding varieties of wheat and rice in developing countries. Amer. Journ. Agric. Econ. 67:1067–1073.

Denham, T.P., S.G. Haberle, C. Lentfer, R. Fullagar, J. Field, M. Therin, N. Porch, and B. Winsborough. 2003. Origins of agriculture at Kuk Swamp in the highlands of New Guinea. Science 301:189–193.

Duvick, D.N. 1984a. Genetic diversity in major farm crops on the farm and in reserve. Econ. Bot. 38:161–178.

Duvick, D.N. 1984b. Genetic contributions to yield gains of U.S. hybrid maize, 1930 to 1980, p. 15–47. *In* W. R. Fehr, ed. Genetic Contributions to Yield Gains of Five Major Crop Plants. CSSA Special Publication No. 7. Crop Science Society of America, American Society of Agronomy, Madison, WI, USA.

Duvick, D.N., R.A. Kleese, and N.M. Frey. 1981. Breeding for tolerance of nutrient imbalances and constraints to growth in acid, alkaline and saline soils. Journal of Plant Nutrition 44:111–129.

Duvick, D.N., J.S.C. Smith, and M. Cooper. 2004. Long-term selection in a commercial hybrid maize breeding program, pp. 109–151. *In* J. Janick, ed. Plant Breeding Reviews, Part 2, Long Term Selection: Crops, Animals, and Bacteria, Vol. 24. John Wiley & Sons, New York.

Edmeades, G.E., and J.A. Deutsch (eds.). 1994. Stress Tolerance Breeding: Maize that Resists Insects, Drought, Low Nitrogen, and Acid Soils, pp. 1–113. CIMMYT, Mexico, D.F.

Edmeades, G.O., M. Bänziger, H.R. Mickelson, and C.B. Peña-Valdivia (eds.). 1997. Developing Drought- and Low N-Tolerant Maize: Proceedings of a Symposium, March 25–29, 1996, CIMMYT, El Batán, Mexico, pp. 1–566. CIMMYT, Mexico, D.F.

Evans, L.T. 1998. Feeding the Ten Billion: Plants and Population Growth. Cambridge University Press, Cambridge, U.K.

Frisvold, G.B., J. Sullivan, and A. Rameses. 2003. Genetic improvements in major US crops: the size and distribution of benefits. Agricultural Economics 28:109–119.

Ghislain, M., R. Nelson, and T. Walker. 1997. Resistance to potato late blight: a global research priority. Biotechnology and Development Monitor 31:14–16.

Gura, T. 1999. New genes boost rice nutrients. Science 285:994–995.

Harlan, J.R. 1992. Crops & Man. 2d ed. American Society of Agronomy, Inc., Crop Science Society of America, Inc., Madison, WI.

Hasegawa, H. 2003. High-yielding rice cultivars perform best even at reduced nitrogen fertilizer rate. Crop Science 43:921–926.

Hayami, Y., and V.W. Ruttan. 1985. Agricultural Development: An International Perspective. The Johns Hopkins Press, Baltimore & London.

Hazell, P., and L. Haddad. 2001. Agricultural research and poverty reduction. Food, Agriculture, and the Environment Disussion Paper 34. International Food Policy Research Institute, Washington, DC.

Jahn, M. 2003. Personal Communication. June 20, 2003.

Kaeppler, H.F. 2000. Food safety assessment of genetically modified crops. Agronomy Journal 92:793–707.

Kaur, Y., A.K. Gupta, and B.R. Jindal. 1999. Socio-economic and psychological impact of migration of farm workers in rural Punjab. J. Agr. Develop. & Policy 11:104–112.

Khush, G.A. 1995. Modern varieties—their real contribution to food supply and equity. GeoJournal 35:275–284.

Lasat, M.M. 2002. Phytoextraction of toxic metals: A review of biological mechanisms. J. Environ. Qual. 31:109–120.

Lee, R.W.H., J. Strommer, D. Hodgins, P.E. Shewen, Y. Niu, and R.Y.C. Lo. 2001. Towards development of an edible vaccine against Bovine Pneumonic Pasteurellosis using transgenic white clover expressing a Mannheimia haemolytica A1 Leukotoxin 50 fusion protein. Infect. Immun. 69:5786–5793.

Leopold, A. 1949. A Sand County Almanac and Sketches Here and There. Oxford University Press, Oxford.

McCarl, B.A., and U.A. Schneider. 2001. Climate Change: Greenhouse gas mitigation in U.S. agriculture and forestry. Science 294:2481–2482.

McLaughlin, S.B., D.G.D.I.T. Ugarte, J.C.T. Garten, M.A. Sanderson, V.R. Tolbert, and D.D. Wolf. 2002. High-value renewable energy from prairie grasses. Environ. Sci. & Technol. 36:2122–2129.

O'Hara, S.L., F.A. Street-Perrott, and T.P. Burt. 1993. Accelerated soil erosion around a Mexican highland lake caused by prehispanic agriculture. Nature 362:48–51.

Ortiz-Monasteriao R., J.I., K.D. Sayre, S. Rajaram, and M. McMahon. 1997. Genetic progress in wheat yield and nitrogen use efficiency under four nitrogen rates. Crop Science 37:898–904.

Persley, G.J. 2003. New Genetics, Food and Agriculture: Scientific Discoveries—Societal Dilemmas. International Council for Science.

RAFI. 1994. Declaring the benefits: The North's annual profit from international agricultural research is in the range of U.S. $4–5 billion. It's time for an accounting. Vol. 1, No. 3. Rural Advancement Foundation International, Ottawa, Ontario, Canada.

Rajaram, S., R.P. Singh, and M.V. Ginkel. 1997. Breeding wheat for wide adaptation, rust resistance and drought tolerance, p. 139–163. Crop Improvement for the 21st Century. Research Signpost, Trivandrum-696 008, India, Trivandrum.

Rosegrant, M.W., X. Cai, and S.A. Cline. 2002. Global Water Outlook to 2025: Averting an Impending Crisis. Food Policy Report. International Food Policy Research Institute and International Water Management Institute, Washington, DC and Colombo, Sri Lanka.

Rosegrant, M.W., M.S. Paisner, S. Meijer, and J. Witcover. 2001. Global Food Projections to 2020: Emerging Trends and Alternative Futures. International Food Policy Research Institute, Washington, DC.

Rudd, J.C., R.D. Horsley, A.L. McKendry, and E.M. Elias. 2001. Host plant resistance genes for fusarium head blight: Sources, mechanisms, and utility in conventional breeding systems. Crop Science 41:620–627.

Ruttan, V.W. 2004. Controversy about agricultural technology: Lessons from the Green Revolution. Int. J. Biotechnology 6:43–54.

Saruul, P., F. Srienc, D.A. Somers, and D.A. Samac. 2002. Production of a biodegradable plastic polymer, poly-ß-hydroxybutyrate, in transgenic alfalfa. Crop Science 42:919–927.

Schmitt, D.A., S.J. Maleki, H.W. Dodo, T. Isleib, and E.T. Champagne. 2003. Screening mutant peanuts for allergen knockouts. J. Allergy Clin. Immunol. 111:S247.

Sedjo, R.A. 1999. Biotechnology and Planted Forests: Assessment of Potential and Possibilities. Discussion Paper 00-06. Resources for the Future, Washington, DC.

Shiva, V. 1996. The seeds of our future. Development 4:14–21.

Specht, J.E., D.J. Hume, and S.V. Kumudini. 1999. Soybean yield potential—a genetic and physiological perspective. Crop Science 39:1560–1570.

Toorchi, M., H.E. Shashidhar, T.M. Gireesha, and S. Hittalmani. 2003. Performance of backcrosses involving transgressant doubled haploid lines in rice under contrasting moisture regimes: Yield components and marker heterozygosity. Crop Science 43:1448–1456.

Villagarcia, M.R., T.E. Carter, Jr., T.W. Rufty, A.S. Niewoehner, M.W. Jennette, and C. Arrellano. 2001. Genotypic rankings for aluminum tolerance of soybean roots grown in hydroponics and sand culture. Crop Science 41:1499–1507.

Waggoner, P.E. 1994. How Much Land Can Ten Billion People Spare for Nature? Task Force Report No. 121. Council for Agricultural Science and Technology.

Waggoner, P.E. 2004. Agricultural technology and its societal implications. Technology in Society 26:123–136.

Waggoner, P.E., and J.H. Ausubel. 2001. How much will feeding more and wealthier people encroach on forests? Population and Development Review 27:239–257.

Walker, J.C. 1966. The role of pest resistance in new varieties, p. 219–242. *In* K. J. Frey, ed. Plant Breeding. The Iowa State University Press, Ames, IA.

Wang, R.R.-C., S.R. Larson, W.H. Horton, and N.J. Chatterton. 2003. Registration of W4909 and W4910 bread wheat germplasm lines with high salinity tolerance. Crop Science 43:746.

Ye, X., S. Al-Babili, A. Klöti, J. Zhang, P. Lucca, P. Beyer, and I. Potrykus. 2000. Engineering the provitamin A (beta-carotene) biosynthetic pathway into (carotenoid-free) rice endosperm. Science 287:303–510.

4

Defining and Achieving Plant-Breeding Goals

A.R. Hallauer, C.F. Curtiss Distinguished Professor in Agriculture (Emeritus), Iowa State University
S. Pandey, Director, Maize Program, CIMMYT (International Maize and Wheat Improvement Center) Mexico

Introduction

Plant breeding is a complex and comprehensive discipline that has played a significant role in the development and improvement of plants for human use. Although significant changes have occurred in most aspects of our lives during the past 40 years (e.g., electronics, computers, automobiles, air travel, space exploration, etc.), plant breeding has never received the same recognition for the changes and improvements that have been made. This lack of recognition exists even though the products derived from plant breeding are very important to sustain the food, feed, fiber, and fuel needs of the global human population. If we include plant improvement in the broader sense (i.e., from plant domestication to present-day improvements), plant breeding is one of our older disciplines. One of the basic human needs is consistent supplies of good-quality food. Plant-breeding activities have occurred since humans recognized that plants have the potential to provide food to sustain individuals, families, and civilizations.

The lack of recognition of the contributions that plant breeding has made to both ancient and modern societies has several possible causes: improvements are incremental from year-to-year and generation-to-generation; incremental changes and improvements are not obvious to the general public; lack of mobility restricts the extent that plants can be included in large exhibitions held in urban areas (e.g., cattle shows, pet shows, state fairs, horse races, etc.); and the plants themselves are far removed and are not obvious in final manufactured products used in modern societies. Although most of the changes are usually not associated directly with plant breeding, the shape or color of a flower has greater recognition within our urban societies than changes in quantity and quality of our food crops. Plant breeding has received more recognition from the public in the past 20 years than during the previous 200 years with the developments in genetically modified organisms (GMOs). Although greater recognition was given, there was not a universal consensus that GMOs were either a desirable or acceptable method to modify and improve our major crop species. Genetic modifications always have been the focus of plant-breeding activities, but it seems to bother some that the modifications are based at the molecular level of the genotype rather on the phenotype, which is also based on genetic changes at the molecular level. Genetic changes and improvements always have been, and will continue to be, the major focus of plant breeding.

Art versus science

Plant breeding is often defined as the art and science of plant improvement (e.g., Allard, 1960; Briggs and Knowles, 1967). Both art and science have contributed significantly to development of cultivars, but the relative importance of art versus science has changed significantly during the past 150 years. Selection was initially emphasized on the phenotype of individuals. Phenotypic selection was practiced among individual plants for ideal phenotype; success depended on ability of the plant breeder to identify phenotypes that would be expressed in the next generation. This ability to identify superior phenotypes would depend on how effectively the plant breeder could visualize what constitutes superior improved cultivars. The original plant breeders were effective in developing cultivated crop plants from their weedy ancestors (Harlan, 1975). These methods also were effective in the improvement of our important crop plants that have specific traits and became more dependent on humans for their survival. Further improvements, based on phenotypic selection, however, became more difficult, exemplified in the attempts to develop improved maize (*Zea mays* L.) cultivars for the United States. During the nineteenth century, phenotypic selection was effective in developing maize cultivars with distinctive plant, ear, and kernel traits and maturity, but grain yield remained relatively unchanged (Hallauer and Miranda Fo, 1988). Score cards were introduced at maize shows that ranked the relative importance of ear and kernel traits that presumably contributed to superior-yielding maize cultivars. Selection of ears that conformed to the score card standards depended on the art—and patience—of the individuals practicing selection. During the early part of the twentieth century, yield trials were conducted to compare cultivars selected on score card standards and for cultivars that did not conform to the score card standards. It was shown that cultivars selected on the basis of specific phenotypic traits were no better, or worse, than those selected without using the criteria of the score cards. The breeder's art was effective in developing cultivars that were phenotypically uniform for specific traits, but the traits selected did not contribute directly to yield improvement. Phenotypic selection was effective for traits having higher heritability but relatively ineffective for traits with lower heritabilities, such as grain yield (Bauman, 1981).

Science became a more important component of plant breeding when the concepts of Darwin (1872), Mendel (Olby, 1966), and Vilmorin (Coons, 1936) developed in the mid-nineteenth century became known, understood, and applied to plant breeding. Although Darwin and Mendel were contemporaries, the concepts of evolution (survival of the fittest) with the genetic principles of segregation and recombination were not integrated until the rediscovery of Mendel's genetic studies in 1900. The rediscovery of Mendel's laws of inheritance generated further genetic studies and provided a basis of interpretation of the data derived from them. For plants, Johannsen in 1903 showed that the phenotypes observed were dependent on genetic and environmental effects (in 1911 he proposed terms genotype and phenotype); Nilson-Ehle in 1908 demonstrated that multiple factors affected inheritance of traits (see Sinnott et al., 1950, for discussion); Shamel (1905), East (1908), and Shull (1908) reported on the effects of inbreeding; and Shull (1909) showed that vigor was restored upon crossing of pure lines, developed by the progeny methods suggested by Vilmorin in 1859. Later, Fisher (1935) presented methods for making valid comparisons between and among treatments. These concepts and studies (and there were many others) provided a scientific basis for plant breeding. During the past 100 years, research has expanded and extended these concepts to the point that science has largely replaced art as the important component of plant breeding. Art, however, still remains as a component in plant breeding because the plant breeder has to make decisions relative to choices of parents to include in crosses, population size for initiating selection, phenotype preferred by the growers (or customers), generation(s) for testing, and the target environments in which the cultivars are to be primarily grown. The importance of phenotypic expression also will vary among horticultural, vegetable, and field crops. Although selection based on phenotype is considered, the effectiveness of phenotypic selection is inversely related to the economic importance of the trait (Bauman, 1981). But these decisions are becoming more scientifically based because of the statistical analyses available to assist plant breeders in the choice of parents, generation of testing, grower desires, and target environments.

Quantitative genetics

The initial genetic studies were conducted on traits that segregated in ratios that conformed to the ratios reported by Mendel [see Sinnott et al. (1950) for translation of Mendel's paper]. It soon became obvious that phenotypic expressions of some traits were not amenable to Mendelian analyses, that is, classified into groups that fit a specific ratio. Other methods were needed to determine the inheritance of more complex traits, which included our more important economic traits such as biomass and grain yield. Fisher (1918), Wright (1921a; 1921b), and Haldane (1932) discussed methods for determining the inheritance of the more complex traits.

These traits were designated as quantitatively inherited traits and integrated the concepts of Mendelian genetics with evolutionary theory (e.g., Wright, 1968). These three individuals developed methods that were heavily dependent on statistical methods of analysis that were not easily understood and accepted (see Provine, 1971, for discussion). Animal breeders accepted the concepts of Fisher, Wright, and Haldane earlier than the plant breeders. Because of the costs and facilities required to maintain large population sizes and the generation intervals for animals, information on sires, dams, and progeny was needed to provide additional information on the inheritance of complex traits, such as carcass weight, milk production, amount and quality of meat, etc. Sewall Wright and his disciples also were effective in demonstrating the use of quantitative analyses in animal breeding. It was not until after World War II, with rapid expansion in plant-breeding programs, the appearance of Mather's book in 1949, and the strong emphasis of the research programs in North Carolina and Cambridge on quantitative genetic studies of plants, that plant breeders recognized that quantitative genetics could have a role in plant-breeding strategies. One other item that generated greater interest in the study of quantitative genetics was the concept of hybrid maize. Double-cross maize hybrids were rapidly replacing the open-pollinated cultivars in the U.S. Corn Belt during the 1940s. The hybrids were superior to the open-pollinated cultivars for grain yield, root and stalk strength, and more uniform in maturity and plant phenotype. The hybrid concept was explicitly described by Shull (1910) and expanded by Jones (1918). Plant vigor, or heterosis, was restored upon crossing pure lines, but the genetic basis of heterosis was not understood. Superior hybrids were identified empirically, based on data collected from yield trial comparisons.

Plant breeders either directly or indirectly recognized the importance of the concepts of quantitative genetics in their breeding programs after 1950. The newer generations of plant breeders had access to courses and texts that presented in a more understandable format the methods presented by Fisher (1918) and Wright (1921a; 1921b) (see Mather, 1949; Kempthorne, 1956; Falconer, 1960; Li, 1976). Studies were initiated to determine the inheritance for different traits, relative importance of additive, dominant, and epistatic effects in trait expression, the relative relations between parents and their offspring for the inheritance of traits, and the correlations for trait expression between parents and their offspring for the same traits and between traits of the same individuals and later generations. A perusal of the literature indicates that quantitative genetic studies were conducted in nearly all important horticultural, vegetable, and field crops.

One topic that received specific attention in quantitative genetic studies was the determination of the genetic basis of heterosis. Because of the popularity and acceptance of double-cross hybrids in maize, research was conducted to develop and test hybrids in horticultural, vegetable, and other field crops. Two general theories were suggested to explain the genetic basis of heterosis: accumulation of dominant favorable alleles, usually designated as additive model, and the nonadditive model, where overdominant and epistatic effects were of greater importance in the expression of heterosis. Specific mating designs and generations were evaluated to determine the relative importance of additive and nonadditive effects and levels of dominance in genetically broad-based populations, F_2 populations developed from crosses of pure lines and different types of hybrids. In the genetically broad-based populations, the estimates of additive genetic variance were greater than the estimates of nonadditive variance with estimates of levels of dominance in the partial to complete dominance range. Within the F_2 populations, estimates of levels of dominance suggested overdominant effects were more important, but it was acknowledged that the estimates could be biased upwards because of repulsion phase linkages. Levels of dominance also were estimated in the same F_2 populations after 5–12 generations of intermating. Estimates of levels of dominance decreased in all instances with increased intermating of the F_2 populations, indicating the original estimates were due to pseudoverdominance because of repulsion phase linkages. Studies were conducted to determine the importance of epistatic effects relative to additive and dominant effects, but in most instances realistic estimates of epistatic variance were not obtained (Silva and Hallauer, 1975). Comparisons of means of different types of hybrids indicated significant nonadditive effects, but the estimates could not be quantified.

Definitive evidence consistently favoring either

of the two theories for the genetic basis of heterosis has not been obtained. Similar to other biological systems, it may not be realistic to expect one general theory will be applicable to all hybrids. Each hybrid is a unique cross between two or more parents that may be inbred, partially inbred, or noninbred. Hence, each hybrid has its own unique combination of alleles from its parents. All types of genetic effects probably have some role in the expression of heterosis. Initial estimates of components of genetic variance in maize populations suggested a preponderance of additive genetic variance relative to nonadditive variance. The importance of additive genetic variance indicated that selection should be effective, but genetic variance estimates obtained from populations suggested that heterosis in maize was primarily due to additive gene effects. To translate the information obtained from populations that were assumed to be randomly mated and in linkage equilibrium to a specific hybrid does not seem reasonable. Each hybrid has its own unique genotype and the relative importance of additive, dominant, and epistatic effects will be different for each hybrid within a plant species and among plant species. Positive, nonadditive gene effects have to be present for the expression of heterosis (Falconer, 1960). The importance of nonadditive genetic effects in different types of hybrids also was shown theoretically by Cockerham (1961). It seems reasonable that the additive accumulation of dominant favorable alleles in combination with interactions of alleles at the same loci (overdominance) and between alleles at different loci (epistasis) all are important to the expression of heterosis. In breeding programs that develop recycled lines to produce hybrids, the development and maintenance of favorable linkage blocks increase the relative importance of epistatic effects in expression of heterosis. It is doubtful that one comprehensive explanation of the genetic basis of hybrids will be obtained.

The contributions of information from theoretical and empirical quantitative genetic studies are greater than generally acknowledged. Because of the different reproduction systems (autogamous, allogamous, vegetatively propagated plant species), different levels of ploidy, annuals and perennials, different generation intervals of reproduction, relative important of different plant traits, and types of progenies (pure lines, half-sibs, full-sibs, testcrosses, hybrids, etc.) that can be evaluated, breeding methods and strategies are developed for each crop species. Quantitative genetic theory has provided guidelines that plant breeders can use in planning their breeding programs. Empirical data from quantitative genetic studies have been reported for most crop species that provide further information for use in planning the desired breeding system. Generally, estimates of genetic variance indicate that selection should be effective for most traits because of the greater importance of the additive genetic variance for plant populations, regardless of the genetic and reproductive systems of the plant species. Nonadditive effects become of greater importance if hybrids are the type of cultivars provided to the growers. If nonadditive effects are either of minor importance or nonexistent, the chances of developing hybrids with better performance than their parents become more difficult.

One important contribution of quantitative genetics to plant-breeding methods is the development of expressions to estimate the relative heritabilities of traits for different breeding systems and types of progenies evaluated (Nyquist, 1991). Because of the different options available for different plant species, it is essential that explicit definitions are used to make valid comparisons among studies, plant populations, crop species, and types of progenies. The classic definition of heritability (h^2) presented by Lush (1945) was $h^2 = \sigma^2_A / \sigma^2_P$, where σ^2_A is additive genetic variance and σ^2_P is the phenotypic variance, has limited relevance to plant breeding unless mass selection of individual plant phenotypes is the main goal. To account for the different systems of mating, breeding methods, and evaluation, the estimates of heritability must include the proper combination of variables to be valid for the proposed method of selection. An important corollary to the estimates of heritability is the prediction of genetic gain (Nyquist, 1991). Eberhart (1970) presented a relation for the prediction of genetic gain that includes variables related to intensity of selection, parental control, types of progenies used to estimate σ^2_A (σ^2_g), phenotypic variance for the testing of the progenies σ^2_P, and number of years to complete a cycle of selection. The prediction equation (Eberhart, 1970) relies on the heritability estimates (Nyquist, 1991) from the evaluation trials, with additional parameters of selection intensity (k), parental control (c), and number of years (y) to

complete one cycle of selection. Valid comparisons for rates of genetic gain (ΔG), different types of progeny, and methods of selection can be determined on a per year basis; that is, $\Delta G = (ck\sigma_g^2)/y\sigma_P$).

The scientific basis of plant breeding has been enhanced with developments in plant quantitative genetics during the past 50 years. The concepts of quantitative genetics certainly contributed to the proper estimation of heritabilities for the different situations that occur in plant-breeding programs. Effectiveness of selection depends on the relative heritabilities of traits considered in selection. If the heritabilities are either not calculated correctly or are inappropriate for the selection method used, effectiveness of selection will be less than anticipated.

Relations between traits are often of interest because they may be useful in selection. Mode and Robinson (1959) reported methods for calculation of genetic and phenotypic correlations between traits that were similar to those used for the estimation of genetic and phenotypic variances but include the components of covariance between traits. Relative heritabilities are different for different traits and may be higher for more easily measured traits (e.g., components of yield) that have some relation to the primary trait (e.g., yield). Because data on the component trait(s) may be easier and less expensive to measure than the primary trait, it is often tempting to collect data on the secondary trait(s) with the goal for increasing the primary trait. Two conditions, however, are necessary for indirect selection (ΔG_{21}) to be more effective than direct selection ($\Delta G_{21} = r_{G_{12}} h_1 h_2 \sigma_{P_2} k_1$): there must be a strong genetic correlation (rG_{12}) between the secondary trait (2) and the primary trait (1), and the heritability of the secondary trait (h_2) must be significantly higher than the heritability of the primary trait (h_1). Usually, direct selection is significantly more effective than indirect selection unless the genetic correlation is near 1 and heritabilities of secondary traits are 0.80 or higher.

The concepts of quantitative genetics are integral components of modern plant-breeding programs, although in most instances, plant breeders may not realize their significance. Extensions of the original concepts have made them applicable to the breeding programs that include crop plants having more complicated genetic systems. The research and teaching of quantitative genetics was passé during the past 20 years because of the interest and expansion in research related to molecular genetics. Quantitative genetics and molecular genetics are polar opposites relative to levels of trait measurements. The research sequence in molecular biology was similar to the research in genetics after the rediscovery of Mendel's laws of inheritance. Initially, research was focused for individual genes at the molecular level for traits that had clear segregation ratios. It also soon became obvious to molecular geneticists that the more important economical traits were complex and that different methods of analyses were needed; that is, quantitative trait loci (QTL). Similar techniques used to determine the inheritance of traits at the phenotypic levels are being used at the molecular level to determine the location and impact of QTLs on trait expression (Lynch and Walsh, 1998). Advances in the study of quantitative genetics will continue with increased knowledge of the genotype. The goal is to determine the most effective breeding methods to genetically improve our important complex traits.

Recurrent selection

Information from the quantitative genetic studies had an impact on developing breeding strategies for the genetic improvement of quantitative traits (Moll, 1974; Hallauer, 1991). Breeding and selection methods, based on both theoretical and empirical information, were suggested and tested to determine the most effective methods for sequential improvement of quantitatively inherited traits. Consistent productivity (grain and biomass) across a series of target environments is the major goal of most plant-breeding programs. It soon became obvious that improvement of consistent productivity could not be achieved with use of the classical Mendelian analyses. Because productivity is usually controlled by a large, unknown number of alleles that are affected by environmental effects, different breeding strategies were needed to ensure consistent genetic improvements; that is, increase the frequency of the alleles that contribute to greater, consistent productivity.

Estimates of genetic components of variance suggested that additive genetic effects were of greater importance than nonadditive genetic effects. But there were exceptions, depending on the population sampled and the methods used in estimation. Interpretations differed, therefore, on what breeding strategy would be more effective. Initially, the differences between the relative importance of additive versus nonadditive effects were interpreted relative to the genetic basis of heterosis in maize. Jenkins (1940) was of the opinion that additive genetic effects were of greater importance and that selection should emphasize general combining ability (GCA). Hull (1945), however, had the opposite view and suggested that selection should emphasize selection for nonadditive effects, or specific combining ability (SCA). Comstock et al. (1949) suggested a selection method that was equally effective for both GCA and SCA. The concepts of GCA and SCA were introduced by Sprague and Tatum (1942), who partitioned the genetic variability among crosses into effects due to primarily either additive (GCA) or nonadditive (SCA) effects. The relative importance of GCA and SCA depended on the extent of previous testing of the parents included in the crosses. The selection methods suggested by Jenkins (1940), Hull (1945), and Comstock et al. (1949) have been designated as recurrent selection.

The basic feature of recurrent selection methods is that they are selection procedures that are conducted in a repetitive manner, or recycling. Because recurrent selection methods are conducted for primarily quantitatively inherited traits, the goal is to increase the frequency of desirable alleles in a consistent manner. The basic premise of natural selection is applied, but plant breeders attempt to manage selection in a more consistent manner for economically important traits for target environments. Because the traits included in recurrent selection tend to have lower heritabilities, they require more testing to determine the breeding values for the progenies tested. Consequently, effective recurrent selection programs are long-term to detect significant genetic improvements. Repetitive cycles of recurrent selection include three important stages (Hallauer, 1985): (1) development of an adequate number of progenies to sample the genetic variability of the population under selection; (2) adequate testing to identify progenies that possess the greater frequency of favorable alleles for the target environments; (3) and intermating the superior progenies to initiate the next cycle of selection. Each stage is important and decisions have to be made that seem most important for the crop species and traits considered in selection (Hallauer, 1985; 1991). Eberhart (1970) presented formulae that include many of the variables that can affect effectiveness of selection.

The first recurrent selection programs were conducted in maize (Hallauer and Miranda Fo, 1988). Grain yield was the more common trait considered in selection. The initial reports were often erratic, and most selection programs were discontinued after a limited number of cycles. For those programs that were continued five or more cycles, significant improvement was generally realized. Pandey and Gardner (1992), for tropical area maize, and Hallauer (1992), for temperate area maize, reported that significant genetic gains for grain yield had been realized. The long-term nature of recurrent selection methods is often discouraging because of seemingly limited genetic progress that is obtained. But the nature of the traits under selection is the primary cause. Because of the complexity of the traits under selection is not amenable to classical Mendelian analyses, recurrent selection methods are, at present, the only alternative available to ensure systematic genetic improvement of complex traits. Separation of the genetic and environmental effects is an important facet of effective recurrent selection methods. Adequate testing of an adequate number (100 or more) of progeny to determine the relative importance of genetic and environmental effects across target environments has time, labor, space, and cost constraints. Adjustments can be made relative to types of progenies tested, types of progenies and methods of intermating, and areas where selection is conducted to reduce time per cycle (Eberhart, 1970).

Although recurrent selection methods were developed initially for grain yield of maize, the methods have been expanded for other traits (disease resistance, stalk and root strength, grain quality, ear traits, etc.) and for most crop species. A perusal of the literature shows that recurrent selection methods have been used for most of the important crop species. Although recurrent selection was initially not considered in autogamous crop species because of the difficulty of intermating, recurrent selection methods have been suggested and used in autogamous crop species (e.g., Gilmore, 1964; Brim and Stuber, 1973; Fehr and Ortiz, 1975; Sorrells and Fritz, 1982; Frey et al., 1988; Diaz-Lago et al., 2002) and for primarily self-pollinated crops (Doggett and Eberhart, 1968). Modifications in selection methods were made for the crop species and traits considered in selection for different autogamous plants to enhance the effectiveness of selection. Similar modifications were made in forage and grass crop species for biomass and disease resistance (e.g., Burton, 1992; Rowe and Hill, 1981).

Recurrent selection methods were originally proposed for the genetic improvement of genetically broad-based populations, which would provide genetic resources for development of cultivars and information on the relative importance of additive and nonadditive genetic effects in selection response. The original genetic resources available to plant breeders were the landraces that were either collected in the wild or had undergone some human selection, such as the different strains of a common cultivar (e.g., different strains of Reid Yellow Dent, an open-pollinated maize cultivar). The different pure lines selected from the landraces were generally better than the original, but they usually had deficiencies for specific traits. The trait deficiencies, however, were not the same for all pure lines. The logical sequence was to intermate pure lines that complemented the strength of the respective pure lines. Selection was practiced among and within the F_2 and subsequent generations to develop cultivars that had the desired traits of the respective parents. For traits with higher heritabilities, one or more backcrosses may have been made to one parent that was superior for the desired trait(s). Superior selections would be released as cultivars and made available to the growers. Although improvements were made, the new cultivars usually had other deficiencies that the breeder wished to correct. Hence, the breeders would cross the newer cultivars to other improved cultivars, either from their own program or other programs to initiate another cycle of pedigree selection. In all instances, pedigree selection and testing were the common breeding strategy. Genetic improvements were gradually made, and improved cultivars were made available to the growers from each cycle of pedigree selection. The concept of recycling pure lines by pedigree selection is similar to the recurrent selection methods suggested for landrace cultivars, that is, increase the frequency of favorable alleles with each cycle of selection (Hallauer, 1985).

Duvick (1977) made a comparison of the rates of gain for pedigree selection versus recurrent selection. Rates of gain were similar for the two selection methods on a per year basis, suggesting the two methods produced the same rate of gain. Duvick (1977), in his calculations, estimated pedigree selection required 13.3 years cycle^{-1} versus 3 years cycle^{-1} for recurrent selection. Because of the changes that have been made in experimental methods since 1977, it is possible to complete one cycle of recurrent selection in 2 years and one cycle of pedigree selection in 6 years; genetic gain would be 1.10 quintels per hectare per year (q ha^{-1} year^{-1}) for recurrent selection versus 1.51 q ha^{-1} year^{-1} for pedigree selection. Rate of gain from Duvick's data would be 37.3% greater by pedigree selection than for recurrent selection with present-day breeding methods. The major difference between pedigree selection and recurrent selection within or between populations would be the types of cultivars developed. Both are important. Pedigree methods emphasize selection within elite line crosses with the major goal of maintaining intact favorable linkage blocks and fine tuning to correct minor deficiencies. The end products are recycled lines that probably have a strong resemblance to one parent. The goal of recurrent selection is to develop improved genetic resources that, depending on the area, are used either directly by the growers or as source germplasm for pedigree selection. Pure line development also can be integrated with recurrent selection methods to develop new, less-related lines that either can be used directly by the growers or provide different genetic variation for pedigree selection programs.

In the broadest context, the principles of recurrent selection are used in all plant-breeding programs. One common activity of plant breeders is choice of parental materials to include in crosses. Based on experience and available information, adjustments in breeding methods are made for crop species, traits considered in selection, and the needs of the growers (e.g., Gardner, 1961; Lonnquist, 1964; Marquez-Sanchez, 1982; Dhillon and Khehra, 1989). Systematic genetic improvement of specific traits is realized by breeding methods that systematically increase the frequency of the desired alleles. This goal can only be achieved by intercrossing superior parents and selecting and testing progenies from the crosses in successive cycles. Repetition is the important element for continued success whether the selection method is within crosses of pure lines to develop recycled pure lines or within populations to develop improved genetic cultivars and/or new pure lines. Ultimate success requires that the breeding method of choice is conducted repetitively during the life of the breeding program.

Advances in plant-breeding techniques

Similar to other disciplines in all areas of science, significant changes and advances were made during the twentieth century for conducting plant-breeding research. The principles of Darwin and Mendel provided a genetic basis for conducting, understanding, and interpreting data from plant research. Without having a clear genetic basis for designing breeding strategies, plant-breeding research would be conducted similar to the previous history of plant breeding. And the results would not be significantly different from those realized during the nineteenth century. Mendelism and the theory of natural selection provided the foundation for the relative heritability of traits and how selection would affect the changes in the frequency of favorable alleles for fitness traits, or greater productivity.

Nongenetic developments during the twentieth century also had significant impacts on the effectiveness and efficiency of plant breeding. Some developments had greater impact than others, but all made important contributions. An all-inclusive list is not possible, but some of the more important contributions would include developments in experimental design and statistical analyses; development of experimental plot equipment to plant and harvest experimental plots, which also include electronic equipment to record data; development of compact, automated equipment with rapid turnaround to analyze quality traits in the laboratory; development of computer hardware and software to collect and analyze data; use of off-season locations to produce crosses, self crosses, seed increases, and advance generations of progenies, which reduce cycle time; and rapid changes in transportation to permit movement of researchers and seed and means of communication among breeders and with producers.

Experimental design and statistical analyses

Developments in experimental design and analyses of data have had a close association with plant research since R.A. Fisher's analyses of data collected while he was at Rothamsted (Mahalanobis, 1964). Researchers in the United Kingdom had a prominent role in showing the importance of randomization, replication, and repetition (growing in different environments) in making valid comparisons among treatments. Because of the nature of plant-breeding research, plant breeders were receptive to the use of proper experimental design and analyses. The basic elements of experimental design and analyses have been extended and refined during the past 50 years to permit options for different situations. These developments also were enhanced with the rapid developments in computer software and hardware, which enhanced the use of more complex designs and greater detail in the analyses of data.

Valid comparisons among genotypes (or cultivars) are important if the plant breeder is to identify correctly the superior genotype. Testing across environments is important if the plant breeder is to have confidence that a cultivar has consistent performance. Earlier experimental studies were conducted to determine the relative interactions of a fixed set of cultivars (G) with locations (L) and/or years (Y). Valid statistical F-tests could be made, but the results were not consistent. It seemed that the significance of the $G \times L$ and $G \times Y$ interactions were dependent on the locations and years included and the general area in which the trials were conducted. In some instances, the second-order interaction ($G \times L \times Y$) was significant, whereas $G \times L$ and $G \times Y$ were not significant; that is, there did not seem to be a consistent combination of factors for the interactions of cultivars with locations and years. Presently, it seems the combination of locations and years is usually designated as a series of environments, unless a common pattern can be determined to partition into locations and years.

Interactions of cultivars with environments are commonly detected in the analyses of variance. For the plant breeder, the detection of interactions of cultivars with environment in the analyses of variance does not specify either which cultivar(s) or how many had significant interactions with environments. Finlay and Wilkinson (1963) and Eberhart and Russell (1966) independently suggested methods to determine the response of each cultivar for the series of environments in which the cultivars were evaluated. The consistent performance (or stability) of a cultivar across a series of environments is very important because the growers rely on the recommendations of the developers in the choice of cultivar for the grower. Reliability of performance is important in making recommendations. Regression analyses were used by Finlay and Wilkinson (1963) and by Eberhart and Russell (1966). Environments were arranged from poor (say, for yield) to good, based on the average yield of all cultivars for each environment. The yield of each cultivar was regressed on the environment yields to estimate the regression value (b) for each cultivar. Finlay and Wilkinson (1963) defined a stable variety as one having $b = 0$; that is, the cultivar had a consistent yield across all environments. Eberhart and Russell's (1966) estimate of b was defined as the response of a cultivar for poor to good environments and the deviations from regression as a measure of stability; that is, the smaller the deviation, the more stable the cultivar performance across environments. Although there have been discussions (e.g., Freeman and Perkins, 1971) about the specific details and assumptions of the stability analyses suggested by Finlay and Wilkinson (1963) and Eberhart and Russell (1966), use of stability analyses is a common practice in large breeding programs.

The stability analyses are one example of extracting additional information from the basic analyses of variance for cultivars tested across a series of environments. As additional information becomes available for climatic factors, soil types, crop management, GPS, and infrared surveys of environments, analyses are adjusted to include these factors either in crop response or altering target environments. The original basic experimental designs are modified when a large number of cultivars is tested. Incomplete block designs were developed originally by Yates (1936), but further modifications have been made for a series of α and β designs. The goal of the incomplete block designs is to reduce experimental error to increase precision of treatment comparisons. The inclusion of a larger set of cultivars in a trial increases the required experimental area and, perhaps, introduces greater variability among plots. Zobel et al. (1988) introduced an analysis, designated as AMMI (additive main effects and multiplicative interactions), in an attempt to present clearer agronomic meaning when cultivars evaluated in different environments have significant interaction components. The extensions and modifications of the designs and analyses of R.A. Fisher have been tested by plant breeders and used where it was found to increase precision of treatment comparisons and to glean additional information from the data. Modifications and extensions in experimental design and analyses will continue, particularly with the availability of modern computers to manage the data. Plant breeders have been in the vanguard in using and adapting experimental designs and statistical analyses; they will continue to do so in the future.

Equipment

Methods of conducting plant-breeding research were labor intensive until the latter part of the twentieth century. Methods for breeding and testing of different crop species were developed after rediscovery of Mendel and remained similar until the 1960s. Methods for making crosses, planting, pest control, harvesting, threshing, data collection and analyses, and preparation of field books and reports required hours of monotonous, repetitive hands-on labor by the principal investigators and their assistants, students, and temporary hires. In the more developed areas of the world, limited sources of laborers and increased costs to hire laborers caused the principal researchers to examine alternative methods to conduct plant-breeding activities in the fields and laboratories. To be effective, plant breeders recognized that adequate numbers of crosses and progenies from the crosses were needed. But the larger numbers required greater investments in producing adequate seed supplies, number of pollinations made in breeding nurseries, land areas required for testing, and greater volume of data to analyze, all of which required more labor. Except for producing the pollinations (either by self- or cross-pollination), which are unique for each crop species, plant breeders started examining the development of experimental plot equipment to reduce costs and enhance the efficiency of plant breeders.

During the 1960s, mobile experimental plot harvesters for small grain crops were being developed and tested in Europe (Gregan, 2003). Previously, small grain test plots had been cut with hand sickles, dried either by solar radiation or commercial dryers, and transported to stationary threshers to separate grain and stover; each step requiring extensive outlays of labor for even a modest breeding program. These early developments in mobile small-grain plot harvesters stimulated interests in other crop species and in other countries. Because of the extensive maize plant-breeding programs conducted by U.S. commercial companies, interest in the development of mobile plot harvesters began in the latter part of the 1960s and during the 1970s. Two approaches were used: custom-made combines designed specifically for small plots and modifications of smaller commercial combines that were adapted to harvest small experimental plots. Both the custom-built and adapted commercial combines were used in subsequent years. Further modifications and refinements were made to increase precision of mobile combines for clean out, loss of grain, gathering ears from stalks, etc. Another component that had a great impact on harvesting was the development of electronic systems that could be installed on combines to record grain yield, grain moisture, and test weight. Continued progress was made by modifying twin-rotor type harvesters that can harvest and record data for two plots simultaneously.

The development and modifications of harvesters for small, experimental test plots have provided maize breeders with equipment that requires only one person. Consider an example from maize-breeding programs that includes an extensive number (30,000–300,000) of test plots. Prior to availability of mobile plot harvesters, eight people could harvest 400–600 plots per day, depending on weather (wet or icy conditions) and crop (lodging and shank strength) conditions. Under these conditions, it required an average of 5–8 minutes (veteran work crew) to harvest, weigh, collect grain-moisture samples (grain moisture later determined in the laboratory), and collect and dispose of harvested ears after ear weight was determined and a moisture sample was collected. Data were analyzed after the completion of harvest. Presently, with a modern state-of-the-art harvester, a plot can be harvested, shelled, and data for grain weight, grain moisture, and test weight can be recorded by one person in 20–30 seconds, or, theoretically, 1,440–960 plots per 8-hour day. Some time is required to empty the grain tank and for combine maintenance, but one person can easily harvest two to three times more plots in one day than eight people could harvest by hand in the same amount of time. If a twin-rotor harvester were used, twice the number of plots would be harvested. In addition to the time element, other advantages of modern maize harvesters include using equipment that resembles equipment used by growers for harvest; reduces harvest variation caused by different individuals hand harvesting and fatigue of hand harvesters, more precise data recorded, electronic data systems to record data, which can be downloaded for immediate analyses, and reduces variation if different hand harvesting crews are used at different locations and years for the same set of cultivars tested.

Similar progress has been made in other phases of plant-breeding research for other crop species. Planters have been developed to accurately sow seeds for small experimental plots for nearly all crop species. The planters also mimic equipment used by growers and ensure more uniform seed depth when planting, seed distribution, and plant stands than with a crew of hand planters. Greater precision of planting has reduced the amount of one labor-intensive task: thinning to have more uniform stands in the field trials. Equipment and devices for recording solar radiation, stalk and root strength, relative maturities, etc., have been made available for plant breeders to use to collect extensive data sets that can assist them in making more effective selection decisions among a larger number of genotypes evaluated over environments.

Data collected from field trials are usually the determining factor for the final release of a new cultivar. But there are, for most crop species, specific minimal quality factors (protein, oil, fatty acids, amino acids, fermentation, baking, taste, etc.) that must be considered in cultivar development. In most instances, these traits are determined in the laboratory. Compared with developing plot equipment for field trials, development of more efficient and accurate laboratory equipment to measure quality traits probably has evolved faster than for field plot equipment. The newer equipment is automated in most instances and integrated with computers to record data. The newer laboratory equipment reduces the time required for collection and analyses of data. The time element is very important to plant breeders because information is often needed at critical times for the breeders to make selections and for meeting planting and harvest schedules.

Off-season nurseries, transportation, and communications

Rapid advances in modes of transportation have affected all aspects of our lives. Rapid and dependable modes of transportation are another nongenetic factor that has increased the efficiency of plant breeding. Use of off-season nurseries has reduced the time required to develop cultivars and also has been very useful for recording classification data, making seed increases of nursery materials, and producing test crosses. Although selection within off-season nurseries usually is not as effective as the home environment, selection for some traits can be done. If local weather catastrophes do not permit seed production scheduled for the growers, off-season seed production can be done. Off-season nurseries are often located great distances from the home stations. But rapid air transportation permits breeders to be present at critical times and for shipment of seed to meet planting and distribution schedules. Although the development of transportation systems may not seem relevant to plant breeding, modern transportation systems have had an important impact.

Another element that has influenced plant breeding is the changes in the communication systems. In earlier years, the postal and telephone systems were our primary methods of communication, and they were not always reliable. Recent developments for use of fax and e-mail have aided plant-breeding activities. Instant communication permits contact at any time and place to permit better supervision and coordination, particularly for large international programs. Exchange of information and data among breeders and supervisors by the modern communication systems permits more effective and timely decision making.

Improvements in the nongenetic factors that impact plant breeding will continue. Plant breeders are innovators and are always receptive to changes that can increase the efficiency of their research. Ideas by plant breeders to increase effectiveness and timeliness of their operations are accepted by technicians and engineers to make their ideas become reality; e.g., adaptation of twin-rotor combines for the harvest of experimental maize plots. Other changes that occur in our societies (e.g., communication and transportation systems) are rapidly accepted and adapted for use in plant breeding in more developed areas. In less-developed areas, the adaptation of more modern technologies will not be as rapid because an adequate labor supply is available to complete the labor-intensive tasks. In these instances, the opportunities to provide employment at the local level may be more important, but the use of the more modern technologies for plant breeding are increasing rapidly in all areas of the world.

One labor-intensive activity that has not been resolved is making the self- and cross-pollinations within the breeding nurseries. Techniques and materials used to make pollinations have improved, but they require individuals to make the actual pollinations. This is an activity that varies widely among crop species, and, in many instances, skilled technicians are needed to produce consistently adequate quantities of seed. It does not seem likely this activity will be amenable to mechanization for most crop species. Significant advances, however, have been made in the commercial production of high-quality seed for the producers. Advances in planting, harvesting, drying, grading, and laboratory analyses for seed diseases and cold and warm germination tests have been made during the past 40 years. In some instances, sterility systems and specialized equipment have been used to reduce labor costs in seed production.

Genetic progress from plant breeding

Grain and biomass yields have increased during the past century for our important crop species. The combination of newer cultivars developed by plant breeders and the improved technologies used by the producers resulted in impressive yield increases. Similar to other aspects of human technological developments during the past century, methods of crop production have changed significantly in highly developed areas and to some extent in the lesser developed areas: plant densities per ha have increased; development of pesticides to reduce the incidence and severity of disease, insect, and weed pressures; development and manufacture of synthetic fertilizers to provide a dependable supply of fertilizers that could be uniformly applied at specific times; continued improvements in equipment to permit timely planting and harvesting of crops; improved transportation systems for movement of seed and harvested crops; improved communications systems for scouting, crop reporting, weather forecasts and patterns, rapid delivery of information to the growers; and, lastly, improved husbandry and management skills of the growers. These factors, and others, have contributed to increased crop yields. The relative importance and impact of the different factors will vary in different areas of the world, but they are being adapted and implemented as quickly as economic and local conditions permit.

Frey (1971) summarized studies that reported comparisons of newer cultivars with either landraces or older cultivars for crop yields of wheat (*Triticum aestivum* L.), alfalfa (*Medicago sativa* L.), rice (*Oryza sativa* L.), oat (*Avena sativa* L.), soybean (*Glycine max* L. Merrill), and maize. In all instances, marked yield improvements through plant breeding were evident in each crop species. Rates of yield improvements varied among crop species: U.S. wheat cultivars had 25–60% greater yielding than cultivars grown 30–70 years previously; maize hybrids grown in Iowa had 50–60% greater yields than the formerly used open-pollinated cultivars; newer rice cultivars had twice the yields of the local rice cultivars in tropical regions; and oat and soybean cultivars grown in Iowa produced 12–14% more grain than cultivars grown 30 years previously. Comparisons generally were made between the original widely grown cultivars and the more recently developed cultivars either in replicated trials or previous data summaries. No attempt was made to determine how much of the yield improvements was because of genetic improvements or because of improved cultural and management skills. The development of "Gaines" wheat cultivar and "IR8" rice cultivar are two examples. Both Gaines and IR8 were developed from different germplasm and were semi-dwarfs, i.e., shorter plant stature than previously grown cultivars. Because of the shorter height, the newer cultivars could be grown at greater plant densities with greater applications of nitrogen fertilizer. Greater yields were realized because newer cultivars had greater resistance to lodging in the husbandry systems required for greater yields; genetic changes were accompanied with husbandry changes, and both were necessary to the realized yield improvements.

The question that was being asked was what portion of increased yields was due to either genetic improvement of the cultivars available to the growers or to the management skills of the growers? Yields from reporting services for the major crops of individual countries and areas have shown trends for consistent yield increases in the highly developed and most-developing countries (Pandey and Gardner, 1992). To determine the relative importance of genetic changes relative to changes in environment and crop management, it required that the cultivars available to the growers for the different years and decades be evaluated in comparative trials that were conducted under similar conditions. These types of trials were not an easy task. Because of changes in crop production, decisions had to be made regarding what type of conditions should be used to make valid comparisons. Usually, both the older and newer crop management styles were included to compare the older and newer cultivars and determine the cultivar response for both types of management practices. Another problem was the availability of viable seed for the older cultivars because some may not have been retained. Compromises were made, but in most instances viable seed of representative cultivars for a specific period desired could be acquired. Types of cultivars available to producers also have changed. In maize, for example, cultivars included open-pollinated varieties, double-cross hybrids, and three-way and single-cross hybrids. Hence, the comparative trials had to consider types of cultivars used, row and plant spacing, response to fertilizers and herbicides, and changing patterns of pest pressures. To reduce possible biases because of seed age and quality, seed of the cultivars included was reproduced at the same time under similar conditions. Most of the trials attempted to include as many variables as feasible in replicated trials repeated across environments to separate and determine the relative importance of genetic and non-genetic effects for yield improvements.

Information from comparative trials have been reported for different crops in different countries. An extensive review will not be presented, but an example will illustrate the genetic improvements for maize in Iowa. Russell (1991) and Duvick (1992) summarized studies that included six decades from the open-pollinated cultivars used in the 1930s and representative hybrids that were widely grown in each decade until 1990. The studies conducted by Russell (1991)and Duvick (1992) were designed to determine the relative portions of the total gain in maize yields that could be attributed to genetic improvements of the cultivars available to the producers and to changes made in the husbandry and management skills of the producers. For the 15 reports included in Russell's summary, average genetic gain was 65.7%. If we separate the five studies reported by Russell and the five studies reported by Duvick, average genetic gain was 67.6 and 70.6%, respectively. Duvick (1992) included 11 studies in his summary, and genetic gain was 70.9%. Russell and Duvick conducted independent studies in central Iowa, and their estimates of average genetic gain were remarkably similar, even though different cultivars, for the most part, were included in the studies (e.g., open-pedigree versus closed-pedigree cultivars). Duvick (1992) also adjusted the differential in yield between research plots and Iowa farm average yield; the estimate of genetic gain was 56% of the total gain in yield. These data suggest that, conservatively, more than 50% of the yield improvements was because of the genetic improvements of the cultivars available to the growers, but it actually may be closer to 66%.

Genetic improvement of cultivar yields is not restricted to maize in Iowa. Similar information has been reported for maize in Yugoslavia (Kojic, 1990), Argentina (Eyherabide et al., 1994), and in the U.S. Corn Belt (Tollenaar et al., 2000). Miller and Kebede (1984) report improved yields for sorghum (*Sorghum bicolar* (L.) Moench), and Gizlice et al. (1994) and Wilcox et al. (1979) have estimated the genetic contributions to yield increases for soybeans. Regardless of the crop species, plant breeders have implemented breeding strategies for effective genetic improvements of successive series of new crop cultivars. Ultimately, yield is the more important trait in plant breeding, but all aspects of crop development must be considered. Genetic changes for traits of a defensive nature (e.g., pest resistance) are necessary if the cultivar is to realize its genetic potential for yield. The matrix of traits that are considered in cultivar development is formidable, but consistent selection for multiple traits must be emphasized if a cultivar is to realize its potential for the known and unknown factors that may occur during any growing season. Information suggests that plant breeders have been effective in making incremental genetic improvements to meet the demands of the growers and consumers.

Future of plant breeding

Every major industry requires a few basic disciplines that provide the foundation for that industry. A basic need of human societies is an adequate daily supply of nutritious foods to maintain our health and contribute to the welfare of our societies. In most areas of the world, industries (growers, processors, refiners, transportation networks, distribution centers, etc.) have developed to provide adequate food supplies. Consumers have choices where they secure food (growers, restaurants, farmers' markets, etc.) and types of foods (vegetables, fruits, meats, that may or may not be grown organically) they wish to consume on a daily basis. Our societies are continually becoming more urban. It is estimated that less than 2% of the U.S. population produces the crops to sustain the food needs for 98% of the U.S. society, primarily urban. At the beginning of the twentieth century, the world was inhabited by 1.5 billion people; the population passed 2 billion in 1927, 3 billion in 1960, 4 billion in 1974, 5 billion in 1987, and 6 billion in 1999. The world's population has increased 75% in the past century. We are adding about 80 million people each year, even with the decreases in birthrates. Some estimate the world will be inhabited by 10 billion people by 2100. Each individual human needs food, and most expect to secure their food needs from local markets. But very few realize the food trail starts with the plant breeder who develops the cultivars that eventually become available to the consumer as food.

The world's population continues to increase, but the area (both land and water) of the world remains the same. The land areas suitable for agriculture generally have been farmed intensively. Other areas of the world are available but have restrictions because the topography limits cultivation, adequate moisture is not available, acid soils limit crop productivity, and social, economic, and political factors affect efficient crop production. Because land areas available for food production will not have significant expansion, it becomes necessary to increase crop production on a unit per area. Increased crop yields have always been an important goal in plant breeding. Emphasis on greater yields per unit land area, however, has been criticized because of local surpluses, exploitation of land, poorer quality, and use of chemicals to reduce pest pressures to enhance yields. But the combination of limited expansion in land use for food production and projected population increases emphasizes the need for greater yields per unit area. Borlaug in 2001 stated that if yields of our major crop species of 40 years ago prevailed today, three times more land in China and the United States and two times more land in India would be needed to meet the cereal demands of these countries. Additional land area is available in each of the three countries, but it is neither amenable nor suitable for crop production. Hence, yield improvement will remain the primary goal in future plant-breeding programs.

Plant breeding and activities related to plant breeding have changed dramatically during the twentieth century, particularly during the past 20 years. After a significant expansion in public plant-breeding programs after World War II, there has been a steady decline in number of public programs and active public plant breeders for cultivar development. The decrease in public breeding programs has been counterbalanced by a significant increase in commercial plant breeding during the past 50 years. Frey (1996) reported that 80% of the plant-breeding scientific years (SYs) was employed by private industry. Among crop species, the percentage of SYs employed by private industry ranged from 27.7% for oat and 41.4% for wheat to 93.5% for dent maize breeding. The changes in emphasis on relative importance of public and private sectors' plant breeding have several possible explanations: (1) development of products that could be commercialized for a profit (e.g., hybrids); (2) legislation was enacted and implemented that protected techniques, equipment, information, germplasm, and cultivars used or developed in breeding programs, which encouraged private investment; (3) the rapid advancements in biotechnology has affected both the public and private sector plant-breeding programs. Major seed companies have been purchased by large agrochemical companies in order to have an avenue to transfer products of biotechnology from laboratory to producers, and significant changes were made in funds available within public institutions to increase support for biotechnology at the expense of public-breeding programs.

Duvick (2002) has addressed several of the issues that will impact plant breeding in the twenty-first century. Plant breeding received more attention in the past 20 years than previously, primarily because it is a vehicle to introduce biotechnology products (transgenes) into elite cultivars (which become genetically modified organisms, or GMOs) that are offered to the producers. These probably will be of greater importance in the 21st century. This is currently an important aspect of plant breeding, but also is minor compared with the overall discipline of plant breeding. Development of elite genotypes (greater yield and greater tolerance to stress and pests to increase stability of yield) will remain a primary goal if the products derived from biotechnology are to achieve the desired, or intended, benefits.

It seems that "classical" plant breeding will always have an important role for developing cultivars with greater yield levels to meet the human food needs. Other disciplines (mutation breeding, plant ideotypes, plant physiology, quantitative genetics, and, currently, molecular genetics) have challenged plant breeding, making it quicker and easier. None of these roles has replaced plant breeding, but each has made information available that enhanced the standard plant-breeding methods, and the newer techniques and information are necessary to maintain systematic, incremental genetic improvements of crop cultivars made available to growers. The extent and sophistications of plant-breeding strategies will range from participatory plant breeders (farmer–breeder) to the large, international companies that can integrate the latest technology with plant-breeding programs. Effective breeding programs are necessarily long term; that is, they need continuity in recycling of germplasm to develop highly efficient genomes that have evolved over time with the proper balance of genetic effects for higher yields. Traditional breeding methods are used in recycling, which is enhanced by the latest developments in genetics, statistical design and analyses, and other non-genetic factors.

One disturbing factor for the future of plant breeding is the elimination and downsizing of publicly supported (e.g., university, governmental, regional, and international) breeding programs. For a few major crop species, cultivar development should not be a major goal of public breeding programs; rather, the goal should be fundamental studies on selection and breeding strategies, pre-breeding to determine genetic potential of germplasm resources, and development and enhancement of germplasm for future use. For other crop species, cultivar development is necessary because commercial interests have determined that the endeavor, either because of limited market or because of inability to produce and control seed supplies, is unprofitable.

Decline of plant-breeding activities in the less-developed countries (LDCs) is especially worrisome. In the LDCs, as well, an increasing amount of plant breeding is carried out by the private sector, which addresses the needs of those farmers who can afford to buy those varieties. The private sector does not develop and promote technologies that they cannot sell and make a profit on. Unfortunately, the majority of the farmers in the LDCs are poor and cannot afford to buy private sector technologies. With declining support for agricultural research, especially plant breeding, by the public sector, a large group of farmers in the LDCs is being left to their own fate. Such farmers have few other alternatives and, therefore, the need for support for public sector plant breeding in the LDCs is far greater than in the developed countries.

Emphasis on breeding goals will vary among public breeding programs, but one important goal should be to educate and train students as plant breeders who can function in field-based breeding programs. Academics are very important, but experience in field research also is very important. Hands-on plant breeders learn what germplasm sources have greater potential, what breeding strategies are more effective, and that the field research is done correctly. Cooperation and interactions with other disciplines have always been important in plant breeding and will be even more important in the future. Students will need strong education and training in basic sciences of genetics, statistics, computers, chemistry, and mathematics, which is the same as in the past, but will also need exposure to the latest developments (e.g., all aspects of genetics). Changes in places of employment have changed significantly during the past 50 years, but a strong demand for well-trained plant breeders will continue in the future.

It is somewhat puzzling why plant breeding is the least recognized part of the food chain, as well as the horticultural and ornamental crops. Plant breeding is the fundamental origin for the development of highly productive and extensively used cultivars (e.g., Gaines wheat, IR8 rice, B73 maize, etc.). But in both public and private organizations, plant breeding is often the first to be downsized because a newer discipline, which is supposed to enhance plant breeding, needs greater support. The methods of recycling are important in plant breeding for making genetic improvements. If programs are eliminated, underfunded, or downsized, the importance and significance of recycling are either reduced or ceased. This is the tragedy when plant-breeding recycling programs are interrupted or stopped; the benefits of recycling are either reduced or lost.

Yields will continue to increase in the future. Yield data obtained from experimental plots and contests among growers are higher than the average yields of the growers. In Iowa, for example, average maize yields are 8.75–10.00 t ha^{-1} compared with 15.00–18.75 t ha^{-1} for contest winners; one winner had more than 25.00 t ha^{-1}. Under excellent husbandry and management practices, the genetic potential for greater yield already exists. Factors that limit our abilities to attain consistently greater yields are those related to the environment: amount and distribution of rainfall, heat stress, disease, insects, weeds, acid soils, topography, etc. Plant breeders will need to develop cultivars that have greater tolerance to those factors that affect yield. Plant biotechnology presents plant breeders with an opportunity to expand the types of cultivars available to the growers. Recently, biotechnology products (transgenics) became available and were inserted in cultivars to increase maize's resistance to pests (e.g., European corn borer, *Ostrinia nubilalis*, Höbner, and corn rootworm, *Diabrotica* sp.), and selective herbicides were developed to reduce the effects of weeds. These types of products may increase in the future with rapid expansion in biotechnology throughout the world. If gene(s) could be identified that reduce the effects of heat, moisture, and soil acidity on crop yields, it would increase crop yields worldwide. Perhaps greater scrutiny of the genome will provide clues to specific regions of the genome that affect yield, either directly or indirectly (e.g., QTLs for greater yields and stress tolerances). Similarly, genes that would enhance nutritional and/or industrial value of crops would have a major effect on farmers, consumers, and economies. These types of information would be of benefit in both developed and lesser-developed areas of crop production. The information and products derived from biotechnology that can be incorporated with plant-breeding strategies in the future are in the development stages, and how much impact they will have is not predictable. But the essential mission of plant breeding is the same: use of cyclical breeding methods to develop improved cultivars and use of the information and products of biotechnology to enhance our mission in ways that are difficult to visualize.

References

Allard, R.W. 1960. Principles of Plant Breeding. Wiley & Sons, Inc., New York, NY.

Bauman, L.F. 1981. Review of methods used by breeders to develop superior inbred lines. Proc. Annu. Corn & Sorghum Ind. Res. Conf. 36:199–208.

Briggs, F.N., and P.F. Knowles. 1967. Introduction to Plant Breeding. Reinhold Publ. Co., New York.

Brim, C.A., and C.W. Stuber. 1973. Application of genetic male sterility to recurrent selection schemes in soybeans. Crop Science 13:528–530.

Burton, G.W. 1992. Recurrent restricted phenotypic selection. p. 101–113. *In* Jules Janick (ed.) Plant Breeding Rev. Vol. 9. John Wiley and Sons, New York, NY.

Cockerham, C.C. 1961. Implications of genetic variances in a hybrid breeding program. Crop Science 1:47–52.

Comstock, R.E., H.F. Robinson, and P.H. Harvey. 1949. A breeding procedure designed to make use of both general and specific combining ability. Agron. J. 41:360–367.

Coons, G.H. 1936. Improvement of the sugar beet. p. 625–656. *In* USDA Yearbook of Agriculture. U.S. Govt. Printing Office, Washington, DC.

Darwin, Charles. 1872. The Origin of Species. 6th ed. Merrill and Baker Publ., New York, NY.

Dhillon, B.S., and A.S. Khehra. 1989. Modified S1 recurrent selection in maize improvement. Crop Science 29:226–228.

Diaz-Lago, J.E., D.D. Stuthman, and T.E. Abadic. 2002. Recurrent selection for partial resistance to crown rust in oat. Crop Science 42:1475–1482.

Doggett, H., and S.A. Eberhart. 1968. Recurrent selection in sorghum. Crop Science 8:119–121.

Duvick, D.N. 1977. Genetic rates of gain in hybrid maize yields during the past 40 years. Maydica 22:187–196.

Duvick, D.N. 1992. Genetic contributions to advances in yield of U.S. maize. Maydica 37:69–79.

Duvick, D.N. 2002. Crop breeding in the twenty-first century. p. 13–14. *In* Manjit Kang (ed.) Crop Improvement: Challenges in Twenty-first Century. The Haworth Press, Inc., Binghampton, NY.

East, E.M. 1908. Inbreeding in corn. p. 419–428. Connecticut Agri. Exp. Stn. Rep. 1907.

Eberhart, S.A. 1970. Factors affecting efficiencies of breeding methods. African Soils 15:669–680.

Eberhart, S.A., and W.A. Russell. 1966. Stability parameters for comparing varieties. Crop Science 6:36–40.

Eyherabide, G.H., A.L. Damilano, and J. C. Colazo. 1994. Genetic gain for grain yield of maize in Argentina. Maydica 39:207–211.

Falconer, D.C. 1960. Introduction to Quantitative Genetics. Ronald Press, New York.

Fehr, W.R., and L.B. Ortiz. 1975. Recurrent selection for yield in soybeans. J. Agric. Univ. Puerto Rico 59:222–232.

Finlay, K.W., and G.N. Wilkinson. 1963. The analysis of adaptation in a plant breeding program. Australian J. Agric. Res. 14:742–754.

Fisher, R.A. 1918. The correlation between relatives on the supposition of Mendelian inheritance. Trans. R. Soc. Edinburgh 52:399–433.

Fisher, R.A. 1935. The Design of Experiments. Hafner Publ. Co., New York, NY.

Freeman, G.H., and J.M. Perkins. 1971. Environmental and genotype-environmental components of variability. VIII. Relation between genotypes grown in different environments and measures of these environments. Heredity 27:15–23.

Frey, K.J. 1971. Improving crop yields through plant breeding. p. 15–58. *In* J. D. Eastin and R.D. Munson (ed.). Moving Off the Yield Plateau. ASA Spec. Publ. 20, ASA, CSSA, and SSSA, Madison, WI.

Frey, K.J. 1996. National Plant Breeding Study-1. Spec. Rep. 98. Iowa Agric. Home Econ. Exp. Stn., Iowa State Univ., Ames, IA.

Frey, K.J., J.K. McFerson, and C.V. Branson. 1988. A procedure for one cycle of recurrent selection per year with spring-grown small grains. Crop Science 28:855–856.

Gardner, C.O. 1961. An evaluation of effects of mass selection and seed irradiation with thermal neutrons on yields of corn. Crop Science 1:241–245.

Gilmore, E.C. 1964. Reciprocal recurrent selection for autogamous species using cytoplasmic and genetic male sterility. Crop Science 4:323–325.

Gizlice, Z., T.E. Carter, and J.W. Burton. 1994. Genetic base for North American public soybean cultivars released between 1947 and 1988. Crop Science 34:1143–1152.

Gregan, M.D. 2003. The effect of plot-to-plot carryover on grain moisture data in corn research yield trials. M.S. Thesis. Iowa State Univ., Ames, IA.

Haldane, J.B.S. 1932. The Causes of Evolution. Longmans and Green, London.

Hallauer, A.R. 1985. Compendium of recurrent selection methods and their application. Critical Rev. Plant Sci. 3:1–33. CRC Press, Inc., Boca Raton, FL.

Hallauer, A.R. 1991. Use of genetic variation for breeding populations in cross-pollinated species. p. 37–67. *In* H. T. Stalker and J. P. Murphy (ed.). Plant Breeding in the 1990s. C.A.B. International, Wallingford, United Kingdom.

Hallauer, A.R. 1992. Recurrent selection in corn. p. 115–179. *In* J. Janick (ed.). Plant Breeding Rev. Vol. 9. Wiley and Sons, New York, NY.

Hallauer, A.R., and J.B. Miranda Fo. 1988. Quantitative Genetics in Maize Breeding. 2nd ed. Iowa State Univ. Press, Ames, IA.

Harlan, J.R. 1975. Crops and Man. ASA and CSSA, Madison, WI.

Hull, H.F. 1945. Recurrent selection for specific combining ability. J. Am. Soc. Agron. 37:134–145.

Jenkins, M.T. 1940. The segregation of genes affecting yield of grain in maize. J. Am. Soc. Agron. 32:55–63.

Jones, D.F. 1918. The effects of and cross breeding upon development. Connecticut Agric. Exp. Stn. Bull. 207:5–100.

Kempthorne, O. 1956. An Introduction to Genetic Statistics. Wiley, New York, NY.

Kojic, L. 1990. Maize improvement in Yugoslavia and Eastern European countries. Natl. Maize Conf., 2nd-Res. Econ. Environ., Grado, Italy.

Li, C.C. 1976. Quantitative Genetics. The Boxwood Press, Pacific Grove, CA.

Lonnquist, J.H. 1964. Modification of the ear-to-row procedure for the improvement of maize populations. Crop Science 4:227–228.

Lush, J.L. 1945. Animal Breeding Plans. Iowa State Univ. Press, Ames, IA.

Lynch, M., and B. Walsh. 1998. Genetics and Analysis of Quantitative Traits. Sinauer Assoc., Inc., Sunderland, MA.

Mahalanobis, P.C. 1964. Professor Ronald Aylmer Fisher. Biometrics 20:238–251.

Marquez-Sanchez, F. 1982. Modifications to cyclic hybridization in maize with single-ear plants. Crop Science 22:314–319.

Mather, K. 1949. Biometrical Genetics. Methuen, London.
Miller, F.R., and Y. Kebede. 1984. Genetic contributions to yield gains in sorghum, 1950 to 1980. p. 1–14. *In* W. R. Fehr (ed.). Genetic Contributions to Yield Gains in Five Major Crop Plants. Spec. Publ. No. 7. CSSA and ASA, Madison, WI.
Mode, C.J., and H. F. Robinson. 1959. Pleiotropism and the genetic variance and covariance. Biometrics 15:518–537.
Moll, R.H. 1974. Quantitative genetics. Empirical results relevant to plant breeding. Adv. Agron. 26:277–313.
Nyquist, W.E. 1991. Estimation of heritability and prediction of selection response in plant populations. Crit. Rev. Plant Sci. 10:235–322.
Olby, Robert. 1966. Origins of Mendelism. 2nd ed. Univ. Chicago Press, Chicago, IL.
Pandey, S., and C.O. Gardner. 1992. Recurrent selection for population, variety and hybrid improvement in tropical maize. Adv. Agron. 48:1–86.
Provine, W.B. 1971. The Origins of Theoretical Population Genetics. Univ. Chicago Press, Chicago, IL.
Rowe, D.E., and R.R. Hill. 1981. Interpopulation selection in alfalfa. Crop Science 21:392–397.
Russell, W.A. 1991. Genetic improvement of maize yields. Adv. Agron. 46:245–298.
Shamel, A.D. 1905. The effect of inbreeding in plants. USDA Yearbook. p. 377–392.
Shull, G.H. 1908. The composition of a field of maize. Am. Breeders' Assoc. Rep. 4:296–301.
Shull, G.H. 1909. A pure line method of corn breeding. Am. Breeders' Assoc. Rep. 5:51–59.
Shull, G.H. 1910. Hybridization methods in corn breeding. Am. Breeders' Mag. 1:98–107.
Silva, J.C., and A.R. Hallauer. 1975. Estimation of epistatic variance in Iowa Stiff Stalk Synthetic of maize. J. Hered. 66:290–296.
Sinnott, E.W., L.C. Dunn, and T. Dobzhansky. 1950. Principles of Genetics. 4th ed. McGraw-Hill Co., Inc., New York, NY.
Sorrells, M.E., and S.E. Fritz. 1982. Application of a dominant male-sterile allele to the improvement of self-pollinated crops. Crop Science 22:1033–1035.
Sprague, G.F., and L.A. Tatum. 1942. General vs. specific combining ability in single crosses of corn. J. Am. Soc. Agron. 34:923–932.
Tollenaar, M., J. Jing, and D.N. Duvick. 2000. Genetic gain in corn hybrids from the Northern and Central Corn Belt. Annu. Corn & Sorghum Res. Conf. 55:53–62.
Wilcox, J.R., J.W.T. Schapaugh, R.L. Bernard, R.L. Cooper, W.R. Fehr, and M.N. Niehaus. 1979. Genetic improvement of soybeans in the Midwest. Crop Science 19:803–805.
Wright, S. 1921a. Systems of mating. Genetics 6:111–178.
Wright, S. 1921b. Correlation and causation. J. Agric. Res. 20:557–595.
Wright, S. 1968. Genetic and Biometric Foundations. Evolution and Genetics of Populations. Vol. 1. Univ. Chicago Press, Chicago, IL.
Yates, F. 1936. A new method of arranging variety trials involving a large number of varieties. J. Agric. Sci. 26:424–455.
Zobel, R.W., M.J. Wright, and H.G. Gauch. 1988. Statistical analysis of a field trial. Agron. J. 80:388–393.

5

Improving the Connection Between Effective Crop Conservation and Breeding

S. Kresovich, A.M. Casa, A.J. Garris, S.E. Mitchell, and M.T. Hamblin, Institute for Genomic Diversity, Cornell University

Introduction

Intuitively, one might assume that a close, coordinated connection exists between effective crop conservation and breeding. Frequently, this assumption could not be further from the truth. Why? There are occasionally divergent goals, different priorities, and constrained resources that impact the connection between curators/conservationists and breeders. Much curatorial work over the past decades has been descriptive and/or retrospective in nature. If the linkage between conservation and breeding is to be improved, curatorial efforts must become more predictive, that is, hypothesizing where new sources of crop diversity can be found. Furthermore, over the past decade curators have become fixated on quantifying and partitioning neutral diversity as determined by the use of anonymous molecular makers. Though this strategy has yielded benefits for conservation through an improved understanding of genetic representation, it has not been effective at building the bridge between conservation and utilization. Based on the great progress in crop genomics, we are now at a point where curators have the ability to move from a focus on neutral diversity to a more "functional" representation of materials they hold in their collections.

In the broadest sense, the goal of conservation activities is the preservation of diversity at the ecosystem, community, and species levels. Conservation of crop genetic resources, with its focus on diversity within species and their wild relatives, differs in that it is inextricably linked to a mandate for utilization. Perhaps the biggest challenge lies in the identification of useful variation not readily assessed at the phenotypic level, due to the complexity of the trait or the masking effects of environment and genetic background (Tanksley and McCouch, 1997). Exploitation of variation in collections has been achieved primarily through phenotypic screens and backcrossing strategies; however, concepts and tools of molecular and population genetics may serve to expedite the identification and deployment of useful alleles. It is our intent to offer some insights into and examples of how understanding diversity is structured allows the generation of data simultaneously useful for both conservation and breeding, enabling us to dissect gene function and consequently assess the predictive value of diversity for crop improvement.

Although the application of strategies of molecular and population genetics holds promise for genetic resources conservation and use, a few caveats should be noted. The methods and examples we will highlight require substantial preliminary data on population structure and appropriate sampling. Patterns of diversity can be influenced not only by selection; the influence of population structure, linkage, and drift must be understood in order to correctly interpret results. While these integrated approaches can identify interesting candidate genes, functional studies still will be required to establish causation. Another key limitation is that some differences that affect phenotype are not coded in DNA, including such phenomena as epigenetics and differential splicing of RNA transcripts. In addition, the importance of regulatory elements in crop domestication and evolution has been demonstrated, indicating that not only structural but also regulatory genes will be critically important in conservation and breeding.

Thoughtful applications of evolutionary and population genetics have the potential to strengthen the link between DNA sequence and phenotype, facilitating conservation and breeding as well as the link between them. Increased access to genomic technologies, in concert with new genetic concepts and improved computational methods for analysis, will make their use both more common and more valuable in maintaining and using crop genetic resources. In the future important and challenging questions will not be constrained by the lack of insightful concepts and appropriate tools, and the allelic diversity in crop collections can be deployed for breeding.

Historical perspectives of the linkage between crop conservation and breeding

In the United States, the genesis of both plant introduction/crop conservation and plant genetics/breeding occurred at approximately the same time. In 1898, the U.S. Department of Agriculture (USDA) established the Section of Seed and Plant Introduction. Both basic and applied plant scientists were also rediscovering the insights of Gregor Mendel in the first decade of the twentieth century. Over the past 100 years, these complementary activities have grown, matured, and been integrated to support advancements in crop agriculture.

At the heart of both good conservation and breeding has been the creation and recognition of a "good" phenotype, whether it was by the earliest plant collectors identifying interesting species of plants in Russia that might be of value as windbreaks in North Dakota or breeders identifying crop ideotypes that fit their particular environment and cropping system. A good eye for phenotyping has been essential for progress. Over time, however, crop conservation and breeding, as complementary activities, have diverged on occasion.

Early crop conservation activities centered primarily on plant introduction and rapid, cursory evaluations of phenotype. The outcome of integrated acquisition, maintenance, and evaluation efforts has been to provide a source of genetic diversity for supporting plant-breeding efforts. Simply put, the coordinated goal was to scour the world for useful phenotypes of crop plants and get them back into the United States for use in agriculture. Whether breeding was an intermediate step to a product (a new cultivar) was irrelevant. In some cases, adaptation studies were the only stage between introduction and use in crop agriculture. The underlying disciplines associated with crop conservation were agronomy and horticulture, with a continued heavy emphasis on identifying useful plant phenotypes. Concurrently, crop breeding was evolving into a practice driven by the disciplines of genetics and statistics. The divergence of crop conservation and breeding was most apparent when developing a vision for "plant genetic resources" management. Conservationists and curators in the first half of the twentieth century focused on *plant* while crop breeders focused on *genetic*. These divergent goals caused the direct linkage between conservation and breeding to become somewhat tenuous at times; however, progress was made. The question is whether this linkage was as effective as it possibly could be.

In the late twentieth century significant changes occurred in crop conservation priorities based on the perception that global biodiversity was being lost at an alarming rate and that these resources represented the raw material for future advances in breeding. The national system evolved from a plant introduction program into an effort addressing long-term conservation of crop genetic resources. Rather than focusing on sheer numbers of holdings, curators began to adopt the concept of genetic representation of collections. That is, curators viewed the quality of their collections based on the genetic breadth and depth of the holdings in relationship to what was known about the crop species in nature. Curators started to address the biological and operational priorities of management by the identification of gaps or redundancies in their collections.

The development of genetic representation of a collection was augmented by the concept of a core collection as proposed by Brown and Clegg (1983). The core collection proposal was based on recognition that there is a need in a large collection for improved access to desirable traits and genes by breeders. Although larger collections could improve the chance that most genes or genotypes are conserved, their large sizes made it difficult to access desirable genes while also maintaining the collection with fixed resources and personnel. An ideal core collection (or subset) within a collection would contain a range of materials that represent the maximum amount of diversity with a minimum of entries. Based on certain assumptions, it was recommended that a core collection would consist of approximately 10% of the whole collection's holdings. Interestingly, in practice, the development of core collections has been more valuable to curators than to the user community. Many curators now focus on genetic representation in their holdings while few breeders screen entire core collections for finding new traits of agronomic importance.

With the incentive to understand and represent genetic diversity in collections of crop species, there has been an increased emphasis over the past 20 years to employ molecular markers and statistical tools to quantify and partition neutral diversity of holdings. While this empirical approach may aid curators in identifying gaps and redundancies, it has done little to improve our ability to understand and maximize functional (trait-based) diversity of a collection. It also has led to a wealth of studies that tended to be descriptive rather than predictive, that is, establishing where to find agronomically or horticulturally useful diversity. Therefore, while great strides were made toward the development of representative collections that could be effectively maintained, the impact on collection utilization was minimal.

Where we are, where we can go: Opportunities and challenges

In the broadest sense, the goal of conservation activities is the preservation of diversity at the ecosystem, community, and species levels. However, conservation of crop genetic resources, whether undertaken in national or international networks, is fundamentally different from classical conservation biology. Conservation of crop genetic resources, with its focus on diversity within species and their wild relatives, differs in that it is inextricably linked to a mandate for utilization. Perhaps the biggest challenge for effective conservation lies in the identification of useful variation not readily assessed at the phenotypic level, due to the complexity of the trait or the masking effects of environment and genetic background (Tanksley and McCouch 1997). As noted previously, exploitation of variation in collections has been achieved primarily through phenotypic screens and subsequent backcrossing once materials moved into breeding programs. However, opportunities now exist to better discover, characterize, evaluate, and use diversity. If thoughtfully approached, coordinated activities may simultaneously benefit both conservation and breeding goals. That is, collections may become more valuable and accessible while breeding efforts may be more efficient at extracting desired genes and genotypes from collections.

Until recently, there has been a conceptual dichotomy between evolutionary disciplines and agriculture. For some reason(s), the disciplines have diverged significantly though agricultural efforts, particularly in crop conservation and breeding, and could benefit greatly by being viewed in an evolutionary context. Whether one studies crop domestication or improvement through breeding, certain unifying principles exist. For example, understanding and predicting the pattern and level of diversity in a genome or population is fundamental to both evolutionary biologists and crop breeders.

Tremendous advances are being made in the basic and applied biological sciences. Progress in the "omics" (genomics, proteomics, metabolomics, etc.) has generated large amounts of data for understanding structural and functional relationships of genes and gene networks in a broad spectrum of species. This trend will only grow in the future as technologies continue to allow increased throughput, reduced unit assay cost, and improved quality of data generated. In tandem, conceptual advances in evolutionary biology, population genetics, molecular genetics, statistical genomics, bioinformatics, and plant breeding greatly increase the value of the data being generated. For example, new approaches to detecting selection show great promise for the discovery of domestically important and agriculturally useful genes or DNA sequences.

In the following sections, we briefly present selected case studies that highlight the integration of evolutionary biology and genomics to better discover unique genes or genotypes that warrant further investigation in an agricultural context. It is likely that if we thoughtfully identify critical agricultural questions, target appropriate populations for comparisons, and employ highly sensitive and statistically rigorous methods to extract biologically relevant information, future conservation and breeding efforts benefit simultaneously. Also, the linkage between conservation and use becomes more tangible.

Linking conservation and use through evolutionary genetics

One possible approach to building the connection from genetic diversity to phenotype (that is, conservation to agricultural use) is linkage disequilibrium mapping, recently proposed as an alternative to traditional methods for mapping traits in plants (Buckler and Thornsberry, 2002). Linkage disequilibrium mapping seeks to identify an ancestral haplotype associated with a phenotype in a sample of gene bank accessions. This ancestral haplotype is detected by the nonrandom association of alleles in a genomic region resulting from their physical linkage; however, population structure, selection, or drift can also give rise to linkage disequilibrium. Estimates of linkage disequilibrium are important as an indicator of how useful linkage disequilibrium-based trait-mapping approaches may be compared to other available methods based on the trade-off between population size and informativeness. If linkage disequilibrium declines rapidly, genome scans will require an excessive marker density, but the testing of candidate genes is feasible. If linkage disequilibrium is too large, resolution may be low, but genome scans may provide a "first cut" at detecting potentially agronomically interesting regions.

In a recent case study in rice (Garris et al., 2003) we provide analysis of linkage disequilibrium in the genomic region containing *xa5*, a bacterial blight-resistance allele that has not been identified or characterized at the molecular level. This study highlights the important role that population structure has had in shaping haplotype diversity in the candidate region for this resistance gene, the extent of linkage disequilibrium in this genomic region in this population, and the complications that arise from genetic heterogeneity.

The analysis of population structure underscores the need for genetic analysis of ecotypic differentiation if linkage disequilibrium and association mapping approaches are to be of value in rice improvement. Population structure in 114 accessions of rice predominantly from Bangladesh and Nepal was examined using 21 simple sequence repeats (SSRs) distributed on the 12 chromosomes. One subpopulation consisted almost entirely of the Bangladeshi *indica* rice ecotype called *aman*. The second group was populated by *aus* and *boro* ecotypes, mainly from Bangladesh and Nepal. Fst values for these two populations showed a high degree of population structure (overall Fst for two populations = 0.89). The population structure data supports a hypothesis of hierarchical levels of divergence within rice, with greater divergence between the *indica* and *aus-boro* groups, and no detectable divergence between the *aus* and *boro* ecotypes at this level of genomic resolution. This population subdivision has a bearing on the distribution of haplotype diversity. Haplotype diversity in the 70-kb candidate region was assessed using single nucleotide polymorphisms (SNPs) in 13 amplicons. In general, each haplotype was found in a single subpopulation, and frequently several closely related haplotypes were found in the same subpopulation. Even in human populations, where levels of Fst are much lower (average Fst = 0.14), population structure is known to confound the association of genotype with phenotype, primarily by increasing the level of false positives (Pritchard et al., 2000). Methods to control for population structure recently applied to maize (Thornsberry et al., 2001; Pritchard, 2001) may overcome this problem in rice, as long as allelic and genetic heterogeneity are not too high (see subsequently).

Linkage disequilibrium in the 70-kb *xa5* region was extensive but potentially informative in reducing the candidate region for *xa5* described in Blair et al. (2003). Linkage disequilibrium, measured as r^2, was significant for the distal 45 kb of the candidate region for resistant accessions from both *indica* and *aus-boro* accessions, a pattern that was not observed in the susceptible groups. Linkage disequilibrium in a larger region encompassing the 70-kb candidate region was assessed with SNPs in an additional five amplicons spanning the proximal 45 kb. Extensive linkage disequilibrium was present; r^2 approaches 0.1 only after 100 kb. This is the same order of magnitude as linkage disequilibrium observed at the *FRIGIDA* flowering time locus in another autogamous organism, *Arabidopsis thaliana*, where significant linkage disequilibrium was detected between pairs of sites up to 250 kb apart (Hagenblad and Nordborg, 2002; Nordborg et al., 2002). As expected, these estimates differ greatly from the limited linkage disequilibrium observed in outcrossing species such as maize, where linkage disequilibrium frequently decays at distances between 100 bp and 1.5 kb (Remington et al., 2001; Thornsberry et al., 2001; Tenaillon et al., 2001). In addition, it is possible that the *xa5* locus is under selection and would therefore be predicted to have more extensive linkage disequilibrium than a locus evolving neutrally.

Analysis of haplotype and phenotype indicate that this theoretically single-gene trait of *xa5* resistance may have underlying allelic or genetic heterogeneity. For example, two highly divergent haplotypes were found in an allele-tested resistant accession, suggesting two origins for this phenotype. In addition, several non-allele-tested, resistant *aman* accessions had haplotypes that differ from the *aus-boro*-resistant haplotype and were nearly identical to some susceptible haplotypes, suggesting that a different locus could be responsible for the resistance in the *aman* population.

A widely held assumption in association studies is that common variants underlie the genetic risk for common phenotypes (Lander, 1996). At this time, little information is available on the distribution of alleles in subpopulations of rice. However, if alleles have arisen after diversification into subpopulations and their isolation has been reinforced by limited gene flow, that assumption is violated.

A similar example of allelic heterogeneity was found for the early flowering *FRIGIDA* locus in *Arabidopsis* (Hagenblad and Nordborg, 2002), and eight independent loss-of-function mutations at this locus conferring early flowering have been identified (Le Corre et al., 2002). Both rice and *Arabidopsis* are predominantly autogamous, and therefore the expectation of a single origin of a phenotype that occurs across subpopulations may be less plausible than in outcrossing species. This has implications for sampling in future linkage disequilibrium or association studies. Isolated populations, as employed in the study of human diseases, may find their plant counterparts in the subpopulations of autogamous crop species that may be more likely to have single origin phenotypes (Shifman and Darvasi, 2001).

From description to prediction: Establishing a framework for identifying agronomically useful variation

The use of population genetics and evolutionary biology principles for identifying potentially useful variation is based on the premise that strong selection can dramatically reduce genetic diversity in the target gene(s) or genomic regions. The extent to which variation will be reduced, however, is not only dependent on the strength of selection but also on the breeding system, the population size, and the levels of recombination observed between the selected site and the molecular marker locus being surveyed. Consequently, knowledge of the distribution of neutral diversity within a genome is a key requirement to build a null hypothesis against which to test levels and patterns of diversity in candidate genes.

Because crop species experienced strong selective pressures during their domestication, they offer a unique opportunity to test the use of neutrally evolving markers in identifying genes controlling traits that have been under selection. Ultimately, implementation of population genetics principles for dissecting molecular diversity will allow both the discovery and characterization of alleles that modify agronomic and developmental traits as well as create community resources for dissecting traits of interest to both conservationists and breeders alike.

Identifying genes of agronomic importance in the grasses

Although a wealth of DNA sequence information is now available for several crop species, our capacity for identifying functionally useful variation is still extremely limited. This is due primarily to the complex nature of many agronomically important traits and to the masking effects of the environment, which in turn hinder our ability to associate genotypes to corresponding phenotypes. Because genome scans of diversity require no a priori knowledge either of the affected trait or of gene function, they have the potential to be used in the identification of adaptive genes.

The first example in which population genetics approaches have been applied to expedite the identification of functionally important alleles in plants comes from maize (Vigouroux et al., 2002). Among the biological characteristics that have made maize an attractive model for this type of study are its high levels of recombination and correspondingly low levels of linkage disequilibrium. Moreover, because historical population sizes for maize were large there is a reasonable expectation that genes neighboring loci under selection will have retained high diversity and can be readily distinguished from those affected by selection.

Vigouroux et al. (2002) compared levels of diversity at SSR loci in cultivated maize to those present in its wild relative teosinte to identify candidate genes involved in domestication. Because the population bottleneck associated with domestication would cause a genome wide loss of diversity, statistical models that incorporated the domestication bottleneck were used. The investigators, therefore, were able to define a threshold above which the loss of diversity was too great to be solely due to the effect of the bottleneck alone. A total of 501 EST-derived SSRs were evaluated, of which 15 exhibited some evidence for selection in maize and 10 showed evidence under stringent criteria. It should be noted, however, that deviation at a SSR locus from neutral expectation is only the first step toward identification of target genes. Incorporation of both DNA sequence diversity and map location of candidate loci will be invaluable for associating candidates with QTLs for traits that were/are under selection. For example, the MADS homologue candidate identified by SSR genomewide scans maps to the short arm of chromosome 1 near a QTL for differences in ear structure between maize and teosinte.

A similar approach for identifying targets of selection also has been employed in sorghum. Sorghum is the fifth most important cereal grown worldwide and is a pillar of food security in the semiarid zones of western and central Africa. The work underway in our laboratory aims to establish a sensitive framework for identifying genomic regions under selection. Genomewide comparisons of diversity in sorghum gene pools (elite inbreds, landraces, and putative wild progenitors) having extensive racial as well as geographic representation were analyzed. A total of 98 SSR loci derived from RFLP clones and small insert genomic libraries were evaluated in a panel of 104 sorghum accessions. SSR data were analyzed for (1) excess of rare alleles (Cornuet and Luikart, 1996); (2) variance in gene diversity (Kauer et al., 2003); and (3) population differentiation based on allele frequencies (Fst; Wright, 1951). A total of 11 loci were flagged as candidates for selection. Because the mutational behavior of an SSR can generate a signal similar to that expected under the selection scenario, follow-up analysis of additional closely linked SSRs (within 100 kb) at three candidate regions was also conducted. Statistical analysis of these new loci indicated that the observed reduction in genetic variation and/or skew in allele frequencies were in some cases also consistent with some type of selection. Acquisition of DNA sequence data for these regions from approximately 30 genotypes, equally representing cultivated and wild sorghum races, will be done next to further test for evidence of selection and to characterize its footprint. Overall, we expect that these data will allow us to quantify the extent and distribution of reduction in diversity across the genome that has accompanied domestication as well as to determine whether reductions in variation are due to strong selection on particular loci or genomewide bottlenecks due to small population sizes. These data also will be valuable for dissecting the influence of breeding system, introgression, and demography on the levels of polymorphism among cultivated and wild sorghums that ultimately will be essential to the discovery, conservation, and utilization of useful alleles.

Although the application of molecular markers holds promise for genetic resources conservation and use, a few caveats should be noted. The methods outlined previously require substantial preliminary data on population structure and appropriate sampling. While these approaches can identify interesting candidate genes, functional studies will be required to establish causation. Increased access to new technologies along with improved computational methods for analysis will make molecular techniques both more common and more useful in maintaining and using crop genetic resources.

Synthesis

In this review, we have attempted to highlight the exciting opportunities and important challenges that now present themselves to better link crop conservation with breeding. Thoughtful applications of molecular and population genetics have the potential to strengthen the bridge between DNA sequence and phenotype. Also, we emphasize the importance of good phenotyping for crop conservation and breeding in the future. As molecular genetic frameworks are established for many crop species, being able to effectively and efficiently document individual phenotypes will greatly enhance the chances for improved crop conservation and improvement. Also, the critical role of phenotyping will engage researchers globally, thus building a true collaborative momentum for future advances.

We also have presented examples of a new and complementary strategy to discovering genes and/or genic regions under selection. In many instances, these candidate genes or genic regions will have adaptive value in natural populations and agronomic value in agricultural settings. Rather than generically using the classic approach to establishing core collections, thoughtful researchers will establish test germplasm arrays based on critical biological questions that simultaneously yield insights to both improved crop conservation and breeding. Again, we emphasize the importance of clearly identifying the biological question and creating an array of biological materials for analysis. Core collections of the future will be based more on functional than neutral diversity.

An improved understanding of natural and breeding population structures (both patterns and amounts of diversity) provides a foundation for prediction and discovery of useful genes and traits. By linking molecular and population genetics with crop conservation and breeding, we will be better able to find useful diversity in the genome and in natural populations.

In the last 20–30 years, the integration of ecological principles into crop agriculture has greatly improved the way we understand and use the environment to feed and shelter people. By analogy, we are entering a period when evolutionary principles can be exploited, through thoughtful integration with crop conservation and breeding efforts, to improve the way we discover and deploy these precious genetic resources that have taken many generations to create.

Acknowledgments

We would like to thank the organizers of the Arnel R. Hallauer International Symposium on Plant Breeding for giving us the opportunity to present this paper.

References

Blair, M.W., A.J. Garris, A.S. Ayer, B. Chapman, S. Kresovich, and S.R. McCouch. 2003. High resolution genetic mapping and candidate gene identification at the *xa5* locus for bacterial blight resistance in rice (*Oryza sativa* L.) Theor. Appl. Genet. 107:62–73.

Brown, A.H.D., and M.T. Clegg. 1983. Isozyme assessment of plant genetic resources. *In* Isozymes: Current Topics in Biological and Medical Research. Vol. 11, M.C. Rattazzi, J.G. Scandalios, and G.S. Whitt, (eds.), 285–295. Liss, NY.

Buckler, E.S., and J.M. Thornsberry. 2002. Plant molecular diversity and applications to genomics. Curr. Opin. Plant Biol. 5:107–111.

Cornuet, J.M., and G. Luikart. 1996. Description and power analysis of two tests for detecting recent population bottlenecks from allele frequency data. Genetics 144:2001–2014.

Garris, A.J., S.R. McCouch, and S. Kresovich. 2003. Population structure and its effect on haplotype diversity and linkage disequilibrium surrounding the *xa5* locus of rice (*Oryza sativa* L.). Genetics 165:759–769.

Hagenblad, J., and M. Nordborg. 2002. Sequence variation and haplotype structure surrounding the flowering time locus *FRI* in *Arabidopsis thaliana*. Genetics 161:289–298.

Kauer MO, D. Dieringer, and C. Schlotterer. 2003. A microsatellite variability screen for positive selection associated with the "out of Africa" habitat expansion of *Drosophila melanogaster*. Genetics 165:1137–1148.

Lander, E.S. 1996. The new genomics: global views of biology. Science 274:536–539.

Le Corre, V.L., F. Roux, and X. Reboud. 2002. DNA Polymorphisms at the FRIGIDA gene in *Arabidosis thaliana*: extensive nonsynonymous variation is consistent with local selection for flowering time. Mol. Biol. Evol. 19:1261–1271.

Nordborg, M., J.O. Borevitz, J. Bergelson, C.C. Berry, J. Chory, et al. 2002. The extent of linkage disequilibrium in *Arabidopsis thaliana*. Nat. Genet. 30:190–193.

Pritchard, J.K. 2001. Deconstructing maize population structure. Nat. Genet. 28:286–289.

Pritchard, J.K., M. Stephens, N.A. Rosenberg, and P. Donnelly. 2000. Association mapping in structured populations. Genetics 155:945–959.

Remington, D.L., J.M. Thornsberry, Y. Matsuoka, L. Wilson, S. Rinehart-Whitt, J.F. Doebley, S. Kresovich, M.M. Goodman, and E.S. Buckler, IV. 2001. Structure of linkage disequilibrium and phenotypic associations in the maize genome. Proc. Natl. Acad. Sci. USA 98:11479–11484.

Shifman, S., and A. Darvasi. 2001. The value of isolated populations. Nat. Genet. 28:309–310.

Tanksley, S.D., and S.R. McCouch. 1997. Seed banks and molecular maps: Unlocking genetic potential from the wild. Science 277:1063–1066.

Tenaillon, M.I., M.C. Sawkins, A.D. Long, R.L. Gaut, J.F. Doebley, et al. 2001. Patterns of DNA sequence polymorphism along chromosome 1 of maize (*Zea mays* ssp. *mays* L.) Proc. Natl. Acad. Sci. USA 98:9161–9166.

Thornsberry, J.M., M.M. Goodman, J.F. Doebley, S. Kresovich, D. Nielsen, and E.S. Buckler, IV. 2001. *Dwarf8* polymorphisms associate with variation in flowering time. Nat. Genet. 28:286–289.

Vigouroux, Y., M. McMullen, C.T. Hittinger, L. Schulz, S. Kresovich, Y. Matsuoko, and J.F. Doebley. 2002. Identifying genes for agronomic importance in maize by screening microsatellites for evidence of selection during domestication. Proc. Natl. Acad. Sci. USA 99:9650–9655.

Wright, S. 1951. The genetical structure of populations. Ann. Eugenics 15:323–354.

6

Breeding for Cropping Systems

E. Charles Brummer, Associate Professor of Plant Breeding, Raymond F. Baker Center for Plant Breeding, Iowa State University

Crop cultivars, the ultimate product of plant breeding, are grown within a cropping system that consists of other crop species and a set of management practices applied to the cultivation of those crops. The breeding process can be conducted in any environment, and it does not need to be done within the cropping system in which the cultivars will be grown. The first objective of this chapter is to assess whether selection in one cropping system will result in the best cultivar for other systems. The second objective is to discuss the case of intercropping, in which two species are grown concurrently in the same field, and describe methods of breeding for this special cropping system. Most breeding is conducted to address particular needs within a single crop species, but considering the overall cropping system productivity, profitability, and ecological value may also alter breeding programs. Thus, the third objective is to address breeding for *total system performance*. A number of excellent reviews covering aspects of these questions have been published previously, and the interested reader is directed to them for further explication of the main points presented here.

Breeding for contrasting cropping systems

The most obvious difference that may accompany a change in cropping systems is that the traits that a particular crop requires in order to be successfully produced may change. For instance, forage crops may be mechanically harvested as hay or silage or harvested directly by livestock as pasture. Alfalfa (*Medicago sativa* L.) breeding programs primarily select under mechanical harvesting conditions, but this procedure does not usually identify plants most tolerant of grazing. As a consequence, breeding directly for grazing tolerance by evaluating genotypes in the presence of animal grazing has been conducted to develop cultivars for pasture systems (Bouton et al., 1991). We compared six cultivars, three of which were selected for grazing and three for biomass yield, in side-by-side experiments to measure mechanically harvested biomass yield and tolerance to grazing by beef cattle (Brummer and Smith, 1999). The three most grazing tolerant cultivars were those that had been selected under grazing; the cultivar with the worst grazing tolerance produced the highest biomass yield (Figure 6.1). Thus, if the trait of most importance differs between systems, then selection obviously must be done for that trait under the conditions in which it will be grown.

More complex is the question of whether selection *for a particular trait* needs to be done in each alternative system in which the cultivar will be grown. For the purposes of this paper, cropping systems can be defined as alternative management strategies, such as conventional tillage versus no tillage, single versus double cropping, irrigated versus rain-fed, inorganic versus organic fertilizer, intercropping versus monocropping, high versus low fertility, and other strategies. The question of breeding in one system for use in another is directly analogous to the case of breeding in "stress" versus "non-stress" environments. The case can be stated as follows: (1) Does the breeder need to select in the target environment in which the cultivar will be grown, and (2) is selection in a highly controlled (e.g., high-input) environment better than selection in a less-controlled (e.g., low-input) environment that is representative of actual farmer conditions. These questions really ask if we need a

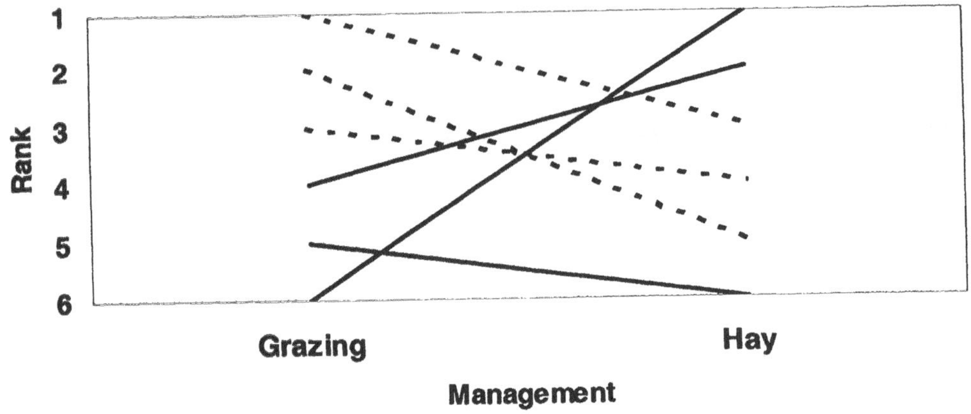

Figure 6.1 Performance of alfalfa cultivars selected under grazing pressure (dotted lines) or under mechanical harvesting (solid lines) when evaluated under either grazing by beef cattle or mechanical harvesting in side-by-side trials at Rhodes, Iowa. A rank of 1 indicates the most grazing-tolerant or the highest-yielding cultivar.

separate breeding program for each system or if breeding can be done in one of them—perhaps the most common program or the one that is the easiest to use—to develop cultivars for all of them.

The general experimental procedure used to address this question is to evaluate a set of cultivars developed under one system (A) in systems A, B, etc. An analysis of variance is used to identify the presence of cultivar by system interactions, which would indicate that cultivar response varies with different systems. Numerous experiments have been conducted to compare different systems; the experiment on tillage methods conducted by Hallauer and Colvin (1985) is representative. They examined 14 maize (*Zea mays* L.) hybrids over five years when grown under four different tillage systems, including fall-plow, strip-till, spring-disk, and no-till. For grain yield, the mean square associated with tillage method was an order of magnitude larger than that for hybrid; both were greater than zero. The hybrid mean square was an order of magnitude larger than the tillage method by hybrid interaction, which was nonsignificant. Similar results were observed for other traits as well. This experiment demonstrated that the response of these cultivars relative to one another is similar regardless of tillage method used.

Results such as these are typically used to argue that breeding programs do not need to be conducted in the different systems under consideration, because the best cultivars under one system are also the best for the alternative system. However, because all the cultivars examined in this and similar experiments were actually selected under only one of the systems being compared, the efficiency of selection within the alternative systems has not been evaluated. The possibility exists that cultivars developed within the alternative system would actually perform better than those developed in the original system. The significance of the method by hybrid interaction does not address this question at all.

Determining if direct selection in a particular system is necessary requires more than an evaluation of cultivars selected under a single set of conditions. The efficiency of direct versus indirect selection of a given trait for a particular system can be ascertained using the following formula (Falconer and Mackay, 1996, p. 322), assuming the selection intensities are the same in both systems:

$$CR_B/R_B = r_G\, h_A/h_B$$

where CR_B is the correlated response of the trait of interest to selection in system A when evaluated in system B, R_B is the direct response to selection in system B when evaluated in system B, r_G is the genetic correlation between systems A and B, and h_A and h_B are the square roots of the heritability of the trait in systems A and B, respectively. The genetic correlation is inversely proportional to the genotype by system interaction variance, and hence, when this variance is large, direct selection will be almost always be more effective (Table 6.1). Indirect selection will be more effective when the ratio of correlated response to direct response

Table 6.1 The ratio of correlated response of indirect selection for a particular trait in system A when grown in system B to the direct response of selection in system B when grown in system B across a range of genetic correlations between systems A and B and a range of heritability ratios measured in systems A and B for the trait under consideration

h^2_A / h^2_B	r_G			
	1.00	0.75	0.50	0.25
	CR_B/R_B			
0.10	0.32	0.24	0.16	0.08
0.25	0.50	0.38	0.25	0.12
0.50	0.71	0.53	0.35	0.18
1	**1.00**	0.75	0.50	0.25
2	**1.41**	0.87	0.58	0.29
4	**2.00**	**1.50**	**1.00**	0.50
10	**3.16**	**2.37**	**1.58**	0.79

Note: The boldfaced numbers indicate those instances in which indirect selection would be equal to or more efficient than direct selection.

Table 6.2 The efficiency of direct versus indirect selection for two soil nutrient levels. Selection made among 116 random F_9-derived oat lines

Environment	Mean yield	h^2	r_G	CR/R
	kg ha^{-1}			
Low P	1140	0.40		0.38
Hi P	2471	0.21	0.52	0.71
Low N	1240	0.32		1.09
High N	2850	0.38	1.08	0.92

Source: Data from Atlin and Frey (1989).

is greater than 1.0. This situation arises when the heritability of the trait is higher in system A than system B, and the genetic correlation is close to 1. However, if the heritability in system B is higher than system A, or if the genetic correlation is low, then direct selection will usually be superior. Progress may still be made by indirect selection, even if the correlated to direct response ratio is less than 1, but it may not be as efficient as direct selection.

The systems most often compared for breeding efficiency have been high- versus low-productivity environments. As an example, Atlin and Frey (1989) assessed the effectiveness of direct versus indirect selection for yield in an evaluation of 116 randomly chosen homozygous oat (*Avena sativa*) lines grown under low or high nitrogen (N) and low or high (P). Their results showed that selection for yield under low-P conditions should be done under low-P conditions, but that selecting for low-N conditions could be just as effectively done under either low- or high-N conditions (Table 6.2). The conventional wisdom holds that because genetic variation is often larger and more accurately estimated in high-productivity environments, selection should be done under those conditions for all environment (Bänziger and Cooper, 2001; Evans, 1993, p. 165, 297). Even if the heritability is larger in the high-productivity environment, selection there cannot be superior to that done directly in the low-productivity environment if the genetic correlation between systems is sufficiently small (Table 6.2). An important observation by Atlin and Frey (1989) was that the heritability of grain yield was actually higher under low-P than high-P conditions, a situation found in other experiments as well (Ceccarelli and Grando, 2002). In this case, direct selection is *always* superior to indirect selection (Table 6.1).

As the difference in productivity between systems under consideration becomes larger, the need for system-specific breeding increases (Bänziger et al., 1997). Selection for productivity across both high- and low-productivity environments may not be compatible with selection for productivity in low production environments only. Although breeding in high-productivity environments may result in improvement for low-productivity environments, the converse is rare (Atlin and Frey, 1990). Regardless, as Simmonds (1991) says: "The sensible response be plant breeders seeking to breed for E_L (low productivity environments) would be to select in E_L; to *select*, be it noted, not merely do trials after selecting in E_H."

Whether alternative systems need distinct selection programs in order to maximize genetic gain can only be assessed by comparative selection programs conducted and evaluated under both systems simultaneously, but this has rarely been done. The need will undoubtedly be context dependent, precluding any simple means to ascertain the correct selection procedure. Breeders need to keep their target environments in mind as they select; potential problems can be avoided or at least minimized by including all target environments (cropping systems) in the breeding program, and subdividing the program into smaller environmental targets as costs permit, if necessary.

Alternative cropping systems have characteristics that suggest that system-specific breeding programs may be needed. As the following examples suggest, different cropping systems may have soils with different biological, chemical, and physical properties. Organic systems have more soil microbes, including mycorrhizae, and higher activities of certain enzymes, such as dehydrogenase and phosphatase, than conventional systems (Mäder et al., 2002). Similarly, a coconut monoculture had less microbial biomass and more organic carbon (C) and total N, P, and Potassium than a complex multistoried coconut-pepper-cacao-pineapple system (Bopaiah and Shetty, 1991). The types and populations of arbuscular mycorrhizal fungi vary due to tillage method and to conventional or low-input management (Galvez et al., 2001). Crop rotations can affect disease prevalence; for example, wheat–pea rotations had less disease than continuous wheat or wheat–fallow systems (Smiley et al., 1996). Generally, increasing agricultural biodiversity improves pest control both through encouraging natural predators and by enhancing the ability of the plants to withstand pests (Gurr et al., 2003). Numerous experiments have demonstrated interrelationships between soil organisms and plant diversity or productivity, suggesting that the soil community affects plant processes and vice versa (Bradford et al., 2002; De Deyn et al., 2003; Wardle et al., 2004). Undoubtedly, all these interactions are affected by fertility; the use of green or animal manures alters the availability of nutrients to the growing crop compared with inorganic fertilizers (Hadas et al., 2004; Pang and Letey, 2000).

Collectively, these results suggest that soil conditions, pest profiles, and other attributes will vary in alternative cropping systems and that more complex cropping systems may have equally complex webs of interacting organisms and different nutrient fluxes than simplified monocultures. The implications of these different system characteristics for cultivar development are not clear because comparative selection programs have not been conducted. Crops bred specifically in more complex cropping systems might be better able to access the services of the more abundant and diverse soil flora and fauna. A tantalizing example that this may be true has been examined in wheat. Breeding over the past 60 years has decreased the dependence of wheat cultivars on mycorrhizal symbioses, which was often observed in cultivars developed prior to 1950 (Hetrick et al., 1993). Thus, as chemical fertilizers became more accessible and wheat crops had less need for mycorrhizal symbioses to access nutrients, wheat cultivars were developed that did not form that dependence. This might suggest that wheat cultivars directly selected within cropping systems in which mycorrhizal symbioses are important for nutrient acquisition, such as those using less synthetic fertilizer, might be expected to perform better than those developed in conventional systems.

The mechanics of breeding in one system or another do not differ; half-sib family-recurrent selection is conducted the same way regardless of the cropping system. However, in different systems at least three aspects of breeding may change. First, the most efficacious method of selection may differ among systems. For example, both mass and full-sib selection were effective at improving maize grain yield under irrigated conditions, but only full-sib selection was successful under dryland conditions (Johnson and Geadelmann, 1989). How often this occurs, and to what extent breeders need to be concerned with this problem, is undoubtedly context dependent and will be determined by the various components of the genetic gain equation (Fehr, 1987). Second, breeding in different systems may be attended by a change in resources, for example, if one large breeding program is cut in half for two separate target systems. If this is the case, the method of breeding may need to be altered to maximize efficiency at that level of resources. Third, different systems will undoubtedly require attention to different traits. We can assume that productivity, in general, will be a common theme, but the spectrum of disease resistances, food or feed quality traits, and other characteristics will likely change, all of which may affect the breeding method. Of particular relevance in this regard are changes, such as elimination of pesticide or herbicide use, that will directly impact the biotic stresses cultivars will encounter.

Breeding for intercropping systems

Intercropping represents a more striking alternative system than those discussed until now, and consequently, it requires more attention. Under intercropping, two or more crop species are grown together in the same field at the same time (Vandermeer, 1989). Mixtures of diverse genotypes of the same species is a conceptually similar idea, often done for similar reasons (Zhu et al., 2000; Zhu et al., 2003). Among the purported benefits of intercrops and mixtures, when compared with monocultures (also called "sole crops"), are higher productivity, better stability of production, and better disease and pest control, although not all intercrops realize these benefits. For our purposes, the reason an intercrop is grown is not as relevant as considerations about breeding crops for intercrops.

Several parameters are of interest in intercrops. First, the relative performance of a crop grown in monoculture can be compared with that in intercrops. Second, within intercrops, the performance of genotypes of the two (or more) species can be ascribed to both general ecological combining ability (GECA) and specific ecological combining ability (SECA) (Harper, 1967; Hill, 1990), terms directly analogous to the commonly used plant-breeding terms *general* and *specific combining ability* (Sprague and Tatum, 1942). In this case, GECA refers to the average performance of a genotype of one species when grown with genotypes of the other; SECA represents the performance of specific combinations of genotypes of the two species. Thus, a general model for intercrop performance for the *i*th entry (e.g., genotype or cultivar) of species A and the *j*th entry of species B can be written as follows:

$$Y(A_iB_j) = \mu + GECA(A_i) + GECA(B_j) + SECA(A_iB_j) + error$$

Comparison of monoculture and intercrop ability is important, particularly in situations where the two intercropped species may also be grown as monocultures. This comparison was made in an experiment analyzing monocultures and intercrops of several cultivars of berseem clover (*Trifolium alexandrinum* L.) and oat (Holland and Brummer, 1999). Oat performance in monoculture was highly positively correlated with performance in intercrops. This result might argue that oat could be selected in monoculture for intercropping situations except for the fact that oat cultivars differentially affected berseem clover performance. Hence, when the selection unit is intercrop performance, selection of oat needs to be done in the presence of the forage legume. For berseem clover, monoculture performance did not reflect intercrop performance, suggesting that clover must be selected in the presence of the small grain if intercrop performance is of interest.

Deciding whether selection needs to be done in intercrop or if it can be as effective under monoculture conditions needs to be evaluated on a case-by-case basis, using a strategy similar to that discussed above. One example will suffice to describe the parameters. Common bean (*Phaseolus vulgaris*) genotypes were evaluated for grain yield and pods per plant both as sole crops and in intercrops with maize (Atuahene-Amankwa and Michaels, 1997). The narrow sense heritability of grain yield was similar between systems (0.35 for sole crop; 0.29 for intercrop); for pods per plant, sole crop heritability was higher (0.35–0.13). The genetic correlation between systems was 0.45 for grain yield and 0.64 for pods per plant. These results lead to a *CR*/R ratio of 0.49 for grain yield and 1.05 for pod per plant, suggesting that direct selection in intercrops is needed to improve intercrop grain yield most efficiently, but that selection in either system was acceptable for pods per plant. Other experiments have shown similar results for bean–maize intercrops, with heritability of grain yield actually higher under intercrops in one of them (Zimmerman et al., 1984).

Interactions of genotypes with cropping systems suggest that the genes controlling a particular trait are different, or act differently, when plants are grown in the systems being compared. Heritabilities and genetic correlations similar to those in the bean–maize intercrop example indicate that breeding in intercrops would be superior to breeding in monoculture. Several other considerations need to be made, however, before concluding that this is true. In particular, breeding in intercrops may be more difficult than monoculture selection; for instance, mechanization may not be possible, or it may be less efficient. If intercrop selection requires smaller population sizes or other alterations in the breeding scheme, then the additional gain in efficiency from direct selection may be eliminated. Monoculture selection may also be useful, particularly at early stages of the process to eliminate poorly performing lines in the most efficacious manner (Davis and Woolley, 1993; Hamblin and Zimmerman, 1986).

Breeding for intercrop performance can be more complex than the familiar selection in monoculture and may be affected in several ways. Most simply, genotypes (clones, families, pure lines, populations, etc.) can be selected for GECA with the companion species by simply growing the

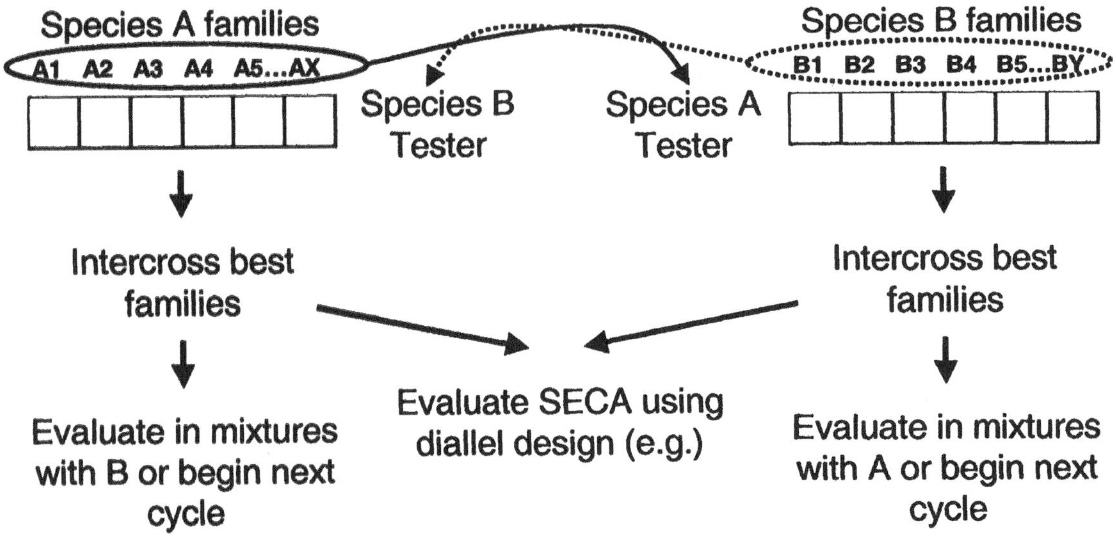

Figure 6.2 General method to improve two species simultaneously for general and specific ecological combining ability (after Hill, 1990, and Francis, 1990).

breeding population in the presence of one or few tester genotypes of the companion species (Davis and Woolley, 1993). The best performing genotypes are selected and intercrossed for the next cycle of selection, or advanced to evaluation trials for potential cultivar release.

Selecting both species for GECA simultaneously can be done using reciprocal recurrent selection for compatibility (Hill, 1990). In this plan, a series of genotypes of species A is tested in combination with a bulk of species B genotypes; similarly, the species A genotypes are bulked to form the tester for species B genotypes (Figure 6.2). In this scheme, the best families are advanced for intercrossing and either continued selection or evaluation for release. Further, a diallel design, in which the best genotypes from A and B are grown in all binary mixtures, can be conducted to select combinations of genotypes for SECA. A more extensive testing for SECA can be carried out under a plan devised by Hamblin et al. (1976). In this method, selection is first made among a series of populations of the two companion species grown in a diallel design; based on performance of populations, lines derived from the populations could be evaluated in a second diallel in the next generation to select good line combinations that are advanced to cultivar status or used to begin another cycle of selection.

A problem of selecting for intercrop performance lies in determining the selection criteria. Yield or productivity of an intercrop does not rely on the productivity of the individual crop, but on the productivity of the combined crops. Clearly, each component must perform adequately, but one component cannot perform in a way that kills, or otherwise unduly limits, the performance of the other. Some method of weighting the value of each crop is needed, and this is not always trivial to obtain. A selection experiment to maximize a maize–bean intercrop used maize yield plus three times bean yield as the selection criterion, based on the economic values of the two crops (O'Leary and Smith, 1999). Further elements besides yield and commodity prices may be incorporated, including ecological values such as improved nutrient cycling or decreased erosion. The value of a green manure crop is more difficult to assess than that of a bushel of grain. A final consideration that needs to be made relative to breeding specifically for intercrop performance—and especially to breeding for SECA—is whether the market for seed is large enough to warrant the effort. The development and marketing of a combination of two cultivars may not be logistically feasible; selection for GECA may result in acceptable cultivars with less effort.

Breeding for sustainable cropping systems

Intercropping shows the importance of considering components simultaneously so that the performance of the whole, rather than the components, is maximized. Practically, this means that the best performing genotype grown in a monoculture may not be the best to maximize the intercrop. I propose to broaden this concept to encompass the entire cropping system of a region, and more importantly, to consider breeding in the context of cropping systems. My argument has two components: first, I will argue that many current cropping systems, focusing on the rotation of maize and soybean (*Glycine max*) monocultures currently occupying much of agricultural land in the midwestern United States, are neither ecologically nor economically sustainable. Second, I will discuss how breeding for whole system performance could improve cultivars within a better cropping system framework.

Over the past 50 years, the typical cropping system in Iowa has been greatly simplified, changing from a complex mixture of species grown in rotation and including some intercrops to a two component serial monoculture of soybean and maize. A quick scan of Iowa agricultural statistics shows that during this time, yield per hectare of both crops has increased (substantially in the case of maize), gross returns in constant dollars per hectacre have diminished, the number of farmers has contracted, and overall farm profitability has declined to the point where government subsidies represent a large portion of farm profit (these data are available at http://www.nass.usda.gov/ia/). A discouraging trend of recurring production problems plague this system, with pests, such as the western corn root worm and soybean cyst nematode, becoming increasingly difficult to control. Further, environmental damage, including contaminated drinking water, hypoxic zones in the Gulf of Mexico, and unacceptably high levels of soil erosion, continues in these systems. Today, maize is overproduced in the United States, and a growing share of the crop is being diverted from food or feed uses to ethanol production for automobile fuel. The "industrial corn-ethanol" cycle is heavily subsidized, to a cost of $3.8 billion (US) in 2004 to produce a fuel that based on whole system accounting is actually *less* sustainable than direct burning of gasoline (Patzek, 2004). I have discussed the context of these problems previously (Brummer, 1998; Brummer, 2004; Keller and Brummer, 2002).

Our current approach to crop improvement is as follows: (1) the performance of system components is individually maximized, (2) problems that arise—for example, a new disease—are addressed by breeding new cultivars to overcome the problem, and (3) increasingly expensive technological methods are employed (or at least, are pursued in the hope they will be employed) in the cultivar development process to overcome the next round of pest problems. Currently, little thought is given to a whole system accounting that questions why problems arise and attempts to prevent them from occurring in the future (Lewis et al., 1997). G.E. Moore in *Principia Ethica* (1903) stated that "the value of the whole bears no regular proportion to the sum of the values of its parts." Thus, while we may be developing good cultivars, they may be grown as part of bad systems.

Current systems have systemic and intractable problems, but fortunately alternative cropping systems that can mitigate these deficiencies exist. Charles Darwin, in *On the Origin of Species* (Darwin, 1991 (1859), p.84) wrote: "Farmers find that they can raise most food by a rotation of plants belonging to the most different orders [.]," and he is neither the first nor last observer outside traditional agricultural circles to realize the value of crop rotation. The benefits of crop rotations extend in many directions, including their ability to interrupt pest cycles (Dick, 1992), decrease weed pressure (Liebman and Gallandt, 1997), efficiently use nutrients (Struik and Bonciarelli, 1997), and diversify the products produced by the farm. Beyond immediate farm management and profitability concerns, rotations have further benefits, not least of which is aesthetic: vast stretches of monoculture may have a minimalist beauty, but they fall short of the idyllic diversified farm with many crops, small woodlands, and a suite of animals. A large literature exists that clearly shows that alternative production systems can be both more sustainable and more productive than current industrial cropping systems typified by the maize-soybean rotation in the U.S. Midwest.

Advances in crop yields have always been made by a synergistic relationship between breeding and agronomic practice (Evans, 1993). We need to think more broadly than our own crop to consider what is best for the entire agricultural system of our region, maximizing total system productivity while minimizing undesirable externalities, such as contaminated water, soil erosion, and low commodity prices. Long-term sustainable systems can only be developed and improved by assessing the ecological resilience of our agricultural systems. Crop species selection and production decisions should be based on sound ecological principles; from there, breeding the crops to make a functioning system better can be undertaken. Thus, selection for constituent crop productivity will be important only to the extent that it is what the overall system needs. Equally important to continued improvement of the major commodity crops is the development of alternative crops that will minimize the disease, pest, and economic pressures that attend simplified crop production systems. In other words, breeding decisions—what crops to breed and what traits to select—need to be informed by the context of the crops. Accounting for externalities is not currently done in most agricultural production systems, and perhaps arguing that it should be done is futile. But if the true cost of agriculture were considered, then the value of crops such as living mulches and green manures would be much more favorable than it currently is.

Breeding for whole-system performance would require a different approach to crop production and improvement than is currently in vogue. First, the private sector plays an important role in developing, producing, and marketing cultivars of the major commodity crops and will continue to do so. It is just this success that has often been used to argue that public plant breeders are no longer needed, or at least, that fewer are needed than formerly. Consideration of the entire cropping system is unlikely to have a major effect on commercial breeding programs until the matrix of government subsidies shifts benefits from a few commodity crops. For the foreseeable future, private breeding programs will likely proceed much as they currently do.

In contrast, it is possible that the public sector will make major strides toward the development and implementation of alternative production systems, and in so doing, make sure that cultivars of the diverse crops that will be needed are available to farmers. I argue that public breeding programs need to be strengthened to address both the major crops, grown in different cropping systems and for different purposes, and alternatives, including forage and bioenergy crops, cover crops, alternative grain and oilseeds, and perennial grains.

One obvious way to limit many of the problems associated with annual crop monocultures is to grow perennial crops instead. Most perennial crops are forages, used for hay or pasture for livestock. Perennial forage crops have been quite profitable to grow in Iowa (and presumably surrounding states) for the past several years, particularly compared with heavily subsidized grain and oilseed crops. Regardless of the benefits of forages, a limited market exists for them and they cannot be used in all farming operations. One alternative is the development of bioenergy crops, such as switchgrass (*Panicum virgatum*); another is the development of perennial grain crops.

Several projects are underway throughout the United States working on perennialization of food crops, with multiple crops being bred at the Land Institute in Salina, Kansas (Cox et al., 2002) and perennial wheat at Washington State University (Scheinost et al., 2001). These efforts are long term in nature, but neither requires a new breeding method to develop new cultivars. A major change from current breeding in annual grain crops, however, needs to be made: multiyear evaluations. This will slow breeding progress, as it has in forage crops, because fewer cycles can be completed in a given time period, consequently slowing gain in yield potential from that which has been observed in annual grains.

To seriously breed for total cropping system performance will require that breeders work on a larger array of crops than is currently done. If only the major crops are continually improved, which is increasingly the situation today, then *of course* other crops will produce at a level that does not warrant their inclusion in the system. Focusing breeding efforts on whole system characteristics—for example, nitrogen cycling ability, profitability, pest suppression—will likely change the dynamics of what crops and which traits are being selected. I am arguing for balance. Maize should continue to be grown and bred in Iowa (although efforts to use surpluses to actually feed the world rather than make products of dubious utility should be enhanced), but it should be within the context of a productive and profitable cropping system that incorporates a diversity of crops for multiple uses.

Conclusion

In conclusion, if current cropping systems result in overproduction of a few commodities with consequent undesirable side effects (e.g., erosion, pest epidemics, low prices), if alternative cropping systems have beneficial properties (e.g., economic and environmental resilience), and if the different biology of alternative systems suggests that the best cultivars need to be developed within those systems, then we should change our cropping systems and change the way we think about breeding. Breeders cannot do all these things, but together with agronomists, economists, rural sociologists, ethicists, and others, they can play a key role in developing an agriculture that is both productive and sustainable.

References

Atlin, G.N., and K.J. Frey. 1989. Predicting the relative effectiveness of direct versus indirect selection for oat yield in three types of stress environments. Euphytica 44:137–142.

Atlin, G.N., and K.J. Frey. 1990. Selecting oat lines for yield in low-productivity environments. Crop Sci. 30:556–561.

Atuahene-Amankwa, G., and T.E. Michaels. 1997. Genetic variances, heritabilities and genetic correlations of grain yield, harvest index and yield components for common bean (*Phaseolus vulgaris* L.) in sole crop and in maize/bean intercrop. Can. J. Plant Sci. 77:533–538.

Bänziger, M., and M. Cooper. 2001. Breeding for low input conditions and consequences for participatory plant breeding: Examples from tropical maize and wheat. Euphytica 122:503–519.

Bänziger, M., F.J. Betran, and H.R. Lafitte. 1997. Efficiency of high-nitrogen selection environments for improving maize for low-nitrogen target environments. Crop Sci. 37:1103–1109.

Bopaiah, B.M., and H.S. Shetty. 1991. Soil microflora and biological-activities in the rhizospheres and root regions of coconut-based multistoried cropping and coconut monocropping systems. Soil Biol. Biochem. 23:89–94.

Bouton, J.H., S.R. Smith, Jr., D.T. Wood, C.S. Hoveland, and E.C. Brummer. 1991. Registration of 'Alfagraze' alfalfa. Crop Sci. 31:479.

Bradford, M.A., T.H. Jones, R.D. Bardgett, H.I.J. Black, B. Boag, M. Bonkowski, R. Cook, T. Eggers, A.C. Gange, S.J.Grayston, E. Kandeler, A.E. McCaig, J.E. Newington, J.I. Prosser, H. Setala, P.L. Staddon, G.M. Tordoff, D. Tscherko, and J.H. Lawton. 2002. Impacts of soil faunal community composition on model grassland ecosystems. Science 298:615–618.

Brummer, E.C. 1998. Diversity, stability, and sustainable American agriculture. Agron. J. 90:1–2.

Brummer, E.C. 2004. Breeding for sustainable cropping systems, p. 63–68. *In* M. Sligh and L. Lauffer, ed., Summit on seeds and breeds for 21st century agriculture. RAFI-USA, Pittsboro, NC http://www.rafiusa.org/pubs/Seeds%20and%20Breeds.pdf.

Brummer, E.C., and M. Smith. 1999. The 1999 Iowa Crop Performance Test—Alfalfa. Extension Bulletin AG-84. Iowa State University Extension, Ames, IA.

Ceccarelli, S., and S. Grando. 2002. Plant breeding with farmers requires testing the assumptions of conventional plant breeding: Lessons from the ICARDA barley program. In D.A. Cleveland and D. Soleri, eds. Farmers, Scientists and Plant Breeding. CAB International, Wallingford.

Cox, T.S., M. Bender, C. Picone, D.L. Van Tassel, J.B. Holland, E.C. Brummer, B.E. Zoeller, A.H. Paterson, and W. Jackson. 2002. Breeding perennial grain crops. Crit. Rev. Plant Sci. 21:59–91.

Darwin, C. 1991 (1859). On The Origin of Species. Prometheus Books, New York.

Davis, J.H.C., and J.N. Woolley. 1993. Genotypic requirement for intercropping. Field Crop. Res. 34:407–430.

De Deyn, G.B., C.E. Raaijmakers, H.R. Zoomer, M.P. Berg, P.C. de Ruiter, H.A. Verhoef, T.M. Bezemer, and W.H. van der Putten. 2003. Soil invertebrate fauna enhances grassland succession and diversity. Nature 422:711–713.

Dick, R.P. 1992. A Review—Long-Term Effects of Agricultural Systems on Soil Biochemical and Microbial Parameters. Agric. Ecosyst. Environ. 40:25–36.

Evans, L.T. 1993. Crop Evolution, Adaptation and Yield Cambridge Univ. Press, Cambridge, UK.

Falconer, D.S., and T.F.C. Mackay. 1996. Introduction to quantitative genetics. 4th ed. Longman, Essex, England.

Fehr, W.R. 1987. Principles of Cultivar Development, Vol. 1: Theory and Technique Iowa State Univ., Ames.

Francis, C.A. 1985. Variety development for multiple cropping systems. Crit. Rev. Plant Sci. 3:133–168.

Francis, C.A. 1990. Breeding hybrids and varieties for sustainable systems, p. 24–54. *In* C. A. Francis, et al., eds. Sustainable agriculture in temperate zones. John Wiley & Sons, New York.

Galvez, L., J. Douds, D.D. Wagoner, and P. Wagoner. 2001. Tillage and farming system affect AM fungus populations, mycorrhizal formation, and nutrient uptake by winter wheat in a high-P soil. Am. J. Alternat. Agric. 16:152–160.

Gurr, G.M., S.D. Wratten, and J.M. Luna. 2003. Multi-function agricultural biodiversity: Pest management and other benefits. Basic and Applied Ecology 4:107–116.

Hadas, A., L. Kautsky, M. Goek, and E.E. Kara. 2004. Rates of decomposition of plant residues and available nitrogen in soil, related to residue composition through simulation of carbon and nitrogen turnover. Soil Biol. Biochem. 36:255–266.

Hallauer, A.R., and T.S. Colvin. 1985. Corn hybrids response to four methods of tillage. Agron. J. 77:547–550.

Hamblin, J., and M.J.O. Zimmerman. 1986. Breeding common bean for yield in mixtures. Plant Breed. Rev. 4:245–272.

Hamblin, J., J.G. Rowell, and R. Redden. 1976. Selection for mixed cropping. Euphytica 25:97–106.

Harper, J. 1967. A Darwinian approach to plant ecology. J. Ecol. 55:247–270.

Hetrick, B.A.D., G.W.T. Wilson, and T.S. Cox. 1993. Mycorrhizal dependence of modern wheat cultivars and ancestors—a synthesis. Can. J. Bot. 71:512–518.

Hill, J. 1990. The three C's—competition, coexistence and coevolution—and their impact on the breeding of forage crop mixtures. Theor. Appl. Genet. 79:168–176.

Holland, J.B., and E.C. Brummer. 1999. Cultivar effects in oat-berseem clover intercrops. Agron. J. 91:321–329.

Johnson, S.S., and J.L. Geadelmann. 1989. Influence of water stress on grain yield response to recurrent selection in maize. Crop Sci. 29:558–564.

Keller, D.R., and E.C. Brummer. 2002. Putting food production in context: Toward a postmechanistic agricultural ethic. BioScience 52:264–271.

Lewis, W.J., J.C. van Lenteren, S.C. Phatak, and J.H. Tumilson, III. 1997. A total system approach to sustainable pest management. Proc. Natl. Acad. Sci. USA 94:12243–12248.

Liebman, M., and E.R. Gallandt. 1997. Many little hammers: Ecological management of crop-weed interactions, p. 291–343. *In* L. E. Jackson, ed. Ecology in agriculture. Academic Press, New York.

Mäder, P., A. Fliebeta bach, D. Dubois, L. Gunst, P. Fried, and U. Niggli. 2002. Soil Fertility and Biodiversity in Organic Farming. Science 296:1694–1697.

O'Leary, N., and M.E. Smith. 1999. Breeding corn for adaptation to two diverse intercropping companions. Am. J. Alternat. Agric. 14:158–164.

Pang, X.P., and J. Letey. 2000. Organic farming: Challenge of timing nitrogen availability to crop nitrogen requirements. Soil Science Society of America Journal 64:247–253.

Patzek, T.W. 2004. Thermodynamics of the corn-ethanol biofuel cycle. Crit. Rev. Plant Sci. 23:519–567.

Scheinost, P.L., D.L. Lammer, X. Cai, T.D. Murray, and S.S. Jones. 2001. Perennial wheat: The development of a sustainable cropping system for the U.S. Pacific Northwest. Am. J. Alternat. Agric. 16:147–151.

Simmonds, N.W. 1991. Selection for local adaptation in a plantbreeding programme. Theor. Appl. Genet. 82:363–367.

Smiley, R.W., H.P. Collins, and P.E. Rasmussen. 1996. Diseases of wheat in long-term agronomic experiments at Pendleton, OR. Plant Dis. 80:813–820.

Smith, M.E., and C.A. Francis. 1986. Breeding for multiple cropping systems, p. 219–249. *In* C. A. Francis, ed. Multiple cropping systems. Macmillan, NY.

Smith, M.E., and R.W. Zobel. 1991. Plant genetic interactions in alternative cropping systems: considerations for breeding methods, p. 57–81. Plant breeding and sustainable agriculture: considerations for objectives and methods, Vol. Special Publication 18. Crop Science Society of America, Madison, WI.

Sprague, G.F., and L.A. Tatum. 1942. General vs. specific combining ability in single crosses of corn. Journal of American Society of Agronomy 34:923–932.

Struik, P.C., and F. Bonciarelli. 1997. Resource use at the cropping system level. Eur. J. Agron. 7:133–143.

Vandermeer, J.H. 1989. The ecology of intercropping Cambridge Univ. Press, Cambridge.

Wardle, D.A., R.D. Bardgett, J.N. Klironomos, H. Setala, W.H. van der Putten, and D.H. Wall. 2004. Ecological linkages between aboveground and belowground biota. Science 304:1629–1633.

Zhu, Y., H. Chen, J. Fan, Y. Wang, Y. Li, J. Chen, J. Fan, S. Yang, L. Hu, H. Leung, T.W. Mew, P.S. Teng, Z. Wang, and C.C. Mundt. 2000. Genetic diversity and disease control in rice. Nature 406:718–722.

Zhu, Y.Y., Y.Y. Wang, H.R. Chen, and B.R. Lu. 2003. Conserving traditional rice varieties through management for crop diversity. Bioscience 53:158–162.

Zimmerman, M.J.O., A.A. Rosielle, and J.G. Waines. 1984. Heritabilities of grain yield of common bean in sole crop and in intercrop with maize. Crop Sci 24:641–644.

7

Participatory Plant Breeding: A Market-Oriented, Cost-Effective Approach

J.R. Witcombe, Centre for Arid Zone Studies, University of Wales
D.S. Virk, Centre for Arid Zone Studies, University of Wales
S.N. Goyal, Gujrat Agricultural University, Main Maize Research Station, Godhra, India
D.N. Singh, Birsa Agricultural University, Kanke, Ranchi, Jharkhand, India
M. Chakarborty, Birsa Agricultural University, Kanke, Ranchi, Jharkhand, India
M. Billore, Jawahar Lal Nehru Krishi Vishav Vidyalaya, Indore, Madhya Pradesh, India
T.P. Tiwari, CIMMYT (International Maize and Wheat Improvement Center)-South Asia, Kathmandu, Nepal
R. Pandya, Maharana Pratap University of Agriculture and Technology, Banswara, Rajasthan, India
P. Rokadia, Maharana Pratap University of Agriculture and Technology, Banswara, Rajasthan, India
A.R. Pathak, Gujrat Agricultural University), Main Maize Research Station, Godhra, India
S.C. Prasad, Gramin Vikas Trust, Ranchi, Jharkhand, India

Introduction

Private-sector and public-sector approaches

Involving farmers more closely in most of the elements of the plant-breeding process makes it more likely that new varieties are adopted by them and meet their needs. The private sector has long used such market research approaches where hundreds of "strip trials," so-called because a strip of land is devoted to each test variety, are grown by hundreds of farmers in dispersed locations.

Public-sector plant breeders, particularly in developing countries, have been more reluctant to involve farmers actively in testing new varieties. Instead, a linear approach to research and extension has been used (Suleman and Hall, 2002) where breeders first breed, test, and release varieties and, after this, extension services promote them. The public sector, when adopting less-linear approaches that are closer to those of the private sector, use terms such as participatory plant breeding (PPB), which have come from social scientists rather than marketing specialists. This vocabulary is related to the concept of empowerment of individuals and communities rather than simply to market research and may well have been unhelpful when encouraging traditional public-sector breeders to adopt more farmer-oriented approaches.

Evolution of conventional programs

PPB programs can easily evolve from conventional breeding programs. They can become more market oriented by testing existing material from those programs for trials in farmers' fields. Usually conventional programs select among the progeny of many crosses, so to take such a program to the field many lines have to be grown by farmers. Hence, scientists who have made existing programs more participatory have adopted the strategy of using relatively few collaborating farmers but have assisted them in the planting of many entries (e.g., Ceccarelli et al., 2001).

New PPB programs

The PPB programs described in this chapter were new breeding programs, thus they could be adapted to the particular advantages and disadvantages of working with farmers: it is relatively more difficult for farmers to grow many entries, but easier for them to grow large populations (they are already cultivating the crop). Hence, a strategy of making few crosses, but advancing large populations from those crosses, has been used. Because only a few crosses are made, much attention must be paid to choosing the parents of the crosses. To increase the probability of getting locally adapted progeny from the cross, at least one of the parents of any cross is locally adapted. Both experimental data and theory support the use of few crosses with large population size (Witcombe and Virk, 2001). However, traditionally, plant breeding has employed a very different strategy of having many crosses with each cross having a small population, and, without a doubt, this method also produces positive results. However, the question is not whether any particular method is feasible but which method is most cost effective. So far the theoretical and recent experimental evidence in favor of few-cross, large-population breeding has not produced a paradigm shift from the traditional approach of many crosses and small population sizes.

In the breeding programs described here, the methods used to select in the populations were kept simple since they sometimes actively involved farmers who had received little training. Mass selection (in maize) or bulk population breeding (in rice) are appropriately simple methods.

Evidence-based adoption of PPB

Testing varieties with farmers using participatory varietal selection (PVS) is now quite widely accepted in developing countries, although regulatory frameworks on varietal release continue to be an obstacle to its widespread adoption. PPB is becoming more widely adopted in the public sector in developing countries, largely as a result of pressure from donors. So far, there has been little published evidence that PPB programs (in contrast to those where farmers select among varieties in PVS) have either been successful or cost effective, so there has been little evidence-based adoption. Recently, such evidence has emerged from PPB programs, and we concentrate here, not on the methods used, but on the impact of these programs and their efficiency compared with more conventional methods.

Methods

PPB in maize

We describe three PPB programs in maize in (1) western India, (2) eastern India, and (3) Nepal, each of which relied on a single composite created predominantly from locally adapted varieties or landraces. In all three cases, crosses were made between yellow- and white-grained varieties, with subsequent selection for the required grain color, and selection was by recurrent-mass selection. Detasseling of 50% of the plants in the population and only advancing generations from detasseled plants avoided selfing that would reduce the efficiency of mass selection.

The first step was participatory varietal selection (PVS) to test a range of already-existing varieties. Hence, farmers were involved in the selection of parents as the best varieties were included as parents of the composite. Farmers were also involved in the recurrent-mass selection for population improvement and in the testing of the resultant varieties in PVS trials that were usually organized in a mother and baby trial system (Snapp, 1999; Witcombe, 2002).

Western India

The collaborating institutes were as follows: the Gramin Vikas Trust (GVT), an Indian NGO (nongovernmental organization); Gujarat Agricultural University (GAU); and the Center for Arid Zone Studies (CAZS), UK. Three white-grained (Gujarat Makka-1, Shweta, and Chandan Makka-2) and three yellow-grained (Mahi Kanchan, Navin, and Chandan Makka-3) varieties, preferred by farmers in PVS trials, were initially crossed to create a composite population (Witcombe et al., 2003). Three open-pollinated varieties, that is, GDRM-186, GDRM 186-1 (a more advanced selection of GDRM 186), and GDRM-187 (GM-6) were derived from this composite but with an increased contribution by backcrossing and pedigree selection of the earlier maturing parent Chandan Makka-2.

Eastern India

GVT, BAU, and CAZS collaborated in this PPB program. Three white-grained (GDRM-187, Shweta, and Rudarpur local) and three yellow-grained (BM-1, Suwan, and Chandan Makka-3) varieties, preferred by farmers in PVS trials, were crossed to create a composite population. One population was derived and improved by recurrent mass selection with farmers (Kumar et al., 2001).

Nepal

The Agricultural Research Station, Pakhribas of the Nepal Agricultural Research Council (NARC) and the University of Wales, Bangor (School of Agriculture and Forest Sciences and CAZS) collaborated in breeding maize for the mid-hill, maize–millet–tree-farming system, where trees grown on the terrace perimeters often shade the maize crop. Crosses were made in the 1999 main season between a yellow-grained variety, CIMMYT (International Maize and Wheat Improvement Center) Pool-21 as female and four white-grained varieties as male parents, that is, Madi local, and three improved varieties adapted to the area: Arun-1 and Manakamana-1, which were bred in Nepal, and Manakamana-3 (CIMMYT Population-22). The pale yellow seeds derived from the crosses were sown for the first random mating. Thereafter, white seeds were selected and the population subjected to recurrent mass selection with two generations a year. From the C5 generation, the entire population was tested as the white-grained variety PM-7 in on-farm, mother-and-baby trials, and in advanced trials conducted by the national coordinated program in Nepal. In addition to the PPB, two released varieties and three varieties in advanced stage of testing were included in a PVS program from 1999 to 2001.

PPB in rice

The rice PPB program was targeted at the rainfed uplands of eastern India. The methods are described in detail in Virk et al. (2003). Few crosses (only five in the first four years of the program) were used that involved carefully chosen parents. The first cross was between an upland (Kalinga III) and a lowland (IR64) variety, both of which were adapted to the target region. Bulk population breeding was used. The F_4 bulk of the cross was grown and selected by farmers in their own fields (defined as collaborative participation), and the result of selection by one farmer, Rajendra Dhan, produced variety Ashoka 200F. In addition, F_4 bulk lines (each derived from all the progeny of a single F_2 plant) were advanced by scientists for selection by farmers in the research station (defined as consultative participation), and this produced variety Ashoka 228. The varieties from this program were tested by on-station and on-farm trials from 1999 to 2001.

Results and discussion

Results in maize

Western India

GDRM-187, a variety from the PPB program, was released in Gujarat state as GM-6 in 2001. It was tested by on-station and on-farm trials from 1997 to 2002. In eight research station trials in Gujarat in 1997 and 1998 it yielded 21% more than the check and was the earliest entry to mature (Table 7.1). It was somewhat lower yielding than conventionally bred variety GM-4 (Table 7.1) but was significantly earlier to mature, and overall, farmers considered that it had the most desirable combination of yield and maturity. In on-farm baby trials from 1998 to 2000 in Gujarat and Rajasthan, GM-6 yielded up to 33% more than the local variety (Figure 7.1). In addition, farmers preferred GM-6 for its good grain and cooking quality, higher market price, and pest resistance due to tight husks.

GDRM-186 differed little from GM-4 for yield and maturity, but further selection produced GDRM-186-1, which was superior to GM-4. In 10 baby trials in 2002 in Jhabua district, MP (Madhya Pradesh), from 1 kg of seed given to each farmer GDRM-186-1 produced on average 48.0 kg of seed, 28% more than GM-4 (37.4 kg) and 56% more than the local variety (30.7 kg). Farmers preferred GDRM-186-1 for its vigorous growth, good cob size, uniform maturity, good taste, good market price, and disease tolerance.

Table 7.1 Mean grain yield (t ha^{-1}) and mean days to 50% silking of PPB-bred varieties (GDRM-186 and GDRM-187) in research station trials in western India compared with conventionally bred variety, GM-4, and the check (GM-1)

Entry	Mean grain yield (t ha^{-1})	Increase over GM-1 (%)[1]	Mean time to 50% silking (days)[1]
GM-4	3.1	37	50
GDRM-186	3.0	31	49
GDRM-187 (GM-6)	2.7	21	45
GM-1 (Check)	2.3	. . .	. . .

[1]Over eight trials; 1997, 1998.

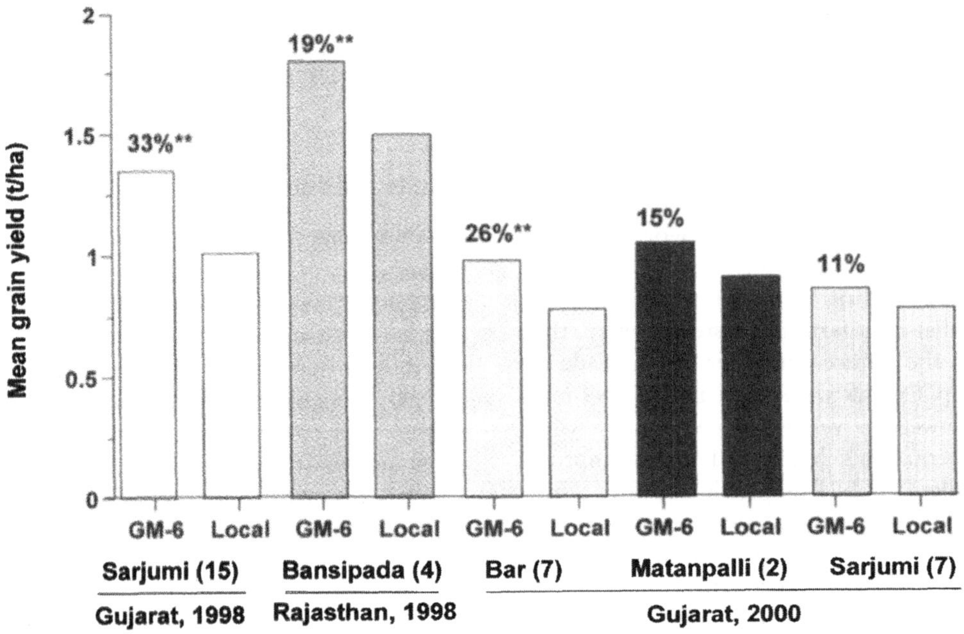

Figure 7.1 Comparative grain yield (t ha^{-1}) of GM-6 and local variety in baby trials conducted by farmers in villages of Gujarat and Rajasthan states in India in 1998 and 2000.

Eastern India

Variety BVM-2 was released by Jharkhand state in 2003. In 8 research trials in Jharkhand from 1999 to 2002 it had a good combination of higher and earlier yield. It yielded 9% more than the recommended check, BM-1, while being 4 days earlier to silk (43 days compared to 47 days) (Table 7.2). Its yield advantage over BM-1 was higher in the poorer environments of farmers' fields mother trials where, in 2000 to 2002 trials, it yielded 20% more than BM-1 and 45% more than the local check in a total of 28 trials (12 trials in 2000, 8 in 2001, and 8 in 2002) in Jharkhand, Orissa, and West Bengal (Table 7.2).

Variety BVM-2 performed well in the All India Coordinated Maize Improvement Project trials of 2000 and 2001. It yielded 12% more than Surya, the early maturing national check, in 38 trials and was as early as Surya.

In focus group discussions with 10 farmers in 2000 in Jharkhand, 90% of farmers preferred BVM-2 for its better taste and higher fodder yield

Table 7.2 Grain yield of BVM-2 in research station, on-farm mother trials in Jharkhand, West Bengal, and Orissa, and in All India Coordinated Maize Improvement Project (AICMIP) trials

	Eight research trials 1999–2002		Twenty-eight mother trials 2000–2002		Thirty-eight AICMIP trials 2000–2001	
Variety	Grain yield (t ha^{-1})	Increase over BM-1 (%)	Grain yield (t ha^{-1})	Increase over BM-1 (%)	Grain yield (t ha^{-1})	Increase over Surya (%)
BVM-2	4.34*	9	3.41**	20	4.65	12
BM-1	3.97*	. . .	2.85	. . .	. . .	
Local	. . .	. . .	2.35	. . .	. . .	. . .
Surya check	. . .	. . .	. . .	. . .	4.15	. . .

*Mean days to 50% silking of 43 days for BVM-2 and 47 days for BM-1; 5 trials from 1999 to 2001.
**Significantly more than BM-1 at 5% level.

than the local variety. In the on-farm trials, farmers opined that its stay-green trait and higher fodder yield made it a desirable variety. They liked it for its good grain and cooking quality, better grain filling, and tight husks for pest tolerance.

Nepal

The PVS program variety Population 22 was liked by farmers for its higher grain yield, stay-green trait, lodging resistance, and desirable grain characteristics. Farmers also liked its tolerance to shade, which was comparable to the local varieties. The National Maize Research Program of the Nepal Agricultural Research Council released this variety as Manakamana-3 in 2002.

Although no variety has yet been released from the PPB program, one variety, PM-7, has reached the advanced stages of testing. In the mother trials in 2001 it yielded significantly more than the local check by 29% (Table 7.3). In the initial yield trials it yielded significantly more than the local check (by 36%) and significantly more than the Manakamana-1 (by 27%). It was taller and later to mature than the local check.

In a survey using household questionnaires, most farmers preferred PM-7 for its long cobs, lodging resistance, stay-green trait, acceptable grain yield, higher palatability and yield of fodder, and improved disease resistance compared with the local variety. There are, as yet, no data on the preference between PM-7 and Manakamana-3, but these two varieties offer a varietal choice to farmers because they differ in important traits such as maturity. Efforts in classical breeding were concentrated on breeding earlier maturing varieties, but farmers found that the medium maturity of PM-7 and Manakamana-3 suited the maize–millet–tree-farming system of the mid-hills.

Results in rice

Performance of PPB varieties

Varieties Ashoka 200F (A 200F) and Ashoka 228 (A 228) were released by Jharkhand state in 2003 for cultivation in the rainfed uplands. In research trials these varieties yielded significantly more (27–28%) than BG 102, a recommended check variety by BAU, and yielded 51–56% more than BG 102 in mother trials (Table 7.4). Both varieties also yielded more than variety Kalinga III, which was identified by farmers as a suitable variety in PVS trials.

Adaptation of Ashoka varieties

Trials in Eastern India

The PPB bred varieties, A 200F and A 228, were as widely adapted to the target environment as those bred by classical breeding (Table 7.5). They had a high mean grain yield and average regression coefficients of 1. They performed better than the check varieties (Kalinga III and BG 102) in all trials; however, they particularly performed well in the poor environments where BG 102 was a poorer performer.

Coordinated project trials

A 200F and A 228 responded less to above-average environments than the check variety Annada in the all India Coordinated trial IVT (Early) in 1999, with 58 varieties tested across 10 sites (Virk et al., 2003). Despite their above-average performance in the below-average sites, the two early-maturing Ashoka varieties were not promoted to advanced all-India trials because of their lower average yields.

Table 7.3 Performance of PM-7 in full-season, hill-zone IYT across three research stations (Lumle, Pakhribas, and Dailekh) in 2002 and in 15 on-farm mother trials in 2001

	Mother trials 2001	IYT 2002		
Variety	Grain yield (t ha^{-1})	Grain yield (t ha^{-1})	Time to 50% silking (days)	Plant height (cm)
PM-7	3.6*	7.5*	77*	273*
Manakamana-1 (check)	3.5*	5.9	74*	237
Local check	2.8	5.5	71*	254
Lsd 5%	0.7	1.0	2	17

* Significantly higher than the local check at 5% level.
IYT, initial yield trials.

Table 7.4 GY and DF of A 200F and A 228 in research trials in Jharkhand and in on-farm mother trials in Jharkhand, Orissa, and West Bengal

	Research trials*				Mother trials**		
Variety	GY(t ha^{-1})	DF	Increase (%) over K III***	Increase (%) over BG 102	GY(t ha^{-1})	Increase (%) over K III***	Increase (%) over BG 102
A 200F	2.58	60	19	28	1.38	19	51
A 228	2.55	67	18	27	1.42	23	56
K III ***	2.16	60	. . .	7	1.16	. . .	27
BG 102	2.01	63	. . .	. . .	0.91	. . .	. . .
LSD	0.03	. . .	. . .	. . .	0.21	. . .	. . .

*6 trials, 1999–2001.
**40 trials, 2000–2001.
***K III = Kalinga III.
GY, grain yield; DF, days to 50% flowering; A 200F, Ashoka 200F; A 228, Ashoka 228.

Table 7.5 Regression parameters for variety mean grain yield (t ha^{-1}) regressed on trial mean for six research trials, Jharkhand, 1999–2001 and five on-farm trials (40 farmer-replications grouped according to village clusters), Jharkhand, Orissa, and West Bengal, 1999–2001

Variety	Overall mean (t ha^{-1})	R^2	$a \pm$ SE	$b \pm$ SE
A 200F	2.10	0.94	0.05 ± 0.19	1.11 ± 0.10
A 228	2.11	0.95	0.16 ± 0.16	1.06 ± 0.08
Kalinga III	1.77	0.96	0.05 ± 0.12	0.94 0.06
BG 102	1.58	0.97	-0.29 ± 0.12	1.01 ± 0.06

Early-maturing entries, which have specific adaptation to lower-yielding environments, tend to be rejected because of their lower overall mean yields.

Trials in western India

A 200F and A 228 performed well in western India even though they were bred in, and targeted for, eastern India. In a nine-variety mother trial, conducted in irrigated and rainfed conditions by 11 farmers in Rajasthan, Gujarat, and MP, the Ashoka varieties and Vandana were the most drought tolerant. Varanideep and Vanprabha, although recommended for uplands, showed the greatest susceptibility to drought, probably because they were late maturing (Figure 7.2). Vanprabha, although higher yielding under irrigated conditions, yielded poorly under rainfed conditions, and Varanideep was the most drought-susceptible variety and did not set seed in any of the rainfed sites. Of these varieties, A 200F, Vandana, and A 228 had the smallest reduction in grain yield in drought conditions compared with irrigated conditions (Figure 7.2).

Farmers in western India ranked the Ashoka varieties very high for their earlier maturity, superior grain quality, better taste, better fodder quality, and higher market value (Table 7.6). They were preferred to the recommended upland variety, Vandana, which has been relatively little adopted because of its poor grain quality.

A crucial finding was that farmers' preference did not agree with simple yield. Rajasthan was the only state in which preference ranking was done for all important traits, and the differences between varieties were significant. Ashoka 200F and Ashoka 228 were greatly preferred for all traits. The overall preference of these varieties was driven by superior taste, cooking quality, fodder yield, and market value of the grain. Higher-yielding entries, such as Vandana and RR453-1, were considered by farmers to be significantly inferior for all these traits. They also had lower ranks, though not significantly, for earlier flowering and grain size.

Adoption and impact of Ashoka 200F and Ashoka 228

In a survey conducted in December 2002 in Jharkhand, Orissa, and West Bengal, the great majority of the farmers preferred A 200F and A 228 for many traits (Figure 7.3) and would continue

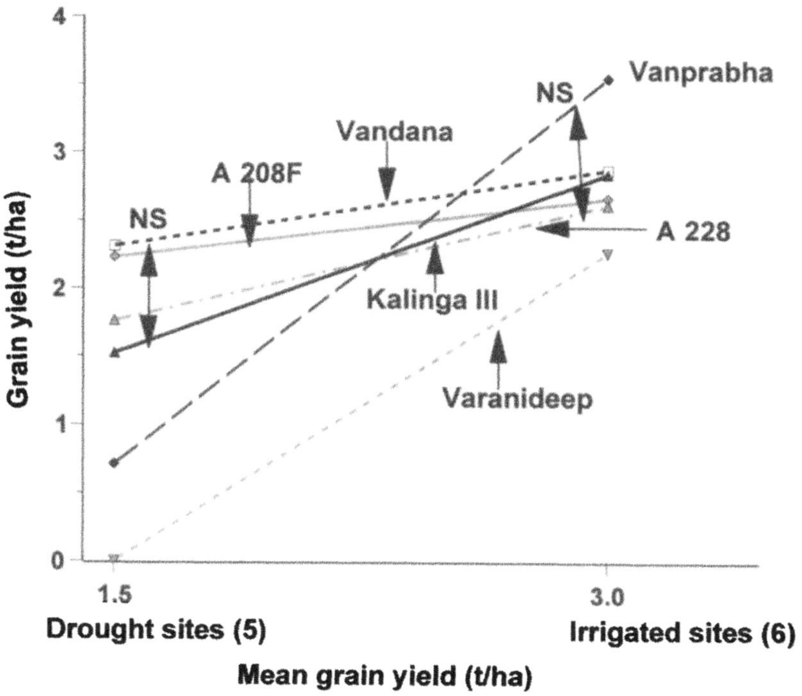

Figure 7.2 Mean grain yield (t ha^{-1}) of six selected rice varieties in six irrigated and five rain-fed sites in Rajasthan, Gujarat, and Madhya Pradesh, 2002.

Table 7.6 Mean preference ranking of selected varieties out of nine tested for traits other than grain yield in Rajasthan (1 = lowest and 9 = highest score)

Variety	Earlier flowering	Grain size	Taste	Cooking quality	Fodder quality	Market value	Overall ranking
A 200F	5.0	5.7	6.0*	6.0*	5.5*	6.0*	6.5*
A 228	5.0	6.0	6.0*	6.0*	5.5*	6.0*	6.0*
RR354-1	3.7	5.0	4.5	5.0	4.5	4.0	4.0
Vandana	3.7	4.7	4.5	4.0	3.5	3.5	3.5
Kalinga III	3.3	3.7	6.0	4.0	4.0	4.0	3.5
Lsd 5%	2.3	1.8	1.2	1.5	1.4	1.7	1.6

*Significantly better than Vandana at 5% level.

growing them from farm-saved seed. However, these data are for the first year, and it is possible, from experience in other studies, that adoption may be somewhat lower in later years. The adoption projected by the interviewed farmers was very high in all three states; sometimes more than 100% of the original upland area. Farmers were either bringing fallow land into cultivation or growing these upland varieties in medium land.

Farmers of smaller- and medium-sized landholdings adopted the two new varieties as much as farmers of medium and large farms. No relationship was found between their area of adoption and the total cultivated land per household (Figure 7.4).

Earlier seed production and dissemination by farmer groups is an important advantage of the nonlinear approach to research and extension. Farmers who tested the A 200F and A 228 in PVS trials saved seed from the harvest for sowing and exchanged it with other farmers. In the 2001–2002 off-season, farmer groups in Orissa produced a total of 56 t seed of the new varieties and in the next off-season 66 t of seed. Most of this was purchased by GVT for dissemination through other NGOs, governmental organizations, and the private sector in three states in western and three states in eastern India. In addition, farmers disseminated seed to other farmers in their villages

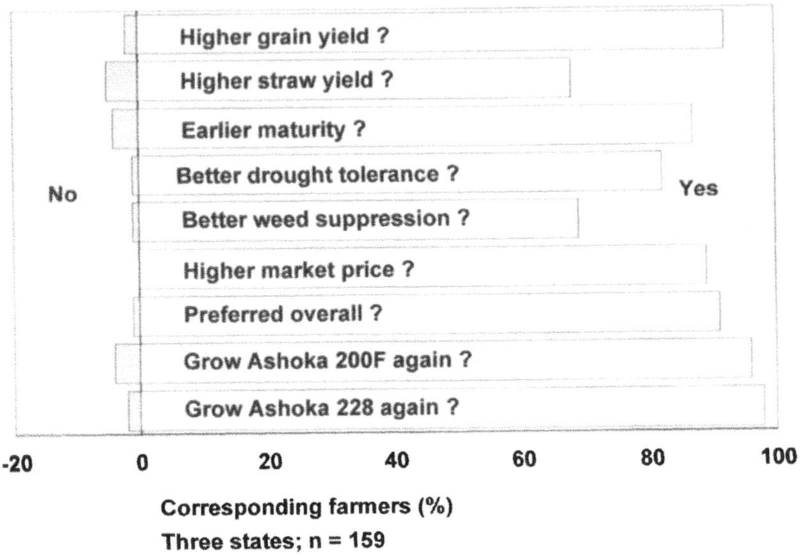

Figure 7.3 Farmers' perception (percentage of farmers) for Ashoka varieties in comparison with the local cultivars. Based on survey in Jharkhand, Orissa, and West Bengal, December 2002.

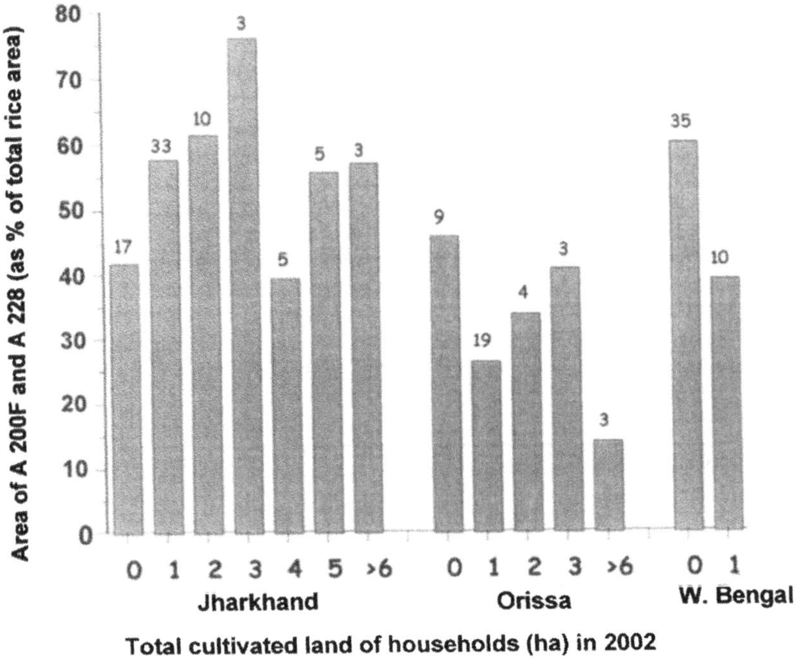

Figure 7.4 Percentage of area of rice land devoted to Ashoka varieties versus total cultivated land of households. Number of farmers in each landholding category indicated above the bars. Based on surveys in three states: Orissa (n = 35 excluding 3 farmers with ≥15 ha), West Bengal (n = 45) and Jharkhand (n = 73 excluding 3 farmers with >15 ha). Survey of December 2002 for adoption in 2002.

and in adjoining villages, some of which were great distances away (Figure 7.5).

Comparison of genetic advance from PPB with classical breeding

Maize

Gains in yield must be considered along with gains in early maturity since increased earliness is normally at the cost of reduced grain yield. The genetic gains in the varieties from the PPB programs in India were higher than those for the most comparable conventionally bred varieties (Table 7.7). The gains from the PPB are greater when the improved earliness of the PPB varieties is also considered.

However, it is important to note that gains for yield per year during the breeding process itself are

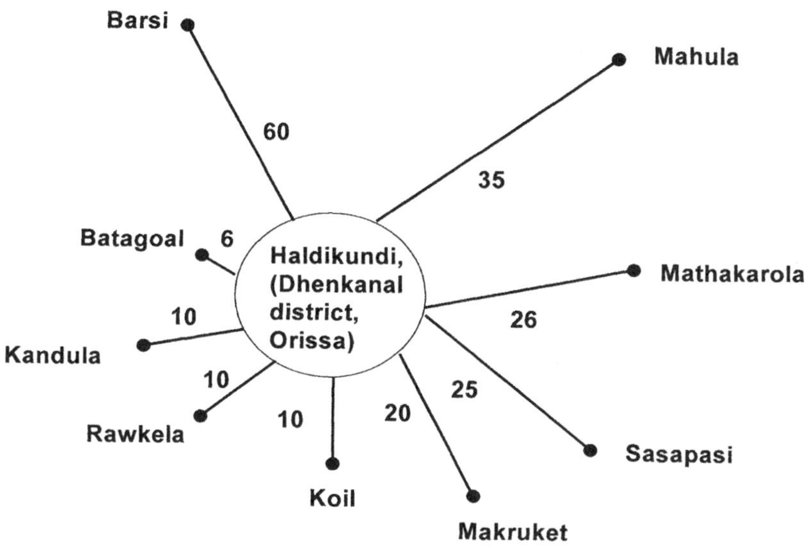

Figure 7.5 An example of seed spread of Ashoka 228 from Haldikundi village; district Dhenkanal, Orissa, to other villages in 2002 (distances shown in kilometers).

Table 7.7 Comparison of genetic gains from PPB and classical breeding in two case studies in maize in India

	Eastern India		Western India	
Basis of comparison	PPB variety (BVM 2)	CB variety (BM 1)	PPB variety (GM-6)	CB variety (GM-4)
Years from cross to completing one year of research trials	3	8	4	4
Years from cross to farmers	3	11	4	12
Yield gains (%) over check on research station	9% over BM-1 in 7 research trials (1999–2001)[1]	35% over Diara in 4 research trials (1990–1993)[2]	21% over GM-1 in 8 research trials (1997–1998)	37% over GM-1 in 8 research trials (1997–1998)
Yield gains (%) over check in farmers' fields	45% over local var. in 28 trials (2000–2002)	21% over local var. in 28 trials (2000–2002)	54% over local var. in 6 trials in MP[1] (2002)	66% over local var. in 6 trials in MP[1] (2002)
Yield gains per year to trials on (a) research station	3.0%	4.4%	5.3%	9.3%
(b) farmers' fields	15.0%	1.9%	13.5%	5.5%

[1]Madhya Pradesh state, India.
CB, classical breeding.

about the same in the two methods. The higher gains in the PPB greatly reflect the reduced time, a savings of eight years, required to reach farmers in a parallel approach to research and extension (Table 7.7). In the conventional model, variety development (breeding) is first completed, and then there is a lag phase for varietal release and seed multiplication. Only after this lag phase is the variety offered to farmers.

Rice

The genetic gains per year from the PPB program in rice were about double of those achieved conventionally (Table 7.8). Again, this did not result from a faster gain during breeding but greatly reflected the reduced time, a savings of 10 years, required for a new variety to reach farmers in a parallel approach to research and extension (Table 7.8). There was a 36% reduction in plant height for the conventionally bred variety but no reduction in plant height for the PPB varieties (Table 7.8). Farmers prefer tall varieties for their higher fodder yield.

Comparison of costs of PPB and conventional breeding

Comparing the costs of PPB and CB (conventional breeding) is not simple. While PPB requires resources to collaborate with farmers, CB involves

Table 7.8 Comparison of genetic gains from participatory plant breeding (PPB) and classical breeding in rice in India

Basis of comparison	PPB variety (Ashoka 200F)	CB variety (BD-101)
Years from cross to completing one year of research trials	4	7
Years from cross to farmers	4	14
Yield gains (%) over check on research station	28% over BG-102 in 6 trials (1999–2001)[1]	18.5% over Brown Gora (LC) in 4 trials (1981–1984)[2]
Yield gains (%) over check in farmers' fields	51% over local variety in 40 trials (2000–2001)	. . .
Yield gains per year in:		
(a) research station trials	7.0%	2.6%
(b) farmers' fields	12.8%	—

[1]0% reduction in plant height for Ashoka 200F but 5% increase of height of A 228 variety over BG-102.
[2]36% reduction in plant height over Brown Gora (Local Collection).

Table 7.9 Cost comparison of PPB and CB in maize in western India and rice in eastern India

Basis of comparison	Maize in PPB variety: GM-6	Western India CB variety: GM-4	Rice in PPB variety: A200F	Eastern India CB variety: BD-101
Cost of method	Simple mass selection One base population	Full-sib progeny testing Many populations	Few crosses, large population Simple pedigree-bulk or bulk methods	Many crosses, small population Costly pedigree method
Staff	1 Research Associate Support from NGO	4 researchers, 2 field assistants (30% of their time) No NGO but supported by extension departments	1 research associate Support from NGO	4 researchers, 6 field assistants (30% of their time) No NGO but supported by extension departments

higher research station costs. To illustrate these difficulties, in the examples considered here, only the PPB used external consultants (they were not needed to conduct traditional breeding), but the influence of consultants, when introducing a new method, may be crucial. Only the PPB required support from an NGO (GVT), although conventional breeding is supported by public-sector extension.

In the maize example from western India, PPB used simple, recurrent selection rather than expensive, full-sib progeny testing used in CB (Table 7.9). PPB also used only a single base population (composite), whereas many composites were used in CB.

In the rice example, the costly pedigree method was used in the CB, but PPB reduced costs by using a few crosses and simple pedigree-bulk or bulk methods. In both cases the CB had more nonstaff costs than the PPB: the recurring budget for CB was six times more than that of PPB.

There are additional cost savings from PPB:

- Farmers, for whom it is only a marginal cost (the increased or reduced yield of the PPB bulk compared with their customary crop), share the costs of growing large populations.
- In rice, early organoleptic testing is cheaper than later replicated yield trials. The work in Nepal and India has demonstrated that it is cheaper and simpler to eliminate varieties for poor grain quality than for low yield.
- In PPB, products reach farmers about 8 to 10 years earlier than for CB.

Economic analysis for rice PPB

A preliminary financial analysis, using the conservative assumptions, has been made for the benefits of A 200F and A 228 in the states of Jharkhand, Orissa, and West Bengal. The benefit/cost ratio of this research was very favorable. From the base year of 2002, the internal rate of return was over

500% by 2010, and, using a 5% discount rate, the net present value was 170 million pounds by 2010 and rose to nearly 500 million pounds by 2015. The model assumed a threefold increase in adoption in each year and an adoption ceiling of 50%. Both these values are lower than actual survey data.

This analysis ignored the additional likely impacts from the adoption of the new varieties in western India (Rajasthan, Gujarat, Madhya Pradesh), which is expected to be substantial, and in other states in eastern India, such as Bihar and Chhattisgarh (formerly eastern MP).

Centralized versus decentralized breeding

Decentralization by breeding for mega-environments

PPB has a greater involvement of farmers. However, it is also decentralization from the research station to the farmers' fields. These two components, participation and decentralization, can be examined separately.

The centralized breeding programs of international agricultural research institutes (IARCs) are decentralized only to the level of mega-environments (Figure 7.6). For example, in the spring wheat-breeding program of CIMMYT, the breeding was mainly conducted in two research stations in Mexico, but varieties bred by the program were tested in multinational, multilocational trials in six mega-environments. This provided a degree of decentralisation (to mega-environment) while having wide adaptation (across the regions of a mega-environment) as a major breeding objective (Rajaram et al., 1995). Many other CGIAR centers follow a similar approach (e.g., Fischer, 1996, for rice in IRRI). In these programs, typically the progeny of hundreds of crosses are advanced to the F_5 or F_6 generation. Entries selected in the environments of the centralized program are then entered into international nurseries for screening by national research programs. Hence, national program breeders select among the finished lines selected outside of the target environment of the country.

In breeding for mega environments, (1) there is a high risk of reduced genetic diversity when a single variety is identified and adopted across countries, and broad regions within countries, within a single mega-environment; (2) it does not directly result in the advantages that can be obtained by more directly involving farmers; and (3) there is often the unintended, but unfortunate, effect of de-emphasising in national programs the role of crossing and selection among segregating generations. When IARCs supply international nurseries or international trials, national program breeders tend to depend on them as sources of new varietal variation for promotion to national-program, multilocational trials. This may be because it is an easy and simple strategy with proven successes, but also because the international trials consume many of the limited available resources.

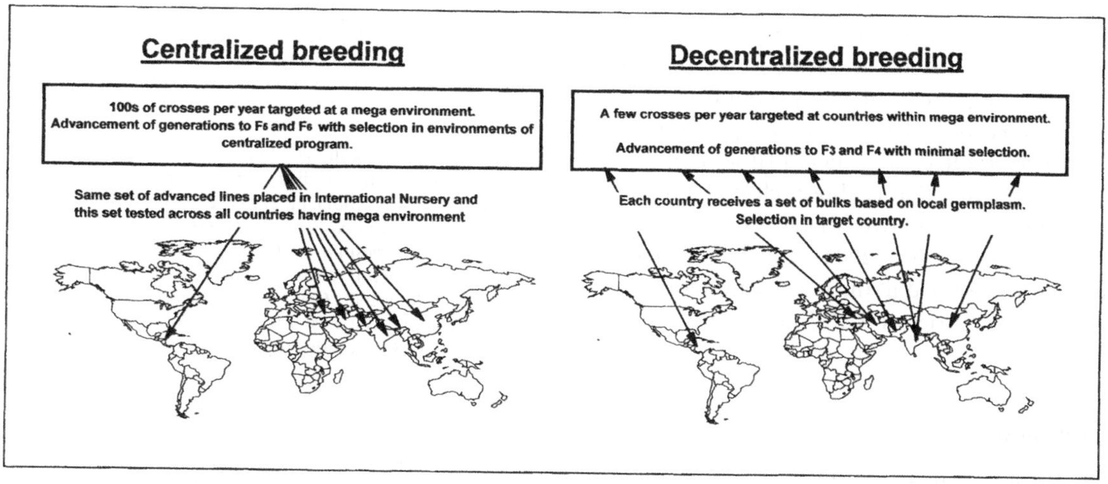

Figure 7.6 Centralized versus decentralized breeding. The decentralized breeding model can involve farmers to a lesser or greater extent in the breeding and varietal testing process.

Solving limited genetic diversity and limited national program involvement

To overcome the problems of the widespread adoption of a few varieties and the limited involvement of national programs can be further decentralized: breeding programs in countries or regions within a mega-environment can produce their own distinct germplasm. The germplasm is based on crosses involving locally adapted genetic material (see, for example, Ceccarelli et al., 2001). IARCs reduce the emphasis on international nurseries and trials and increase support to targeted crossing programs in which national program breeders are actively involved.

Reaping the benefits of greater involvement of farmers

The limited benefits of decentralization can be enhanced by greater involvement of farmers. Selection of parents of crosses becomes more precise as the traits important to farmers are better identified. Selection is carried out in the actual target environment of farmers' fields under farmer management, and the entire process of breeding and testing is shortened. The germplasm distributed to national programs needs to be that most suitable for PPB. Hence, in inbreeding crops, advanced generations produced by bulk population breeding methods would be distributed that are ideal for incorporation into PPB programs. Such a model is being followed in, for example, a collaborative project between CIMMYT, CAZS, and national program partners in Nepal, India, and Bangladesh. The program uses a PVS and PPB approach with a range of germplasm from international and national breeding programs. These programs are targeted at the more marginal wheat-growing areas of these countries where baseline surveys had revealed that past breeding efforts had resulted in farmers primarily growing obsolete varieties, that is, those released at least 20 years previously. In two wheat-growing seasons (2002–2003 and 2003–2004), the PVS programs have already resulted in the large-scale adoption of more modern varieties (CIMMYT, 2003).

Conclusions

One possible criticism of PPB is that it is not cost effective. Results have shown in both maize and rice that genetic gains can be as high, or higher, than conventional breeding. Farmers also like the varieties since they have been involved in the setting of goals for the breeding program and in the selection of material very early in the breeding process. Fewer resources are required since the PPB programs rely on fewer populations and crosses than conventional breeding. The results of PPB can be widely applied if the target environment is widely distributed. In rice, this is illustrated by the wide adaptability of Ashoka 200F and Ashoka 228. In maize, GM-6 is adapted to a large area of maize in at least three states in western India.

Greater adoption of PPB would help solve the problem of farmers in marginal environments having little varietal choice. However, it is a valuable supplement to more centralized breeding programs and not a replacement. The PPB programs here have all used parental material from more centralized programs.

References

Ceccarelli, S., S. Grando, A. Amri, F.A. Asad, A. Benbelkacem, M. Harrabi, M. Maatougui, M.S. Mekni, H. Mimoun, R.A. Rl-Einen, M. El-Felah, A.F. El-Sayed, A.S. Shreidi, and A. Yahyaoui. 2001. Decentralized and participatory plant breeding for marginal environments. pp 115–135. *In* H.D. Cooper, C. Spillane, and T. Hodgkin, eds. *Broadening the genetic base of crop production*. CABI Publishing, FAO and IPGRI.

Ceccarelli, S., S. Grando, E. Bailey, A. Amri, M. El-Felah, F. Nassif, S. Rezgui, and A. Yahyaoui. 2001. Farmer participation in barley breeding in Syria, Morocco and Tunisia. Euphytica 122:521–536.

CIMMYT. (2003). Farmers and researchers find common ground in South Asia. CIMMYT in 2002–2003. *In* Innovation for development. Mexico, D.F.: CIMMYT. pp. 40–41.

Fischer, K.S. 1996. Research approaches for variable rainfed systems—thinking globally, acting locally. pp 25–38. *In* Cooper, M., and Hammer, G.L., eds. *Plant Adaptation and Crop Improvement*. CAB International, Wallingford, United Kingdom.

Kumar, A., D.K. Ganguli, S.C. Prasad, and J.S. Gangwar. 2001. Participatory plant breeding in maize for the Chhotanagpur plateau of eastern India. pp 229–232. *In An exchange of experience from South and South East Asia: Proceedings of the international symposium on participatory plant breeding and participatory plant genetic resource enhancement*, Pokhara, Nepal, 1–5 May 2000. Cali, Colombia: Participatory Research and Gender Analysis Program, Coordination Office; International Center for Tropical Agriculture, 2001.

Rajaram, S., M. van Ginkel, and R.A. Fischer. 1995. CIMMYT's wheat breeding mega-environments (ME). pp. 1101–1106. *In* Z.S. Lin and Z.Y. Xin, Eds. *Proceedings of the Eighth International Wheat Genetics Symposium*. Vol 2. China Agriculture Scientech Press, Beijing, China.

Snapp, S. 1999. Mother and baby trials: A novel trial design being tried out in Malawi. *In* TARGET. The Newsletter of the Soil Fertility Research Network for Maize-Based Cropping

Systems in Malawi and Zimbabwe. Jan. 1999 issue. CIMMYT, Zimbabwe.

Suleman, V.R., and A. Hall. 2002. Beyond technology dissemination: Reinventing agricultural extension. Outlook on Agriculture 31:225–233.

Virk, D.S., D.N. Singh, S.C. Prasad, J.S. Gangwar, and J.R. Witcombe. 2003. Collaborative and consultative participatory plant breeding of rice for the rainfed uplands of eastern India. Euphytica 132:95–108.

Witcombe, J.R. 2002. A mother and baby trial system. pp 79–89. *In* Witcombe, J.R., Parr, L.B., and Atlin, G.R., eds. Breeding rainfed rice for drought-prone environments: Integrating conventional and participatory plant breeding in South and Southeast Asia. Proceedings of a DFID Plant Sciences Research Programme/IRRI Conference, 12–15 March 2002, IRRI, Los Banos, Laguna, Philippines.

Witcombe, J.R., and D.S. Virk. 2001. Number of crosses and population size for participatory and classical plant breeding. Euphytica 122:451–462.

Witcombe, J.R., A. Joshi, and S.N. Goyal. 2003. Participatory plant breeding in maize: A case study from Gujarat, India. Euphytica 130:413–422.

8

Plant Breeding Education

Elizabeth A. Lee, Department of Plant Agriculture University of Guelph
John W. Dudley, Department of Crop Sciences, University of Illinois Champaign-Urbana

"The best tool that a breeder can have is a prepared mind." (anonymous)

Introduction

Plant-breeding education is vital to continued plant improvement. Without being overly dramatic, we could say that plant-breeding education is vital to the continued survival of the human race. Without well-educated plant breeders, progress in food, fiber, and forage production will not keep up with the increasing human population. In addressing the topic of plant-breeding education, we used three main sources of information: (1) John Dudley's insight gathered over more than 40 years as a professor of plant breeding; (2) the National Plant Breeding Survey-IV (Frey, 2000); and (3) an e-mail survey sent to approximately 50 plant breeders (breeders represented a diverse array of species, countries, organizations, and generations and were asked to pass this survey on to colleagues who might be interested in responding). We asked the breeders what they thought were the important issues regarding educating future plant breeders. We asked them to think about this in terms of educating future professors of plant breeding versus private company plant breeders versus plant breeders for international centers. We asked them to think in terms of educating breeders for developed countries versus emerging countries versus third world countries and we asked them what they thought we were doing well as educators; what they felt needed to be changed in the educational experience; and how it should be changed. And, finally, we asked them how they would attract students to the profession. Additional insight and information were gleaned from several published papers on the general subject of changes in plant breeding (Schillinger, 1994; Coffman, 1998; Coors, 2001).

What is a plant breeder?

Before we discuss how to educate plant breeders, we need to understand the discipline of plant breeding. Plant breeding has been defined as "the application of genetic analysis to development of plant lines better suited for human purposes" (Miglani, 1998); "the art and science of the genetic improvement of plants" (Fehr, 1987); "the systematized attempts to develop plants better suited to satisfying man's needs" (Allard, 1960); and "the art and science of changing and improving the heredity of plants" (Poehlman, 1979). All of these definitions encompass two key elements: (1) the final objective is improvement of the plant's phenotype to better suit human needs, and (2) the approach is firmly grounded in the science of genetics.

By definition, a plant breeder practices the discipline of plant breeding. But applying genetics to plant improvement does not make one a plant breeder. Yes, a plant breeder does this, but the plant breeder differs from other plant geneticists in two fundamental areas. The plant breeder's methodology is firmly based on the principles underlying evolution: recombination to create novel genetic variation followed by selection to identify the superior genotypes. And, the plant breeder applies the equation $P = G + E + G \times E +$ error when practicing selection. Where P is the plant's phenotype, G is the plant's genotype, E is the environment in which the plant is growing, and $G \times E$ is the interaction that occurs between the plant's genotype and the environment. It is the reliance on recombination to create novel G's, the requirement to work with the E and $G \times E$ components of this equation, and the need to cope with the quantitative nature of P that sets plant breeders apart from other plant scientists that apply genetics to plant improvement. The plant breeder has always been a systems biologist. It is the nature of both the methodology and the equation, $P = G + E + G \times E +$ error that forces the plant breeder to approach genetic improvement "whole-istically." Systems biology is being hailed by some as the wave of the future (Begley, 2003). The focus of systems biology is on understanding the system's dynamics and structure (Kitano, 2002; Chong and Ray, 2002) rather than on understanding the components of the system, as important as that understanding may be. Plant breeders are systems integrators, integrating and applying the disciplines of genetics (classical, molecular, quantitative, evolutionary, developmental, and population), statistics and experimental design, agronomy and soil science, molecular biology, weed science, crop and plant physiology, plant pathology, and entomology. Plant breeders should be keen observers. And finally, they should possess people, communication (verbal and written), and fiscal skills. Most plant breeders practice their profession in the corporate world (two of every three scientist years devoted to plant breeding in the early 1990s were in private companies; Frey, 1996) where people and communication skills are vital, and those that work in the public sector will find these skills to be highly beneficial as well. The most asked questions by industry people when seeking references for potential plant breeders are "Can this individual work with people?" and "Can he/she communicate?"

Attracting graduate students

The first obstacle in plant-breeding education is attracting graduate students to the discipline. Demographics of future plant breeders have changed. There are far fewer undergraduate students choosing conventional plant breeding as a profession. Instead, they tend to be attracted to the more glamorous molecularly oriented laboratory careers. The result is a smaller qualified applicant pool. The remaining applicant pool still contains quality students, but at a lower frequency, and some of those prospective students lack a basic agricultural background. How do we educationally compensate for this deficit in basic agricultural knowledge? Historically, most North American plant breeders came from agriculture backgrounds, but this was not the case for other parts of the world. In North America this change in demographics poses an educational challenge, but in many cases it is the desire of the student, not the background that is important. Yes, we do need to create opportunities that can compensate for some deficits in agricultural knowledge, but it is not the most pressing matter facing us. Persuading the most talented and brightest undergraduates to become interested in agriculture, and more specifically plant breeding, so that they choose it as a career is by far our greatest challenge.

One of the underlying causes of this trend is the confusion of *tools* with *science*. Very few students who chose to study in tissue culture labs in the 1970s did so because they were interested in understanding the biology and genetics of cell cycle regulation and totipotency. Instead, many students were drawn there because of the potential that tissue culture held for *improving genotypes*. In the 1980s, again very few students who chose to study in molecular marker labs did so because they were interested in the basis of DNA sequence variation. Instead they were attracted to the promise that marker-assisted selection (MAS) held for *improving genotypes* through more efficient selection. In the 1990s, students began flocking to structural genomics laboratories. But, like their predecessors, they were drawn there not because they wanted to understand genome organization; they were mesmerized by the prospect that knowing all of the G's, A's, T's, and C's held for *improving genotypes*. Likewise in the early twenty-first century students are still confusing *tools* with *science*. They continue to be drawn to the newest tool on the block, functional genomics, metabolomics, transcriptomics, bioinformatics, and the hope that these *tools* hold for *improving genotypes*. Who is to blame for all of this confusion? It is a three-slice pie of blame. We as educators and researchers, granting agencies, and companies all have a share of the blame. In addition, it is human nature to assume that newer, more high-tech science must be better science. In reality, plant breeders practice the most complex, high-tech "omics" of them all, phenomics (Bernardo, personal communication), but do it using very traditional *tools*. As educators we need to be proactive in preaching our "omics" to granting agencies and prospective students. We need to be charismatic, dynamic scientists who are excited by our discipline. We need to build educational equality into our graduate programs. Employers need to demand that molecularly oriented students be exposed to the same curriculum that field-oriented students are exposed to and vice versa.

Because the agricultural population is becoming smaller in the developed countries, it is becoming more difficult to find students with interests in agriculture. Thus, it is imperative that students from urban areas be recruited to enter colleges and universities offering education in agricultural disciplines. Everyone in the plant-breeding community can help with this by identifying bright high school students with an interest in science and suggesting to them the value of working in plant breeding. To accomplish this task, as well as others that will be discussed later, will require the collaboration of the public, private, and international sectors. As an example, some commercial corn-breeding companies in the United States offer internships to undergraduates who express an interest in plant breeding or the seed industry. If the internship is well planned and provides the student a real learning experience (as opposed to being a way to get another pair of hands for pollinating), it may well spark the student's interest in plant breeding as a profession and, if all goes well, lead to a job for the student in the future.

Educational philosophy

Once we have attracted qualified students, how should we educate them? We have several education biases.

1. We need to *educate*, as opposed to *train*, plant breeders. They need to understand the theory and principles behind what they do, not just how to perform the task. For many, the words *educating* and *training* are synonymous, but in terms of educational philosophy differences between them, while subtle, are substantial. *Training*, by definition, is directing a student toward a desired objective or outcome and is generally achieved through drilling. On the other hand, *educating*, by definition, is to provide with information (i.e., inform) and supervised practice in a skill, trade, or profession (Merriam-Webster, 2003). In essence, a *trained* student does not understand the theory underlying the task that they are routinely performing. This may not be a problem until the task needs to be altered, improved upon, or have new technologies and approaches integrated into it. For the task to evolve or for new technologies to be adopted, understanding the fundamental theories underlying the task is required.
2. We need to *educate* them as "plant" breeders as opposed to "corn" or "soybean" or "canola" breeders. They need to develop an appreciation for the uniqueness of each organism: how differences in mode of pollination, mode of propagation, generation time, final product, etc. influence breeding strategies and objectives. Most U.S. private-sector plant-breeding effort is focused on eight plant species. In the early 1990s, 25% of the plant-breeding scientist years (SYs) were devoted to maize, 7% were devoted to soybean, 6% each were devoted to cotton and wheat, while only 2–5% of the SYs were devoted to tomato, alfalfa, sorghum, and potato (Frey, 1996). Do these statistics imply that if you want to be a plant breeder in the United States you should develop an interest in one of these crops? To a certain extent the answer is yes. You should be aware of the uniqueness of each species and how that influences breeding strategies and objectives.
3. Plant-breeding students must work in the field. This creates an opportunity for gaining experiences that cannot be acquired through other means. It gives them the opportunity to observe the equation $P = G + E + G \times E +$ error in action. The plant's phenotype (P), how genetic variation (G) influences P, the environment (E), and how the plant interacts with the environment ($G \times E$) can only be seen in the field.

Plant-breeding curriculum—A new paradigm is needed

What should the academic experience involve? What courses do plant-breeding graduate students need to take? The educational experience and curriculum needs are defined by our goal: to produce masters and doctorate candidates who are capable of implementing the equation, $P = G + E + G \times E$ + error. Due to the breadth of knowledge required to accomplish this, the list of courses can be quite extensive. However, a basic list would consist of, but not be limited to: graduate level plant breeding, quantitative genetics, experimental design, statistics (analysis of variance,regression), graduate level plant genetics, biochemistry, molecular biology, and population genetics. Students can only take so many courses. Thus, our challenge as educators is to design courses that are both fundamental in terms of teaching the concepts (theories) and that are novel in that they cross over disciplines to integrate concepts. These courses should challenge convention, incorporate discussion of landmark and current research papers, and, when possible, develop assignments that allow the students to apply the concepts to "real world" data sets. We need to encourage learning outside of the classroom, such as in journal clubs, as well as the importance of being familiar with the literature outside the student's immediate realm of research.

Besides working in the advisor's public-sector breeding program, how can graduate students acquire other breeding experiences? One way to introduce graduate students to the private sector, a major source of employment, is by having them work with a breeder from a private company on their thesis problem. There are many possible degrees of involvement between the private breeder and the student. A model, which has worked well, is for the company to fund a research assistantship and provide some in-kind or actual dollar support for the research. The company contribution can take many forms. In some cases, the company provides molecular marker data and/or locations for testing. Whatever the financial contribution, a most important contribution is the willingness of the private breeder to become an integral part of the research and the education of the student. In some cases, the private breeder serves on the examining committee for the student and advises on analysis of the data. A most valuable part of a program of this sort, where it is practical, is the opportunity for the student to spend an extended period of time (from two weeks to a semester) working in the laboratory of the private breeder. This allows the student to get a good idea of the nature of a commercial breeding operation and the private breeder to assess the potential of the student as a future employee.

Retaining academic plant-breeding positions

The number of academic plant breeders has been on a slow and steady decline in the United States (Collins and Phillips, 1991; Frey, 1996). Private-sector plant-breeding companies were built upon the products of academic plant breeders. While the private sector may no longer heavily rely upon academic plant breeders for elite germplasm, there is still a dependence upon academics for education, basic research, germplasm enhancement, and integration of emerging technologies into the discipline. Can we educate future plant breeders without academic plant-breeding positions? No.

> A sad consequence of increased private sector plant breeding has been that funding organizations (both public and private granting agencies) are less likely to properly fund university plant breeding departments, the base for educating future breeders. These agencies have the impression that "industry can do it all." But industry cannot do it all. And furthermore, what the industry requires is a constant output of well-educated, creative, imaginative plant breeders from university plant breeding departments. (D. Duvick, personal communication)

Is this a reflection of the demand for plant breeders? No. Plantbreeding research and development in the public sector decreased 2.5 SYs per year from 1990 to 1994. During the same 5-year period, however, private sector SYs increased at a rate of 32 per year (Frey, 1996). Instead, the reduction in public-sector plant breeders is a reflection of the economics of academia. The academic environment in which many plant breeders were educated has fundamentally changed. At many land-grant universities, retiring plant-breeding faculty are not being replaced. Instead, these positions are being replaced by molecularly oriented faculty that can attract large grants and publish high-profile papers (Knight, 2002). The economics of academia have led to the erosion of our academic knowledge base in many agronomy departments. And it is this knowledge base that includes many of the courses that plant-breeding graduate students should take.

How do we combat the erosion of our academic

knowledge base? In the long-term, there needs to be greater advocacy from the private sector in protecting core agronomy academic positions, for the erosion will only become worse. Can we completely undo the damage that has been incurred in many agronomy departments? No. We need to accept this and arrive at a solution that will still allow us to achieve a critical mass from which to mount graduate degrees. How can we achieve this critical mass? One possibility is simply consolidating academic resources across universities. In the age of distance education courses through the Web and teleconferencing, it is possible, for example, to mount a cross-institution collaborative effort in plant-breeding education. This would involve a joint effort to develop on-line teaching materials, incorporate examples and data sets from the wide range of species that we collectively work on, along with some kind of on-line system to put students in contact with each other, with breeding professors at different universities, and with practicing plant breeders in government and industry. The approach would allow a graduate student in Saskatoon to gain some level of experience with maize, strawberry, asparagus, and soybean breeding from Guelph. At the same time, a student in Guelph can gain some level of familiarity with lentil, chickpea, and flax breeding from Saskatoon. Keeping in mind that plant-breeding students need a fairly broad knowledge base, this would involve more than just plant-breeding courses and plant-breeding faculty; it would include statistics, experimental design, crop physiology, etc. Faculty members, departments, and universities have different strengths and expertise that could be drawn upon to create a virtual Canadian plant-breeding institute. While not as striking as in Canada, differences in faculty and departmental strengths exist within and between regions in the United States.

Accompanying our eroding academic knowledge base is a fundamental change in the activities of many of the remaining academic plant-breeding programs. We have previously argued that plant-breeding students must work in the field that we need to educate them as "plant" breeders who have an appreciation for the uniqueness of each organism. One of Frey's (2000) recommendations is that comprehensive plant-breeding educational programs need to be reestablished. These programs need to have the resources available to them to permit a full-spectrum of plant-breeding research and development (R&D) activities. The R&D activity that is most often lacking from the existing academic programs is cultivar development (Frey, 1994, 2000). The obstacle hindering public-sector cultivar development in the major crop species, finances aside, is working with relevant germplasm. Most of the elite germplasm sources for the major crops are proprietary. It is not academically useful to practice cultivar development in historically important, but currently irrelevant, germplasm. An alternative to this scenario would be to develop an educational partnership between universities and private plant-breeding companies. We have alluded to the changing role that industry needs to play in educating future plant breeders, and this is just another opportunity for industry to be involved. By working with a breeder from a private company on their thesis problem, students would be exposed to elite germplasm and cultivar development within the elite germplasm pool.

Continuing the educational experience beyond graduate school

How do plant breeders obtain new knowledge after graduate school? The third and final need driving Frey's (2000) call for a new paradigm in plant-breeding education was continuing education. An example of a continuing education effort, which has provided continuing updates in knowledge applied to plant breeding, is the Illinois Corn Breeders' School. This school, which celebrates its 40th anniversary in 2004 and which attracts over 100 commercial corn breeders each year, was designed to provide information about the latest scientific advances, as well as practical information useful to practicing corn breeders. It also provides an opportunity for interaction among commercial breeders. There are a number of specialized short courses available in more specific areas. For example, there is the Iowa State University short course on molecular markers, the Illinois course on statistical analysis of molecular marker data, and the North Carolina State University series of short courses on statistical analysis of genetic data. Learning is a life-long endeavor, and the state of knowledge and scope of disciplines that need to be integrated into plant breeding are constantly changing. Both as educators and practicing plant breeders we need to recognize this need and address it. We need to create opportunities that permit hands on learning experiences and foster discussion of science and technology in the context of plant breeding. To accomplish this will again require the joint efforts of the plant-breeding industry and academia. The Illinois Corn Breeders School grew out of an informal conversation between a commercial breeder and an academic back in 1964. Where will the twenty-first century version of that conversation be held? This brings us to our final point, how do we regain our "culture"?

Regaining our lost "culture"

Throughout this chapter we have been building a case for a new paradigm in plant-breeding education, one that relies more on private sector involvement than ever before. Paramount to achieving Frey's (2000) new paradigm is the need for frequent, informal interactions between academic and private-sector plant breeders. Somehow over the years we have lost our culture as plant breeders, that is, the opportunity to interact with one another on a regular and an informal basis and to freely exchange ideas. These interactions stimulated discussions that led to new ideas, presented opportunities for graduate students to interact with practicing breeders and potential employers, kept the academically oriented plant breeders in touch with the industry, and finally created opportunities for continuing education. How do we regain our culture? We, as academic and private-sector breeders, need to participate in scientific meetings, in company and university field days, in tradeshows. We also need to make sure that our graduate students participate in these meetings and that there is funding set aside to do this. What scientific meetings should we be participating in? In Canada there are meetings called the Expert Committee Meetings on (crop name here), in the United States they are called the Agricultural Experiment Station Regional Committees (e.g., NCR-167 is the North Central Region Committee on corn breeding and quantitative genetics). These meetings tend to be fairly small and focused. Large gatherings such as the American Society of Agronomy (ASA) annual meetings are not always conducive to free-flowing and informal discussions. The ASA, committed to increasing member involvement, needs to consider changing their meeting format. They need to reexamine why people attend meetings and what comprises good, productive meetings, perhaps creating a format where one of the five days of meetings is devoted to specialties within each of the divisions. For example, within the Crop Science Society of America (CSSA) the C1 division is devoted to crop breeding, genetics, and cytology. During this day, the C1 division could hold workshops by crop (e.g., maize, soybean, etc.) or by subject area (e.g., estimating genetic variances). The programs could be a mixture of workshops and lectures (rather than scientific talks) geared to specific topics.

There are a number of obstacles to the redevelopment of these interactions. The recent emphasis on intellectual property rights, both in the private and public sectors, has reduced communication among breeders from different companies, between breeders in the public and private sectors, and, to a lesser extent, between breeders in the public sector. The plant-breeding community is subdivided by species to the point that informal interactions among breeders of different species of the major agronomic crops are rare except for those that take place within an academic institution or a company. For the horticultural crops where a single breeder may work on several species and there may only be a few breeders for each species, this lack of opportunity for communication among breeders of different species is not as true. Breeders in academia spend more and more time seeking funding to maintain their programs, while company breeders, with the advent of year-round nurseries and transgenics, find themselves constantly trying to keep up with their workload. In both academia and the commercial sector, the advent of research by committee requires more and more time in meetings focused on what needs to be done today.

Although we have listed a number of obstacles to informal communications among plant breeders, none of them is insurmountable. The organizers of this symposium are to be congratulated not only for honoring one of the giants in the plant-breeding field, but also for providing a venue for promoting interaction among plant breeders from the public and private sectors, from many different countries working on many different species. As we enhance public–private collaborations to educate plant breeders, communications among breeders will also be enhanced.

Epilogue

We have built upon the theme of Frey's (2000) survey on plant breeding: There needs to be a new paradigm in plant-breeding education. That new paradigm needs to involve the private sector in all aspects of the endeavor. We need to revitalize existing academic plant-breeding and agronomic programs so that they have the resources necessary to effectively educate future generations of plant breeders and practicing plant scientists. We need to put a halt to the erosion of academic knowledge that is occurring within many universities, and to offset the loss of this knowledge base at any one institution we need to consider creating virtual multi-university institutes. And finally, and perhaps most importantly, we need to regain our lost culture of informal interaction and scientific discourse. Through this interaction, continuing education opportunities will arise, awareness of changes occurring within the academic sector will be more evident, and it will strengthen the links both between the public and private sector and among universities.

Acknowledgments

We would like to thank the steering committee for giving us an opportunity to share our educational philosophy at the Arnel R. Hallauer International Symposium on Plant Breeding. We are grateful to all who responded to our email survey: Arnel Hallauer, Don Duvick, Randy Holley, Deon Stuthman, Levi Mansur, Thomas Hoegemeyer, Graham Scoles, Istvan Rajcan, Mike Listman, Duane Falk, Lyn Kannenberg, James Anderson, Richard Trethowan, Lee Stromberg, Jon Geadelmann, Tim Welbanks, Diane Mather, Larry Darrah, Dave Mies, Jean Beigbeder, Pat Byrne, Margaret Smith, and Bruce Hunter. And finally, we are indebted to Arnel Hallauer, not only for his academic achievements in plant breeding, but also for being an outstanding role model and mentor for legions of plant breeders.

References

Allard, R.W. 1960. Principles of Plant Breeding. John Wiley and Sons, New York.

Begley, S. 2003. Science Journal, Wall Street Journal, Feb. 21.

Coffman, W.R. 1998. Future of plant breeding in public institutions. Proc. of the 53rd Annual Corn and Sorghum Research Conference. 53:69–80.

Chong, L., and L.B. Ray. 2002. Whole-istic biology. Science. 295:1661.

Collins, W.W., and R.L. Phillips. 1991. Plant breeding training in public institutions in the United States. USDA Rep. 591 (Rev.). National Plant Genetic Resources Board. Office of Under Secretary for Science and Education, USDA. U.S. Gov. Print. Office, Washington, DC.

Coors, J.G. 2001. Changing role of plant breeding in the public sector. Proc. of the 56th Annual Corn and Sorghum Research Conference. 56:48–66.

Fehr, W.R. 1987. Principles of Cultivar Development, vol. 1. Theory and Technique. MacMillan Publishing Co., New York, NY.

Frey, K.J. 1996. National Plant Breeding Study-I: Human and financial resources devoted to plant breeding research and development in the United States in 1994. Special Report 98. Iowa Agricultural and Home Economics Experiment Station, Iowa State University, Ames, IA.

Frey, K.J. 2000. National Plant Breeding Study-IV: Future priorities for plant breeding. Special Report 102. Iowa Agricultural and Home Economics Experiment Station, Iowa State University, Ames, IA.

Kitano, H. 2002. Systems biology: A brief overview. Science. 295:1662–1664.

Knight, J. 2002. A dying breed. Nature. 421:568–570.

Merriam-Webster OnLine. 2003. http://www.m-w.com/home.htm.

Miglani, G.S. 1998. Dictionary of Plant Genetics and Molecular Biology. Food Products Press, New York, NY.

Poehlman, J.M. 1979. Breeding Field Crops, 2nd edition. Avi Publishing Co., Inc., Westport, CT.

Schillinger. 1994. ASTA Proceedings.

9

Theoretical and Biological Foundations of Plant Breeding

J.B. Holland, USDA-ARS Plant Science Research Unit, North Carolina State University

Of what use is theory for plant breeding?

The value of quantitative genetic theory to practical plant breeding is debatable. Among the more pessimistic views, Simmonds (1984) suggested that quantitative genetics "helped to interpret what has already been done . . . [but] had little impact on the actual practice of breeding." Baker (1984) agreed with this sentiment, but only if breeding is taken in a very strict sense: "One might ask if the understanding and application of quantitative genetic principles can be expected to enhance crop improvement efforts. If one takes the narrow view that plant breeding consists primarily of generating variability and subsequent selection of superior segregants, the answer is probably no." However, Baker (1984) suggested that quantitative genetics principles were key to maximizing the efficiency of plant-breeding programs by aiding *a priori* comparisons between selection schemes and guiding decisions on allocation of testing resources and on population sizes needed to maintain long-term selection gains. Similarly, Dudley (1997) suggested that quantitative genetic theory had immediate practical uses in choosing appropriate parents for breeding crosses, for weighting among-line and within-line selection during inbreeding, for designing efficient recurrent selection schemes, and for appropriately weighting DNA marker information in marker-assisted selection programs.

Historically, some people with no formal scientific training have been successful plant breeders; indeed the daunting task of domesticating crop species from their sometimes phenotypically very distinct progenitors was accomplished thousands of years ago, well before the publication of Allard's (1960) textbook! Furthermore, Darwin (1872) deduced the effectiveness of selection in modifying phenotypes over generations, even without a correct understanding of genetics. However, many recent refinements in breeding methods are due to application of quantitative genetics theory and would not have been accomplished easily without help from theory.

Some ways in which quantitative genetic and population genetic theory has been useful to plant breeding include:

1. estimation of the relative importance of genotypic, $G \times E$, and environmental effects on phenotype;.
2. estimation of heritability and prediction of gain from selection;
3. estimation of genetic correlations and prediction of correlated changes under selection;
4. design of efficient evaluation and selection schemes based on optimal allocation of resources;
5. understanding changes in partitioning of genetic variance among and within lines at different levels of inbreeding;
6. understanding the effects of population size and mating system on inbreeding and genetic drift; and
7. understanding of the effects of different methods of population maintenance on genetic variability in germplasm samples.

An application of theory that is only recently bearing practical fruit is the use of best linear unbiased prediction (BLUP) in plant breeding.

Although the basic theory was developed by Henderson in 1974, BLUP was only recently introduced into the plant-breeding literature (Bernardo, 1996; Panter and Allen, 1995). One reason was that the intensive computational resources needed to solve the BLUP equations were not widely available until recently. A second reason is that plant breeders likely did not see the value of a theory that was initially developed to compare breeding values of animals in very badly unbalanced experiments; after all, an advantage of plant breeding is the ability to save seed and put all of one's best experimental cultivars in a single head-to-head evaluation that can be replicated over environments in a balanced manner. However, Panter and Allen (1995) demonstrated that use of BLUP could assist in the choice of breeding parents, which has traditionally been a breeding step that involved more art and less science. Having a numerical method to rank potential breeding parents could be a great asset to breeders who cannot possibly evaluate progeny from all possible breeding crosses. Similarly, Bernardo (1996) demonstrated the use of BLUP to predict the genotypic values of untested hybrids based on the phenotypic values of tested hybrids and their genetic relatedness to the untested hybrids. A method that permits breeders to initially test only a subset of their hybrids and to use that to identify hybrids with no phenotypic data that have a good chance to be good performers would obviously be a great benefit to hybrid breeding programs. In the absence of theory, however, it is highly unlikely that breeders could have come up with anything like the BLUP equation:

$$\begin{bmatrix} X'R^{-1}X & X'R^{-1}Z \\ Z'R^{-1}X & Z'R^{-1}Z+G^{-1} \end{bmatrix} \begin{bmatrix} \hat{\beta} \\ \hat{u} \end{bmatrix} = \begin{bmatrix} X'R^{-1}y \\ Z'R^{-1}y \end{bmatrix}$$

(Henderson, 1974).

Nor is it likely that breeders would imagine that they could pick better hybrids in the absence of empirical data on those hybrids and have any chance at being successful. Theory permitted development of the prediction scheme and also gave (at least a few) breeders confidence that the scheme was worth testing.

These accomplishments were primarily achieved with what I will refer to as the "standard theory" of quantitative genetics, which consists of equations that most breeders are likely to understand and might actually use. The standard theory is laid out in textbooks such as Falconer and Mackay (1996) and includes important generic equations such as the covariance of relatives ($Cov(X,Y) = r\sigma_A^2 + u\sigma_D^2$), heritability ($h^2 = \sigma_A^2 / \sigma_P^2$), and response to selection ($R = i\sigma_P h^2$).

In addition to its practical applications to breeding, another function of theory in plant breeding is as a basis of scientific explanation. Theory should improve understanding of the relationships between disparate phenomena such as selection response and the mechanisms of regulation of gene expression. The biological basis of standard quantitative genetic theory is based on the assumptions of many genes, each with relatively small effects, that act primarily in an independent fashion (i.e., epistasis is ignored). Typically, linkage is often ignored, and if inbreeding is considered, then dominance is ignored. These assumptions are made to make the equations tractable and intelligible, but the extent to which this framework supports explanation of quantitative trait phenotypes as outputs of gene expression mechanisms is in question. To investigate this issue, our current understanding of gene regulation mechanisms follows.

Current understanding of the regulation of gene function

Substantial evidence from molecular biology indicates that gene expression is affected by complex, interconnected pathways of regulation. At the molecular level, genes typically interact with products of the "genetic background" (i.e., portions of the rest of the genome). To illustrate the ubiquity of the interconnectedness of gene function, examples of molecular interaction at different levels of expression (from transcription to phenotype) follow. Abundant evidence exists for these interactions; only a brief example will be presented to illustrate each.

Gene regulation by transcription factors

For gene expression to occur, RNA polymerases must first initiate contact with DNA in the promoter region of a gene. This interaction is mediated by transcription factors that can specifically regulate a single locus or a suite of loci. Since the transcription factor itself is the product of a separate locus, the physical interaction between the transcription factor protein and the promoter region of the gene being regulated occurs each time a gene is transcribed into RNA. For example, production of anthocyanin in maize aleurone tissue requires a complex of transcription factors. The transcription factor complex induces anthocyanin production by directing the transcription of some structural genes encoding catalytic steps of anthocyanin biosynthesis. A combination of one of the *myc*-like B or R proteins and one of the *myb*-like Pl or C1 proteins is required to induce transcription (Figure 9.1). Specifically, transcription of the duplicate structural genes encoding chalcone synthase (CHS), *colorless2* (*c2*) and *white pollen1* (*whp1*), and of the dihydroflavanol reductase-encoding (DFR-encoding) structural gene, *anthocyaninless1* (*a1*), is mediated by the *myc-myb* transcription factor complex. Products of *c2*, *whp1*, and *a1* catalyze steps in a pathway that converts malonyl-CoA and *p*-coumaroyl-CoA to visible anthocyanin pigments (Holton and Cornish, 1995; Figure 9.1). Different allelic combinations at genes encoding transcription factors and structural enzymes naturally give rise to the phenotypic variations that are the visible display of epistatic interactions. The simplest of these observable interactions directly correspond to classical definitions of complementary and duplicate gene interactions (Mather and Jinks, 1977, p. 102–104).

Complementary gene interactions occur between genes affecting a common biochemical or signaling pathway. In the anthocyanin pathway example (Figure 9.1), DFR and at least one of the CHS gene products (from either *c2* or *whp1*) are required to produce anthocyanins. If the structural enzymes catalyzing any of these steps are missing, then anthocyanins are not made. Thus, homozygosity for a nonfunctional allele at *a1* or both *c2* and *whp1* will mask variation due to segregation at other loci of the anthocyanin pathway. Complementary interactions also occur among regulatory genes or between regulatory genes and structural genes (Figure 9.1).

Duplicate gene interactions occur between loci that serve similar functions. For example, since both *c2* and *whp1* encode CHS (Figure 9.1), one of the two loci can be "knocked out" without necessarily causing an observable phenotype. Similarly, *b1* and *r1* represent one duplicated gene pair, and *pl* and *c1* represent another (Figure 9.1). Genome duplication and genetic redundancy are common in plants (Pickett and Meeks-Wagner, 1995), as whole genome sequencing has revealed in even the simple genome of the model plant, *Arabidopsis* (The Arabidopsis Genome Initiative, 2001). Therefore, duplicate gene interactions are expected to be common in plants (Holland, 2001).

Gene duplication also provides raw material for evolution to operate on, as duplicated loci are freer to evolve new functions or to become subfunctionalized (Dias et al., 2003; Lynch et al., 2001). For example, *b1* and *r1* are not entirely identical, as *b1* does not regulate *a1*, whereas *r1* does (Holton and Cornish, 1995). Also, the duplicate CHS genes *whp1* and c2 have retained identical functions, but are regulated by different transcription factors and by different posttranscriptional regulatory genes in some tissues (Franken et al., 1991; Szalma et al., 2002). Whereas the exon sequences of *whp1* and *c2* are highly homologous, their noncoding sequences have diverged (Franken et al., 1991). The evolution of similar, but not identical, functions by paralogous loci expands the possibilities for epistatic interactions beyond simple duplicate interactions into more subtle forms, in which combinations of mutations at similar loci result in unpredictable phenotypes (Holland, 2001; Martienssen, 1999).

Metabolic interactions

When different biochemical pathways or different branches of a common pathway compete for a limited pool of precursors, the competition creates antagonistic interactions between the enzymes and regulatory gene products involved in the different pathways. For example, flavanone is a precursor of both flavones (such as maysin) and 3-deoxyanthocyanins that are found in silk tissue, as well as for 3-hydroxyanthocyanin pigments (Figure 9.1). In silk tissue, *a1* is required for the production of 3-deoxyanthocyanins and is regulated by a transcription factor encoded by the *pericarp color* (*p*) locus (Figure 9.1, McMullen et al., 2001). The P transcription factor also regulates other genes of the flavanoid synthesis pathway, including the gene encoding the enzyme flavone synthase, which catalyzes the conversion of flavanone to flavone (McMullen et al., 1998). Thus, P can affect the accumulation of both flavones and 3-deoxyanthocyanins in silk tissue by its regulation of both pathways.

The flow of biochemical intermediates between

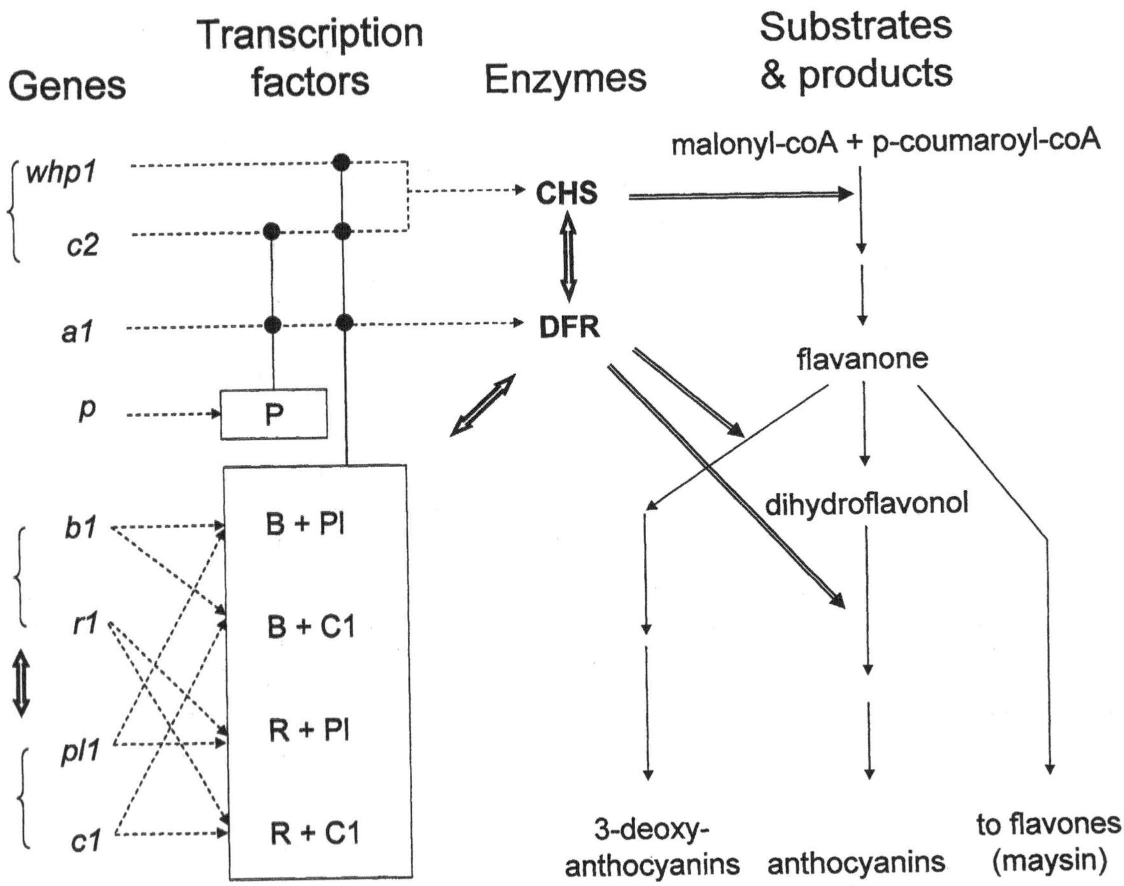

Figure 9.1 Biochemical pathway for anthocyanin production in maize, adapted from Holton and Cornish (1995) and McMullen et al (2001). CHS is chalcone synthase; DFR is dihydroflavonol reductase. ——▶ represents a biochemical reaction, ----▶ represents transcription and translation of genes into transcription factors or enzymes, ⟹ represent a steps catalyzed by a specific enzyme, ——● represents the regulation of structural gene transcription by transcription factors, ⟺ represents a complementary epistatic interaction mediated by interactions among or between transcription factors and enzymes, and brackets represent duplicate gene pairs that cause duplicate epistatic interactions. Anthocyanins (specifically, 3-hydroxyanthocyanins) are produced in many tissue types, including aleurone, whereas 3-deoxyanthocyanins have been found only in pericarp and silk tissue and maysin only in silk tissue.

neighboring pathway branches is influenced by the activity of the first catalytic step of a pathway branch. The DFR structural enzyme encoded by *a1* is the first catalytic step in 3-deoxyanthocyanin biosynthesis at the branch-point between the anthocyanin and flavone pathways (Figure 9.1). It is possible for *a1* to be a major quantitative trait loci (QTL) for both 3-deoxyanthocyanin and flavone accumulation in maize silks, even though it is not required for flavone synthesis, by its competitive effect on substrate needed for both the flavone and 3-deoxyanthocyanin pathways (McMullen et al., 2001). Changes in the flux of biochemical intermediates through one pathway branch occur through reduced or increased activity of the first catalytic step of a pathway branch and result in the increased or decreased flux of biochemical substrates shared with neighboring pathway branches, respectively (McMullen et al., 2001; Szalma et al., 2002).

Further, *a1* and *p* interact epistatically for both flavone and 3-deoxyanthocyanin production in silks (McMullen et al., 2001). The complementary epistatic interaction between the two loci observed for anthocyanin production occurs because *p* regulates *a1* (both genes must be functional). The epistatic interaction observed for flavone production occurs because the competition for substrate by DFR has no effect on flavone production in genotypes that lack a functioning *p* allele to activate the flavone pathway (McMullen et al., 2001). Therefore, competition for precursors between enzymes in alternative, interconnected biochemical pathways can be a biological basis of epistasis as well as pleiotropy (McMullen et al., 2001).

Direct protein–protein interactions

Some gene products function by direct physical interaction with other proteins. For example, chaperonins are a class of proteins that function to promote the folding of other proteins into specific tertiary structures required for their proper activity. The chaperonin coded by the *Hsp90* locus in *Arabidopsis* plays an important role in promoting the proper maturation of a wide variety of proteins, as demonstrated by the aberrant phenotypes that tend to occur when *Hsp90* proteins are inactivated by chemical treatment (Queitsch et al., 2002). When the *Hsp90* chaperonin was inactivated in recombinant inbred lines of *Arabidopsis*, different mutant phenotypes were observed in different lines, and in some cases, the mutant phenotypes were observed only at higher temperatures (Queitsch et al., 2002). This suggests that natural populations maintain cryptic variation at numerous loci that is normally epistatically masked by the function of *Hsp90*. Only when *Hsp90* is inactivated do these variants cause observable phenotypic changes. Queitsch et al. (2002), following Rutherford and Lindquist (1998), suggested that *Hsp90* acts as a capacitor for morphological evolution; that is, numerous variants at key enzymatic loci can be maintained in a population because their potentially deleterious effects are masked by the function of *Hsp90*. This may permit the accumulation of alleles that individually would have deleterious effects, but may yet result in novel, favorable allele combinations. If the favorable effects of such combinations are sufficiently large, they may be expressed phenotypically, even in the presence of active *Hsp90* products. Queitsch et al. (2002) suggested this as a possible mechanism for populations to develop new, adaptive combinations of alleles at multiple loci, without necessarily first suffering declines in fitness.

Gene regulation in complex regulatory networks

Development is controlled by networks of regulatory genes, in which multiple transcription factors interact by binding to each other's *cis*-regulatory regions (including promoters). The interactions among the transcription factor genes and their products create positive and negative feedback loops that are ultimately integrated to regulate expression of the developmental differentiation genes (Davidson et al., 2002).

General features of regulatory networks are likely to be (1) that transcription factor genes are more numerous than the downstream structural or differentiation genes, and that most of the regulation occurs in the *cis*-regulatory elements of the network transcription factors (Davidson et al., 2002); and (2) that the effects of various transcription factors binding to the *cis*-regulatory region of a regulatory gene can be represented as a set of programming commands, that typically involves Boolean logical operators (Yuh et al., 1998). These features, combined with the facts of genetic redundancy in plants, suggest that such regulatory networks are complex systems that will exhibit robustness and the maintenance of system function despite external or internal component variations, and that will rarely suffer catastrophic failure of the entire system (Carlson and Doyle, 2002).

Summary of gene expression control

In summary, our current understanding of the nature of gene expression, which is the key intermediary step between genotype and phenotype, includes the following highlights: (1) many genes may be involved in influencing the final product of developmental or metabolic pathways or gene expression regulatory networks; (2) genomic duplication (with some variation among paralogous genes) is common in plant genomes; (3) genes and gene products are inherently interactive; and (4) the expression level of genes depends greatly on the status of the rest of the genome at any given moment. The last point suggests the possibility that the allelic variation in regulatory genes, rather than in structural genes, may be the molecular basis of many QTL. Also, allelic variation in nonstructural coding regions of genes (such as promoters, introns, and 5′ untranslated regions) that affect the tissue specification, timing, and quantity of gene transcription may also be important contributors to quantitative variation in typical breeding populations. Although many major-effect mutations may be due to coding gene knockouts, they may be under such strong selection pressure that they are either fixed or eliminated from most breeding populations and are not important contributors to quantitative trait variation in breeding populations. Evidence for this view is accumulating from sequence comparisons of alleles at the few QTL that have been sequenced to date: *teosinte branched 1* in maize (Wang et al., 1999) and *fw2.2* (Frary et al., 2000) and *Brix9-2-5* (Fridman et al., 2000) in tomato. In each case, the differences between QTL alleles with different phenotypic effects are located in promoter or intron regions.

Reconciling the biological basis of gene expression and quantitative genetics theory

A drawback to the standard quantitative genetics theory is that it is difficult to relate to the underlying biological reality of multigenic control of traits. Of the four key features of gene expression control listed above, only the first (that many genes may affect a trait) is common between standard quantitative genetic theory and our current understanding of biological reality.

Does this deficiency necessarily invalidate the standard theory? For if it meets the needs of practitioners by always providing good predictions of selection response, we can consider the theory successful as a tool, although it does not satisfy the criterion of providing scientific explanation of observed phenomena. The issue of reconciling it with biological reality in such a case may be largely an academic issue of little interest to most plant breeders. However, if the standard theory does not always make good predictions, perhaps theory can be improved by better grounding it in the known features of gene expression.

Experimental studies generally bear out the predictions of standard theory. For example, substantial evidence indicates that selection response is greater for traits with higher heritabilities than those with lower heritabilities. However, experimental studies also sometimes produce surprising results. Examples of such surprises can be found in the literature on the Iowa State University maize-breeding program and include (1) the frequently lower realized gains from selection compared with predicted gains from selection; (2) the poor response to S_2 recurrent selection programs in maize; and (3) unexpectedly good response to selection in highly related crosses and the possibly related phenomenon of increased genetic variance in some populations with very small population sizes. Such surprises are summarized below, and in each case, modifications to standard theory are considered that might clarify the observations.

Realized gains from selection are often lower than predicted gains

Weyhrich et al. (1998b) observed that estimated heritabilities for grain yield over four cycles of six different recurrent selection methods in maize ranged from 47 to 87%, but that realized heritabilities ranged from 9 to 26%. Such observations are not limited to maize; for example, Holland et al. (2000) found that heritabilities for grain yield in an oat recurrent selection population ranged from 34 to 49% across three cycles of selection, but that realized heritability was only 19%. Weyhrich et al. (1998b) noted that heritabilities estimated in the selection trials could be upwardly biased by genotype-by-year interactions (since the selection trials for each cycle were performed in only one year each). Another source of bias occurs because the numerator of heritability estimated as the ratio of the family variance component to the phenotypic variance of family means ($\hat{\sigma}^2_F / \hat{\sigma}^2_P$) is not equal to the covariance between selection units and response units, which is the covariance that actually predicts the response to selection (Holland et al., 2003; Lamkey and Hallauer, 1987). Standard theory predicts this problem for full-sib families because a portion of the dominance variance is included in the family variance component but not in the covariance between selected families and their intermated progenies. However, many of the recurrent selection methods used by Weyhrich et al. (1998b) and Holland et al. (2000) involved inbred progeny, in which case standard theory (e.g., Falconer and Mackay, 1996; Mather and Jinks, 1977) ignores the full ramifications of dominance. With inbred progeny, dominance can contribute to important discrepancies between heritability estimated based on the family variance component and realized heritability.

When dominance is important and inbreeding occurs in the pedigrees involved in the selection program, covariances between relatives and predicted responses to selection can be determined using theory developed by Cockerham (1983) and Cockerham and Matzinger (1985).

Ignoring epistasis, the expected covariance between two individuals, *X* and *Y*, is:

$$E[Cov(X, Y)] = 2\theta_{XY}\sigma_A^2 + 2\delta_{X+Y}\sigma_D^2 + 2(\gamma_{X+Y} + \gamma_{XY})D_1 + \delta_{XY}D^*_2 + (\Delta_{X\cdot Y} - F_X F_Y)H^*$$

This equation involves higher-order identity by descent measures developed by Cockerham (1971; 1983) and three additional genetic components that are absent from the covariance of noninbred relatives. The additional parameters are D_1, the covariance between additive effects and their respective homozygous dominance deviation effects; D_2^*, the variance of homozygous dominance effects; and H^*, the sum of squared inbreeding depression effects (Cockerham, 1983; Weir and Cockerham, 1977).

As an example of how this theory helps to distinguish the estimate of heritability from the heritability function that predicts response to selection among inbred progenies, consider $S_{1:2}$ selection (Figure 9.2). Applying the equation to the pedigree diagram in Figure 9.2, in which the selection unit is the family composed of individuals $X_{11}, X_{12}, \ldots, X_{1n}$, and the response unit is Z_i, the expected variance component due to families is equal to the expected covariance between any two individuals in that family, say X_{11} and X_{12}:

$$E[\hat{\sigma}_F^2] = E[Cov(X_{11}, X_{12})] = C_{122} = \frac{3}{2}\sigma_A^2 + \frac{1}{8}\sigma_D^2 + \frac{5}{2}D_1 + \frac{9}{16}D_2^* + \frac{1}{16}H^*$$

(Cockerham, 1983). The expected response to $S_{1:2}$ family selection is twice the covariance between X_{11} and Z_1, which is:

$$E[R] = 2\cdot E[Cov(X_{11}, Z_1)] = 2C_{122\infty} = \frac{3}{2}\sigma_A^2 + \frac{5}{4}D_1$$

(Cockerham and Matzinger, 1985). Thus, if any of the nonadditive components of covariance are dif-

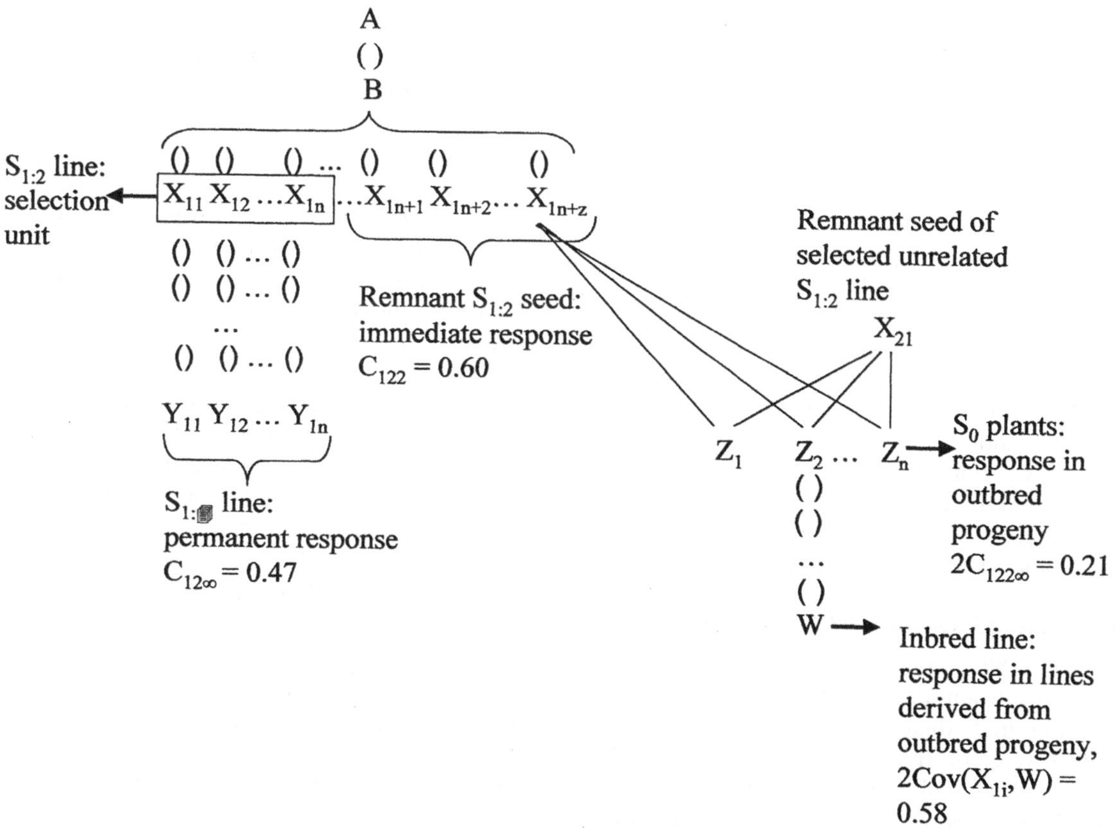

Figure 9.2 Pedigree diagram representing selection among $S_{1:2}$ families and response to selection measured in various response units. Expected covariances ($C_{tgg'}$) between evaluation units and response units are presented based on genetic parameter estimates from the BS13(C0) population made by Edwards and Lamkey (2002).

Table 9.1 Genetic parameter estimates for grain yield in the BS13(C0) maize population (Edwards and Lamkey, 2002) and their coefficients in numerators of heritability functions based on selection among $S_{1:2}$ families measured in different generations of offspring.

	Estimated genetic components of covariances of inbred relatives (± standard errors)				
	σ^2_A	σ^2_D	D_1	D^*_2	H^*
	0.29 ±0.05	0.32 ±0.09	−0.18 ±0.06	0.85 ±0.19†	1.55 ±0.48
	Parameter coefficients in numerators of heritability functions				
Response unit	$2\theta_Y$	$2\delta_{\ddot{X}+\ddot{Y}}$	$2(\gamma_{\ddot{X}Y}+\gamma_{X\ddot{Y}})$	$\delta_{\ddot{X}\ddot{Y}}$	$\Delta_{\ddot{X}\cdot\ddot{Y}}-F_XF_Y$
Remnant S1:2 seed (C_{122})	1.5	0.125	2.5	0.056	0.063
Outbred progeny ($2C_{122\infty}$)	1.5	0	1.25	0	0
Homozygous inbreds derived from remnant seed ($C_{12\infty}$)	1.5	0	2.75	0.625	0
Homozygous inbreds derived from outbred progeny	1.5	0	1.88	0.056	0

†Error in standard error of D_2* in original publication corrected (K.R. Lamkey, pers. comm.).

ferent from zero, they can contribute to bias in the heritability estimated from the family variance component.

To gauge how important such biases might actually be, I used estimates of the components of covariance for yield of inbred relatives recently obtained by Edwards and Lamkey (2002) from the BS13(C0) maize population. Using the estimated values of the covariance parameters as true values, we find that the typical estimator of heritability overestimates the expected response to selection by about three times in this population (C_{122} = 0.60, vs. $2C_{122\infty}$ = 0.21; Table 9.1,; Figure 9.2).

The equations presented above ignore epistatic components of variance, which could also be a contributing factor to the upward bias in heritability estimators. For example, if we include only additive-by-additive epistatic variance in the equations, we find that the coefficient of this component in the numerator of the heritability estimator is 9/4, whereas it is 9/8 in the numerator of the predicted response to selection.

The failure of S_2 selection methods in maize

Lamkey (1992) reported that although seven cycles of selection among half-sib families (created by crossing individual plants with a double-cross tester) were effective in improving mean yield of the Iowa Stiff Stalk Synthetic population, an additional six cycles of S_2 selection gave no response. Reasons for this lack of response to inbred selection may include genetic drift, increased selection for traits other than yield, and a negative correlation between inbred and outbred genotypic values in this population (Lamkey, 1992).

Genetic drift can counteract gains from selection by increasing the inbreeding level of the population. Helms et al. (1989) demonstrated that the S_2 selection program was effective at increasing the frequency of favorable alleles, but also resulted in decreased population mean yield caused by genetic drift.

Selection for agronomic type may have reduced gains in yield because of negative correlations between yield and other agronomic traits. Greater selection pressure seems to have been placed on characters other than grain yield in the S_2 selection program compared with the half-sib program (Lamkey, 1992). Thus, although selection differentials were positive among the $S_{1:2}$ lines that were tested for grain yield (Lamkey, 1992), this does not necessarily imply that the selection differentials were positive with respect to the population of random potential $S_{1:2}$ families.

Under a completely additive model of inheritance, S_2 selection should be more effective than half-sib selection per cycle (Lamkey, 1992). When dominance is important, however, it is difficult to predict response from selection among inbred progenies, because the covariance between selection and response units involves the additional components of covariance of inbred relatives discussed previously.

The component D_1 is a covariance between additive and homozygous dominance deviations and, as such, can be negative or positive. If D_1 is negative and sufficiently large, it can cancel much of the response expected based on the additive component of genetic variance. Edwards and Lamkey (2002) reported a large negative estimate of D_1 in BS13 (Table 9.1), and this negative correlation between dominance deviations of inbreds and their breeding value in random-mated population is partially responsible for the surprisingly limited response to S_2 selection in this population. Coors (1988) also estimated a negative D_1 component in the Golden Glow maize population, so this situation is not necessarily unique to BS13.

Response to selection conceivably can be measured in homozygous inbred progeny, either formed by selfing directly from the selected $S_{1:2}$ lines or by first intermating the selected lines, then by selfing. In the former case, the covariance between selection and response units is $C_{12\infty}$ (Table 9.1), with an expected value of 0.47 in the BS13(C0) population (Figure 9.2). In the latter case, there is no previously published notation, and the covariance is expected to be 0.57 in the BS13(C0) population (Table 9.1; Figure 9.2). Thus, when dominance is important, and nonadditive components of the covariance of relatives are important, inbred selection units tend to be more highly correlated with inbred than noninbred response units. Therefore, we can predict that inbreds developed from the later cycles of S_2 selection would show more improvement compared with original cycle inbreds, even if little improvement was observed in noninbred plants of later cycles. This hypothesis is testable and would serve as a check on theory, although Lamkey (1992) already suggested that inbred lines derived from the S_2 selection program in BS13 performed poorly as lines *per se*.

In contrast to the lack of response to S_2 selection in the BS13 population, Weyhrich et al. (1988) reported that S_2 selection in the BS11 population resulted in greater response than other selection methods, most of which did not involve inbreeding. A difference between the two populations is that additive genetic variance is substantially greater than dominance genetic variance in the BS11, but not in the BS13, population. Thus, selection response in the BS11 population is reasonably well predicted by standard theory (Weyhrich et al., 1998b). This suggests that breeders could first estimate additive and dominance variance components (or effects) using relatively simple mating designs (Hallauer and Miranda, 1988), and if additive variance is predominant, then standard theory will probably suffice. If dominance variance is similar to or greater than the additive variance, however, and if inbreeding is to be used during selection or measurement of response, then breeders should be concerned about nonadditive components of variance including D_1, D_2^*, and H^*, if they plan to theoretically compare selection methods that involve inbreeding. Estimating the genotypic components of variance between inbred individuals is substantially more difficult than estimating additive and dominance variance in noninbred progenies, so this implies significant additional empirical work before making predictions about selection response.

Continued genetic gain and genetic variance in small populations

Another surprising result of some recent work in the Iowa State maize-breeding program was the finding that five cycles of recurrent selection for grain yield in the BS11 population did not reduce the genetic variance for yield, even when only five S1 lines were selected and intermated each cycle (Guzman and Lamkey, 2000). This result was particularly surprising, since the population's per se yield decreased when the small population size was used (Weyhrich et al., 1998a), indicating significant inbreeding depression associated with genetic drift. Standard theory predicts that genetic drift will cause a decrease in genetic variance (Falconer and Mackay, 1996). Possible explanations include (1) low initial frequencies of favorable alleles at most of the genes controlling yield, and (2) epistasis (Guzman and Lamkey, 2000). Epistasis can cause temporary increases in additive genetic variance when genetic drift occurs (Cheverud and Routman, 1996; Holland, 2001), although this effect is ignored in the standard theory. Rasmusson and Phillips (Rasmusson and Phillips, 1997) also suggested that epistasis permitted continued gains from selection in genetically narrow barley populations.

Reconciling quantitative genetics theory and biological knowledge of gene expression

The preceding examples illustrate that experimental quantitative genetic studies are important as checks on theory and can highlight deficiencies in theory. They can serve as impetus to refine theory, which in turn should improve the predictive power of later theoretical investigations. Two modifications to standard theory were proposed to reconcile the surprising findings of empirical research with quantitative genetics theory. One was the use of the complete single-locus theory for the covariance of inbred relatives, which involves three additional genetic components that do not appear in the covariance of noninbred relatives. This modification is not based on a more detailed understanding of gene regulation, however; it is almost a purely statistical improvement that is based on the increased frequency of homozygous dominance deviations under inbreeding. The second modification was the consideration of epistatic variance as an important component of the genetic variance, which has strong motivation in our current understanding of gene expression.

Given the importance of molecular-level interactions on gene expression, it seems logical to conclude that we need more realistic models of quantitative genetic inheritance that incorporate gene interactions (epistasis) at a fundamental level. Holland (2001) reviewed the evidence for and against epistasis controlling quantitative traits in plants and concluded the following: (1) epistasis tends to be detected more frequently in self-pollinating species than in outcrossing species, and (2) epistatic gene *effects* are more frequently found to be important than epistatic *variances*. Epistatic variances are rarely found to be near the magnitude of additive variance.

The paradox of interactions at molecular level and additivity at the phenotypic level

We are left, then, with a seeming paradox. Genes and gene products are highly interactive at the level of gene expression and metabolism. Yet, the evidence for epistasis that can be observed with quantitative genetics experiments on traits like yield is underwhelming. Why is there such a seemingly disjointed behavior of genes at these different levels of observation?

As Omholt et al. (2000) stated: "We by no means have a clear mechanistic understanding of the underlying causes . . . of gene action at the molecular genetic level, how these are related, and in which way they generate and contribute to the various variance components.. . . It remains to be explained why and how gene regulatory networks and signal transduction pathways, with all their nonlinear interactions and hierarchical organization, behave in such a way that the linear 'bean bag model' of quantitative genetics has such a predictive power when implemented within a statistical methodological apparatus."

One simple answer is that quantitative genetics models are inherently biased toward additivity (Lynch, 2000). The additive genetic variance is a function of squared average statistical effects, not of additive gene action effects. The average statistical effect for an allele incorporates not only its additive gene action effect, but also its dominance and epistatic interaction effects (Cheverud and Routman, 1995; Holland, 2001). In contrast, the additive-by-additive statistical effect of an allele pair is affected only by epistatic gene action effects (Holland, 2001). Thus, additive variance components tend to be greater than epistatic variance components, even if the epistatic effects are important.

Another answer is that the standard theory works well as a first approximation to the true behavior of genes on phenotypes and that by extending the theory to better explain the data implies a diminishing return on adding complexity to the models. Difficulties in applying more complex quantitative genetic theory to plant breeding include more complex experimental designs required to estimate additional model parameters and substantially more complicated equations required to relate key concepts such as response to selection to model parameters. The standard theory's wide applicability in practice is based, in part, on the fact that it can be easily understood and related to breeding. As an example of what happens when the assumptions are relaxed and more realistic conditions are modeled, consider Weir and Cockerham's (1977) equation 6 for the variance of an inbred population when epistasis, dominance, inbreeding, linkage, and linkage disequilibrium are simultaneously considered. The equation required two pages to write out, and Weir and Cockerham (1997) concluded: "The complexity of the expressions . . . [is a] negative feature. As it stands, the result is of little use." Thus, although the standard theory relies on numerous untenable assumptions, the simplification permitted by these assumptions results in a robust theory.

Finally, a more philosophical answer to this apparent paradox is that when many genes are involved in controlling a trait, trait-level phenomena are distinct from phenomena occurring at the underlying gene level. In some complex systems, the greater the underlying complexity, the more robust the system is to individual deviations in the components (Carlson and Doyle, 2002). The trait, then, is an emergent phenomenon that must be analyzed as itself and cannot be easily broken down into its parts. Mayr (1982, p. 76) suggested that one of the principles of a philosophy of biology would be to acknowledge "that the patterned complexity of living systems is hierarchically organized and that higher levels in the hierarchy are characterized by the emergence of novelties." The result of this organization is that, "processes at the higher hierarchical level are often largely independent of those at the lower levels" (Mayr, 1982, p. 60). Thus, it may not be possible to predict quantitative trait expression based on even a complete knowledge of the underlying gene interactions and environmental influences, just as it may be impossible to derive the effect of a transcription factor on a promoter from the known properties of atoms. This implies that (1) searching for all of the causal connections between gene expression and quantitative traits will be at best extremely difficult, and perhaps will be fruitless; and (2) most of the current theory as a basis for explanation is sufficient if one is willing to accept as "explanation" that additivity is an emergent property of complex underlying interconnections among genes.

This is not to say that genetic analysis of quantitative traits is hopeless. What it does suggest, however, is that the reason the standard quantitative genetics theory is robust is that it deals with the aggregate effects of many genes taken together, rather than attempting to model all the possible gene interactions that underlie a phenotype. The standard theory deals with quantitative genetics at the level of the trait, which is, indeed, the appropriate level to deal with.

Conclusions

Should we be content with ignoring epistasis and the higher-order components of the covariances of inbred relatives? There are several reasons why we should not. First, where careful (and difficult) experiments have been conducted to accurately measure these components, they have sometimes been found to be quite important (Cockerham and Zeng, 1996; Edwards and Lamkey, 2002). Second, even if epistatic effects do not cause large epistatic variances, identifying those situations in which epistasis affects the response to selection may be an important contribution of theoretical and empirical studies, because the consequences in terms of response to selection differ when traits are under primarily additive or strongly epistatic control. Importantly, epistasis may cause the following otherwise unexpected phenomena (even in some cases when epistatic variance is hard to measure): (1) the increase of additive genetic variance following genetic bottlenecks discussed above, (2) temporary response to selection that can be captured as heterosis, and (3) rugged fitness landscapes in which selection drives gene frequencies toward local rather than global optima (Holland, 2001).

Given that quantitative genetic models that include epistasis and higher-order components of the covariance of inbred relatives quickly become unwieldy and that these parameters are hard to estimate, making theoretical models difficult to fine-tune, is there any hope for making predictions about the likelihood of the implications of epistasis and inbreeding effects if they cannot be measured very well? Three approaches seem to have merit in this regard: (1) developing models of phenotypic outputs of gene regulatory outputs based on observed properties of such systems, (2) using computer simulations to understand the effects of selection in populations segregating for components of such regulatory networks, and (3) continued empirical investigation of selection response and genetic architecture of quantitative traits, combining field experiments with genomics information.

Models that posit realistic behavior of regulatory gene networks have been constructed to determine how variation in their components segregating in populations might give rise to observable phenotypic variation. Most such models to date have been relatively simple, compared with what is known about some actual developmental networks (Davidson et al., 2002). Nevertheless, early studies (Gibson, 1996; Omholt et al., 2000) suggest that additive, dominant, and epistatic gene actions result naturally from simple network feedback regulatory mechanisms. Some caution is in order in interpreting these studies, however, due to their simplicity and the conflation of epistatic gene action with pleiotropy in the Omholt et al. (2000) study. Ultimately, models should be developed that are based first on interactions between gene promoters and transcription factors, second on the effects of complex networks of genes, and finally on the effects of enzyme activity of gene products that exist in the same or competing biochemical pathways.

Other studies have increased the complexity of the networks considered by modeling generic networks of many segregating genes and studying their evolution under selection by stochastic computer modeling studies. Typically, such models are based on Kauffman's (1993) *N*:*K* model. The *N*:*K* model is explicitly epistatic; it permits one to specify the number (*N*) of genes in the network and the level of interactivity (*K*, the number of other genes with which each locus interacts). Instead of modeling specific regulatory mechanisms based on empirical molecular biology findings, however, a generic Boolean network is posited, in which each gene has one of two states (on or off, represented by 0 or 1), and the "phenotype" is some mathematical function of the gene states (Frank, 1999). Podlich and Cooper's (1998) Qu-Gene software implements such a network specifically in a plant-breeding framework. Cooper and Podlich's (2002) initial studies with the *N*:*K* modeling approach verify that response to selection is generally reduced as epistasis and genotype-by-environment interactions increase in importance. This simulation platform should provide a way to compare breeding methods for their long-term effectiveness, assuming varying levels of epistasis.

The use of such computer modeling and simulation studies should help identify those circumstances in which epistatic or inbreeding effects influence the response to selection. These results can then guide empirical studies to determine if such effects are observed in plant populations. One major difficulty with this research program is that the effects may be observed only in long-term breeding programs, in which case breeders may be forced to choose between breeding methods that are known to be effective in the short term and strategies that simulation studies suggest will be better in the long run, without any empirical evidence on which to base the choice.

In summary, the dilemma facing theoreticians in the post-Hallauer era is between developing a theory that is too far abstracted from biological reality and developing a theory that is bogged down in too much biological detail. Descriptions of the dangers of these two extremes can be found in the nonscientific literature. Tolstoy wrote in *War and Peace* of a character, Pfuel, who clung to a theory of battle despite its frequent failure to accurately predict the outcomes of wars:

> [Pfuel] had a science—the theory of oblique movements deduced by him from the history of Frederick the Great's wars, and all he came across in the history of more recent warfare seemed to him absurd and barbarous—monstrous collisions in which so many blunders were committed by both sides that these wars could not be called wars, they did not accord with the theory, and therefore could not serve as material for science.
>
> In 1806 Pfuel had been one of those responsible, for the plan of campaign that ended in Jena and Auerstadt, but he did not see the least proof of the fallibility of his theory in the disasters of that war. On the contrary, the deviations made from his theory were, in his opinion, the sole cause of the whole disaster, and with characteristically gleeful sarcasm he would remark, "There, I said the whole affair would go to the devil!" Pfuel was one of those theoreticians who so love their theory that they lose sight of the theory's object—its practical application. His love of theory made him hate everything practical, and he would not listen to it. He was even pleased by failures, for failures resulting from deviations in practice from the theory only proved to him the accuracy of his theory (Tolstoy, 1962, p. 50, translated by L. Maude and A. Maude).

In contrast, Borges (1998) wrote of mapmakers whose zeal for detail drove them to create a map so intricate as to be useless:

> In that Empire, the Art of Cartography attained such Perfection that the map of a single Province occupied the entirety of a City, and the map of the Empire, the entirety of a Province. In time, those Unconscionable

> Maps no longer satisfied, and the Cartographers' Guilds struck a Map of the Empire whose size was that of the Empire, and which coincided point for point with it. The following Generations, who were not so fond of the Study of Cartography as their Forbears had been, saw that the vast Map was Useless, and not without some Pitilessness was it, that they delivered it up to the Inclemencies of Sun and Winters. In the Deserts of the West, still today, there are Tattered Ruins of that Map, inhabited by Animals and Beggars. . . . (Borges, 1946, translated by A. Hurley, 1998).

The challenge will be to navigate between the Scylla and Charybdis of these two extremes to develop a theory that captures the key facts of biological reality (so as to serve as an explanatory tool) and also proves useful to practical plant breeders.

Acknowledgments

Thanks to Dr. Steve Szalma for helpful comments on this chapter.

References

Allard, R.W. 1960. Principles of plant breeding. Wiley, New York.

Baker, R.J. 1984. Quantitative genetic principles in plant breeding. p. 147–176. Proceedings of the 16th Stadler Genetics Symposium. University of Missouri Agricultural Experiment Station, Columbia.

Bernardo, R. 1996. Best linear unbiased prediction of maize single-cross performance. Crop Sci. 36:50–56.

Borges, J.L. 1998. On exactitude in science. p. 325. *In* A. Hurley (ed.), Collected Fictions. Viking, New York.

Carlson, J.M., and J. Doyle. 2002. Complexity and robustness. Proc. Nat. Acad. Sci. 99:2538–2545.

Cheverud, J., and E. Routman. 1996. Epistasis as a source of increased additive genetic variance at population bottlenecks. Evol. 50:1042–1051.

Cheverud, J.M., and E.J. Routman. 1995. Epistasis and its contribution to genetic variance components. Genet. 139: 1455–1461.

Cockerham, C.C. 1971. Higher order probability functions of identity of alleles by descent. Genet. 69:235–246.

Cockerham, C.C. 1983. Covariances of relatives from self-fertilization. Crop Sci. 23:1177–1180.

Cockerham, C.C., and D.F. Matzinger. 1985. Selection response based on selfed progenies. Crop Sci. 25:483–488.

Cockerham, C.C., and Z.-B. Zeng. 1996. Design III with marker loci. Genet. 143:1437–1456.

Cooper, M., and D.W. Podlich. 2002. The E(NK) model: Extending the NK model to incorporate gene-by-environment interactions and epistasis for diploid genomes. Complexity 7:31–47.

Coors, J.G. 1988. Response to four cycles of combined half-sib and S1 family selection in maize. Crop Sci. 28:891–896.

Darwin, C. 1872. The origin of species by means of natural selection. 6th Ed. John Murray, London.

Davidson, E.H., J.P. Rast, P. Oliveri, A. Rasnick, C. Calestani, C.-H. Yuh, T. Minokawa, G. Amore, V. Hinman, C. Arenas-Mena, O. Otim, C.T. Brown, C.B. Livi, P.Y. Lee, R. Revilla, A.G. Rust, Z.J. Pan, M.J. Schilstra, P.J.C. Clarke, M.I. Arnone, L. Rowen, R.A. Cameron, D.R. McClay, L. Hood, and H. Bolouri. 2002. A genomic regulatory network for development. Science 295:1669–1678.

Dias, A.P., E.L. Braun, M.D. McMullen, and E. Grotewold. 2003. Recently duplicated maize R2R3 *myb* genes provide evidence for distinct mechanisms of evolutionary divergence after duplication. Plant Physiol. 131:610–620.

Dudley, J.W. 1997. Quantitative genetics and plant breeding. Advances in Agronomy 59:1–23.

Edwards, J.E., and K.R. Lamkey. 2002. Quantitative genetics of inbreeding in a synthetic maize population. Crop Sci. 42:1094–1104.

Falconer, D.S., and T.F.C. Mackay. 1996. Introduction to quantitative genetics, 4th ed. Longman Technical, Essex, U.K.

Frank, S.A. 1999. Population and quantitative genetics of regulatory networks. J. Theor. Biol. 197:281–294.

Franken, P., U. Niesbach-Klosgen, U. Weydemann, L. Marechal-Drouard, H. Saedler, and U. Wienand. 1991. The duplicated chalcone synthase genes *C2* and *Whp* (*white pollen*) of *Zea mays* are independently regulated; evidence for translational control of *Whp* expression by the anthocyanin intensifying gene *in*. EMBO J. 10:2605–2612.

Frary, A., T.C. Nesbitt, A. Frary, S. Grandillo, E. van der Knapp, B. Cong, J. Liu, J. Meller, R. Elber, K.B. Alpert, and S.D. Tanksley. 2000. *fw*2.2: A quantitative trait locus key to the evolution of tomato fruit size. Science 289:85–87.

Fridman, E., T. Pleban, and D. Zamir. 2000. A recombination hotspot delimits a wild-species quantitative trait locus for tomato sugar content to 484 bp within an invertase gene. Proc. Nat. Acad. Sci. 97:4718–4723.

Gibson, G. 1996. Epistasis and pleiotropy as natural properties of transcriptional regulation. Theor. Pop. Biol. 49:58–89.

Hallauer, A.R., and J.B. Miranda. 1988. Quantitative genetics in maize breeding. 2nd Edition. Iowa State Univ. Press, Ames, IA.

Helms, T.C., A.R. Hallauer, and O.S. Smith. 1989. Genetic drift and selection evaluated from recurrent selection programs in maize. Crop Sci. 29:602–607.

Henderson, C.R. 1974. General flexibility of linear model techniques for sire evaluation. J. Dairy Sci. 57:963–972.

Holland, J.B. 2001. Epistasis and plant breeding. Plant Breed. Rev. 21:27–92.

Holland, J.B., Å. Bjørnstad, K.J. Frey, M. Gullord, D.M. Wesenberg, and T. Buraas. 2000. Recurrent selection in oat for adaptation to diverse environments. Euphytica 113:195–205.

Holland, J.B., W.E. Nyquist, and C.T. Cervantes-Martinez. 2003. Estimating and interpreting heritability for plant breeding: An update. p. 9–111. *In* J. Janick (ed.), Plant Breeding Reviews. Wiley, New York.

Holton, T.A., and E.C. Cornish. 1995. Genetics and biochemistry of anthocyanin biosynthesis. Plant Cell 7:1071–1083.

Kauffman, S.A. 1993. The origins of order. Oxford Univ. Press, Oxford, U.K.

Lamkey, K.R. 1992. Fifty years of recurrent selection in the Iowa stiff stalk synthetic maize population. Maydica 37:19–28.

Lamkey, K.R., and A.R. Hallauer. 1987. Heritability estimated from recurrent selection experiments in maize. Maydica 32:61–78.

Lynch, M. 2000. The limits to knowledge in quantitative genetics. p. 225–237. *In* M.T. Clegg, M.K. Hecht, and R.J.

MacIntyre (ed.), Evolutionary Biology. Vol. 32. Kluwer Academic/Plenum Publishers, New York.

Lynch, M., M. O'Hely, B. Walsh, and A. Force. 2001. The probability of preservation of a newly arisen gene duplicate. Genet. 159:1789–1804.

Martienssen, R. 1999. Copying out our ABCs: The role of gene redundancy in interpreting genetic hierarchies. Trends Genet. 15:435–437.

Mather, K., and J.L. Jinks. 1977. Introduction to biometrical genetics. Cornell Univ. Press, Ithaca, NY.

Mayr, E. 1982. The growth of biological thought. Harvard University Press, Cambridge, MA.

McMullen, M.D., P.F. Byrne, M.E. Snook, B.R. Wiseman, E.A. Lee, N.W. Widstrom, and E.H. Coe. 1998. Quantitative trait loci and metabolic pathways. Proc. Nat. Acad. Sci. (USA) 95:1996–2000.

McMullen, M.D., M. Snook, A.E. Lee, P.F. Byrne, H. Kross, T.A. Musket, K. Houchins, and E.H.J. Coe. 2001. The biological basis of epistasis between quantitative trait loci for flavone and 3-deoxyanthocyanin synthesis in maize (Zea mays L.). Gen. 44:667–676.

Omholt, S.W., E. Plahte, L. Oyehaug, and K. Xiang. 2000. Gene regulatory networks generating the phenomena of additivity, dominance, and epistasis. Genet. 155:969–980.

Panter, D.M., and F.L. Allen. 1995. Using best linear unbiased predictions to enhance breeding for yield in soybean: I. Choosing parents. Crop Sci. 35:397–405.

Pickett, F.B., and D.R. Meeks-Wagner. 1995. Seeing double: Appreciating genetic redundancy. Plant Cell 7:1347–1356.

Podlich, D.W., and M. Cooper. 1998. QU-GENE: A simulation platform for quantitative analysis of genetic models. Bioinformatics 14:632–653.

Queitsch, C., T.A. Sanger, and S. Lindquist. 2002. Hsp90 as a capacitor of phenotypic variation. Nat. 417:618–624.

Rasmusson, D.C., and R.L. Phillips. 1997. Plant breeding programs and genetic diversity from de novo variation and elevated epistasis. Crop Sci. 37:303–310.

Rutherford, S.L., and S. Lindquist. 1998. Hsp90 as a capacitor for morphological evolution. Nat. 396:336–342.

Simmonds, N.W. 1984. Gene manipulation and plant breeding. p. 637–654. Proceedings of the 16th Stadler Genetics Symposium. University of Missouri Agricultural Experiment Station, Columbia.

Szalma, S.J., M.J. Snook, B.S. Bushman, K.E. Houchins, and M.D. McMullen. 2002. Duplicate loci as QTL: The role of chalcone synthase loci in flavone and phenylpropanoid biosynthesis in maize. Crop Sci. 42:1679–1687.

The Arabidopsis Genome Initiative. 2001. Analysis of the genome sequence of the flowering plant *Arabidopsis thaliana*. Nat. 408:196–815.

Tolstoy, L. 1962. War and Peace, Vol. 2. Heritage Press, New York.

Wang, R.-L., A. Stec, J. Hey, L. Lukens, and J. Doebley. 1999. The limits of selection during maize domestication. Nature 398:236–239.

Weir, B.S., and C.C. Cockerham. 1977. Two-locus theory in quantitative genetics. p. 247–269. *In* E. Pollack, O. Kempthorne, and T.B. Bailey, Jr. (ed.). Proc. First International Conference on Quantitative Genetics, Ames, IA. August 16–21, 1976. Iowa State Univ. Press, Ames, IA.

Weyhrich, R.A., K.R. Lamkey, and A.R. Hallauer. 1998a. Effective population size and response to S1-progeny selection in the BS11 maize population. Crop Sci. 38:1149–1158.

Weyhrich, R.A., K.R. Lamkey, and A.R. Hallauer. 1998b. Responses to seven methods of recurrent selection in the BS11 maize population. Crop Sci. 38:308–321.

Yuh, C.-H., H. Bolouri, and E.H. Davidson. 1998. Genomic cis-regulatory logic: Experimental and computational analysis of a sea urchin gene. Science 279:1896–1902.

10

Integrating Breeding Tools to Generate Information for Efficient Breeding: Past, Present, and Future

M. Cooper, O.S. Smith, R.E. Merrill, L. Arthur, D.W. Podlich, C.M. Löffler
Pioneer Hi-Bred International Inc.

Introduction

Breeding tools have come in a diversity of forms, and the needs of breeding programs have changed over time. Effective integration of tools for use in a breeding program should be examined in relation to the creation and flow of sources of trait genetic–phenotypic data and information that are relevant to the breeding objectives of the program, the goal being (1) to create gene-to-phenotype trait knowledge for breeding objectives, and (2) to use that knowledge in product development and deployment. We consider this topic from the perspective of the Pioneer corn-breeding program. This is a large multinational breeding program that has evolved into its current form over a period spanning approximately 80 years. It is important to appreciate that while the motivations for considering breeding tool development for such a program may have many scientific issues in common with smaller breeding programs, many of the issues related to developing tools that can be applied at the scale of a multinational program are potentially quite different. Where appropriate, we will comment on some relevant distinctions.

Hallauer and Miranda (1988) discussed the importance and interdependencies of setting appropriate short, intermediate, and long-term breeding goals for a successful breeding program. Assuming breeding objectives are clearly defined, the effective development and application of breeding tools requires an understanding of

1. the processes by which gene-to-phenotype information is created for traits,
2. the processes used to characterize new inbreds and hybrids for traits,
3. the flows of gene-to-phenotype information through the breeding process,
4. the limits on access to this trait knowledge, and
5. any utilization bottlenecks that exist in the cyclical process of the breeding program.

For the purposes of this chapter, "information" is defined as any gene-to-phenotype knowledge created by research on the traits that are the targets for improvement in the breeding program. Using this broad definition, it is possible to consider development of tools for breeding strategies that focus on using either phenotypic (including environmental characterization), genotypic (germplasm), gene, or DNA sequence data or some combination of these sources of information. Following general considerations of the cyclical nature of the breeding process we consider past, present, and likely future views of gene-to-phenotype information and their roles in effective breeding.

Cyclical nature of a corn-breeding program

The large commercial corn-breeding programs that operate today have evolved from smaller programs that were designed to improve the grain yield (and/or silage yield) and yield stability of hybrids for a geographically defined target popula-

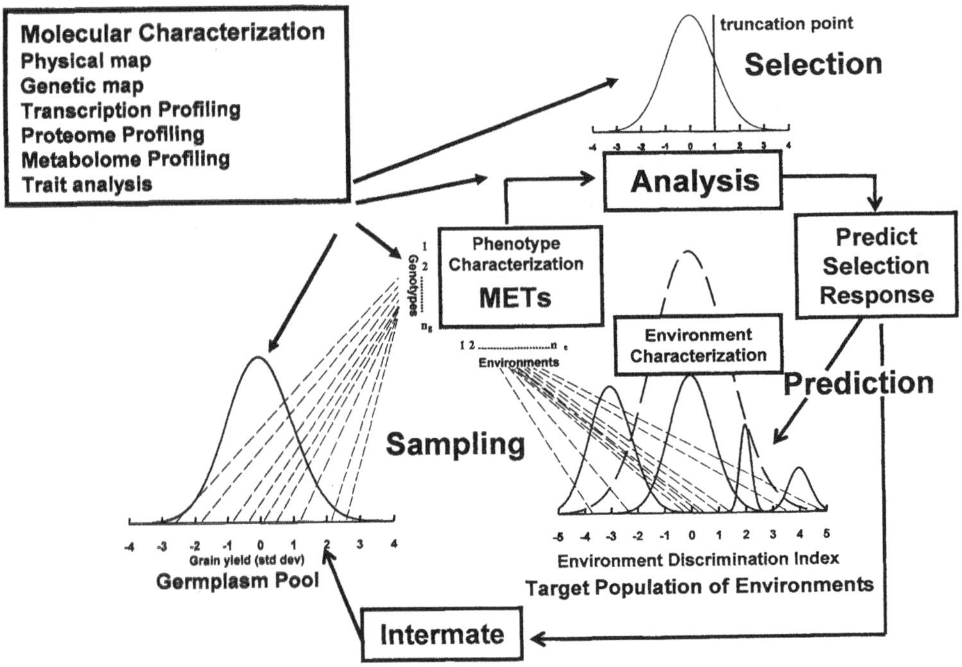

Figure 10.1 Schematic representation of key phases in the cyclical process of a breeding program.

tion of environments (TPEs). In general, the core germplasm improvement program will consist of some long-term cyclical process that includes the basic components of a recurrent selection process:

1. **Evaluation**: measurement of the phenotypic performance of new genotypes (depending on the stage of testing, inbreds in testcross combination for early stages and candidate commercial hybrids for later stages) in a sample of environments;
2. **Selection**: quantitative process for sorting the candidate inbred lines and hybrids into a selected group that is the germplasm to be retained for further use and a group of rejected genotypes discarded from the breeding program; and
3. **Utilization and Prediction**: two principal outcomes of selection, and as such there is commercialization of those hybrids that demonstrate superior performance relative to the current commercial hybrids and intercrossing within the enriched pool of new and old inbreds to create a new set of genotypes for the next round of inbred–hybrid development and evaluation (Figure 10.1).

The details of the breeding process applied within the general breeding cycle shown in Figure 10.1 will differ among programs. For the Pioneer corn-breeding program, pedigree breeding has historically been, and continues to be, the core breeding strategy (Duvick et al., 2004). Initially (1920s to 1960s) the breeding program operated effectively as a single population improvement program. During this early period, commercial products were predominantly double-cross (four-parent) hybrids. With the introduction of single-cross hybrids in the 1960s and the formation of key Stiff-Stalk (SS) and Non-Stiff-Stalk (NSS) heterotic groups, the program evolved into a large reciprocal recurrent selection breeding process with pedigree breeding operating within each of the heterotic groups. Throughout this long-term breeding process grain yield has undergone continual improvement for the range of favorable and less-favorable environmental conditions encountered in the North American Corn Belt (Figure 10.2a).

In association with the increases in grain yield and yield stability over the history of the breeding program (Duvick et al., 2004), secondary traits and combinations of traits have changed with time (Figure 10.2). During the early stages of the pro-

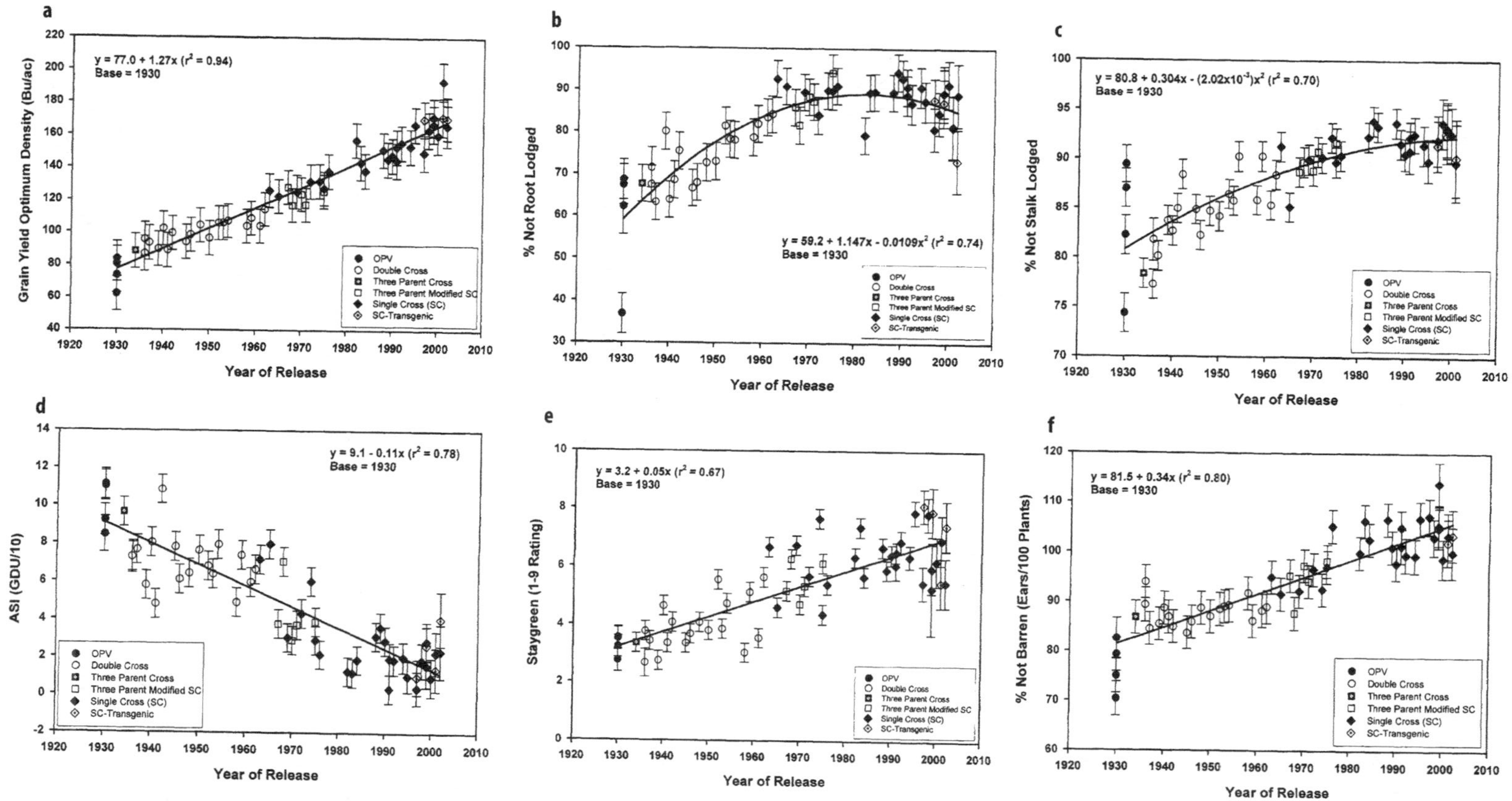

Figure 10.2 Grain yield and secondary trait values for successful hybrids developed by the Pioneer breeding program plotted against the year of hybrid release.

gram, when double-cross hybrids were developed, there were strong contributions to yield improvement through improvements in crop standability, with an increase in the percentages of plants not root-lodged (Figure 10.2b) and not stalk-lodged (Figure 10.2c). There were also changes in the other secondary traits such as reduced anthesis to silking interval (ASI) (Figure 10.2d), increased staygreen (Figure 10.2e), and reduced barrenness (Figure 10.2f). Duvick et al. (2004) discusses these and other trends in more detail.

Views of gene-to-phenotype information

At present, the large commercial corn-breeding programs are in the middle of a transition from what were historically successful conventional breeding programs, which largely selected new genotypes based on some direct measurements of the phenotypic performance for target traits, to molecular-enhanced breeding programs, that focus on selection of superior genotypes directly on genetic variation at the DNA sequence level (Koornneef and Stam, 2001). Molecular breeding strategies involve the use of transgenic approaches (single transgenes and transgene stacks) and a broad family of marker assisted selection (MAS) methods based on marker trait associations.

The differing areas of focus among researchers will naturally emphasize different views of the trait gene-to-phenotype information continuum. These different views will in turn emphasize different research and breeding tools. For example, for researchers involved in developing new inbreds, the pedigree relationships among the inbreds and the breeding values (BVs) of the inbreds represent key pieces of information. Important tools are a database and graphical tools to make clear the pedigree relationships and software to implement the statistical genetic methods used to compute BVs. Researchers involved in characterizing the performance of hybrids in multienvironment trials (METs) will emphasize the multiple-trait phenotypes of the hybrids as key information. In this case, important tools are a database to manage the data from a large number of experiments, powerful software algorithms to analyze the data, and graphical tools to visualize and understand the results of the METs. Researchers involved in studying the genetic architecture of traits (trait mapping) emphasize the genetic map locations of candidate genes and quantitative trait loci (QTL) and the effects of different QTL or gene alleles as key pieces of information. Important tools are a database to manage the results of the studies and powerful algorithms to map the traits. More recently, with the availability of high-throughput genomics, the alignment of genetic maps with physical sequence and sequence diversity information, for example, in the form of single nucleotide polymorphism (SNP) haplotypes, represents key information. Again, important tools are a database and the necessary mining tools for sequence analysis and comparisons between genetic, physical, and sequence diversity maps. From the perspective of designing improved breeding strategies and developing tools to support the implementation of the breeding strategies, one or all of these information sources and associated analysis tools may be relevant.

From the above considerations on views of gene-to-phenotype information and breeding tools, a central tool for any commercial breeding program is a database to store and manipulate the diverse data types that are generated by breeding programs. To this end Pioneer has developed its own proprietary database and data management systems to meet the needs of the current breeding program and to provide a platform for future development.

Information flow and information management in a breeding program

The views of information from the perspective of individuals and disciplines, as discussed above, should be distinguished from the concept of information flow in the breeding process. Usually, the view of gene-to-phenotype information from the perspective of a researcher is founded from the perspective of the data generation and interpretation process. In most cases, the information generated by the researcher has to be moved from the realm of the researcher who acquired the data and created the information to a broader audience involved in other parts of the product development process. It is this movement of the information from its point of origin to the relevant user community throughout the breeding process that is of concern when we consider information flow. This is an aspect that differs significantly between small and large breeding programs. In the case of small programs conducted by 1 to 5 corn breeders, those involved in generating the data and gene-to-phenotype information are likely to represent the entire user community. Whereas in a large commercial program the user community may be 10s to 100s of researchers that do not have any direct involvement in the data generation, quality control, and its interpretation. Examples that would be familiar to most that could differ with size of the breeding program include

1. the collection of trait phenotypic data from small plot METs and product advancement decisions,
2. the generation of molecular marker fingerprints of inbred lines and selection of new breeding crosses, and
3. the insertion and initial efficacy assessment of a transgene and its wide area evaluation in different hybrid combinations.

If we distinguish between the research that is undertaken to generate the gene-to-phenotype information for traits on the one hand and the needs to move the information to the relevant components of the breeding program at the correct times on the other hand, we can recognize that two different types of breeding tools are required. In the former case we are concerned with the discovery tools that support the genetics research process and in the latter case we are concerned with the deployment tools to support information flow in harmony with the decision making stages within the annual cycle of the breeding program (Figure 10.1). Developing appropriate tools that support both of these aspects of a large commercial breeding program is important to the continued success of the program. Understanding the information needs throughout the breeding process and the timing of critical decisions within the breeding program provides a basis for prioritizing the development and integration of breeding tools.

In the following sections we consider past breeding tool development and the current areas of investigation and investment and then speculate on some likely trends for the not too distant future.

Breeding tools: Past

Without attempting to distinguish between the recent and remote past, there are a number of features that can be used to characterize the commercial breeding programs of the past. These early breeding programs can be represented by the stage when individual breeders were required to be involved in all aspects of design and implementation of the breeding program cycle. The predominant data types used to make selection decisions were trait phenotypes measured on the individuals and their relatives, tested in hybrid combinations.

Breeding tools that were considered important during this stage include many aspects of field experiment mechanization and, more recently, computerization. Advances in both areas enabled the breeder to scale up the number of genotypes (inbreds–hybrids) that could be tested in small plot experiments and the number of environments in which they could be tested.

Mechanization had two major effects on the Pioneer breeding programs. First, it allowed a breeder or breeding station to test a much larger number of genotypes at more locations, that is, better sample both the genetic variation and the TPE. Second, it allowed collection of phenotypic data, that is, grain yield, in a more timely manner such that breeders could start using data-driven off-season nurseries. Off-season nurseries had been used in conventional breeding programs for some time before mechanical harvesting methods were implemented; however, full advantage could not be made of these nurseries until data-driven decisions about the germplasm to be sent could be implemented.

Population and quantitative genetic theory was developed and applied to study the efficiency of alternative corn-breeding strategies (e.g., Hanson and Robinson, 1963; Comstock, 1977; Hallauer and Miranda, 1988). However, in the absence of empirical data on the genetic architecture of traits, much of the theoretical framework was developed from assumed finite locus models. For these assumed polygenic models, statistical prediction equations were developed and used as a framework for (1) the design of single-population and dual-population breeding strategies (e.g., Comstock et al., 1949), and (2) to understand how the selection methods implemented in breeding strategies exploited sources of additive and nonadditive genetic variation (e.g., Hallauer and Miranda, 1988; Comstock, 1996).

In parallel, many quantitative traits were studied using a variety of mating designs to quantify the extent of additive and nonadditive sources of genetic variation. From this work the concepts of general combining ability (GCA) and specific combining ability (SCA) emerged (Sprague and Tatum, 1942).

More recently, statistical mixed model methodology was applied to explicitly incorporate information from relatives and to compute BVs of inbred lines based on the theoretical framework advanced by Henderson (Henderson, 1975). Routine use of BVs in Pioneer to predict performance of breeding crosses and experimental products began in 1996. This was possible because of the development of efficient proprietary algorithms, coupled with an increase in computational power (release of the Alpha computer chip by Digital Corporation in 1993). These two developments facilitated the computation of additive and dominance relationships between all pairs of parents, as well as the solution of a variance–covariance matrix for all observed crosses, which typically includes several million elements.

Computation of BVs required extraction of information from two different sources, historical pedigree information, methods to electronically store this information for a large multinational program and algorithms that could rapidly retrieve and compute coefficients of coancestry for very large numbers of genotypes, and a database of the phenotypic data, where experimental design information could be combined with the phenotypic trait data to make maximum use of this in the mixed linear model analyses. These analyses represented the next step in the process of summarizing data collected on breeding METs.

Independently of these quantitative genetic investigations and the empirical development of the inbred–hybrid breeding strategies, detailed molecular genetic studies were conducted on a range of qualitative traits. While these investigations contributed a lot of basic knowledge on the genetics of traits in corn, this work had little direct impact on the design and conduct of breeding programs.

Breeding tools: Present

Here we identify the present as the period beginning from around the early 1990s, where testing of large numbers of hybrids in wide-area METs has become commonplace (e.g., Figure 10.3) and the technologies of molecular genetics are beginning to be used in breeding programs. At this time many of the selection decisions for quantitative traits in the core breeding program are still based largely on phenotypic data. However, in contrast with the breeding programs of the past, the volume of data generated has now increased to a scale that requires teams of researchers to be involved in data collection, quality control, database management, analysis, and interpretation. The advancements in information technologies have made this feasible.

Selection of hybrids that demonstrate broad adaptation and superior yield performance across large geographical areas has been a major focus of wide-area testing programs used in commercial corn breeding. The importance of genotype-by-environment (G × E) interactions for hybrid grain yield and the variable contributions of some key secondary traits to hybrid yield stability have been recognized and quantified.

The initial stages of molecular breeding strategies began to take shape through the latter half of the 1990s and into the 2000s. Transgenic hybrids have been developed and deployed for insect resistance genes and herbicide resistance. The suitability of a number of molecular-marker technologies for studying the genetic architecture of traits has been evaluated. The cost of DNA-sequencing technologies has declined, and a range of structural and functional genomic technologies can now be routinely applied to discover the genes responsible for determining traits and to study genetic variation in breeding populations at the DNA sequence level. While there is a large investment in genomics, at present the majority of the applications of these technologies to the study of gene-to-phenotype relationships for traits are still highly descriptive. However, from this early molecular characterization work two major research trends have emerged in plant breeding: (1) the use of high-throughput genotyping to study the molecular genetic diversity of the germplasm created by the breeding program, and (2) advancements in molecular biology technologies that have enabled the compilation of a knowledge base of the genetic architecture of traits for a range of organisms. The outcomes from the research to date are starting to create a picture that indicates that the details of the genetic architecture of the target traits of a breeding program is a continuum extending from simple to complex genetics. We are currently in the early exploratory phase of this continuum.

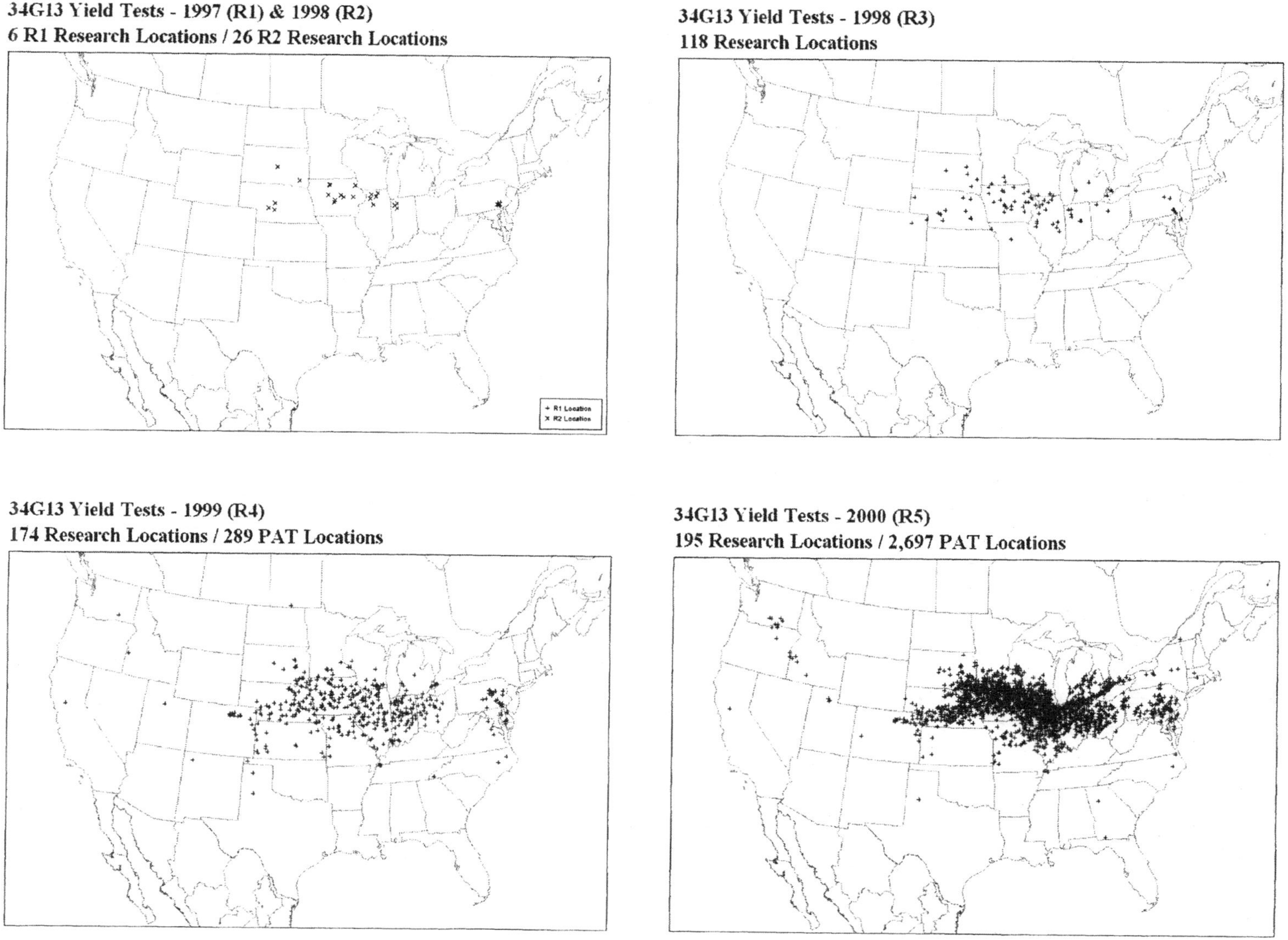

Figure 10.3 The sequence of trials conducted over the period from 1997 to 2000 that comprised the set of multienvironment trials (METs) in which the hybrid released as 34G13 was tested in North America.

Many traits have been genetically mapped, using both low- and high-density molecular marker maps. The results from this body of work have identified many candidate regions in the form of QTL. These QTL provide a basis for either further investigation and/or manipulation using MAS. Depending on the trait and the reference population in which the mapping was conducted, QTLs have been found to show a mixture of effects on traits that are sometimes consistent across genetic backgrounds and environments, and sometimes genotype and environment specific. Complementary work has also been conducted to examine the contributions of a range of candidate genes to the phenotypic variation for traits among elite lines. A part of the picture that emerges from these empirical investigations is as we may have expected. Many of the results of the mapping studies for important traits are context dependent. The detail of the genetic architecture of the standing natural variation for traits that have been subjected to intense selection over breeding cycles is likely to depend on the history of the breeding. Thus, mapping of traits in order to realize benefits from MAS in a commercial breeding program will need to be conducted within the context of the elite germplasm pool of the breeding program as the reference population (e.g., Jansen et al., 2003).

Breeding tools: Future

It is expected that the large commercial corn breeding programs will continue to invest heavily in molecular technologies in the future. Thus, future breeding tools will be focused around those necessary to support and realize benefits from molecular enhanced breeding strategies. Perhaps somewhat paradoxically this large investment in molecular technologies is driving the need for a parallel investment into appropriate technologies for phenotyping in ways that are different from the high throughput phenotyping that is conducted today and that which was used in the past. Two relevant examples can be considered in this context: (1) Mapping drought tolerance as a component of grain yield stability, and (2) studying the regulation of gene expression.

Water-deficit (drought) is an important component of the TPE for corn in North America. Severe drought that causes significant yield loss occurs in about one year in four to five years in the Western Corn Belt. Genetic variation for drought tolerance and genotype-by-environment (G × E) interactions for grain yield are observed under variable drought regimes. Under drought conditions, grain yield measured in traditional breeding small-plot METs generally has a low heritability. The importance of this environmental component of the TPE and its effects on selection and realized gains for improved yield stability has focused efforts to: (1) create suitable managed drought environments for high-throughput phenotyping of inbreds and hybrids, and (2) study genetic variation for key component drought tolerance traits to evaluate their contributions to yield under drought. The creation of dedicated drought breeding facilities itself creates specific information management requirements to integrate genetic, environmental, and phenotypic data sources for inbred and hybrid evaluation that are different from the needs for traditional breeding METs.

A number of investigations of the genetic basis of variation for quantitative traits in model and crop species have indicated the importance of variation in both coding and regulatory components of genes. This has created considerable interest in studying the basis of variation in gene regulation and genetic regulatory networks. The application of RNA-profiling technologies within a trait-mapping context (e.g., to recombinant inbred lines) demonstrates a more fine-grained level of defining phenotypes than most breeders have previously considered (e.g., Jansen and Nap, 2001; Jansen, 2003; Schadt et al., 2003). Molecular breeding will become more than the application of markers to a breeding program. It will include RNA, protein, and metabolite profiling. The challenge in understanding the genetic regulation of a trait remains knowing which level(s) to collect phenotypic data; it is now feasible to include the molecular level. Initially, the development of breeding tools required for this level of integration will involve modifications of existing QTL mapping tools. As we understand more about how to leverage these tools, databases and analyses can be designed to integrate molecular and field phenotypes. Certainly, one of the most significant potential impacts that high-throughput, low-cost molecular technologies will have on crop breeding will be less dependence on the extrapolation of gene-to-phenotype models from model organisms such as yeast and *Arabidopsis*. The tools and cost will be such that the experiments can be carried out with the crop itself.

It is important to appreciate that in a relatively short space of time the plant-breeding community has moved from a state of studying the genetic architecture of traits within a quantitative genetics framework of assumed simple statistical models to a stage where a large number of candidate QTL regions and candidate genes are now examined in terms of the details of their DNA sequence and the regulation of their expression and contributions to traits. The next step is to understand the functional properties of allelic variation for the important genes and the impact of this variation on trait variation in elite germplasm.

As the gene-to-phenotype knowledge for multiple traits accumulates, many new opportunities for breeding are unfolding. Perhaps one of the key differences that will likely exist between the gene-to-phenotype knowledge for traits we seek today and the tools that will be increasingly required in the future is in the area of knowledge integration for multigenic, multitrait gene-to-phenotype investigations. It is highly likely that trait knowledge sharing in the future will be in the form of dynamic gene-to-phenotype models for traits rather than static lists of QTLs, DNA sequence, and phenotypic data. The level of detail required in such models will differ among traits and will depend on the genetic complexity that regulates phenotypic variation for the trait variation. As the gene-to-phenotype models for traits improve, these models will provide a robust foundation for *in silico* breeding.

In the future there will be greater emphasis on predicting the performance characteristics of hybrids in the TPE. To improve our ability to predict the expected performance distributions of commercial hybrids across geographical areas and years and to determine how to quickly develop new hybrids that overcome the limitations of the current hybrids, characterizing the incidence of important G × E interactions and understanding their causes will become more important (e.g., yield variation due to genetic variation for drought tolerance and environmental variation for drought incidence). A key step is achieving a better understanding of the distribution of environmental conditions that influence these interactions and their frequencies of occurrence within the TPE. Developing such higher-resolution views of the TPE, and how METs can be designed to represent the TPE, identifies opportunities for targeted breeding for specific traits and continual refinement of wide-area testing across all stages of the breeding program.

At Pioneer, we characterize the environments for all trial sites. Performance information is reported, with breakouts by environment, thus phenotype measures are linked to environment type. We assess the relevance of the MET to the TPE, and advancement decisions can be made with some understanding of the scope of inference of our performance data. Figure 10.4 illustrates the dramatic shifts in distribution of environmental conditions that could occur from year to year using an environmental characterization based on features of the environment associated with repeatable G × E interactions for grain yield. Five environmental types are identified and are labeled high latitude, subtropical, temperate, temperate-dry and temperate-humid; Loffler et al. (2003) give more detail on these environment types. The geographical distribution of the dominant environment types for the period 1952–2002 is shown in Figure 10.4a. Examples of the interyear variation are shown in Figures 10.4b–d, which contrast 2002 (Figure 10.4d) with 1988, an extremely dry year (Figure 10.4b), and 1993, an extreme wet year (Figure 10.4c).

Grain yield and yield stability are expected to remain the primary selection criteria for corn for the foreseeable future. Validation of new hybrids for their performance characteristics in the TPE will remain a critical component of any molecular-enhanced breeding strategy. However, a trend that can be anticipated for the future is a move away from grain yield being the dominant trait for selection decisions. With improved gene-to-phenotype knowledge and high-throughput molecular characterization it becomes feasible to select for different traits at different stages of the breeding process. As the gene-to-phenotype knowledge base expands, the range of molecular tools broadens, and their utility for breeding is validated, there will be a continuation of the coevolution of information management requirements and breeding tool development. We have seen this process unfold as the first generation transgenic products were developed, evaluated, and scaled up to commercial production levels.

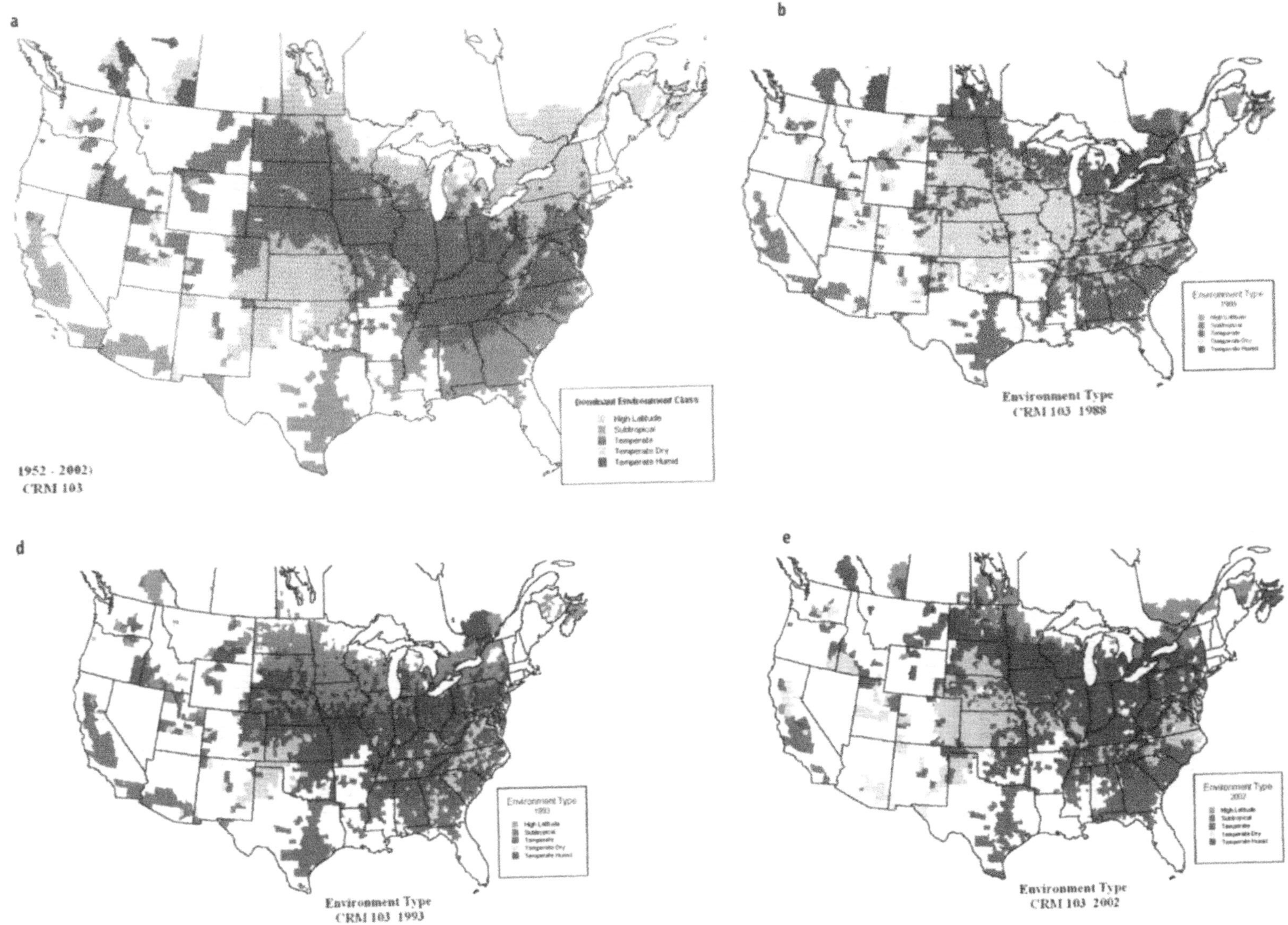

Figure 10.4 Geographical distribution of five environment types (high latitude, Subtropical, temperate, temperate dry, and temperate humid), for the North American Corn Belt characterized using the methodology described by Löffler et al. (2003); (a) the dominant environment types over the period 1952–2002, (b) 1988, (c) 1993, and (d) 2002.

Modeling breeding strategies

A third trend, that goes with the other two listed above, is the need for an extension of the current quantitative genetics framework that builds on our growing knowledge of the genetic architecture of traits. Our early theoretical framework, out of necessity, assumed that many genes acted additively and independently in determining the phenotypic variation for traits. While it was always recognized that these model assumptions were likely to have limited applicability in the details of applied breeding, we now have a growing body of empirical evidence that demonstrates the importance of biological interactions between genes and interactions between genes and environmental variables for important traits. Thus, it is important to consider the implications of these features of the genetic architecture of traits for short-term and long-term response to selection.

Kempthorne (1988) discussed the need to move our theoretical framework beyond the simple models used in classical quantitative genetics. He indicated the importance of computer simulation as a path forward. Developments in this area have occurred (Podlich and Cooper, 1998; Cooper and Podlich, 2002), and it is now feasible to use computer simulation methods to model the power of conventional and molecular breeding strategies for gene-to-phenotype models that can be defined to include more of the trait-specific details than the assumed models of classical quantitative genetic theory (Cooper et al., 2002; Wang et al., 2003). Integration of such a modeling framework with the empirical corn gene-to-phenotype databases will enable strategic and tactical investigations and *in silico* design of robust breeding strategies. Some of the preliminary concepts and components for such *in silico* breeding strategy design and computer breeding have been developed and or discussed (Cooper et al., 2000; Eagles et al., 2001; Peleman and van der Voort, 2003).

The application of MAS is an example of an area where high throughput *in silico* research tools can be used to better identify and understand the factors that influence trait improvement. Here, the power of many different approaches to MAS can be evaluated across a broad range of gene-to-phenotype models in a relatively short period of time. For example, Figure 10.5 shows the average improvement in trait performance for a simulated reciprocal recurrent selection breeding program using (1) phenotypic information only, and (2) a combination of phenotypic and molecular marker information. On average, it can be seen that MAS outperformed conventional phenotype selection for the scenarios considered here. Results such as those presented in Figure 10.5 give us an impression of how a given breeding strategy will perform "on average" across a broad range of genetic models. However, for any single realization of the breeding process there will be variation around this average result. For example, Figure 10.6 shows the relative improvement in performance (if any) achieved by MAS over phenotypic selection for a broad range of genetic models, where each point on the figure represents a different trait architecture. Clearly, the effectiveness of MAS is going to differ among traits as demonstrated by the results shown here. Hence, as our understanding of the genetic architecture of traits improves, we will use *in silico* solutions to help refine our breeding strategies to reflect the specific complexities of the traits we are trying to manipulate.

Discussion

The requirements for effective integration of tools to enhance breeding efficiency have always been the same; develop the tool and support its application when value to the breeding program has been demonstrated. In the past, and also today for small breeding programs, deciding how to prioritize these activities was simpler for breeding programs when only a few breeders and technicians were involved in all aspects of the program. However, today the diverse range of technologies that is available for use in the study of gene-to-phenotype relationships for traits and the range of opportunities that exist for implementing molecular-enhanced breeding have contributed to greater complexity in the breeding process and the need to prioritize the efforts in breeding tool development.

Current needs for complex information flows can be met due to the widespread use of computers and advances in information technologies. Powerful, high performance databases are an essential part of the commercial breeding program today.

Around the core information management process there is a continual need for an exploratory research process that evaluates the needs and technical solutions for new breeding tools.

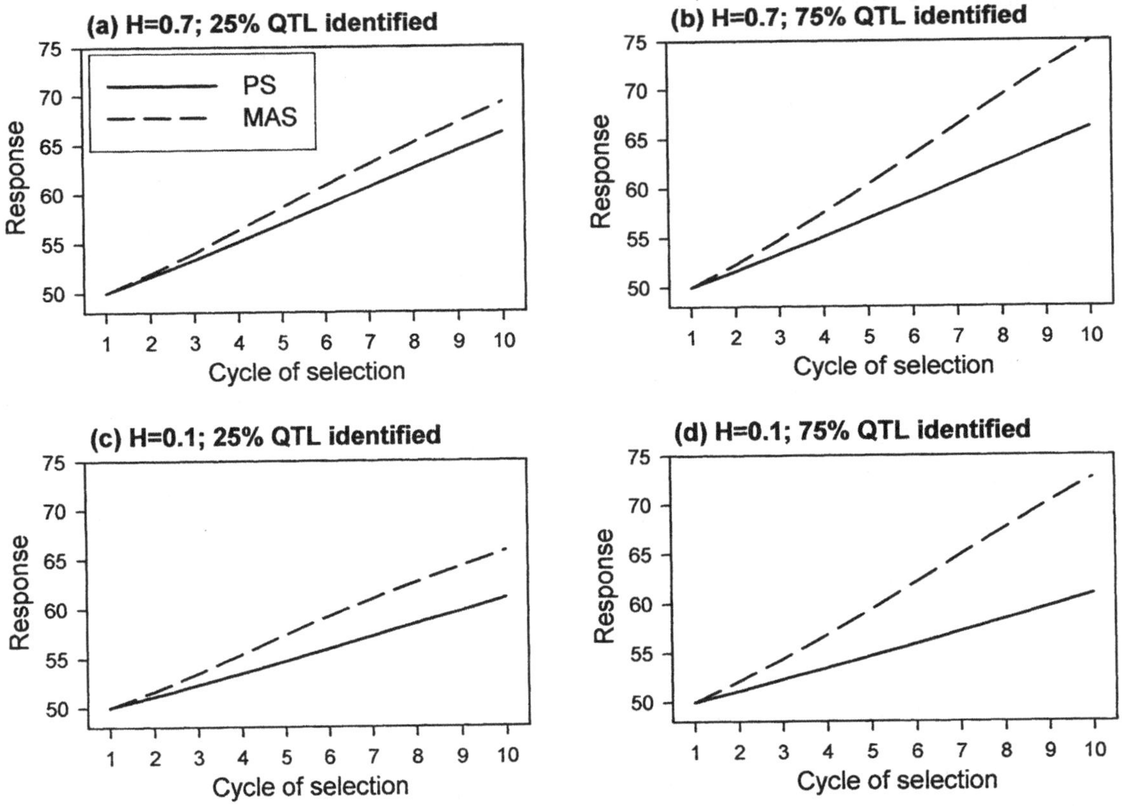

Figure 10.5 Average response to selection for a simulated reciprocal recurrent selection breeding program using (1) phenotype information only (PS), and (2) combination of phenotypic and marker information (MAS). Results are shown for two levels of heritability (H) and for scenarios with different levels of genetic variation explained by markers associated with quantitative trait loci (QTL).

Major challenges and areas for further research

Given the current efforts into the use of genomics technologies to study the genetics of traits for model species and commercially important crops, it is tempting to move quickly to the assumption that we will soon have a clear understanding of the genetic architecture of many of the key traits and the functional basis of natural allelic variation in elite breeding material and thus be able to define the opportunities to use molecular methods to manipulate all traits (e.g., Peleman and van der Voort, 2003). However, there are still likely to be many surprises as we build gene-to-phenotype models for traits that apply within the TPE of a breeding program. For example, the finding by Fu and Dooner (2002) of the lack of microcolinearity in the DNA sequence in the genic regions between two corn inbreds raises many structural and functional questions about the organization of the genetic architecture of traits within a plant genome. Currently, the *Arabidopsis* community articulated in the *Arabidopsis* 2010 vision that it would have a comprehensive understanding of the function of all genes for this model organism (Somerville and Dangl, 2000). While this will be a valuable resource, comparative genomics investigations have already shown us that much of the gene-to-phenotype knowledge for traits developed in *Arabidopsis* will have to be refined and in some cases investigated *de novo* in corn.

We argue that building an appropriate gene-to-phenotype knowledge base for the germplasm pool of a breeding program, and in particular for the elite inbreds and hybrids, will be an area of intense research for the foreseeable future. Success in these activities will provide the foundation for implementation of targeted molecular breeding efforts. In association with these gene-to-phenotype and molecular breeding research efforts there will be continual refinement of information manage-

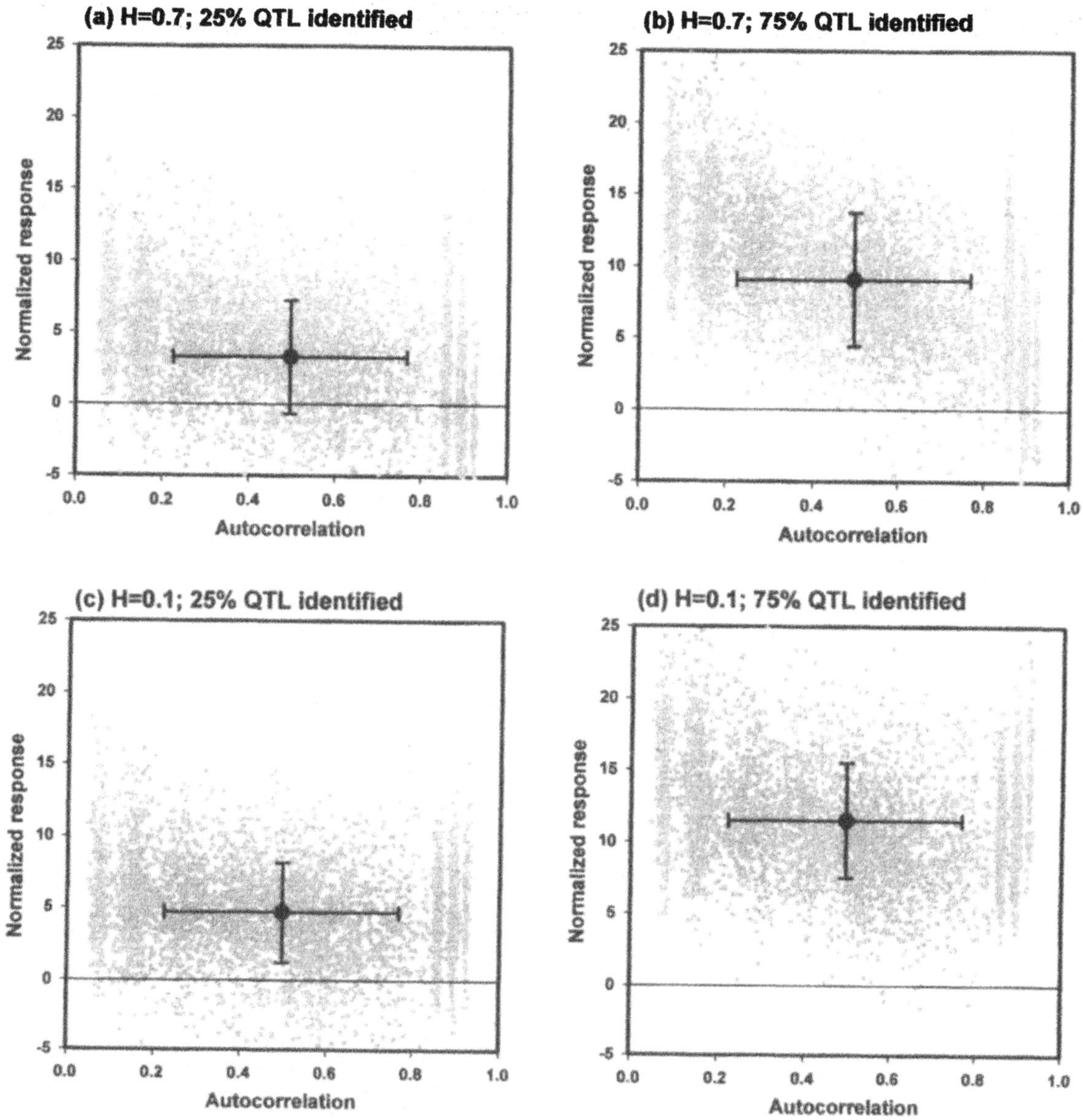

Figure 10.6 Performance of Marker Assisted Selection (MAS) relative to phenotype selection (PS) at cycle 10 (MAS − PS = normalized response) plotted against the complexity of the genetic models considered in the experiment (as measured in terms of an autocorrelation value). Each point on the figure represents a different genetic model. Simple genetic models generally reside on the right side of each figure panel, and more complex genetic models generally reside on the left side. The average performance across all points is shown in Figure 10.5.

ment needs and appropriate breeding tool development, implementation, and support. One trend that is expected to continue is the increase in the rate of data generation in exploratory and production research. Thus, there is always a need to increase capacity and to balance flexibility in the information management for the exploratory research phase and stability of the high-throughput measurement processes contributing to product development. Striking the appropriate balance is a continual challenge, given the rapid pace of advances in gene-to-phenotype knowledge and the current early stage of evolution of molecular breeding strategies.

Acknowledgments

We thank Tim Fast for his assistance in preparing Figure 10.3.

References

Comstock, R.E. 1977. Quantitative genetics and the design of breeding programs. pp. 705–718. *In* E. Pollack, O. Kempthorne, and T.B. Bailey, Jr., eds., Proceedings of the International Conference on Quantitative Genetics, August 16–21, 1976, Iowa State University Press, Ames, IA.

Comstock, R.E. 1996. Quantitative Genetics with Special Reference to Plant and Animal Breeding. Iowa State University Press, Ames, IA.

Comstock, R.E., H.F. Robinson, and P.H. Harvey. 1949. A procedure designed to make maximum use of both general and specific combining ability. Agron. Jour. 41:360–367.

Cooper, M., and D.W. Podlich. 2002. The *E(NK)* model: Extending the *NK* model to incorporate gene-by-environment interactions and epistasis for diploid genomes. Complexity 7:31–47.

Cooper, M., D.W. Podlich, K.P. Micallef, O.S. Smith, N.M. Jensen, S.C. Chapman, and N.L. Kruger. 2002. pp. 143–166. Complexity, quantitative traits and plant breeding: A role for simulation modelling in the genetic improvement of crops. In M.S. Kang, ed. Quantitative Genetics, Genomics and Plant Breeding. CABI Publishing, Wallingford, United Kingdom.

Cooper, M., D.W. Podlich, and S.C. Chapman. 2000. Computer simulation linked to gene information databases as a strategic research tool to evaluate molecular approaches for genetic improvement of crops. pp. 162–166. *In* J.-M. Ribaut and D. Poland eds., Molecular Approaches for the Genetic Improvement of Cereals for Stable Production in Water-Limited Environments. A Strategic Planning Workshop held at CIMMYT, El Batan, Mexico, 21–25 June 1999. Mexico D.F., CIMMYT.

Duvick, D.N., J.S.C. Smith, and M. Cooper. 2004. Long-term selection in a commercial hybrid maize breeding program. pp. 109–151. In *Plant Breeding Reviews 24(Part 2), Long Term Selection: Crops, Animals, and Bacteria*. J. Janick (Ed.) John Wiley & Sons, New York, NY.

Eagles, H.A., M. Cooper, R. Shorter, and P.N. Fox. 2001. Using molecular information for decision support in wheat breeding. pp. 285–298. *In* R.J. Henry (ed.), Plant Genotyping the DNA Fingerprinting of Plants. CABI Publishing, Wallingford, United Kingdom.

Fu, H., and H.K. Dooner. 2002. Intraspecific violation of genetic colinearity and its implications in maize. Proc. Nat. Acad. Sci. 99:9573–9578.

Hallauer, A.R., and J.B. Miranda, Fo. 1988. Quantitative Genetics in Maize Breeding, Second Edition. Iowa State University Press, Ames, IA.

Hanson, W.D., and H.F. Robinson, eds. 1963. Statistical Genetics and Plant Breeding. Publication 982, National Academy of Sciences—National Research Council, Washington, DC.

Henderson, C.R. 1975. Best linear unbiased estimation and prediction under a selection model. Biometrics 31:423–447.

Jansen, R.C. 2003. Studying complex biological systems using multifactorial perturbation. Nature Reviews Genetics 4:145–151.

Jansen, R.C., J.-L. Jannink and W.D. Beavis. 2003. Mapping quantitative trait loci in plant breeding populations: Use of parental haplotypes sharing. Crop Science 43:829–834.

Jansen, R.C., and J.-P. Nap. 2001. Genetical genomics: The added value from segregation. Trends in Genetics 17: 388–391.

Kempthorne, O. 1988. An overview of the field of quantitative genetics. pp. 47–56. In B.S. Weir, E.J. Eisen, M.M. Goodman, and G. Namkoong, eds., Proceedings of the Second International Conference on Quantitative Genetics. Sinauer Associates, Inc., Sunderland, MA.

Koornneef, M., and P. Stam. 2001. Changing paradigms in plant breeding. Plant Physiology 125:156–159.

Löffler, C.M., J. Wei, T. Fast, and R. Merrill. 2003. Classification of maize environments using crop simulation and GIS. pp. 242–243. *In* Book of Abstracts: Arnel Hallauer International Symposium on Plant Breeding. 17–22 August 2003, Mexico City, Mexico, D.F.

Peleman, J.D., and J.P. van der Voort. 2003. Breeding by design. Trends in Plant Science 8:330–334.

Podlich, D.W., and M. Cooper. 1998. QU-GENE: A simulation platform for quantitative analysis of genetic models. Bioinformatics 14:632–653.

Schadt, E.E., S.A. Monks, T.A. Drake, A.J. Lusis, N. Che, V. Collnayo, T.G. Ruff, S.B. Milligan, J.R. Lamb, G. Cavet, P.S. Linsley, M. Mao, R.B. Stoughton, and S.H. Friend. 2003. Genetics of gene expression surveyed in maize, mouse and man. Nature 422:297–302.

Somerville, C., and J. Dangl. 2000. Plant biology in 2010. Science 290:2077–2078.

Sprague, G.F., and L.A. Tatum. 1942. General vs. specific combining ability in single crosses of corn. J. Am. Soc. Agron. 34:923–932.

Wang, J., M. van Ginkel, D. Podlich, G. Ye, R. Trethowan, W. Pfeiffer, I.H. DeLacy, M. Cooper, and S. Rajaram. 2003. Comparison of two breeding strategies by computer simulation. Crop Science. 43:1764–1773.

11

Genotype by Environment Interaction—Basics and Beyond

Fred van Eeuwijk
Wageningen University, Department of Plant Sciences, Laboratory of Plant Breeding

Introduction

In plant breeding, genotype by environment interaction is usually described as a problem that occurs whenever the most basic quantitative genetic model fails to describe the relation between the phenotype on the one hand and the genotype plus the environment on the other hand. This most basic model states that the phenotype, P, is simply the sum of a genotypic contribution, G, and an environmental contribution, E: $P = G + E$. A bit more sophisticated, the phenotype for genotype i ($i = 1 \ldots I$) in environment j ($j = 1 \ldots J$) is written as $\underline{P}_{ij} = \mu + G_i + E_j + \underline{e}_{ij}$, where G_i and E_j are defined as parameters that are centered around the general mean μ, and $\underline{e}_{ij}$ is an error term that is assumed to be normally distributed with zero mean and constant variance. For notational clarity, we adopt the convention to underline random terms.

Thinking about the processes governing development, whether physiological, genetic or otherwise, it would be peculiar if the simple additive model, $P = G + E$, could provide an adequate description of the phenotypic responses of a set of genotypes. The additive model assumes that differences between genotypes remain the same across environments. In the light of the heavy nonlinear developmental processes underlying plant growth, the additive model can be expected to provide only exceptionally an adequate description of phenotypic responses. The additive model might be satisfactory for a relatively homogeneous set of genotypes over a relatively short environmental range. Outside these limited conditions, the additive model will not be satisfactory, and some extension of the additive model will be required. The most common extension adds in one term indexed by both genotypes and environments, GE_{ij}. This extra term is often referred to as genotype by environment interaction (GEI), a description that is somewhat deceptive because the concept of GEI is better reserved for a wider class of phenomena (see below). Including the extra term GE_{ij} produces the model $\underline{P}_{ij} = \mu + G_i + E_j + GE_{ij} + \underline{e}_{ij}$. In comparison with the additive model, the GE_{ij} term can also be interpreted as adapting the additive model to more complicated data by the introduction of a nonadditivity term. Note that this nonadditivity term still enters the model in an additive sense, that is, GE_{ij} is included as another additive model term.

The phenotype is the cumulative result of causal interactions between the genetic makeup of a plant and the environment over time. In classical breeding, the genetic makeup of a plant is tacitly taken to be equivalent to the genotype. However, this equivalence masks the crucial fact that the set of active genes and the intensity with which the active genes create gene products, changes in its dependence on environmental stimuli (nutrients, minerals, etc.) and developmental time. Furthermore, the history of individual plants will determine which genes will become active. Just as the genotype changes over time, so does the environment. The required nutrients, minerals, and stimuli depend on a plant's history and developmental age. Plants will differ with respect to the efficiency and adequacy with which they convert environmental input and stimuli into desirable products or appropriate and adaptive responses using their idiosyncratic genetic software. Environments will differ with respect to the quality and quantity of the resources and cues they present to plants differing in genetic makeup. The way in which the phenotype materializes, the series of complex interactions between gene networks and environmental stimuli sets, must lead to many occurrences of GEI, that is, phenotypic differences between genotypes being conditional on the environmental circumstances.

The equivalence of GEI and nonadditivity in linear statistical models is rather restrictive. It is fruitful to look at GEI in a wider modeling context. In such a context, genotypes will be understood loosely as labels for the levels of a genotypic factor that merely categorizes entities known to differ in smaller or larger parts of their genetic constitution, without these differences being made more explicit initially. Similarly, environments are just labels for the levels of an environmental factor, again, initially without explicit reference to further environmental descriptions. Statistical modeling of GEI is, firstly, trying to find a model for the differential (mean) phenotypic expression of individual genotypes across environments in terms of nonparallel responses, taking into account genotypic and environmental factors and variables that can cause GEI. Simultaneously, the observed pattern of genotypic and environmental variances and correlations, the variance–covariance structure (VCOV), should be modeled, again, when possible, in relation to genotypic and environmental factors and variables.

Preferentially, mean responses and VCOV are modeled simultaneously to adequately represent the data and as a prerequisite for valid inference on the mean responses. The classical analysis of variance (ANOVA)/linear model framework must then be generalized to the more appropriate linear mixed model framework (Searle et al., 1992). The latter allows more elaborate modeling of patterns of variation.

Within a mixed-model framework, choices must be made about individual model terms being fixed or random. These choices will affect the interpretation of the results. Many papers have been written on this issue, but no consensus has been reached. In this chapter, the pragmatic attitude is taken that a model term is taken to be random when it is necessary for the research question at hand. A condition that must be fulfilled is that there is enough information (enough independent observations or degrees of freedom, for example more than 10) in the data to allow a sufficiently precise estimate of the variance and covariance parameters related to the specific random term. Furthermore, the estimates for the individual random effects should look as if they could have come from a normal distribution (although this is in practice difficult to verify). We consider a few simple examples to illustrate the choice of terms as fixed and random. Later, some of these examples will be further elaborated. For the estimation of genetic correlations between environments, genotypic main effects must be taken to be random, otherwise the genetic correlations are by definition zero. When genotypic stability variances are of interest, the GEI must be random, whereas the genotypic main effects are often taken to be fixed. In experiments with many genotypes and environments, simple convenience could dictate the choice of genotypes and environments as random, because, when these terms are random, fewer parameters need to be estimated than when these terms are fixed.

As remarked above, the mean, and to a lesser extent the VCOV, should be modeled as much as possible in relation to genotypic and environmental covariables. Candidate genotypic covariables for inclusion in our models follow from physiology and developmental genetics (Cooper and Hammer, 1996); genotypic factors/variables underlying GEI are developmental stage/maturity, adaptability, sensitivity to environmental stimuli, stability, resistance, and tolerance against diseases and abiotic stresses, vigor, etc. The list of candidate genotypic covariables can be extended by quantifications of pedigree relationships, absence or presence of quantitative trait loci and molecular markers, gene expression in a microarray, etc. For the environmental factor in GEI analyses, candidate covariables underlying differential responses are incidence and intensity of biotic and abiotic stress factors and limiting factors.

The structure of the rest of this chapter is as follows. First, modeling of VCOVs will be treated. Logically, after that, modeling of the mean receives attention. Two examples should further elucidate practical and theoretical aspects of modeling means and VCOVs. The first example deals with modeling the differential expression of a quantitative trait locus (QTL) in relation to the environment, where aspects of modeling the mean and variance are important. The second example is directed principally at investigating mean structures by graphical displays.

Modeling

Modeling of VCOVs

The mixed model framework was introduced as a suitable framework for studying GEI. Mixed models consist of a model for the mean and a model for the VCOV. Standard linear models with normally distributed independent errors of constant variance are contained within the class of linear mixed models. Standard linear models are mixed models with a simple VCOV based on only one parameter, the error variance. For example, the two-way (fixed) additive model $P_{ij} = \mu + G_i + E_j + e_{ij}$, has a model for the mean (fixed part of mixed model), $\mu_{ij} = \mu + G_i + E_j$, while the model for the VCOV (random part of mixed model) is $\text{var}(\underline{P}_{ij}) = \sigma^2_e$, for the variance of individual genotype by environment means. The assumption of independence implies that the covariance between different genotype by environment means should be zero, $\text{cov}(P_{ij}, P_{i^*j^*}) = 0$.

Within mixed models, GEI can appear in various guises. It appears as nonadditivity in the model for the mean and as heterogeneity of variance or correlation in the model for the VCOV. Extensive discussions on the modeling of VCOVs in the context of GEI can be found in Denis et al. (1997), Piepho (1997), Piepho and van Eeuwijk (2002), Smith (1999), Smith et al. (2001), and van Eeuwijk et al. (2001).

A recommended strategy to fitting mixed models, using restricted maximum likelihood, or REML, is to start with elaborate models for the fixed and random part, that is, models that are certainly not underparameterized, thus with rather large models for mean and VCOV (Verbeke and Molenberghs, 2000). Subsequently, first the model for the VCOV is made more parsimonious. Next, given the VCOV, the model for the mean is inspected again to see whether some simplification is possible. A deviance statistic, -2 times the log likelihood ratio, can be used for comparing nested VCOVs. Under the null hypothesis of two nested VCOVs giving equivalent fits, the distribution of the deviance is approximately a $\chi 2$ distribution with the number of degrees of freedom equal to the difference in the number of parameters between the two models. After having chosen a VCOV, a Wald test can be used to test for dropping fixed terms from the model for the mean (Schabenberger and Pierce, 2002; Verbeke and Molenberghs, 2000). Model building is an iterative process. The cycle of successive checking of the VCOV and mean model may have to be repeated a few times before settling on a definite model for VCOV and mean.

Modeling of the VCOV will be illustrated by some examples. Assume genotype by environment data are available and interest centers on estimating genotype-dependent stability variances. This state of affairs forms the starting situation for a description of GEI in terms of heterogeneity of genotypic variance. A well-known model in this context was proposed by Shukla (1972); $\underline{P}_{ij} = \mu + G_i + \underline{E}_j + \underline{GE}_{ij} + \underline{e}_{ij}$, with $\underline{E}_j$, $\underline{GE}_{ij}$, and $\underline{e}_{ij}$ being random terms for environmental main effects, nonadditivity, and error, distributed as normal variates with zero means and variances σ^2_e, σ^2_{GEi}, and σ^2_e, respectively. The stability variances, σ^2_{GEi}, thus depend on the genotype. It will be obvious that the interaction terms, $\underline{GE}_{ij}$, should be chosen randomly if interpretations of GEI are wanted in terms of heterogeneity of stability variances. In practice, the error, $\underline{e}_{ij}$, will represent an average plot and experimental error across replications. Parameter estimates for variance components and fixed and random terms can be obtained by REML. The question, of course, is whether another model related to the stability variance model would have provided a better description of the data. Two options exist, either the heterogeneity of genotypic variance model is too simple and more parameters would be required to describe the data, or, alternatively, the heterogeneity of variance model is too complex and a simpler model would have been enough. Which related models spring to mind and how can we test for which model deserves our preference? For the two-way genotype by environment mean write $\underline{P}_{ij} = \mu_i + \underline{e}_{ij}$, where μi represents the fixed part of the model and $\underline{e}_{ij}$ the random part. The model for the mean states that $\mu_{ij} = \mu_i$, while the VCOV model for the random terms is

$$\begin{bmatrix} \sigma^2_E + \sigma^2_{GE_1} & & \\ \sigma^2_E & \sigma^2_E + \sigma^2_{GE_2} & \\ \sigma^2_E & \sigma^2_E & \sigma^2_E + \sigma^2_{GE_3} \end{bmatrix},$$

restricting ourselves to the first three genotypes. The VCOV for the heterogeneity of variance model has 1 (genotypic main effects variance) + I (stability variances) parameters. A more elaborate model is the unstructured model (Wolfinger, 1996), with VCOV $\begin{bmatrix} \sigma_1^2 & & \\ \sigma_{21} & \sigma_2^2 & \\ \sigma_{31} & \sigma_{32} & \sigma_3^2 \end{bmatrix}$, where each genotype has its own variance, σ_i^2, and each pair of genotypes i and i* (i $\neq$ i*) has its own covariance, σ_{ii}*. In the unstructured model there are $I(I + 1)/2$ parameters. To test whether the goodness of fit for the heterogeneity model differs from that of the unstructured model, a deviance test can be employed. Under the null hypothesis that both models are equivalent, the difference in deviance is approximately χ^2 distributed with degrees of freedom equal to $I(I + 1)/2 - (1 + I)$.

If it is suspected that the heterogeneity model is too complex, the compound symmetry model with VCOV $\begin{bmatrix} \sigma_E^2+\sigma_{GE}^2 & & \\ \sigma_E^2 & \sigma_E^2+\sigma_{GE}^2 & \\ \sigma_E^2 & \sigma_E^2 & \sigma_E^2+\sigma_{GE}^2 \end{bmatrix}$ could do. For the compound symmetry model, only two variance parameters have to be estimated, σ_E^2 and σ_{GE}^2. To test for the necessity of genotype-dependent stability variances in comparison with a common genotype by environment interaction variance, a deviance test can be performed, with the test statistic having approximately a χ^2 distribution with $(1 + I) - 2 = I - 1$ degrees of freedom.

A popular class of models for VCOVs that combines high flexibility with parsimony in parameters is the class of factor analytic or multiplicative models (Gogel et al., 1995; Piepho, 1997; Smith, 1999; Smith et al., 2001). For example, a model with just one multiplicative term (parameters $\lambda 1 \ldots \lambda I$) and a residual genotype dependent heterogeneity, $\sigma_{d_i}^2$, leads to $\begin{bmatrix} \lambda_1\lambda_1+\sigma_{d_1}^2 & & \\ \lambda_2\lambda_1 & \lambda_2\lambda_2+\sigma_{d_2}^2 & \\ \lambda_3\lambda_1 & \lambda_3\lambda_2 & \lambda_3\lambda_3+\sigma_{d_3}^2 \end{bmatrix}$ This VCOV is of intermediate complexity (number of parameters); for larger genotype-by-environment tables, it can be located in between the simple heterogeneity of variance model with constant covariance between genotypes (σ_e^2) and the unstructured model. This factor analytic model with $2I$ parameters can be tested by deviance tests against the heterogeneity of a genotypic variance model on the one hand and the unstructured model on the other hand.

Looking at genotype by data from the perspective of the environments, the models in the last section remain valid, but the interpretation changes from genotypic stability and genotypic covariance/correlation to their environmental counterparts. For example, to model the (genetic) correlation between environments, environments should be chosen as fixed and genotypes as random, leading to the mixed model $\underline{P}_{ij} = \mu_j + e_{ij}$, with $\mu_{ij} = \mu_i$ and the VCOV (J $\times$ J) for $\underline{e}_{ij}$ in its simplest form, being $\begin{bmatrix} \sigma_G^2+\sigma_{GE}^2 & & \\ \sigma_G^2 & \sigma_G^2+\sigma_{GE}^2 & \\ \sigma_G^2 & \sigma_G^2 & \sigma_G^2+\sigma_{GE}^2 \end{bmatrix}$, for compound symmetry. The compound symmetry model for the environmental VCOV implies that the genetic correlation is constant across pairs of environments, $\frac{\sigma_G^2}{\sigma_G^2+\sigma_{GE}^2}$, a not very credible model. At the other end of the spectrum, the unstructured model prescribes a different genetic correlation for each pair of environments, a model that is often too complicated given the customary limited amounts of available data. A factor analytic structure may provide a good approximation to an unstructured VCOV at the cost of fewer parameters. An illustration on the use of different VCOV models for modeling genetic correlations between environments in relation to indirect response theory is given in van Eeuwijk et al. (2001), while Malosetti et al. (2004) compare the same range of VCOV models to model residual polygenic variation in QTL modeling.

Modeling of the mean

In the section on modeling VCOVs, it was assumed that during the comparisons of different models for the VCOV, the model for the mean remained the same. This constancy of the VCOV model allowed the use of deviance tests for comparing VCOV models. It is advisable to start with a model for the mean that is on the large side, containing possibly too many parameters (for example, too many factorial interactions). After having determined an adequate model for the VCOV, the next step in the modeling process involves reduction of the model for the mean. In this section, we will concentrate on finding models for the mean. In standard linear models, we can use F tests in our search for a parsimonious model for the mean. In mixed models, Wald tests can be used in very much the same way as F tests (Schabenberger and Pierce, 2002). Below, we will treat the modeling of the mean in the simplest setting, namely within the standard linear model context. An important reason to do so is notational simplicity. For balanced data, results carry over easily to mixed models, the only adaptation being the construction of slightly more complicated error terms for testing proposed reductions in the mean structure of the model. A second reason to deal with the modeling of mean under the assumption of the simplest model for the error structure is that the majority of the literature on GEI refers to that situation.

In the modeling of the mean we attempt to explain the phenotypic variation as much as possible in terms of model parameters that are indexed by either genotype or environment. We try to avoid the inclusion of double-indexed model terms, because the latter do not allow us to reduce the complexity of the table of genotype by environment means, which has undesirable consequences for both prediction and interpretation. In line with the strategy to describe the phenotype as a function of single-indexed parameters, multiplicative models for interaction can be used to replace the nonadditivity, GE_{ij}, by one or more product terms of the type a_i b_j: $\mu_{ij} = \mu + G_i + E_j + a_i b_j$. The genotypic parameter, a_i, represents a genotypic sensitivity to the environmental characterization, b, with value b_j in environment j. Two main types of multiplicative models can be distinguished depending on whether a_i, b_j, or both are known constants (measured variables) or parameters. Multiplicative models containing explicit genotypic and/or environmental information are called factorial regression models (Denis, 1988, 1991; van Eeuwijk et al., 1996). To give an example, when environmental characterizations such as temperature or rainfall are measured, genotypic sensitivities to these environmental variables can be estimated. The model $\mu_{ij} = \mu + G_i + E_j + a_i \, \mathrm{Rain}_j$ is an ordinary regression model, and estimation and testing proceed as dictated by standard regression theory. On the other hand, when genotypic characteristics, such as earliness and water-use efficiency, (WUE) are measured, environmental characterizations can be estimated, again within the standard regression context, for example, $\mu_{ij} = \mu + G_i + E_j + \mathrm{WUE}_i b_j$. It is also possible that measurements are available for both genotypes and environments. In that case a proportionality constant remains to be estimated, for example, $\mu_{ij} = \mu + G_i + E_j + k\mathrm{WUE}_i \mathrm{Rain}_j$.

In factorial regression models GE_{ij} is replaced by regression(s) on genotypic covariables, $\sum_{s=1}^{S} x_{si}\rho_{sj}$, and/or environmental covariables, $\sum_{t=1}^{T} \beta_{ti} z_{tj}$, with ρ_{sj} being an environmental characterization matching the genotypic covariable x_s, with values x_{si} and β_{ti} being a genotypic sensitivity to an environmental covariable z_t with values z_{tj}.

Genotypic and environmental covariables in factorial regression can be both quantitative and qualitative (see van Eeuwijk et al., 1996). Introduction of qualitative covariables introduces group structure in genotypes and/or environments. Quantitative genotypic covariables that are always candidates for explicit description of GEI are phenological characterizations, resistance and tolerance indicators to stress factors, and measures for vigor. Less obvious, but equally important for inclusion in factorial regression models, are probabilities for QTL genotypes, calculated from flanking markers. These QTL genotype probabilities can be included in models for QTL by environment interaction (QEI), so that models for QEI become nested within factorial regression models for GEI. In a later section, we will elaborate an example of factorial regression for the description of QEI. Qualitative genotypic covariables that are candidates for incorporation in factorial regression models for GEI are maturity classes, geographical origin, marker alleles, and marker genotypes, etc. Environmental covariables that merit consideration are climatological and soil characterizations, with an emphasis on indicators of biotic and abiotic stress factors. When, in addition to the response that is modeled, other phenotypic response variables have been measured in the same trials, genotypic and/or environmental means of these other responses can act as covariables, too. For example, measurements on precocity might be averaged across environments and introduced as a genotypic covariable in a model for the GEI in yield.

Factorial regression models are useful for verifying ideas about genotypic and environmental variables underlying GEI. Depending on the VCOV, testing individual covariables for incorporation in the model may proceed by Wald tests or F tests. The problem is how to select covariables from a large set of possible candidates. Some answers to this question are given by variable subset selection procedures (Denis, 1988). Another possibility is to screen covariables by including them passively in biplots of GEI for response variables such as yield, see below. The best approach is to avoid statistical choices between covariables by using physiological knowledge (Voltas et al., 1999a, 1999b). However, this last option has only limited applicability, because prior knowledge on variables underlying GEI is usually restricted. Therefore, the graphical approach via biplots seems the most general and promising approach. This approach is tightly connected to another class of multiplicative models for interaction, the bilinear models (Denis and Gower, 1994, 1996).

In contrast to factorial regression models, where either genotypic sensitivities or environmental characterizations need to be estimated, in bilinear models both these elements require estimation. One way of understanding bilinear models is in terms of hypothetical environmental variables that are constructed such as to make genotypes differ maximally in sensitivity to these variables. Bilinear models derive their name from the fact that upon fixation of the row parameters, the models become linear in the column parameters, while upon fixation of the column parameters, the models become linear in the row parameters. Bilinear models are especially useful in the exploratory analysis of GEI, thanks to their close connection with biplots, graphical displays of interaction patterns. Factorial regression models can test more firmly hypotheses generated by the application of bilinear models. (Note that the random counterpart of the bilinear models for the mean in the current section are the factor analytic models for the VCOV in the previous section.)

A popular linear–bilinear model is $\mu_{ij} = \mu + G_i + E_j + \sum_{m=1}^{M} \gamma_{mi}\delta_{mj}$, where the nonadditivity, GE_{ij}, is replaced by a sum of products of genotypic sensitivities and hypothetical environmental variables. Introduced by Gollob (1968), and later elaborated by Gabriel (1978), Gauch (1988) popularized the model and introduced the acronym AMMI, from additive main effects and multiplicative interaction effects model. Besides the AMMI model, another well-known bilinear model is the row regression, or regression on the mean model (Yates and Cochran, 1938; Finlay and Wilkinson, 1963), $\mu_{ij} = \mu + G_i + \beta_i E_j$, where the environmental main effect, E_j, represents a biological measure for the environment, and β_i a genotypic sensitivity. The phenotypic expressions are expressed as a set of converging, diverging, and intersecting straight lines with intercepts $\mu + G_i$ and slopes β_i. GEI is represented in the row regression model by heterogeneity of the slopes. In an additive model all slopes are equal to one. A slight reformulation of the regression on the mean model, imposing an additional, and unnecessary, sum-to-zero constraint on the slopes, gives $\mu_{ij} = \mu + G_i + E_j + \beta^*_i E_j$. From this formulation, it is easily seen how the regression on the mean model is nested within the AMMI model. The row regression model can be extended to have more than one multiplicative term: $\mu_{ij} = \mu + G_i + \sum_{m=1}^{M} \gamma_{mi}\delta_{mj}$.

The environmental equivalent of the regression on the mean model is known as sites regression model (Crossa and Cornelius, 1997), $\mu_{ij} = \mu + E_j + \sum_{m=1}^{M} \gamma_{mi}\delta_{mj}$. A last category of bilinear models, the shifted multiplicative models, consists of an intercept term, μ, followed by M product terms: $\mu_{ij} = \mu + \sum_{m=1}^{M} \gamma_{mi}\delta_{mj}$. Site regression models and shifted multiplicative models are frequently used to cluster environments (Crossa and Cornelius, 2002).

Assuming independent normal errors with constant variance, estimation of parameters in bilinear models for complete genotype by environment tables is relatively easy. The estimation procedure consists of two steps. First, for row and column regression models, and AMMI models, row and/or column main effects can be fitted in the standard least-squares sense, that is, calculating row and/or column means and then deducing the general mean. The table of residuals from the main effect(s) model is then decomposed by a singular value decomposition to obtain estimates for the multiplicative parameters (Gollob, 1968; Gabriel, 1978). For shifted multiplicative models, an exhaustive search algorithm works best, that is, to deduce a range of intercept values from the genotype by environment data and then decompose the residual table with a singular value decomposition for a given number of multiplicative terms. The estimate for the intercept term is then given by the value that minimizes the residual sum of squares (Denis and Gower, 1994, 1996).

When particular combinations of the genotype-by-environment table are missing, the above estimation procedure breaks down. An alternative estimation procedure that works for both complete and incomplete tables uses the property that fixation of the row parameters reduces the bilinear model to a standard linear model for the column parameters and vice versa. By fixing the row parameters G_i and $\gamma_{1i}\ldots\gamma_{Mi}$ in, for example,

$$\mu_{ij}=\mu+G_i+\hat{E}_j+\sum_{m=1}^{M}\gamma_{mi}\hat{\delta}_{mj}$$

, the only parameters requiring estimation are the column main effects, E_j, and the column scores, $\delta_{1j}\ldots\delta_{Mj}$. Alternatively, by fixing the column parameters E_j, and $\delta_{1j}\ldots\delta_{Mj}$, the row parameters G_i and $\gamma_{1i}\ldots\gamma_{Mi}$ become the parameters to be estimated:

$$\mu_{ij}=\mu+\hat{G}_i+E_j+\sum_{m=1}^{M}\hat{\gamma}_{mi}\delta_{mj}$$

. Starting with arbitrary values and iterating between row and column regression, least-squares estimates will be found for all the parameters. This scheme works not only for continuous data assumed to be normally distributed with constant variance, but also for data having other distributions within the exponential family, with the variance depending on the mean, like counts having a Poisson distribution. For such situations, generalized bilinear models are appropriate,

$$g(\mu_{ij})=\mu+G_i+E_j+\sum_{m=1}^{M}\gamma_{mi}\delta_{mj}$$

, in which not the mean, but a function of the mean, $g(\mu_{ij})$, is linear–bilinear in the parameters. The estimation algorithm then consists of alternating generalized linear regressions (van Eeuwijk, 1995b; Gabriel, 1998).

An important question in the application of bilinear models is the assessment of the number of multiplicative terms to retain. A simple test procedure is to translate the eigen values (squared singular values = amount of variation explained by a multiplicative term) to mean squares, by dividing the eigen values by an approximate number of degrees of freedom. Various suggestions have been made for the degrees of freedom. Gollob (1968) proposed $(I-1)+(J-1)-\mu\,(2m-1)$ degrees of freedom for the m-th multiplicative term in an AMMI model. Gollob's procedure works well when structure is easily separable from noise, that is, when the first multiplicative terms stand out from the later ones with respect to explained variation. For more sophisticated testing procedures, Crossa and Cornelius (2002) offer a detailed overview of methods for assessing rank in the various bilinear models.

In addition to testing for the number of multiplicative terms to retain within one type of bilinear model, one can test for one type of bilinear model being more appropriate than another type. For example, the regression on the mean model is nested within an AMMI model with one multiplicative term; that is, the latter model reduces to the first model when $\delta_j = E_j$. An approximate F test would test the difference in residual sum of squares between regression on the mean model and AMMI model divided by $J-2$ degrees of freedom against an error estimate that could be based on the deviations from AMMI model. See van Eeuwijk et al. (1996) and Crossa and Cornelius (2002) for a more general treatment of this problem.

Biplots

A major application of bilinear models in plant breeding involves the graphical exploration of similarities and dissimilarities between genotypes and environments with respect to GEI in biplots (Kempton, 1984; van Eeuwijk, 1995a). Biplots help in grouping genotypes with similar (parallel) responses across environments, as well as grouping environments that elicit similar responses in genotypes.

The relation between bilinear models and (planar) biplots stem from the equivalence between the expression for the approximation of the nonadditivity term GE_{ij} by a sum of multiplicative terms, $\gamma_{1i}\delta_{1j} + \delta_{2i}\sigma_{2j}$, and the inner product between the two-dimensional score vectors $\gamma_I = \begin{pmatrix}\gamma_{1i}\\ \gamma_{2i}\end{pmatrix}$, and $\sigma_j = \begin{pmatrix}\delta_{1j}\\ \delta_{2j}\end{pmatrix}$, for genotype i and environment j, respectively (Gabriel, 1971). The inner product $(\gamma_i, \delta_j) = |\gamma_i||\delta_j|\cos(\gamma_i,\delta_j)$ = sign $|\gamma_i||$ projection of δ_j on $\gamma_i|$ = sign$|\delta_j||$ projection of γ_i on $\delta_j|$, with sign as +1 for vectors γ_i and σ under acute angles and −1 under obtuse angles. $|\ |$ stands for length (square root out of the sum of the squared elements), while cos is cosine. The bilinear fit to the nonadditivity, GE_{ij}, can be read off from a biplot as the projection of the vector for genotype i on the vector for environment j times the length of the latter vector.

When a set of I genotypes are evaluated in two environments, the relation between the performance in both environments can be visualized by a scatter plot, with genotypes portrayed as points and environments determining the axes. The performance of individual genotypes can be found through orthogonal projection of genotypic points on the environmental axes. For evaluations in J environments, the biplot is the natural multidimensional generalization of the ordinary two-dimensional scatter plot. Genotypes are still depicted as points in a plane, with coordinates $(\gamma_{1i}, \gamma_{2i})$, and their performance in individual environments is still given by orthogonal projection on environmental biplot axes, whose direction is given by $(\delta_{1j}, \delta_{2j})$. To help interpretation, scale marks can be added to the environmental biplot axes (Gower and Hand, 1996, chapter 2; Graffelman and van Eeuwijk, 2005).

Interpretation rules for a biplot are as follows: Genotypes that appear close together exhibited similar behavior over the test environments, while genotypes far apart exhibited dissimilar behavior. Genotypes close to the origin behaved in an additive fashion, that is, they showed no adaptive behavior to any specific environment, although they may be broadly adapted. In contrast, genotypes that are located far from the origin were relatively well adapted to some environments and badly adapted to other environments. The amount of interaction, an indication for stability, is proportional to the distance from the origin. The cosine of the angle between genotypic vectors approximates the "correlation" between the responses of the genotypes. The extremes are zero degree angles, or coinciding vectors, reflecting perfect positive correlations; 90-degree angles, or orthogonal vectors, reflecting zero correlations; 180-degree angles, or opposed vectors, reflecting perfect negative correlations. Because this type of analysis departs from a fixed model, the term correlation should be interpreted loosely.

The interpretation of the behavior of environments with respect to interaction is similar to that of genotypes described above. For example, environments whose markers are close together elicited similar adaptive responses. Environments far apart elicited dissimilar responses, etc.

A GEI biplot gives a rank two approximation to the nonadditivity, GE_{ij}. It is useful to know how accurate the approximation is for the whole of the table and for individual genotypes and environments. For the whole of the table, one can look at the ratio of the first two eigen values to the total sum of squares for interaction. For individual genotypes and environments, a measure for the quality of representation is the ratio of the sum of squares in the biplot plane, $\gamma_{1i}^2+\gamma_{2i}^2$ and $\delta_{1i}^2+\delta_{2i}^2$, respectively, to the total sum of squares for that genotype or environment, $\sum_{k=1}^{K}\gamma_{ki}^2$ and $\sum_{k=1}^{K}\delta_{kj}^2$, respectively, with K the minimum of $I-1$ and $J-1$.

Biplots can be enriched with directions of greatest change for additional, explicit genotypic and environmental covariables. The imposition of additional variables is a special case of what is called the addition of passive or supplementary points, points that have zero weight in the analysis (Gower and Hand, 1996). Use the coefficients of the regression of genotypic covariable, x with values x_i, on the genotypic scores, γ_1 and γ_2, with values γ_{i1} and γ_{i2}, respectively, to define the coordinates for a point representing that genotypic covariable. Projection of a genotypic point on the line determined by the covariable point approximates the value of the projected genotype on the portrayed genotypic covariable. For environmental covariables an analogous procedure can be followed, starting with regressing an environmental covariable, z, on the environmental scores, σ_1 and σ_2. Examples of the use of biplots for the screening of measured genotypic and environmental covariables with respect to a possible role in GEI are given by Crossa et al. (1999), Vargas et al. (1999), and Voltas et al. (2002).

The exploration of GEI by biplots is closely linked to the AMMI model with two multiplicative terms. Another interesting biplot is connected to the use of the sites regression model. The biplot corresponding to the sites regression model is equivalent to the classical principal components biplot for the situation where we interpret the environments as variables and the genotypes as objects or experimental units (Gabriel, 1971). Instead of exploring GEI, we explore genotypic main effects plus GEI, the part of the phenotypic response of most importance to breeders. For the possibilities of this type of biplot, which is also called GGE biplot, see Yan and Kang (2003).

Example 1: Mixed factorial regression for modeling QTL and QTL × E—yield in barley

Factorial regression models do provide a convenient framework for modeling QTL expression across environments. A genome scan for QTL main effects could consist of fitting the following factorial regression model at a number of evaluation positions along the chromosomes:

$\underline{P}_{ij} = \mu + x_i\rho + \underline{G}_i^* + E_j + \underline{GE}_{ij} + \underline{e}_{ij}$. The term x_i stands for a genetic covariable or predictor that is obtained as a linear function of QTL–genotype probabilities. These QTL–genotype probabilities are calculated from flanking marker genotypes. The parameter ρ is the corresponding putative QTL main effect. $\underline{G}_i^*$ is a random residual genotypic effect. The remaining terms have their familiar meaning.

The QTL main effect and the environmental main effect are chosen fixed. For the QTL main effect, only one parameter is fitted, and, for that reason, it seems preferable to interpret this term as a fixed term. In most QTL studies involving multiple environments, the number of environments is modest (<10), which favors a "fixed" interpretation of the environmental main effect. The other model terms pertain to enough parameters or levels to make a random interpretation attractive.

As a test statistic for QTL detection, the Wald statistic can be used. Some form of multiple-testing correction will be necessary, whether in the form of type I error control by a Bonferroni procedure or in the form of control of the false discovery rate (Storey and Tibshirani, 2003). The most complicated part in the application of mixed factorial regression for QTL analysis may seem to be the construction of the genetic predictors, x_i. To this purpose, Jiang and Zeng (1997) developed an algorithm that covers all common QTL designs. For some standard designs, the textbook by Lynch and Walsh (1998) gives sufficient details.

Screening for QEI in addition to screening for QTL main effects can be accomplished by extending the above model with the QEI term $x_i\rho_j$, where x_i is the genetic predictor described earlier and ρ_j represents the deviation from the main effect of QTL expression in environment j. The mixed factorial regression model will read

$\underline{P}_{ij} = \mu + x_i\rho + \underline{G}_i^* + E_j + x_i\rho_j + \underline{GE}_{ij}^* + \underline{e}_{ij}$. Similar to the partitioning of the genetic main effect, G_i, into a part due to QTL main effect expression, $x_i\rho$, and residual genetic variation, $\underline{G}_i^*$, $\underline{GE}_{ij}$, is partitioned into a part due to QEI, $x_i\rho_j$, and a residual, $\underline{GE}_{ij}^*$. Another Wald statistic can be used to test for QEI.

To achieve higher transparency and flexibility in modeling, the QTL main effect and QEI are often combined into one fixed term, which effectively means that for each environment a separate QTL effect is estimated at a given evaluation position. In the same vein, the residual genotypic main effect and GEI are modeled together as one random term with a VCOV model that allows heterogeneity of variance and correlation between environments. The factor analytic model, with one multiplicative term and residual heterogeneity of variance dependent on the environment, is often a good choice (Malosetti et al., 2004). More formally, for the environments j and j^*, the VCOV for the compound random term $\underline{G}_i^* + \underline{GE}_{ij}^* = (\underline{G} + \underline{GE})_{ij}^*$ has entry $\lambda_j\lambda_{j^*}$ for $j \neq j^*$, and $\lambda_j^2 + \sigma_{d_j}^2$ for $j = j^*$.

A further step in modeling QEI is to regress the environment-specific QTL effects, ρ_j, on an environmental covariable, z with values z_j, leading to the model $\underline{P}_{ij} = \mu + x_i\rho + \underline{G}_i^* + E_j + \kappa x_i z_j + x_i\underline{\rho}_j^* + \underline{GE}_{ij}^* + \underline{e}_{ij}$ (van Eeuwijk et al, 2002; Malosetti et al., 2004). The term $\kappa x_i z_j$ reflects the part of the QEI that can be described by the regression on the environmental covariable z, while $x_i\rho_j^*$ stands for residual QEI that cannot be ascribed to a dependence on the environmental factor z. The proportionality constant, κ, provides an immediate opportunity to predict GEI for new environments conditional on knowledge of QTL (marker) alleles and measurements on the environmental variable z (van Eeuwijk et al., 2004).

QEI was studied for yield data from the North American Barley Genome project. One hundred fifty doubled haploid lines developed out of a cross, Steptoe × Morex, were evaluated in 10 trials across the United States and Canada in the years

1991 and 1992. Malosetti et al. (2004) report on analyses of QEI using the above introduced mixed factorial regression models. They compiled meteorological data for the trials from records of nearby weather stations. The environmental covariables that were finally screened for incorporation in the model for QEI were minimum and maximum temperature, temperature range, rainfall, ratio of evapo-transpiration to rain, and days of growing degrees for three growth stages, referred to as vegetative, heading, and grain filling. An initial genome scan was done using the model $\underline{P}_{ij} = \mu + x_i \rho_j + E_j + \underline{e}_{ij}$, where the random term $\underline{e}_{ij}$ contained residual genotypic main effects, residual GEI, and plot error. The analysis was conducted on a table of genotype by environment means. As a VCOV model for $\underline{e}_{ij}$, the factor analytic model described above was chosen. A Wald test for QTL main effects plus QEI was performed jointly. The initial genome scan was equivalent to a simple interval-mapping exercise. As a result, a number of chromosome positions with putative QTLs were identified. Restricted composite interval mapping was then carried out by scanning the genome again under the inclusion in the model of genetic predictors for QTLs at chromosomes (that is, cofactors) other than the chromosome under evaluation; $\underline{P}_{ij} = \mu + \sum_{c \in C} x_c \rho_{cj} + x_i \rho_j + E_j + \underline{e}_{ij}$, with $\sum_{c \in C} x_c \rho_{cj}$ representing the cofactor set. Various QTLs were found. The QTL on chromosome 2 (2H) showed strong QEI, which could be explained by a regression on the temperature range during heading (Figure 11.1). Individual Steptoe alleles increased yield by 0.112 ton/ha (SE = 0.018) for every degree Celsius with which the temperature range at heading increased. These Steptoe alleles apparently were involved in an adaptive response against extreme daily temperature variations. The QTL main effect at this position was less important than the QEI, because it offered an advantage of only 0.067 ton/ha (SE 0.043) for a Steptoe allele. The covariable explained almost all of the QEI, since inclusion of the covariable left only one environment with significant residual QEI.

Example 2: Graphical analyses of genotype by environment data—yield in rice

The biplot is a powerful graphical tool to visualize the main features of genotype by environment data. The biplot is most easily understood as a multivariate generalization of the bivariate scatter plot. The utility of biplots, but also that of standard scatter plots, for uncovering data features related to adaptation and stability will be illustrated on rice yield data from The Gambia (Manneh, 2004). As part of a QTL mapping experiment in the year 2000, 104 recombinant inbred lines (RILs) of rice were exposed to four types of environmental conditions, a factorial combination of freshwater (S1) versus salt (S2) water with low (N1) versus high (N2) nitrogen input. A split-split plot design with three replicates was used, with salinity levels as main plots, nitrogen levels as subplots, and RILs as sub-subplots.

For an analysis of the phenotypic responses in a plant-breeding context, only that part of the ANOVA table that involves the RILs is relevant. Table 11.1 shows that all terms involving RILs were significant, that is, the main effects and all genotype by environment interactions. Even the three-way interaction, RIL by salinity by nitrogen, was significant. These test results point in the direction of complicated interaction patterns. Still, taking the sums of squares as criterion, the RIL main effect and the RIL by salinity interaction dominated. Table 11.2 gives a summary of the data in terms of environmental variances and correlations. According to expectation, presence of salt reduced phenotypic variation. Correlations of some magnitude existed only between the phenotypic expressions at low and high nitrogen input for freshwater and between low and high nitrogen input for salt water. The GGE biplot of Figure 11.2 replaces a matrix of scatter plots. The acute angle between S1N1 and S1N2, and the acute angle between S2N1 and S2N2, immediately show the stronger correlations for those pairs of environments. The biplot axes for the four environments contain scale marks (Gower and Hand, 1996; Graffelman and van Eeuwijk, 2005). Projecting the RILs on the environmental biplot axes emphasizes the larger variation in the freshwater environments, S1N1 and S1N2.

The biplot axes define zones that can be used to classify RILs. RILs having positive projections on all four environmental biplot axes, zone I in Figure 11.2, can be called widely adapted; they did well everywhere (area between parts above the origin of the biplot axes for S1N1 and S2N2). In contrast, RILs projecting on the negative halves of

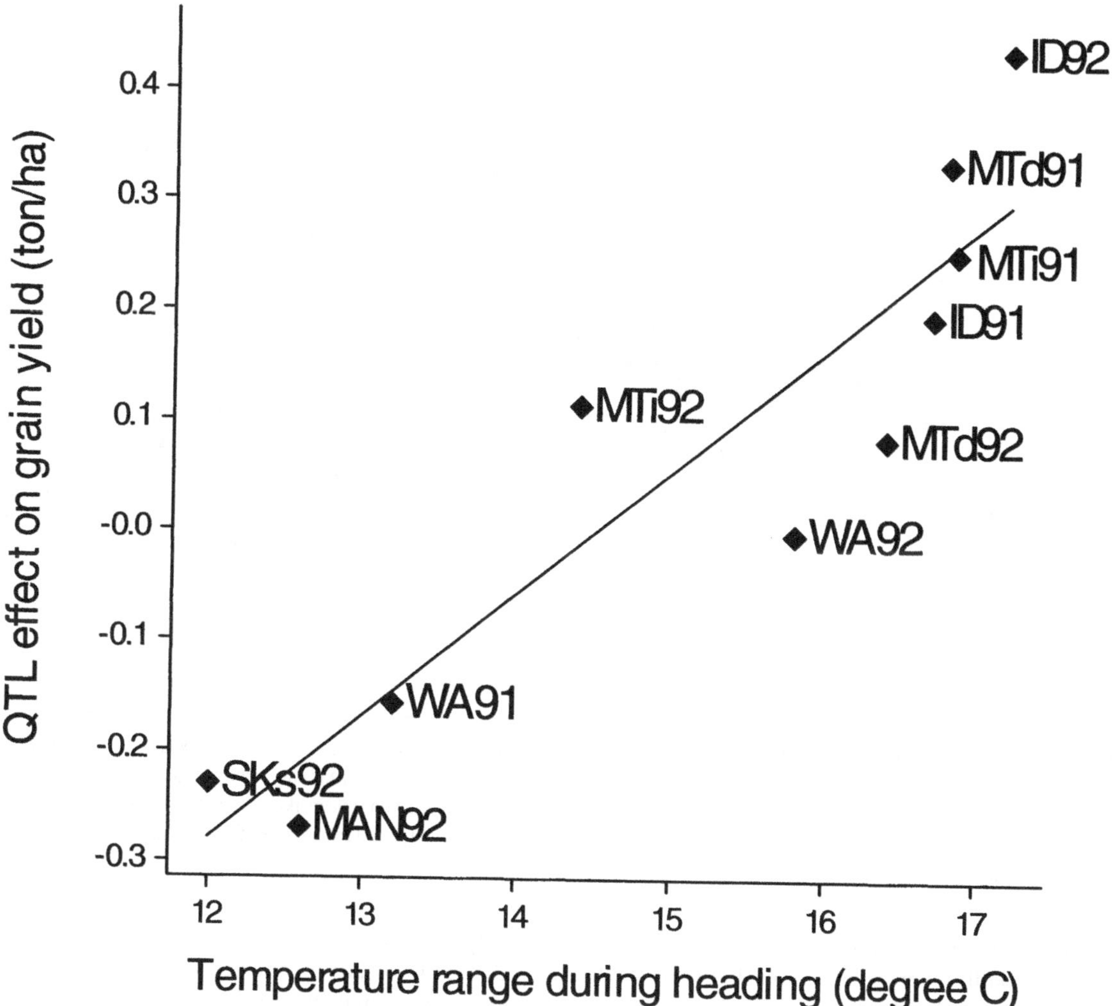

Figure 11.1 QTL effects on chromosome 2 (2H) for yield in barley against temperature range during heading for a series of trials (evaluation environments) stemming from the North American Barley Genome Project.

Table 11.1 Part of analysis of variance table for yield data of rice RILs under salinity and nitrogen stress

Source	DF	SS	MS	F	p
RIL	103	217.3	2.11	9.59	<.001
RIL.salinity	103	140.4	1.36	6.20	<.001
RIL.nitrogen	103	54.5	0.53	2.41	<.001
RIL.salinity.nitrogen	103	39.4	0.38	1.74	<.001
Residual	821(3)	180.6	0.22		

Table 11.2 Phenotypic variances and correlations for the four environments of the rice experiment

Variance			
		N1	N2
	S1	0.732	1.014
	S2	0.346	0.365
*** Correlation matrix ***			
s1n2	**0.683**		
s2n1	0.129	0.047	
s2n2	0.334	0.300	**0.524**
	s1n1	s1n2	s2n1

S1, freshwater; S2, salt water; N1, low nitrogen; N2, high nitrogen.

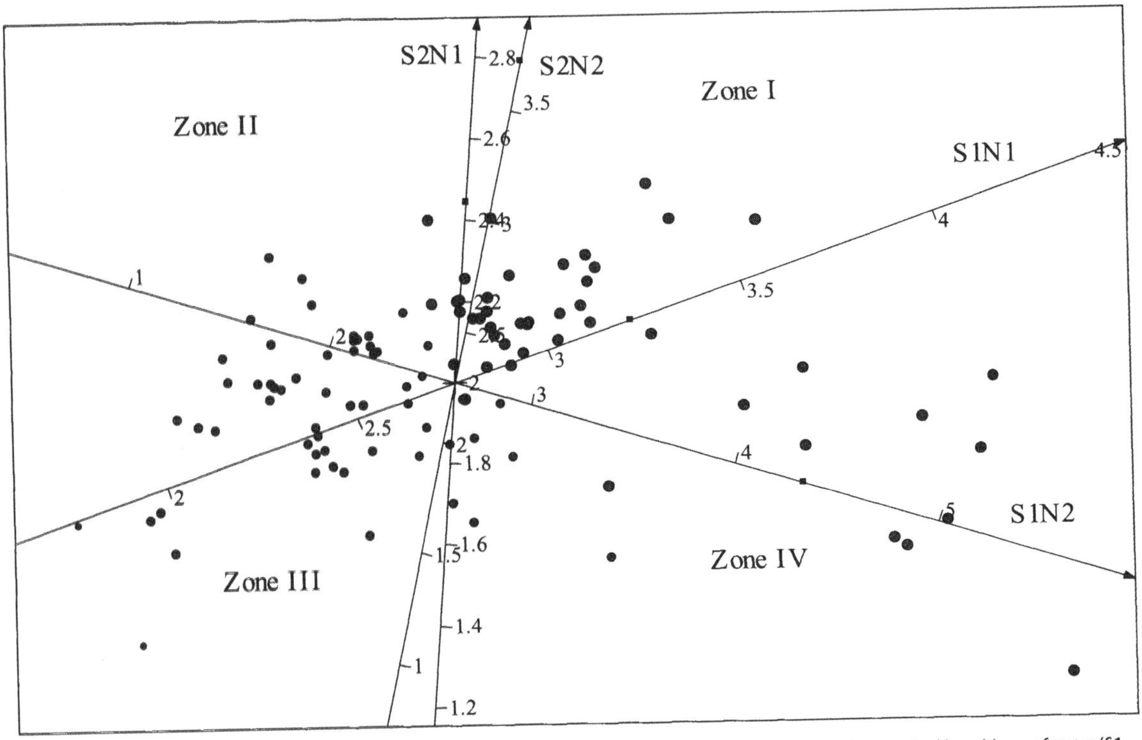

Figure 11.2 Biplot (GGE) showing yield responses of 104 rice RILs in four environments. The environments are characterized by saltiness of water (S1 = freshwater, S2 = salt water) and level of nitrogen input (N1 = low, N2 = high). Symbol size for RILs (circles) and environments (squares) is proportional to mean yield for that particular RIL or environment. Biplot axes for the environments are enriched with scale marks to facilitate interpretation. The zones I–IV represent groups of RILs with a specific adaptation pattern (see text). By means of example, for each zone some RILs with an adaptation pattern typical for that zone have been encircled.

the environmental biplot axes, zone III in Figure 11.2, did badly everywhere (area between parts below the origin for biplot axes S1N1 and S2N2). Specifically adapted RILs can be found in zone II and performed relatively well under saltwater conditions and relatively poorly under freshwater conditions (area between the part above the origin for biplot axis S2N1 and the part below the origin for biplot axis S1N2), while zone IV contains RILs that were relatively good under freshwater conditions and bad under saltwater conditions (area between part above origin for biplot axis S1N2 and part below origin for biplot axis S2N1). (In this setting, adaptation is little more than a shorthand for doing relatively well in one or more environments.) Comparing RILs from zones II and IV with each other can lead to the identification of cross-over interactions, where the superiority of one RIL compared with another is conditioned by the environment.

The biplot of Figure 11.3 is equal to that of Figure 11.2, except for the inclusion of an extra genotypic covariable and a rotation. In Figure 11.3, the biplot axis for the genotypic performance in an average environment is added, that is, an approximation to the genotypic main effect. The direction for this average environment biplot axis is most simply obtained by averaging the directions of the four individual environmental biplot axes. Alternatively, one could have regressed the genotypic main effects vector on the two genotypic score vectors, after which the regression coefficients would have provided the direction. The whole of the biplot of Figure 11.2 was rotated to make the direction of the biplot axis representing the genotypic performances in an average environment coincide with the horizontal axis. Figure 11.3 retains all interpretational features of Figure 11.2, but, in addition, the horizontal axis provides a direct means for assessing wide adaptation and the vertical axis allows an appreciation of stability (eco-valence; Wricke, 1962). The farther to the right a RIL is located, the higher the average performance of that RIL, while the farther from the

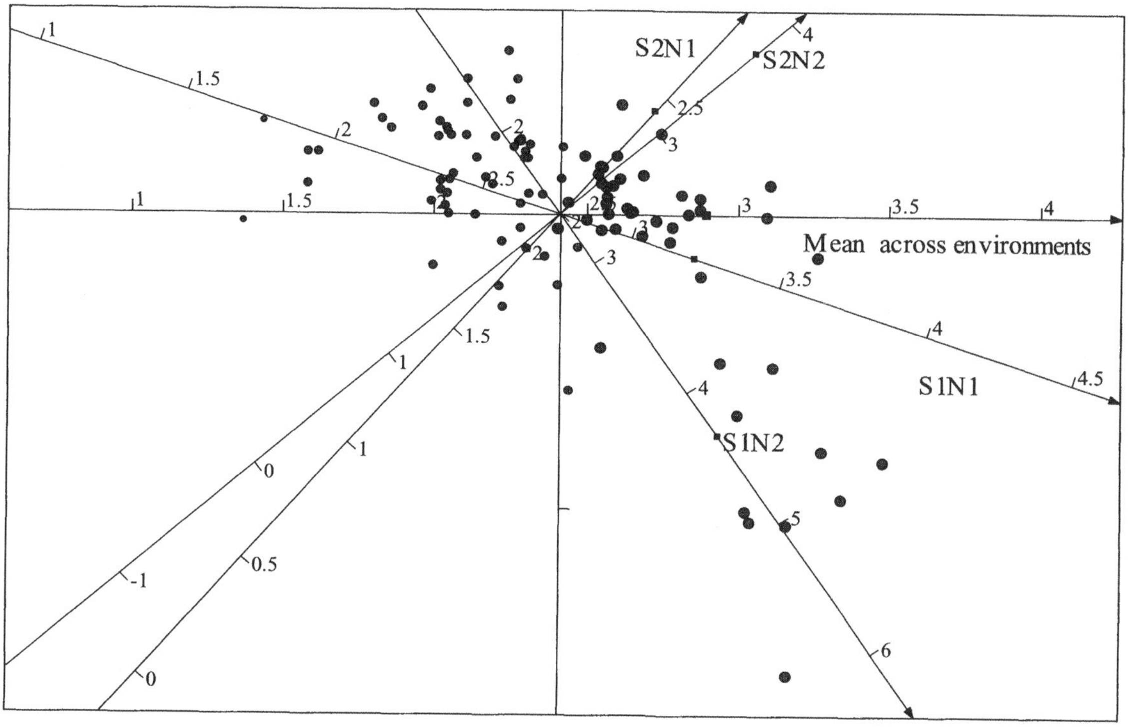

Figure 11.3 Biplot (GGE-type) for 104 rice RILs in four environments. As Figure 11.2, but with an additional genotypic covariable, mean yield across environments, along the horizontal axis. From left to right mean yield increases. The vertical axis can be interpreted in terms of stability, the farther a RIL is from the horizontal axis, the larger the genotype by environment interaction and the lower the stability.

center in the vertical direction a RIL is found, the larger the nonadditivity for that RIL.

Although the biplot of Figure 11.3 conveys the main adaptation and stability patterns underlying GEI, for this specific data set a more illuminating graphical analysis would have been possible. The ANOVA indicated that the dominant interaction was the line by salinity interaction. A scatter plot of average yield under saline conditions, averaged across low and high nitrogen input, versus average yield under freshwater conditions points to a curvilinear trend (Figure 11.4). One group of lines fared poorly under both freshwater and saline water. A second group was adapted to freshwater, but did not have tolerance to saline water conditions. Finally, a group of lines did relatively well in saline conditions, while doing reasonably well in freshwater conditions. A scatter plot illustrating the smaller nitrogen by line interaction is given in Figure 11.5, where it is apparent that growth is generally reduced under low-nitrogen conditions. The minor three-way interaction originated in the larger dispersion of the phenotypic responses at high nitrogen input that interfered lightly with the curvilinear trend of saltwater responses versus freshwater responses, that is, curvilinearity, was slightly less for high nitrogen input.

Conclusion

A philosophy for modeling phenotypic expression across environments should take into account aspects of adaptability and stability. Adaptability refers to the capacity of genotypes to react, on average, to changes in the environment. The breeding term *adaptability* is thus closely related to the mean in statistical models. Stability refers to deviations from average responses, for which no straightforward control mechanism can be envisioned. The breeding term *stability* is thus closely related to the variance in statistical models. This chapter described a framework for modeling mean and VCOV for genotype-by-environment data and tried to present tools for a better interpretation of GEI. Two examples were given to elucidate the theory. The first example showed how QTL, by environment interaction, could be described by regres-

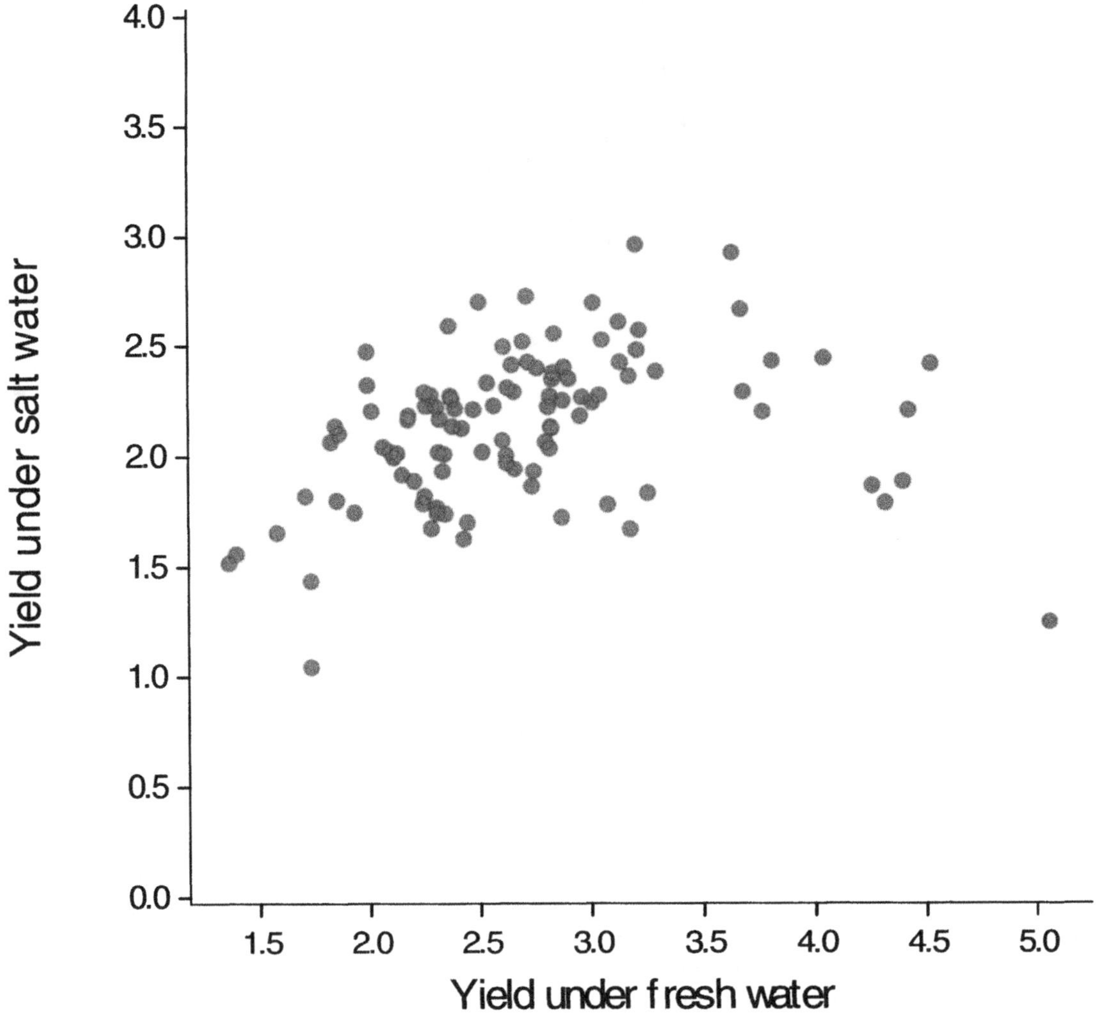

Figure 11.4 Scatter plot showing yield under saltwater conditions versus yield under freshwater conditions for 104 rice RILs. Size of RIL dots is proportional to mean yield across all four environments.

sion in a mixed-model context, where differential QTL expression was directly linked to environmental variables. This example demonstrated aspects of modeling of mean and variance. The second example served to give an impression of the power of graphical displays in analyzing GEI. The second example concentrated strongly on modeling the mean.

Acknowledgments

I would like to thank Mark Cooper (Pioneer), Hans-Peter Piepho (University of Hohenheim), and Martin Boer (Biometris) for their critical remarks on draft versions of this chapter. Marcos Malosetti and Baboucarr Manneh provided the examples and helped with insightful discussions on the interpretation of the results.

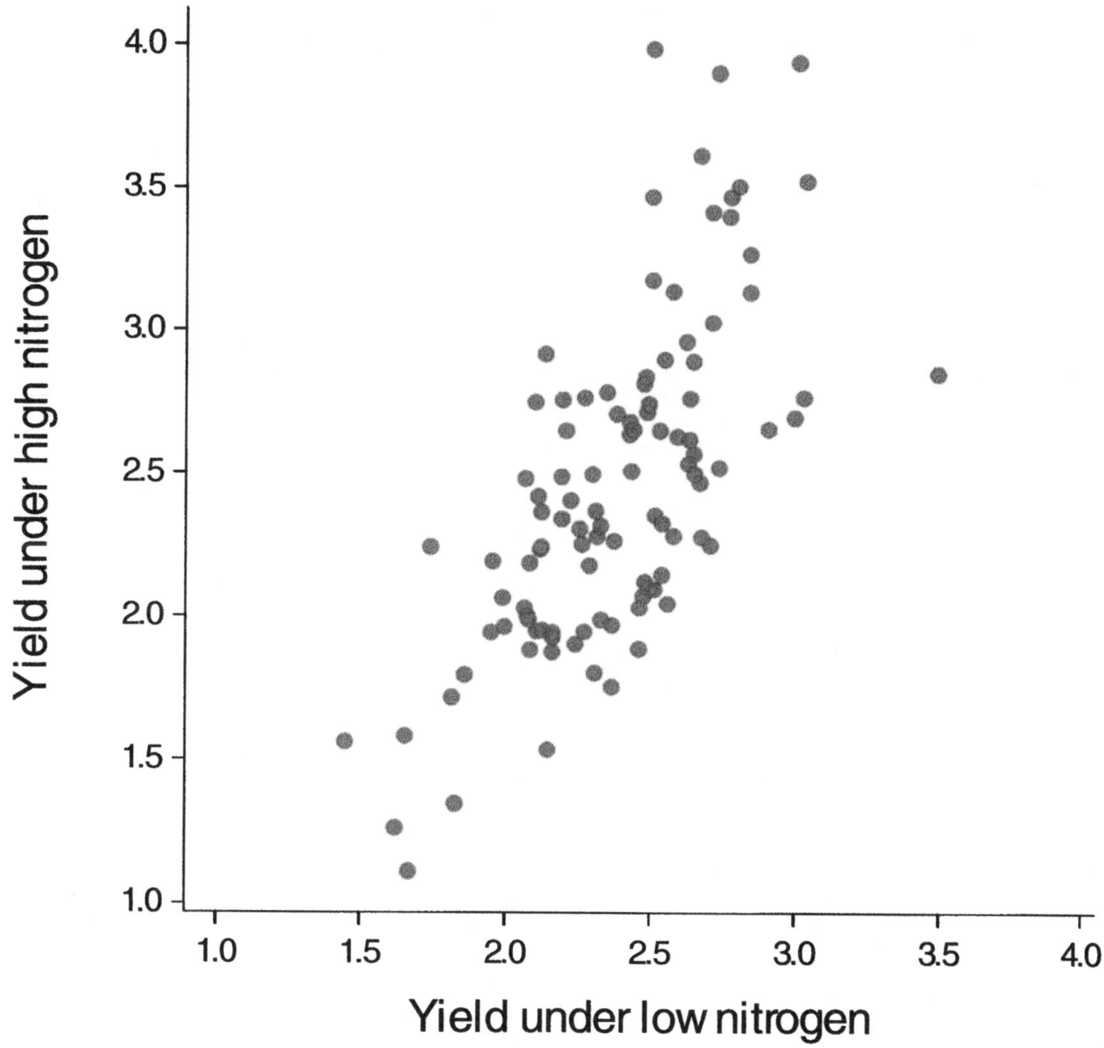

Figure 11.5 Scatter plot showing yield under high-nitrogen conditions versus yield under low-nitrogen conditions for 104 rice RILs. Size of RIL dots is proportional to mean yield across all four environments.

References

Cooper M., and G.L. Hammer. 1996. Plant adaptation and crop improvement. CAB International, Wallingford, UK.

Crossa, J., and P.L. Cornelius. 1997. Sites regression and shifted multiplicative model clustering of cultivar trials sites under heterogeneity of error variance. Crop Science 37:405–415.

Crossa, J., and P.L. Cornelius. 2002. Linear-bilinear models for the analysis of genotype-environment interaction. p. 305–322. *In* M.S. Kang ed., Quantitative genetics, genomics and plant breeding. CAB International, Wallingford.

Crossa, J., M. Vargas, F.A. van Eeuwijk, C. Jiang, G.O. Edmeades, and D. Hoisington. 1999. Interpreting genotype × environment interaction in tropical maize using linked molecular markers and environmental covariables. Theoretical and Applied Genetics 99:611–625.

Denis, J.-B. 1988. Two-way analysis using covariates. Statistics 19:123–132.

Denis, J.-B. 1991. Ajustement de modèles linéaires et bilinéaires sous contraintes linéaires avec données manquantes. Revue de Statistique Appliquée 39:5–24.

Denis, J.-B., and J.C. Gower. 1994. Asymptotic covariances for the parameters of biadditive models. Utilitas Mathematica 46:193–205.

Denis, J.-B., and J.C. Gower. 1996. Asymptotic confidence regions for biadditive models: Interpreting genotype-environment interactions. Applied Statistics 45:479–492.

Denis, J.-B., H.P. Piepho, and F.A. van Eeuwijk. 1997. Modelling expectation and variance for genotype by environment data. Heredity 79:162–171.

Finlay, K.W., and G.N. Wilkinson. 1963. The analysis of adaptation in a plant breeding programme. Australian Journal of Agricultural Research 14:742–754.

Gabriel, K.R. 1971. The biplot graphic display of matrices with applications to principal components analysis. Biometrika 58:453–467.

Gabriel, K.R. 1978. Least squares approximation of matrices by additive and multiplicative models. Journal of the Royal Statistical Society, Series B 40:186–196.

Gabriel, K.R. 1998. Generalised bilinear regression. Biometrika 85:689–700.

Gauch, H.G. 1988. Model selection and validation for yield trials with interaction. Biometrics 44:705–715.

Gogel, B.J., B.R. Cullis, and A.P. Verbyla. 1995. REML estimation of multiplicative effects in multi-environment variety trials. Biometrics 51:744–449.

Gollob, H.F. 1968. A statistical model which combines features of factor analysis and analysis of variance techniques. Psychometrika 33:73–115.

Gower, J.C., and D.J. Hand. 1996. Biplots. Chapman and Hall, London.

Graffelman, J., and F.A. van Eeuwijk. 2005. Multivariate scatter plots for exploratory analysis of relations within and between sets of variables in genomic research. Submitted.

Jiang, C., and Z.-B. Zeng. 1997. Mapping quantitative trait loci with dominant and missing markers in various crosses for two inbred lines. Genetica 101:47–58.

Kempton, R.A. 1984. The use of biplots in interpreting variety by environment interactions. Journal of Agricultural Science, Cambridge 103:123–135.

Lynch, M., and B. Walsh. 1998. Genetics and the analysis of quantitative traits. Sinauer Assoc., Inc., Sunderland.

Malosetti, M., J. Voltas, I. Romagosa, S.E. Ullrich, and F.A. van Eeuwijk. 2004. Mixed models including environmental variables for studying QTL by environment interaction. Euphytica 137:139–145.

Manneh, B. 2004. Genetic, physiological and modelling approaches towards tolerance to salinity and low nitrogen supply in rice. Ph.D. thesis, Wageningen University.

Piepho, H.P. 1997. Analyzing genotype-environment data by mixed models with multiplicative effects. Biometrics 53: 761–766.

Piepho, H.-P., and F.A. van Eeuwijk. 2002. Stability analysis in crop performance evaluation. p. 315–351. *In* M.S. Kang, ed., Crop improvement: Challenges in the 21st century. Food Products Press, Binghampton.

Schabenberger, O., and F.J. Pierce. 2002. Contemporary statistical models for the plant and soil sciences. CRC Press, Boca Raton.

Searle, S.R., G. Casella, and C.E. McCulloch. 1992. Variance components. Wiley, New York, NY.

Shukla, G.K. 1972. Some statistical aspects of partitioning genotype-environmental components of variability. Heredity 29:237–245.

Smith, A.B. 1999. Multiplicative mixed models for the analysis of multi-environment trial data. Ph.D. thesis Dept. of Statistics, The University of Adelaide, Adelaide, South Australi, 2005, Australia.

Smith, A.B., B.R. Cullis, and R. Thompson. 2001. Analyzing variety by environment data using multiplicative mixed models and adjustments for spatial field trend. Biometrics 57:1138–1147.

Storey, J.D., and R. Tibshirani. 2003. Statistical significance for genomewide studies. Proceedings of the National Academy of Sciences 100:9440–9445.

van Eeuwijk, F.A. 1995a. Linear and bilinear models for the analysis of multi-environment trials: I. An inventory of models. Euphytica 84:1–7.

van Eeuwijk, F.A. 1995b. Multiplicative interaction in generalized linear models. Biometrics 51:1017–1032.

van Eeuwijk, F.A., M. Cooper, I.H. DeLacy, S. Ceccarelli, and S. Grando. 2001. Some vocabulary and grammar for the analysis of multi-environment trials, as applied to the analysis of FPB and PPB trials. Euphytica 122:477–490.

van Eeuwijk, F.A., J. Crossa, M. Vargas, and J.-M. Ribaut. 2002. Analysing QTL by environment interaction by factorial regression, with an application to the CIMMYT drought and low nitrogen stress programme in maize. p. 245–256 *In* M.S. Kang ed., Quantitative genetics, genomics and plant breeding. CAB International, Wallingford.

van Eeuwijk, F.A., J.B. Denis, and M.S. Kang. 1996. Incorporating additional information on genotypes and environments in models for two-way genotype by environment tables. p 15–50. *In* M. S. Kang & H. G. Gauch, eds., Genotype-by-environment interaction. CRC Press, Boca Raton.

van Eeuwijk, F.A., M. Malosetti, X. Yin, P.C. Struik, and P. Stam. 2004. Modelling differential phenotypic expression. *In* "New directions for a diverse planet." Proceedings of the 4th International Crop Science Congress, 26 Sep–1 Oct 2004, Brisbane, Australia. Published on CDROM. Web site www.cropscience.org.au

Vargas, M., J. Crossa, F.A. van Eeuwijk, M.E. Ramírez, and K. Sayre. 1999. Using AMMI, factorial regression, and partial least squares regression models for interpreting genotype × environment interaction. Crop Science, 39:955–967.

Verbeke, G., and G. Molenberghs. 2000. Linear mixed models for longitudinal data. Springer, New York.

Voltas, J., F.A. van Eeuwijk, A. Sombrero, A. Lafarga, E. Igartua, and I. Romagosa. 1999a. Integrating statistical and ecophysiological analysis of genotype by environment interaction for grain filling of barley in Mediterranean areas. I. Individual grain weight. Field Crops Research 62:63–74.

Voltas, J., F.A. van Eeuwijk, J.L. Araus, and I. Romagosa. 1999b. Integrating statistical and ecophysiological analysis of genotype by environment interaction for grain filling of barley in Mediterranean areas. II. Grain growth. Field Crops Research 62:75–84.

Voltas, J., F.A. van Eeuwijk, E. Igartua, L.G. del Moral, J.L. Molina-Cano, and I. Romagosa. 2002. Genotype by environment interaction and adaptation in barley breeding: Basic concepts and methods of analysis. *In* Barley science: Recent advances from molecular biology to agronomy of yield and quality, pp. 205–241, G. Slafer, ed., Food Products Press, Binghampton.

Wolfinger, R.D. 1996. Heterogeneous variance-covariance structures for repeated measures. Journal of Agricultural, Biological, and Environmental Statistics 1:205–230.

Wricke, G. 1962. Über eine Methode zur Erfassung der ökologischen Streubreite in Feldversuchen. Zeitschrift für Pflanzenzüchtung 47:92–93.

Yan, W., and M.S. Kang. 2003. GGE biplot analysis: A graphical tool for breeders, geneticists, and agronomists. CRC Press, Boca Raton.

Yates, F., and W.G. Cochran. 1938. The analysis of groups of experiments. Journal of Agricultural Science 28:556–580.

12

Applications of Comparative Genomics to Crop Improvement

Mark E. Sorrells, Professor of Plant Breeding, Department of Plant Breeding and Genetics, Cornell University

Introduction

Comparative genomics is a broad field of research that uses sequence and map-based tools to estimate structural and functional similarity among living organisms at some level of organization. In recent years, comparative genomics has received a great deal of attention, and advances in this field have dramatically changed research strategies. The genomes of Arabidopsis and rice have already been sequenced, and plans are underway for the sequencing of several other plant genomes over the next five to seven years. Consequently, plant research is now strongly model organism-oriented and is increasingly driven by questions that can be addressed by whole genome sequence analyses and related technologies such as large-scale reverse and forward genetics, whole genome transcript/proteomics analysis, and large-scale genotyping. Plant species for which there is little genomic sequence available will likely be anchored to such model species, using comparative genomics methodologies. Elucidation of gene and genome structure–function relationships is most efficient in model species, and efficient methods of transferring that information to other species are vitally important for crop species with large complex genomes or less research support.

It is the complementarity of information available from different species that lends power to comparative analyses. This is because different species have evolved different alleles, genes with different functions, and differential gene expression. In addition, humans have shaped the evolution of some species to their benefit and in the process focused on certain traits that often differ between species. However, there are a number of traits in common across species that are fundamental to domestication, and comparative analysis of the variation in those genes can reveal much about structural and functional relationships.

Comparative genomics research has several goals: (1) to compare the organization of related genomes and infer the basic processes of genome evolution, (2) to transfer information from model species to related organisms, and (3) to integrate information on gene location and expression across species. Comparative maps based on anchor probes mapped in multiple species are a critical tool for information transfer among species. Also, consensus maps have been useful for amplifying the number of markers available for comparison, especially for species with low polymorphism. Trait dissection, integration of information about metabolic pathways, gene expression, and chromosome location facilitate the rational selection of candidate genes. Assessments of allelic diversity and relative value are required for the identification of superior alleles for genes of economic importance. This information can be used by plant breeders to assemble the best alleles into superior crop varieties. This review is intended to be a general review of comparative mapping studies and will focus on principles of comparative genomics, examples of applications to cereal crops, impacts on our concepts of genome evolution, and the use of comparative genomics for cross-species gene cloning.

Anchor probes

The purpose of using anchor probes is to identify orthologous loci across genomes of multiple species or genomes within a polyploidy species. Typically, they are cDNA clones that are sufficiently conserved that they will hybridize well to genomic DNA from species that belong to different genera or families. Single- or low-copy clones work best because they reduce the likelihood of mapping a paralogous locus. Anchor probes are developed by screening anonymous cDNA clones on "garden blots" containing DNA of the species to be used for comparative mapping. Those clones that hybridize well to the species on the garden blot are then screened for polymorphism using DNA from parents of mapping populations. Because of attrition, hundreds of clones must be screened to identify those that can be mapped in multiple species and give good genome coverage. Van Deynze et al. (1998) screened 1,800 probes on garden blots containing DNA of rice, maize, sorghum, sugarcane, wheat, barley, and oats, and 153 of them were selected as anchors. The number mapped in each species will then depend on the polymorphism between the parents of each of the populations. The cDNAs derived from etiolated leaf libraries used by Van Deynze et al. (1995) mostly coded for proteins indicative of heterotrophic activity involved in the TCA cycle or the glycolytic pathway. A comparison of the frequency of cross-hybridization on DNA from five species for cDNAs derived from barley, maize, oat, and rice showed that the cDNAs derived from oats were much more useful than the other three species (Van Deynze et al. 1998). The rice cDNAs hybridized to the fewest species even though the clones had been previously evaluated and mapped in rice.

Southern hybridization using anchor probes has long been the method of choice for evaluation of relationships among species and genera, especially for comparative mapping (Van Deynze et al., 1998). This is because PCR-based fragment amplification may be an all or none reaction (dominant), may amplify nonorthologous loci, or may inadequately sample sequence variation because of the specificity of the primers. To be useful for comparative mapping, a molecular marker must identify orthologous loci in two or more species and exhibit a sufficient level of polymorphism within a species to facilitate determination of map location. It is apparent that for PCR-based markers these criteria are in direct conflict because DNA sequence variation is essential for polymorphism whereas conservation of DNA sequence is essential for designing primers that function within and across species. However, recently expressed sequence tags (ESTs) containing simple sequence repeats (EST-SSRs) have been recognized as a valuable source of molecular markers that can identify orthologous loci across species (Scott et al. 2000; Kantety et al. 2002). DNA sequences containing conserved regions of a gene that flank a hypervariable region are most useful for designing PCR-based markers that can amplify orthologous gene fragments across species. Although microsatellite markers derived from genomic libraries are more polymorphic than those from expressed genes (Cho et al., 2000; Eujayl et al. 2001; Eujayl et al. 2002; Scott et al., 2000), genomic microsatellite markers generally will not amplify loci in a species other than the one from which it originated. Yu et al. (2003) used sequence similarity analysis to identify 156 cross-species superclusters and 138 singletons for developing primer pairs that were then tested on the genomic DNAs of four grass species: barley, maize, rice, and wheat. Primer pairs for 141 superclusters and 128 singletons produced PCR amplicons, and 228 primers amplified DNA from two or more species. Like anchor probes, EST-SSRs can also be useful for identifying orthologous loci in the different genomes of polyploids. Mapping multiple loci from a probe or EST-SSR aids in the identification of homoeologous chromosomes.

Consensus maps

Consensus maps are important components of a comparative mapping strategy. Consensus maps bring together information on locus order from multiple maps for a particular species or from multiple genomes within a polyploid into a single comprehensive map for each chromosome (Nelson et al. 1995a, 1995b; barley consensus). This greatly amplifies the number of loci that can be used for comparison with other species maps, thus increasing the resolution of the comparative map. Construction of consensus maps can present challenges, and their reliability is directly proportional to the number of loci in common among the maps included. Computer programs have been developed to create consensus maps (e.g., Qi-X et al., 1996); however, if there are sufficient loci for alignment, two maps can be merged manually. Consensus maps are particularly useful for species with low polymorphism because parents of different populations are, to some extent, complementary as to which probes are polymorphic. In addition, polyploid species such as wheat often have genomes that are closely related, and the individual genome maps can be merged into a consensus map for the species. Consensus maps are complicated by probes identifying multiple loci within a genome, and different loci may be polymorphic in different populations. The order of loci on a consensus map that are situated between common loci is, at best, an estimate. However, for low-resolution comparative maps, small differences in locus order on a consensus map are inconsequential.

Comparative maps of Gramineae

For the domesticated grasses, the conserved linkage blocks and their relationships with rice linkage groups have led to hypotheses about the basic organization of the ancestral grass genome (e.g., Gale and Devos, 1998; Moore et al., 1995; Wilson et al., 1999) and have provided impetus for subsequent investigations examining conservation in more detail. Comparative maps, like many biological phenomena, become more complex as we delve further using more sophisticated tools and techniques. Comparisons of genetic linkage maps are limited in their resolution by the number of orthologous loci and by population sizes. Early comparative maps (e.g., Ahn and Tanksley, 1993; Hulbert et al. 1990; Gale and Devos, 1998) greatly underestimated the complexity of genome relationships. Later studies using higher-density maps (Wilson et al., 1999) and large-scale genomic sequencing (Chen et al., 1997; Tikhonov et al., 1999) revealed more rearrangements. Wilson et al. (1999) described a higher-resolution rice and maize (*Zea mays* L.) comparative map that detailed more than 20 rearrangements, including chromosome duplications, inversions, and translocations. Those maps were based almost entirely on RFLP analyses, and their resolution was well below what is required for microcolinearity assessment.

It has become apparent that the use of DNA sequence-based comparative genomics for evolutionary studies and for transferring information from model species to related large-genome species is critical for molecular genetics and crop-improvement strategies. The use of comparative sequence analysis methods to cross-reference genes between species makes it possible to greatly enhance the resolution of comparative maps, study gene evolution patterns, identify conserved regions between the genomes, and facilitate interspecies gene cloning. Devos et al. (1999) compared rice ESTs with the Arabidopsis genome sequence and assessed the colinearity between these model representatives of the monocot and dicot subclasses of flowering plants. Their comparisons of two regions of up to 3 cM on Arabidopsis chromosome 1 and rice ESTs with homology to Arabidopsis genes from 10 bacterial artificial chromosome (BAC) clones revealed little conservation, even from regions containing closely linked genes in one of the species. Sorrells et al. (2003) compared 4485 ESTs that were physically mapped in wheat chromosome bins to the public rice genome sequence data from 2251 ordered BAC/PAC (P1-derived artificial chromosome) clones using BLAST (Basic Local Alignment Search Tool). A rice genome view of homoeologous wheat genome locations based on comparative sequence analysis revealed numerous chromosomal rearrangements that will significantly complicate the use of rice as a model for cross-species transfer of information in nonconserved regions (Figure 12.1).

The structural relationships between the genomes indicate that for most individual rice chromosomes there is a preponderance of wheat genes from one or two wheat homoeologous groups. Most of these genome relationships were apparent from earlier RFLP-based comparative maps (Kurata et al. 1994; Van Deynze et al., 1995a, 1995b, 1995c; Sarma et al., 2000). Features of the rice–wheat genome relationship revealed by this analysis compared with the RFLP-based maps include a high frequency of breakdown in colinearity throughout the genomes, and localized homoeology between the genomes not previously reported. The inverse view showing the relationship between the wheat deletion map and rice genomic sequence location revealed partial conservation of gene content and order at the resolution conferred by the chromosome deletion breakpoints in the wheat genome (Table 12.1). Examples of the most- and least-conserved chromosome relationships between wheat and rice are illustrated by wheat chromosomes 3 and 5, respectively. Wheat chromosome 3 and rice chromosome 1 have many genes in common. However, using only single-copy genes, even deletion bins in the

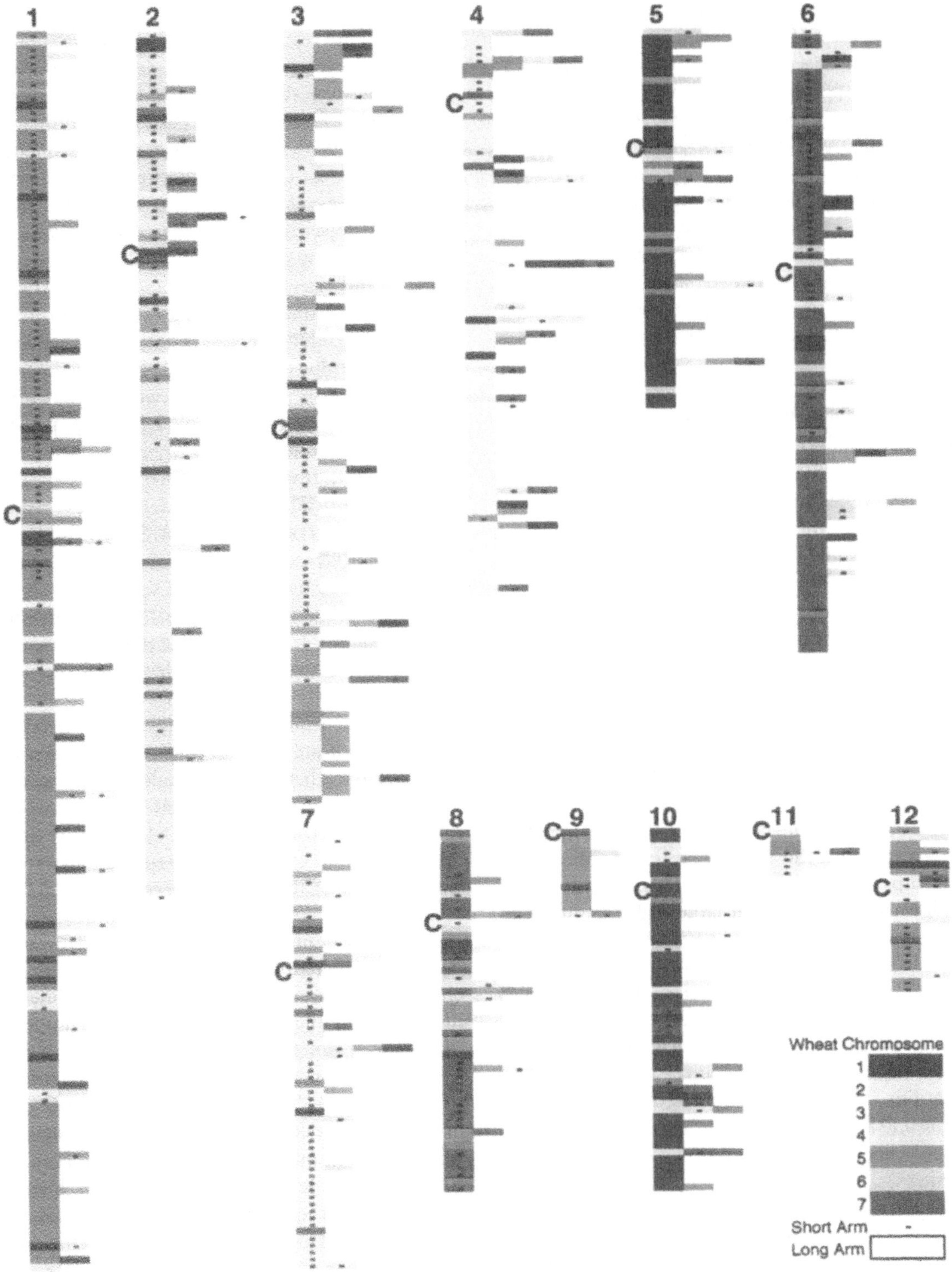

Figure 12.1 Rice–wheat genome relationships. Rice genome view showing the wheat chromosome arm location for the most similar wheat gene sequences. Each colored box represents a rice–wheat gene sequence match at ≥ 80% identity. When the wheat EST mapped to more than one wheat chromosome, the other color-coded locations are positioned adjacent to the first. Homoeologous wheat chromosome locations are grouped together. Rice BAC/PAC sequences that did not match any wheat sequence, as well as redundant matches, are omitted. The rice centromere location is indicated by "C."

Table 12.1 Wheat–rice genome relationships for wheat chromosomes 3 and 5

Wheat deletion bin name	Rice Chromosome 1	2	3	4	5	6	7	8	9	10	11	12	Number of ESTs with no Signif- cant hits	Total number of ESTs mapped
3AS4-0.45-1.00	28	5	1	1		2	2						52	91
3AS2-0.23-0.45	7		2									1	11	21
C-3AS2-0.23	6										1		4	11
C3A	1												3	4
C-3AL3-0.42	29						1						16	46
3AL3-0.42-0.78	36		1	1	1	1	1	1		1			25	68
3AL5-0.78-1.00	22		1			3	3			1	1		52	83
3BS8-0.78-1.00	7	1	1			2							26	37
3BS9-0.57-0.78	12	2		2									26	42
3BS1-0.33-0.57	17	1	1									1	21	41
C-3BS1-0.33	14		2								1		24	41
C3B	1												1	2
C-3BL2-0.22	30			1			1	1					19	52
3BL2-0.22-0.50	25			1		1		1		1			28	57
3BL10-0.50-0.63	11		1		1		1				1		11	26
3BL7-0.63-1.00	31		1			1	2			2			55	92
3DS6-0.55-1.00	21	3		1		1							42	68
3DS3-0.24-0.55	25	1	3									1	35	65
C-3DS3-0.24	6		1			1	1				1		7	17
C3D	2												1	3
C-3DL2-0.27	34						1	1					16	52
3DL2-0.27-0.81	54		1	1		1		1		1	1		55	115
3DL3-0.81-1.00	9		1			2	1			1			39	53
5AS7/10-0.98-1.00													1	1
5AS3-0.75-0.98	3	2		1						1		4	28	39
5AS1-0.40-0.75												8	13	21
C-5AS1-0.40	1	1		1				1		1		2	14	21
C5A													4	4
C-5AL12-0.35								1					15	16
5AL12-0.35-0.57	1	2	1			3		2	7		1	1	29	47
5AL10-0.57-0.78	1		8	1			3	1	2	1	1		44	62
5AL17-0.78-0.87			7				2						26	35
5AL23-0.87-1.00	1		6	1	2	1		4					23	38
5BS6-0.81-1.00	3	1								1			15	20
5BS5-0.71-0.81		1											12	13
5BS8-0.56-0.71		1										3	9	13
5BS4-0.43-0.56									1			6	19	26
C-5BS4-0.43										1			10	11
C5B													3	3
C-5BL6-0.29						2			1				14	17
5BL6-0.29-0.55		1						1				2	17	21
5BL1-0.55-0.75		3	1				1	2	6	1	1	1	24	40
5BL14-0.75-0.76							1		1				17	19
5BL9-0.76-0.79		1	12	1			1			2	1		32	50
5BL16-0.79-1.00	4		14	1	1		3	3		1			50	77
5DS2-0.78-1.00	4	1											18	23
5DS5-0.67-0.78		1											4	5
5DS1-0.63-0.67												3	4	7
C-5DS1-0.63		1		1						1		7	25	35
C5D		1											4	5
C-5DL1-0.60		1		1		3		3	6	1		4	58	77
5DL1-0.60-0.74	2	1		1			2	1	8				30	45
5DL9-0.74-0.76			1						1		1		7	10
5DL5-0.76-1.00	2		23	2			4	2		5			79	117

most conserved regions often contained sequences from more than one rice chromosome. This suggested that there has been an abundance of rearrangements, insertions, deletions, and duplications that, in many cases, will complicate the use of rice as a model for cross-species transfer of information in nonconserved regions. While occasional artifacts may arise from using "the best hits" between wheat ESTs and rice genomic sequence, the high stringency used in their study tends to reduce such errors. This DNA sequence-based comparative map effectively increases the resolution by 30- to 40-fold versus RFLP-based maps and provides a much better estimate of the shortest conserved evolutionary unit sequence (SCEUS) (O'Brien et al., 1993). Although it appears that different regions of the grass genomes evolve at different rates (e.g., Akhunov et al., 2003), earlier estimates of the average number of structural changes (0.14) per million years of divergence (Paterson et al., 2000) may have been an underestimate. However, the enhanced resolution afforded by comparative DNA sequence analysis for wheat and rice, especially in conserved regions, will facilitate the selection of markers for saturation mapping a wheat chromosome region and for selecting candidate genes, both of which are important for developing functional molecular markers and for understanding triticeae gene evolution.

The sequence-based comparisons between wheat and rice genomes described above, as well as recent studies of the Indica and Japonica rice subspecies (Feng et al., 2002) and maize inbreds (Fu and Dooner, 2002), indicate that grass genomes may be more labile than previously thought. Gaut (2002) recently reexamined the evolution of grass genomes with respect to their phylogeny. In the review, he reanalyzed published comparative map data as well as comparative sequence analyses. Gaut (2002) also used a phylogenetic analysis of the grasses (Kellogg, 2000) to illustrate the conceptual problems with considering rice as an ancestral genome despite its small size and simple structure. He concluded that grass genomes are evolutionarily labile for many characteristics, including genome size and chromosome number, and that the current colinearity paradigm is in need of reassessment. Feng et al. (2002) analyzed DNA sequence alignments between 2.3 Mb of three contiguous segments of chromosome 4 from the two rice subspecies, Indica and Japonica. Although there was extensive sequence colinearity, they identified 9056 single-nucleotide polymorphisms (1 per 268 bp) and 63 and 138 insertion/deletions (many in coding regions) for the Indica and Japonica sequences, respectively.

Fu and Dooner (2002) sequenced over 100 kb from the *bz1* genomic region of two different maize lines and found substantial differences between them. Retrotransposon clusters and genetic complement differed markedly between the two lines, demonstrating that genetic microcolinearity can be violated within the same species. Their results relate to the underlying genetic basis of hybrid vigor in maize, the meaning of "allelism," and the assessment of genetic distances. The implications are that lines lacking different genes would complement one another and exhibit hybrid vigor, but lines lacking most of the same genes would not complement and thus would be placed in the same heterotic group. These studies suggest that there are genomewide mechanisms effecting frequent rearrangements that characterize these genomes at the megabase level.

Domestication genes

Domestication genes are those that control traits that are important for a plant to survive in nature but that limit the value of plants to humankind. Such traits include seed shattering; inflorescence or fruit size; free threshing; lodging; certain plant, seed, or fruit colors; and others. Lin et al. (1995) reported that quantitative trait loci (QTLs) with major effects on height and flowering in sorghum have counterparts in homoeologous segments of the rice, wheat, barley, and maize genomes. Tillering in maize is one of the important traits that differentiates it from its ancestor teosinte. Doebley et al. (1995) discovered that one of the QTLs associated with tillering was involved with the control of lateral branch length and floral development. Paterson et al. (2000) proposed that QTLs affecting seed size, nonshattering of grain, and photoperiod sensitivity are likely to be orthologous in sorghum, rice, and maize based on correspondence of map position across these genera. However, there are examples of traits apparently controlled by genes that are not orthologous across species. Paterson et al. (2000) detected an occasional lack of correspondence for domestication-related QTLs across taxa. Gale and Devos (1998) hypothesized that different species could have unique genes that allow adaptation to the specific environmental conditions to which they are exposed.

Applications

Marker-assisted selection

While there are a number of theoretical papers on marker-assisted selection (MAS) in the literature (e.g., Lande and Thompson, 1990; Hospital et al., 1992; Dudley, 1993), there are few reports of successful variety or germplasm releases where molecular markers were used for selection. This may be due to the paucity of public plant-breeding programs, the high cost of MAS, or other limitations. Reports describing the use of MAS include the development of isolines or special genetic stocks for lepidopteran resistance in soybeans (Walker et al., 2002), bacterial blight resistance in rice (Singh et al., 2001), malting quality in barley (Han et al., 1997), and heading date in rice (Lin et al., 2000).

Implementation of MAS requires polymorphic markers that are tightly linked to the allele of interest. To be cost-effective, the markers should be adaptable to high-throughput detection systems. Because the molecular marker maps of most crop species are low resolution, the number of markers available is usually quite limited. This, combined with the low polymorphism typical of elite germplasm, results in markers being the limiting factor in most programs. Low-resolution comparative maps can be used to identify homoeologous regions in the model species, and DNA sequences from that species can then be evaluated for marker development. With relatively high-resolution maps, it may be possible to identify candidate genes that are responsible for the trait of interest. In any case, the low polymorphism is a limitation that is difficult to overcome without considerable time and expense. One approach to circumventing the lack of polymorphism is to clone DNA sequences close to or in the gene being transferred. The sequences from the parents are compared, and primers specific to the DNA sequence differences between the parents are designed. In the case of polyploids, this can be somewhat challenging due to the multiple copies of homoeologous regions and duplicated segments of the genome. This is because the polymorphism must be unique, not only between the parents but also genome specific. The ideal marker is a PCR-based, allele-specific assay where the primers are designed to amplify only the sequence responsible for the desired phenotype.

Another limitation of MAS is related to the fact that most traits of economic importance are controlled by several genes with small effects. These QTL are often influenced by the genetic background of the parent and interact with the environment. In general, using today's technology, if more than half of the variation for a trait can be explained by three or fewer genes, then MAS has some value. If it is controlled by more than three genes or they explain less than half of the variation in the trait, then conventional breeding and selection techniques will likely be more efficient. One strategy for countering the limitation of selecting QTL with small effects is to screen germplasm accessions for alleles that have larger effects. Essentially, this is what transformation attempts to do, that is, introduce a gene with a large beneficial effect. If the gene(s) controlling the trait are known, then accessions can be screened for variation and classified according to their alleles for association analysis. A successful search for an allele with a major effect on a quantitative trait can have a large impact on a breeding program and could facilitate the use of the marker or even direct phenotypic screening.

Positional cloning

One of the anticipated benefits of the complete rice genome sequence is its use for generating markers that can be used for high-resolution comparative mapping to facilitate positional gene cloning in grass species. Positional cloning depends on identifying markers whose genetic distance to the gene is within a large-insert clone so that a library can be screened with that marker (Tanksley et al., 1995). Thus, conservation of gene content and order at the megabase level (as well as a large-insert library in the target or closely related species) is essential for efficient use of a model species for this purpose. However, assessments of microcolinearity between rice and the Triticeae have revealed both conservation (Dubcovsky et al., 2001; Dunford et al., 1995; Yan et al., 2003) and intergenic breakages and segmental translocations (Kilian et al., 1995; Ramakrishna et al., 2002; Bennetzen and Ramakrishna, 2002). Yan et al. (2003) used accessions of *Triticum monococcum* differing in vernalization response to fine map and clone the *Vrn1* gene. They reported almost perfect colinearity for rice, sorghum, and wheat for genes in the *Vrn1* region spanning approximately 0.1 cM. Gene composition and order were also found to be conserved in the *adh1* region of maize and sorghum, but not in rice (Tikhonov et al., 1999; Tarchini et al., 2000). Duplications of loci separated by large genetic distances in different regions of the same chromosome can complicate comparative mapping, especially when polymorphism levels limit the number of fragments mapped in a given population. Gene duplication followed by sequence divergence and small translocations of single genes (Tarchini et al., 2000), multigene families (Dubcovsky and Dvorak, 1995), and the rapidly evolving nature of certain genes, such as disease-resistance genes (Leister et al., 1998; Keller and Feuillet, 2000) can all lead to rapid rearrangement of resistance-like genes and nonsyntenic distribution in cereal genomes (Leister et al., 1998). Although macrocolinearity does not always predict microcolinearity, the level of conservation can be assessed simultaneously with fine mapping. The need to evaluate microcolinearity for most situations complicates the use of model species because of the time and labor required for phenotyping and mapping a large population for fine-scale analysis.

Integration of different sources of information

Online databases (e.g., Graingenes: http://wheat.pw.usda.gov; ZMDB: http://www.zmdb.iastate.edu) are a wealth of information for germplasm, genes, and maps, and they are presenting the information in novel ways that facilitate interpretation and utilization. One of the most exciting prospects is the integration of information about genes, their expression, metabolic pathways, and agronomic phenotypes. Most metabolic pathways have been elucidated using microorganisms and model species; however, much of the information is applicable to a broad range of species. The Kyoto Encyclopedia of Genes and Genomes (KEGG: http://www.kegg.com/) has an impressive array of genomic information for microbes that is linked to metabolic pathways, regulatory pathways, and molecular assemblies. New visualization tools for cross-species analyses facilitate the transfer of information to crop species by comparative genetics. Ultimately, linking gene to phenotype is our goal.

Figure 12.2 illustrates the integration of various sources of information that allow us to identify the genes controlling a trait of interest and eventually understand their function. Given a trait of interest, we first need to know how many important genes control the trait and where they are located in the genome. QTL mapping is still the most common approach to acquiring that information, although analysis of various kinds of mutants and association analysis also contribute to that knowledge base. Once we know the approximate location of the genes, we next want to learn their function. Knowing something about the metabolic pathway that might be involved may make it possible to select a subset of candidate genes that have been previously located to that region of the genome. These may be cloned and characterized genes or ESTs that have been assigned a putative function. Supporting evidence for the candidate gene may be obtained from gene expression data or developmental specificity. Finally, once there is ample evidence for the role of a particular gene, the final but most important step is to characterize allelic variation in the gene. It is critical that superior alleles be identified for variety improvement; however, locus × environment and locus × locus interactions are likely to complicate this process.

Conclusion

As we move to the genomics model for biology, in which one starts with sequence and then proceeds to function, the key to understanding will be the ability to execute high-throughput genotyping and precision-phenotyping experiments. Genomics research has emphasized structural aspects in recent years; however, the focus is shifting to determining the functional role of genes and the mechanisms of evolutionary change that have resulted in the diversity of living organisms we see today. Methods for genomewide gene-expression studies are developing rapidly and are critical to our understanding of protein structure–function relationships that are necessary for predicting gene function and for rationally engineering genes. Bioinformatics will play an increasingly important role in the integration of information from different species and sources through the use of novel approaches to analysis and visualization of complex data. Struc-

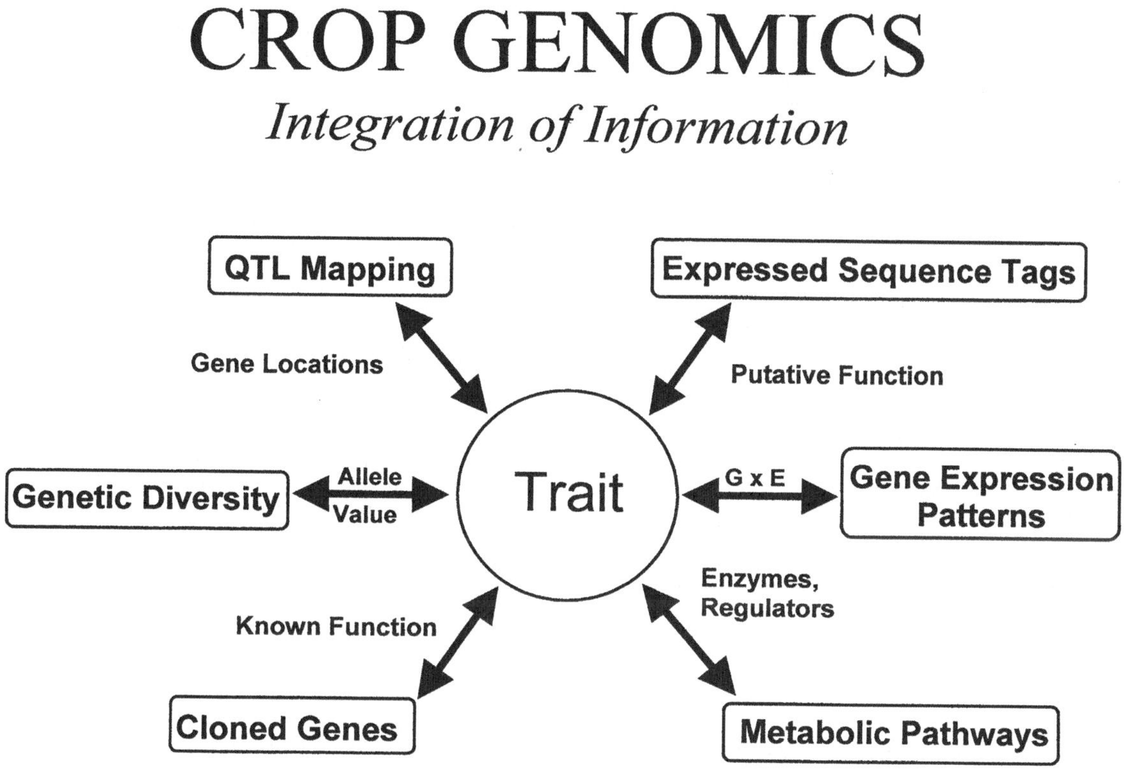

Figure 12.2 Integration of genomic information can facilitate gene discovery and characterization for trait improvement.

tural genomic research linking genes and genomes across species benefits all species but is especially important for large-genome species as well as those that receive less funding. We have already gained a great deal of knowledge about biological systems and their never-ending complexity. Sorting out the components that can be easily and reliably manipulated is the challenge.

Breeders and geneticists must not lose sight of our long-term goal, which is crop improvement. Breeding progress depends on (1) discovery and generation of genetic variation for agronomic traits, (2) development of genotypes with new or improved attributes due to superior combinations of alleles at multiple loci, and (3) accurate selection of rare genotypes that possess the new improved characteristics. Consequently, efficient methods are needed for identifying and evaluating allelic effects on a large scale so that desirable alleles can be assembled in superior varieties. This can be facilitated by integration of genetic information across species, identification of superior alleles, and by focusing on the most important genes and traits for the species of interest.

References

Ahn, S., and S.D. Tanksley. 1993. Comparative Linkage Maps of the Rice and Maize Genomes. Proceedings of the National Academy of Sciences 90:7980–7984

Akhunov et al. 2003. PNAS. 100:10836–10841.

Bennetzen, J.L., and W. Ramakrishna. 2002. Numerous small rearrangements of gene content, order, and orientation differentiate grass genomes. Plant Mol. Biol. 48:821–827.

Chen, M., P. San Miguel, A.C. de Oliveira, S.S. Woo, H. Zhang, R.A. Wing, and J.L. Bennetzen. 1997. Microcolinearity in sh2-homologous regions of the maize, rice, and sorghum genomes. Proceedings of the National Academy of Sciences 94:3431–3435.

Cho, Y.G., T. Ishii, S. Temnkh, X. Chen, L. Lipovich, S.R. McCouch, W.D. Park, N. Ayer, and S. Cartinhour. 2000. Diversity of microsatellites derived from genomic libraries and GenBank sequences in rice (Oryza sativa). Theor. Appl. Genet. 100:713–722.

Devos, K.M., J. Beales, Y. Nagamura, and T. Sasaki. 1999. *Arabidopsis*-Rice: Will Colinearity Allow Gene Prediction Across the Eudicot-Monocot Divide? Genome Research 9:825–829.

Doebley, J., A. Stec, and C. Gustus, 1995. *teosinte branched1* and the origin of maize: Evidence for epistasis and the evolution of dominance. Genetics 141:333–346.

Dubcovsky, J., and J. Dvorak. 1995. Ribosomal RNA loci: Nomads in the Triticeae genomes. Genetics 140:1367–1377.

Dubcovsky, J., W. Ramakrishna, P. SanMiguel, C.S. Busso, L. Yan, B.A. Shiloff, and J.L. Bennetzen. 2001. Comparative sequence analysis of colinear barley and rice BACs. Plant Physiology 125:1342–1353.

Dudley, J.W. 1993. Molecular markers in plant improvement: Manipulation of genes affecting quantitative traits. Crop Science 33:660–668.

Dunford, R.P., N. Kurata, D.A. Laurie, T.A. Money, Y. Minobe, and G. Moore. 1995. Conservation of fine-scale DNA marker order in the genomes of rice and the Triticeae. Nuc. Acid Res. 23:2724–2728.

Eujayl, I., M.E. Sorrells, M. Baum, P. Wolters, and W. Powell. 2001. Assessment of genotypic variation among cultivated durum wheat based on EST-SSRS and genomic SSRS. Euphytica 119:39–43.

Eujayl, I., M.E. Sorrells, M. Baum, P. Wolters, and W. Powell. 2002. Isolation of EST-derived microsatellite markers for genotyping the A and B genomes of wheat. Theor. Appl. Genet. 104:399–407.

Feng, Q., et al. 2002. Sequence and analysis of rice chromosome 4. Nature 420:316–320.

Fu, H., and H.K. Dooner. 2002. Intraspecific violation of colinearity and its implications in maize. Proc. Natl. Acad. Sci. 99:9573–9578.

Gale, M.D., and K.M. Devos. 1998. Comparative genetics in the grasses. Proc. Natl. Acad. Sci. 95:1971–1974.

Gaut, B.S. 2002. Evolutionary dynamics of grass genomes. New Phytologist 154:15–28.

Han, F., I. Romagosa, S.E. Ullrich, B.L. Jones, P.M. Hayes, and D.M. Wesenberg. 1997. Molecular marker-assisted selection for malting quality traits in barley. Molecular Breeding 3:427–437.

Hospital, F., C. Chevalet, and P. Mulsant. 1992. Using markers in gene introgression breeding programs. Genetics 132:1199–1210.

Hulbert, S.H., T.E. Richter, J.D. Axtell, and J.L. Bennetzen. 1990. Genetic mapping and characterization of sorghum and related crops by means of maize DNA probes. Proc. Natl. Acad. Sci. 87:4251–4255.

Kantety, R.V., M. La Rota, D.E. Matthews, and M.E. Sorrells. 2002. Data Mining for Simple Sequence Repeats in Expressed Sequence Tags from Barley, Maize, Rice, Sorghum and Wheat. Plant Mol. Biol. 48:501–510.

Keller, B., and C. Feuillet. 2000. Colinearity and gene density in grass genomes. Trends Plant Sci. 5:246–251.

Kellogg, E.A. 2000. The grasses: A case study in macroevolution. Annu. Rev. Ecol. Syst. 31:217–238.

Kilian, A., D.A. Kudrna, A. Kleinhofs, M. Yano, N. Kurata, B. Steffenson, and T. Sasaki. 1995. Rice-barley synteny and its application to saturation mapping of the barley Rpg1 region. Nucleic Acids Research 23:2729–2733.

Kurata, N., G. Moore, Y. Nagamura, T. Foote, M. Yano, Y. Minobe, and M. Gale. 1994. Conservation of genomic structure between rice and wheat. Bio Technol. 12:276–278.

Lande, R. and R. Thompson. 1990. Efficiency of marker-assisted selection in the improvement of quantitative traits. Genetics 124:743–756.

Leister, D., J. Kurth, D.A. Laurie, M. Yano, T. Sasaki, K. Devos, A. Graner, and P. Schulze-Lefert. 1998. Rapid reorganization of resistance gene homologues in cereal genomes. Proc. Natl. Acad. Sci. 95:370–375.

Lin, H.X., Yamamoto-T, Sasaki-T, and Yano-M. 2000. Characterization and detection of epistatic interactions of 3 QTLs, Hd1, Hd2, and Hd3, controlling heading date in rice using nearly isogenic lines. Theor. Appl. Genet. 101: 1021–1028.

Lin, Y.R., K.F. Schertz, and A.H. Paterson. 1995. Comparative analysis of QTLs affecting plant height and maturity across the Poaceae, in reference to an interspecific sorghum population. Genetics 141:391–412.

Moore, G., K.M. Devos, Z. Wang, and M.D. Gale. 1995. Grasses, line up and form a circle. Curr. Biol. 5:737.

Nelson, J.C., A.E. Van Deynze, E. Autrique, M.E. Sorrells, Y.H. Lu, M. Merlino, M. Atkinson, and P. Leroy. 1995a. Molecular mapping of wheat. Homoeologous group 2. Genome 38:517–524.

Nelson, J.C., A.E. Van Deynze, E. Autrique, M.E. Sorrells, Y.H. Lu, S. Negre, M. Bernard, and P. Leroy. 1995b. Molecular mapping of wheat. Homoeologous group 3. Genome 38:525–533.

O'Brien, S.J., J.E. Womack, L.A. Lyons, K.J. Moore, N.A. Jenkins, and N.G. Copeland. 1993. Anchored reference loci for comparative genome mapping in mammals. Nat. Genet. 3:103–112.

Paterson, A.H., J.E. Bowers, M.D. Burow, X. Draye, C.G. Elsik, C. Jiang, C.S. Katsar, T. Lan, Y. Lin, R. Ming, and R.J. Wright. 2000. Comparative genomics of plant chromosomes. Plant Cell 12:1523–1539.

Qi-X, P. Stam, and P. Lindhout. 1996. Comparison and integration of four barley genetic maps. Genome 39:379–394.

Ramakrishna, W., J. Dubcovsky, Y.-J. Park, C.S. Busso, J. Emberton, P. SanMiguel, and J.L. Bennetzen. 2002. Different types and rates of genome evolution detected by comparative sequence analysis of orthologous segments from four cereal genomes. Genetics 162(3):1389–1400.

Sarma, R.N., L. Fish, B.S. Gill, and J.W. Snape. 2000. Physical characterization of the homoeologous Group 5 chromosomes of wheat in terms of rice linkage blocks, and physical mapping of some important genes. Genome 43:191–198.

Scott, K.D., P. Eggler, G. Seaton, M. Rossetto, E.M. Ablett, L.S. Lee, and R.J. Henry. 2000. Analysis of SSRs derived from grape ESTs. Theor. Appl. Genet. 100:723–726.

Singh, S., J.S. Sidhu, N. Huang, Y. Vikal, Z. Li, D.S. Brar, H.S. Dhaliwal, and G.S. Khush. 2001. Pyramiding three bacterial blight resistance genes (xa5, xa13 and Xa21) using marker-assisted selection into indica rice cultivar PR106. Theor. Appl. Genet. 102:1011–1015.

Sorrells, M.E., M.L. Rota, C.E. Bermudez-Kandianis, R.A. Greene, R. Kantety, J.D. Munkvold, M. Miftahudin, A. Mahmoud, X. Ma, P.J. Gustafson, L.L. Qi, B. Echalier, B.S. Gill, D.E. Matthews, G.R. Lazo, S. Chao, O.D. Anderson, H. Edwards, A.M. Linkiewicz, J. Dubcovsky, E.D. Akhunov, J. Dvorak, D. Zhang, H.T. Nguyen, J. Peng, N.L.V. Lapitan, J.L. Gonzalez-Hernandez, J.A. Anderson, K. Hossain, V. Kalavacharla, S.F. Kianian, D.W. Choi, T.J. Close, M. Dilbirligi, K.S. Gill, C. Steber, M.K. Walker-Simmons, P.E. McGuire, and C.O. Qualset. 2003. Comparative DNA sequence analysis of wheat and rice genomes. Gen Res. 13:1818–1827

Tanksley, S.D., M.W. Ganal, and G.B. Martin. 1995. Chromosome landing: A paradigm for map-based gene cloning in plants with large genomes. Trends in Genetics 11:63–68.

Tarchini, R., P. Biddle, R. Wineland, S. Tingey, and A. Rafalski. 2000. The complete sequence of 340 kb of DNA around the rice Adh1-Adh2 region reveals interrupted colinearity with maize chromosome 4. The Plant Cell 12:381–391.

Tikhonov, A.P., P.J. SanMiguel, Y. Nakajima, N.M. Gorenstein, J.L. Bennetzen, and Z. Avramova. 1999. Colinearity and its exceptions in orthologous adh regions of maize and sorghum. Proc. Natl. Acad. Sci. 96:7409–7414.

Van Deynze, A.E., J. Dubcovsky, K.S. Gill, J.C. Nelson, M.E. Sorrells, J. Dvorak, B.S. Gill, E.S. Lagudah, S.R. McCouch, and R. Appels. 1995a. Molecular-genetic maps for group 1 chromosomes of Triticeae species and their relation to chromosomes in rice and oat. Genome 38:45–59.

Van Deynze, A.E., J.C. Nelson, L.S. O'Donoughue, S.N. Ahn, W. Siripoonwiwat, S.E. Harrington, E.S. Yglesias, D.P. Braga, S.R. McCouch, and M.E. Sorrells. 1995b. Comparative mapping in grasses. Oat relationships. Mol. Gen. Genet. 249:349–356.

Van Deynze, A.E., J.C. Nelson, E.S. Yglesias, S.E. Harrington, D.P. Braga, S.R. McCouch, and M.E. Sorrells. 1995c. Comparative mapping in grasses. Wheat relationships. Mol. Gen. Genet. 248:744–754.

Van Deynze, A.E., M.E. Sorrells, W.D. Park, N.M. Ayres, H. Fu, S.W. Cartinhour, E. Paul, and S.R. McCouch. 1998. Anchor probes for comparative mapping of grass genera. Theor. Appl. Genet. 97:356–369.

Walker, D., H.R. Boerma, J. All, and W. Parrott. 2002. Combining cry1Ac with QTL alleles from PI 229358 to improve soybean resistance to lepidopteran pests. Molecular Breeding 9:43–51.

Wilson, W.A., S.E. Harrington, W.L. Woodman, M. Lee, M.E. Sorrells, and S.R. McCouch. 1999. Can we infer the genome structure of progenitor maize through comparative analysis of rice, maize, and the domesticated panicoids? Genetics 153:453–473.

Yan, L., A. Loukoianov, G. Tranquilli, M. Helguera, T. Fahima, and J. Dubcovsky. 2003. Positional cloning of the wheat vernalization gene VRN1. Proc. Nat. Acad. Sci. 100:6263–6268.

Yu, J.K., S. Singh, T.M. Dake, D. Benscher, B.S. Gill, and M.E. Sorrells. 2003. Development and mapping of EST-derived simple sequence repeat (SSR) markers for hexaploid wheat. Genome 47:805–818.

13

Perspectives on Finding and Using Quantitative Disease Resistance Genes in Barley

P.M. Hayes, Department of Crop and Soil Science, Oregon State University
L. Marquez-Cedillo, Department of Crop and Soil Science, Oregon State University
C.C. Mundt, Department of Botany and Plant Pathology, Oregon State University
K. Richardson, Department of Crop and Soil Science, Oregon State University
M.I. Vales, Department of Botany and Plant Pathology, Oregon State University

Genetic resistance is the most economical and environmentally appropriate strategy for disease control in plants. Plant disease resistance can be classified as qualitative or quantitative, based on the inheritance of the resistance and the degree of symptom expression. Qualitative resistance facilitates genetic analysis and selection, but it is likely to be nondurable due to the evolution of virulence in the pathogen population. Quantitative resistance is more complicated, due to complex inheritance, but a large body of theory and empirical data indicate that it is more likely to be durable. Stripe rust (caused by *Puccinia striiformis* West. f.sp. *hordei*) is an important disease of barley throughout the world, and it has emerged as a major threat in the Americas. Over 10 years ago we initiated a collaborative program with Dr. H. Vivar (ICARDA/CIMMYT) to map and use genes conferring quantitative resistance to stripe rust. We determined the number and genome location of both quantitative and qualitative resistance genes in several accessions and proceeded to assemble these resistance genes, in various configurations, in elite breeding lines. Current dimensions of this research, in cooperation with Dr. Flavio Capettini of ICARDA and Dr. Sergio Sandoval-Islas of the Colegio de Postgraduados, include continued breeding for quantitative resistance; assessment of the role of mapping population size in estimation of resistance QTL number, effect, and interaction; development and characterization of the structure and function of near-isogenic lines (NILs) of individual resistance QTL and combinations of QTL; and physical mapping and characterization of quantitative resistance genes.

We reviewed our collaborative stripe rust resistance efforts (Hayes et al., 2001) in the context of the symposium held in honor of Dr. Vivar's retirement. That information, which is also available on the Internet at www.barleyworld.org, is still current and summarizes, in a comprehensive fashion, the rationale for our efforts, our strategies, the results of mapping multiple resistance genes in multiple germplasm accessions, and the introgression of these resistance genes into germplasm adapted to the Pacific Northwest of the United States. In the context of this symposium, we feel it may be more relevant and interesting if we address some broader issues raised by these efforts and explore some of the lessons learned in the "school of hard knocks" in resistance breeding. First, it may be useful to provide some perspectives on barley, barley stripe rust, and quantitative resistance.

Barley has long been a model for quantitative resistance genetics and breeding, and the characterization of host plant resistance to disease has remained a substantial controversy ever since Vanderplank (1963, 1968) suggested the dichotomy between vertical and horizontal resistance. At one extreme have been those who consider vertical and horizontal resistance to be qualitatively different traits, and that horizontal resistance will be permanent, owing to lack of interaction with pathogen genotypes (Robinson, 1976; Vanderplank, 1982). At the other extreme have been those who consider all resistance genes to be potentially race specific, but that this specificity can be masked by factors such as precision of measurement and interactions with host genetic background (Ellingboe, 1976; Nelson, 1978; Parlevliet and Zadoks, 1977). This confusion prompted Johnson (1981) to coin the term *durable resistance*, defined as resistance that "remains effective while a cultivar possessing it is widely cultivated" and "includes no statement or implication about the genetic control of the resistance, its mechanism, its degree of expression, or its race specificity." Though the durable resistance concept has been highly useful in attaining the immediate goals of resistance breeding programs, a more precise characterization of resistance is also needed so as to develop improved strategies for attaining durability in the future.

In this presentation, we will use the term "qualitative resistance" to designate Mendelian genes of large effect that clearly interact on a gene-for-gene basis with the pathogen. We consider quantitative resistance (QR) to designate resistance that shows continuous variation and is usually incomplete in expression. We consider the race specificity of QR to be a question that is yet unresolved, but we accept the assumption that quantitative resistance is more likely, on average, to be more durable than is qualitative resistance.

Parlevliet and Zadoks (1977) demonstrated through modeling approaches that gene-for-gene interactions may occur in QR (and, hence, allow for pathogen adaptation) but be particularly difficult to demonstrate with traditional analysis-of-variance approaches. Quantitative host genotype × pathogen genotype interactions have sometimes been detected experimentally (e.g., Jenns et al., 1982; Latin et al., 1981; Parlevliet, 1977). Interactions with environment, however, may cause these interactions to be irrelevant to pathogen adaptation (Jenns et al., 1982; Kulkarni and Chopra, 1982). Adaptation of pathogen populations to QR can be demonstrated in greenhouse and growth chamber evaluations (e.g., Ahmed et al., 1995; Ahmed et al., 1996; Caten, 1974; Clifford and Clothier, 1974; Jeffrey et al., 1962; Jinks and Grindle, 1963; Kolmer and Leonard, 1986; Lehman and Shaner, 1997; Leonard, 1969), but it is more difficult to study in the field. A wheat (*Triticum aestivum*) cultivar quantitatively resistant to *Mycosphaerella graminicola* (causal agent of *Septoria tritici* blotch) eroded very substantially over a 10-year period in the Willamette Valley of Oregon (Mundt et al., 2002). On the other hand, Vanderplank (1978) presented data suggesting that potato (*Solanum tuberosum*) cultivars with quantitative resistance to *Phytophthora infestans* remained stable for more than 30 years.

Vanderplank (1978; 1982) argued that horizontal resistance can be controlled by a very small number of genes and that the stability of horizontal relative to vertical resistance is due primarily to qualitative differences in mechanism and not to gene number. Until the relatively recent advent of QTL (quantitative trait locus) analysis, estimates of gene number for quantitative traits were derived from statistical analyses of phenotypic data collected from segregating crosses. Geiger and Heun (1989) provided an excellent review of this topic and concluded that the number of effective factors controlling QR ranges from 2 to 10, a range lower than had been predicted in earlier years. However, they also noted that model assumptions (equal effects of genes, no linkage, no epistasis, etc.) could result in these estimates greatly underestimating the number of genes controlling QR. They concluded that there is "much uncertainty over the number of genes involved."

The components of QR are defined relative to the important life history traits of the pathogen: infection frequency, latent period, lesion size, sporulation rate, and infectious period (Parlevliet, 1979). The epidemiological importance of these resistance components varies among pathogens and is affected by epidemic speed (e.g., Leonard and Mundt, 1984). The components of QR are often highly correlated (Das et al., 1993; Parlevliet, 1989; Parlevliet, 1986). Association among resistance components could, of course, be caused by either pleiotropy or linkage. Parlevliet (1986) derived strong genetic evidence for pleiotropic control of latent period and infection efficiency for barley (*Hordeum vulgare*) leaf rust (caused by *Puccinia hordei*), though tight linkage could not be completely excluded as an explanation for the correlation.

Overall, we know relatively little about ontogenic changes in expression of QR genes. Quantitative resistance may be expressed at any or all plant growth stages. For some diseases, for example, the cereal rusts, durable resistance with partial expression is sometimes expressed exclusively or at higher levels in the adult plant stage (Quayom and Line, 1985; Hulbert, 1997; Hulbert et al., 2001). Though such adult plant resistance may be quantitatively inherited, there are also some examples of control by single, dominant genes, for example, the *Lr34* gene for resistance to wheat leaf rust (e.g., Kolmer, 1996; Singh and Gupta, 1992). Even when QR is expressed at both seedling and adult stages, it is still possible that one or more different genes are expressed differentially at different growth stages. By no means can seedling and adult plant resistance be considered independent traits as a general rule; selection of QR on seedlings in the greenhouse, however, is often positively correlated with QR in adult plants in the field (e.g., Parlevliet, 1989).

The characterization of plant resistance genes at the molecular level has provided information upon which to develop models involving signal detection, signal transduction, and response (Beynon, 1997). These studies (Buschges et al., 1997; Martin et al., 1993; Salmeron et al., 1994; Schulze-Lefert et al., 1997; Zhou et al., 1995) have provided molecular evidence confirming hypotheses based on whole plant data (summarized by Ellingboe, 1976; Gabriel and Rolfe, 1990), indicating that "monogenic" gene-for-gene relationships are recognition processes that turn on multiple genes in a resistance pathway. At the same time, QTL analysis procedures have facilitated dissection of quantitative disease resistance. In some cases, a significant proportion of the total variance in the expression of quantitative traits may be attributable to one locus or a few loci (Chen et al., 1994; Hayes et al., 1996a; Michelmore, 1995; Young, 1996), supporting the classical quantitative genetic studies discussed above. Thus, the overall picture would seem to be one of converging lines of evidence supporting complexity in some qualitative models and simplicity in some quantitative models.

The QTL concept has represented an important step forward in understanding traits showing quantitative variation as described in greater detail below for barley stripe rust. More broadly, as Robertson (1985) observed nearly 20 years ago, "qualitative and quantitative traits may be the result of different types of variation of DNA at the loci involved." In other words, quantitative variation may be attributable to certain alleles and qualitative variation to other alleles at the same locus, or loci. As elegantly demonstrated in rice and tomato, with sufficient resources and the appropriate genetic stocks, QTLs can be identified as Mendelian loci and cloned (Yano et al., 1997, 2000; Yamamoto et al., 1998; Frary et al., 2000). More recently, human medical genetics efforts have also entered the "land between Mendelian and multifactorial inheritance," as so elegantly described by Burghes et al. (2001).

Molecular analyses have shown that many resistance genes in plants are found in clusters (Hulbert, 1997; Hulbert et al., 2001; Michelmore, 1995; Kanazin et al., 1996; Ellis et al., 1998). We have found qualitative and quantitative resistance genes conferring resistance to fungal and viral pathogens in proximity (Toojinda et al., 2000). Genes conferring qualitative and quantitative resistance to a range of fungal, bacterial, and viral pathogens have been mapped in the barley genome, and there are certainly patterns of association of multiple resistance loci (Hayes et al., 2003a, and posted on the Internet at www.barleyworld.org). In the case of barley and powdery mildew (caused by *Erysiphe graminis* [= *Blumeria graminis*] f. sp. *hordei*), a particularly well-studied system, the *Mla* (powdery mildew) resistance cluster is an excellent example of local clustering of multiple specificities in a short physical region (Wei et al., 1999).

A 1996 review of quantitative resistance QTLs summarized 11 studies, incorporating substantial diversity among causal agents (Young, 1996). This summary showed that the number of identified QTLs associated with QR ranged from 2 to 11, with a mean of 5.2, a median of 3.8, and a mode of 3. The percentage of QR variation explained by the identified QTLs ranged from 14 to 82%. A more recent survey (Kover and Caicedo, 2001) included 85 QTL studies of disease or insect resistance, incorporating 100 mapping population × pathogen (or insect) combinations. The number of identified QTLs ranged from 0 to 18, with a mean of 4.6, a median of 4.2, and a mode of 2. On average, the identified QTLs accounted for 51 and 67.5% of the phenotypic and genotypic variance, respectively. The authors of both reviews (Young, 1996; Kover and Caicedo, 2001) noted that estimates of QTL numbers are biased downward owing to small population sizes. Indeed, the issue of small population size is one that has plagued studies of nearly all quantitative traits. QTL analyses depend on relating molecular marker polymorphisms with phenotypic variants in a structured population. In the case of a "small" (e.g., $n \leq 100$) population, the chances of recovering recombination events between marker and target loci are limited. Thus, the limited population sizes used in many QTL detection studies have led to underestimation of QTL number, overestimation of QTL effects, and a failure to quantify QTL interactions (Beavis, 1998; Jansen and Stam, 1994; Kaeppler, 1997; Melchinger et al., 1998; Utz et al., 2000; Zeng, 1994). These concerns regarding population size, in fact, prompted the first phase of our barley stripe rust QTL mapping project, the development of a large doubled haploid mapping population (the Oro population), focus on a quantitative trait of high heritability (resistance to barley stripe rust), and a fruitful collaboration with Drs. Schoen and Utz at the University of Hohenheim regarding statistical analyses of QTL data.

In some cases, there appears to be evidence for the existence of epistasis among QTLs, plant developmental effects, and host resistance QTL × pathogen race interactions (reviewed in Kover and Caicedo, 2001; Young, 1996). The importance of these effects is unclear in many cases, however. For example, Leonards-Schippers et al. (1994) identified potato QTLs that interacted with two races of *P. infestans*. However, this interaction was not repeatable among trials, which differed in both time and developmental stage of the test plants. Thus, as with classical studies of QR, there may be putative host × pathogen interactions that are really host × pathogen × environment interactions that are not repeatable.

Very little is known about QTLs associated with different components of QR. Though studying components of QR was not a goal of their study, Wang et al. (1994) noted that "many" of the ten QTLs for lesion number that were identified also affected lesion size, though only two of the lesion size QTLs were statistically significant. The correlation was further clouded by the fact that these data were derived from polcyclic tests and by competition among lesions, which is well known to cause a negative correlation between lesion size and number.

Barley is an excellent species for genome mapping and map-based cloning. This diploid ($2n = 14$) species has seven cytologically distinct chromosomes containing approximately 5.3×10^9 bp DNA (Bennett and Smith, 1976). Although barley is an autogamous species, there is sufficient DNA-level diversity for efficient linkage map construction in populations derived from crosses between related genotypes (Becker et al., 1995; Graner et al., 1991; Hayes et al., 1997; Kasha et al., 1995; Kleinhofs et al., 1993). The North American Barley Genome Project (NABGP) has focused on building maps in elite germplasm in order to facilitate the direct application of these maps to plant breeding (reviewed by Hayes et al., 1997). Several thousand loci have been placed on these maps, providing a comprehensive catalogue of markers. Higher throughput markers, such as amplified fragment length polymorphisms, have been used for barley map construction (Becker et al., 1995; Hayes et al., 1997). Microsatellite polymorphism has been demonstrated (Saghai-Maroof et al., 1994; Ramsay et al., 2000) and used for barley germplasm characterization and map construction (Becker and Heun, 1995; Powell et al., 1996; Russell et al., 1997; Toojinda et al., 2000). The Scottish Crop Research Institute (SCRI) has a very productive simple sequence repeat (SSR) development program (http://www.scri.sari.ac.uk/SSR/). We are currently cooperating with the SCRI in an international barley SSR characterization effort, and we are systematically mapping the SCRI SSRs on NABGP populations. Recently, this effort has expanded to mapping single nucleotide polymorphisms (SNPs) and expressed sequence tags (ESTs).

Barley stripe rust (BSR) was first reported in South America in 1975 and in the United States in 1991 (Marshall and Sutton, 1995). In the late 1980s, we initiated a program to transfer quantitative resistance to barley varieties adapted to the Pacific Northwest before the arrival of the pathogen in this region. Barley germplasm developed by the ICARDA/CIMMYT program in Mexico allows limited symptom development when exposed to the spectrum of virulence encountered in field tests in South America, Mexico, and the United States. The fact that this germplasm has remained resistant over a 17-year period may be grounds for describing it as durable (Johnson, 1981). Sandoval-Islas et al. (1998) provided additional evidence for the quantitative and durable nature of the resistance of genotypes in the ICARDA/CIMMYT program.

A collaborative effort was initiated to use molec-

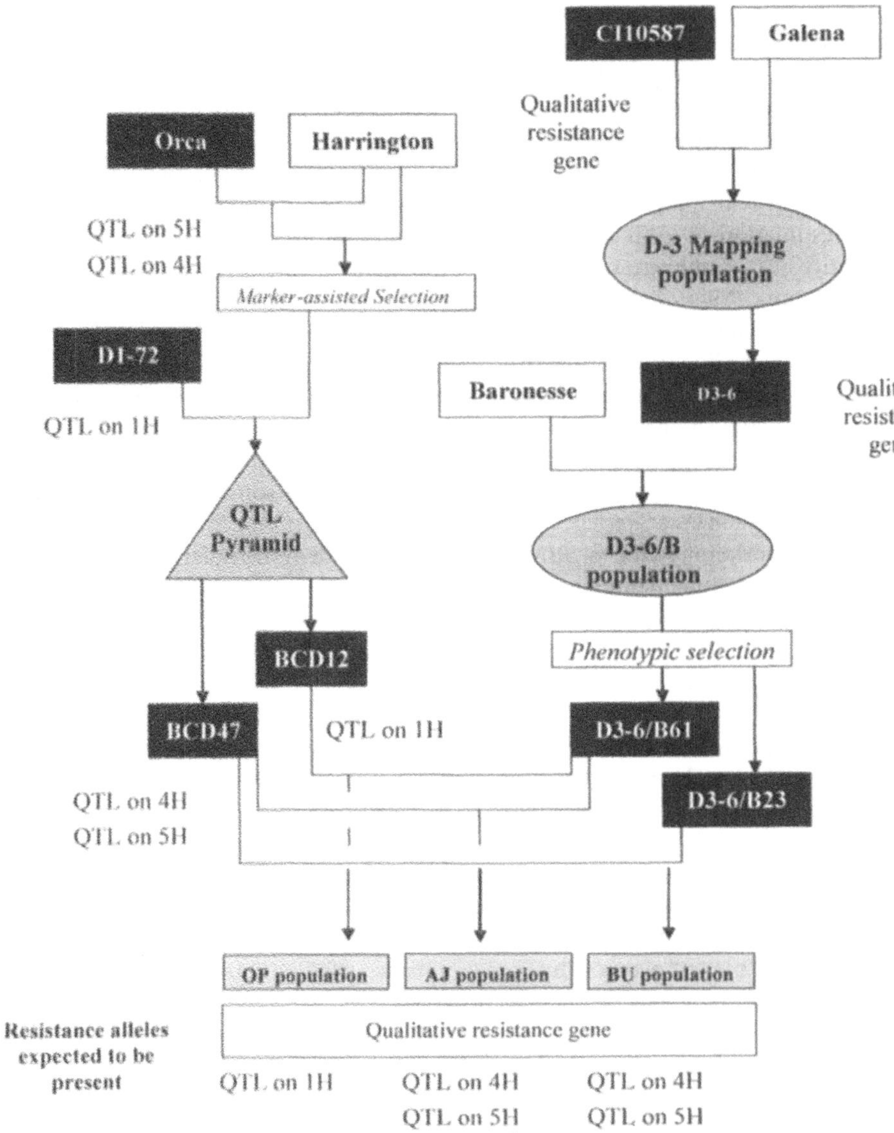

Figure 13.1 Schematic showing the development of the qualitative/quantitative disease-resistance populations. Black boxes represent resistance sources and white boxes represent susceptible parents.

ular markers for resistance QTL mapping and marker-assisted selection (reviewed by Hayes et al., 2000b), and the germplasm derivation process is illustrated in Figure 13.1. We mapped two QTLs for BSR resistance to barley chromosomes 4H and 5H of the resistance source "Calicuchima-sib" (Chen et al., 1994). Toojinda et al. (1998) described marker-assisted introgression of these resistance QTLs into the cultivar Steptoe, resulting in the release of the cultivar Tango. Using the resistance source Shyri, we identified BSR resistance QTLs on chromosomes 1H, 2H, 3H, and 6H (Toojinda et al., 2000). No QTLs were detected on chromosomes 4H and 5H, suggesting that Calicuchima and Shyri have different BSR resistance QTLs. Based on this assumption, we developed a complex population combining the QTLs on chromosomes 4H and 5H from Calicuchima with the resistance QTL on chromosome 1H from Shyri and confirmed the QTL effects in the new genetic background (Castro et al., 2003b) Additional BSR resistance QTLs have been mapped by Thomas et al. (1995) on chromosomes 1H, 5H, and 7H.

BSR resistance QTLs appear to be clustered within the barley genome. Thomas et al. (1995) mapped BSR resistance QTLs on chromosomes 1H, 5H, and 7H. The QTL on chromosome 1H is located in the same region as the QTL detected by Toojinda et al. (2000), and the QTL on chromosome 5H is in the same region as the QTL detected by Chen et al. (1994). The chromosome 1H QTL identified by Thomas et al. (1995) maps to the same region as *Yr4*, a qualitative BSR resistance gene mapped by Von Wettstein-Knowles (1992). We recently mapped another qualitative BSR resistance gene on chromosome 7H (Hayes et al., 1999; Castro et al., 2003a, 2003b, and 2003c). Five of the seven BSR QTLs that we have identified thus far map to areas containing qualitative genes for resistance to powdery mildew. The other two BSR resistance QTLs map to the same regions as genes conferring resistance to other barley diseases.

Our mapping and introgression experiments have been focused on adult plant field resistance, based on the experience and success of the ICARDA/CIMMYT barley program and perspectives on durable resistance obtained in the Pacific Northwest with wheat stripe rust (Milus and Line, 1986). However, it is also of interest to determine if there is growth-stage specificity associated with BSR resistance QTLs. Hayes et al. (1996) mapped seedling resistance QTLs in Calicuchima-sib to chromosomes 4H and 6H based on artificial inoculation with a defined isolate of *P. striiformis* f. sp. *hordei*. The chromosome 4H QTL mapped to the same region as the adult plant QTLs. Using three defined isolates with divergent virulence patterns, we recently mapped genes conferring resistance at the seedling stage in the Shyri × Galena population in which we had previously mapped adult plant resistance QTLs (Castro et al., 2002). The infection type data for each of the three isolates fit a 3:1 (susceptible\resistant) ratio, which is the expected ratio in a doubled haploid population if two genes are required for resistance. QTL effects and significance were estimated using three different procedures. In all cases, two resistance QTLs (one on chromosome 1H and one on chromosome 6H) were detected for each isolate, and in all cases Shyri contributed the resistance alleles. The two seedling resistance QTLs map to the same regions of the genome as two of the four adult plant resistance QTLs. Interestingly, however, all of the adult plant QTLs that we have identified show additive effects, while the two seedling QTLs do not. Thus, preliminary data suggest only a partial association between seedling and adult plant resistance and that gene action may depend on plant growth stage. However, these results are confounded by differences in methodology that have traditionally been used to study resistance at the seedling and adult stages. Seedling resistance was evaluated in a monocyclic inoculation test in the greenhouse, with disease reaction being measured as reaction type on a one-to-nine scale. In contrast, adult plant QTLs were identified from polycyclic tests in the field, with disease reaction being measured as percentage of leaf area covered by stripe rust lesions. Clearly, associations and differences among seedling and adult plant QTLs need to be evaluated using the same methodology.

Stripe rust in barley has not been the subject of extensive quantitative genetics research, as have leaf rust and mildew, presumably because in Europe it has not been a disease of major importance, it is not an ideal pathogen for controlled experimental research, and it can usually be held in check by the extensive use of fungicides. There may have been some accumulation of minor genes for resistance because we have found that most European barley varieties, and North American varieties with European germplasm in their pedigrees, are somewhat tolerant of stripe rust under the epidemic conditions that generally prevail in North America. However, in South America and Mexico, this tolerance is not apparent or sufficient. Whether this is due to pathogen virulence or environment, or both, is an open question. What is certain is that the land race varieties of the Andean region, which trace to post-Conquest introductions, are highly susceptible, and the disease, as a consequence, was devastating. The effects of stripe rust were also apparent on North American six-row malting barley germplasm. To put it bluntly, this germplasm is a "stripe rust magnet."

In the remainder of this chapter we will highlight some key points that have direct bearing on stripe rust resistance breeding in barley and broader implications for other resistance breeding efforts.

Resistance comes in many forms

The phenotypic frequency distributions for stripe rust severity (in percentage) in several doubled

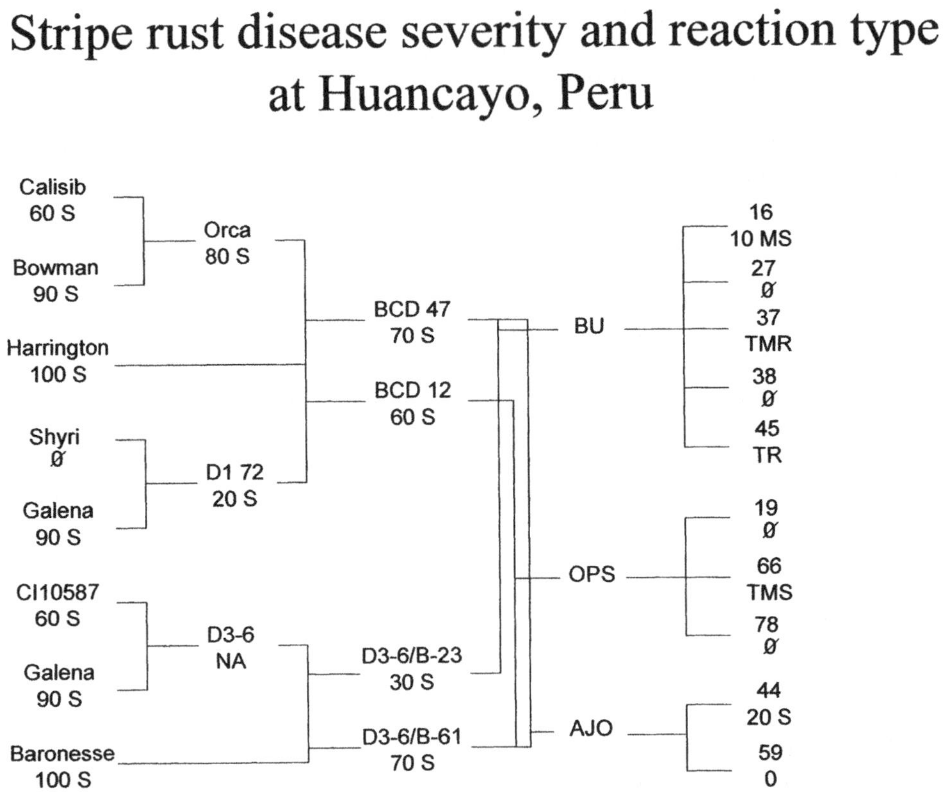

Figure 13.2 Phenotypic frequency distributions for stripe rust disease severity (%) in six doubled haploid-mapping populations.

haploid-mapping populations are shown in Figure 13.2. Several points can be made from these figures. Resistance shows inheritance patterns ranging from Mendelian to bell-shaped curves typical of quantitative traits. Many of the frequency distributions show patterns that could, with grouping of severity values, lead to groupings that fit one or two gene ratios. Others defy classification. The standard errors on these phenotype values are very low when based on assessments at Toluca, Mexico. Heritability in these environments is consistently high; for example, for the Oro (BCD47 × Baronesse) population the heritability is 95.7%, based on the average of three experiments in Toluca, Mexico.

The Toluca environment has proven to be where we have consistently been able to achieve high heritability for stripe rust resistance showing different inheritance patterns. In other environments, such as in the Pacific Northwest at Mt. Vernon, Washington, there is less spread in resistance phenotype values, heritabilities are lower, and fewer QTLs are detected (Tables 13.1 and 13.2).

A plethora of resistance genes?

As summarized in Hayes et al. (2001), Castro et al. (2003a), and in this chapter, we and others have mapped multiple resistance genes in multiple accessions. Surely, there are yet uncharacterized sources of resistance—it is a matter of securing the available resources to document what is available in world collections; the ICARDA/CIMMYT program would clearly be an excellent starting point, because this program has consistently accumulated resistance genes from multiple sources by a process of cyclic introgression.

Tango lessons

Once we had mapped several stripe rust resistance QTL (in the Calicuchima × Bowman mapping population) and had molecular markers identified that bracketed the stripe rust resistance QTL, we were ready to attempt marker-assisted introgression of stripe rust resistance genes into adapted germplasm. As described in detail by Toojinda et al. (1998) and summarized by Hayes et al. (2003a), we performed one cycle of marker-assisted back-

Table 13.1 Summary of stripe rust resistance QTLs detected in the BCD47/Baronesse DH population in Toluca, Mexico, using composite interval mapping

		Average of three Mexico experiments		
Closest marker	Chromosome	LOD[a]	R^2 (%)[b]	Additive effect[c]
Bmac093	2H	10.0	6.6	6.45
Bmag225	3H	17.6	14.0	−9.36
Bmag606	3H	17.2	12.6	−8.88
EBmac788	4H	47.9	33.9	−14.58
Bmag337	5H	5.2	2.6	−4.03
GMS001	5H	4.0	2.3	3.80
Bmac316	6H	8.3	5.4	−5.85
HVCMA	7H	8.4	4.8	5.51

[a]LOD is the log-likelihood at the position.
[b]R^2 is the percentage of phenotypic variation explained by the QTL.
[c]Negative and positive values indicate that BCD47 and Baronesse, respectively, contributed the resistance QTL allele.

Table 13.2 Summary of stripe rust resistance QTLs detected in the BCD47/Baronesse DH population in Washington state, using composite interval mapping.

		Average of two Washington state experiments		
Closest marker	Chromosome	LOD[a]	R^2 (%)[b]	Additive effect[c]
HvML03	4H	19.0	14.0	−3.96
Bmag337	5H	4.3	3.0	−1.81
Bmac316	6H	10.5	8.9	−3.11
HvWaxy4a	7H	7.7	6.5	2.66
Bmag507	7H	10.0	7.5	2.88

[a]LOD is the log-likelihood at the position.
[b]R^2 is the percentage of phenotypic variation explained by the QTL.
[c]Negative and positive values indicate that BCD47 and Baronesse, respectively, contributed the resistance QTL allele.

crossing into the variety Steptoe (at the time, the most popular variety in the Pacific Northwest of the United States), produced doubled haploids from the progeny, and generated the variety Tango. As described in the variety release (Hayes et al., 2003b), the good news is that the experiment worked, and we rapidly had a stripe rust–resistant version of Steptoe available for Pacific Northwest growers. The bad news is that, as experience and theory have demonstrated, with backcrossing the agronomic performance level is set by the recurrent parent. In fact, in the absence of disease pressure, Tango never quite matched Steptoe for agronomic performance traits, producing from 5–20% lower grain yield in the absence of disease. In summary, the Tango development and release process taught us several important lessons:

1. Stripe rust resistance QTLs are real, and their introgression into a susceptible background can lead to resistance.
2. Molecular marker–assisted breeding and doubled haploid production can accelerate variety development time.
3. Backcrossing is indeed a conservative breeding strategy and the choice of recurrent parent is a very important decision.
4. Lesson 1 will be a recurring theme throughout this chapter. Lesson 2 is a rather obvious one and confirms that with sufficient resources and hard work one can meet serious breeding challenges. Lesson 3 is taught in Introductory Plant Breeding, but apparently refresher courses are sometimes necessary.

Three additional lessons we are still studying are that

1. there may be costs to using resistance genes from exotic germplasm,
2. genetic background may have unexpected and profound effects,
3. estimates of QTL number and effect are subject to revision.

These concepts will be explored in greater detail in the remainder of this chapter.

What is the cost of disease-resistance genes?

Stripe rust is a recurring threat to commercial barley production in California. Fortunately, the disease has not emerged as a consistent economic threat to barley production in other Western U.S. barley states (Colorado, Idaho, Montana, Oregon, Washington, and Wyoming). In general, all of our stripe rust–resistant varieties and potential varieties are lower yielding than the local check in disease-free environments. An example comparing resistance gene pyramids with check varieties (Baronesse and Harrington) is shown in Table 13.3. In a practical sense, a grower choosing a variety needs to consider the probability of disease occurring versus the yield penalty paid by growing a resistant variety. The million-dollar question, of

Table 13.3 Grain yield at Pendleton, Oregon (expressed as % of "Baronesse"), of stripe rust resistance QTL pyramid lines in 2001 compared with their stripe rust disease severities at Huancayo, Peru, in the same year

Selection	Yield (% of Baronesse)	BSR Peru (% severity)	Other positive traits
BU16	90	10	Scald, BYDV
BU27	79	0	Scald, BYDV
Bu 37	84	Trace	Russian Wheat aphid, scald
Bu 38	80	0	Russian wheat aphid, Scald, BYDV
Bu45	82	Trace	None
Ajo44	87	20	Scald
Ajo59	89	0	Scald
Ops19	87	0	Scald, BYDV
Ops66	82	Trace	Russian wheat aphid
Ops78	90	0	Russian wheat APHId

Note: Other positive traits are notable resistance to scald (caused by *Rhynchosoporium secalis*), BYDV, and the Russian wheat aphid (*Diuraphis noxia*). Scald and BYDV were assessed at Davis, California (in cooperation with L. Jackson), and Russian Wheat Aphid was assessed at Stillwater, Oklahoma (in cooperation with D. Mornhinweg).
BYDV, barley yellow dwarf virus.

course, is the genetic basis of this "resistance insurance premium": is it incurred by linkage, pleiotropy, or both? At this point we simply don't know, although linkage drag seems a plausible explanation. Until recently, there were simply insufficient markers, and costs were too high, to limit linkage drag by multiple marker selection: We simply used flanking markers coupled with phenotypic selection for agronomic and quality phenotypes. As will be described under the heading, What Are Quantitative Resistance Genes?, the development of QTL NILs should help to resolve the question of why we have seen associations of poor agronomic performance with resistance.

An additional point is that the stripe rust–resistance pyramid lines also have resistance to multiple diseases (Table 13.3), which may in part compensate for the yield penalty. This germplasm is also resistant to what may be a new race that was detected in a nursery in Huancayo, Peru, in 2001 (Figure 13.2). This multiple resistance is testimony to the resistance gene accumulation strategies of the ICARDA/CIMMYT program.

What about genetic background?

Tango was a success, at least as measured by publication. The Colter conversion, a simultaneous effort, was not a success by publication or variety release, but it has certainly taught us a lesson in terms of genetic background. The same two QTLs that reduced stripe rust severity in a Steptoe background had no significant effect in another North American six-row variety, Colter. This result was, of course, quite disappointing and perplexing. Possible explanations were errors in genotyping, unaccounted for gene partners in epistatic interactions, and resistance suppressors. We went so far as to produce a doubled haploid-mapping population from the cross of a Steptoe-derived selection with resistance QTL alleles according to both marker genotype and resistance phenotype (Tango) and a Colter-derived selection (CR30-3) with resistance QTL alleles according to marker genotypes. Our expectation was that since both parents had the same alleles at the target QTL, if we saw phenotypic segregation in the progeny it would be due to whatever loci were leading to a resistance phenotype in Tango and a susceptible phenotype in the Colter-derived selection. We did indeed observe segregation for resistance in the doubled haploid progeny tested under severe disease pressure in Mexico. Unfortunately, we could not persuade grant reviewers that the project was worth funding, and so the seed sits on the shelf. Cooperators interested in exploring this phenomenon are welcome!

How big is big?

The size of the mapping population is an important issue in QTL detection, as alluded to in the introductory section. We accordingly created and are studying the Oro population. This population of 409 doubled haploids, derived from the cross of BCD47 × Baronesse, has afforded us the luxury of empirically assessing population size and its role in estimates of QTL number, location, effect, and interaction. The phenotypic frequency distributions for disease severity, based on the whole population, are shown in Figure 13.3. Based on the parents of the cross, we hypothesized that we would find QTLs on chromosomes 4H and 5H. In fact, additional loci are detected using the whole population, including resistance QTL with small effects, where the susceptible parent contributes resistance alleles.

QTL analysis using the average of three Mexico experiments identified BSR QTLs on chromosomes 3H, 4H, 5H, and 6H, where the resistant parent contributed favorable alleles, and QTLs on chromosomes 2H, 5H, and 7H, where the susceptible parent contributed favorable alleles (Table 13.1). Cumulatively, the QTLs detected in Mexico account for nearly 60% of the variation in phenotypic expression. Using the Washington data we identified QTLs on 4H, 5H, and 6H, where the resistant parent contributed favorable alleles, and on chromosome 7H, where the susceptible parent contributed favorable alleles (Table 13.2). Cumulatively, the QTLs detected in Washington account for nearly 40% of the variation in phenotypic expression. QTL × environment interaction was detected in Toluca, Mexico, for chromosome 4H and in Washington for chromosome 1H, due to slight changes in magnitude of the QTL effects. Two locus epistatic interactions were not detected in either the Mexico or Washington state experiments. Preliminary analyses using subsets of the Oro population of 50, 100, 200, and 400 lines indicate that population size does have an effect on QTL estimates. The number of QTLs detected increased as the population size increased (Figure 13.2). The proportion of the total variance explained by the QTLs, the background markers, and any explanatory variables decreased as the population size increased, confirming that small population sizes overestimate the percentage of genetic variance explained by the QTLs (Beavis, 1998). Since the amount of variance explained by the QTLs detected with 200 DH lines and 400 DH lines is the same in the Mexico environments (60%), we conclude that a population size of 200 DH would be appropriate for further evaluations. Nevertheless, in an environment where the expression of the disease is not optimum, a larger population size maximizes the identification of the QTLs and its effects. Of greater importance, however, is the test environment. Using the full population, the number and effect of QTLs was greater in the Mexico than in the U.S. tests (Table 13.1, Table 13.2), underscoring the necessity of maximizing heritability and data quality in QTL-mapping experiments. In other words, a good screen on a small population is likely to be more useful than a large population under less-effective screening procedures.

How much is a QTL allele worth?

One of the most important lessons we have learned in our stripe rust–mapping efforts is the variation in allele value depending on genetic background. In addition to genetic background, other possible causes include bias in estimation of QTL effect due to limited population size and variation in pathogen virulence in different test environments. As an example of this difference in allele value, the effects of QTL on chromosomes 4H, 1H, and 5H are shown in Table 13.4. The chromosome 1H and 5H QTL are of lower value than expected, and the 4H QTL is of higher value than expected. However, we have recently seen an apparent reversal of this trend. In crosses involving Tango and Orca as donors of the resistance QTL alleles on 4H and 5H to susceptible Midwestern germplasm (Stander and Excel), chromosome 5H has again emerged as an important resistance locus. A set of recombinant inbred lines (called the Stander/Orca/Tango/Excel, or SOTE, lines) was evaluated for stripe rust resistance at Toluca in 2001 and 2002. We found equally significant allele effects at the 4H and 5H QTLs, as well as significant QTL × QTL interaction (Table 13.4). The reasons for the renewed importance in the 5H QTL are not known. Perhaps there is a relationship with the germplasm base in question. The 5H QTL has always had a larger effect in six-row than in two-row germplasm.

The harder they come, the harder they fall

As documented in two recent papers (Castro et al., 2003a and 2003b) and illustrated in Figure 13.1, we have used marker information to define the allelic architecture of lines tracing to crosses involving multiple resistance donors. The message, as shown in Figure 13.4, is that in the case of quantitative resistance, multiple QTL alleles confer more resistance than single QTL alleles. In the case of pyramids involving qualitative and quantitative re-

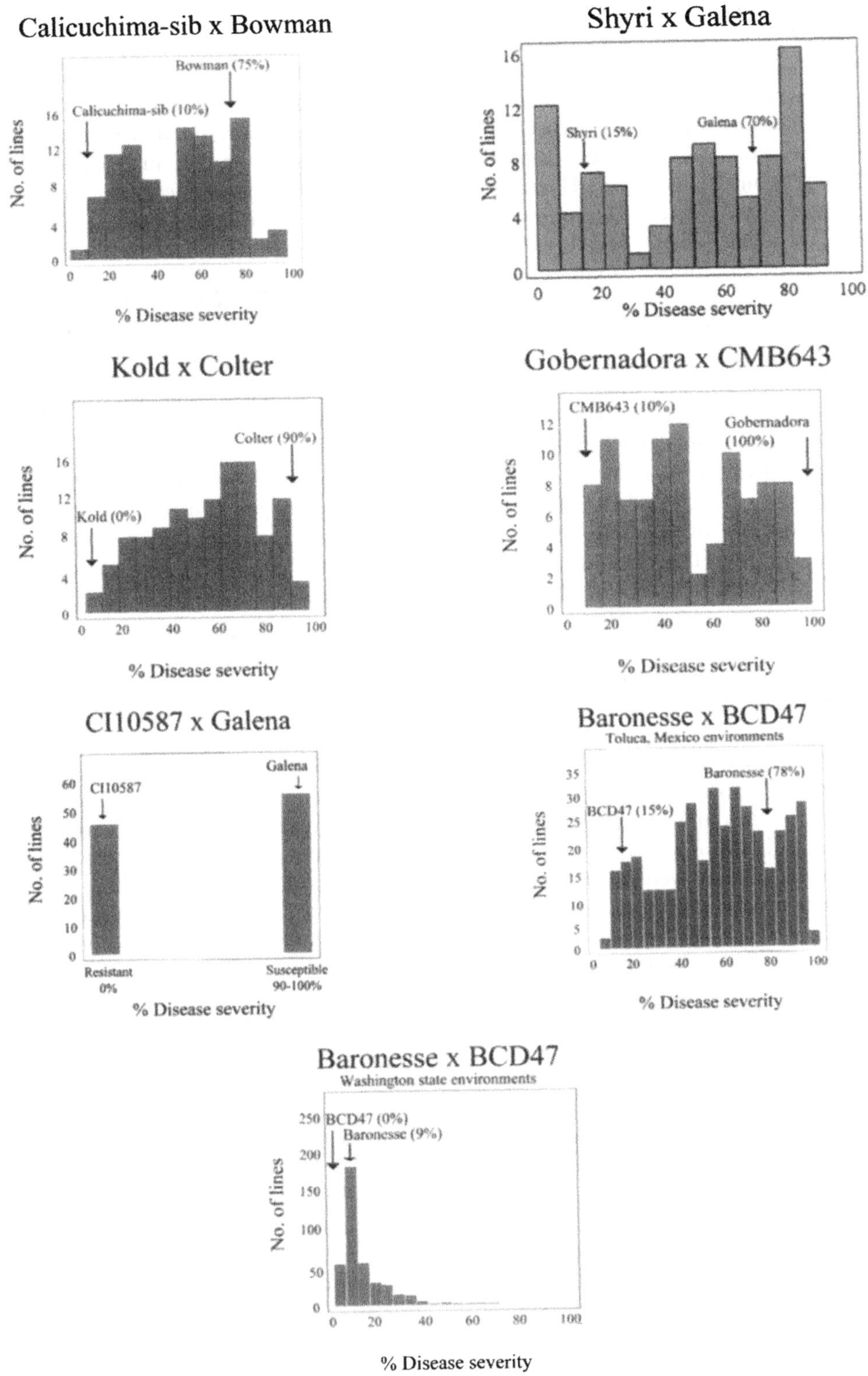

Figure 13.3 Stripe rust disease severities (%) and reaction types for germplasm tested in the USDA cooperative 2001 stripe rust screening nursery conducted at Huancayo, Peru.

Table 13.4 Comparison of the amount of genotypic variance for disease severity (σ_G^2) explained by the QTL effects in the original data sets in the MAS-derived QR pyramid population, and in the SOTE

	Original report			Pyramid population		SOTE	
QTL location	Population	% σ_G^2	*p* value	% σ_G^2	p value	% σ_G^2	*p* value
Chromosome (1H)	Shyri/Galena	91	<0.0001	22	<0.0001		
Chromosome (4H)	Cali/Bowman	2	0.0089	56	<0.0001	26	<0.0001
Chromosome (5H)		73	<0.0001	16	<0.0001	26	<0.0001

Source: Chen at al., 1994; Toojinda et al., 2000; Castro et al., 2003a; and unpublished data.

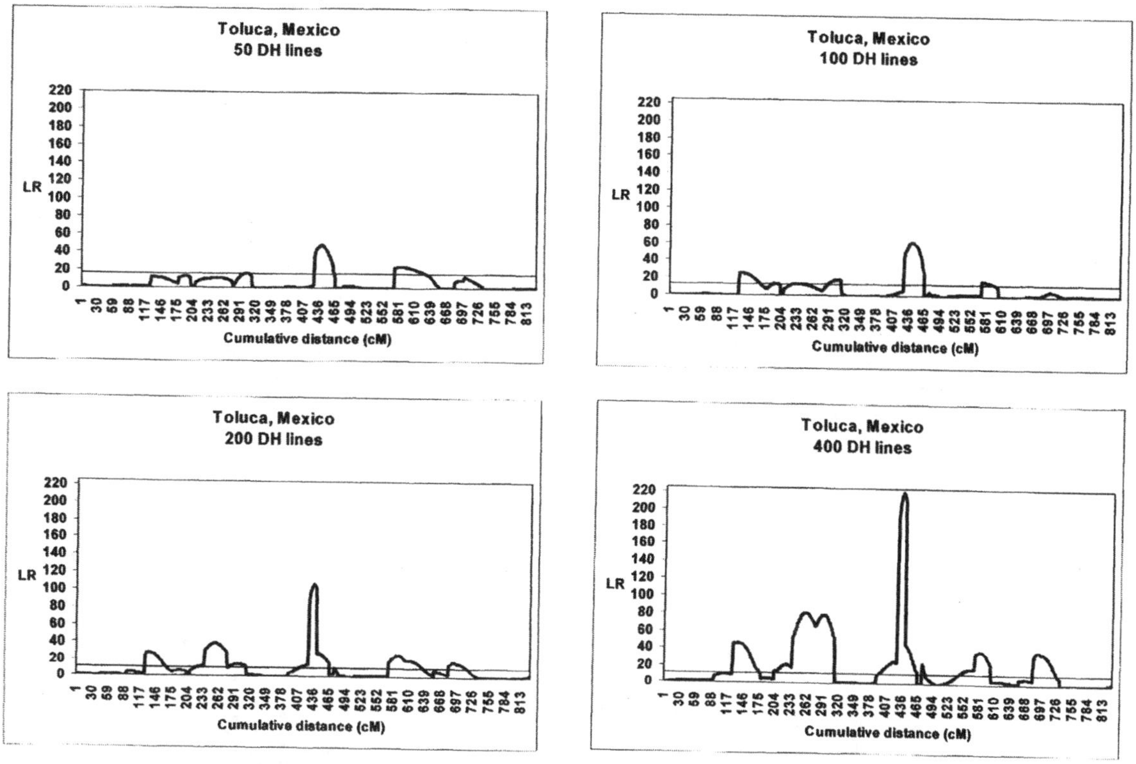

Figure 13.4 Effect of population size on the detection of QTLs for barley stripe rust severity in Toluca, Mexico.

sistance loci, the effects of the qualitative resistance locus overshadow the effects of the quantitative resistance loci (Figure 13.5). Fortunately, we have not yet encountered a race of stripe rust in Mexico or the United States that will allow us to determine if the assemblage of multiple resistance genes is indeed more effective than deploying individual genes. However, some interesting preliminary data were generated in the U.S. Department of Agriculture cooperative 2001 stripe rust screening nursery conducted at Huancayo, Peru. In this test, the qualitative resistance gene is defeated and the quantitative resistance sources all show higher disease levels than in Mexico, but the resistance gene pyramids show very low disease severities. Cooperators are welcome to help us pursue this line of research.

What are quantitative resistance genes?

As reviewed in the introductory section, it has long been a question of interest as to whether qualitative and quantitative are due to the same or differ-

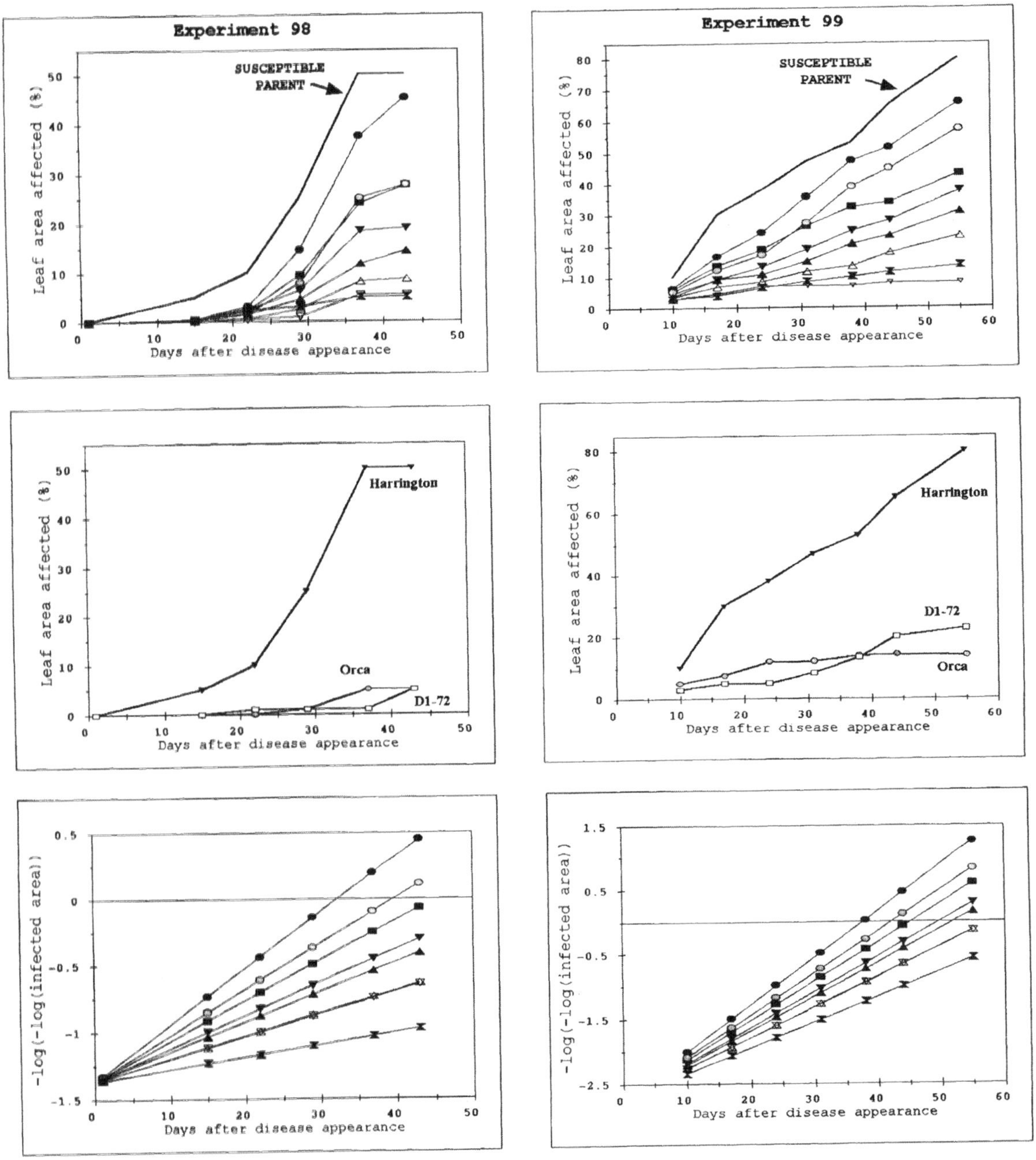

Figure 13.5 Average disease progress curves for doubled haploid lines with different combinations of resistance QTL alleles in two of the six experiments where this phenotype was measured. Figures on the left (98) correspond to the third planting date in 1998. Figures on the right (99) correspond to the first planting date in 1999. Figures in the first panel show results based on untransformed data; figures in the center show the parental lines; and figures in the lower panel show the results from the model adjusted with the Gompertz transformation, which was used to calculate the infection rate.

ent genes. One outcome of mapping and cloning qualitative resistance genes is that these genes tend to cluster in plant genomes. Many of the quantitative stripe rust resistance loci we have mapped are found in such regions; for example, on 4H the major QTL effect maps near the *Mlo* locus, a cloned mildew resistance gene, and we are in the process of further characterizing the physical structure of this region, capitalizing on barley and rice genetic resources. One important implication of clustered resistance genes is that even with extensive sequence information at hand, it may be difficult to actually determine which candidate locus is actually the determinant of quantitative resistance.

From a breeding standpoint, as long as there are not repulsion linkage issues, it will be simpler and advantageous to move multiple resistance factors in large blocks than individual genes one at a time. From the standpoint of genetic analysis and understanding quantitative resistance mechanisms, however, it will be useful to work with one genome region at a time.

If quantitative resistance genes are to be used efficiently, we need to understand their effects and interactions with each other and with genes determining other economically important, quantitatively inherited phenotypes. More precise genetic characterization of quantitative resistance will aid in the development of improved selection methodologies. If all, or at least most, of the genes controlling quantitative resistance can be identified and tagged, the corresponding regions of the genome can be tracked and incorporated into new genotypes by marker-assisted selection. Information on markers defining quantitative resistance regions is also essential for pyramiding resistance QTLs, since based on phenotype alone, it may not be possible to distinguish intervals with different numbers and combinations of resistance genes. Also, understanding the genetic basis of quantitative resistance is critical in order to predict how pathogen populations may respond to deployment of such resistance. We are currently developing a series of NILs, called the BISON (barley isogenic lines) population, for some of the stripe rust–resistance QTL regions detected in the Oro population. The NILs will be completely homozygous genotypes, each one representing an approximately 20-cM insertion of resistance donor genome in a Baronesse genetic background. Figure 13.7 shows a flow chart of the germplasm derivation and the NIL development for the BISON population. The F_1 four-way cross generation (n = 237) from the cross of Baronesse × BCD47 and Baronesse × BCD12 was screened using 11 SSRs. These SSRs were previously mapped to locations that flank the QTL regions on chromosomes 1H, 4H, and 5H. BCD47 contains the favorable allele for the BSR resistance QTL on chromosomes 4H

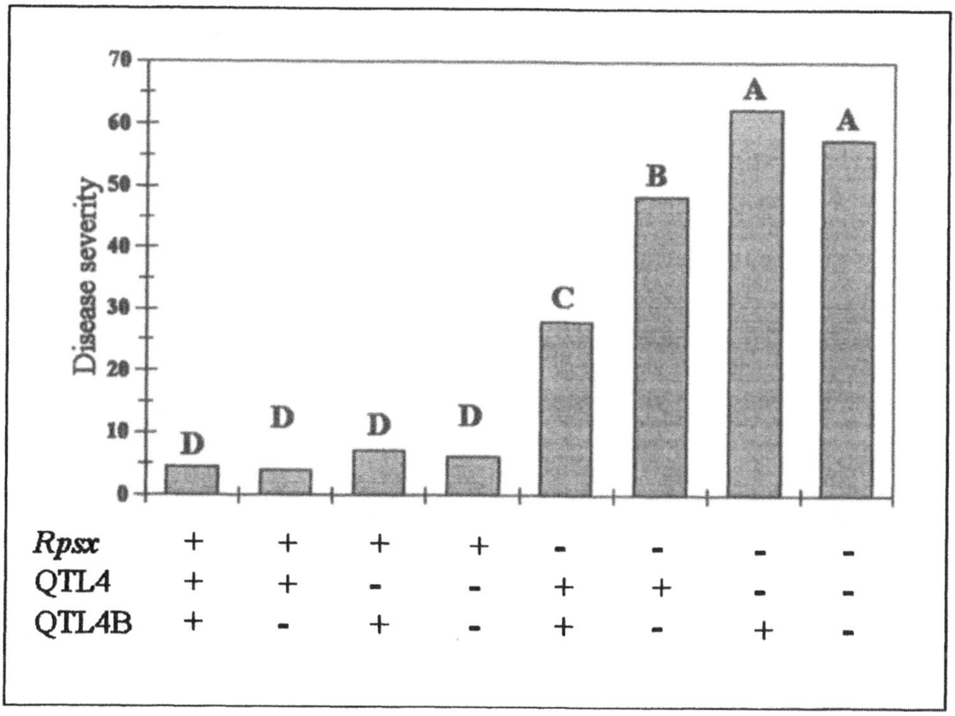

Figure 13.6 Least squares means of disease severity in DH lines of the AJ and BU populations classified according to the presence or absence of the resistance alleles at *Rpsx*, QTL4, and QTL4B QTL regions. Bars with the same letter are not significantly different ($p < 0.05$) based on pairwise comparisons.

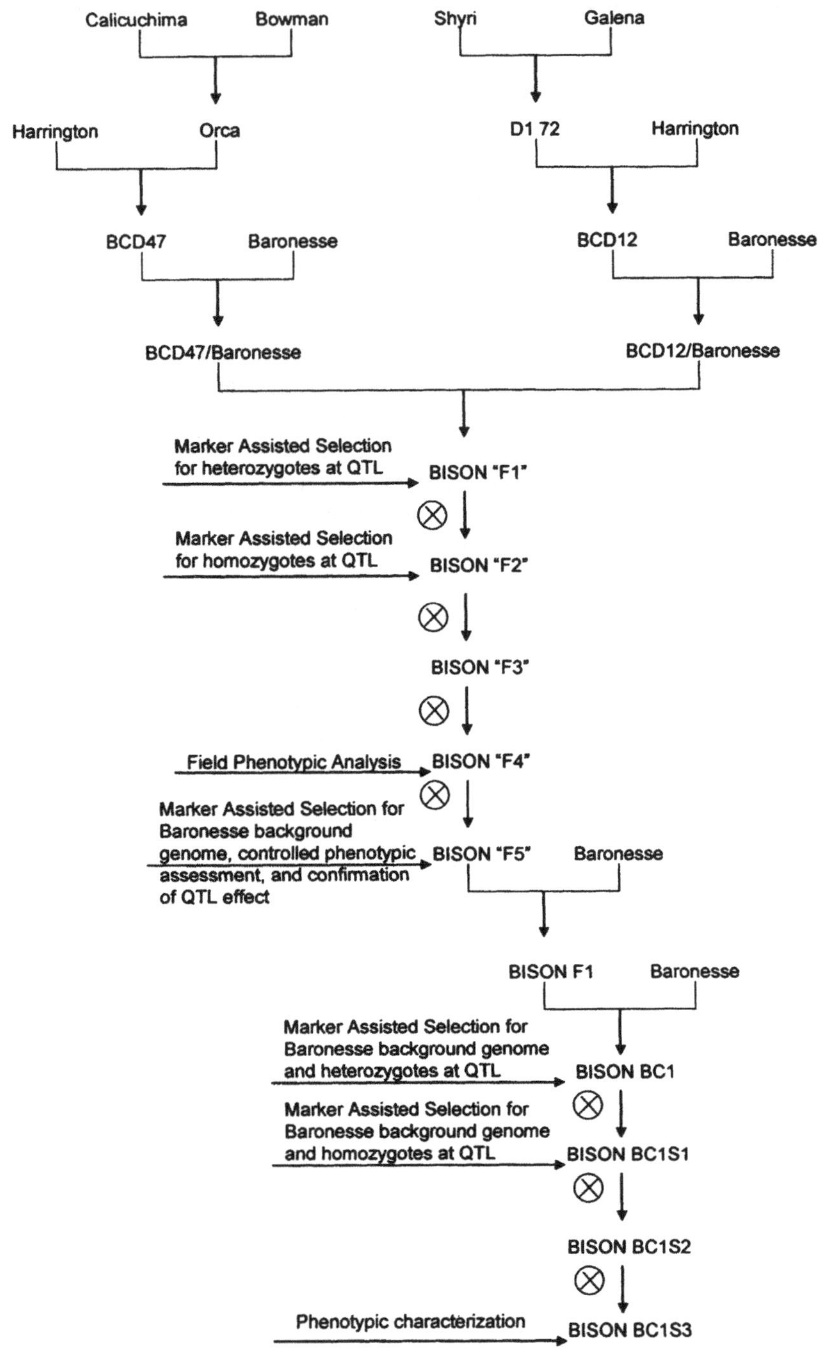

Figure 13.7 Flow chart showing the derivation of the stripe rust resistance QTL near-isogenic lines (BISON).

and 5H, and BCD12 contains the favorable allele for the BSR resistance QTL on chromosome 1H. We have now developed lines with single-resistance QTL alleles and lines with all possible combinations of QTL alleles and are currently phenotyping and genotyping this germplasm.

As well as the BISON, we are also creating BISON chromosome 7H lines. D36/B23 is the source of the favorable allele for the BSR resistance qualitative gene on chromosome 7H, and these lines should be of particular use in addressing the question of linkage drag associated with the introgression of this resistance allele from the land race CI10587.

Of model organisms, synteny, and resistance

One of the more exciting developments in plant biology is the power of comparative genetic analysis. In the case of disease resistance in general, and stripe rust in particular, there are two interesting and practical applications. The first and most obvious is the integration of the extensive stripe rust genetics research effort in wheat, with the resources available in barley. Many qualitative stripe rust resistance genes are described in wheat, but only a subset are mapped. The polyploid nature of wheat has complicated such efforts, although the recent development and characterization of deletion lines with mapped ESTs (http://wheat.pw.usda.gov/wEST/) should facilitate this effort. Singh et al. (2000) reported the map location of several stripe rust resistance genes in wheat, including *Yr28* on chromosome 4DS, and minor genes on chromosomes 7DS, 3BS, 3DS, and 5DS. With the abundant EST resources now available in wheat and barley, it should be possible to efficiently integrate disease resistance–mapping efforts in these homoeologous genomes.

Going a bit further afield, genetically speaking, the relationship of rice blast resistance genes in barley and rice has been the subject of recent investigations. Rice blast is a major disease of rice. Barley and this pathogen have not coevolved, yet some barley varieties show resistance to rice blast. This research was prompted by occasional reports in the literature and recent efforts to integrate rice and barley in rotations. Sato et al. (2001) reported rice blast resistance QTLs on chromosomes 4H and 5H of barley, and it is of considerable interest to us that the locations of these QTLs are coincident with those for stripe rust. The rice blast QTLs were mapped in a different mapping population from the stripe rust QTLs, and the blast-mapping population parents are both susceptible to stripe rust, which suggests that these regions of the genome may harbor multiple resistance genes conferring resistance to multiple pathogens. Chen et al. (2003) provided further evidence for syntenous clusters of disease-resistance genes in a comprehensive study involving the same barley-mapping population used by Sato et al. (2001), using three rice blast isolates and a rice-mapping population. This allowed for direct alignment of syntenous QTL regions in the barley and rice. The 4H and 5H regions reported by Sato et al. (2001) were confirmed, as well as 8 additional blast-resistance QTLs in barley and 12 in rice. The barley 4H and 5H QTLs have been the subject of frequent discussion throughout this report. Unfortunately, the short arm of chromosome 3 in rice, which is syntenous to the long arm of 4H in barley, is one of the last regions of the rice genome to be completely sequenced, but the initial results are quite interesting.

Conclusions

In summary, our collaborative stripe rust resistance efforts have been both rewarding and productive. We have mapped multiple resistance loci and demonstrated that they can be introgressed into susceptible varieties where they will confer resistance. Along the way, we have learned important lessons regarding the effects of alleles in different genetic backgrounds and the importance of reducing linkage drag. The germplasm resources exist, or are under development, to further understand the role of genetic background. We have demonstrated that resistance genes can be pyramided, and we have preliminary data that such pyramids may confer resistance to a new race (or races) that are virulent on individual resistance genes or simple combinations of resistance genes. Again, the germplasm resources exist to further characterize this phenomenon. The availability of many more cost-effective markers from the barley and Triticeae EST programs should allow us to be more efficient in locating and introgressing resistance genes. There are exciting opportunities to capitalize on synteny to better understand and manipulate durable resistance. In short, technology should make us more efficient, but molecular breeding is a tool—of paramount importance are genetic resources. In this era of "pay as you go" and "patent first" we need to find ways to ensure continued open exchange of germplasm and data.

Acknowledgments

We would like to thank all the colleagues who have made so many contributions to the stripe rust resistance breeding effort. These include Bill Brown,

Ann Corey, Flavio Capettini, Ariel Castro, Xianming Chen, Tanya Filichkin, Sergio Sandoval-Islas, Mareike Johnston, Andy Kleinhofs, Diane Mather, Jayne Osborne, Doris Prehn, Carlos Rossi, Kaz Sato, Chris Schoen, Theerayut Toojinda, and Hugo Vivar.

References

Ahmed, H.U., C.C. Mundt, and S.M. Coakley. 1995. Host-pathogen relationship of geographically diverse isolates of *Septoria tritici* and wheat cultivars. Plant Pathol. 44:838–847.

Ahmed, H.U., C.C. Mundt, M.E. Hoffer, and S.M. Coakley. 1996. Selective influence of wheat cultivars on pathogenicity of *Mycosphaerella graminicola* (anamorph *Septoria tritici*). Phytopathology 1986:454–458.

Beavis, W.B. 1998. QTL analyses: Power, precision, and accuracy. *In* A.H. Patterson (ed.), Molecular dissection of complex traits CRC Press, Boca Raton.

Becker, J., and M. Heun. 1995. Barley microsatellites: Allele variation and mapping. Plant Mol. Biol. 27:835–845.

Becker, J., P. Vos, M. Kuiper, F. Salamini, and M. Heun. 1995. Combined mapping of AFLP and RFLP markers in barley. Mol. Gen. Genet. 249:65–73.

Bennett, M.D., and J.B. Smith. 1976. Nuclear DNA amounts in angiosperms. Phil. Trans. R. Soc. Lond. Biol. 274:227–274.

Beynon, J.L. 1997. Molecular genetics of disease resistance: An end to the gene-for-gene concept? *In* I.R. Holub and I.J. Burdon (ed.). The gene-for-gene relationship in plant-parasite interactions. CAB International, Wallingford, United Kingdom.

Burghes, A.H.M, H.E.F. Vaessin, and A. de la Chapelle. 2001. The land between mendelian and multifactorial inheritance. Science (21)293: 2213–2214.

Buschges, R., K. Hollricher, R. Panstruga, G. Simons, M. Wolter, A. Frijters, R. van de Daelen, T. van de Lee, P. Diergaarde, J. Groenendijk, S. Topsch, P. Vos, F. Salamini, and P. Schulze-Lefert. 1997. The barley *Mlo* gene: A novel control. Cell. 8:695–705

Castro, A., P.M. Hayes, T. Filichkin, and C. Rossi. 2002. Update of barley stripe rust resistance QTL in the Calicuchima-sib × Bowman mapping population. Barley Genetics Newsletter 32:1–12.

Castro, A.J., X. Chen, P.M. Hayes, S.J. Knapp, R.F. Line, T. Toojinda, and H. Vivar. 2002b. Coincident QTL which determine seedling and adult plant resistance to stripe rust in barley. Crop Science 42:1701–1708.

Castro, A., F. Capettini, A. Corey, T. Filichkina, P.M. Hayes, A. Kleinhofs, D. Kudrna, K. Richardson, S. Sandoval-Islas, C. Rossi, and H. Vivar. 2003a. Mapping and pyramiding of qualitative and quantitative resistance to stripe rust in barley. Theor. Appl. Genet. In press.

Castro, A., X. Chen, A.E. Corey, T. Filichkina, P.M. Hayes, C. Mundt, K. Richardson, S. Sandoval-Islas, and H. Vivar. 2003b. Pyramiding quantitative trait locus (QTL) alleles determining resistance to barley stripe rust: Effects on adult plant resistance. Crop Science 43:651–659.

Castro, A.J., X. Chen, P.M. Hayes, S.J. Knapp, R.F. Line, T. Toojinda, and H. Vivar. 2003c. Coincident QTL which determine seedling and adult plant resistance to stripe rust in barley. Crop Science 42:1701–1708.

Caten, C.E. 1974. Intra racial variation in *Phytophthora infestans* and adaptation to field resistance for potato late blight. Ann. Appl. Biol. 77:259–270.

Chen, F., D. Prehn, P.M. Hayes, D. Mulrooney, A. Corey, and H. Vivar. 1994. Mapping genes for resistance to barley stripe rust (*Puccinia striiformis* f. sp. *hordei*). Theor. Appl. Genet. 88:215–219.

Chen, H., S. Wang, Y. Xing, C. Xu, P.M. Hayes, and Q. Zhang. 2003. Comparative analysis of genomic locations and race specificities of loci for quantitative resistance to *Pyricularia grisea* in rice and barley. Proc. Natl. Acad. Sci. 100:2544–2549.

Clifford, B.C., and R.B. Clothier. 1974. Physiologic specialization of *Puccinia hordei* on barley hosts with nonhypersensitive resistance. Trans. Brit. Mycol. Soc. 63:421–430.

Das, M.K., S. Rajaram, W.E. Kronstad, C.C. Mundt, and R.P. Singh. 1993. Associations and genetics of three components of slow rusting in leaf rust of wheat. Euphytica 68:99–109.

Ellingboe, A.H. 1976. Genetics of host-parasite interactions. In R. Heitefuss and P.H. Williams, (ed). Physiological plant pathology, Springer-Verlag, Berlin.

Ellis, J.G., G.J. Lawrence, W.K. Peacock, and A.J. Pryor. 1998. Approaches to cloning plant genes conferring resistance to fungal pathogens. Annu. Rev. Phytopathol. 26:245–263.

Frary, A., T.C. Nesbitt, A. Frary, S. Grandillo, E. van de Knaap, B. Cong, J. Liu, J. Meller, R. Elber, K.B. Alpert, and S.D. Tanksley. 2000. *fw2.2*: A quantitative trait locus key to the evolution of tomato fruit size. Science 289:85–88.

Gabriel, D.W., and B.G. Rolfe. 1990. Working models of specific recognition in plant-parasite interactions. Annu. Rev. Phytopathol. 28:365–391.

Geiger, H.H., and M. Heun. 1989. Genetics of quantitative resistance to fungal diseases. Annu. Rev. Phytopathol. 27:317–341.

Graner, A., A. Jahoor, J. Schondelmaier, H. Siedler, K. Pillen, G. Fischbeck, G. Wensel, and R.G. Herrmann. 1991. Construction of an RFLP map of barley. Theor. Appl. Genet. 83:250–256.

Hayes, P.M., D. Prehn, H. Vivar, T. Blake, A. Comeau, I. Henry, M. Johnston, B. Jones, and B. Steffenson. 1996a. Multiple disease resistance loci and their relationship to agronomic and quality loci in a spring barley population. J. Quant. trait loci. http://www.probe.nalusda.gov:8000/otherdocs/jqtl/index.htm.

Hayes, P.M., F.Q. Chen, A. Kleinhofs, A. Kilian, and D. Mather. 1996b. Barley genome mapping and its applications. *In* P.P. Jauhar (ed.). Methods of genome analysis in plants, CRC Press, Boca Raton, FL.

Hayes, P.M., J. Cerono, H. Witsenboer, M. Kuiper, M. Zabeau, K. Sato, A. Kleinhofs, D. Kudrna, A. Kilian, M. Saghai-Maroof, D. Hoffman, and N.A.B.G.M.P. 1997. Characterizing and exploiting genetic diversity and quantitative traits in barley (*Hordeum vulgare*). J. Quant. Trait Loci. http://www.probe.nalusda.gov:8000/otherdocs/jqtl/jqtl1997-02/.

Hayes, P.M., X. Chen, A. Corey, M. Johnston, A. Kleinhofs, J. Korte, D. Kudrna, T.T. Toojinda, and H. Vivar. 1999. A summary of barley stripe rust mapping efforts. Plant and Animal Genome VII.

Hayes, P.M., A.E. Corey, R. Dovel, R. Karow, C. Mundt, K. Rhinart, and H. Vivar. 2000a. Registration of Orca barley. Crop Sci. 40:849–851.

Hayes, P.M., A. Castro, A. Corey, L. Marquez-Cedillo, B. Jones, D. Mather, I. Matus, C. Rossi, and K. Sato. 2000b. A summary of published barley QTL reports. http://www.css.orst.edu/barley/nabgmp/qtlsum.htm.

Hayes, P.M., A. Castro, A.E. Corey, T. Filichkin, C. Rossi, J.S. Sandoval, M.I. Vales, H.E. Vivar, and J. von Zitzewitz. 2001. Collaborative stripe rust resistance gene mapping and deployment efforts. *In* H.E. Vivar and A. McNab (eds.). Breeding barley in the new millenium: Proceeding of an international symposium, Ciudad Obregon, Sonora, Mexico, D.F.: CIMMYT.

Hayes, P.M., A. Castro, L. Marquez-Cedillo, A. Corey, C. Henson, B.L. Jones, J. Kling, D. Mather, I. Matus, C. Rossi, and K. Sato. 2003a. Genetic diversity for quantitatively inherited agronomic and malting quality traits. *In* R. Von Bothmer, H. Knupffer, T. van Hintum, and K. Sato (ed.). Diversity in Barley. Elsevier Science Publishers, Amsterdam.

Hayes, P.M., A.E. Corey, C. Mundt, T. Toojinda, and H. Vivar. 2003b. Registration of Tango barley. Crop Sci. 43:729–731.

Hulbert, S.H. 1997. Structure and function of the *rp1 complex* conferring rust resistance in maize. Annu. Rev. Phytopathol. 39:285–312.

Hulbert, S.H., C.A. Webb, S.M. Smith, and Q. Sun. 2001. Resistance gene complexes: evolution and utilization. Annu. Rev. Phytopathol. 39:285–312.

Jansen, R.C., and P. Stam. 1994. High resolution mapping of quantitative traits into multiple loci via interval mapping. Genetics 136:1447–1455.

Jeffrey, S.I., B. Jinks, J. L, and M. Grindle. 1962. Intraracial variation in *Phytophthora infestans* and field resistance to potato late blight. Genetica 32:323–328.

Jenns, A.E., K.J. Leonard, and R.H. Moll. 1982. Variation in the expression of specificity in two maize diseases. Euphytica 31:269–279.

Jinks, J.L., and M. Grindle. 1963. Changes induced by training in *Phytophthora infestans*. Heredity 18:245–264.

Johnson, R. 1981. Durable resistance: Definition of, genetic control, and attainment. Phytopathology 71:567–568.

Kaeppler, S. 1997. Power analysis for quantitative trait locus mapping in populations derived by multiple backcrosses. Theor. Appl. Genet. 95:618–621.

Kanazin, V., L.F. Marex, and R.C. Shoemaker. 1996. Resistance gene analogs are conserved and clustered in soybean. Proc. Natl. Acad. Sci. 93:11746–11750.

Kasha, K.J., A. Kleinhofs, A. Kilian, M.S. Maroof, G.J. Scoles, P.M. Hayes, F.Q. Chen, X. Xia, X.Z. Li, R.M. Biyashev, D. Hoffmann, L. Dahleen, T.K. Blake, B.G. Rossnagel, B.J. Steffenson, P.L. Thomas, D.E. Falk, A. Laroche, W. Kim, and S.J. Molnar. 1995. The North American barley genome map on the cross HT and its comparison to the map on cross SM. *In* K. Tsunewaki (ed.). Plant genome and plastome: Their structure and evolution Kodansha Scientific LTD, Tokyo Japan.

Kleinhofs, A., A. Kilian, M.A.S. Maroof, R.M. Biyashev, P. Hayes, F.Q. chen, N. Lapitan, A. Fenwick, T.K. Blake, V. Kanazin, E. Ananiev, L. Dahleen, D. Kudrna, J. Bollinger, S.J. Knapp, B. Liu, M. Sorrells, M. Heun, J.D. Franckowiak, D. Hoffman, R. Skadsen, and B.J. Steffenson. 1993. A molecular, isozyme and morphological map of barley *(Hordeum vulgare*). Theor. Appl. Genet. 86:705–712.

Kolmer, J.A. 1996. Genetics of resistance to wheat leaf rust. Annu Rev. Phytopathol. 34:435–455.

Kolmer, J.A., and K.J. Leonard. 1986. Genetic selection and adaptation of *Cochliobolus heterostrophus* to corn hosts with partial resistance. Phytopathology 76:774–777.

Kover, P.X., and A.L. Caicedo. 2001. The genetic architecture of disease resistance in plants and the maintenance of recombination by parasites. Mol. Ecol. 10:1–16.

Kulkarni, R.N., and V.L. Chopra. 1982. Environment as the cause of differential interaction between host cultivars and pathogen races. Phytopathology 72:1384–1386.

Latin, R.X., D.R. MacKenzie, and H.J. Cole. 1981. The influence of host and pathogen genotypes on the apparent infection rates of potato late blight epidemics. Phytopathology 71:82–85.

Lehman, J.S., and G. Shaner. 1997. Selection of populations of *Puccinia recondita* f. sp. *tritici* for shortened latent period on a partially resistant wheat cultivar. Phytopathology 87:170–176.

Leonard, K.J. 1969. Selection in heterogeneous populations of *Puccinia graminis* f. sp. *avenae*. Phytopathology 59: 1851–1857.

Leonard, K.J., and C.C. Mundt. 1984. Methods for estimating epidemiological effects of quantitative resistance to plant diseases. Theor. Appl. Genet. 67:219–230.

Leonards-Schippers, C., W. Gieffers, R. Pregl-Schafer, S.J. Knapp, F. Salamini, and C. Gebhardt. 1994. Quantitative resistance to *Phytophthora infestans* in potato: A case study for QTL mapping in an allogamous plant species. Genetics 137:67–77.

Marshall, D., and R.L. Sutton. 1995. Epidemiology of stripe rust, virulence of *Puccinia striiformis* f. sp. *hordei*, and yield loss in barley. Plant Dis. 70:732–737. Barley Genetics VII Proc. of Intl. Symp., Saskatoon, Canada.

Martin, G.B., S.H. Brommonschenkel, J. Chunwongse, A. Frary, M.W. Ganal, R. Spivey, T. Wu, E.D. Earle, and S.D. Tanksley. 1993. Map-based cloning of a protein kinase gene conferring disease resistance in tomato. Science 262:1432–1436.

Melchinger, A.E., H.F. Utz, and C.C. Schon. 1998. Quantitative trait locus (QTL) mapping using different testers and independent population samples in maize reveals low power of QTL detection and large bias in estimates of QTL effects. Genetics 149:383–403.

Michelmore, R.W. 1995. Molecular approaches to manipulation of disease resistance genes. Annu. Rev. Phytopathol. 33:393–428.

Milus, E.A., and R.F. Line. 1986. Number of genes controlling high-temperature, adult plant resistance to stripe rust in wheat. Phytopathology 76:93–96.

Mundt, C.C., C. Cowger, and K.A. Garrett. 2002. Relevance of integrated disease management to resistance durability. Euphytica 124:245–252.

Nelson, R.R. 1978. Genetics of horizontal resistance to plant diseases. Annu. Rev. Phytopathol. 16:359–378.

Parlevliet, J.E. 1977. Evidence of differential interaction in the polygenic *Hordeum vulgare Puccinia hordei* relation during epidemic development. Phytopathology 67:776–778.

Parlevliet, J.E. 1979. Components of resistance that reduce the rate of epidemic development. Annu. Rev. Phytopathol. 17:203–222.

Parlevliet, J.E. 1986. Pleiotropic association of infection frequency and latent period of two barley cultivars partially resistant to barley leaf rust. Euphytica 35:267–272.

Parlevliet, J.E. 1989. Identification and evaluation of quantitative resistance. McGraw-Hill, New York, NY.

Parlevliet, J.E., and J.C. Zadoks. 1977. The integrated concept of disease resistance; a new view including horizontal and vertical resistance in plants. Euphytica 26:5–21.

Powell, W., E. Baird, A. Booth, P. Lawrence, M. MacAulay, N. Bonar, G. Young, W.T.B. Thomas, J.W. McNicol, and R. Waugh. 1996. Single and multi-locus molecular assays for barley breeding and research. Barley Genetics VII Proc. of Intl. Symp., Saskatoon, Canada.

Quayom, A., and R.F. Line. 1985. High-temperature, adult-plant resistance to stripe rust of wheat. Phytopathology 75:1121–1125.

Ramsay, L., M. Macaulay, S.D. Ivanissevich, K. MacLean, L. Cardle, J. Fuller, K.J. Edwards, S. Tuvesson, M. Morgante, A. Massari, E. Maestri, N. Marmiroli, T. Sjakste, M. Ganal, W. Powell, and R. Waugh. 2000. A simple sequence repeat-based linkage map of barley. Genetics 156:1997–2005.

Robertson, D.S. 1985. A possible technique for isolating genic DNA for quantitative traits in plants. J. Theor. Biol. 117:1–10.

Robinson, R.A. 1976. Plant pathosystems, Springer-Verlag, New York.

Russell, J., J. Fuller, G. Young, B. Thomas, G. Taramino, M. Macaulay, R. Waugh, and W. Powell. 1997. Discriminating between barley genotypes using microsatellite markers. Genome 40:442–450.

Saghai-Maroof, M.A., R.M. Biyashev, G.P. Yang, Q. Zhang, and R.W. Allard. 1994. Extraordinarily polymorphic microsatellite DNA in barley: Species diversity, chromosomal locations, and population dynamics. Proc. Natl. Acad. Sci. 91: 5466–5470.

Salmeron, J.M., S.J. Barker, F.M. Carland, A.Y. Mehta, and B.J. Staskawicz. 1994. Tomato mutants altered in bacterial disease resistance provide evidence for a new locus controlling pathogen recognition. The Plant Cell 6:511–520.

Sandoval-Islas, J.S., L. Broers, H. Vivar, and K. Osada. 1998. Evaluation of quantitative resistance to yellow rust in ICARDA/CIMMYT's barley breeding programme. Plant Breed. 117:127–130.

Sato, K., T. Inukai, and P.M. Hayes. 2001. QTL analysis of resistance to the rice blast pathogen in barley (*Hordeum vulgare*). Theor. Appl. Genet. 102:916–920.

Schulze-Lefert, C. Peterhaensel, and A. Freialdenhoven. 1997. The gene-for-gene relationship in plant-parasite interactions, CAB International, Wallingford, United Kingdom.

Singh, R.P., and A.K. Gupta. 1992. Expression of wheat leaf rust resistance gene *Lr34* in seedlings and adult plants. Plant Dis. 76:489–491.

Singh, R.P., J.C. Nelson, and M.E. Sorrells. 2000. Mapping *Yr28* and other genes for resistance to stripe rust in wheat. Crop Sci. 40:1148–1155.

Thomas, W.T.B., W. Powell, R. Waugh, K.J. Chalmers, U.M. Barua, P. Jack, V. Lea, B.P. Forster, J.S. Swanston, R.P. Ellis, P.R. Hanson, and R.C.M. Lance. 1995. Detection of quantitative trait loci for agronomic, yield, grain and disease characters in spring barley (*Hordeum vulgare* L.). Theor. Appl. Genet. 91:1037–1047.

Toojinda, T., E. Baird, A. Booth, L. Broers, P. Hayes, W. Powell, W. Thomas, H. Vivar, and G. Young. 1998. Introgression of quantitative trait loci (QTLs) determining stripe rust resistance in barley: An example of marker-assisted line development. Theor. Appl. Genet. 96:123–131.

Toojinda, T., E. Baird, L. Broers, X.M. Chen, P.M. Hayes, A. Kleinhofs, J. Korte, D. Kudrna, H. Leung, R.F. Line, W. Powell, and H. Vivar. 2000. Mapping quantitative and qualitative disease resistance genes in a doubled haploid population of barley. Theor. Appl. Genet. 101:580–589.

Utz, H.F., A.E. Melchinger, and C.C. Schön. 2000. Bias and sampling error of the estimated proportion of genotypic variance explained by quantitative trait loci determined from experimental data in maize using cross validation and validation with independent samples. Genetics 154:1839–1849.

Vanderplank, J.E. 1963. Plant diseases: Epidemics and control, Academic Press, New York.

Vanderplank, J.E. 1968. Disease resistance in plants, Academic Press, New York.

Vanderplank, J.E. 1978. Genetic and molecular basis of plant pathogenesis, Springer-Verlag, Berlin.

Vanderplank, J.E. 1982. Host pathogen interactions in plant disease, Academic Press, New York.

Von Wettstein-Knowles, P.V. 1992. Cloned and mapped genes: Current status CAB International. Wallingford, United Kingdom.

Wang, G.L., D.J. Mackill, J.M. Bonman, S.R. McCouch, M.C. Champoux, and R.J. Nelson. 1994. RFLP mapping of genes conferring complete and partial resistance to blast in a durably resistant rice cultivar. Genetics 136:1421–1434.

Wei, F., K. Gobelman-Werner, S.M. Morrol, J. Kurth, L. Mao, R. Wing, D. Leister, P. Schulze-Lefert, and R.P. Wise. 1999. The *Mla* (powdery mildew) resistance cluster is associated with three NBS-LRR gene families and suppressed recombination within a 240-kb DNA interval on chromosome 5s (1HS) of barley. Genetics 153:1929–1948.

Yamamoto, T., Y. Kuboki, S.Y. Lin, T. Sasaki, and M. Yano. 1998. Fine mapping of quantitative trait loci *Hd-1*, *Hd-2* and *Hd-3*, controlling heading date of rice, as single Mendelian factors. Theor. Appl. Genet. 97:37–44.

Yano, M., Y. Harushima, Y. Nagamura, N. Kurata, Y. Minobe, and T. Sasaki. 1997. Identification of quantitative trait loci controlling heading date in rice using a high-density linkage map. Theor. Appl. Genet. 95:1025–1032.

Yano, M., Y. Katayose, M. Askikari, U. Yamanouchi, L. Monna, T. Fuse, T. Baba, K. Yamamoto, and T. Sasaki. 2000. http://www.intlpag.org/pag/8/abstracts/lpag8819.htm.

Young, N.D. 1996. QTL mapping and quantitative disease resistance in plants. Ann. Rev. Phytopathol. 34:479–501.

Zeng, Z.B. 1994. Precision mapping of quantitative trait loci. Genetics 136:1457–1468.

Zhou, J., Y.Y. Loh, R.A. Bressan, and G.B. Martin. 1995. The tomato gene *Pti1* encodes a serine/threonine kinase that is phosphorylated by *Pto* and is involved in the hypersensitive response. Cell 83:925–935.

14

Breeding for Resistance To Abiotic Stresses in Rice: The Value of Quantitative Trait Loci

David J. Mackill, International Rice Research Institute (IRRI), Philippines

Abstract

While rice breeders have been very successful in developing cultivars that are widely accepted by farmers and consumers, many challenges remain, in particular, improving complex traits such as yield and tolerance for abiotic stresses. This paper discusses how the mapping of quantitative trait loci (QTLs) has provided an important tool for improving rice for quantitative traits. With present technology, marker-assisted selection (MAS) will most likely be effective for traits controlled by a few number of QTLs with large effects. In rice, such QTLs have been identified for tolerance to abiotic stresses, including submergence, salinity, P deficiency, low temperature, Fe toxicity, and Al toxicity. Some putative drought-resistance QTLs also appear promising. In addition, there are a number of cultivars that are widely grown in South and Southeast Asia, where abiotic stresses frequently limit rice production. Incremental improvement of these cultivars by marker-assisted backcrossing (MAB) is a viable strategy to develop new and improved varieties. Effective use of MAB will ensure that newly deployed QTLs will be in a genetic background acceptable to farmers in these regions. The practice of MAB with major QTLs is available with existing technology. The advances of functional genomics and further cost reductions of marker technology will allow MAS for QTLs to be more widely integrated into conventional rice-breeding programs.

Introduction

Rice breeding in the latter half of the twentieth century has been a remarkable success. Modern rice varieties have been developed and have spread to large areas, enabling the countries of Asia to meet the food needs of their expanding populations. After achieving the breakthrough in yield potential with the semidwarf varieties, rice breeders successfully incorporated early maturity, improved cooking quality, and resistance to insects and diseases. It is now true that traditional varieties are more the exception than the rule (Khush, 1995). One of the key achievements during this period was the development and spread of rice varieties with resistance to diseases and insect pests (Khush, 1984; Bonman et al., 1992). Newer varieties are continually being developed that have improved grain quality as well as resistance to new disease races or insect biotypes. However, many challenges remain for rice breeders, in particular, the improvement of complex traits such as yield, nutritional quality, and resistance to abiotic stresses.

Improvement of rice for quantitative traits, which includes most agronomic traits such as yield, resistance to abiotic stresses, and partial resistance to biotic stresses, has depended on the standard breeding methods for self-pollinated crops, in particular the pedigree method of breeding. Studies using classical methods of quantitative genetics have provided information on heritability and gene action of many quantitative traits, but have had little practical impact in rice-breeding programs. It was only the availability of DNA markers in the late 1980s that allowed the identification of QTLs underlying the inheritance of these traits. Rice breeders developed a number of advanced populations using fixed lines that could be used for QTL mapping of multiple traits in multiple environments. These included recombinant inbred lines (RILs), doubled-haploid lines (DHs), and backcross-inbred lines (BILs). The RILs of a cross between the *indica* cultivar CO39 and the tropical *japonica* cultivar Moroberekan was the first such population used to map a number of quantitative traits, including partial resistance to blast disease (Wang et al., 1994) and drought-related traits (Champoux et al., 1995). Many QTLs for additional traits in a range of rice populations were mapped (for reviews see McCouch and Doerge, 1995; Yano and Sasaki, 1997; Li, 2001; Xu, 2002). QTL mapping has been proceeding at an accelerating pace over the last decade, but this technology has not yet been generally adopted for rice breeding. This paper discusses the application of QTL mapping and MAS to breeding for tolerance to abiotic stresses, which are promising target traits for this technology in rice.

Why breeders have been slow to use QTLs

Despite increasing information on QTLs for economically important quantitative traits, breeders have not been able to take advantage of this technology to develop improved cultivars. Constraints to the use of MAS for quantitative traits include the following:

- **Poor resolution of QTLs.** In most cases the QTL position can only be estimated in a fairly large chromosomal region (more than 20 cM on the genetic map).
- **Small effects of QTLs.** For many traits, many QTLs of relatively small effect control the trait, which would require a cumbersome process of selection for multiple QTLs in a MAS program.
- **Interaction of QTLs with environment or genetic background.** The effects of QTLs are not consistent across populations or in different environments.
- **Use of mapping populations not relevant to breeding objectives.** In most cases, these populations have been selected for sufficient polymorphism and divergence of the traits to allow mapping.
- **Expense of genotyping.** While costs are decreasing, the use of MAS in the large breeding populations used by breeders is still problematic.

Breeders would be much more likely to use MAS if QTLs were of relatively large effect and were independent of genetic background (i.e., would be expressed in a wide range of genotypes). In addition, traits that are more difficult to measure would offer an attractive target for MAS. While mapping these QTLs would be expensive, the investment would pay off in a better screen once effective QTLs were identified. The problems of poor resolution of QTLs and inconsistent occurrence across trials result in part from the low precision of many QTL mapping experiments. A high broad-sense heritability (H, the proportion of the phenotypic variance explained by genotype) is necessary to reliably detect QTLs. This high H can be achieved by use of carefully designed screens and adequate replication within and among trials. While traits that normally exhibit low H are thought to be appropriate subjects for QTL analysis, the H should be maximized in mapping experiments for accurate QTL detection.

The value of QTLs for breeding can be assessed by several statistics. These include the LOD (log-odds) score, which is a likelihood score for the presence of the QTL; the R^2 value, which is the proportion (or percentage) of the phenotypic variation explained by the QTL; and the effect of the QTL, which is the additive value of an allele at the locus. In the following discussion, QTLs with high LOD scores (usually above 6) or R^2 values (usually above 20%) are considered "major" QTLs. The value and reliability of each of these statistics is greatly affected by the design and precision of the QTL-mapping experiment. The expected value of R^2 is the product of H and the proportion of the genetic variance explained by the QTL. Therefore, even QTLs with rather large effects are difficult to detect in experiments with low levels of replication or imprecise phenotyping. High grain yield is an essential requirement for rice varieties and is the most important quantitative trait, along with superior grain quality. Yield QTLs have been mapped by a number of researchers in rice (14.1). In these studies, QTLs have relatively low LOD scores; in the studies cited, only one QTL contributed 15.7% to phenotypic variation, and all others were 12%

or lower. If these QTLs actually represented a 12% yield increase above the existing levels, they would be highly promising. However, these levels are usually measured in relation to segregating populations and should not be considered as a direct add-on to present yields. Furthermore, the identification of QTLs is specific to the genetic background of the population used and the location of testing. They may not be observed in other locations or other populations.

A recent report (Hittalmani et al., 2003) summarized QTLs identified in a single population (the DH population of IR64/Azucena) in nine Asian locations. Three QTLs were identified for grain yield; however, these were identified in three, two, or one location only, and percentage of variation explained ranged from 7.7 to 15.1 per locus. The highest LOD score observed was 3.62. These results indicate that yield is not currently a suitable trait for manipulation by MAS in rice. Considering the requirements for effective use of MAS for QTLs, abiotic stresses that can be realistically imposed in precise screens seem to offer unique opportunities in the application of markers.

Abiotic stresses in rice

Climatic and soil factors often result in unfavorable growing conditions to the rice plant. Excess or deficits of water, extremes of temperature, and mineral deficiencies or toxicities are the common abiotic stresses affecting rice. The abiotic stresses have been a long-term objective of rice-breeding programs. However, progress in developing cultivars with tolerance to these specific stresses has been slow for several reasons, including the quantitative nature of their inheritance, the difficulty of devising accurate screens, the undesirable traits of the best donor cultivars, and the presence of multiple stresses in many target areas. Progress in identifying QTLs for these traits is briefly described below.

Water stresses

Drought

Drought is the most widespread and damaging of abiotic stresses and has also attracted the most interest for QTL mapping. Breeding for drought resistance has been hampered by a low level of genetic variability, and complex inheritance of the trait. Probably the most serious constraint to improving drought resistance is the difficulty of measuring the trait (phenotyping) accurately. Early work on drought resistance focused on symptoms such as leaf death and leaf rolling observed under vegetative stage stress. However, it has become increasingly clear that these evaluations of drought resistance were not usually related to the most important trait, yield under stress, or yield in the target environment (which would include yield under stress as well as yield potential without stress). These types of measurements are more expensive and difficult to obtain, but the effort is needed for accurate phenotyping.

Early QTL studies focused on secondary traits thought to be related to drought resistance, such as root depth, thickness, and volume; root penetration ability; osmotic adjustment; and leaf rolling or death. This has resulted in accumulation of considerable QTL data. The following populations have been used to map drought-related QTLs (I = *indica* subspecies, J = *japonica* subspecies, L = lowland, U = upland):

- CO39(I,L)/Moroberekan(J,U) RILs (Champoux et al., 1995; Lilley et al., 1996; Ray et al., 1996; Zheng et al., 2000)
- CT9993(J,U)/IR62266(I,L) DHLs (Tripathy et al., 2000; Zhang et al., 2001; Babu et al., 2003)
- Azucena(J,U)/Bala(I,U) RILs (Price and Tomos, 1997; Price et al., 1997; Price et al., 2000; Price et al., 2002a; Price et al., 2002b; Price et al., 2002c)
- IR64(I,L)/Azucena(J,U) DHLs (Courtois et al., 2000; Hemamalini et al., 2000; Zheng et al., 2000; Shen et al., 2001; Lafitte et al., 2002; Venuprasad et al., 2002)
- IR20(I,L)/63-83(J,U) F_2 (Quarrie et al., 1997)
- IR58821-23-B-1-2-1(I,L)/IR52561-UBN-1-1-2(I,L) RILs (Ali et al., 2000)

Very few studies have mapped QTLs related to the most important trait, yield under drought. A field study of the population CT9993/IR62266 was conducted in India by Babu et al. (2003). They observed a region on chromosome 4 that contained major QTLs for plant height, grain yield, and number of grains per panicle under stress. There were also QTLs in this region for root traits. A few QTLs were identified that had relatively strong effects, although LOD scores were generally modest.

At the moment, the use of MAS for any of these QTLs would be problematic. It is not clear if the QTLs would have a detectable effect in different genetic backgrounds or under varied types of drought stress. One approach might be to introgress some of the promising QTLs into a suitable genetic background and determine if they contribute to drought resistance. For example, Shen et al. (2001) introduced four QTLs for deeper roots from the upland cultivar Azucena into the lowland high-yielding variety IR64 by backcrossing, although yield performance was not reported. However, these introgressions did not have a consistent effect on root system size or depth in the introgressed lines (Shen et al., 2001), and recent agronomic evaluations have not shown them to have consistent effects on yield under stress (IRRI, unpublished data).

At IRRI, the focus on QTL analysis for drought tolerance has shifted from the genetic dissection of secondary physiological traits to the evaluation of grain yield under stress. A range of populations derived from crosses between highly resistant and highly susceptible parents have been generated for this purpose. This approach is designed to identify alleles with major effects on yield under stress. Although there are large differences among rice lines in yield in drought-stressed environments, it remains to be seen if these differences are due to genes with effects large enough to be useful in MAB.

Flooding

Three traits related to flooding tolerance are (1) tolerance to short-term submergence, (2) rapid internode elongation ability (to escape deep water), and (3) germination under anaerobic (flooded) conditions. Genetic studies have not been reported for the last trait. QTLs for internode elongation ability were mapped in a RIL population from IR74 (lowland) crossed with the deepwater variety Jalmagna (Sripongpangkul et al., 2000). A major locus near or at the semidwarf locus *sd1* was responsible for plant height and increase in plant height and internode length in response to rising water level, with less-important loci on other chromosomes. Toojinda et al. (2003) found alleles for shoot elongation under submergence on other chromosomes.

Submergence tolerance is the most useful of these survival mechanisms for tolerance to flooding. Submergence tolerant cultivars can survive periods up to two weeks under water. Most cultivars are severely damaged within a week of flooding. The most widely used source of submergence tolerance is the Indian cultivar FR13A. A major QTL was shown to control submergence tolerance in this trait (Xu and Mackill, 1996). FR13A also has additional QTLs that contribute to its tolerance (Nandi et al., 1997; Toojinda et al., 2003).

The major QTL from FR13A, designated *Sub1*, has been fine-mapped to an interval of less than 0.5 cM (Xu et al., 2000). Simple sequence repeat markers closely linked to this locus have been used to transfer it into different genetic backgrounds, including both *japonica* and *indica* varieties (Siangliw et al. 2003; Xu et al. 2004). Therefore, this QTL is an excellent candidate for application of MAS.

Temperature stresses

Most studies have focused on tolerance to low temperature, which is a common stress in both temperate and subtropical regions and in high-elevation areas of the tropics. *Japonica* cultivars are more tolerant than *indica* cultivars at both the vegetative and reproductive stages. For tolerance at the booting stage, Saito et al. (2001) identified two QTLs that were transferred by backcrossing from a tropical *japonica* cultivar into a Japanese termperate *japonica* rice. Andaya and Mackill (2003b) identified QTLs in a cross between a temperate *japonica*, M-202, and an *indica*, IR50. In general, these QTLs at the booting stage had a relatively small effect.

In contrast to the booting stage, cold tolerance at the vegetative stage is controlled by both major and minor QTLs. Major QTLs for wilting and chlorosis were identified by conventional genetic studies (Kwak et al., 1984; Nagamine, 1991). Major QTLs were also identified in the M-202/IR50 population for cold-induced wilting and tolerance to necrosis (Andaya and Mackill 2003a). These QTLs might have an advantage for improving the cold tolerance of *indica* cultivars that are preferred in tropical environments. This trait is required where a dry-season crop of rice is seeded from November to January in higher latitudes, for example, the boro crop in India or Bangladesh, or in high-elevation tropical locations. QTLs for tolerance to low-temperature at the germination stage were identified by Misawa et al. (2000) in an *indica*/*japonica* cross.

Soil-related stresses

Phosphorus deficiency

Phosphorus (P) deficiency is a widespread problem in many rice-growing areas where farmers often do not have access to phosphate fertilizers, and rice soils frequently have a high P-fixing capacity. Two QTL-mapping studies have been conducted in rice. Wissuwa et al. (1998) used a backcross inbred population with the recurrent parent Nipponbare (*japonica*, sensitive) and the variety Kasalath (*indica*, tolerant). In addition to some minor QTLs, a major QTL on chromosome 12 was identified for P uptake, P use efficiency, dry weight, and tiller number. For P uptake, this QTL had a LOD score of 10.74 and explained 27.9% of the phenotypic variation. Ni et al. (1998) found a similarly strong QTL in the same location on chromosome 12 using RILs from a cross of IR20 (tolerant) with IR55178-3B-9-3 (sensitive). They measured relative tillering ability, relative shoot dry weight, and relative root dry weight. When this chromosome 12 locus, designated *Pup1*, was transferred by three backcrosses into the variety Nipponbare, the resulting lines showed 170% increase in P uptake and 250% increase in yield when grown under low-P conditions (Wissuwa and Ae, 2001b). The NILs with the *Pup1* allele from Kasalath had increased root growth under low-P conditions, but the differences in root growth and P uptake were not observed under anaerobic soil conditions (Wissuwa and Ae, 2001a).

Salinity

Koyama et al. (2001) used a RIL population of a cross between IR4630-22-2-5-1-3 (tolerant) and IR15324-117-3-2-2 (sensitive) to map QTLs for salt tolerance. QTLs were identified for a number of traits, but all had relatively low R^2 values. Similarly, Prasad et al. (2000), using a doubled haploid population of IR64/Azucena, observed a number of seedling-stage QTLs with low LOD scores and R^2 values. A major QTL for salt tolerance, named *saltol*, was mapped on rice chromosome 1 using recombinant inbred lines in a cross between the tolerant cultivar Pokkali and the susceptible IR29 (Gregorio, 1997). The gene had a LOD score of 14.5 and explained up to 80% of the phenotypic variation. This locus has been fine-mapped (Bonilla et al., 2002) and is being transferred by MAS into improved cultivars (G. Gregorio, personal communication).

Aluminum toxicity

A number of QTLs have been identified using two *indica/japonica* crosses. In the cross IR1552 (*indica*)/Azucena(*japonica*), four QTLs were identified, although none explained over 20% of the variation (Wu et al., 2000). In the cross CT9993 (*japonica*)/IR62266(*indica*), 10 QTLs for root-length ratio (stress/control) were identified, including major QTLs on chromosomes 1 and 8 (Nguyen et al., 2002). Two regions on chromosomes 1 and 9 appeared to be the same in both crosses, so the locus on chromosome 1 would seem to be a particularly important one, explaining up to 19% of the variation in IR1552/Azucena and 24% of the variation in CT9993/IR62266. In analysis of a population using the wild species *Oryza rufipogon* as a source of tolerance (Nguyen et al., 2003), seven QTLs were identified, with those on chromosomes 3 and 7 being particularly important. These loci could be very significant, because a progeny from this population has been released as a cultivar in Vietnam and has spread to over 100,000 ha of cultivation in the acid sulfate areas (D.S. Brar, personal communication).

Iron toxicity

Wu et al. (1997) measured iron toxicity tolerance in a doubled haploid population of Azucena and IR64. An Azucena allele for a QTL on chromosome 1 explained 32% of the variation in the population.

Target QTLs for MAS

Nearly all the studies on mapping abiotic stress genes cited above have used visual symptoms of plants as the measurement. However, in many cases, these symptoms correspond to the actual damage that is observed under field conditions, particularly when plant survival is the trait measured, as opposed to quantitative yield reduction. This would give confidence that the measurements are relevant to producing higher yields under stress, with the exception of drought resistance as described above. QTL candidates for marker-assisted selection should have a relatively large effect, be expressed in different genotypic backgrounds (or at least the background of the cultivar that should be improved), and have closely linked markers that can be used to select for the trait. The

Table 14.1 Target QTLs for marker-assisted backcrossing of abiotic stress resistance in rice

Trait	Chr	Markers	LOD	R^2	Population[a]	Reference
Highest priority QTLs						
Al toxicity (RRL)	3	CD01395-RG391	8.4	24.9	IR64/***O. rufipogon***	Nguyen et al., 2003
Submergence	9	C1232	36.0	69.0	**IR40931-26**/PI543851 (japonica)	Xu and Mackill, 1996
P deficiency	12	G2140-C443	10.7	27.9	Nipponbare/**Kasalath**	Wissuwa et al., 1998
Salt tolerance	1	C52903S-C1733S	6.7	43.9	**Pokkali**/IR29	Bonilla et al., 2002
Second priority QTLs						
Al toxicity (RRL)	7	RZ629-RG650	5.4	22.5	IR64/***O. rufipogon***	Nguyen et al., 2003
Al toxicity	1	CD0345	8.1	24.1	**CT9993**/IR62266	Nguyen et al., 2002
Al toxicity	8	C1121	8.2	28.7	**CT9993**/IR62266	Nguyen et al., 2002
Al toxicity	12	RG9	6.8	20	IR1552/**Azucena**	Wu et al., 2000
Fe toxicity	1	C955-C885	3.2	20.5	**Nipponbare**/Kasalath	Wan et al., 2003
Plant elongation (flooding)	1	RG109-*sd1*	21.5	29.6	IR74/**Jalmagna**	Sripongpangkul et al., 2000
Root length (29 d) (drought)	11	RG2 + 24 cM	6.9	29.8	Bala/**Azucena**	Price and Tomos, 1997
Submergence	6	AFLP markers	4.7	26.5	IR74/**FR13A**	Nandi et al., 1997
Submergence	7	AFLP markers	3.6	23.4	IR74/**FR13A**	Nandi et al., 1997
Submergence	5	R1553	11.2	34.1	**IR49830**-7/CT6241	Toojinda et al., 2003
P deficiency	6	AFLP markers	7.8	33.6	**IR20**/IR55178-3B-9-3	Ni et al., 1998
P deficiency	12	RG9-RG241	16.5	54.0	**IR20**/IR55178-3B-9-3	Ni et al., 1998
Cold tolerance	4	RM335-RM261	8.4	20.8	**M-202**/IR50	Andaya and Mackill, 2003a
Cold tolerance	12	RM101-RM292	18.5	41.7	**M-202**/IR50	Andaya and Mackill, 2003a
Cold tolerance	7			22.1	Akihikari/**Koshihikari**	Takeuchi et al., 2001
Drought–cell membrane stability	3	RZ403	12.1	42.1	CT9993/**IR62266**	Tripathy et al., 2000
Drought–cell membrane stability	9	RZ698-RM219	10.4	37.4	**CT9993**/IR62266	Tripathy et al., 2000
Drought–cell membrane stability	8	RG598	7.5	29.4	**CT9993**/IR62266	Tripathy et al., 2000
Cold-booting stage	4	R2737	28		Norin PL8 (introgression from Silewah into Hokkai241)	Saito et al., 1995; Saito et al,. 2001
Yield/drought	4	RG476-RG939	4.7	15.8	**CT9993**/IR62266	Babu et al., 2003
Basal root thickness	4	RG476-RG939	14.0	37.6	**CT9993**/IR62266	Zhang et al., 2001
Grain yld drought	12	AFLP	7.5	22.3	**CT9993**/IR62266	Babu et al., 2003

[a]Parent in bold is considered source of the desirable allele.

effect of the QTL should be sufficient to make a measurable difference in performance of a rice variety under farmers' field conditions. Some of the best candidate QTLs identified so far are listed in Table 14.1. It can be seen that these loci compare very favorably to those identified for yield (Table 14.2). This would suggest that measurable advances could be obtained by transferring these loci into elite genotypes.

Target cultivars for MAS

In general, rice breeding has been a highly successful enterprise, with the result that improved varieties have spread to most rice farmers. However, the release and adoption of new cultivars varies greatly from country to country. For simplicity, we can consider three general situations of variety adoption by farmers:

- Farmers are growing improved varieties, and newly released varieties that are superior to existing varieties in one or more important traits spread rapidly. This is common in more favorable areas of rice cultivation where there is good infrastructure, extension services, and seed multiplication capabilities.
- Farmers are growing unimproved varieties, and newly developed varieties are generally not adopted by farmers. This is found in unfavorable growing environments, especially where abiotic stresses limit the potential of improved varieties and farmers use low levels of inputs.
- Farmers are growing improved varieties, but newly released varieties are not widely adopted by the farmers. This can occur in favorable or unfavorable growing environments.

In the first case, farmers cultivate a number of varieties that include older varieties, where the area

Table 14.2 QTL mapping studies for grain yield in rice

Population	Environment	No. of QTLs (LOD > 2.5)	R^2 of strongest QTL	Reference
CT9993/IR62266 DHL	Tamil Nadu, India	None		Babu et al., 2003
Zhenshang 97/Minghui 63 RIL	Los Banos, Philippines	2	12.2	Cui et al., 2003
Zhenshan 97B/Milyang 46 RIL	Hangzhou, China	6	4.4	Zhuang et al., 2002
IR64/Azucena DH	Bangalore, India	1	15.7	Venuprasad et al., 2002
Zhenshang 97/Minghui 63 RIL	Wuhan, China 1997	3	7.2	Xing et al., 2002
	Wuhan, China 1998	4	10.0	
IR64/Azucena DH	Punjab, India	1	11.6	Hittalmani et al., 2002
Zhenshang 97/Minghui 63 F2/F3	Wuhan, China 1994	5	11.7	Yu et al., 1997
	Wuhan, China 1995	6	10.2	
O. rufipogon/V20 BC2	Hunan, China	7	5.2	Xiao et al., 1998
TSA/CB F_2	Hangzhou, China	5	11.4	Zhuang et al., 1997

of cultivation tends to be declining over time, and newly developed varieties whose area is increasing over time. In these areas, the life of a single variety may be short-lived, because newer, improved varieties are rapidly adopted. In the second case, farmers cultivate a large number of varieties that are genetically diverse. These varieties are suitable to the unfavorable conditions that predominate in these areas.

In the third situation, farmers have adopted improved varieties, but they are reluctant to adopt newer varieties. This situation is relatively common in favorable or mildly unfavorable areas of South and Southeast Asia. Limited adoption of new varieties in these areas may be due to deficiencies of these varieties in characteristics that were not evident during the evaluation process. Most commonly, these deficiencies may be related to grain quality or lack of tolerance to abiotic stresses. In other cases, new varieties may not be adopted even if they are clearly superior to the older varieties. A highly successful cultivar creates its own standard to which all new competitors are compared. The farmers and processors may have adopted practices optimized for this particular cultivar, and they are therefore reluctant to switch to cultivars with a different grain type or handling characteristics, even if consumers would find them perfectly acceptable. Another factor that may limit adoption of new cultivars is the inadequacy of public-sector testing programs. Most programs do not have resources to evaluate new breeding lines with sufficient number of well-managed yield trials to obtain reliable information about a line's merits.

There are a handful of rice varieties that are widely grown throughout tropical Asia. Improved varieties such as Swarna, Mahsuri, Samba Mahsuri, IR64, and BR11 are each grown on millions of hectares. In addition to these, several million hectares of rice in Thailand are cultivated to the variety Khao Dawk Mali 105 or the derived cultivar RD6, which are pure line selections from traditional land races. Most of these varieties were released over 10 or even 20 years ago, and they have been difficult to replace with new varieties. It is clear that these varieties have characteristics that make them highly preferred by many rice farmers. These varieties were spread rapidly through farmer-to-farmer contact and have not required major promotion efforts by the government or NGOs.

In circumstances where one or a few varieties are widely grown, a strategy of incremental improvement of these varieties by backcrossing is a viable approach for their improvement. This is particularly the case if there are one or a few genes/QTLs that would have a significant impact on the performance of the variety. Despite this situation, rice breeders have been reluctant to use the backcrossing method. One reason is that backcrossing is labor intensive, involving production of a large number of crossed seeds. Another reason is that the backcrossing method is considered a conservative approach that will not result in multiple improvements over the existing cultivar. An additional concern is that the recurrent parent may be obsolete by the time the new line is available. Finally, in a backcrossing program without the use of markers, a large chromosome fragment is introduced along with the gene of interest (Stam and Zeven, 1981), and genes on other chromosomes will also be carried along by random drift.

Despite these disadvantages, the backcrossing approach has many appealing features. As indicated above, there are many varieties that are widely accepted by rice farmers. Using these varieties as a basis for introducing valuable QTLs would have the advantage that the breeder would be more confident of their acceptability with the farmers.

Tanksley and Nelson (1996) advocated an approach termed advanced backcross QTL (ABQTL) for practical utilization of QTLs in plant breeding. While most mapping studies used F_2 or derived populations, the ABQTL approach used sequential backcrosses to transfer chromosomal regions into a superior cultivar. Because useful QTLs are identified in an elite cultivar, they can be immediately exploited for further breeding or varietal release.

The advantages of using molecular markers for introducing QTLs by backcrossing (i.e., MAB) include (1) the ability to rapidly remove the donor chromosomal segments not associated with the trait of interest, (2) the ability to select for recombination on either side of the gene being introduced, thus removing the effect of linkage drag, and (3) the improvement in efficiency of selection for recessive traits or those that are difficult to accurately measure, and (4) the fewer number of plants that need to be genotyped compared with the standard breeding approaches like the pedigree method.

The graphical genotype (Young and Tanksley, 1989) displays the parental origin of the chromosomal segments of segregating plants. When a relatively large number of backcross plants are generated, these plants can be genotyped using markers scattered over the genome. Plants with fewer markers from the donor can be selected for backcrossing. In this way, the recurrent parent genotype at markers unlinked to the gene being introduced can be recovered with fewer backcrosses and also reduce the number of marker data points needed (Frisch et al., 1999a).

An additional strategy is to select for recombination near the gene being introduced, so that unnecessary donor genes are not introduced in the MAB. For flanking markers that are less than 5 cM away from the gene, relatively large population sizes are needed to recover recombinants (Frisch et al., 1999b). For example, Chen et al. (2000) selected for recombinants within 0.8 and 3.0 cM flanking the bacterial blight resistance gene *Xa21* being introduced into a hybrid rice parent. This scheme will require hundreds of backcross F_1 plants to obtain close recombinants as well as to select against unlinked donor segments. In some cases this selection for flanking marker-recombination may not be necessary, because the chromosomal segment does not carry unfavorable genes. However, the main advantage of the MAB scheme is that the recurrent genotype is reliably reconstituted, and this reduction of the chromosome segment length transferred may be crucial to the success of the undertaking. One of the reasons that conventional breeding has not been successful in replacing these widely grown varieties is that they represent unique and rare accomplishments. In practice, it is observed that backcrosses by the conventional method fail to produce lines that are sufficiently like the recurrent parent to be used in their place.

The third advantage of the MAB approach is the ability to transfer traits that are difficult to screen for by selection for phenotype. This is the case for most quantitative traits, which require careful, replicated measurements to achieve high levels of H. Of course this makes accurate mapping of these QTLs a challenge, but once the investment has been made to do the mapping, the markers can be reliably used in many populations. Tolerance to abiotic stresses is often difficult to measure, but this will be one of the major advantages for MAB with these traits.

Rice-breeding programs can vary greatly in size, but most would grow a substantial number of plants for selection purposes. A large breeding program would grow upwards of one million F_2 plants. Selection of plants grown in progeny rows over subsequent generations would narrow this to a few hundred uniform lines (F_7) that would be ready for more intensive evaluation. With the existing expense of the technology, application of molecular markers would have to be very selective and highly targeted (Koebner and Summers, 2003). With the MAB approach, the population sizes are more amenable to genotyping.

Future prospects and conclusions

The above discussion presents a clear case for the use of the MAB scheme for introducing QTLs for abiotic stress tolerance into widely adapted rice cultivars. The development of such improved lines would be appropriate to add some stability to the production in these areas, where farmers suffer from fluctuations in growing conditions, as well as allow the adoption of these superior cultivars in areas where the prevalence of abiotic stresses has prevented their cultivation. Table 14.2 gives a list of the some of the primary targets for this approach in rice. In the first group are QTLs with sufficient information on their efficacy in the main genetic backgrounds needed so that they could be incorporated into such a strategy directly. In the second group are some highly promising QTLs that may need further study before they could be used with confidence. However, MAB can also be used as an approach for fine-scale mapping as well as validation of a QTL effect in the appropriate genetic background. Thus, once QTLs are identified, they can be backcrossed into a major target cultivar as a means of assessing their potential.

From the foregoing, it is clear that QTL–MAS is presently a reality in rice, and it is being used now to transfer the most significant QTLs into the standard rice cultivars. This approach is a promising strategy to add value to the predominant cultivars that are still popular with the farmers. It is not, however, an approach that would be used exclusively in addressing the problems of unfavorable growing environments. Undoubtedly, new and improved varieties can still be produced through conventional breeding methods, and these varieties may be improved for a number of traits not currently amenable to selection by markers. The success rate for these programs, in terms of (1) number of superior lines developed, (2) number of cultivars released, and (3) number of cultivars adopted by farmers in large areas, is low. However, even limited success will easily compensate for the cost of these activities, which would be relatively low compared with the benefits.

New advances in genomics have the potential to extend the application of molecular methods further into the conventional breeding process. There are a wealth of available resources in rice, including complete genome sequence for the two subspecies (Goff et al., 2002; Yu et al., 2002), functional genomics tools (Kishimoto et al., 2002), and a high-density microsatellite map (McCouch et al., 2002). Large-scale gene identification will occur in rice, and a major objective will be to determine what genes are responsible for the QTLs underlying economic traits. Desirable alleles at these loci will result in markers that can be used to select for multiple QTLs. High-throughput methods of DNA extraction and marker assays will enable screening of thousands of early-generation lines for any chromosome fragment or QTL of interest (Mackill and McNally, 2004). In the future, we can expect to see much more use of molecular markers as a substitute for detailed phenotyping in early generation populations, although the cost of the assays must be further reduced. While careful visual selection would continue to be applied, marker genotypes would predict performance for traits not evident on a single-plant basis or in a single-selection environment. The availability of these candidate gene markers will allow the use of MAS even for traits without major QTLs, such as grain yield. A potentially more powerful outcome of functional genomics is the ability to directly introduce genes from other sources or modify existing genes to achieve traits not available in the germplasm. Indeed the transgenic approach is compatible and in fact equivalent to the MAB approach outlined above.

Acknowledgments

The author is grateful for the valuable comments of the following IRRI colleagues: Gary Atlin, Abdel Ismail, Glenn Gregorio, and Renee Lafitte.

References

Ali M.L., M.S. Pathan, J. Zhang, G. Bai, S. Sarkarung, H.T. Nguyen. 2000. Mapping QTLs for root traits in a recombinant inbred population from two indica ecotypes in rice. Theor. Appl. Genet. 101:756–766.

Andaya V.C., D.J. Mackill. 2003a. Mapping of QTLs associated with cold tolerance during the vegetative stage in rice. J. Exp. Bot. 54:2579–2585.

Andaya V.C., D.J. Mackill. 2003b. QTLs conferring cold tolerance at the booting stage of rice using recombinant inbred lines from a *japonica* X *indica* cross. Theor. Appl. Genet. 106:1084–1090.

Babu R.C., B.D. Nguyen, V. Chamarerk, P. Shanmugasundaram, P. Chezhian, P. Jeyaprakash, S.K. Ganesh, A. Palchamy, S. Sadasivam, S. Sarkarung, L.J. Wade, H.T. Nguyen. 2003. Genetic analysis of drought resistance in rice by molecular markers: Association between secondary traits and field performance. Crop Science 43:1457–1469.

Bonilla P., J. Dvorak, D.J. Mackill, K. Deal, G. Gregorio. 2002. RFLP and SSLP mapping of salinity tolerance genes in chromosome 1 of rice (*Oryza sativa* L.) using recombinant inbred lines. Philipp. Agric. Sci. 85:68–76.

Bonman J.M., G.S. Khush, R.J. Nelson. 1992. Breeding rice for resistance to pests. Annu. Rev. Phytopathol. 30:507–528.

Champoux M., G. Wang, S. Sarkarung, N. Huang, D.J. Mackill, J.C. O'Toole, S.R. McCouch. 1995. Locating genes associated with root morphology and drought avoidance in rice via linkage to molecular markers. Theor. Appl. Genet. 90:969–981.

Chen S., Lin X.H., C.G. Xu, Q.F. Zhang. 2000. Improvement of bacterial blight resistance of "Minghui 63," an elite restorer line of hybrid rice, by molecular marker-assisted selection. Crop Science 40:239–244.

Courtois B., G. McLaren, P.K. Sinha, K. Prasad, R. Yadav, L. Shen. 2000. Mapping QTLs associated with drought avoidance in upland rice. Mol. Breed. 6:55–66.

Cui K.H., S.B. Peng, Y.Z. Xing, S.B. Yu, C.G. Xu, Q. Zhang. 2003. Molecular dissection of the genetic relationships of source, sink and transport tissue with yield traits in rice. Theor. Appl. Genet. 106:649–658.

Frisch M., M. Bohn, A.E. Melchinger. 1999a. Comparison of selection strategies for marker-assisted backcrossing of a gene. Crop Science 39:1295–1301.

Frisch M., M. Bohn, A.E. Melchinger. 1999b. Minimum sample size and optimal positioning of flanking markers in marker-assisted backcrossing for transfer of a target gene. Crop Science 39:967–975.

Goff S.A., D. Ricke, T.H. Lan, G. Presting, R.L. Wang, M. Dunn, J. Glazebrook, et al. 2002. A draft sequence of the rice genome (*Oryza sativa* L. ssp. *japonica*). Science 296:92–100.

Gregorio G.B. 1997. Tagging salinity tolerance genes in rice using amplified fragment length polymorphism (AFLP). Ph.D. thesis, University of the Philippines, College, Laguna.

Hemamalini G.S., H.E. Shashidhar, S. Hittalmani. 2000. Molecular marker assisted tagging of morphological and physiological traits under two contrasting moisture regimes at peak vegetative stage in rice (*Oryza sativa* L.). Euphytica 112:69–78.

Hittalmani S., N. Huang, B. Courtois, R. Venuprasad, H.E. Shashidhar, J.Y. Zhuang, K.-L. Zheng, G.-F. Liu, G.-C. Wang, J.S. Sidhu, S. Srivataneeyakul, V.P. Singh, P.G. Bagali, H.C. Prasanna, G. McLaren, G.S. Khush. 2003. Identification of QTL for growth- and grain yield-related traits in rice across nine locations of Asia. Theor. Appl. Genet. 107: 679–690.

Hittalmani S., H.E. Shashidhar, P.G. Bagali, N. Huang, J.S. Sidhu, V.P. Singh, G.S. Khush. 2002. Molecular mapping of quantitative trait loci for plant growth, yield and yield related traits across three diverse locations in a doubled haploid rice population. Euphytica 125:207–214.

Khush G.S. 1984. Breeding rice for resistance to insects. Prot. Ecol. 7:147–165.

Khush G.S. 1995. Modern varieties—their real contribution to food supply and equity. GeoJournal 35:275–284.

Kishimoto N., J.S. Yazaki, F. Fujii, K. Shimbo, T. Ohta, Z. Shimatani, Y. Nagata, A. Hashimoto, S. Kikuchi. 2002. Rice cDNA microarray: A powerful tool for transcriptome analysis in rice functional genomics. p. 49–59. *In* Recent Research Developments in Plant Biology. Vol 2: Recent Research Developments in Plant Biology. Research Signpost, Trivandrum 695008.

Koebner R.M.D., R.W. Summers. 2003. 21st century wheat breeding: plot selection or plate detection? Trends Biotechnol. 21:59–63.

Koyama M.L., A. Levesley, R.M.D. Koebner, T.J. Flowers, A.R. Yeo. 2001. Quantitative trait loci for component physiological traits determining salt tolerance in rice. Plant Physiol. 125:406–422.

Kwak T.S., B.S. Vergara, J.S. Nanda, W.R. Coffman. 1984. Inheritance of seedling cold tolerance in rice. SABRAO J. 16:83–86.

Lafitte H.R., B. Courtois, M. Arraudeau. 2002. Genetic improvement of rice in aerobic systems: progress from yield to genes. Field Crop Res. 75:171–190.

Li Z. 2001. QTL mapping in rice: A few critical considerations. p. 153–171. *In* G.S. Khush, D.S. Brar, B. Hardy (eds), Rice genetics IV. International Rice Research Institute, Los Baños, Philippines.

Lilley J.M., M.M. Ludlow, S.R. McCouch, J.C. O'Toole. 1996. Locating QTL for osmotic adjustment and dehydration tolerance in rice. J. Exp. Bot. 47:1427–1436.

Mackill D.J., K.L. McNally. 2004. A model crop species: molecular markers in rice. P. 39–54. *In* Lörz H, Wenzel G (eds), Molecular marker systems in plant breeding and crop improvement. Vol. 55. Biotechnology in Agriculture and Forestry. Springer Verlag, Heidelburg.

McCouch S.R., R.W. Doerge. 1995. QTL mapping in rice. Trends Genet. 11:482–487.

McCouch S.R., L. Teytelman, Y. Xu, K.B. Lobos, K. Clare, M. Walton, B. Fu, R. Maghirang, Z. Li, Y. Zing, Q. Zhang, I. Kono, M. Yano, R. Fjellstrom, G. DeClerck, D. Schneider, S> Cartinhour, D. Ware, L. Stein. 2002. Development and mapping of 2240 new SSR markers for rice (*Oryza sativa* L.). DNA Res. 9:199–207.

Misawa S., N. Mori, S. Takumi, S. Yoshida, C. Nakamura. 2000. Mapping of QTLs for low temperature response in seedlings of rice (*Oryza sativa* L.). Cereal Res. Commun. 28:33–40.

Nagamine T. 1991. Genic control of tolerance to chilling injury at seedling stage in rice. Jpn. J. Breed. 41:35–40.

Nandi S., P.K. Subudhi, D. Senadhira, N.L. Manigbas, S. Sen-Mandi, N. Huang. 1997. Mapping QTLs for submergence tolerance in rice by AFLP analysis and selective genotyping. Mol. Gen. Genet. 255:1–8.

Nguyen B.D., D.S. Brar, B.C. Bui, T.V. Nguyen, L.N. Pham, H.T. Nguyen. 2003. Identification and mapping of the QTL for aluminum tolerance introgressed from the new source, *Oryza rufipogon* Griff., into indica rice (*Oryza sativa* L.). Theor. Appl. Genet. 106:583–593.

Nguyen V.T., B.D. Nguyen, S. Sarkarung, C. Martinez, A.H. Paterson, H.T. Nguyen. 2002. Mapping of genes controlling aluminum tolerance in rice: comparison of different genetic backgrounds. Mol. Genet. Genomics 267:772–780.

Ni J.J., P. Wu, D. Senadhira, N. Huang. 1998. Mapping QTLs for phosphorus deficiency tolerance in rice (*Oryza sativa* L.). Theor. Appl. Genet. 97:1361–1369.

Prasad S.R., P.G. Bagali, S. Hittalmani, H.E. Shashidhar. 2000. Molecular mapping of quantitative trait loci associated with seedling tolerance to salt stress in rice (*Oryza sativa* L.). Curr. Sci. 78:162–164

Price A.H., J.E. Cairns, P. Horton, H.G. Jones, H. Griffiths. 2002a. Linking drought-resistance mechanisms to drought avoidance in upland rice using a QTL approach: progress and new opportunities to integrate stomatal and mesophyll responses. J. Exp. Bot. 53:989–1004.

Price A.H., K.A. Steele, J. Gorham, J.M. Bridges, B.J. Moore, J.L. Evans, P. Richardson, R.G.W. Jones. 2002b. Upland rice grown in soil-filled chambers and exposed to contrasting water-deficit regimes II. Mapping quantitative trait loci for root morphology and distribution. Field Crop Res. 76:25–43.

Price A.H., K.A. Steele, B.J. Moore, P.B. Barraclough, L.J. Clark. 2000. A combined RFLP and AFLP linkage map of upland rice (*Oryza sativa* L.) used to identify QTLs for root-penetration ability. Theor. Appl. Genet. 100:49–56.

Price A.H., A.D. Tomos. 1997. Genetic dissection of root growth in rice (*Oryza sativa* L.). 2. Mapping quantitative trait loci using molecular markers. Theor. Appl. Genet. 95:143–152.

Price A.H., J. Townend, M.P. Jones, A. Audebert, B. Courtois. 2002c. Mapping QTLs associated with drought avoidance in upland rice grown in the Philippines and West Africa. Plant Mol. Biol. 48:683–695.

Price A.H., E.M. Young, A.D. Tomos. 1997. Quantitative trait loci associated with stomatal conductance, leaf rolling and heading date mapped in upland rice (*Oryza sativa*). New Phytol. 137:83–91.

Quarrie S.A., D.A. Laurie, J.H. Zhu, C. Lebreton, A. Semikhodskii, A. Steed, H. Witsenboer, C. Calestani. 1997. QTL analysis to study the association between leaf size and abscisic acid accumulation in droughted rice leaves and comparisons across cereals. Plant Mol. Biol. 35:155–165.

Ray, J.D., L. Yu, S.R. McCouch, M.C. Champoux, G. Wang, H.T. Nguyen. 1996. Mapping quantitative trait loci associated with root penetration ability in rice (*Oryza sativa* L.). Theor. Appl. Genet. 92:627–636.

Saito K., K. Miura, K. Nagano, Y. Hayano-Saito, H. Araki, A. Kato. 2001. Identification of two closely linked quantitative trait loci for cold tolerance on chromosome 4 of rice and their association with anther length. Theor. Appl. Genet. 103:862–868.

Saito, K., K. Miura, K. Nagano, Y. Hayanosaito, A. Saito, H. Araki, A. Kato. 1995. Chromosomal location of quantitative trait loci for cool tolerance at the booting stage in rice variety 'Norin-PL8'. Breeding Sci. 45:337–340.

Shen, L., B. Courtois, K.L. McNally, S. Robin, Z. Li. 2001. Evaluation of near-isogenic lines of rice introgressed with QTLs for root depth through marker-aided selection. Theor. Appl. Genet. 103:75–83.

Siangliw, M., T. Toojinda, S. Tragoonrung, A. Vanavichit. 2003. Thai jasmine rice carrying QTLch9 (SubQTL) is submergence tolerant. Ann. Bot. 91:255–261.

Sripongpangkul, K., G.B.T. Posa, D.W. Senadhira, D. Brar, N. Huang, G.S. Khush, Z.K. Li. 2000. Genes/QTLs affecting flood tolerance in rice. Theor. Appl. Genet. 101:1074–1081.

Stam, P., A.C. Zeven. 1981. The theoretical proportion of the donor genome in near-isogenic lines of self-fertilizers bred by backcrossing. Euphytica 30:227–238.

Takeuchi, Y., H. Hayasaka, B. Chiba, I. Tanaka, T. Shimano, M. Yamagishi, K. Nagano, T. Sasaki, M. Yano. 2001. Mapping quantitative trait loci controlling cool-temperature tolerance at booting stage in temperate japonica rice. Breed Sci. 51:191–197.

Tanksley, S.D., J.C. Nelson. 1996. Advanced backcross QTL analysis: a method for the simultaneous discovery and transfer of valuable QTLs from unadapted germplasm into elite breeding lines. Theor. Appl. Genet. 92:191–203.

Toojinda, T., M. Siangliw, S. Tragoonrung, A. Vanavichit. 2003. Molecular genetics of submergence tolerance in rice: QTL analysis of key traits. Ann. Bot. 91:243–253.

Tripathy, J.N., J. Zhang, S. Robin, T.T. Nguyen, H.T. Nguyen. 2000. QTLs for cell-membrane stability mapped in rice (*Oryza sativa* L.) under drought stress. Theor. Appl. Genet. 100:1197–1202.

Venuprasad, R., H.E. Shashidhar, S. Hittalmani, G.S. Hemamalini. 2002. Tagging quantitative trait loci associated with grain yield and root morphological traits in rice (*Oryza sativa* L.) under contrasting moisture regimes. Euphytica 128:293–300.

Wan, J.L., H.Q. Zhai, J.M. Wan, H. Ikehashi. 2003. Detection and analysis of QTLs for ferrous iron toxicity tolerance in rice, *Oryza sativa* L. Euphytica 131:201–206.

Wang, G.L., D.J. Mackill, J.M. Bonman, S.R. McCouch, M.C. Champoux, R.J. Nelson. 1994. RFLP mapping of genes conferring complete and partial resistance to blast in a durably resistant rice cultivar. Genetics 136:1421–1434.

Wissuwa, M., N. Ae. 2001a. Further characterization of two QTLs that increase phosphorus uptake of rice (*Oryza sativa* L.) under phosphorus deficiency. Plant Soil 237: 275–286.

Wissuwa, M., N. Ae. 2001b. Genotypic variation for tolerance to phosphorus deficiency in rice and the potential for its exploitation in rice improvement. Plant Breed. 120: 43–48.

Wissuwa, M., M. Yano, N. Ae. 1998. Mapping of QTLs for phosphorus-deficiency tolerance in rice (*Oryza sativa* L.). Theor. Appl. Genet. 97:777–783.

Wu, P., C.Y. Liao, B. Hu, K.K. Yi, W.Z. Jin, J.J. Ni, C. He. 2000. QTLs and epistasis for aluminum tolerance in rice (*Oryza sativa* L.) at different seedling stages. Theor. Appl. Genet. 100:1295–1303.

Wu, P., A. Luo, J. Zhu, J. Yang, N. Huang, D. Senadhira. 1997. Molecular markers linked to genes underlying seedling tolerance for ferrous iron toxicity. Plant Soil 196:317–320.

Xiao, J.H., J.M. Li, S. Grandillo, S.N. Ahn, L.P. Yuan, S.D. Tanksley, S.R. McCouch. 1998. Identification of trait-improving quantitative trait loci alleles from a wild rice relative, *Oryza rufipogon*. Genetics 150:899–909.

Xing, Y.Z., Y.F. Tan, J.P. Hua, X.L. Sun, C.G. Xu, Q. Zhang. 2002. Characterization of the main effects, epistatic effects and their environmental interactions of QTLs on the genetic basis of yield traits in rice. Theor. Appl. Genet. 105:248–257.

Xu, K., R. Deb, D.J. Mackill. 2004. A microsatellite marker and a co-dominant PCR based marker for marker-assisted selection of submergence tolerance in rice. Crop Science 44:248–253.

Xu, K., D.J. Mackill. 1996. A major locus for submergence tolerance mapped on rice chromosome 9. Mol. Breeding 2:219–224.

Xu, K., X. Xu, P.C. Ronald, D.J. Mackill. 2000. A high-resolution linkage map in the vicinity of the rice submergence tolerance locus *Sub1*. Mol. Gen. Genet. 263:681–689.

Xu, Y. 2002. Global view of QTL: Rice as a model. p. 109–134. *In*, Kang, M.S. (ed), Quantitative genetics, genomics and plant breeding. CAB International, Wallingford, United Kingdom.

Yano, M., T. Sasaki. 1997. Genetic and molecular dissection of quantitative traits in rice. Plant Mol. Biol. 35:145–153.

Young, N.D., S.D. Tanksley. 1989. Restriction fragment length polymorphism maps and the concept of graphical genotypes. Theor. Appl. Genet. 77:95–101.

Yu, J., S.N. Hu, J. Wang, G.K.S. Wong, S.G. Li, B. Liu, Y.J. Deng, et al. 2002. A draft sequence of the rice genome (*Oryza sativa* L. ssp *indica*). Science 296:79–92.

Yu, S.B., J.X. Li, C.G. Xu, Y.F. Tan, Y.J. Gao, X.H. Li, Q. Zhang, M.A.S. Maroof. 1997. Importance of epistasis as the genetic basis of heterosis in an elite rice hybrid. Proc. Natl. Acad. Sci. USA 94:9226–9231.

Zhang, J., H.G. Zheng, A. Aarti, G. Pantuwan, T.T. Nguyen, J.N. Tripathy, A.K. Sarial, S. Robin, R.C. Babu, B.D. Nguyen, S. Sarkarung, A. Blum, H.T. Nguyen. 2001. Locating genomic regions associated with components of drought resistance in rice: comparative mapping within and across species. Theor. Appl. Genet. 103:19–29.

Zheng, H.G., R.C. Babu, M.S. Pathan, L. Ali, N. Huang, B. Courtois, H.T. Nguyen. 2000. Quantitative trait loci for root-penetration ability and root thickness in rice: Comparison of genetic backgrounds. Genome 43:53–61.

Zhuang, J.Y., Y.Y. Fan, Z.M. Rao, J.L. Wu, Y.W. Xia, K.L. Zheng. 2002. Analysis on additive effects and additive-by-additive epistatic effects of QTLs for yield traits in a recombinant inbred line population of rice. Theor. Appl. Genet. 105:1137–1145.

Zhuang, J.Y., H.X. Lin, J. Lu, H.R. Qian, S. Hittalmani, N. Huang, K.L. Zheng. 1997. Analysis of QTL × environment interaction for yield components and plant height in rice. Theor. Appl. Genet. 95:799–808.

15

The Phenotypic and Genotypic Eras of Plant Breeding

Michael Lee, Department of Agronomy, Iowa State University

Introduction

The composition of the phenotype, the observable properties of an organism (Johannsen, 1909) is simply expressed as the outcome of three major sources of variation: the genotype, the environment, which includes all factors external to the plant that affect development and growth, and interactions of all kinds. Recently, the observable properties of organisms have expanded to include their primary DNA sequences and several categories of molecular phenotypes (e.g., metabolomics, proteomics, and transcriptomics). The availability of such information, technology, and material have significant implications for the strategies and tactics of plant breeding.

Plant breeding is the genetic adaptation of plant species to the desires of human societies and the demands of nature in the context of agriculture. From the early stages of crop plant domestication thousands of years ago, through most of the twentieth century, plant breeding has succeeded by selecting at the level of the phenotype. Despite the addition of the progeny test, and various genetic and statistical methods that have been developed to identify genetic components of phenotypic variation (Hallauer and Miranda Fo, 1981; Simmonds and Smartt, 1999), the genetic improvement of crops through the mid-1990s has been based almost entirely on phenotypic selection, a method that will always be vital.

The phenotypic era of plant breeding has had mixed results and some limitations. Genetic gains from phenotypic selection have been assessed for many plant species and environments, and the progress has varied widely (Duvick, 1984, 1986; Volenec et al., 2002). When there has been progress, it is important to note that the rate of improvement may be low by contemporary expectations but steady. Despite instances of spectacular success, phenotypic selection has revealed little about the fundamental basis of progress achieved by plant breeding. Retrospective analyses have shown that resistance to biotic and abiotic stress and shifts in photo-assimilate distribution have been important, but they reveal scant information about options or expectations for crop improvement in a changing and more challenging world.

The genotypic era of plant breeding includes fundamentally new approaches to crop improvement. The advent of genetic transformation of crop species, complete genome sequences, and large-scale assessments of gene products and their putative pathways have inspired suggestions that direct analysis, selection, and manipulation of the genome is the next important source of variation for crop improvement and a new paradigm for plant breeding (Conway and Thoenniessen, 1999; Koornneef and Stam, 2001). It is obvious that basic knowledge of genetics, allied disciplines, and some new biotechnologies will have a greater role in crop improvement, but that is true primarily because they have actually contributed very little in a direct manner to genetic improvement of crop species through the mid-1990s. Besides providing a rudimentary understanding of meiosis, recombination, gametogenesis, and polyploidy, the principles and knowledge of basic genetics have had a very limited impact on actual advancements and genetic gain from plant breeding. While there have been a few widely adapted products from the first phase of genomics, such as transgenic soybeans (herbicide resistance) and cotton and maize (both for insect resistance), subsequent developments have yet to deliver significant advances, a few of which might be delayed by concerns related to transgenic crops (i.e., genetically modified organisms, GMOs). Also, some initial observations of model and crop genomes, the focus of this chapter, suggest that they are more complex than might have been imagined; thus, some aspects of the genotypic era of crop improvement might be more difficult than predicted or promised.

Comparative genetics and models systems

A major goal of the genotypic era of crop improvement is understanding the connection(s) between genotype and phenotype; a relationship mediated by the redundancy of the plant genomes and physiology, many interactions among genetic and environmental factors, and the seemingly random nature of developmental processes as well as the timing, intensity, and combination of signals from the external environment (Pickett and Meeks-Wagner, 1995; Mayr, 1997). The ultimate goal of such understanding is determining the physiologically significant role(s) of a gene, a serious challenge and presumably one that is fundamental to realizing the potential of this era.

Two of the hallmarks of the genotypic era have been the emergence of comparative genomics and model systems for plants. Based on observations that groups of sexually isolated species exhibit a surprising degree of genome conservation with regard to gene content and order, comparative genomics has provided a basis for systematic compilation of information that may facilitate technology development in an unprecedented manner (Gale and Devos, 1998). For example, previously cloned and annotated maize genes related to viviparary may be used to identify homologous DNA sequences in wheat associated with preharvest sprouting, thereby accelerating a marker-assisted selection strategy in wheat. While not a panacea, this approach will create new strategies and tactics for crop improvement, especially so when gene functions are more clearly defined.

Presumably, gene function will be most rapidly and comprehensibly accomplished in the first model systems for plants Arabidopsis and Oryza. Model systems, with their relative simplicity and rapid experimental cycle needed for hypothesis formation and reformation, have been vital to the process of understanding gene functions, but one should not underestimate the complexity of that task. For example, perhaps the first model system for molecular biologists, bacteriophage lambda, has been under intense scrutiny since the 1950s. The complete DNA sequence of that virus was known in 1982; yet, functional analysis of that genome of a few dozen genes continues to the present (Ptashne, 1987). So, the expectations for the analyses of relatively large and complex plant genomes in dynamic environments should be adjusted accordingly.

The initial drafts of the Arabidopsis and Oryza genomes were complete in 2000 and 2002 (TAGI 2000; Yu et al., 2002; Goff et al., 2002). The Arabidopsis genome was estimated to contain nearly 26,000 genes, of which 69% could be assigned to one or more broad functional categories (e.g., metabolism), using the best tools of computational biology; thus, 31% of the genes were complete mysteries. Despite its relatively small size (125 Mb) and disomic inheritance patterns, roughly 58% of the genome consists of segmental duplications at the molecular level (Vision et al., 2000). The Oryza genome initiatives and the slightly larger genome (420–466 Mb) resulted in a more complicated initial assessment of gene content and organization, with the predicted gene number between 32,000 and 65,000; 16 to 70% of the genes assigned to a functional category; and 77% of the genome duplicated. Despite their scientific stature as model systems, the actual functions of only a small proportion of their genes have been determined through direct experimentation; less than 10% of the Arabidopsis genes (Somerville and Dangl, 2000; Breyne and Zabeau, 2001; Van Montagu, 2002) and less than 100 rice genes (Cyranoski, 2003). The proportion is smaller if one expects to know the physiologically significant substrates, the relevant contexts of gene function (i.e., at the subcellular, cellular, tissue, organismal, and community levels), and secondary functions of the gene product(s). Gene products often have secondary functions, but the problem, or bias, of science is that we usually detect only that which we seek. Of course, knowledge of gene function increases each day, and a complete understanding of the genes and their interactions is not essential as exemplified throughout the history of the phenotypic era. But, such understanding should help maximize the progress and minimize the unfortunate surprises that may result from uninformed manipulation of genomes. Obviously, much annotation remains for the draft sequences of the model species.

As good and important as they are, the draft sequences fail to capture some potentially important information regarding the status and information content of the primary DNA sequence. In Angiosperms, 20–30% of the cytosines are methylated, often in the context of CG or CnG sequences. Such epigenetic modification is reversible, heritable, often related to transcriptional regulation, and may exhibit a developmental gradient within a plant (Richards, 1997). The functional significance and consequences of methylation are not readily predicted and are revealed through direct experimentation. However, the potential significance may be ubiquitous because an "exceptionally high" CG content was observed in an exon in almost every rice gene and the content was 25% higher at 5′ end (Yu et al., 2002).

The number of gene products per gene and thereby the information content of the draft sequence may also be grossly underestimated. In humans, nearly 50% of the genes produce more than one transcript through alternative splicing mechanisms (Modrek and Lee, 2002) and functions are gradually being assigned to many of the alternative splice products. Also, nearly 3200 human genes produce an antisense transcript with functions related to editing, nuclear retention, gene silencing, and chromatin status (Carmichael, 2003). Similar assessments have not been conducted in any plant species thus far, although computational comparisons of expressed sequence tags and genomic DNA sequence of Arabidopsis indicates that alternative splicing occurs with a small proportion of the genes (i.e., 1.5%; Zhu et al., 2003).

Perhaps the most significant omission of the draft DNA sequences of model plant species is the limited mention of sequences for noncoding RNA species (i.e., RNA that does not function as ribosomal, transfer or messenger RNA). Today, several types of noncoding RNAs have been identified in plants (e.g., sRNA, ncRNA, stRNA, miRNA, siRNA) and their list of functions includes the regulation of the timing and fate of developmental processes such as organ formation and flowering (Carrington and Ambros, 2003). Noncoding RNA is a relatively new, dynamic and challenging area (e.g. some of the noncoding RNAs move systemically) of investigation and so it is difficult to appreciate the potential effects of such gene products on phenotypes. At this point, we only know that RNA is much more interesting than previously imagined.

While it is clear that we have much to learn during this genotypic era, some early derivatives have become standard tools and options in plant breeding. They include DNA markers such as single nucleotide polymorphism for detailed DNA fingerprinting, haplotype analysis, association mapping, genetic linkage analysis, and marker-assisted selection. The advent of large-scale transposon-based insertional mutagenesis systems, sequence-based mutant detection (e.g., TILLING), physical mapping, and modest improvements in transformation methods have enabled investigations to proceed from the phenotype to the gene, or the reverse, in more reasonable lengths of time and at reasonable cost. Knowledge gained through such investigations will eventually enable plant-breeding programs to create phenotypes by design as exemplified by the development of "golden rice," a product that could never fulfill the intended social agenda yet represented a significant scientific and technical milestone (Ye et al., 2000).

The model systems and comparative genomics have much to offer to the genotypic era of crop improvement. However, there are always limits to which one may reasonably extrapolate and transfer information from the model to the more recalcitrant crop species. For example, disease-resistance genes seem to be among the more elusive targets of comparative genomics (e.g., Gale and Devos, 1998). Also, genomes such as Arabidopsis and Oryza may have too many fundamental differences in comparison with larger genomes, such as maize, to serve as informative models for some areas of investigation; such differences include the proportion of repetitive DNA, a larger proteome, and limited matches between the respective sequences of expressed sequence tags and proteins (Brendel et al. 2002).

Some Observations from the Maize Genome

Analyses of the adh1, b1, and bz regions of the maize genome have provided some clear examples of the potential complexity that awaits the genotypic era of plant breeding. Diagnostic sequencing of the intergenic region of the adh1 gene revealed that most of the intergenic DNA consisted of complete or fragmented long terminal repeat (LTR)-retrotransposons inserted into each other. Such sequence composition and arrangement is ubiquitous in the maize genome and may represent 60% or more of the entire genome (San Miguel et al., 1996). The long terminal repeats (i.e., LTRs) contain enhancer, promoter, and termination signal sequences that are recognized by the host cell transcriptional components. In addition, the intergenic DNA contains sequences that function as matrix-attachment regions in binding assays (Avramova et al., 1995). So, the so-called "junk" DNA is actually full of information that should affect transcription of adjacent genic DNA, native or transgenic, and the content of the intergenic DNA is highly variable among genotypes (Fu and Dooner, 2002).

Likewise, the content of genic DNA may also vary substantially among maize genotypes. A comparison of the regions flanking the bz gene in two inbred lines revealed that they shared only one of the 12 or 13 families of LTR-retrotransposons detected in each inbred and that one inbred contained four "genes" totally lacking in that region of the other inbred. In other words, with respect to the transposon DNA and the four genes, the two inbreds were hemizygous and not at all collinear. Such observations will certainly be made for many other regions of the maize genome, and the genomes of other plants species as insertion-deletion polymorphism (indels) are reported with greater frequency (now that we know to look for them).

Analysis of the b1 region with respect to paramutation has revealed a clear example of long-distance regulatory sequences that are more common in mammalian species. At b1, the sequences involved in the meaningful methylations related to paramutation were discovered in two regions, more than 10 and 100 kb pairs from the start of transcription (Stam et al., 2002a, 2002b). The extent to which such long-distance control and epigenetic phenomenon, such as paramutation, are significant for other plant genes is not known. Summaries from plant transformation experiments suggest that key regulatory sequences are often found within 2-kb pairs of the transcribed sequences. In large genomes, such as maize and wheat, the potential for long-distance control is high because many genes are separated by long tracts of repetitive DNA.

Summary and forecast

Our understanding of plant genomes is very much like the proverbial "tip of the iceberg," and many discoveries and annotation await us. The subsurface portion of the iceberg of knowledge is even larger, since this assessment ignored the complexities of the proteome, transcriptome, and metabolome, as well as the problems and severe limitations of transformation technology. So, one certainty at this time is that we are in the primordial stages of the genotypic era of plant breeding and only the simplest strategies and tactics have been tested and deployed. As always, we will make mistakes, and we will hope that they are relatively low-cost and reversible.

Knowing that we are in an early stage of this new era, it would seem wise to avoid condemnation and irrational commitments; yet, it is easy to observe examples of them on a regular basis. Perhaps human nature changes more slowly than plant-breeding methods, if it changes at all. Another certainty is that the phenotypic era of plant breeding is endless and irreplaceable. The real challenge is how to enable phenotypic selection and to make it more effective. "The development of the phenotype involves many stochastic processes that preclude a one-to-one relation between genotype and phenotype. This is, of course, precisely why we must accept the phenotype as the object of selection rather than the genotype" (Mayr, 1997).

References and suggested readings

Arumuganathan, K., and E.D. Earle. 1991. Nuclear DNA content of some important plant species. Plant Mol. Biol. Rep. 9:208–219.

Attwood, T.K. 2000. The Babel of bioinformatics. Science. 290:471–473.

Avramova, Z., P. SanMiguel, E. Georgieva, and J.L. Bennetzen. 1995. Matrix attachment regions and transcribed sequences within a long chromosomal continuum containing maize *Adh*1. The Plant Cell 7:1667–1680.

Baulcombe, D. 2002. An RNA microcosm. Science. 297: 2002–2003.

Bevan, M. 2003. Surprises inside a green grass genome. Science. 300:1514–1515.

Bennetzen, J.L. 2001. Arabidopsis arrives. Nat. Gen. 27:3–5.
Bennetzen, J.L. 2002. Opening the door to comparative plant biology. Science. 296:60–63.
Bochner, B. 2003. New technologies to assess genotype-phenotype relationships. Nat. Rev. Genet. 4:309–314.
Brendel, V., S. Kurtz, and V. Walbot. 2002. Comparative genomics of Arabidopsis and maize: prospects and limitations. Genome Biol. 3:1–6.
Breyne, P. and M. Zabeau. 2001. Curr. Opin. Plant Biol. 4:42.
Breyne, P., Rombaut, D., van Gysel, A., van Montagu, M., and Gerats, T. 1999. AFLP analysis of genetic diversity within and between *Arabidopsis thaliana* ecotypes. Mol. Gen. Genet. 261, 627–634.
Carrington, J.C., and V. Ambros. 2003. Role of microRNAs in plant and animal development. Science. 301:336–338.
Cardon, L.R., and J.I. Bell. 2001. Association study designs for complex diseases. Nat. Rev. Genet. 2001. 2:91–99.
Chory, J., J.R. Ecker, S. Briggs, et al. 2000. National science foundation-sponsored workshop report: "The 2010 Project." Plant Phys. 123:423–425.
Conway, G., and G. Thoenniessen. 1999. Feeding the world in the twenty-first century. Nat.(Supplement). 402:55–58.
Cremer, T., and C. Cremer. 2001. Chromosome territories, nuclear architecture and gene regulation in mammalian cells. Nat. Rev. Genet. 2:292–301.
Cyranoski, D. 2003. A Recipe for revolution? Nat. 422: 796–798.
Duvick, D.N. 1984. Progress in conventional plant breeding. p. 17–31. *In* J.P. Gustafson (ed.), Gene Manipulation in Plant Improvement, 16th Stadler Genetics Symposium. Plenum Press, NY.
Duvick, D.N. 1986. Plant breeding: Past achievements and expectations for the future. Econ. Bot. 40:289–97.
Feuillet, C., and B. Keller. 2002. Comparative genomics in the grass family: Molecular characterization of grass genome structure and evolution. Ann. Bot. 89:3–10.
Finnegan, E.J., R.K. Genger, W.J. Peacock, and E.S. Dennis. 1998. DNA methylation in plants. Annu. Rev. Plant Physiol. Plant Mol. Biol. 49:223–247.
Fu, H., and H.K. Dooner. 2002. Intraspecific violation of genetic colinearity and its implications in maize. Proc. Natl. Acad. Sci. USA. 99:9573–9578.
Gale, M.D., and K.M. Devos. 1998. Plant comparative genetics after 10 years. Science. 282:656–659.
Gaut, B.S., and J.F. Doebley. 1997. DNA sequence evidence for the segmental allotetraploid origin of maize. Proc. Natl. Acad. Sci. USA 94:6809–6814.
Goff, S.A, D. Ricke, T.-H. Lan, et al. 2002. A draft sequence of the rice genome (Oryza sativa L. ssp. japonica). Science 296:92–100.
Gottlieb, T.M., M.J. Wade, and S.L. Rutherford. 2002. Potential genetic variation and the domestication of maize. BioEssays. 24:685–689.
Hallauer, A.R., and J.B. Miranda Fo. 1981. Quantitative Genetics in Maize Breeding. Iowa State University Press, Ames, Iowa, USA.
Hedges, S.B. 2002. The origin and evolution of model organisms. Nat. Rev. Genet. 3:838–849.
Johannsen, W.L. 1903. Ueber erblichkeit in populationen und in reinen leinen. Gustav Fischer, Jena.
Johannsen, W.L. 1909. Elemente der exakten Erblichkeitslehre. Gustav Fischer, Jena.
Koornneef, M., and P. Stam. 2001. Changing paradigms in plant breeding. Plant Phys. 125:156–159.
Kumar, A., and J.L. Bennetzen. 1999. Plant retrotransposons. Annu. Rev. Genet. 33:479–532.
Mann, C.S. 1999. Crop scientists seek a new revolution. Science. 283:310–314.
Mazur, B., E. Krebbers, and S. Tingey. 1999. Gene discovery and product development for grain quality traits. Science. 285:372–375.
Mayr, E. 1997. The objects of selection. Proc. Natl. Acad. Sci. USA. 94:2091–2094.
Modrek, B. and C. Lee. 2002. A Genomic view of alternative splicing. Nat. Genet. 30: 13–19.
Pennacchio, L.A., and E.M. Rubin. 2001. Genomic strategies to identify mammalian regulatory sequences. Nat. Rev. Genet. 2:100–109.
Pickett, F.B., and D.R. Meeks-Wagner. 1995. Seeing double: Appreciating genetic redundancy. The Plant Cell 7:1347–1356.
Ptashne, M.A. Genetic switch gene control and phage lambda. 1987. Blackwell Scientific Publications and Cell Press. Palo Alto, CA.
Remington, D. L., J. M. Thornsberry, Y. Matsuoka, L. M. Wilson, S.R. Whitt, J. Doebley, S. Kresovich, M.M. Goodman, and E.S. Buckler IV. 2001. Structure of linkage disequilibrium and phenotypic associations in the maize genome. Proc. Natl. Acad. Sci. USA. 98:11479–11484.
San Miguel, P., A. Tikhonov, Y.-K. Jin, N. Motchoulskaia, D. Zakharov, A. Melake-Berhan, P.S. Springer, K.J. Edwards, M. Lee, Z. Avramova, and J. L. Bennetzen. 1996. Nested retrotransposons in the intergenic regions of the maize genome. Science. 274:765–768.
San Miguel, P., and J.L. Bennetzen. 1998. Evidence that a recent increase in maize genome size was caused by the massive amplification of intergene retrotransposons. Ann. Bot. 82:37–44.
Stam, M., C. Belele, J.E. Dorweiler, and V.L. Chandler. 2002a. Differential chromatin structure within a tandem array 100 kb upstream of the maize b1 locus is associated with paramutation. Genes & Dev. 16:1906–1918.
Stam, M., C. Belele, W. Ramakrishna, J.E. Dorweiler, J.L. Bennetzen and V.L. Chandler. 2002b. The regulatory regions required for B′ paramutation and expression are located far upstream of the maize b1 transcribed sequences. Gen. 162:917–930.
Schauer, S.E., S.E. Jacobson, D.W. Meinke, and A. Ray. 2002. DICER-LIKE1: Blind men and elephants in Arabidopsis development. Trends in Plant Sci. 7:487–491.
Simmonds, N.W., and J. Smartt. 1999. Principles of Crop Improvement. Blackwell Science Ltd., London.
Somerville, C., and J. Dangl. 2000. Plant biology in 2010. Science. 290:2077–2078.
Storz, G. 2002. An expanding universe of noncoding RNAs. Science 296:1260–1263.
The Arabidopsis Genome Initiative. 2000. Analysis of the genome sequence of the flowering plant *Arabidopsis thaliana*. Nat. 408:796–815.
The Rice Chromosome 10 Sequencing Consortium (Wing, R., et al.). 2003. In-depth view of structure, activity, and evolution of rice chromosome 10. Science. 300:1566–1569.
Vision, T.J., D.G. Brown, and S.D. Tanksley. 2000. The Origins of genomic duplications in Arabidopsis. Science. 290: 2114–2117.
Richards, E.J. 1997. DNA methylation and plant development. Trends in Genet. 13:319–323.
Volenec, J.J., S.M. Cunningham, D.M. Haagenson, W.K. Berg, B.C. Joern and D.W. Wiersma. 2002. Physiological genetics of alfalfa improvement: Past failures, future prospects. Field Crops Res. 75:97–110.
Weatherall, D.J. 2001. Phenotype-genotype relationships in monogenic disease: Lessons from the thalassemias. Nat. Rev. Genet. 2:245–255.
Ye, X., S. Al-Babili, A. Kloti, J. Zhang, P. Lucca, P. Beyer and I. Potrykus. 2000. Engineering the provitamin A (β-carotene) biosynthetic pathway into (carotenoid-free) rice endosperm. Science. 287:303–305.

Yelin, R., D. Dahary, R. Sorek, E.Y. Levanon, O. Goldstein, A. Shoshan, A. Diber, S. Biton, Y. Tamir, R. Khosravi, S. Nemzer, E. Pinner, S. Walach, J. Bernstein, K. Savitsky, and G. Rotman. 2003. Widespread occurrence of antisense transcription in the human genome. Nat. Biotech. 21:379–386.

Yu, J., S. Hu, J. Wang et al. 2002. A Draft sequence of the rice genome (Oryza sativa L. ssp. indica). Science 296:79–92.

Zhang, M.Q. 2002. Computational prediction of eukaryotic protein-coding genes. Nat. Rev. Genet. 3:698–709.

Zhu, W., S.D. Schlueter and V. Brendel. 2003. Refined annotation of the Arabidopsis genome by complete expressed sequence tag mapping. Plant Phys. 132:469–484.

16

The Historical and Biological Basis of the Concept of Heterotic Patterns in Corn Belt Dent Maize

W.F. Tracy and M.A. Chandler
Department of Agronomy, College of Agricultural and Life Sciences, University of Wisconsin-Madison

Abstract

Currently, the concept of heterotic patterns is fundamental to maize-breeding theory and practice, especially in temperate regions. As the use of hybrids increases in tropical maize and in other crop species, plant breeders apply the lessons of Corn Belt Dent (CBD) heterotic patterns. However, the origin and development of the concept of CBD heterotic patterns have not been critically examined. CBD heterotic patterns were created by breeders, and are *not* the result of historical or geographical contingencies. While the phenomenon of hybrid vigor (heterosis) and its effects on various traits have been known since the early 1900s, the concept of heterotic patterns developed in the 1960s and 1970s. Academic interest in heterotic patterns increased in the late 1980s, stimulated by the availability of DNA-based markers and attempts at using markers to identify heterotic patterns. For CBD open-pollinated varieties and first cycle inbreds it would *not* have been possible to identify heterotic groups using molecular markers, had markers been available. CBD heterotic patterns were created by breeders through trial and error from a single race of corn. The application of the current concept of heterotic patterns in a hybrid breeding program results in increased divergence between the groups.

Introduction

Currently, the concept of heterotic patterns is an integral component of hybrid maize-breeding theory and practice. Heterotic patterns simplify germplasm management and organization. Heterotic patterns inform the breeder when choosing parents for crosses for inbred development and inbred testers to evaluate combining ability of newly developed inbreds. Usually, there are two groups in a heterotic pattern, and there may be subgroups within the two main groups. The current concept of heterotic patterns suggests that the parents of populations for inbred development should come from the same group and testers for newly developed inbreds come from the opposite group.

Melchinger and Gumber (1998) define a heterotic group as "a group of related or unrelated genotypes from the same or different populations, which display similar combining ability and heterotic response when crossed to genotypes from other genetically distinct heterotic groups." A heterotic pattern is a specific pair of two heterotic groups.

As the use of hybrid cultivars increases in tropical maize and in other crop species, plant breeders apply the lessons of CBD heterotic patterns to those crops. However, the origin and development of the concepts underlying our ideas on CBD heterotic patterns have never been critically examined.

In the 1970s, when I (W.F. Tracy) first became involved in maize breeding, I was told the story of CBD heterotic groups (I'm not sure those exact words were used). I was told of the origin of CBD in which Southern Dents from the southeastern United States and Northern Flints from the northeast were carried by pioneers across the Appalachians into the Northwest Territory (lands bordered by the Great Lakes and the Mississippi and Ohio Rivers). The two races of corn intermated, at first accidentally and then deliberately by farmers, creating a new maize race, Corn Belt Dent. As part of the story I was told that the most important maize hybrids were made by crossing inbreds derived from the open-pollinated variety (OPV) Reid Yellow Dent (Reid) with inbreds derived from Lancaster Surecrop (Lancaster), another OPV. Perhaps most importantly, I learned that these two facts were connected. Reid was developed in Illinois and Iowa and was mostly Southern Dent. Lancaster was developed in Pennsylvania, in relative isolation from Reid and the rest of the Corn Belt cultivars, and had a higher percentage of Northern Flint in its background. This geographic and phylogenetic history was the basis for the excellent combining ability between Reid and Lancaster inbreds.

To distill the story to its essence: The major heterotic pattern in Corn Belt Dent is based on geographic/phylogenetic distance of the source germplasm and therefore the Reid– Lancaster pattern was waiting to be discovered. The whole story made perfect sense and, for me at least, became the orthodox or canonical (to use Stephen J. Gould's term; Gould, 2002) story of the biological basis of heterotic groups. From lectures and writings of other plant breeders (e.g., Havey, 1998; Melchinger and Gumber, 1998; Cheres et al., 2000), I believe this has become the canonical story regarding CBD heterotic groups for many breeders.

The late Stephen J. Gould, evolutionist and essayist, often wrote about canonical stories in biology (Gould, 2002). He believed that canonical stories, like folk tales, teach important lessons in a simplified, easily remembered way. He also believed that canonical stories could get in the way of our understanding of complex biological systems. Gould wrote a number of essays on canonical stories, explaining what the intended message was and explaining what we as scientists were missing due to the oversimplification of complex systems. Put simply, if a story is too good to be true, it probably isn't. And so it is with the Reid-by-Lancaster heterotic groups.

The main message of the Reid–Lancaster story for novice plant breeders is clear, simple, and important. Genetic diversity is needed for high levels of heterosis. However, for many, the canonical story of CBD heterotic patterns may be misleading. For example, if we accept that Reid-by-Lancaster heterosis is due to a historical contingency, the geographic isolation of these two varieties, and that all important hybrids are based on this pattern, we would draw certain logical conclusions. We might conclude that when beginning a hybrid breeding program, one should search for maximum diversity, create groups by dividing the germplasm along the lines of maximum diversity, and develop hybrids by making crosses between inbreds derived from different groups. The validity of such a conclusion, however, depends on the factual basis of the canonical story. Specifically, are most important CBD hybrids based on the famous Reid–Lancaster pattern? And are high levels of heterosis due to the geographically diverse origins of Reid and Lancaster? Our intent is to look more deeply at the historical and biological basis of the concept of CBD heterotic groups and see if the canonical story leads to a misunderstanding of the process.

In this chapter we will address four questions: (1) What issues confronted early hybrid corn breeders? (2) When did the concept of heterotic groups develop in the Corn Belt? (3) What was the actual role of Lancaster? (Was geographical isolation of Lancaster required for the success of hybrid corn?) (4) How did CBD heterotic groups develop?

Methods

We reviewed corn-breeding literature focusing primarily on the United States. There are many excellent corn-breeding reviews and books, and these were examined, but, whenever possible we went to the primary literature. We looked for articles covering heterosis or combining ability in corn in the indices of all volumes of *Crop Science* and in the *Agronomy Journal* between 1920 and 1980. We reviewed the table of contents of all the Proceedings of the American Seed Trade Association Corn and Sorghum Research Conferences and the Proceedings of the Illinois Corn Breeders School. We reviewed the minutes from all the meetings of the North Central Region Corn Improvement Conferences up until 1985. Potentially rich sources of primary literature we did not review are the numerous experiment station bulletins, reports, and circulars.

A brief history of hybrid corn

Hybrid corn was such a major technological and economic event that a number of excellent histories have been written (Crabb, 1942; Wallace and Brown, 1956; Hayes, 1963; Hallauer et al, 1988; Hallauer, 1999).

E.M. East (1908), G.H. Shull (1908), and others experimented on inbreeding corn in the early 1900s. In the first decade of the twentieth century, Shull (1908, 1909, 1952) made three key observations: (1) individual plants in a normal corn OPV were hybrids, (2) by inbreeding, hybrids could be reduced to true breeding strains (inbreds), and (3) uniform hybrids could be produced by crossing two inbreds. East was in the audience when Shull first publicly discussed his results and recognized the importance of Shull's discovery and the relevance of his own work to corn improvement (Shull, 1952; Singleton, 1963).

Shull moved on to other genetic research, but East and his students continued research on inbreeding and crossbreeding corn (Hayes, 1963). East's students became the leading corn breeders and geneticists of the next generation and were instrumental in the development of hybrid corn (Peterson and Bianchi, 1999).

Despite the scientific interest surrounding Shull's discovery, hybrid corn did not appear economically viable (Baker, 1984). This was because the inbreds developed directly from OPVs were very weak and could not produce quantities of seed at prices farmers would pay. D.F. Jones overcame this problem with the invention of the double-cross hybrid (Jones, 1918). A double cross is created by making two single-cross hybrids (A × B) and (C × D) and then crossing the two single crosses the following season. The seed sold to farmers was from this second cross. The male and female parents in a double cross are vigorous F_1 hybrids, and the female parent produces large quantities of high-quality seed. Double-cross hybrids were first sold in the Midwest in the 1930s and were rapidly accepted by Midwestern farmers. By 1943 nearly all of Iowa corn acreage was planted to hybrid corn, and by 1960 virtually all U.S. corn was hybrid (Hallauer and Miranda Fo, 1981; Sprague, 1983). While hybrids yielded 10 to 20% more than open-pollinated cultivars, other traits of the new hybrids also played a role in the rapid acceptance of hybrids. Hybrids came on the scene as hand harvesting was being replaced by mechanical harvesting, and hybrid traits such as increased uniformity and decreased lodging were highly valued.

The first inbreds were derived by self-pollinating plants from the numerous OPVs. Later generations of inbreds were developed by intermating existing inbreds and then selfing in a pedigree-breeding program. As a result of selection and recombination, later cycle inbreds were more vigorous and higher yielding. Some of the improved inbreds could produce economic levels of hybrid seed directly. In the 1960s single-cross hybrids began to replace double crosses and by the mid-1980s nearly all new hybrids were single crosses (Hallauer et al., 1988).

Issues confronting early hybrid corn breeders

The number one issue confronting early hybrid corn breeders was the poor agronomic quality of the first generation of inbreds derived directly from OPVs. In 1984 Raymond Baker (1984) wrote

> Just keeping those early inbreds from open-pollinated corn alive was an art... Most practical breeders predicted that hybrid corn would never succeed because of these weak rooted first cycle inbreds.

George Sprague (1984) recalled that in the Iowa breeding program, Lancaster made good inbreds (combining ability), but all were so weak rooted that only two were named and released, L289 and L317.

The invention of the double cross (Jones, 1918) allowed these weak inbreds to be used commercially, but breeders wanted to develop improved inbreds.

Early hybrid corn breeders were developing theory as they developed new inbreds and hybrids. The relationship between genetic divergence and combining ability was initially unclear and required 20–30 years of research before the relationship was firmly established. Since this relationship was unclear and breeders needed to improve inbred performance, they made breeding crosses among elite inbreds. The crosses were designed so that weaknesses in one inbred were compensated by strengths in the other. Less attention was given to maintaining diversity.

Richey (1927) suggested the breeding scheme called *convergent improvement* to test the dominance theory of heterosis. Convergent improvement is a double-backcross program in which the F_1 (A × B) is backcrossed to each parent A and B. Richey (1927) hypothesized that if heterosis was due to dominance, it should be possible to improve the performance of the inbreds by accumulating favorable dominant alleles in A′ and B′ without altering the performance of the hybrid. Experiments by Richey and Sprague (1931) and Hayes and students at Minnesota (Murphy, 1942) supported this approach, but the method was not widely used by corn breeders.

Convergent improvement is, in fact, a program to breed for decreased diversity between the parents of the hybrid. Furthermore, since its purpose is to improve inbred performance without altering hybrid performance, if effective it will result in decreased heterosis. Sprague (1955b) and Sprague and Eberhart (1977) devoted considerable space to convergent improvement in corn-breeding chapters in the first and second editions of *Corn and Corn Improvement*. In the third edition, Hallauer et al. (1988) briefly mention convergent improvement, but say that it is not widely used. While never much used, the persistence of convergent improvement in the literature indicates that the imperative of inbred improvement outweighed the need for maintaining or increasing diversity.

In 1950 Richey (1950) wrote, "This would lead to the expectation that crosses between inbreds from different varieties would tend to be more productive than crosses between inbreds of the same variety. This expectation has been justified by the general experience of corn breeders."

A few years later Griffing and Lindstrom (1954) wrote, "Corn breeders have frequently suggested that the degree of heterosis is to some extent proportional to the genetic divergence of the parent inbreds. If this hypothesis is correct . . .."

These quotes from leading corn breeders and geneticists indicate that the relationship between diversity and combining ability was still not settled in the 1950s. The work of Lonnquist, Moll, and collaborators (Lonnquist and Gardner, 1961; Moll et al., 1962; Paterniani and Lonnquist, 1963) finally settled the issue more than 40 years after hybrid corn breeding began.

While corn breeders in the 1940s and 1950s began to establish the relationship between diversity and its role in combining ability, the need to develop improved inbreds was an overriding concern. Thus, many second- and third-cycle inbreds were derived from crosses between parents from what we now consider to be opposite heterotic groups. This was especially true with Lancaster germplasm, which had relatively poor root and stalk quality. As a result, most second-cycle Lancaster inbreds were, by pedigree, 50% or less Lancaster (Gerdes and Tracy, 1993) (Table 16.1.)

As the number of publicly developed inbreds proliferated, corn breeders were confronted with another problem: how to organize the inbreds to make breeding programs more efficient? In 1947 G.S. Stringfield, corn breeder at the Ohio Agricultural Experiment Station, raised this issue at the annual meeting of corn breeders from the North Central Region, the North Central Regional Corn Improvement Conference. Following is a direct quote from the minutes of the 1947 meeting.

> G.H. Stringfield discussed the advisability of grouping lines for breeding purposes. He urged that crosses for the improvement of lines then should be made only among lines of the same group. The object would be to maintain genetic diversity and avoid relationships among lines that later are used in the production of hybrids. (Anon., 1947)

A committee was formed to study the situation and suggest such a program of operation for the Corn Belt. The conference did not meet in 1948, but at the 1949 meeting the Committee on Grouping of Inbred Lines for Breeding Purposes presented the following report (Anon., 1949).

> The committee recommends that the inbred lines of the North Central Corn Improvement Conference be divided into two groups, which are to be kept distinct in breeding advance cycle lines. This means that no crosses for breeding purposes are to be made except between lines belonging to the same group.
>
> Each group should contain inbreds representing widely diverse maturities and desirable plant characters. As an arbitrary division, the committee recommends that the lines having odd entry numbers in the 1948 uniform tests of inbreds be tentatively assigned

Table 16.1 Background and percentage of Lancaster contribution to background of 27 inbreds classified in the Lancaster heterotic group ranked in order of decade of release

Inbred	Background	% Lancaster (by pedigree)	Decade of release
L289	Lancaster OP	100	1920
L317	Lancaster OP	100	1920
C103	Lancaster OP	100	1940
Oh43	Oh40B × W8	50	1940
Mo17	C.I. 187-2 × C103	50	1960
A619	(A171 × Oh43)Oh43	37.5	1960
Pa375	CH22 × C103	50	1970
H95	Oh43 × C.I.90A	25	1970
Va26	Oh43 × K155	25	1970
Va35	(C103 × T8)T8	25	1970
B70	M14 × C103	50	1970
Oh570	Oh07 × C103	50	1980
Oh572	Oh07 × C104	50	1980
A682	[(AS-D × Mo17)Mo17(2)]	40.6	1980
A683	[(AS-D × Mo17)Mo17(2)]	40.6	1980
H108	(Mo17 × H99)Mo17	40.6	1980
H109	(Mo17 × H99)Mo17	40.6	1980
N197	(Mo17 × Early Krug line)Mo17	37.5	1980
N198	(Mo17 × Early Krug line)Mo17	37.5	1980
Pa869	75F-5 × Pa83	25	1980
Pa870	75F-5 × Oh43	25	1980
T167	Mo17 × C.I.66	25	1980
H107	(H99 × H98)H99	15.6	1980
CM555	(Mo17 × MAG)MAG	12.5	1980
NC258	Complex pedigree	12.5	1980
NC260	(Mo44 × Mo17)Mo44	3.1	1980
B93	(B70 × H99)H99	21.9	1990

Source: Gerdes and Tracy (1993).

> to Group A and that those having even entry numbers be tentatively assigned to Group B. In cases of known relationship between inbreds, the originating station shall be responsible for shifting lines to provide for maximum genetic diversity between groups.
>
> It is recommended that each station submit a revised list to the committee in order that a permanent grouping may be presented at the next meeting of the conference.

This report was moved and passed by the conference and became the policy of the committee through the late 1980s.

Given a current perspective. it appears that this plan was the beginning of what we now call heterotic groups. But notice the way in which the lines were assigned, odd numbers in group A and even in group B. The next sentence did address the issue of relationship among inbreds and maximum genetic diversity between groups. But a review of the lists shows that the breeders' understanding of relationship did not reflect the canonical story (Reid–Lancaster). The first list released in 1950 shows that most states followed the odd/even scheme, and closely related lines ended up in both groups, for example, inbreds I205 (Iodent), L317 (Lancaster), and Os420 (Osterland Reid) were in group A, while I159 (Iodent), L289 (Lancaster), and Os426 (Osterland Reid) were in group B. B10, one of the first inbreds developed from Iowa Stiff Stalk Synthetic (BSSS), was assigned to group B (Anon., 1950).

BSSS, a source of many important inbreds, is a 16-line synthetic developed by George Sprague in the 1930s. Seventy-five percent of BSSS background traces back to improved strains of Reid Yellow Dent (RYD) (Troyer, 2000b). Therefore BSSS inbreds are usually classified as a subgroup of the Reid heterotic group (Troyer, 2000a).

The second list was published in 1953 (Anon., 1953). New inbreds were added, and some inbreds were moved to the opposite group to better reflect their origin. But there were still cases of inbreds from the same OPV assigned to both groups. Most of the important Reid and Lancaster inbreds were assigned to group A, clearly indicating that in 1953 the leading corn breeders did not recognize the Reid–Lancaster heterotic pattern of the canonical story. All of the BSSS inbreds were in group B. This arrangement persisted through the 1980s when the committee for grouping inbred lines was discontinued.

In 1971, there was some discussion regarding the usefulness of the groups. In that discussion Dr. Steve Eberhart was quoted as follows:

> Heterosis depends on differences in gene frequencies and dominance effects so that on the average, greater heterosis is observed between divergent groups. Since the A and B groups were originally established on the basis of heterosis between Midland and Reid types, the grouping has and could continue to serve its purpose . . . . (Anon., 1971)

While there is now no dispute with the first sentence, a thorough review of the committee minutes from 1947 to 1971 found no evidence that the groups were originally formed around a Midland–

Reid heterotic pattern. Indeed, the early lists had few, if any, Midland lines (Anon., 1950, 1953).

History of the concept of heterotic patterns

Today, the concept of heterotic patterns seems fundamental to our ideas on breeding hybrid crops. But the concept and terminology were expressed in modern terms 40–50 years after the beginning of hybrid corn breeding. George Sprague (1984) wrote, "In retrospect it appears that the concept of heterotic patterns was slow in developing."

Since CBD heterotic patterns developed empirically (Hallauer and Miranda Fo, 1981; Hallauer, 1999; Troyer, 2000a), yield testing of many hybrids had to be done before any patterns could become obvious (Hallauer, 1999; Hallauer et al., 1988). Ideas and observations underlying the concept of heterotic patterns (combining ability, grouping of inbreds, relationship between diversity and heterosis, and recognition of the importance of specific OPVs) needed to develop prior to the development of our current concepts. Published observations on the importance of inbreds derived from Reid, Krug, and Lancaster began in the 1940s (Anderson, 1944). Krug is an improved strain of Reid (Gracen, 1986).

Public and private breeders began grouping inbreds in the 1950s (Smith et al., 1999). Dr. D. Duvick, retired research director of Pioneer Hibred, recalled that groups were initially created based on whether the inbred was an acceptable seed parent or pollen parent. B37, a public inbred used by Pioneer in the 1950s, was a good seed parent but was not a good pollen producer. It became part of the "female" group (D. Duvick, pers. comm.). B37 was derived from BSSS, and other BSSS-related inbreds were also eventually placed in the female pool. Inbreds that combined well with BSSS were placed in the male pool. In publications by Pioneer researchers, the Pioneer female pool is also called stiff stalk (SS), and the male pool is designated non-stiff stalk (non-SS) (Smith et al., 2000; Romero-Severson et al., 2001; Casa et al., 2002; Duvick et al., 2004).

The first mention of the term *heterotic pattern* (or *heterotic group*) that we could find in the literature was in 1972 by B. Tsotsis (1972), then director of corn breeding with Dekalb Agresearch Inc. Tsotsis (1972) discussed the Reid–Lancaster heterotic pattern and research designed to identify new heterotic patterns. Tsotsis (1972) attributed the research to unpublished work of C.W. Crum of Dekalb Agresearch in 1970. Thus, it is clear that our current concept of heterotic patterns was familiar to some corn breeders at least by the late 1960s. The work by the Dekalb group, Crum, Kaufman, and Tsotsis, focused on developing new heterotic patterns (Tsotsis, 1972; Crum, 1973; Kaufman et al., 1982). Their methodology and experimental design were similar to the earlier work of Lonnquist and Moll and collaborators (Lonnquist and Gardner, 1961; Moll et al., 1962; Paterniani and Lonnquist, 1963). But these workers did not discuss their work in terms of identifying heterotic patterns or groups.

Discussions on heterotic patterns are then found in corn-breeding literature, for example, *Proceedings of the ASTA Corn and Sorghum Research Conference*, *Proceedings of the Illinois Corn Breeders School*, minutes of the North Central Regional Corn Improvement Conference, occasionally in the 1970s and early 1980s (Crum, 1973; Beil, 1975; Kannenberg, 1976; Kaufman et al., 1982). Hallauer and Miranda Fo (1981) discuss heterotic patterns in *Quantitative Genetics in Maize Breeding*. Hallauer et al. (1988) devote five pages to the topic of heterotic patterns in the corn-breeding chapter of the third edition of *Corn and Corn Improvement* (Sprague and Dudley, 1988).

Perhaps more revealing is where the terminology did not appear. No mention of heterotic groups or patterns are found in the books *The Hybrid Corn Makers* (Crabb, 1942), *Corn and Its Early Fathers* (Wallace and Brown, 1956), *A Professor's Story of Hybrid Corn* (Hayes, 1963), *Corn* (Manglesdorf, 1974), or numerous important chapters about corn breeding (Anderson and Brown, 1952; Jenkins, 1978; Russell and Hallauer, 1980; Zuber and Darrah, 1987; Sprague, 1983). Some of these books and book chapters did mention Reid–Lancaster hybrids and/or groups or families of inbreds, but they did not use the terms *heterotic groups* or *patterns*. George Sprague, one of the leading corn breeders and corn-breeding theoreticians of the twentieth century, edited all three editions of *Corn and Corn Improvement* (Sprague, 1955a, 1977; Sprague and Dudley, 1988). He also wrote the corn-breeding chapters in the first and second editions (Sprague, 1955b; Sprague and Eberhart, 1977). In the first edition Sprague

(1955b) does not mention heterotic groups or patterns nor does he mention Reid or Lancaster. In the second edition, Sprague and Eberhart (1977) mention the importance of Reid, Lancaster, and Krug germplasm. They do not mention heterotic groups or patterns. In sharp contrast, just seven years later in a lecture to the Illinois Corn Breeders School, Sprague (1984) said, "The single most important element of a breeding program is the recognition and utilization of heterotic patterns, this recognition simplifies and increases the efficiency of all subsequent operations." Clearly, the concept of heterotic groups grew from a minor point to a major concept during the 1970s and early 1980s.

The terms "heterotic group," "heterotic groups," "heterotic pattern," and "heterotic patterns" were seldom used in literature included in databases such as Agricola (http://agricola.nal.usda.gov/) and CAB (http://www.cabi-publishing.org/) until the late 1980s (Table 16.2; Figure 16.1). Searching different databases resulted in different numbers of citations and different year of first use. But the overall pattern is quite consistent (Table 16.2). The CAB database resulted in more citations than Agricola, and the largest number of citations, 122, resulted from searching the term "heterotic groups." In contrast, searching the CAB database for "combining ability" resulted in 9039 citations going back to 1972, the earliest year of the CAB database. The earliest citation using any of the terms related to heterotic groups was a 1978 abstract by Mishra and Geadelmann (1978). The earliest refereed publications to use "heterotic groups" were in 1986, with one paper on wheat (Murphy et al., 1986) and another on corn (Smith, 1986).

The terms dealing with heterotic groups or patterns were seldom used prior to the late 1980s, and then use increased dramatically (Figure 16.1). Many of the papers referring to heterotic patterns from 1987 through 2003 dealt with the use of DNA-based markers for sorting germplasm into

Table 16.2 Number of citations and the year first cited for the key words: heterotic group, heterotic groups, heterotic pattern, heterotic patterns, heterotic pool, heterotic pools, and combining ability

	Database			
	Agricola		CAB	
Key words	Number of citations	Year first cited	Number of citations	Year first cited
Heterotic group	17	1987	43	1986
Heterotic groups	42	1986	122	1978
Heterotic pattern	17	1990	53	1980
Heterotic patterns	26	1988	68	1984
Heterotic pool	1	1998	2	1998
Heterotic pools	0	—	7	1987
Combining ability	2062	1967	9039	1973

Note: Inclusive years were 1967–2002 for Agricola and 1973–2002 for CAB.

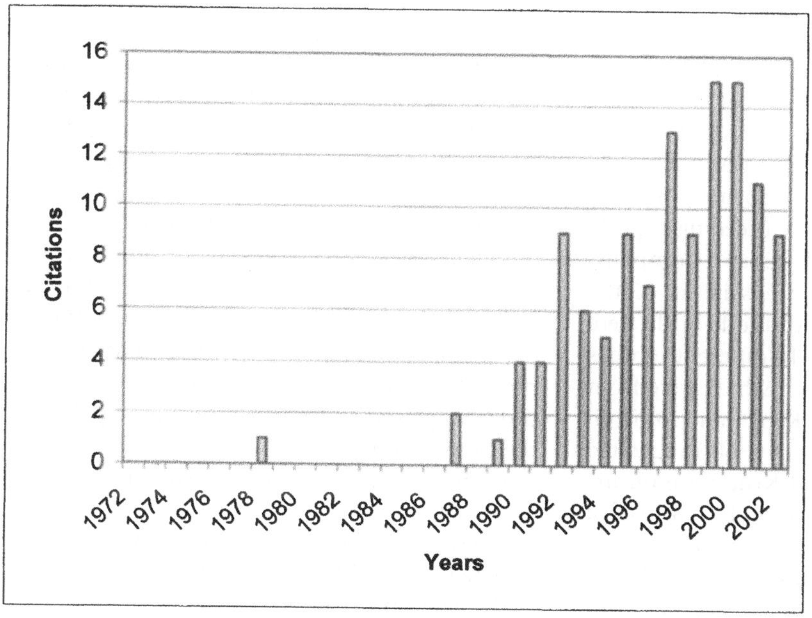

Figure 16.1 Number of citations per year from a search for the key words "heterotic groups" on the CAB database. Database for 1972–2002 inclusive. Search done in August 2003.

heterotic groups. The first papers describing RFLPs for use in maize breeding and genetics appeared in the mid-1980s (Helentjaris et al., 1985; Evola et al., 1986). In 1987, Walton and Helentjaris (1987) presented a paper on the use of RFLP technology in maize breeding at the ASTA corn and sorghum research conference. The first use they listed was "organization of germplasm" (Walton and Helentjaris, 1987).

In summary, our current concept of heterotic patterns crystallized in the late 1960s and early 1970s and became widely recognized and accepted in the 1970s and early 1980s. It is unclear why the concept of heterotic groups developed when it did. Many of the ideas underlying the concept were developed earlier, and the importance of Reid and Lancaster was recognized much earlier. It may be that the change to single-cross hybrids in the 1960s and the importance of Reid (Wf9, B14, B37) and Lancaster inbreds (C103, C123, Oh43, Mo17) in these early single crosses made the concept of heterotic patterns quite clear and useful.

What was the actual role of Lancaster?

A key feature of the canonical story of CBD heterotic patterns is that Lancaster germplasm was uniquely important, and by implication, geographical isolation of Lancaster was required for the success of hybrid corn in the Corn Belt. The current and historical importance of Reid inbreds, especially in the form of BSSS inbreds, is very clear, but what was the actual role of Lancaster?

The excellent combining ability of three Lancaster inbreds, L289, L317, and LDG was noted by Edgar Anderson (1944). Anderson was not a corn breeder, and he received this information from Raymond Baker, manager of the breeding department of Pioneer Hi-Bred Corn Company. Anderson reported on the inbreds in six of the most widely grown hybrids. All the hybrids were double crosses. He noted that 18 different inbreds were used in these crosses, 12 of which were from Reid Yellow Dent, 3 from Krug, and 3 from Lancaster (Table 16.3). Anderson (1944) suggested that contributions of Reid and Krug were unsurprising due to the importance and wide use of these OPVs in the Corn Belt. He was very surprised that Lancaster, an obscure OPV from Pennsylvania, had such a significant impact. He wondered, "If Lancaster Surecropper is really an effective source of good inbreds is there anything in its history to suggest why this might be so?"

Table 16.3 The number of double-cross hybrids and the inbred background of the four parents of the hybrids discussed by Anderson (1944)

	Number and background of inbreds		
Number of hybrids	Reid	Lancaster	Krug
3	3	1	0
1	2	1	1
1	2	0	2
1	4	0	0

Of the six double-cross hybrids studied by Anderson (1944), four had one Lancaster inbred (Table 16.3). The remaining two had no Lancaster contribution. One was all Reid and the other was 50% Reid and 50% Krug (Reid).

Other authors noted the importance of Lancaster inbreds in CBD hybrids (Crabb, 1942; Anderson and Brown 1952; Wallace and Brown, 1956). Interestingly Crabb (1942) in The Hybrid Corn Makers briefly mentions Lancaster inbreds L289 and L317, but does not include a discussion of Lancaster or its developers, while there were lengthy discussions on Reid and Krug. The 1992 reprinting of his book has significantly more information on Lancaster and its developer Isaac Hershey (Crabb, 1992). Clearly awareness of Lancaster increased over time.

The initial observations of the importance of Lancaster inbreds were based on their contributions to important hybrids of the 1930s, such as US13 [(Wf9 × 38-11) × (HY × L317)] and Iowa 939 [(I205 × L289) × (Os420 × Os426)], two of the most popular hybrids in history. George Sprague (1964, 1984) discussed the importance of Lancaster inbreds and mentioned that their uniqueness was very apparent in the Iowa program. Later, Oh43 and Mo17 (both 50% Lancaster) and their derivatives became important in the 1960s and 1970s. B73 × Mo17, the most important public single cross of the 1970s and 1980s, probably played an important role in popularizing the canonical story.

Raymond Baker (1984) pointed out that

> typically modern hybrids have one inbred parent derived from Iowa Stiff Stalk Synthetic. The other side of the cross usually has

some Lancaster in its origin. Usually, the non-Stiff Stalk parent is only half Lancaster with Reid or some northern variety like Minnesota 13 as the other half.

Zuber and Darrah (1981) reported that in 1979 39% of the U.S. germplasm was related to Lancaster and 42% to Reid. Zuber and Darrah grouped Oh43 and Mo17 as 100% Lancaster (Darrah, pers. comm.). Since both these inbreds and all their derivatives are no more than 50% Lancaster by pedigree, 39% Lancaster is an overestimate and 42% is an underestimate for Reid. Five years later, the Lancaster contribution had dropped to 12% with 44% Reid and 24% Iodent (Darrah and Zuber, 1986).

It is clear that public sector breeders recognized the importance of Lancaster in CBD hybrids, but much of the written record on Lancaster's importance was retrospective, after Mo17 and Oh43 had significant impacts. A closer look at the both the historical and current impact of Lancaster tells a different story.

Of the six hybrids discussed by Anderson (1944), four had one Lancaster line (Table 16.3). Of the remaining two hybrids, one had four Reid inbreds and the other two Reid and two Krug. Since Krug is an improved strain of Reid, on a percentage basis Reid constituted 83.3% of these six hybrids and Lancaster 16.7%.

Russell (1974) studied the contribution of breeding to increased corn yields by comparing hybrids from different decades in replicated yield trials. He chose four hybrids from each decade. The hybrids were chosen because they were among the most popular and widely grown in central Iowa. The contribution of Reid and Lancaster was calculated as the percentage of contribution by pedigree totaled over the four hybrids from each decade (Figure 16.2). These numbers were calculated based on the estimated pedigree contribution; for example, Mo17 is 50% Lancaster and 50% Krug (Reid). All inbreds derived from BSSS were grouped with Reid. The four most popular hybrids all had similar pedigrees in the 1930s. Each had three Reid inbreds and one Lancaster (Figure 16.2). Since Anderson (1944) wrote on sources of important germplasm in 1944, these and similar hybrids would have formed the data set he used. What would Anderson have written if he had based his conclusions on hybrids of the 1940s or 1950s rather than the 1930s? In the 1940s Reid accounted for 82.5% of the hybrids, Lancaster only 6.25% (Figure 16.2). In the 1950s there was no Lancaster germplasm in the four hybrids. In the 1960s Lancaster's share increased to 4.69%. With the advent of the single cross, the Lancaster contribution increased to 25%, the historical high in the 1970s. Clearly, Lancaster was not required for successful commercial hybrids. Recently published

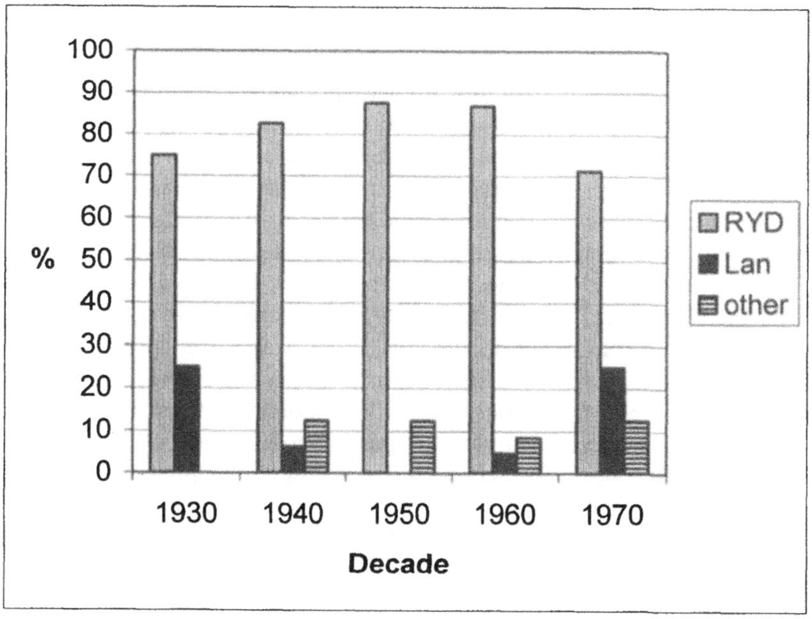

Figure 16.2 Percentage of contribution of Reid Yellow Dent (RYD), Lancaster Surecrop (Lan), or other OPVs to the inbred parents of four popular Iowa hybrids per decade from the Russell era studies (Russell, 1974). Reid Yellow Dent includes Stiff Stalk Synthetic inbreds.

papers documenting the contribution of germplasm to Pioneer Hi-Bred's commercial hybrids support this conclusion (Smith et al., 1999, 2004; Romero-Severson et al., 2001; Casa et al., 2002; Duvick et al., 2004). In Pioneer's program, Lancaster contribution peaked in the 1940s (16.9%) and has since declined to historic lows in the 1990s (2.9%) (Smith et al., 1999). Duvick et al. (2004) put the current Lancaster contribution to a series of successful hybrids for the west-central Corn Belt at 3.45%. Smith et al. estimated the Lancaster contribution at 4.9% and that of Lindstrom Long Ear at 2.9%. Troyer (2004) has suggested that Lindstrom Long Ear was derived from Lancaster. If this is so, it would raise the contribution of Lancaster to 8% in current Pioneer germplasm. While significant, clearly Lancaster is not a major component of Pioneer's successful breeding program.

Smith et al. (1999) suggest that Lancaster was more important in the public sector than it was for Pioneer. However, the proportion of Lancaster in public-sector Lancaster inbreds has decreased with each cycle of breeding (Table 16.1) (Gerdes and Tracy, 1993). First-cycle inbreds such as L317 and C103 were 100% Lancaster. However, with each cycle of breeding the proportion of Lancaster was reduced 50%. While foundation-seed companies still group families with names such as Mo17 or C103 (Anon., 1995), it is clear based on the morphology of these newer inbreds that they have substantial amounts of non-Lancaster germplasm. Williams and Hallauer (2000) wrote that the primary guide used to classify lines as Lancaster was because they exhibited good combining ability with lines from BSSS.

What was the true role of Lancaster? With a few exceptions successful hybrids were never more than 25% Lancaster. Many successful modern hybrids have no Lancaster by pedigree, and for those that do, the percentage of Lancaster is probably less than 12.5% (Gerdes and Tracy, 1993; Troyer, 2000a; Romero-Severson et al., 2001, Casa et al., 2002). Troyer (1999) indicated that Lancaster contributed approximately 4% by pedigree to commercial germplasm. The declining influence of Lancaster can be seen in the terminology used in the literature. The heterotic pattern was first described as Reid–Lancaster (Tsotsis, 1972; Hallauer and Miranda 1981). Later, as the contributions of BSSS became clear, the pattern was usually called SS-Lancaster (Geadelmann, 1984; Dudley, 1984). Today, knowledgeable writers discuss SS–NSS (Casa et al., 2002; Duvick, 2004).

If the role of Lancaster is overstated, what is the origin of the canonical story? Anderson's paper in 1944 hinted at a unique place of Lancaster in the success of hybrid corn. But the Anderson and Brown (1952) paper in the book *Heterosis* (Gowan, 1952) explicitly stated, ". . . sources of good combining inbreds are open-pollinated varieties with a stronger infusion of Northern Flint than was general in the Corn Belt. This is particularly true of Lancaster Surecrop. . . ."

This is the essence of the canonical story and probably its original written source. The 1950 Heterosis Symposium was influential, and the resulting book, *Heterosis* (Gowan, 1952), was widely read and cited. It is clear from later writings that Dr. William Brown, later president of Pioneer Hi-Bred, was convinced of the importance of the Northern Flint germplasm in determining Lancaster's combing ability (Brown, 1953; 1967; Weatherspoon, 1973).

How did CBD heterotic patterns develop?

CBD heterotic patterns developed empirically by trial and error, based on crosses among inbreds initially derived from the available OPVs (Hallauer et al., 1988; Hallauer, 1999; Troyer, 2000a). Hallauer (1999) wrote, "Heterotic groups do not evolve naturally except for being genetically dissimilar for allele frequencies."

When hybrid breeding began, breeders had a number of OPVs available to them, but the choice of patterns was not systematic (Hallauer et al., 1988). Later, breeders, including Tsotsis (1972), Crum (1973), and Kaufman (1982), attempted to identify new CBD heterotic patterns by systematic crossing. Once heterotic groups have been established and improved by breeding, however, it is difficult to develop competitive new patterns (Hallauer et al., 1988; Melchinger, 1999).

The number and choice of heterotic groups are arbitrary decisions. Some breeders prefer a large number of specific groups (Troyer, 2000a), while others prefer to arrange their program based on two large, diverse groups (Hallauer et al., 1988; Hallauer, 1999). When two main groups are used, there are usually subgroups within the main groups (Hallauer et al, 1988). SSS—Lancaster is the best known CBD heterotic pattern because it fits well with grain requirements for the main type of corn (No. 2 yellow), is adapted to the central Corn Belt, and was developed by the public sector. White corn was more important in Tennessee and Kentucky, and the consumer would not tolerate yellow kernels in the corn. Thus a number of the successful early double-cross hybrids consisted of four inbreds from the same white OPV (Hayes, 1963; Jenkins, 1978). Troyer (2000a) lists a number of alternate heterotic groups that fit the northern Corn Belt better than SSS–Lancaster. If the main part of the Corn Belt was 200 miles north, the most famous heterotic group may have been Reid–Minnesota 13. It is now apparent that the most important heterotic pattern for Pioneer Hi-Bred in the central Corn Belt is not SS–Lancaster, but instead SS–non-SS. Pioneer non-SS has a large contribution from Iodent and Minnesota 13, with a smaller contribution from Lancaster (Romero-Severson et al., 2001, Casa et al., 2002).

All inbreds within CBD heterotic groups are not necessarily related (Geadelmann, 1984; Casa et al., 2002; Duvick et al., 2003). What the inbreds have in common is high combining ability with inbreds from the opposite heterotic group (Hallauer et al., 1988; Williams and Hallauer, 2000). When an inbred unrelated to either group in a heterotic pattern combines equally well with inbreds from the two groups, the breeder must choose the group into which the inbred is incorporated. Geadelmann (1984) wrote that when such a situation occurred, he essentially "flipped a coin" in assigning the germplasm to a group. Of the two main CBD groups, SS and non-SS, it appears that the SS group is more homogeneous than the non-SS (Casa et al., 2002; Liu et al., 2004).

Heterotic groups are not constant, nor absolute (Hallauer et al., 1988; Gerdes and Tracy, 1993; Smith et al., 1999). The genetic composition changes over time. The public Lancaster group has become less Lancaster and more Reid with each cycle of breeding (Gerdes and Tracy, 1993).

Pioneer Hi-Bred has documented the change in its germplasm (Smith et al., 1999, 2004). This information makes an excellent case study on heterotic groups and especially the aspect of the canonical story that phylogenetic distance based on geography or some other historical contingency is needed for the creation of successful heterotic groups.

Pioneer began grouping inbreds in the 1950s, roughly the same time that the public sector began to do so (Anon., 1950; Smith et al., 1999; Duvick et al., 2004; Duvick, pers. comm.). The initial criteria for grouping included whether the inbreds made good seed or pollen parents (Duvick, pers. comm.). B37, a good seed producer and a poor pollen shedder, was placed in the female pool, which evolved into the SS group. Inbreds that were good pollen shedders, unrelated to SS, and combined well with SS, were placed in the non-SS group. The groups were formalized between 1960 and 1989 (Duvick et al., 2004). Pioneer was never as dependent on Lancaster inbreds as was the public sector and some other companies (Smith et al., 1999). Pioneer's highest use of Lancaster was in the 1940s, with about 15%, and the 1970s with 8.6%. In the 1990s, Lancaster contributed approximately 3% to Pioneer commercial hybrids (Smith et al., 1999).

If Lancaster is not the main constituent of the non-SS group, what is? A major constituent of the Pioneer non-SS pool is Pioneer Iodent (Romero-Severson et al., 2001; Casa et al., 2002). Iodent OPV was an early-maturing strain of Reid selected at Iowa State College (Troyer, 1999). However, Pioneer Iodent inbreds are not pure Iodent (Hallauer and Miranda, 1981; Troyer, 1999; Romero-Severson et al., 2001). While the highest contribution is Iodent, they have a diversity of germplasm sources including other Reid strains and northern and southern germplasm. The most consistent component following Iodent is Minnesota 13. Of the germplasm backgrounds of five Iodent inbreds revealed in Romero-Severson et al. (2001), only one has any Lancaster. A number of these inbreds do have Lindstrom Long Ear in their pedigrees, and if Lindstrom Long Ear was derived from Lancaster, as Troyer (2004) suspects, most of these inbreds would have some Lancaster. When the germplasm sources are totaled, most of the five inbreds are 50% or more Reid (Iodent plus Reid sources) (Romero-Severson et al., 2001). Thus, both groups in the Pioneer pattern are more than 50% Reid. This does not fit the canonical story. The Pioneer heterotic groups are not derived from material that was geographically or phylogenetically distant. Substantial portions of Pioneer's SS and non-SS are derived from the same cultivar. The predominance of Reid in Pioneer's current heterotic groups was summarized by Smith et al.

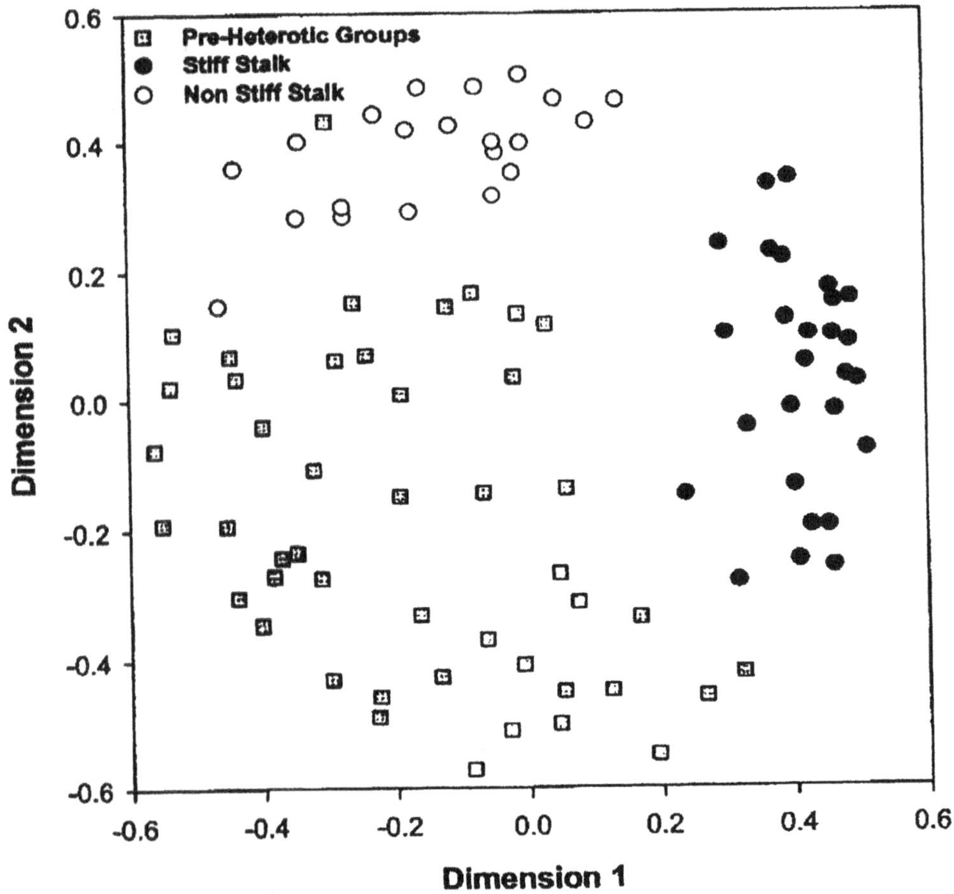

Figure 16.3 Scores for 94 inbreds contributing to the era hybrids on the first two dimensions of the multidimensional scaling analysis of the SSR polymorphism data for 298 SSR loci ($R2 = 0.45$ for the two-dimension model) (Duvick et al. 2004).

(2004) as follows: "...a performance potential that was previously latent in RYD has been realized as evidenced by the combining ability of lines developed from BSSS (largely Reid) when crossed to lines that are predominantly Iodent, a strain of Reid."

Duvick et al. (2004) examined SSR polymorphisms among the inbred parents of the hybrids in the Pioneer era studies. In those studies, widely grown hybrids from different decades are compared to determine the changes that have occurred over time and the proportion of change that is due to genetics and breeding (Duvick et al., 2004). Duvick et al. (2004) found that the inbreds from the preheterotic group era formed one large cluster with no clear groupings (Figure 16.3). On the other hand the modern SS and non-SS inbreds form discrete groups divergent from one another and the pre-heterotic group cluster. Duvick et al. (2004) wrote, "The SSR polymorphism data indicate a clear divergence between the allele profiles of the inbreds created by pedigree breeding in the SS and the non-SS heterotic groups."

The divergent groups were created by breeders. They did not exist in the original germplasm.

Summary

Heterotic patterns are useful tools for increasing the efficiency of breeding programs, but breeders should be wary of adhering to the canonical story too rigidly. Corn breeding abounds with examples of successful breeders who developed very important inbreds from crosses between parents from the two heterotic groups. Indeed this may be a crucial factor in inbred development. If there is a group of elite inbreds with many excellent characteristics but deficient in some character such as root quality, the best source of improved roots may be inbreds from the opposite group. Because there are usually numerous subgroups within groups it is possible to make intergroup crosses and still develop excellent inbreds that have excellent combining ability.

The first breeders of hybrid corn were confronted by a number of problems, perhaps the most important of which was that first-cycle inbreds were extremely weak and difficult to maintain. These breeders needed to develop improved inbreds, and they did it by crossing the best inbreds available with relatively little regard for relationships among the inbreds. An excellent example is Mo17, which was derived from a cross between a Lancaster inbred and a Krug (Reid) inbred.

A second problem confronted by early breeders was one of organization; how to organize the breeding program with the flood of inbreds being developed by public breeders. Breeders in the 1940s chose to do this by splitting the inbreds into two groups and creating the groups in an apparently arbitrary way with little attention to phylogeny. While this may seem surprising, there is theoretical, experimental, and empirical support for this approach. Cress (1967), based on the results of computer simulations, suggested that the way to make the most gain in a reciprocal recurrent selection program is to form one pool with the available germplasm and then arbitrarily split the pool into groups. Genetic drift will create an initial divergence of allele frequencies, and the selection program will enhance those differences (Cress 1967). Butruille et al. (2004) did exactly this in an experimental population. After six cycles of recurrent selection, they detected a significant increase in the yield of the population cross. Allele frequencies diverged, but it was not possible to determine if the changes were due to drift or selection (Butruille and Coors, 2004). The empirical date from the Pioneer breeding program as shown in the Era experiments also lends support to the actions of the breeders in the 1950s. Pioneer breeders started with inbreds derived from CBD OPVs. Using SSRs it was not possible to determine any population structure among these progenitor lines (Duvick et al., 2004). After 60 years of selection, distinct heterotic groups were detected using molecular markers (Duvick et al., 2004). In a separate study on the population structure of CBD OPVs, Labate et al. (2003) found no evidence of two broad groupings of Reid and Lancaster.

Conclusion

Parts of the canonical story are incorrect; CBD heterotic patterns were created by breeders, and are *not* the result of historical or geographical contingencies. The canonical story originated from an article by Anderson and Brown (1952), based on successful hybrids of the 1930s. The concept of heterotic patterns developed in the 1960s and 1970s. Academic interest in heterotic patterns increased in the late 1980s. Academic interest was stimulated by the availability of DNA-based markers and attempts at using markers to identify heterotic patterns. Such examinations have shown that it would not have been possible to identify heterotic groups for CBD, OPVs, and first cycle inbreds using molecular markers, had they been available in the early years of hybrid corn breeding (Labate et al., 2003; Duvick et al., 2004). If breeders had been able to identify Lancaster in the 1930s and tried to keep the Lancaster group pure, breeding progress would have been greatly impeded (poor agronomics of Lancaster).

CBD heterotic patterns were created by breeders through trial and error from a single race of corn. Using heterotic groups as a tool in a hybrid breeding program results in divergent heterotic groups.

Acknowledgments

We thank Don Duvick and Pioneer Hi-Bred International, Inc., for allowing us to reprint Figure 16.3, and Don Duvick, Jim Coors, Fritz Behr, Terry Foley, Everett Gerrish, Scott Johnson, and Forrest Troyer for comments on the manuscript. We acknowledge funding from the College of Agricultural and Life Sciences, University of Wisconsin-Madison.

References

Anderson, E. 1944. Sources of effective germplasm in hybrid maize. Ann. Bot. Gard. 31:355–361.

Anderson, E., and W.L. Brown. 1952. 124–148. *In* J.W. Gowan (ed.), Heterosis. Iowa State College Press. Ames, Iowa.

Anon. 1947. Minutes of the meeting of the North Central Corn Breeding Technical Committee. USDA. Beltsville, Maryland.

Anon. 1949. Minutes of the meeting of the North Central Corn Breeding Technical Committee. USDA. Beltsville, Maryland.

Anon. 1950. Minutes of the meeting of the North Central Corn Breeding Technical Committee. USDA. Beltsville, Maryland.

Anon. 1953. Minutes of the meeting of the North Central Corn Breeding Technical Committee. USDA. Beltsville, Maryland.

Anon. 1971. Minutes of the meeting of the North Central Corn Breeding Technical Committee. USDA. Beltsville, Maryland.

Anon. 1995. MBS Inc. Genetics Handbook. 22nd edition. MBS. Inc. Ames, Iowa.

Baker. R. 1984. Some of the open-pollinated varieties that contributed the most to modern hybrid corn. 16:1–20. 1984 Illinois Corn Breeders School. 16:20–31. University of Illinois at Urbana-Champaign.

Beil, G.M. 1975. Selection and development of inbred material for use in early maturing corn hybrids. P.131–149. *In* Doris Wilkinson (ed.) Proceedings of the 30th Annual Corn and Sorghum Research Conference. American Seed Trade Association. Washington D.C.

Brown, W.L. 1953. Sources of germplasm for hybrid corn. P.11–16. *In* Doris Wilkinson (ed.), Proceedings of the 8th Annual Corn and Sorghum Research Conference. American Seed Trade Association. Washington D.C.

Brown, W.L. 1967. Results of non-selective inbreeding in maize. Sonderabdruck aus. . .Der Zuchter 37(4)155–159.

Butruille, D.V., H.D. Silva, S.M. Kaeppler, and J.G. Coors. 2004. Response to Selection and Genetic Drift in Three Populations Derived from the Golden Glow Maize Population. Crop Sci. 44:1527–1534.

Casa, A.M., S.E. Mitchell, O.S. Smith, J.C. Register III, S.R. Wessler, and S. Kresovich. 2002. Evaluation of Hbr (MITE) markers for assessment of genetic relationships among maize inbred lines. Theor. Appl. Genet. 104:104–110.

Cheres, M.T., J.F. Miller, J.M. Crane, and S.J. Knapp. 2000. Genetic distance as a predictor of heterosis and hybrid performance within and between heterotic groups in sunflower. Theor. Appl. Genet. 100:889–894.

Crabb, A.R. 1942. The Hybrid Corn Makers: Prophets of Plenty. Rutgers University Press. New Brunswick, New Jersey.

Crabb, A.R. 1992. The Hybrid Corn Makers: Prophets of Plenty. The Golden Anniversary Edition. West Chicago Publishing Co. West Chicago, Illinois.

Cress, C.E. 1967. Reciprocal recurrent selection and modifications in simulated populations. Crop Sci. 7:561–567.

Crum, C.W. 1973. The role of source populations in a plant breeding-corn breeding program. Illinois Corn Breeders School. 5:1–11. University of Illinois at Urbana-Champaign.

Darrah, L.L., and M.S. Zuber. 1986. 1985 U.S. farm maize germplasm base and commercial breeding strategies. P. 40. *In* Minutes of the meeting of the North Central Corn Breeding Technical Committee. USDA. Beltsville, Maryland.

Dudley, J.W. 1984. Identifying parents for use in a pedigree breeding program. 176–188. *In* Doris Wilkinson (ed.), Proceedings of the 39th Annual Corn and Sorghum Research Conference. American Seed Trade Association. Washington D.C.

Duvick, D.N., J.S.C. Smith, and M. Cooper. 2004. Long term selection in a commercial hybrid maize program. Plant Breed. Rev. 24:109–151.

East, E.M. 1908. Inbreeding in corn. Conn. Agric. Expt. Stn. Rep. 1907. pp.419–428.

Evola, S.V., F.A. Burr, and B. Burr. 1986. The suitability of restriction-fragment-length-polymorphisms as genetic-markers in maize. Theor. Appl. Genet. 71:765–771.

Geadelmann, J.L. 1984. Using exotic corn to improve northern corn. 98–110. *In* Doris Wilkinson (ed.), Proceedings of the 39th Annual Corn and Sorghum Research Conference. American Seed Trade Association. Washington D.C.

Gerdes, J.T., and W.F. Tracy. 1993. Pedigree diversity within the Lancaster Surecrop Heterotic group of maize. Crop. Sci. 33:334–337.

Gowan, J.W. 1952. Heterosis. Iowa State College Press. Ames, Iowa.

Gould, S.J. 2002. Jim Bowie's letter and Bill Buckner's legs. 54–70. *In* I Have Landed. Harmony Books. New York.

Gracen, V.E. 1986. Sources of temperate maize germplasm and potential usefulness in tropical and subtropical environments. Advances in Agronomy 39:127–172.

Griffing, B., and E.W. Lindstrom. 1954. A study of the combining of corn inbreds having varying proportions of Corn Belt and non-Corn Belt germplasm. Agron. J. 46:545–552.

Hallauer, A.R. 1999. 483–500. *In* J.G. Coors and S. Pandey (eds.), The Genetics and Exploitation of Heterosis in Crops. Crop Science Society of America. Madison, Wisconsin.

Hallauer, A.R., W.A. Russell, and K.R. Lamkey. 1988, Corn breeding. *In* G.F. Sprague an J.W. Dudley (eds.) Corn and Corn Improvement, third edition. Crop Science Society of America. Madison, Wisconsin.

Hallauer, A.R., and J.B. Miranda Fo. 1981. Quantitative Genetics in Maize Breeding. Iowa State University Press. Ames, Iowa.

Havey, M.J. 1998. Molecular analysis and heterosis in the vegetables: Can we breed them like maize? 29–44. *In* K.R. Lamkey and J.E. Staub. (eds.), Concepts and Breeding of Heterosis in Crop Plants. Crop Science Society of America. Madison, Wisconsin.

Hayes, H.K. 1963. A Professor's Story of Hybrid Corn. Burgess Publishing Co., Minneapolis, Minnesota.

Helentjaris. T., G. King, M. Slocum, C. Siedenstrang, and S. Wegman. 1985. Restriction fragment polymorphisms as probes for plant diversity and their development as tools for applied plant-breeding. Plant Molecular Biology 5:109–118.

Jenkins, M.T. 1978. Maize breeding during the development and early years of hybrid maize. 13–28. *In* D.B. Walden (ed.), Maize Breeding and Genetics. Wiley-Interscience New York.

Jones. D.F. 1918. The effects of inbreeding and crossbreeding upon development. Conn. Agric. Expt. Stn. Bull. 207:5–100.

Kannenberg, L.W. 1976. An open-ended hierarchical procedure for the development of heterotic breeding populations. 11. Minutes of the meeting of the North Central Corn Breeding Technical Committee. USDA. Beltsville, Maryland.

Kaufman, K.D., C.W. Crum, and M.F. Lindsay. 1982. Exotic germplasm in a corn breeding program. Illinois Corn Breeders School. 14:23–38. University of Illinois at Urbana-Champaign.

Labate, J.A., K.R. Lamkey, S.E. Mitchell, S. Kresovich, H. Sullivan, and J.S.C. Smith. 2003. Molecular and historical aspects of Corn Belt Dent diversity. Crop Sci. 43:80–91.

Liu, K.M. Goodman, S. Muse, J.S.C. Smith, E. Buckler, and J. Doebley. 2004. Genetic Structure and diversity among inbred lines as inferred from DNA microsatellites. Submitted.

Lonnquist, J.H., and C.O. Gardner. 1961 Heterosis in intervarietal crosses in maize and its implications in breeding procedures. Crop Sci. 1:179–183.

Manglesdorf, P.C. 1974. Corn: Its Origin, Evolution, and Improvement. Belknap Press. Cambridge, Massachusetts

Melchinger, A.E. 1999. Genetic diversity and heterosis. 99–118. *In* J.G. Coors and S. Pandey (eds.), The Genetics and Exploitation of Heterosis in Crops. Crop Science Society of America. Madison, Wisconsin.

Melchinger, A.E., and R.K. Gumber. 1998. Overview of heterosis and heterotic groups in agronomic crops. 29–44. *In* K.R. Lamkey and J.E. Staub. (eds.), Concepts and Breeding of Heterosis in Crop Plants. Crop Science Society of America. Madison, Wisconsin.

Mishra, T.N., and J.L. Geadelmann. 1978. 57. Classification of maize populations into heterotic groups. Agronomy Abstracts. American Society of Agronomy. Madison, Wisconsin.

Moll, R.H., W.S. Salhuana, and H.F. Robinson. 1962. Heterosis and genetic diversity in variety crosses of maize. Crop Sci. 2:197–198.

Murphy, R.P. 1942. Convergent improvement with four inbred lines of corn. Jour. Amer. Soc. Agron. 34:138–150.

Murphy, J.P., T.S. Cox, and D. M. Rodgers. 1986. Cluster analysis of red winter wheat cultivars based upon coefficients of parentage. Crop Sci. 26:672–676.

Paterniani, E., and J.H. Lonnquist. 1963. Heterosis in interracial crosses of corn. Crop Sci. 3:504–507.

Peterson, P.A., and A. Bianchi. 1999. Maize Genetics and Breeding in the 20th century. World Scientific Publishing Co. Singapore.

Richey, F.D. 1927. The convergent improvement of selfed lines of corn. Amer. Naturalist 61:430–449.

Richey, F.D. 1950. Corn breeding. Advances in genetics 3:159–192.

____, and G.F. Sprague. 1931. Experiments on hybrid vigor and convergent improvement in corn. U.S. Dept. Agr. Tech. Bull. 1354:1–18.

Romero-Severson, J., J.S.C. Smith, J. Ziegle, J.L. Hauser, and G. Hookstra. 2001. Pedigree analysis and haplotype sharing within diverse groups of *Zea mays* L. inbreds. Theor. Appl. Genet. 103:567–574.

Russell, W.A. 1974. Comparative performance of maize hybrids representing different eras of maize breeding. 81–101. *In* Doris Wilkinson (ed.), Proceedings of the 29th Annual Corn and Sorghum Research Conference. American Seed Trade Association. Washington D.C.

Russell, W.A., and A.R. Hallauer. 1980. Corn. 299–309. *In* W.R. Fehr and H.H. Hadley (eds.), Hybridization of Crop Plants. Crop Science Society of America. Madison, Wisconsin.

Shull, G.H. 1908. The composition of a field of maize. Amer. Breeders' Assoc. Rep 4:296–301.

Shull, G.H. 1909. A pure line method of corn breeding. Amer. Breeders' Assoc. Rep 5:51–59.

Shull, G.H. 1952. Beginnings of the heterosis concept. 14–49. *In* J.W. Gowan (ed.) Heterosis. Iowa State College Press. Ames, Iowa.

Singleton, W.R. 1963. Handbook of hybrid corn. Journal of Heredity 54(5) 205–206.

Smith, J.S.C., D.N. Duvick, O.S. Smith, A. Grunst, and S.J. Wall. 1999. Effects of hybrid breeding on genetic diversity of maize. 119–126. *In* J.G. Coors and S. Pandey (eds.), The Genetics and Exploitation of Heterosis in Crops. Crop Science Society of America. Madison, Wisconsin.

Smith, J.S.C., D.N. Duvick, O.S. Smith, M. Cooper, and L. Feng. 2004. Changes in pedigree backgrounds of pioneer brand maize hybrids widely grown from 1930 to 1999. Crop Sci. 44:1935–1946.

Smith, O.S. 1986. Covariance between line *per se* and testcross performance. Crop Sci. 26:540–543.

Smith, O.S., H. Sullivan, B. Hobart, and S.J. Wall. 2000. Evaluation of a divergent set of SSR markers to predict F1 grain yield performance and grain yield heterosis in maize. Maydica 45:235–241.

Sprague, G.F. 1955a. Corn and Corn Improvement, 1st edition. American Society of Agronomy. Madison, Wisconsin.

Sprague, G.F. 1955b. Corn breeding. In G.F. Sprague (ed.) Corn and Corn Improvement, 1st edition. American Society of Agronomy. Madison, Wisconsin.

Sprague, G.F. 1964. Estimates of genetic variations in two open-pollinated varieties of maize and their reciprocal F1 hybrids. Crop Sci. 3:332–334.

Sprague, G.F. 1977. Corn and Corn Improvement, 2nd edition. American Society of Agronomy. Madison, Wisconsin.

Sprague, G.F. 1983. Heterosis in maize: Theory and practice. 47–70. *In* R. Frankel (ed.), Heterosis. Springer-Verlag, Berlin.

Sprague, G.F. 1984. Organization of breeding programs. 1984 Illinois Corn Breeders School. 16: 20–31. University of Illinois at Urbana-Champaign.

Sprague. G.F., and J.W. Dudley, 1988. Corn and Corn Improvement, 3rd edition. American Society of Agronomy. Madison, Wisconsin.

Sprague, G.F., and S.A. Eberhart. 1977. Corn breeding. 305–362. *In* G.F. Sprague (ed.), Corn and Corn Improvement, 2nd edition. American Society of Agronomy. Madison, Wisconsin.

Troyer, A.F. 1999. Background of U.S. Hybrid Corn. Crop Sci. 39:601–626.

Troyer, A.F. 2000a. Temperate corn—background, behavior, and breeding. 393–466. In A.R. Hallauer (ed.), Specialty Corns. CRC Press. Boca Raton, Florida.

Troyer, A.F. 2000b. Origins of modern corn hybrids. 27–42. *In* Doris Wilkinson (ed.) Proceedings of the 55th Annual Corn and Sorghum Research Conference. American Seed Trade Association. Washington D.C.

Troyer, A.F. 2004. Background of U.S. Hybrid Corn II: Breeding, Climate, and Food. Crop Sci. 44:370–380.

Tsotsis, B. 1972. Objectives of industry breeders to make efficient and significant advances in the future. 93–107.*In* Doris Wilkinson (ed.), Proceedings of the 27th Annual Corn and Sorghum Research Conference. American Seed Trade Association. Washington D.C.

Wallace, H.A., and W.L. Brown. 1956. Corn and Its Early Fathers. Michigan State University Press. East Lansing, Michigan.

Walton, M., and T. Helentjaris 1987. Application of restriction fragment length polymorphism (RFLP) technology to maize breeding. 48–75. *In* Doris Wilkinson (ed.), Proceedings of the 42nd Annual Corn and Sorghum Research Conference. American Seed Trade Association. Washington D.C.

Weatherspoon, J.H. 1973. Usefulness of recurrent selection schemes in a commercial corn breeding program. 137–143. *In* Doris Wilkinson (ed.), Proceedings of the 28th Annual Corn and Sorghum Research Conference. American Seed Trade Association. Washington D.C.

Williams, T.R., and A.R. Hallauer. 2000. Genetic diversity among maize hybrids. Maydica 45:163–171.

Zuber, M.S., and L.L. Darrah. 1981. 1979 U.S. corn germplasm base. *In* Minutes of the meeting of the North Central Corn Breeding Technical Committee. USDA. Beltsville, Maryland.

Zuber. M.S., and L.L. Darrah. 1987. Breeding, genetics, and seed corn production. 31–52. *In* S.A. Watson and P.E. Ramstad. Corn: Chemistry and Technology. American Association of Cereal Chemists. St. Paul, Minnesota.

17

Hybrid and Open-Pollinated Varieties in Modern Agriculture

Kevin V. Pixley, International Maize and Wheat Improvement Center (CIMMYT), Mexico

Introduction

Recent rates of increase in cereal production are insufficient to meet the expected demand for basic cereals in coming decades. This threatened food crisis has led scientists and policy makers to rethink agricultural research and policy priorities and strategies. This chapter considers the potential roles and implications of using hybrid and open-pollinated varieties in addressing the future demand for production of four important cereal crops: rice, sorghum, pearl millet, and maize. Emphasis is given to the perspective of developing countries, for this is where the greatest challenges and opportunities lie, including greatest anticipated production shortfalls and largest potential to increase productivity.

The potential of hybrids to contribute to increased agricultural productivity and production gains has been demonstrated for several crops and was the subject of a recent international symposium (Coors and Pandey, 1999). The question remains, however, whether hybrid technology is viable and likely to meet the growing food and feed demand in developing countries, where crop yields generally lag far behind those achieved in many developed countries. A corollary question is whether open-pollinated varieties (OPVs) have a significant contributing role in modern agriculture toward meeting cereal production demand. This chapter examines technical merits of hybrids and OPVs, social preferences ascribed to each, and uses experimental data to model economic considerations influencing the suitability of hybrid relative to OPV maize technology. Much of the general discussion centers on maize but addresses issues of wider relevance.

The next section briefly reviews literature citing the potential of hybrids relative to improved varieties and traditional or local varieties for rice, sorghum, pearl millet, and maize. Reasons influencing adoption, or lack of widespread adoption, of hybrids are summarized for each crop to elicit insights about circumstances and conditions under which hybrids are likely to succeed. The section entitled Stability of Hybrids and OPVs reviews evidence about stability of hybrids relative to OPV and traditional varieties' performance across a wide range of environmental conditions to address questions about increased risk or vulnerability sometimes associated with adoption of hybrid technology.

A case study for maize in southern Africa is developed in Choosing Between Hybrid and Open-Pollinated Maize Varieties, using simple assumptions to predict profitability of growing hybrids relative to OPVs. The study includes performance data for maize hybrids and OPVs in environments where average yield ranged from 1 to 10 t ha^{-1} and includes data indicating the consequences of planting farmer-saved (F_2) grain instead of F_1 seed.

Selected economic and philosophical considerations are then raised, before reaching some overarching conclusions and recommendations for future research.

Potential of hybrids

Rice

Rice (Oriza sativa L.) was produced on more than 150 M ha during 2000. Asian countries grew 136.3 M ha of rice with average milled yield of 2.6 t ha^{-1}, Africa grew 7.3 M ha at 1.5 t ha^{-1}, Latin

America 5.7 M ha at 2.5 t ha^{-1}, North America 1.3 M ha at 4.8 t ha^{-1}, and Europe grew 0.6 M ha at 3.4 t ha^{-1} (Coats, 2003). Aggregate demand for rice in Asia by 2025 will exceed 1990 consumption by 50–60% (Pingali et al., 1997), and there is therefore considerable concern about declining annual percentage growth rates of rice production (4.35, 2.59, 3.24, 1.25), rice area harvested (1.34, 0.60, 0.31, 0.10) and rice yield (2.70, 1.88, 2.86, 1.06) in Asia during recent decades (1960s, 1970s, 1980s, and 1990s, respectively) (Van Tran, 2001).

The best, new, improved rice cultivars today achieve maximum yield similar to that of "IR8" in the 1960s (9-10 Mg ha^{-1}), suggesting that a yield plateau has been reached (Peng et al., 1999). And, there is general consensus that significant closure of the gap between yield potential and actual yield achieved by farmers will not be profitable or economically sensible for many rice farmers (Herdt, 1996; Pingali et al., 1997; Van Tran, 2001). The International Rice Research Institute (IRRI) has aggressively pursued a rice ideotype, dubbed New Plant Type or NPT rice, characterized by few tillers, large panicles, more and heavier grains per panicle, and strong stems. The NPT rice has to date failed to significantly raise the yield ceiling set by IR8 in the 1960s (Peng et al., 1999). The emerging conclusion is that hybrid cultivars may be the most effective technology to achieve significant yield increase in rice within a short to medium time frame.

Rice is normally self-pollinating, with an average of only 1–4% natural outcrossing (Moldenhauer and Gibbons, 2003); therefore, modern varieties are mainly inbred lines. The first commercial hybrid rice cultivar was released in China in 1976, using cytoplasmic male sterility in a three-line system (male sterile line, maintainer line, and a restorer line) (Pingali et al., 1997). The commonly cited yield advantage of best hybrid over best inbred rice cultivars is 15–20%, although several reports range between 9 and 33% (Andrews, 2001; Horie, 2001; Janaiah and Hossain, 2001; Lin, 1994; Peng et al., 1999; Virmani, 2001). There is speculation that this yield advantage of hybrids may be doubled by use of NPT lines in hybrid combinations and/or use of inter-subspecific crosses (e.g., among *O. indica, O. japonica, and O. javanica*) (Peng et al., 1999; Virmani, 2001). Importantly, in addition to higher yield, hybrid cultivars offer a higher marginal rate of return to labor investment than inbred cultivars (Lin, 1994). This largely relates to reduced seeding rates for hybrid cultivars (about one-third relative to inbred cultivars), which translate into less labor required for transplanting seedlings.

Successful hybrid rice production requires higher use of fertilizer and crop protectants (e.g., fungicides and pesticides) relative to production of inbred cultivars (Janaiah and Hossain, 2001; Lin, 1994; Pingali et al., 1997; Virmani, 2001). Table 17.1 lists hindrances or constraints to adoption of hybrid rice in India, China, the United States, and in general. Many of the technical constraints, such as higher susceptibility to pests and diseases, lesser stability across stressed environments, and inferior grain quality, can largely be overcome through breeding efforts, but highlight the enormity of the task of developing a new breeding program (i.e., the three-line system). The social constraints, however, reflect a combination of neglect of vari-

Table 17.1 Hindrances to adoption of rice hybrids

General (Virmani, 2001)
Very high expectations by farmers
Inconsistent performance of the first set of released hybrids
Inadequate understanding of agronomic management of hybrids
Inadequate availability of pure seeds of parental lines and hybrids
Poor grain quality of hybrids compared with premier-quality rice varieties
Inadequate level of disease/insect resistance in released hybrids
Inconsistent seed yields
High cost of hybrid seeds
Traditional habit of rice farmers of using their own seed
General (Pingali et al., 1997)
Knowledge intensive technology (therefore not readily accessible)
India (Janaiah and Hossain, 2001)
Inferior cooking quality
Inferior storage quality
Inferior taste
Inferior stickiness after cooking
Unpleasant smell after cooking
Higher cost of production (seed and crop protectants)
Price penalty for hybrid grain
Unavailability of pure hybrid seed
Formation of sterile grains in the productive tillers
Unstable yield
China (Lin, 1994)
High risk: few hybrids were available, yet environments were many and diverse
Lack of stability across environments, particularly adverse environments
Seed must be purchased every year
Complicated seed production and distribution system
Later maturity of hybrids meant difficulty achieving two crops
Low cooking quality of hybrid rice (later overcome)
Hybrid rice yields best with higher fertilizer rate than other modern varieties
United States (Andrews, 2001)
Expensive seed
Lower grain quality results in discounted price

ous characters (e.g., cooking quality and taste preferences) by breeders and inherent nature of the hybrid technology (e.g., higher cost of seed, loss of seed self-sufficiency through planting saved grain, and resultant increased household vulnerability).

China is the only country where hybrid rice is widely grown, covering about 15 million ha or 50% of the rice area (Eizenga and Rutger, 2003). Other countries where commercial hybrid rice production exists include Vietnam (280 k ha), India (150 k ha), Philippines (5 k ha), and Bangladesh (250 k ha) (Virmani, 2001). Factors positively associated with adoption of rice hybrids in China include size of landholding, level of education, capital endowment, and government-imposed production quotas (Lin, 1994). In general, policies promoting national food security by maximizing total grain production should favor adoption of rice hybrids. By contrast, policies or market forces favoring specialty and value-added products would probably not lead to increased use of hybrid rice cultivars.

Sorghum

Sorghum (*Sorghum bicolor* L.) was produced on more than 43 M ha during 1998. African countries grew 23.0 M ha of sorghum with average grain yield of 0.87 t ha^{-1}, Asia grew 12.7 M ha at 1.15 t ha^{-1}, Latin America grew 4.0 M ha at 3.13 t ha^{-1}, the United States grew 3.1 M ha at 4.23 t ha^{-1}, and Europe grew 0.2 M ha at 4.16 t ha^{-1} (Smith, 2000).

Grain sorghum is predominantly self-pollinating, but outcrossing occurs naturally at an average of about 6% (range of 2–35%) (Rooney and Smith, 2000). Consequently, traditional or landrace varieties are heterogeneous populations maintained and modified by individual plant selections. Many modern sorghum varieties are pure-line (inbred) cultivars, but hybrid cultivars account for 100% of sorghum area in Argentina, Australia, Mexico, and the United States (Toure et al., 2002). Hybrids are possible in sorghum through various nuclear and cytoplasmic-nuclear sterility systems, but all commercial hybrids use cytoplasmic-nuclear sterility, and most use the A1 cytoplasm (Maunder, 2000; Rooney and Smith, 2000). Other cytogenetic-nuclear male sterility-inducing cytoplasms have been described (Rooney and Smith, 2000), and their agronomic usefulness has been investigated (e.g., Moran and Rooney, 2003) due to concerns about extensive use of the A1 system. Many first-generation sorghum hybrids (i.e., the first sorghum hybrids developed by any program) are/were topcross hybrids using a cytoplasmic male sterile inbred line as female for a popular cultivar as male.

House et al. (1997) summarized literature indicating −44–180% hybrid superiority over improved varieties and 49–185% hybrid superiority over local land race cultivars. Kapran et al. (2002) reported 80% yield advantage of hybrid NAD-1 over best local sorghum varieties in Niger. Hageen Dura-1 produced 58% more grain than the best local cultivar under irrigation and 52% more under rain-fed conditions in Sudan (House et al., 2000). The first sorghum hybrid released in India in 1964, CSH-1, produced 40% more grain than local varieties (House et al., 2000). In the United States, first sorghum hybrids in the late 1950s had average superiority of about 33% over standard cultivars (Maunder, 2000). Duvick (1999a) estimates the average superiority of hybrid over best nonhybrid sorghum varieties at the time of first hybrid introduction has been about 40%.

Doggett (1967) summarized results from trials in East Africa, Rhodesia, and the United States and concluded that none of the six studies (each with 32–128 trials) gave "any support to the idea that heterosis is expressed more with good than with poor farming." In fact, both Doggett and Jowett (1966) and Haussmann et al. (1998) found highest hybrid superiority over their parent lines or male OPV at the most severely stressed sites. Yield of hybrids in such stressed environments was nevertheless low, making hybrids economically viable mainly in the higher-yielding environments.

Inbreeding depression in sorghum can be severe, and yield from planting F_2 grain may be less than the mid-parent value for many hybrids (Table 17.2) (Liang et al., 1972). This result raises concerns about the likely consequences of recycling hybrid sorghum seed.

Reasons for rapid and intensive adoption (95–100%) of hybrid sorghum in the United States included (Maunder, 2000) (1) the availability of associated agronomic technologies (e.g., fertilizer, herbicide, narrower row spacing, irrigation) that dramatically increased the total productivity gain, and (2) the existence of well-established hybrid (maize) seed industry. Rapid adoption of sorghum hybrids in India (currently about 40% according to Toure et al. [2002)), was facilitated by the avail-

Table 17.2 Mid-parent, F_1 and F_2 values, high-parent heterosis, and inbreeding depression for 10 bi-parental sorghum crosses among five inbred sorghum lines

	Mid-parent	F_1	F_2	High-parent Heterosis	Inbreeding Depression
				%	%
Grain yield (g/plant)	62.3	70.2	60.0	6.0 (−0.4, 16.3)[a]	14.1 (5.4, 29.0)[a]
Kernel weight (g/1000)	29.1	29.6	28.3	−1.4 (−24.7, 25.3)	4.0 (−6.9, 17.9)
Days to flower (d)	67.9	66.1	67.1	−0.9 (−7.6, 3.5)	−1.6 (−5.4, 5.3)

Source: (Liang et al., 1972).
[a]Average (low, high value).

ability of excellent public hybrids and strong governmental policy support to the seed sector, including empowering each state to establish seed certification agencies and creating the "truthfully labeled" category of seed that made certification voluntary (House et al., 2000). Ahmed et al. (1996) indicate that in sub-Saharan Africa most new sorghum cultivars have been adopted, not because of greater yield potential, but because their greater earliness forms part of a risk-avoidance strategy.

Despite these advantages, sorghum hybrids have not been widely adopted in much of Africa; for example, Nigeria and Niger are the only two out of seventeen sorghum-growing West and Central African countries that have released sorghum hybrids (Toure et al., 2002). Reasons for slow or little adoption of sorghum hybrids include the following:

1. The first hybrids released and promoted often were not good enough! For example, the first released hybrid in India (CHS-1) had poor grain quality relative to local varieties (House et al., 1997); in West and Central Africa, where hybrid sorghum yields were high, increased plant lodging, head bugs, and grain mold often cancelled the advantage, and use of unadapted exotic (e.g., from India or USA) male sterile genotypes as the female parents of the hybrids led to severe leaf disease problems, greater susceptibility to Striga, lesser grain quality, and photoperiod insensitivity (unlike most local varieties that flower at the end of the rainy season ensuring grain maturation under favorable, dry conditions) (Toure et al., 2002).
2. The absence of viable seed production and distribution mechanisms has been a major factor discouraging hybrid adoption in several countries (House et al., 1997).
3. In Zimbabwe, farmers prefer the white-grained cultivars to higher-yielding hybrids (Mangombe and Mushonga, 1996).
4. The inability to recycle grain for use as seed for subsequent crops (Haussmann et al., 1998; Mangombe and Mushonga, 1996) has been cited as an impediment to adoption of hybrid sorghum cultivars.

Most scientists will agree that hybrid sorghum is a viable technology for increasing overall sorghum grain production and productivity. However, most will also agree that seed production and marketing structures capable of reliably supplying good quality seed are an essential precondition to the successful adoption of hybrid technology, and are largely deficient in Africa (House et al., 2000; Murty, 2002; Toure et al., 2002).

Pearl Millet

Pearl millet (*Pennisetum glaucum* L.) was grown on about 15–19 M ha in 1992 (Dendy, 1995). The following figures are inexact because production statistics for pearl millet are typically combined with total millet production. Some facts seem clear, however, such as greater than 99% of all pearl millet is produced in developing countries, primarily in India (42%), Nigeria (23%), Niger (9%), Mali (5%), Senegal (4%), Burkina Faso (3%), Sudan (3%), and Pakistan (2%) (estimated from Dendy, 1995). Global average yield of pearl millet is around 0.75 t ha^{-1}, with Sudan (0.22 t ha^{-1}) and perhaps Nigeria (0.82 t ha^{-1}) illustrating the extremes. Pearl millet is grown primarily as a subsistence crop in areas with poor soils and inconsistent or low rainfall (e.g., < 300 mm).

Like sorghum, pearl millet is characterized by large panicles consisting of numerous perfect flowers. Pearl millet is primarily cross-pollinating because it is protogynous; stigmas emerge and are receptive (typically one to three days) before anthers emerge and release pollen (House et al., 1995). Hence, most landrace, traditional, and modern varieties are OPVs. Kumar et al. (2002) listed five types of hybrids that are possible for pearl millet:

1. Pro-hybrid: Fertile inbred female × OPV male (uses protogyny)
2. Single-cross: Cytoplasmic male sterile (CMS) inbred female × fertile (restorer) inbred male
3. Top-cross hybrid: CMS inbred female × OPV male
4. Three-way hybrid: (CMS × unique maintainer line) as female × fertile (restorer) inbred male
5. Variety cross hybrid: OPV × OPV (uses protogyny

Developing appropriate pearl millet hybrids is complicated immensely by the facts that pearl millet is primarily grown as a subsistence crop in the harshest of environments, and it is grown as a dual-purpose crop for which both grain and stover are highly valued. Estimates of heterosis and hybrid superiority relative to OPVs are heavily dependent on the evaluation environment, often becoming negative at driest or most-stressed sites. Variety cross hybrids using local OPVs in West Africa yielded up to 59% more than best local cultivars (Kumar et al., 2002). Mid-parent heterosis for grain yield was 15–215% in Mali, although none of the variety cross hybrids was significantly better than the best local OPVs (Hanna, 2001). Mid-parent heterosis for variety cross hybrids in a nine-parent diallel ranged from −14 to 30% at drought stressed sites, and from −9 to 17% at favorable sites in India (Presterl and Weltzien, 2003). Monyo et al. (1996) reported variety cross hybrid yielded 7–47% more than the best OPV in trials in Zimbabwe (19–87% high-parent heterosis), while another set of variety cross hybrids in Tanzania outyielded the best OPV check by 33.3 to 89.8% (high-parent heterosis 13–99%).

Topcross hybrids offer greater yield advantage than variety crosses. Twenty topcross hybrids evaluated for two years showed on average 73% higher yield than their OPV male parents and were 30% higher yielding than the best OPV check (Kumar et al., 2002). Mahalakshmi et al. (1992) also reported higher grain yield (but similar biomass yield) of topcross hybrids relative to their OPV males; however, local landraces and their topcrosses were higher yielding than improved cultivars and their topcrosses at the lowest yielding sites (< 1 t ha^{-1}). Bidinger et al. (1994) evaluated 32 topcrosses using 16 landraces as males for two CMS lines and reported the following: (1) landraces outyielded OPV checks for grain in the arid environment only, but for fodder at all sites; (2) the "grain-type" topcrosses outyielded landraces for grain at all sites and yielded less fodder than landraces at two sites; and (3) "dual-purpose" topcrosses outyielded landraces for grain and fodder at all sites. An important constraint to use of topcrosses in Africa has been the disease susceptibility (e.g., downy mildew and grain smut) of the available, exotic male-sterile lines, but this issue is being addresses through breeding efforts to convert local germplasm to male sterility (Kumar et al., 2002).

Between 1965 and 1992 in India, the area of pearl millet grown to hybrids and improved OPVs grew from 5 to 55%, and average yield rose from 0.36 to 0.65 t ha^{-1} (Govila et al., 1997; Rai et al., 1997). Rapid adoption occurred because farmers value the higher yield potential when good rains occur and the greater Striga ("witch weed") resistance of hybrids (Tripp and Pal, 1998). In 1995, there were more than 30 private companies marketing about 50 pearl millet hybrids in India (Govila et al., 1997), a fact that indicates viability of the seed system and enabled widespread adoption of hybrids.

Reasons for low rates of adoption of pearl millet hybrids in various regions are summarized in Table 17.3. Several authors, approximating a consensus, have proposed that the most appropriate pearl millet hybrids for Africa may be topcross hybrids using male-sterile or fertile (using protogyny) inbred lines as female seed parent for locally adapted OPVs or landraces. Such hybrids are higher yielding than OPVs, possess better disease resistance, have better general adaptation, and are easier and cheaper to produce than single cross hybrids.

Maize

Maize (*Zea mays* L.) is grown on more than 140 M ha worldwide (Aquino et al., 2000). Although aver-

Table 17.3 Hindrances to adoption of pearl millet hybrids (and/or improved varieties)[a]

General (Mahalakshmi et al., 1992)
- Poorer adaptation to environmental stress in arid regions
- Lesser stability

India (Kelley et al., 1996)
- Belief that traditional yields more straw than improved varieties in dry years
- Low grain yield in dry years
- Low straw yield in dry years
- Poor grain quality
- Poor straw quality
- Problems with seed availability
- Perceived to be riskier than traditional cultivars

India (Tripp and Pal, 1998)
- Lower fodder yield and quality
- Inferior food quality (e.g., for making roti)
- Lack of knowledge about improved varieties
- Lack of trust toward seed and input traders

India (vom Brocke et al., 2002)
- Lower yield under stress conditions
- Lower stover productivity
- Lesser tillering habit (tillers offer yield security against stress)

Africa (Govila et al., 1997)
- Nonavailability of suitable parental (CMS) lines
- Downy mildew and ergot susceptibility of CMS lines

Sub-Saharan Arica (Ahmed et al., 1996)
- Greater fertilizer requirement
- Lack of sufficient, high-quality seed
- Lack of policies making inputs more accessible and production more profitable
- Lack of alternative grain uses to avoid price collapses in surplus production years

West Africa (Kumar et al., 2002)
- Lack of seed production infrastructure
- Lack of assured food grain markets (and diversified end uses for grain)

Niger (Ndjeunga and Sidi, 2002)
- Unreliable seed sources, with high transaction costs

[a]It was not always possible to distinguish between factual and perceived factors, but all were given as important hindrances to adoption of hybrids or modern varieties.

Table 17.4 Maize production statistics, 1997–1999

Region	Harvested Area ('000 ha)	Yield t ha^{-1}
East and southern Africa	15,436	1.5
Africa excluding South Africa	11,745	1.3
West and Central Africa	9,223	1.2
South Asia	8,147	1.7
Southeast Asia	8,185	2.4
East Asia	25,592	4.8
Mexico, Central America & Caribbean	9,601	2.2
Andean Region	2,082	1.9
Southern Cone, South America	15,501	3.2
East Europe and former Soviet Union	9,577	3.8
West Europe & North America	34,543	8.3
World Total	**140,182**	**4.3**

Source: Aquino et al. (2000).

age yield in the United States and other high-income countries is 8.3 t ha^{-1}, most of Africa and large areas of Asia and South America achieve yield below 2 t ha^{-1} (Table 17.4). Demand for maize is projected to increase from 1995 levels 50% globally and 93% in sub-Saharan Africa by 2020 (International Food Policy Research Institute, as cited by Pingali and Pandey, 2001). Given these facts, it is alarming that annual rate of growth of maize yield between 1956 and 1995 was less than 2% across all less-developed countries and was less than 1% in sub-Saharan Africa (Pingali and Heisey, 1999).

Maize is a cross-pollinating crop, which implies that landrace and traditional cultivars are highly heterozygous and heterogenous. Because female and male flowers are physically separate from each other (on the ear and tassel, respectively), control of parentage and deliberate formation of improved OPVs and hybrids are readily possible. The added facts that each maize plant typically yields 250–500 seeds, and only 20,000–50,000 seeds ha^{-1} are commonly planted for maize production, make commercial use of hybrids economically feasible under many circumstances.

Several types of hybrids, spanning a broad range of heterogeneity, are used commercially in maize. Paliwal (2000) summarized data indicating that average yield advantage of hybrids relative to OPVs was 46% for single cross, 30% for three-way, 23% for double, 37% for double topcross, 28% for topcross, and 17% for variety cross hybrids. Hybrids involving one or more noninbred parents are cheaper and easier to produce (and are therefore usually sold at lower price), but typically offer lower yield potential than hybrids using inbred parents. As expected, because inbreds are lower yielding than OPVs, hybrids formed using inbred parents have much higher heterosis (e.g., 90–300%) than hybrids among OPVs (Hallauer and Miranda Fo, 1981; Melchinger and Gumber, 1998).

Duvick (1999a) reported that the first hybrids released in the United States (in the 1920s) were 15% higher yielding than the best OPVs. Similarly, Fakorede et al. (2001) found yield advantage of first hybrids introduced in Nigeria over best OPVs was 29% for white and 15% for yellow hybrids. Chiduza et al. (1994) compared five commercial

Table 17.5 Maize area planted to improved OPVs and hybrids in developing countries, late 1990s

	Total maize area (million ha)	Area planted using farm-saved seed (%)	Area planted using commercial seed		
			OPVs (%)	Hybrids (%)	All MVs (%)
Latin America	27.1	55.1	5.0	39.9	44.9
excluding Argentina	24.5	59.6	5.3	35.1	40.4
Sub-Saharan Africa	23.3	53.3	16.1	30.6	46.7
excluding South Africa	19.2	63.9	18.9	17.2	36.1
East, South & Southeast Asia	19.6	33.7	22.0	44.3	66.3
All regions	70.0	48.5	13.5	38.0	51.5
All nontemperate regions	63.3	52.9	14.6	32.5	47.1

Source: Morris (2001).

hybrids and five elite OPVs at eight sites in Zimbabwe and found 16% hybrid advantage for unfertilized and 19% hybrid advantage for fertilized trials. Menkir and Akintunde (2001) evaluated 30 hybrid, 30 OPV, and 30 landrace varieties in Nigeria and found hybrid superiority over OPVs was 9% under well-watered conditions and 7% under drought stress, whereas superiority over landraces was 76% and 67% for hybrids and 62% and 56% for OPVs, under well-watered conditions and drought stress, respectively. Pixley and Bänziger (2004) reported 18% average hybrid advantage over elite OPVs at 16 sites with mean grain yield from 1.8 to 7.3 t ha^{-1}. A second study, including 78 trials across Zimbabwe, again found 18% average advantage of hybrids over OPVs (Pixley and Bänziger, 2004). It seems reasonable to conclude that, in regions where both types of maize variety have been actively or recently developed, superiority of best hybrids over best OPVs may average 15–20%.

The extent of inbreeding depression in maize is highly relevant to the majority of maize farmers in the nontemperate world, who regularly plant farm-saved grain ("recycled" seed) (Table 17.5). Yield reduction of hybrids due to inbreeding depression is generally inversely proportional to the number of inbred lines involved in each parent (Kiesselbach [1933], as cited by Morris et al., [1999]). Morris et al. (1999) used simulation models and estimated average F_1 to F_2 inbreeding depression of 33% for single-cross, 17% for three-way, and 8% for double-cross hybrids. In highland Mexico, Beck and Torres (2003) measured 36% inbreeding depression for single-cross, 31% for three-way, and 15% for double cross hybrids. Pixley and Bänziger (2004) found 32% (significant, $P < 0.01$), 16% (significant, $P < 0.01$), and 5% (not significant) yield loss from planting recycled seed of inbred-parent hybrids (mostly three-way, but also a few single-cross hybrids), double-topcross hybrids, and OPVs (F_2 versus F_3 seed) across five locations in Zimbabwe (Table 17.6). It is also of great relevance to note that inbreeding depression is less for "synthetic" maize populations (e.g., OPVs) formed using lines with good tolerance to inbreeding than for populations maintained by random cross-pollination among full-vigor plants (Hallauer and Miranda Fo, 1981; Lima et al., 1984; Miranda Fo, 1999; Paliwal, 2000).

Adoption of maize hybrids in the United States reached 95% within 15 years of their introduction in the 1930s (Duvick, 1999a). Reasons for this impressive rate of adoption included higher grain yield; improved resistance to root and stalk lodg-

Table 17.6 Comparison of variety types across generations across five sites in Zimbabwe

Variety Type	Generation of seed planted	Mean yield (t ha^{-1})	Yield loss (%)	Days to male flowering
Hybrid	F_1	6.12 a	32.4	72.5 b
Hybrid	F_2	4.14 e		74.2 a
OPV	F_2	4.66 c	4.9	67.7 d
OPV	F_3	4.43 cd		67.7 d
Topcross	F_1	5.08 b	15.8	69.3 c
Topcross	F_2	4.28 de		69.9 c
LSD		0.22		0.8

Means followed by the same letter are not significantly different from each other (DMRT), $P = 0.05$.

Source: Pixley and Bänziger (2004).

ing, which facilitated mechanical harvest; improved drought tolerance (Duvick, 1999b); increased availability of chemical inputs, mechanization, and increasing value of labor (Tomes, 1998); and various subsidies, including the public research system. Hybrids are currently planted on only 33% of nontemperate maize-growing area, however, and improved OPVs cover only an additional 15% (Table 17.5). Use of hybrids is actually declining in some areas of southern Africa, where removal of subsidies to inputs has decreased the competitiveness of local with respect to imported maize. Some reasons cited by farmers for not adopting maize hybrids (or improved OPVs) are listed in Table 17.7.

Table 17.7 Hindrances to adoption of maize hybrids (and/or improved varieties)

Sub-Saharan Africa (Pixley and Bänziger, 2004)
Expensive seed cost
Cash constraint at planting time
Poor availability of seed at local shops
Need to also purchase fertilizer if growing hybrids
Small or no yield difference compared to local varieties
Lack of adaptation (poor performance in local environment)
Poor storability
Poor processing quality
Sub-Saharan Africa (Abalu, 2001)
Inferior performance to farmers' wants and needs
Limited access to improved seed and inputs (e.g., fertilizer)
Malawi, Tanzania, Zambia, and Zimbabwe (Phiri et al., 2003)
Expensive seed
Fertilizer not available or not affordable
Low value (price) for maize grain
Poor storage qualities
Credit not available
Poor local processing quality (or yield)
Confusion about or unfamiliarity with names of varieties
Poor access to seed (long distance to travel)
Poor taste in traditional foods
Malawi (Smale et al., 1995)
Prefer local maize for home consumption
Large maize requirement for household consumption
Lack of experience with hybrid maize
Kaduna, Nigeria (Akpoko et al., 2001)
Lack of access to desired quantity of inputs (fertilizer too expensive)
Unconvinced about yield advantage
Unconvinced about profitability
Poor availability (timeliness) of inputs
Unconvinced about types (specific varieties) of hybrids sold
Chiapas and Oaxaca, Mexico (Bellon et al., 2003)
Mistrust or lack of confidence in "packaged" seed
Improved varieties are riskier; less well known
Prefer local varieties for own consumption (processing and local foods)

General merits of hybrids and OPVs

The average yield advantage of hybrids relative to best OPVs has generally been 15–40% for the cereals considered herein. This advantage has often been less in marginal environments, where little research effort has been devoted to developing hybrid technology. Many of these environments may remain unattractive to private research investment for many reasons (see Some Economic Considerations). Given this scenario, it is important to consider whether to invest short- and medium-term public research efforts aimed at these environments into hybrid or OPV development. Coors (1999) summarized results of nearly 130 long-term population improvement (recurrent selection) studies and found that gains from selection have been comparable to those achieved through hybrid breeding in the United States (Table 17.8). This certainly suggests that substantial and sustained gains in productivity can be achieved with OPV technology. Other important considerations are discussed below.

Stability of hybrids and OPVs

There are many contradictory reports in the literature about the value of heterogeneity and associated "population buffering" in contributing to yield stability for maize, sorghum, and pearl millet. Theory and most academic studies using random geno-

Table 17.8 Summary of published results for gains from selection for maize grain yield for long-term recurrent selection studies

		Realized	Predicted[a]
Method	Studies No.	--------kg ha^{-1} yr^{-1}--------	
Hybrids	11	96	
Mass sel'n	16	82	129–192
Mod. ear to row	25	83	157–234
Half sib (HS)	12	50	91–135
Full sib (FS)	27	87	178–265
Mean intra-pop'n	80	79	149–221
HS recip. rec. sel'n	15	82	125–186
FS recip. rec. sel'n	14	116	197–292
Mean inter-pop'n	29	98	160–237
S1	13	93	154–229
S2	6	43	78–116
Mean inbred	19	77	130–193

Source: Coors (1999).

[a]Annual response for 1% (left) and 0.01% (right number) selection intensity.

types conclude that heterogeneity contributes to yield stability across a broad range of environments (e.g., Eberhart and Russell, 1969; Pixley and Bjarnason, 2002; Reich and Atkins, 1970; Schnell and Becker, 1986). On the other hand, citing several studies that conclude that stable hybrids can be developed by selecting and using stable inbreds, Janick (1999) has argued that pure stands of best (adapted) homogenous cultivars can be expected to outyield heterogeneous mixtures. The two schools almost meet at very low-yielding sites, where Ceccarelli and Grando (1996) and other authors maintain that heterogeneity within cultivars and diversity of crops is key to risk aversion and long-term stability of production. These environments are typical of one-third to one-half of the world's farmers and are the primary focus of this discussion.

In addition to heterogeneity, heterozygosity and associated heterosis (hybrid vigor) have been demonstrated to contribute to stability for grain yield of sorghum (e.g., Doggett, 1967; Reich and Atkins, 1970) and maize (e.g., Schnell and Becker, 1986; Tomes, 1998). Schnell and Becker (1986) compared stability of various population structures (varied for heterozygosity and heterogeneity) of sorghum and maize. For sorghum, they found that both heterozygosity and heterogeneity contributed to increased yield stability, and their contribution was of similar importance or magnitude. For maize, however, although both contributed to increased yield stability, heterozygosity was of much greater importance than heterogeneity.

Literature abounds with reports of superior yield stability of hybrids relative to OPVs and reports of exactly the opposite. The former situation is generally the case for experiments that do not include low-yielding environments. For studies including severely stressed or marginal environments, results are often strongly affected by the quality or suitability of the test cultivars (e.g., exotic cytoplasmic male sterile seed parents may contribute disease susceptibility, making hybrids inferior to local OPVs). Where little breeding effort has been invested for extremely stressed environments, it is not realistic to expect new hybrids or OPVs to consistently outperform local varieties.

Pixley and Bänziger (2004) compared performance of commercial and elite CIMMYT experimental maize hybrids with elite CIMMYT OPVs across a broad range of environments in Zimbabwe. The first experiment evaluated 10 OPVs and 4 commercial hybrids (all genotypes were of similar maturity) at 16 sites with average yield from 1.8 to 7.3 t ha^{-1} (Figure 1). The hybrids were more responsive than OPVs to favorable environments (b = 1.09 and b = 0.96, respectively) and were on average 18% higher yielding than the OPVs. A second experiment compared 3 hybrids with 4 OPVs in trials at 78 locations in Zimbabwe and also found 18% average yield superiority of hybrids over OPVs (b = 1.08 and b = 0.97, respectively). The authors concluded that good hybrids consistently yielded about 18% more grain than good OPVs across low- to high-yielding sites.

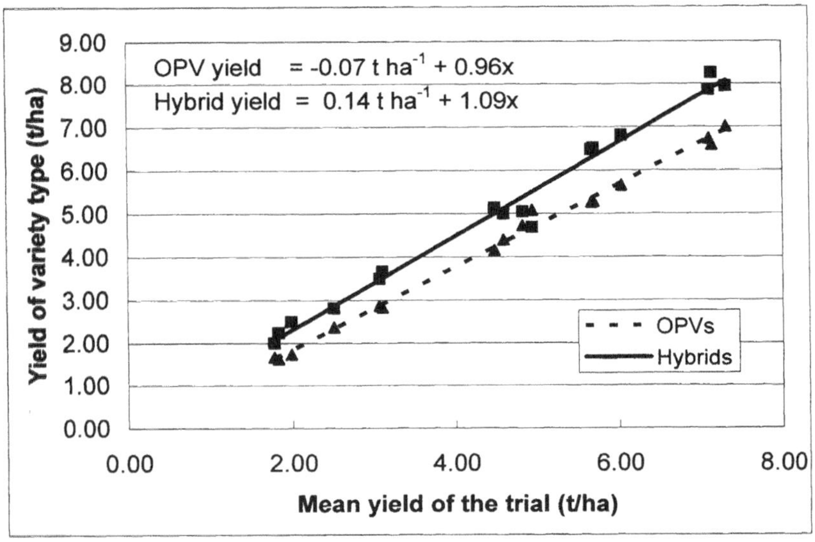

Figure 17.1 Average grain yield of elite OPVs and hybrids evaluated at 16 locations in Zimbabwe to establish the effect of management level on the productivity of hybrids and OPVs (Source: Pixley and Bänziger, 2004).

Following are a few conclusions regarding the role of stability in the debate over suitability of hybrid versus open-pollinated cultivars:

- Evidence for the value of heterogeneity in conferring stability certainly exists, but becomes counterbalanced in uniform environments by theory and evidence that increasingly homogeneous cultivars have greater yield potential, provided they are "ideal" genotypes. Such spatial and temporal uniformity is not typical of low-input agriculture in stress-prone environments.
- Heterozygosity and associated heterosis is similarly expressed across productivity range, and several studies suggest it is largest (in percentage) in more-stressed environments (e.g., Haussmann et al., 1998; Tomes, 1998; Virmani, 2001).
- Stability (especially temporal) of local varieties has not happened by chance; it has been achieved through generations of deliberate and natural selection. Several reports have concluded that the most effective approach to develop suitable cultivars (e.g., pearl millet and sorghum hybrids) for marginal environments is using local varieties either as source germplasm or as parents in topcross hybrids (e.g., Bidinger et al., 1994; Kumar et al., 2002; Mahalakshmi et al., 1992; Monyo et al., 1996; Presterl and Weltzien, 2003).
- Whereas modern plant breeding generally seeks to develop cultivars that are widely adapted (with big market or wide geographical impact potential), successful cultivars for severely stressed environments likely need to be specifically developed (Ceccarelli and Grando, 1996) and surely need to be deliberately tested in the stress environments where they will be deployed.
- Stable, good performance can only be achieved with excellent, locally adapted cultivars. Too often, hybrids (and improved OPVs) have been promoted without being "good enough."

Choosing between hybrid and open-pollinated maize varieties[1]

Hybrid maize varieties were released in sub-Saharan Africa more than 40 years ago, yet adoption of hybrids (Table 17.5) and average grain yield (Table 17.4) remain low. Hybrid seed is generally available in areas where the private seed sector can operate profitably; conversely, it is generally not available in remote rural areas with poor infrastructure and where farmers have low purchasing power. In addition to unavailability of seed, farmers cite a variety of reasons for not adopting hybrid varieties (Table 17.7). These circumstances and arguments beg the question whether hybrids indeed are a preferred alternative to open-pollinated or local varieties under resource-poor farmers' conditions characterized by insecure seed availability, low input use, and substantial risk of crop failures (Pixley and Bänziger, 2004).

Table 17.9 summarizes some of the benefits and opportunity costs to farming communities that grow primarily hybrid, OPV, or local maize varieties. While many farmers may be unaware of the benefits (e.g., uniform crop stand) of chemically treated, high-quality seed, or from securing the presence of private research efforts, it is also likely that many scientists and seed growers do not appreciate the value that resource-poor farmers place on seed security (independence from relying on availability or access [cash] to seed). In their study, Pixley and Bänziger (2004) quantified the genetic advantage and examined the relative profitability of growing hybrids relative to OPVs across a range of maize-growing conditions typical for southern and eastern Africa, both when first or second generation (recycled) seed is used.

Two field studies estimated the relative grain yield of best available commercial or experimental hybrids relative to best commercial or experimen-

Table 17.9 Types of benefits available from maize seed and relative access to these benefits if farmers grow hybrid, improved open-pollinated or local varieties

Type of benefit	Hybrids	Benefit from Improved OPVs	Local maize
Access to genetic gain	High	Medium	Low
Benefits from seed treatment and seed quality control	High	Only when purchased as certified seed	No benefits
Presence of a viable seed sector that continues to provide access to new genetic gains	Likely	Questionable	Unlikely
Independence of farming communities	Low	Medium	High

Source: Pixley and Bänziger (2004).

[1]Adapted from Pixley and Bänziger (2004).

tal OPVs (Pixley and Bänziger, 2004). The authors concluded that hybrids can be expected to outyield OPVs by about 18% across low- to high-yielding production environments (see Figure 17.1 and associated discussion, above). A second study examined the yield reduction incurred by planting recycled relative to fresh seed of three-way hybrids (mostly, although a very few single crosses were included), double-topcross hybrids, or OPVs. Across five locations, the effect of planting recycled seed was negligible for OPVs, severe for hybrids (>30% loss), and intermediate for topcross hybrids (about 16%) (Table 17.6) (Pixley and Bänziger, 2004). These results are similar to Duvick's (1999a) estimate of 15% superiority of the first commercial hybrids over best OPVs in the United States, and with estimates by Morris et al. (1999) for yield loss from recycling various types of maize cultivars (see discussion, above).

Using these results, Pixley and Bänziger (2004) calculated two scenarios with the following parameters:

1. Elite hybrids produce 18% more grain than elite OPVs.
2. Recycled hybrid seed produces 32% less grain than fresh F1 hybrid seed.
3. Recycled OPV seed produces 5% less grain than fresh OPV seed.
4. Seeding rates are 20 kg ha^{-1}.

The authors acknowledged that these assumptions deny the seed quality benefits (e.g., chemical treatment) of commercially purchased hybrid seed. The simplification, that "there is no further inbreeding depression (beyond the 32 and 5% yield reduction for hybrids and OPVs, respectively) from second and subsequent recycling of seed. . .should favor the hybrids, as theory predicts they will suffer additional inbreeding depression" (Pixley and Bänziger, 2004).

In their *constant management* scenario Pixley and Bänziger (2004) assumed that farmers apply the same crop management (i.e., fertilizer application, weeding, planting date, etc.) regardless of the variety used. The cost of hybrid seed was assumed greater than for OPV, recycled OPV, and recycled hybrid seed. Considering expected differences in grain yield for each type, the market prices for seed and grain determined the relative profitability of types at any given management level. When the authors assumed a realistic price ratio of 1:7:14 for grain/OPV seed/hybrid seed, recycling OPV seed was the most profitable option at the 1 t ha^{-1} level and purchase of hybrid seed became the more profitable option at or above 2 t ha^{-1}. A similar conclusion was reported by Mekuria and Siziba (2003), who found that hybrids must outyield OPVs by more than 30% to repay the added cost of hybrid seed at maize yield levels below 1 t ha^{-1} in Zimbabwe. If one applies Duvick's (1999b) "rule of thumb" that "first time use of hybrid seed should enable the farmer to earn an extra profit equal to at least three times the added cost of seed," this model indicates that purchase of hybrid seed becomes advantageous over purchase of OPV seed followed by recycling (three times) only at management level around 4.5 t ha^{-1}. In general, Pixley and Bänziger (2004) found that recycling of seed became less profitable as management level increased, and recycling of hybrid seed was the least profitable option.

In their *constant investment* scenario, Pixley and Bänziger (2004) assumed that farmers will invest a fixed amount of cash for crop inputs (restricted to seed and nitrogen fertilizer, in their example) and that savings on seed purchase will be used to buy additional nitrogen fertilizer. The scenario was developed using a realistic price ratio of 1:7:14:11 for grain/OPV seed/hybrid seed/nitrogen fertilizer, and assuming that each kilogram of nitrogen would result in yield increase of 20 kg (Muza et al., 2004) for all maize crops. "OPVs, whether purchased or recycled, were the most profitable option at the 1 and 2 t ha^{-1} management levels" (Pixley and Bänziger, 2004). Fresh hybrid seed became the most profitable option at 3 t ha^{-1}, and recycling of hybrid seed remained the least profitable option at all yield levels. If we apply Duvick's (1999b) suggestion that use of hybrid seed should result in extra profit equal to at least three times the added cost of seed, then purchase of OPV seed (and recycling three times) with use of small amount of nitrogen fertilizer within the constant investment scenario would remain advantageous over purchase of hybrid seed even at 5 t ha^{-1} management level.

Pixley and Bänziger (2004) make two additional comments that are widely applicable and pertinent to this debate throughout sub-Saharan Africa. First, when bumper harvests occur, as they did in Uganda during 2001, maize grain prices generally decline sharply at local and national level, and the yield at which purchase and use of hybrid seed is more profitable than use of OPVs may climb as high as 3.5 t ha^{-1}. The second comment came from Chiduza et al. (1994), who reported that because hybrid seed at two remote rural communities in Zimbabwe costs five times the price of the same seed in the capital city (Harare), use and recycling of OPV seed together with modest use of fertilizer gave the highest net benefit, followed by use of hybrid seed with fertilizer.

Pixley and Bänziger (2004) concluded that economic analyses of returns to farmers' investments are not adequate, on their own, to determine which variety type is best for a given farmer, community or area. "Aspects such as (i) access to the benefits from research investments in genetic improvement of new varieties, (ii) access to the benefits of seed treatment and seed quality control as is typical for certified seed, (iii) the [likelihood of a] continued presence of a viable seed sector, and (iv) the livelihood strategies of resource-poor maize farmers, must [also] be considered" (Pixley and Bänziger, 2004). The authors concluded that improved OPVs represent a valuable option for maize farmers under some circumstances that are common in eastern and southern Africa.

Some economic considerations

Griliches (1960) studied the adoption of hybrid maize (replacing OPVs) in the United States and published what are considered classic analyses about technology adoption in general. He noted that hybrid yield advantage over OPVs was 15–20% across a wide range of environments. Adoption of hybrids occurred first in areas with highest yield ("good" maize-growing areas), most mechanical harvesters, largest total maize area, largest proportion of farmland devoted to maize, and, in summary, greatest overall profit potential. Farmers generally began by planting only 20–30% of their maize area to hybrid cultivars and took several years to reach 100% adoption. Sociological variables, such as farmers' personality, level of education, economic status, and social environment were unsuccessful in explaining hybrid adoption patterns. Griliches (1960) predicted that areas with low and variable yields (e.g., western Nebraska, South Dakota, and Kansas) had reached an equilibrium level of hybrid adoption at about 30–60%. In a later paper, Griliches (1980) explains that the eventual development of suitable hybrids, combined with gradual unavailability of the old technology (OPVs), raised the adoption rate of hybrids even in these marginal production environments. He concluded that hybrid corn was more profitable and first adopted in the "good" areas, which is illustrative of a tendency for technological change to accentuate regional disparities in levels of prosperity.

Once hybrid (and associated) technology was adopted in the United States, impressive results ensued: maize yields climbed from below 2 t ha^{-1} to almost 8 t ha^{-1}, total production nearly quadrupled, while cultivated maize area declined about 20% within 50 years (Kloppenburg, 1988). Surprisingly, however, farmers' profits declined from 38 to 17%, while expenditures on farm inputs climbed from 44 to 68% of gross farm income (Strange, 1988). The increased production prompted the government to implement four broad types of subsidies (subsidized loans, payments to reduce production, purchase of surplus production, and price-support payments to producers) designed to prevent commodity prices from plunging (Cochrane, 1979). Small family farms gave way to large commercial farms as an increasingly large area was needed to remain viable given declining profit margins. Thus, while agriculture in the United States is impressively productive, it has generally not benefited most farmers.

The U.S. agricultural model provides useful lessons, but it is not directly applicable or an entirely desirable model for developing countries seeking to increase their agricultural production. As described in earlier sections of this chapter, many farmers in Africa achieve very low yields, farm small land areas, have limited access to input and output markets, and may not realize an immediate profit from purchasing and planting hybrid seed; they do not fit the profile of Griliches' (1960) new technology adopter.

Byerlee and Eicher (1997) reviewed maize research and development activities in Africa over the past 30 years and identified four key issues for designing appropriate food-production strategies. First, they recommend focusing resources on one or two crops and on favorable production areas. Second, they recommend focusing on smallholder agriculture, stating that small farms can be competitive with large commercial farms if they are provided with appropriate technology and economic incentives. Third, they recommend aggressive promotion of use of external inputs and focus on hybrids (because the private sector lacks interest in OPVs). Finally, they highlight the importance of the question, but offer no answer, on how to assist smallholder farmers in areas of marginal production potential and poor infrastructure.

Most economists agree that the route to national, regional, or global food security is to focus on intensive agriculture in good production environments. "Trickle-down" effects, including lower food prices and increased labor opportunities, should provide some relief to farmers in marginal areas. Heisey and Edmeades (1999), however, warned that anticipated maize demand will not be met without at least maintaining current production in drought-stressed environments. They conclude that demand will only be met through large increases in productivity in favorable areas, small increases in marginal areas, and some growth in total maize area.

Philosophical perspectives

Several philosophical arguments contribute to the debate about appropriateness or suitability of hybrid and open-pollinated cultivars in modern agriculture. The following are gross simplifications that undoubtedly will raise objections from each of the proponent groups, but they are presented here in the belief that each holds an important message.

Socialists argue that agribusiness, and most agricultural research efforts (which are inevitably strongly influenced by capital and agribusiness), emphasize creating an agriculture dependent on inputs that it provides. Adoption of this production system deprives farmers of their autonomy and compromises their livelihood security. Hybrid seed is a favorite example of this because it is a modification of nature that deprives farmers of their previous ability to be self-reliant for seed while providing "a form of economic protection that is more effective than the patent system" (Kloppenburg, 1988). Some authors have asked whether OPVs could compete with hybrids if comparable investment in their development were made (Cleveland, 2001; Kloppenburg, 1988; Lewontin and Berlan, 1990).

Ecologists warn that commercial agriculture is based on modifying or simplifying the environment to fit the requirements of a plant ideotype (high-yielding hybrid) that is incapable of thriving and is inappropriate in nature. This modification of the environment requires an unsustainable and ecologically undesirable extensive use of fossil fuels and chemicals (e.g., herbicides, pesticides). Their recommendation is to develop technologies suited to farmers' environments and to maximize productivity and profitability of ecologically sensible and sustainable farming systems (Kirschenmann, 2003; Scott, 1998). This view should be neutral on the issue of hybrids and OPVs.

Sociologists explain that modern technologies have frequently failed to achieve their intended impact because they have not understood and met the complex needs and preferences of resource-poor farmers. Appropriate technologies can and should be developed by involving farmers in their development and/or verification (prior to dissemination). This view would encourage provision of hybrids and OPVs, enabling farmers to choose when to use either, neither, or both.

As discussed in the previous section, economists recommend focusing efforts on the higher productivity environments, using high-input, high-tech (e.g., hybrids) strategies (Byerlee and Eicher, 1997; Griliches, 1960; Pingali, 1999). If this approach is followed, they conclude that the private sector will be a strong ally that will ensure continued research investment and reliable input and output markets, all of which will result in sustainable development. Resource limitations regrettably require that little or no attention can be directed to farmers in marginal areas. This view favors hybrids for their greater production potential relative to OPVs.

Finally, plant breeders are confident that appropriate cultivars can be developed to meet the growing global agricultural production requirements. We realize this will not be easy and would not like to discard any available technology that might assist us to achieve this objective more effectively or efficiently. We are perhaps too quick to dismiss other philosophies as naive, complacent, and perilously incapable of meeting the critical food production demand. There is an awareness, particularly among breeders working in and for marginal environments, of the complementary roles of OPVs and various types of hybrids in providing useful options to farmers. It is probably less common, however, for breeders to ponder the possibility that OPVs may be capable of competing with hybrids, as suggested by data compiled by Coors (1999). This possibility may be especially intriguing in marginal areas where relatively little investment in formal plant-breeding research has occurred.

Conclusions

There is no single technological intervention that will quickly increase productivity to meet anticipated cereal demand, secure livelihoods, and eliminate poverty for resource-poor households in Africa and elsewhere. Their agricultural environment (e.g., soils and climate) is complex and characterized by extreme variation that necessitates and has evolved risk-averting technologies and strategies. Their economic and social circumstances, plus the often weak infrastructure and markets, also affect their ability to accept the risk of experimenting with new technologies. Fortunately, there are many technological options that can be made available to farmers. The above discussion for sorghum, pearl millet, and maize (see Potential of Hybrids), the general discussion on stability of hybrids and OPVs (see Stability of Hybrids and OPVs and Choosing Between Hybrid and Open-Pollinated Maize Varieties about choosing between hybrids and OPVs), all identified an important range of heterogeneity and heterozygosity among available types of hybrids, such that a transition from OPVs to hybrids need not be drastic. It is also encouraging that extensive evidence indicates that even the poorest of farmers do adopt, or often modify and partially adopt, technologies proven useful to them. Interesting examples of literature on this process include Bellon et al.'s (2003) description of the "creolization" (essentially, converting to "local") of improved maize germplasm by farmers in Mexico and vom Brocke et al.'s (2002) description of farmers' seed-management practices for pearl millet in India.

It is clear that most agricultural production will continue to occur in favorable environments. Development and promotion of homogeneous hybrids will be profitable in these environments (see Stability of Hybrids and OPVs), and the expectation is that most of the required research will be conducted by the private sector. The complementary role of public research may be to leverage pressure for ecologically sustainable production through leadership in the development and advocacy of appropriate technologies, including suitable varieties.

Plant-breeding research for marginal environments, particularly those lacking infrastructure and markets to attract private investments, should develop improved varieties with little inbreeding depression when recycled. This should include development of synthetic open-pollinated cultivars by recurrent selection, using schemes that involve selection among inbred lines. It should be exciting to test the limits of performance attainable with synthetic varieties (OPVs). Secondly, this should include development of heterogeneous hybrids, such as double topcross and double-cross hybrids in maize, that allow exploitation of heterosis with relatively little inbreeding depression when recycled and that will generally offer better stability of yield than expected from homogeneous hybrids (particularly when research budget limits the testing program).

Technologies for resource-poor farmers in stress-prone environments must offer options that fit within their risk-management strategies. This is probably less an issue than one might initially imagine, because ample evidence (e.g., Tables 17.1, 17.3, and 17.7) suggests farmers will not adopt technologies that do not meet this criterion. On the other hand, development of varieties that meet the expectations and needs of resource-poor farmers will have to include farmers in the evaluation and selection of varieties for promotion. Promotion of new varieties that do not meet this standard of usefulness has in past resulted in nonadoption and will further increase farmers' mistrust and hesitance to adopt new technologies.

Two crucial, nontechnical questions remain unanswered: First, how will these heterogeneous improved varieties be supplied to farmers? In southern Africa, David Rohrbach (ICRISAT agricultural economist, personal communication) has noted "there is a common assumption that there is no payoff to marketing non-hybrid seed—except for the residual (though consistent) demand of relief programs. Correspondingly, there is no investment in developing retail marketing networks for these crops." The consequence is that superior improved open-pollination (and heterogeneous hybrid) cultivars are rarely produced and made available to farmers. The second question is How will the required plant-breeding research be funded? There has been a declining trend of funding for public research, and, concomitantly, many public sector breeders are having to generate income through their research efforts. With this mandate to generate income, the research agenda has naturally gravitated toward serving the "good" agricultural areas, where new technology is most likely to be adopted. Answering these questions has become part of the job of public sector plant breeders and will be essential to substantially and sustainably increasing productivity while improving livelihoods of farmers in less-favorable agricultural environments.

Acknowledgments

Numerous colleagues provided ideas that influenced this review. I am particularly thankful to Marianne Bänziger, Robert Tripp, Mauricio Bellon, Richard Jones, Alex Phiri, Abebe Menkir, David Rohrbach, Frederick Kirschenmann, Kendall Lamkey, Augustine Langyntuo, Miloje Denic, Mike Lee, David Beck, and Shephard Siziba. Opinions expressed in this paper are not necessarily those of CIMMYT.

References

Ahmed, M.M., J.H. Sanders, and W.T. Nell. 1996. New sorghum and millet cultivar introduction in sub-Saharan Africa: Impacts and policy implications. In: Leuschner, K., and C.S. Manthe (eds.), Drought-tolerant crops for southern Africa. Proc. of the SADC/ICRISAT regional sorghum and pearl millet workshop, 25–29 July 1994, Gaborone, Botswana. ICRISAT, Patancheru, India.

Andrews, R.D. 2001. The commercialization and performance of hybrid rice in the United States. In: Peng, S., and B. Hardy (eds.). Rice research for food security and poverty alleviation. Proceedings of the international rice research conference, 31 March–3 April 2000, Los Banos, Philippines. Los Banos, Philippines: IRRI. 692 p.

Aquino, P., F. Carrion, R. Calvo, and D. Flores. 2000. Selected maize statistics. In: Pingali, P.L. (ed.), CIMMYT 1999–2000 world maize facts and trends. Meeting world maize needs: Technological opportunities and priorities for the public sector. Mexico, D.F.: CIMMYT. 60 p.

Beck, D., and J.L. Torres. 2003. Inbreeding depression in recycled seed of highland hybrids sown under optimal and nitrogen stressed conditions. Arnel R. Hallauer International Symposium on Plant Breeding, 17–22 August 2003, Mexico City, Mexico. Abstr.

Bellon, M.R., M. Adato, J. Becerril, and D. Mindek. 2003. The impact of improved maize germplasm on poverty alleviation: The case of Tuxpeno-derived material in Mexico.

Bidinger, F.R., E. Weltzien, R.V. Mahalakshmi, S.D. Singh, and K.P. Rao. 1994. Evaluation of landrace topcross hybrids of pearl millet for arid zone environments. Euphytica 76:215–226.

Byerlee, D., and C.K. Eicher. 1997. Introduction: Africa's food crisis. In: Byerlee, D., and C.K. Eicher (eds.), Africa's emerging maize revolution, Lynne Rienner Publ., Inc., Boulder, CO, USA. 301 p.

Ceccarelli, S., and S. Grando. 1996. Drought as a challenge for the plant breeder. Plant Growth Regul. 20:149–155.

Chiduza, C., S.R. Waddington, and I.K. Mariga. 1994. Grain yield and economic performance of experimental open-pollinated varieties and released hybrids of maize in a remote semi-arid area of Zimbabwe. Zimbabwe J. Agric. Res. 32(1):33–43.

Cleveland, D.A. 2001. Is plant breeding science objective truth or social construction? The case of yield stability. Agriculture and Human Values 18:251–270.

Coats, B. 2003. Global rice production. In: Smith, C.W., and R.H. Dilday (eds.), Rice: Origin, history, technology and production. John Wiley & Sons, Inc., New Jersey. 642 p.

Cochrane, W.W. 1979. The development of American agriculture: A historical analysis. Univ. Minn. Press., Minneapolis, MN. 464 p.

Coors, J.G. 1999. Selection methodology and heterosis. In: Coors, J.G., and S. Pandey (eds.), The genetics and exploitation of heterosis in crops. Amer. Soc. Agron., Madison, WI. 524 p.

Coors, J.G., and S. Pandey (eds.). 1999. The genetics and exploitation of heterosis in crops. Amer. Soc. Agron. Inc., Madison, WI. 524 p.

Dendy, D.A.V. 1995. Sorghum and the millets: Production and importance. In: Dendy, D.A.V. (ed.), Sorghum and millets: Chemistry and technology. American Ass. Cereal Chemists, Inc., St. Paul, MI. 406 p.

Doggett, H. 1967. Yield increase in sorghum hybrids. Nature 216:798–799.

Doggett, H., and D. Jowett. 1966. Yields of maize, sorghum varieties and sorghum hybrids in the east African lowlands. J. Agric. Sci. Camb. 67:31–39.

Duvick, D.N. 1999a. Heterosis: Feeding people and protecting natural resources. In: Coors, J.G., and S. Pandey (eds.), The genetics and exploitation of heterosis in crops. Amer. Soc. Agron., Madison, WI. 524 p.

Duvick, D.N. 1999b. Commercial strategies for exploitation of heterosis. In: Coors, J.G., and S. Pandey (eds.), The genetics and exploitation of heterosis in crops. Amer. Soc. Agron., Madison, WI. 524 p.

Eberhart, S.A., and W.A. Russell. 1969. Yield and stability for a 10-line diallel of single-cross and double-cross maize hybrids. Crop Sci. 9:357–361.

Eizenga, G.C., and J.N. Rutger. 2003. Genetics, cytogenetics, mutation and beyond. In: Smith, C.W. and R.H. Dilday (eds.), Rice: Origin, history, technology and production. John Wiley & Sons, Inc., New Jersey. 642 p.

Fakorede, M.A.B., J.M. Fajemisin, S.O. Ajala, J.G. Kling, and A. Menkir. 2001. Hybrid maize and hybrid seed production in Nigeria: Lessons for other west and central African countries. In: Badu-Apraku, B., M.A.B. Fakorede, M. Ouedraogo, and R.J. Carsky (eds.), Impact, challenges and prospects of maize research and development in west and central Africa, Proceedings of a regional maize workshop, 4–7 May, 1999, IITA-Cotonou, Benin. Ibadan, Nigeria: IITA. 525 p.

Govila, O.P., K.N. Rai, K.R. Chopra, D.J. Andrews, and W.D. Stegmeier. 1997. Breeding pearl millet hybrids for developing countries: Indian experience. In: Proceedings of the international conference on genetic improvement of sorghum and pearl millet. Lubock, TX. Sept. 22–27, 1996. USAID and ICRISAT. 708 p.

Griliches, Zvi. 1960. Hybrid corn and the economics of innovation. Science 132:275–280.

Griliches, Zvi. 1980. Hybrid corn revisited: A reply. Econometrica 48:1463–1465.

Hallauer, A.R., and J.B. Miranda Fo. 1981. Heterosis. Chapter 10, In: Quantitative genetics in maize breeding. Iowa State University Press, Ames, IA. 468 p.

Hanna, W.W. 2001. Breeding pearl millet with improved performance and stability. In: Grain sorghum/pearl millet collaborative research support program (CRSP), INTSORMIL 2001 annual report. INTSORMIL, Univ. Nebraska, Lincoln, NE. 178 p.

Haussmann, B.I.G., A.B. Obilana, A. Blum, P.O. Ayiecho, W. Schipprack, and H.H. Geiger. 1998. Hybrid performance of sorghum and its relationship to morphological and physiological traits under variable drought stress in Kenya. Plant Breeding 117:223–229.

Heisey, P.W., and G.O. Edmeades. 1999. Maize production in drought-stressed environments: technical options and research resource allocation. Part 1 of CIMMYT 1997/98 world maize facts and trends; maize production in drought-stressed environments: technical options and research resource allocation. Mexico, D.F.: CIMMYT.

Herdt, R.W. 1996. Summary, conclusions and implications. In: Evenson, R.E., R.W. Herdt, and M. Hossain (eds.), Rice research in Asia: Progress and priorities. CAB International, Wallingford, United Kingdom. 418 p.

Horie, T. 2001. Increasing yield potential in irrigated rice: breaking the yield barrier. In: Peng, S., and B. Hardy (eds.). Rice research for food security and poverty alleviation. Proceedings of the international rice research conference, 31 March–3 April 2000, Los Banos, Philippines. Los Banos, Philippines: IRRI. 692 p.

Hossain, M. 1996. Recent developments in the Asian rice economy: Challenges for rice research. In: Evenson, R.E., R.W. Herdt, and M. Hossain (eds.), Rice research in Asia: Progress and priorities. CAB International, Wallingford, United Kingdom. 418 p.

House, L.R., M. Gomez, D.S. Murty, Yi Sun, and B.N. Verma. 2000. Development of some agricultural industries in several African and Asian countries. In: Smith, C.W., and R.A. Frederiksen (eds.), Sorghum: origin, history, technology and production. John Wiley & Sons, Inc., New York, NY. 824 p.

House, L.R., M. Osmanazi, M.I. Gomez, E.S. Monyo, and S.C. Gupta. 1995. Agronomic principles. In: Dendy, D.A.V. (ed.), Sorghum and millets: chemistry and technology. American Ass. Cereal Chemists, Inc., St. Paul, MI. 406 p.

House, L.R., B.N. Verma, G. Ejeta, B.S. Rana, I. Kapran, A.B. Obilana, and B.V.S. Reddy. 1997. Developing countries breeding and potential of hybrid sorghum. In: Proceedings of the international conference on genetic improvement of sorghum and pearl millet. Lubock, TX. Sept. 22–27, 1996. USAID and ICRISAT. 708 p.

Janaiah, A., and M. Hossain. 2001. Adoption of hybrid rice technology in India: An economic assessment of early farm-level experiences. In: Peng, S. and B. Hardy (eds.), Rice research for food security and poverty alleviation. Proceedings of the international rice research conference, 31 March–3 April 2000, Los Banos, Philippines. Los Banos, Philippines: IRRI. 692 p.

Janick, J. 1999. Exploitation of heterosis: Uniformity and stability. p. 319-333. In: Coors, J.G., and S. Pandey (eds.), The genetics and exploitation of heterosis in crops. Amer. Soc. Agron., Madison, WI.

Kapran, I., M. Adamou, G. Ejeta, and J.D. Axtell. 2002. Review of sorghum hybrid research in Niger. In: Proceedings of the west African hybrid sorghum and pearl millet seed workshop. Sep. 28–Oct. 2, 1998, Niamey, Niger. USAID. 279 p.

Kelley, T.G., P. Parthasarathy Rao, E. Weltzien R., and M.L. Purohit. 1996. Adoption of improved cultivars of pearl millet in an arid environment: Straw yield and quality considerations in western Rajasthan. Expl. Agric. 32:161–171.

Kirschenmann, F. 2003. New seeds and breeds for a new revolution in agriculture. Paper presented at: The summit on seed and breeds for the 21st century, Sept. 6–8, 2003, Washington, D.C.

Kloppenburg Jr., J.R. 1988. First the seed: The political economy of plant biotechnology, 1492–2000. Cambridge Univ. Press, New York, NY. 349 p.

Kumar, K.A., B. Ouendeba, S. Boureima, S.C. Gupta, and S. Ouattara. 2002. Current status of hybrid pearl millet in West Africa. In: Proceedings of the west African hybrid sorghum and pearl millet seed workshop. Sep. 28–Oct. 2, 1998, Niamey, Niger. USAID. 279 p.

Lewontin, R.C., and J.P. Berlan. 1990. The political economy of agricultural research: The case of hybrid corn. In: Carroll, C.R., J.H. Vandermeer and P. Rosset (eds.), Agroecology. McGraw-Hill Publ. Co., NY. 641 p.

Liang, G.H., C.R. Reddy, and A.D. Dayton. 1972. Heterosis, inbreeding depression, and heritability estimates in a systematic series of grain sorghum genotypes. Crop Sci. 12:409–411.

Lima, M., J.B. Miranda Fo., and P. Boller Gallo. 1984. Inbreeding depression in Brazilian populations of maize (Zea mays L.). Maydica 29:203–215.

Lin, J.Y. 1994. The nature and impact of hybrid rice in China. In: David, C.C. and K. Otsuka (eds.), Modern rice technology and income distribution in Asia. Lynne Rienner Publ., Inc., Boulder, CO. 475 p.

Mahalakshmi, V., F.R. Bidinger, K.P. Rao, and D.S. Raju. 1992. Performance and stability of pearl millet topcross hybrids and their variety pollinators. Crop Sci. 32:928–932.

Mangombe, N., and J.N. Mushonga. 1996. Sorghum and pearl millet on-farm research work in Zimbabwe. In: Leuschner, K., and C.S. Manthe (eds.), Drought-tolerant crops for southern Africa. Proc. of the SADC/ICRISAT regional sorghum and pearl millet workshop, 25–29 July 1994, Gaborone, Botswana. ICRISAT, Patancheru, India.

Maunder, A.B. 2000. History of cultivar development in the United States: From "Memoirs of A.B. Maunder—Sorghum breeder." In: Smith, C.W., and R.A. Frederiksen (eds.), Sorghum: Origin, history, technology and production. John Wiley & Sons, Inc., New York, NY. 824 p.

Mekuria, M., and S. Siziba. 2003. Economics of maize seed types: Options for Zimbabwe smallholder farmers. IAAE 25th international conference, 16–22 August, 2003, Durban, South Africa. Poster.

Melchinger, A.E., and R.K. Gumber. 1998. Overview of heterosis and heterotic groups in agronomic crops. In: Lamkey, K.R., and J.E. Staub (eds.), Concepts and breeding of heterosis in crop plants. CSSA Special Publ. No. 25, Crop Sci. Soc. Amer., Madison, WI. 127 p.

Menkir, A., and A.O. Akintunde. 2001. Evaluation of the performance of maize hybrids, improved open-pollinated and farmers' local varieties under well-watered and drought stress conditions. Maydica 46:227–238.

Miranda Fo., J.B. 1999. Inbreeding and heterosis. In: Coors, J.G., and S. Pandey (eds.), The genetics and exploitation of heterosis in crops. Amer. Soc. Agron., Madison, WI. 524 p.

Moldenhauer, K.A.K., and J.H. Gibbons. 2003. Rice morphology and developments. In: Smith, C.W., and R.H. Dilday (eds.), Rice: Origin, history, technology and production. John Wiley & Sons, Inc., New Jersey. 642 p.

Monyo, E.S., S.A. Ipinge, S. Mndolwa, N. Mangombe, and E. Chintu. 1996. The potential of local landrace varieties in pearl millet improvement. In: Leuschner, K., and C.S.

Manthe (eds.), Drought-tolerant crops for southern Africa. Proc. of the SADC/ICRISAT regional sorghum and pearl millet workshop, 25–29 July 1994, Gaborone, Botswana. ICRISAT, Patancheru, India.

Moran, J.L., and W.L. Rooney. 2003. Effect of cytoplasm on the agronomic performance of grain sorghum hybrids. Crop Sci. 43:777–781.

Morris, M.L. 2001. Assessing the benefits of international maize breeding research: An overview of the global maize impacts study. In: Pingali, P.L. (ed.), CIMMYT 1999–2000 world maize facts and trends. Meeting world maize needs: Technological opportunities and priorities for the public sector. Mexico, D.F.: CIMMYT. 60 p.

Morris, M.L., J. Risopoulos, and D. Beck. 1999. Genetic change in farmer-recycled maize seed: a review of the evidence. CIMMYT economics working paper No. 99-07. Mexico, D.F.: CIMMYT.

Murty, D.S. 2002. Hybrid sorghum in West Africa. In: Proceedings of the west African hybrid sorghum and pearl millet seed workshop. Sep. 28–Oct. 2, 1998, Niamey, Niger. USAID. 279 p.

Muza, Lucia, Stephen R. Waddington, and Marianne Bänziger. 2004. Preliminary results on the response of "nitrogen use efficient" OPV and hybrid maize to N fertilizer on smallholder fields in Zimbabwe. In: Friesen, D.K., and A.F.E. Palmer (eds.), Integrated approaches to higher maize productivity in the new millennium: Proceedings of the seventh Eastern and Southern Africa Regional Maize Conference, 5-11 February 2002, Nairobi, Kenya. CIMMYT (International Maize and Wheat Improvement Center) and KARI (Kenya Agricultural Research Institute). p. 245–249.

Ndjeunga, J., and A. Sidi. 2002. Pearl millet seed systems in Niger. In: Proceedings of the west African hybrid sorghum and pearl millet seed workshop. Sep. 28–Oct. 2, 1998, Niamey, Niger. USAID. 279 p.

Paliwal, R.L. 2000. Hybrid maize breeding. In: Paliwal, R.L., G. Granados, H.R. Lafitte, and A.D. Violic (eds.), Tropical maize improvement and production. FAO, Rome, Italy. 363 p.

Peng, S., K.G. Cassman, S.S. Virmani, J. Sheehy, and G.S. Khush. 1999. Yield potential trends of tropical rice since the release of IR8 and the challenge of increasing rice yield potential. Crop Sci. 39:1552–1559.

Phiri, M.A.R., M. Mekuria and M. Bänziger. 2003. Assessment of smallholder farmers' utilization of improved maize seed in the SADC region: A study of Malawi, Tanzania, Zambia and Zimbabwe. Southern African Drought and Low Fertility Project (SADLF), Harare, Zimbabwe: CIMMYT.

Pingali, P.L. 1999. Role of heterosis in meeting world cereal demand in the 21st century. In: Coors, J.G., and S. Pandey (eds.), The genetics and exploitation of heterosis in crops. Amer. Soc. Agron., Madison, WI. 524 p.

Pingali, P.L., and P.W. Heisey. 1999. Cereal crop productivity in developing countries. CIMMYT Economics Paper 99-03. Mexico D.F.: CIMMYT.

Pingali, P.L., M. Hossain, and R.V. Gerpacio. 1997. Asian rice bowls: The returning crisis? CAB International, Wallingford, United Kingdom. 341 p.

Pingali, P.L., and S. Pandey. 2001. Meeting world maize needs: technological opportunities and priorities for the public sector. In: Pingali, P.L. (ed.), CIMMYT 1999-2000 world maize facts and trends. Meeting world maize needs: Technological opportunities and priorities for the public sector. Mexico, D.F.: CIMMYT. 60 p.

Pixley, K., and M. Bänziger. 2004. Open-pollinated maize varieties: A backward step or valuable option for farmers? In: Friesen, D.K., and A.F.E. Palmer (eds.), Integrated approaches to higher maize productivity in the new millennium: Proceedings of the seventh Eastern and Southern Africa Regional Maize Conference, 5–11 February 2002, Nairobi, Kenya. CIMMYT (International Maize and Wheat Improvement Center) and KARI (Kenya Agricultural Research Institute). p. 22–28.

Pixley, K.V., and M.S. Bjarnason. 2002. Stability of grain yield, endosperm modification and protein quality of hybrid and open-pollinated quality protein maize (QPM) cultivars. Crop Sci. 42:1882–1890.

Presterl, T., and E. Weltzien. 2003. Exploiting heterosis in pearl millet for population breeding in arid environments. Crop Sci. 43:767–776.

Rai, K.N., K. Anand Kumar, D.J. Andrews, S.C. Gupta, and B. Ouendeba. 1997. Breeding pearl millet for grain yield and stability. In: Proceedings of the international conference on genetic improvement of sorghum and pearl millet. Lubock, TX. Sept. 22–27, 1996. USAID and ICRISAT. 708 p.

Reich, V.H., and R.E. Atkins. 1970. Yield stability of four population types of grain sorghum, Sorghum bicolor (L.) Moench, in different environments. Crop Sci. 10:511–517.

Rooney, W.L., and C.W. Smith. 2000. Techniques for developing new cultivars. In: Smith, C.W., and R.A. Frederiksen (eds.), Sorghum: origin, history, technology and production. John Wiley & Sons, Inc., New York, NY. 824 p.

Schnell, F.W., and H.C. Becker. 1986. Yield and yield stability in a balanced system of widely differing population structures in *Zea mays* L. Plant Breeding 97:30–38.

Scott, J.C. 1998. Taming nature: An agriculture of legibility and simplicity. Chapter 8, In: Scott, J.C. Seeing like a state: How certain schemes to improve the human condition have failed. Yale Univ. Press, New Haven, USA.

Smith, C.W. 2000. Sorghum production statistics. In: Smith, C.W., and R.A. Frederiksen (eds.), Sorghum: Origin, history, technology and production. John Wiley & Sons, Inc., New York, NY. 824 p.

Strange, M. 1988. Family farming: A new economic vision. Univ. Nebraska Press, Lincoln, NE. 311 p.

Tomes, D.T. 1998. Heterosis: performance stability, adaptability to changing technology, and foundation of agriculture as a business. In: Lamkey, K.R., and J.E. Staub (eds.), Concepts and breeding of heterosis in crop plants. CSSA Special Publ. No. 25, Crop Sci. Soc. Amer., Madison, WI. 127 p.

Toure, A., H.F.W. Rattunde, and I. Akintayo. 2002. Heterosis and hybrid sorghum seed production in west Africa. In: Proceedings of the west African hybrid sorghum and pearl millet seed workshop. Sep. 28–Oct. 2, 1998, Niamey, Niger. USAID. 279 p.

Tripp, R., and S. Pal. 1998. Information exchange in commercial seed markets in Rajesthan. Overseas Development Institute (ODI), Agric. Res. and Ext. Network (AgREN), Network Paper No. 83.

Van Tran, D. 2001. Closing the rice yield gap for food security. In: Peng, S., and B. Hardy (eds.). Rice research for food security and poverty alleviation. Proceedings of the International Rice Research conference, 31 March–3 April 2000, Los Banos, Philippines. Los Banos, Philippines: IRRI. 692 p.

Virmani, S.S. 2001. Opportunities and challenges of developing and using hybrid rice technology in the tropics. In: Peng, S., and B. Hardy (eds.). Rice research for food security and poverty alleviation. Proceedings of the International Rice Research Conference, 31 March–3 April 2000, Los Banos, Philippines. Los Banos, Philippines: IRRI. 692 p.

vom Brocke, K., T. Presterl, A. Christinck, E. Weltzien, and H.H. Geiger. 2002. Farmers' seed management practices open up new base populations for pearl millet breeding in a semi-arid zone of India. Plant Breeding 121:36–42.

18

Breeding Vegetatively Propagated Crops[1]

Rodomiro Ortiz, Carine Dochez, Robert Asiedu, Francis Moonan
International Institute of Tropical Agriculture (IITA), Nigeria

Introduction

The most important vegetatively propagated food crops are potato, cassava, sweet potato, yam, plantain/banana, sugar cane, and fruit trees. Other crops with asexual propagations are some ornamentals, grasses, and forages. Cross-breeding methods for vegetatively propagated crops rely on sexual hybridization, that is, seeds are needed for producing new genotypes after crossing selected parents. The main goal of breeding clones will be to obtain genotypes that are phenotypically uniform (homogeneous), but often highly heterozygous, particularly if nonadditive gene action controls the commercial trait(s) of interest. Nonadditive gene action may arise from intra- or interallelic (epistasis) interactions.

The conventional plan for breeding clones consists of (1) selecting appropriate parents for crossing schemes, (2) early or late selection in clonal generations, which will be determined by the heritability of the targeted trait(s), and (3) adequate environmental sampling (i.e., number of locations and years) for testing advanced breeding materials leading to cultivar development. Innovative approaches include genetic manipulations of complete chromosome sets, which are called ploidy manipulations or scaling up and down chromosome numbers of a species within a polyploid series. Chromosome sets are manipulated with haploids, 2n gametes, and through interspecific-interploidy crosses. Analytical breeding schemes rely mainly on ploidy manipulations to "capture" diversity from exotic (wild or nonadapted) germplasm and use 2n gametes to incorporate this genetic diversity through unilateral (USP; n × 2n or 2n × n) or bilateral (BSP; 2n × 2n) polyploidization. Haploids are propagules with the gametophytic chromosome number (n), and 2n gametes possess the sporophytic chromosome number of the parental source.

This review provides an update on breeding the five most important vegetatively propagated starchy food crops; namely potato (the model system), *Musa* (banana and plantain), cassava, sweet potato, and yams. The review includes both conventional and new methods for manipulating the genomes in each of these crops.

Potato: The model system

Potato is an Andean tuber crop (*Solanum tuberosum* L.) that was originally domesticated in South America and started its worldwide dissemination after Columbus's voyages. Today, the potato is one of the five most important crops and the most important staple starchy food in the world. Potato yields on average more food energy and protein per unit of land than cereals (Horton, 1988). The potato crop should be seen not only as an important food (fresh or processed, but also as the raw material for the starch-processing industry, as feed because its vines are fed to animals, and as a potential resource for medicine because of the compounds in its true seed.

The cultivated tetraploid potato (2n = 4*x* = 48 chromosomes) was the result of crosses of two diploid tuber-bearing *Solanum* species (2n = 2*x* = 24 chromosomes) producing 2n gametes and with two EBNs (endosperm balance numbers) (Ortiz

[1]Keynote in International Arnel Hallauer Plant Breeding Symposium (Mexico, 17–22 August, 2003).

and Ehlenfenldt, 1992). In the Andean highlands of South America (2000–4000 m). Gametes with the plant chromosome number are 2n gametes, while EBN refers to endosperm development in interploidy crosses, both intraspecific and interspecific. Under this theory, normal endosperm development occurs only in some angiosperm species when a balance of two EBNs from the female parent matches with one EBN from the male parent in the resulting endosperm. Any deviations from this two maternal/one paternal EBN ratio leads to faulty endosperm and lack of normal seed. EBN is a unifying concept for predicting endosperm function in intraspecific, interploidy, and interspecific crosses characteristic of a species and is more important for predicting the success of a cross and determining the ploidy in the offspring (Ehlenfenldt and Ortiz, 1995). EBN also provides a means for understanding the role of endosperm in evolution and its proper use in breeding methods for gene transfer. For example, through the right ploidy and EBN manipulations, genes from all tuber-bearing and other *Solanum* species are incorporated into the cultivated germplasm pool (Ortiz, 1998a).

Scaling up and down chromosome sets

Ploidy manipulations with haploids (plants with gamete chromosome number), 2n gametes and wild species, still remain today as one of the most impressive and exciting crop germplasm enhancement methods ensuing from cytogenetics research. Indeed, the genetic enhancement strategy for potato germplasm involves species, haploids, 2n gametes, and EBN, in which species are the source of genetic diversity, haploids provide a method for "capturing" the diversity, and 2n gametes and EBN are involved in an effective and efficient method of transmitting diversity to cultivars.

There are two main methods for ploidy manipulations in potato: unilateral sexual polyploidization (4*x*-n gametes × 2*x*-2n gametes or vice versa) and bilateral sexual polyploidization (ensuing from crosses between 2*x*-2n gamete-producing parents). For these breeding schemes the diploid progenitors ensue from crosses between potato haploids and tuber-bearing diploid species. Maternal haploids are easily extracted through parthenogenesis from most tetraploid cultivars and crossed with diploid species for breeding at the diploid level. The locally adapted haploid species hybrids are selected because they possess 2n gametes, acceptable tuber characteristics, and additional desired attributes, for example, disease or pest resistance. Most of the hybrids ensuing from sexual polyploidization in the potato are tetraploids because of a strong triploid block in potato. With this knowledge, Prof. Stanley J. Peloquin (University of Wisconsin) and his "school"—particularly in Italy, Poland, and the Centro Internacional de la Papa (CIP; Lima, Peru)—were able to develop new potato genotypes that combine high and stable yield plus disease or pest resistance, which also allow the widening of potato growing in areas of the world that were previously unsuitable for this crop (Ortiz et al., 2005). The potato genotypes ensuing from their work are amenable to both fresh table markets and chipping industry.

Most potato cultivars are tetrasomic polyploids in which purebred lines are very rare because of the high outcrossing rates in this species whose hybrid vigor for high tuber yield appears to be maximized by multiallelism per locus. Hence, the best diploid parents for tetraploid–diploid crosses are those producing first division restitution (FDR) 2n gametes because they have greater heterozygosity than gametes produced from second division restitution (Watanabe et al., 2004). Not surprisingly, tetraploid hybrids derived from diploid progenitors producing FDR 2n pollen often outyield their half-sib tetraploid hybrids from intermating tetraploid progenitors. Specific crosses are recommended for cultivar development after testing specific combining ability between locally selected parents.

Broadening the genetic base and true potato seed

Analysis of allele frequency in isozyme loci in excess of 2400 farmer-selected tetraploid potato cultivars held in the CIP gene bank and in excess of 100 North American cultivars allowed us to monitor allozyme frequency changes between cultivar pools. This helped us understand the manipulation of the potato genome by plant breeders in North America since the nineteenth century, that is, South American farmers who still grow old landraces (Ortiz and Huamán, 2001). Results from this research indicate that allozyme frequency changes resulted from directional selection of isozyme marker linked to quantitative trait loci (QTL) affecting agronomic or quality characteristics. Furthermore, there were allozymes in some North American cultivars that were not observed in the Andean farmers' selections, which confirms that plant breeders already incorporated genes from wild species or other primitive cultivars into this gene pool. Genetic erosion in North American cultivars was observed mostly for rare alleles in the South American cultivars. A genetic bottleneck as determined by allozyme number per locus was observed for the chromosome bearing a QTL for the glycoalkaloid solanine in potato.

Those results reveal the need for targeted broadening of the genetic base for specific chromosomes (or chromosome regions), which may be facilitated by genetic research. Indeed, progress in plant breeding depends on the availability of genetic variation in the reference population whose original genetic variation, as well as the selection method, may influence genetic gains in further cycles of crop improvement. For example, the most important sources of the original CIP lowland breeding population were clones from the potato groups Tuberosum, Neotuberosum (or Andigena selected under long-days), and 4x-2x hybrids (DTO) between Tuberosum and Phureja. Selection for adaptation was made with earliness for tuber bulking under heat as the primary target trait and high yield per se, plus resistance to bacterial wilt, potato viruses Y (PVY), and X, as well as acceptable tuber quality. A few clones were selected from thousands because they met those standards in the humid lowland tropics of Peru (0.06% selection intensity). After some cycles of recurrent selection, selection intensity increased to 10%, and this material became one of the sources of CIP breeding population for producing tubers from true potato seed (TPS).

TPS refers to true potato seed or commercial potato production from true (sexual) seed. The Incas used this propagation system in the Andes. TPS appears to be very promising in warm tropical environments, where potato growers are affected by high cost of seed tubers and lack of clean planting materials because of high pest pressure. TPS lowers production costs, reduces the incidence of pests such as viruses that are not transmitted by true seed, and allows true seed to be a source of planting material even if parental plants are diseased. TPS technology enables low-income small landholders in the developing world to grow potatoes, thereby expanding the geographic range of this crop worldwide, especially in locations where transport and cold storage of seed tubers are not feasible (Ortiz, 1997a).

CIP assembled an original TPS breeding population from various genetic sources, whose diversity was previously tested for tuber yield and its stability of performance across lowland tropic locations. Years later, breeding at CIP required new sources of genetic variation and a heterogeneous TPS breeding population was assembled using introductions from Europe and North America in crosses with selected CIP lowland breeding materials (hereafter referred as intermediate stage). An advanced TPS breeding population ensued from the second cycle of recurrent selection ensuing from the intermediate breeding cycle stage.

Variance components and heritability were calculated in the source population, as well as intermediate and advanced breeding stages of the TPS breeding populations (Figure 18.1). Heritability was higher in the intermediate stage than in the source population for plant survival but lower in the advanced breeding stage owing to the high percentage of survival in this population after some cycles of recurrent selection. These results suggest that CIP breeders were able to keep enough genetic

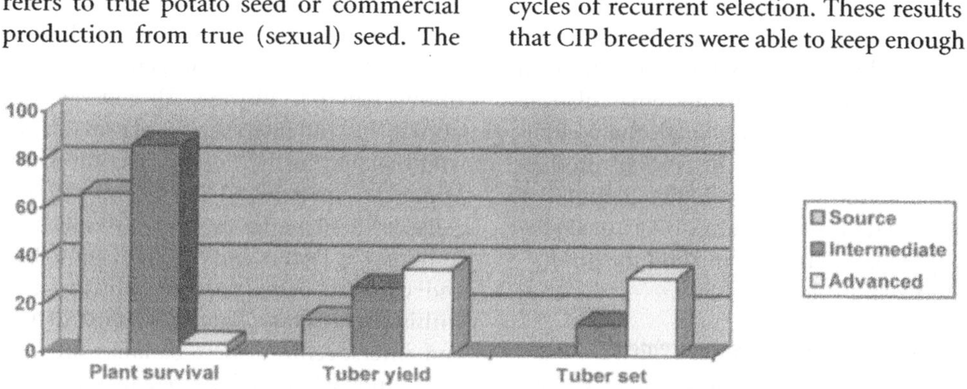

Figure 18.1 Heritability in source population, intermediate and advance breeding stages of true potato seed (after Ortiz and Golmirzaie, 2002).

variation for most important characteristics for potato production from true seed in their intermediate breeding materials by adding new sources of variation to the original breeding population. Similarly, heritability for tuber yield and tuber set in the advanced selection stage was higher than in the intermediate stage or source population, which suggests that recombination through more cycles of recurrent selection brought untapped variation for both characteristics in this breeding material.

Synthetic TPS cultivars may result from polycrosses (Golmirzaie and Ortiz, 2002b), which may be further bred through local adaptive testing and early selection of most promising genotypes, according to their seedling vigor (Golmirzaie and Ortiz, 2002a). A new method of producing inexpensive tetraploid hybrid true potato seed may also ensue from bilateral sexual polyploidization and natural insect pollination. It consists of using unrelated, locally adapted diploid haploid-species hybrids as parents, according to their combining ability, with profuse flowering, attractiveness to bumblebees (natural pollinators), and other desired attributes. The diploid male parent has high male fertility, and very high frequency of (or almost only) first division restitution 2n pollen and a heterozygous monogenic dominant marker tightly linked to the centromere. The female parent combines male and female fertility, self-incompatibility, very high frequency 2n egg production, and lacks the dominant marker (recessive genotype). Both diploid haploid-species parents are grown using an interplanting field designed to allow natural pollination and gene flow between them. TPS are harvested only from female parents and their seedlings grown in a nursery to eliminate those showing poor vigor and lacking the dominant marker phenotype of the male progenitor. Hence, the tuber harvest of this offspring will include mostly (if not only) tetraploid hybrids for potato production in the field. With this TPS scheme, emasculation, pollen collection, and hand pollination are eliminated, thereby saving 50% of the costs of producing hybrid tetraploid seed. It would be desirable to select diploid parents that are able to set 10,000 hybrid seeds per plant with this method for producing TPS (Ortiz and Peloquin, 1991).

Biotechnology tools for genetic enhancement

Potato breeders can incorporate their genetic knowledge and crop improvement methods for the twenty-first century with wild, landrace, or exotic germplasm whose breeding may be further facilitated by recent advances in gene technology. For example, there are transgenic potatoes with resistance to Colorado potato beetle or viruses such as PVY and potato leaf roll virus. On average reported profits in the United States (in U.S. dollars) are $22.40 per acre for *Bt*-potato (Gianesi and Carpenter, 1999), and ex-ante analysis suggests a profit of $288.80 for virus-resistant potato in Mexico (Qaim, 1998).

Advances in potato genomics ensued from the extensive genetic mapping using diploid stocks in the 1990s (Tanksley et al., 1992). These genetic maps may assist in the marker-assisted incorporation or introgression of *Solanum* genetic resources into the tetraploid breeding populations. So far, molecular-aided genetic analysis allowed the dissection of complex quantitative characteristics into their discrete genetic factors or confirmed early hypotheses about transmission of heterozygosity through 2n gametes and helped to elucidate the mode of 2n gamete formation in diploid parents (Ortiz, 1998a, and references therein).

Musa: Genetic knowledge leads new breeding schemes

Banana and plantain (*Musa* spp.) are not trees but giant herbs that can grow up to 15 m tall, making them the largest perennial herb worldwide, and they are vegetatively propagated through suckers. The most important bananas and all plantains are triploid, with a few, almost sterile, cultivars (Ortiz, 2005), which evolved mostly through somatic mutations as determined by descriptors (Ortiz, 1997b), quantitative variation (Ortiz et al., 1998a), and molecular markers (Crouch et al., 2000), confirming the low level of gene flow via pollen among triploid cultivars. All cultivars originated from intra- and interspecific crosses of two diploid wild bananas in the *Eumusa* section of the genus *Musa*: *M. acuminata* Colla. and *M. balbisiana* Colla., which are the sources of A and B genomes, respectively. Desert and east African highland beer and cooking bananas from intraspecific crosses within *M. acuminata* are *Musa* spp. AAA; plantain and interspecific dessert bananas are *Musa* spp. AAB; and Asian cooking bananas from interspecific crosses are *Musa* spp. ABB.

Plantains and cooking bananas are major food crops in developing countries, and desert bananas are also an important export crop (annual exports above $5 billion). Their fruits are highly nutritious, containing large amounts of carbohydrates and minerals, such as phosphorus, calcium, and potassium, as well as vitamins A and C. They are also important sources of revenue for many small-scale farmers. About 90% of the world's bananas and plantains are grown on small farms and consumed locally. The fruits can be fried, baked, or roasted and are also sold in pulp form, as chips, and in confectionery. In some countries they are used to produce alcohol. The leaves and pseudostem are also often used, for example for wrapping food, for thatching, and in textiles. The fruits can also be used as animal feed.

Breeding for host plant resistance to pests and diseases

Black Sigatoka is now pantropic and has become a major constraint to expanding cultivation of edible *Musa* (Craenen and Ortiz, 2003). The causal pathogen of black Sigatoka, *Mycosphaerella fijiensis* Morelet, is a fungus that attacks the leaves. Black Sigatoka has spread rapidly to all major banana and plantain growing areas and the spread is still continuing. Chemical control strategies exist, but they are environmentally unsound and socioeconomically inappropriate, particularly within the framework of the resource-poor farmers that grow the crop in Africa (Craenen et al., 2000, and reports therein).

Genetic manipulations through interspecific and interploidy hybridization (Vuylsteke et al., 1997) or genetic engineering (May et al., 1995; Sagi et al., 1995) appears to offer the only means for broadening the genetic diversity in triploid banana and plantain farming systems, which should not be regarded as close to extinct irrespective of recent false claims suggesting that "the world's favorite fruit could disappear forever in 10 years' time," (*New Scientist*, 2003). In the early 1990s, a team from the International Institute of Tropical Agriculture (IITA), led by the late Dirk R. Vuylsteke (Ortiz, 2001), was able to rapidly (in about five years) develop improved plantain–banana hybrid germplasm with resistance to black Sigatoka using a range of conventional and innovative approaches, such as interspecific hybridization, ploidy manipulation, embryo culture, rapid in vitro multiplication, field testing, and selection (Vuylsteke and Ortiz, 1995; Vuylsteke et al., 1993, 1995). On average, it took 1000 seeds produced from hand pollination of 200 plants (0.12 ha) to obtain one selected tetraploid hybrid per year. This result is a noteworthy achievement, considering that programs elsewhere required decades of breeding before *Musa* hybrids became available. The potential impact of using black Sigatoka-resistant plantains shows a cost-benefit impact of 10:1 over fungicides during periods of adequate production in rural southeastern Nigeria, while this advantage may reduce to 5.5:1 during periods of scarcity in plantain production and dramatically influence the prices of plantain fruit (Ortiz et al., 1997a). Owing to its pioneering research for development on breeding hybrid plantains resistant to black Sigatoka and for advances made in the genetics of *Musa,* IITA received from the Consultative Group on International Agricultural Research (CGIAR) the King Baudouin Award in 1994.

More tetraploid hybrids with heavy bunch weight and stable yield across environments were selected after multilocational testing across sub-Saharan Africa (Ortiz, 1998b) and shared with local researchers first in Africa (Ortiz, 1997c), and more recently in other tropical locations around the world. PITA 14 (or TMPx 7152-2; Ortiz and Vuylsteke, 1998b) appears to be one of the most promising IITA plantain hybrids in Nigeria because of its early fruiting, high bunch weight, and big fruits. While detailed analysis of the acceptability of PITA 14 in southeastern Nigeria is underway, it is noteworthy that several farmers have established sucker multiplication plots and are selling suckers to other farmers (CGIAR/TAC, 2001). Likewise, the cooking banana, BITA 3 (or TMBx 5295-1; Ortiz and Vuylsteke, 1998a), seems to be the preferred hybrid in India owing to the taste of its slender starchy fruits. Because of this early success, in 2001 IITA started large-scale introduction (on-farm) of hybrids with black Sigatoka resistance in the farming community in 11 Nigerian states of the plantain belt, and in 2002 began new projects (in partnership with the International Plant Genetic Resources Institute and local researchers) in Cameroon, Ghana, Mozambique, Tanzania, and Uganda.

Nematodes successfully colonize a greater variety of habitats than any other group of multicellular animals. Worldwide, bananas are attacked by a complex of endoparasitic nematodes, of which

Radopholus similis, *Pratylenchus goodeyi*, *Pratylenchus coffeae*, and *Helicotylenchus multicinctus* are the most important. Banana nematodes feed, multiply, and migrate in roots, resulting in a necrotic and reduced root system. Nematode-infested plants have reduced ability to uptake water and nutrients, which may result in a delay in flowering and ratooning and in reduced yield. Also, plant anchorage is affected, resulting in plant toppling, especially at bunch filling and when strong winds prevail. A promising way of controlling nematodes is the development of hybrids with resistance to nematodes. The first step in breeding for nematode resistance is identifying sources of resistance (Tenkouano et al., 2003), which can then be included in the breeding programs to develop new hybrids with resistance to these nematodes. Two sources of resistance to *R. similis* are widely confirmed: Pisang Jari Buaya and Yangambi Km 5. All plant material (wild bananas, landraces, and hybrids) needs to be tested using reliable screening methods. An early screening method uses individual root inoculation (Dochez et al., 2000; De Schutter et al., 2001). Promising genotypes selected through this early screening method are further tested in pot trials to verify their host plant resistance, before including them in field testing. However, nematode resistance may be effective to only a single nematode species or even a pathotype. The resistance may not be durable if the target nematode species has a high level of genetic variability. Breeding efforts should focus, therefore, on the most pathogenic nematode population, whose reproductive fitness should be assessed as a function of time and inoculum level. DNA fingerprinting can also assist to define distinct nematode populations.

Diploid breeding stocks are regarded as the best sources of genes for breeding at other ploidy levels in *Musa*. Inheritance research suggests that most traits of economic importance are more predictably inherited from diploid sources than from parents with higher ploidy (Tenkouano et al., 1998b). Hence, *Musa* breeding programs worldwide invest significantly in genetic betterment of diploid stocks (Ortiz and Vuylsteke, 1996). Two IITA diploid banana hybrids, TMB2x 5105-1 and TMB2x 9128-3, show resistance to black Sigatoka and breeding potential in *4x-2x* crosses when included in crossing blocks for secondary triploid hybrids (Tenkouano et al., 2003).

Evolutionary crop breeding

A new approach was proposed for further genetic gains in the *Musa* crop (Ortiz, 1997c). In this evolutionary breeding scheme, resulting from the genetic knowledge accumulated during the conventional cross-breeding of plantains (Ortiz, 2000), heterozygous triploid landraces are the sources of allelic diversity, which is released after crossing the landraces with diploid accessions showing desired traits—particularly resistance to pests and diseases. High-yielding primary tetraploid hybrids are selected according to their specific combining ability in the segregating population for new crosses with selected diploid breeding stocks to obtain improved secondary triploid hybrids (Ortiz et al., 1998), which may result from artificial hand pollination or through polycrosses among selected tetraploid and diploid parents (Ortiz and Crouch, 1997). The parental sources of the polycrosses are selected according to combining ability tests after artificial tetraploid-diploid crosses. Synthetic populations derived from the polycrosses can be tested in other locations to identify promising offspring for cultivar development because the genotype-by-environment interaction affects significantly bunch weight across environments (Ortiz, 1998b).

The promise of biotechnology in genetic enhancement

New *Musa* germplasm with enhanced adaptation, particularly in pest-prone environments, will assist sustainable and perennial *Musa* farming systems. This breeding of the *Musa* gene pool can be enhanced by adding genetic variation with molecular markers that may accelerate the process of recurrent selection on this crop species. Furthermore, genetic markers linked to fruit parthenocarpy may account independently or jointly for significant variation observed in segregating plantain–banana hybrid populations for fruit and bunch characteristics (Ortiz and Vuylsteke, 1995), and a major gene for resistance to black Sigatoka can account for most variation for this characteristic (Craenen and Ortiz, 1997) or fruit traits (Craenen and Ortiz, 1996).

Researchers at IITA and partners elsewhere have been working since the mid-1990s to identify genetic markers for fruit parthenocarpy and other traits (Crouch et al., 1998b; Ortiz et al., 1997b) so they can select at the seedling stage in hybrid populations of a giant perennial plant where the first bunch emerges between 12 and 18 months after planting (Ortiz, 2004). In the process, they identified RAPD markers for A and B genomes in *Musa* species (Pillay et al., 2000) and adapted a fluorescent in situ hybridization technique to determine distinct *Musa* genomes (Osuji et al., 1997). They also assessed variation in *Musa* germplasm with many DNA marker systems (Crouch et al., 1998b, 1999; Pillay et al., 2001; Ude et al., 2002a,b) or used micro-satellites for genetically aided analysis (Crouch et al., 1998a, 1999a,b) and cultivar registration (Ortiz et al., 1998). Nonetheless, *Musa* breeders at IITA have tried without great success to predict heterosis with micro-satellites (Tenkouano et al., 1998a), but their research indicated that pedigree-based analysis might still prove useful for selecting parents of prospective *Musa* hybrid populations (Tenkouano et al., 1999a,b). More of their recent work led to the finding of an amplified fragment length polymorphism (AFLP) band likely to be associated with fruit parthenocarpy, but they still rely on field testing and selection for getting new, elite plantain and banana hybrids (Tenkouano et al., 1998a). Perhaps the main public good from this investment in *Musa* genomics at IITA will be the abundant knowledge gathered. However, this promise of *Musa* genomics to assist plantain and banana breeding still remains, and therefore at present "the jury is out."

IITA and other researchers are assessing for *Musa* genomics a new technology known as rapid analysis of gene expression (RAGE) to identify and manipulate multiple genetic components whose quantitative and qualitative expression influence the phenotype. RAGE is a form of gene expression profiling that provides allelic fingerprinting of expressed RNA transcript sequences from a particular tissue and treatment and, to some degree, also includes an assay of the steady-state RNA levels of those fingerprinted transcripts. The produced gene expression profiles are compared pairwise or in multiples. Statistical correlations with particular morphological, physiological, or biochemical traits are analyzed in relation to both peak quantitative levels, as well as peak location, in a particular gene-profiling chromatogram. The result is the identification of new alleles whose gene expression levels or allelic diversity statistically correlate with expression of a particular trait of interest. The fingerprints found in these profiles are produced as cDNA fragment libraries made with real-time polymerase chain reaction (PCR) and AFLP technologies, and are analyzed as chromatograms of these fragments. Figure 18.2 illustrates a typical fingerprint-generating process. To make these diagnostic fragments for RAGE, independent cDNA amplicon library pools are produced from RNA from expressed tissue of different plant genotypes or treatments by real-time PCR. These library pools are then used with AFLP-like generating methods and with real-time PCR to produce an extensive library of bi-tagged cDNA fragments, which may be subsequently profiled into a single chromatogram. The amplification can be targeted to specific genes of known sequence, be generated by random restriction enzyme sequence tags, or be amplified using conserved contextual target sequences, such as the translation initiation context sequences that show conservation (but differences) with monocot and dicot plants. Accordingly, the chromatograms can profile either narrow allelic diversity or broad allelic diversity. The chromatograms function as assays of not only allelic diversity of expressed genes, but also function to some extent, as an assay of relative steady-state RNA levels for the fragments from which the individual cDNA fragments are derived. This point of the impact of gene regulation on phenotype expression is often ignored in traditional breeding, but clearly is important, particularly as one attempt to breed for traits of lower heritability.

Cassava: From a poor man crop to a source of cash in rural areas

Among the most important vegetatively propagated food crops for dry land agriculture is cassava (*Manihot esculenta*), which became the most important food crop in sub-Saharan Africa, which accounts for most of the root harvest worldwide, followed by Asia and Latin America—the center of origin for *Manihot* species. In Africa and Latin America, cassava is mostly used for human consumption, while in Asia and parts of Latin America it is also used commercially for the production of animal feed and starch-based products. Roots are processed into granules, pastes, and flours, or eaten fresh, boiled or raw. The leaves are also eaten in Africa and some Asian locations as a green vegetable, which provides protein and vitamins A and B.

Cassava is regarded as a crop adapted to

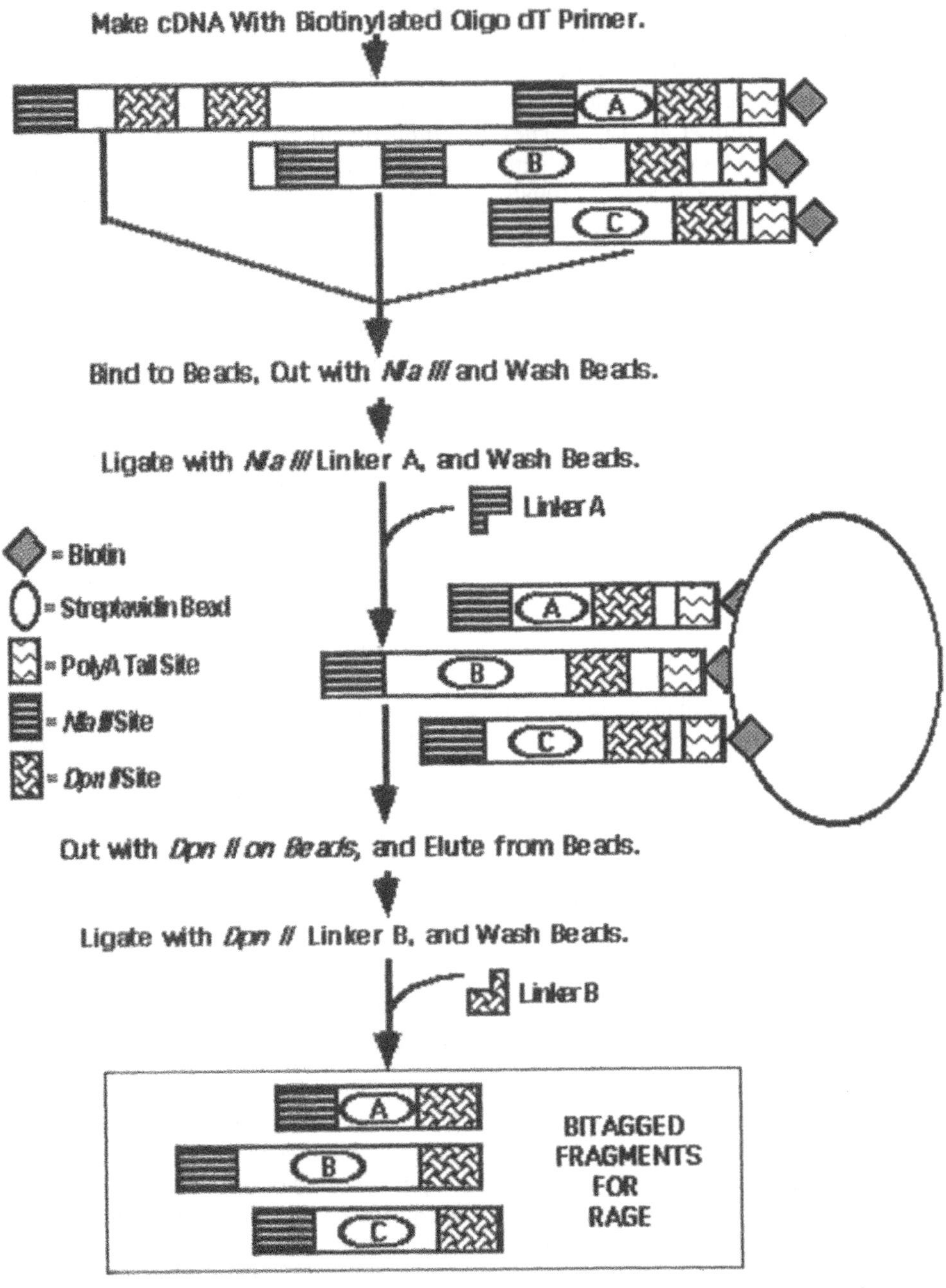

Figure 18.2 An example of the production of bitagged cDNA fragments for gene expression profiling of individual crop plants by a single set of restriction enzymes and corresponding linkers A and B comprising sites of said restriction enzymes. PCR amplification with fluorescent-tagged primers would be done prior to chromatographic separation on a fluorescent DNA-sequencing instrument.

drought-prone environments, where cereals and other crops do not grow well, and it also grows well in poor soil (Cock, 1985; Nassar, 2001a). Under drought stress, the cassava plant reduces water use by following an avoidance strategy of stomatal closure and leaf area reduction. After the stress, recovery of cassava leaf area occurs (Connor and Cock, 1981; Connor and Palta, 1981; Connor et al., 1981), which, of course, influences root yield in cassava depending on the developmental stage of the crop and the environment where it grows (Baker et al., 1989). Because cassava roots can be stored in the ground for up to 24 months, and some varieties for up to 36 months, harvest may be delayed until market, processing, or other conditions are favorable.

Wild species for germplasm enhancement

There are about 100 wild *Manihot* species, which provide an important genetic endowment for cassava breeding (Nassar, 2000a, 2001a), though some of the wild *Manihot* species are under threat of extinction in natural habitats in South and Central America (Nassar, 2000b). Polyploidy and apoximis could participate in the evolution of cassava within the genus *Manihot*. Polyploidy enhances genetic variation, whereas the apomictic mechanism, which appears to be favored by natural selection, offers an escape from lethality and provides a means for perpetuating the genotype (Nassar, 2001b). Sexual polyploidization occurs in cassava because of 2n gametes, and some polyploid cassava genotypes show high plant vigor during growth and root yield (Hahn et al., 1990). Polyploids arising from sexual polyploidization from intermating diploid stocks may assist to incorporating desirable traits from wild species to the cultivated gene pool of cassava. For example, tetraploid hybrids arising from diploid interspecific crosses showed high vigor and root yield, and one of them outyielded the best cultivar in trials undertaken in Nigeria (Hahn et al., 1990). Furthermore, cassava roots are well known for containing very low levels of proteins, but the wild species *Manihot tristis* stores large amounts of protein in the roots. Hence, IITA in the 1990s started a crossing program aimed at hybrids that would possess the food quality characteristics of cassava and the high protein content *of M. tristis*. However, the retention of the protein during processing and the nutritional quality of these proteins still need to be assessed. Similar germplasm enhancement approaches may be useful with other wild *Manihot* species as well as for targeting micronutrient breeding (or biofortification).

Population improvement impacts

Although tetraploid hybrids resulting from 2*x*-2*x* crosses appear as an effective method of incorporating desired traits from wild species into cassava, the breeding methods of this crop still rely on conventional approaches that benefit from the array of genetic variation generated by farmers throughout the cultivation of this crop. Professional cassava breeding started in the twentieth century and was spurred by increasing population demands, which were affected by the limited supply of energy food crops, particularly in sub-Saharan Africa. International aid resources have been given since the late 1960s for cassava research for development in Latin America and the Caribbean, Asia, the Pacific, to Centro Internacional de Agricultura Tropical (Cali, Colombia), and to IITA for Africa. Both organizations undertook in-depth crop research that included its variation worldwide as well as its genetic enhancement through plant breeding. The two international centers of the CGIAR included improving yield per unit area as well as incorporating new areas, particularly in marginal or pest-prone environments, among their breeding goals (Jennings and Iglesias, 2002). In their breeding strategy the target was broad-based breeding populations that would be further selected by national researchers and local partners, according to their needs (Hahn et al., 1979; Hershey and Jennings, 1992). Hence, crosses among local cultivars were high in their breeding agenda as was incorporating exotic germplasm into the desired gene complexes, but minimizing inbreeding and restoring heterozygosity to escape from inbreeding depression. The improved cassava germplasm was sent for testing across locations and for breeding populations.

The most notable results from cassava breeding are seen today in Africa (Nweke et al., 2002), where cassava was transformed from being a poor man's crop to an urban food, from being a subsistent crop to an industrial cash crop owing to long-term research by IITA and partners that led to the development of improved, high-yielding Tropical *Manihot* Selection cultivars that increased cassava yields 40% without the use of fertilizer. Breeding for pests such as cassava mosaic disease (Hahn et al., 1980b), which benefited from early work in East Africa (Jennings, 1957, 1994), and bacterial blight (Hahn et al., 1980a) account for this breeding success. About 206 releases of cassava cultivars from IITA germplasm are recorded in 20 African nations. In the 1990s African programs incorporated IITA-bred materials in 80% of their cassava-bred germplasm, which led to 50% gains in cassava yields on average (Manyong et al., 1999). The improved cultivars raised per capita output 10% continent-wide, benefiting 14 million farmers.

The national research capacity available in Africa

and support from IITA provide a means to deal with new threats affecting this crop on the continent (Legg, 1999; Otim-Nape et al., 2001). For example, the total benefits from the cassava multiplication research-for-development partnership project between NARO (Uganda) and IITA to combat the cassava mosaic disease pandemic in six districts was approximately $36 million (U.S.) over four years (1998–2001) for an initial investment of $0.8 million. Partnerships between National Agricultural Research Systems and IITA were key for this and other successes in the genetic enhancement of cassava in Africa. Through the introduction of more productive cultivars that are resisting prevailing pests, and the effective biological control of the cassava mealybug (Herren and Neuenschwander, 1991) and other pests, large-scale famine was avoided in sub-Saharan Africa; that is, cassava production would be 50% or less, or over 13 million t $year^{-1}$ of dry cassava, enough to meet the calorie requirements of 65 million African people. Furthermore, governments in Ghana and Nigeria are taking steps for promoting agro-processing (e.g., through cassava starch industry) as a major vehicle for job creation and poverty reduction in rural areas. Cassava exports could be capable of generating an income of $1.5 billion (U.S.) within two years in Nigeria alone. Moreover, the low labor requirements for cassava and its potential in drought-prone environments provide an alternative to other African nations with the same constraints.

Utilization of new genetic variation effectively requires extensive intercrossing and selection, particularly for a crop such as cassava in which many pathogens still take their toll and occasionally in epidemic proportions in both traditional and new areas of production, for example, cassava brown streak disease in the coasts of East and Southern Africa. Likewise, the wide range of target agro-ecologies, farming systems, utilization patterns, and consumer preferences for cassava worldwide demands that a number of different recombining populations are handled in partnerships with local researchers. Back-up special populations are essential as feeders of agro-ecologically broad-based populations and as sources of specific trait populations. Hence, ongoing research at CIAT (International Center for Tropical Agriculture) and IITA includes analysis of genotype-by-environment interaction for selection of testing sites and distribution areas. Recently, a double-haploid breeding approach capitalizing on combining ability was advocated for further genetic gains in cassava (Ceballos et al., 2002). The new cassava-breeding domains are included in Table 18.1.

Table 18.1 New domains for cassava breeding

Domain
Decentralizing cassava breeding with farmer participatory methods
End-user and market-driven breeding
Broadening the genetic endowment with genes adapted to harsh environments
Biotechnology for cassava genetic enhancement to reach farming systems
Improving protein content with genes from wild *Manihot* species

Genetic enhancement through biotechnology

Tissue culture for micropropagation of healthy planting materials and *in vitro* gene banking, genetic engineering, and DNA markers are among the biotechnology methods for improving cassava. Fregene and Puonti-Kaerlas (2002) provide the most recent update of the potential of cassava biotechnology, which so far provides a better understanding of the cassava genome through gene mapping, and the potential for a more rapid and efficient improvement through genetic transformation and marker-aided breeding. For example, micro-satellites are being useful to study diversity in *Manihot* gene pools and to identify duplicates in the gene bank collection, and recently some researchers in Africa, America, and Australia started developing expressed sequence tags and micro-array techniques for cassava. The most promising results are from interval mapping with RFLP and micro-satellite markers (Fregene et al., 1997, Mba et al., 2000). Two markers flanking a dominant gene may provide new sources of resistance to cassava mosaic diseases (Akano et al., 2002), and new research may include the cloning and sequencing of this putative "gene." Likewise, research at the Royal Veterinary and Agricultural University showed that two RAPD markers linked to genome regions coding for the enzyme catalyzing the first committed step of the biosynthesis of cyanogenic glucosides were tested positively in 46 African genotypes.

Sweet potato: Tapping vitamin A in the fight against blindness

This hexaploid ($2n = 6x = 90$) crop (*Ipomoea batatas*) ranks third among root and tuber staples worldwide, after potato and cassava, with China accounting for 65% of total acreage. A high density for this crop occurs in the farming systems of the Central African highlands, whereas in other locations the crop shows a low density but appears to be grown elsewhere in the tropics. The crop originated in Latin America, and the greatest variation occurs in northwest South America and some locations in Central America, where the wild species *I. trifida* complex appears to be abundant (mostly as diploid or tetraploid). Gene flow seems to occur with the *I. trifida* complex that includes a ploidy series (from diploids to hexaploids). Cytogenetic research suggests that sweet potato may be a polysomic polyploid species but with a high degree of genomic duplication (Nishiyama et al., 1975; Shiotani, 1988; Ukoskit and Thompson, 1997). Owing to the occurrence of tetravalents and hexavalents, two inheritance models (hexasomic, tetra-disomic) were tested in sweet potato (Kumagai et al., 1990). Likewise, the ratios of linkage in the coupling phase to linkage in repulsion phase, and of non-simplex to simplex markers, suggest mostly polysomic inheritance with some degree of preferential pairing (Kriegner et al., 2003). However, Jones (1967) classified the segregation ratios observed in this crop as quantitative. Sweet potato cultivars are affected by self-incompatibility, which prevents inbreeding depression and which seems to be a deleterious trait in a highly heterozygous crop. Sweet potato breeders rank epistasis, that is, a unique gene assortment rather than unique genes *per se*, as very important for outstanding clones.

Polycross breeding and ploidy manipulations

Polycrosses are the main conventional population improvement method for sweet potato, which require an environment that enhances flowering of all clones included in the polycross. Farmers in Papua, New Guinea are well known for taking advantage of natural polycrosses, as shown by the many distinct clones in the same field. The goal in the genetic enhancement of sweet potato is to increase the frequency of favorable alleles in the population, from which outstanding clones may be selected for cultivar development (Martin and Jones, 1986). Because of the high heritability for most traits (Jones, 1986), recurrent mass selection (a long-term goal) followed by polycross mating (a short-term goal) proved to be a very effective method for sweet potato breeding (Jones et al., 1976). Through mass selection in a population of 3000 seedlings, about 30 selected clones are included (after field-testing) in polycrosses, using open pollination in isolated fields. Promising clones for cultivar release may result after selecting genotypes for multilocational testing over years. Each successive cycle of recurrent mass selection starts with the remaining seed of selected clones rather than the clone itself to preserve the gene complex rather than the selected genotype.

The main breeding targets, arranged by tropical regions worldwide, are listed in Table 18.2. Although sweet potato breeders in the southern United States developed better cultivars in the twentieth century, yield gains in Africa and Latin America recorded negative rates from the 1960s to the 1990s. Yield rates were only positive (almost

Table 18.2 Main breeding targets in sweet potato according to tropical regions worldwide

Target	Latin America	South and West Asia	East Southeast Asia and Pacific	Africa
Weevil	a	a	a	
Virus				a
Drought	a	a	a	a
Dry matter (starch)		a		a
Foliage	b	b		b
Non-sweet cultivars				(West)[a]
Storability	a	a	a	a

[a]The main breeding target.
[b]Potential demand for forage only.
Source: After CIP (1995).

doubling) in Asia, a continent in which wild species of *I. trifida* were successfully incorporated in sweet potato breeding (Sakamoto, 1970). The Japanese cultivar Minamiyutaka has one-eighth *I. trifida* genes, which provide resistance against root-lesion nematode, high starch content, and high yield through heterosis (Sakamoto, 1976). Hence, new methods were needed for the genetic enhancement of this crop; it may be further broadened with exotic or wild germplasm, which does not produce storage roots. In this regard, the CIP began in the 1980s (and until the mid 1990s) ploidy manipulations in this crop (Iwanaga et al., 1991) through creating tetraploid storage root producers (from crosses between sweet potato and diploid *I. trifida*) to assess wild species root traits, particularly within the *I. trifida* complex (Orjeda 1989), and to search for 2n gametes (Orjeda et al., 1990), because most of *I. trifida* accessions are diploids and tetraploids. Freyre et al. (1991) reported 2n pollen in triploid hybrids ensuing from 4*x*-2*x* crosses within the *I. trifida* complex. They were further included in crosses with sweet potato to transfer genes from the wild species to the crop gene pool. They also reported that synthetic hexaploids (after chromosome doubling with colchicine) showed poor vigor owing to inbreeding depression, which may also limit transmitting heterozygosity from *I. trifida* to sweet potato. Their results suggest that FDR 2n pollen from triploid *I. trifida* offers the most attractive path for producing high-yielding offspring due to maximum heterozygosity.

Biofortification

Sweet potatoes possess beta-carotene as the predominant carotenoid (Takahata, 1995; Takahata et al., 1993), which serves as the main source of provitamin A in the roots and which is converted by the human body into the essential nutrient vitamin A (Woolfe, 1992). CIP breeders selected some sweet potato clones with deep orange flesh (under the acronym VITAA). CIP, IITA, and local researchers are using farmer-participatory breeding methods to provide such genetic resources to the farming systems in eastern and southern Africa, where rural poor people are more vulnerable to vitamin A deficiency, leading to blindness. About 50 million African children (below age 6 years) may benefit by eating VITAA cultivars daily. Local entrepreneurs in eastern and southern Africa are also processing (e.g., flour) VITAA cultivars for urban end-users, who need products to address vitamin A deficiency.

The potential of biotechnology in genetic enhancement

Protocols for genetic engineering of sweet potato are available (Okada et al., 2001; Wambugu, 2001) and provide another means for the genetic enhancement of the crop. For example, researchers at the Kenya Agricultural Research Institute—through a grant from the United States Agency for International Development and in collaboration with Monsanto—incorporated in the 1990s genes for resistance to a virus in eight sweet potato cultivars. Field-testing started in 2001 with materials already screened under containment in the greenhouse (Wambugu, 2001). *Ex-ante* analysis suggests a benefit ranging from $42.31 to $101.12 (U.S. dollars) per acre (Qaim, 1998; Marra, 2001). Likewise, RAPD and AFLP markers within respective genetic linkage maps are available in sweet potato (Kriegner et al., 2003, Ukoskit and Thompson, 1997), but so far molecular markers provided only a means for determining polysomic inheritance in the crop. The best results from marker-aided analysis in sweet potato refer to assessing genetic variation in cultivars or gene bank accessions with restriction analysis of chloroplast DNA, microsatellites, or RAPD (Buteler et al., 1999; Huang and Sun 2000; Jarret and Austin, 1994; Jarret and Bowen, 1994; Zhang et al., 1998, 2000, 2001)

Yams: The ploidy puzzle

Yams (*Dioscorea* spp.) are important tuber crops in humid and sub-humid tropics, particularly in West Africa, which accounts for 93% of the world production (Asiedu et al., 1997); Nigeria alone produces 70% of the total (Hahn et al., 1979). The tubers are processed into pounded yam, boiled yam, roasted or grilled yam, fried yam slices, yam balls, mashed yams, yam chips, and yam flakes. Fresh yam tubers are also peeled, chipped, dried, and milled into flour that is used to prepare a dough called *amala* or *telibowo*.

The *Dioscorea* species shows a wide ploidy polymorphism (Dansi et al., 2001), but most important cultivars accounting for 99% of all food yams among 600 *Dioscorea* species belong to the white yam (*D. rotundata*; 2n = 40, 80), yellow yam (*D. cayenensis*; 2n = 36 – 140), and trifoliate yam (*D. dumetorum*; 2n = 36 – 54), water yam (*D. alata*; 2n = 20 – 80), and Chinese yam (*D. esculenta*; 2n = 30 – 100), aerial yam (*D. bulbifera*; 2n = 30 – 100), and cush-cush yam (*D. trifida*, 2n = 54 – 81) (Hahn, 1995; Dansi et al., 1999). Domestication appears to occur still in West Africa, thereby leading to the establishment of new genotypes in the farming systems (Dumont and Vernier, 2000). The white and yellow yam are the most important in West Africa because of their tuber taste after cooking. Farmers also grow water yam there, the second more important per its total production and with the widest spread in the tropics. The breeding objectives for this crop are high and stable yield of marketable tubers with acceptable quality (i.e., dry matter content, cooking texture, taste, dormancy, and rate of enzymatic browning). The steps in yam genetic improvement at IITA, one of the largest programs for this crop worldwide, are in Table 18.3.

IITA and national or local partners generated several new cultivars of water and white yams with high and stable yield of tubers (50–100% superior to popular local cultivars), as well as good storability and food quality attributes through breeding and selection. High levels of host plant resistance bred into the cultivars against the two most important diseases of the crop, that is, yam anthracnose disease caused by *Colletotrichum gloeosporioides* and yam mosaic virus (YMV), contribute significantly to the high level and stability of field performance. With the aim of limiting production cost, the improved yam cultivars were selected for good performance in the absence of external inputs of fertilizer or staking (in the moist savanna zone), and emphasis was placed on tuber shapes that facilitate harvesting. Many of these new cultivars were assessed at multiple sites in the yam-producing locations of West Africa for suitability in local farming and food systems in comparison with popular indigenous cultivars and with active participation of potential farmers. Three IITA-bred cultivars of white yam were formally released by Nigeria in 2001. Several others are in the pipeline in the other major producing countries in the sub-region.

Table 18.3 Breeding steps for improving yams

1. Assessment of variation of accessions held in gene bank and from other sources
2. Selection of accessions according to breeding objectives
3. Trials[a] to assess selected accessions towards cultivar development or for potential parental sources
4. Full-sib and OP half-sib seeds from selected parental sources in isolated crossing blocks or OP seeds from nursery trials or farmers' fields[b]

Ensuing seedlings included in nursery trials and selected genotypes start again from step 2 onward.

[a]Trials start with preliminary nonreplicated plots including variable number of clones and stands per clone, followed by clonal assessment in hill trials of same size and preliminary yield plots with two replications. Advanced and uniform trials are for next testing stages in randomized complete block designs with three to six replications. Cooking and processing tests begin after the fifth year, that is, with materials from advanced or uniform trials.

[b]Many white yams do not flower in fields, and water yam normally does not flower owing to genetic, physiological, and environmental factors, thereby preventing their gene flow through crosses. Furthermore, flowering synchronization between male and female cultivars can also hamper hybridization.

Water yam, a species introduced to Africa from Asia, deserves special mention. It is generally superior to the indigenous white yam in yield potential (especially under low to average soil fertility), ease of propagation (production of bulbils and reliability of sprouting), early vigor for weed suppression, and storability of tubers. Indeed, it has superior characteristics for sustainable production. Its major limitation in the field is the susceptibility of most cultivars to anthracnose. The tuber quality of most cultivars of this species is inferior to that of white yam in the preparation of West African food. IITA researchers bred new water yam cultivars with much improved food quality, resistance to anthracnose, and high tuber yield, which are now under multisite testing with partners in Nigeria and Côte d'Ivoire. Already one of the key parental sources, earlier introduced from Puerto Rico has gained very wide acceptance in West Africa. Introduction to farmers in Ebonyi State of Nigeria has led to a rapid spread in that state and neighboring ones. A technological package has been developed for the production of fried yam chips suitable for use in the fast-food industry to meet the demands of an increasingly urban population. This package includes suitable yam cultivars (with defined physical and chemical characteristics), partial frying of the chips followed by freezing, and final frying to get the finished product. Processing parameters were established at the bench scale, and these will soon be adapted to commercial scale.

Methods for rapid production of large quantities of healthy seed tubers were developed and disseminated by IITA and partners elsewhere because of limited availability and cost of seed tubers (about 50% of total production cost), which are major considerations in the productivity of yam cultivation. The crop appears to be in high demand in Africa, as shown by the annual growth rates in acreage, production, and yield from the early 1960s to 2000: 3.81%, 4.06%, and 0.23%, respectively, which today allow the crop to be second in cultivation (after cassava) on this continent. These methods include the use of minisetts from field-grown tubers and major biotechnology advances in pathogen (especially virus) diagnostics, micropropagation, and seed disinfestations.

Advances in DNA mapping and potential of marker-aided breeding

Other biotechnology options in this crop are gene mapping and assessment of germplasm variation with DNA markers. For example, genetic diversity and phylogenetic relationships in *Dioscorea* were determined with isozymes or DNA markers (Asemota et al., 1996; Dansi et al., 2000; Hamon and Touré, 1990; Mignouna et al., 1998, 2002a; Ramser et al., 1996, 1997; Terauchi et al., 1992; Zoundjihékpon et al., 1994). Today, AFLP maps of the water yam include 338 markers on 20 linkage groups (1055 cM) (Mignouna et al., 2002b), whereas there are 107 markers in 12 linkage groups (585 cM) for male and 13 linkage groups (700 cM) for female in the white yam (Mignouna et al., 2002c).

Three QTL and one QTL, with effects on resistance to YMV, the most widespread and economically important viral disease affecting white yam, were identified on the maternal and paternal linkage maps, respectively (Mignouna et al., 2002c). Segregation ratios from early research indicated that a single dominant gene in a simplex condition governs the resistance in clone TDr 89/01444, while the resistance in clone TDr 93-2 is associated with the presence of a major recessive gene in duplex configuration (Mignouna et al., 2001b). Likewise, one AFLP marker (*E-14/M52-307*) on linkage group 2 appears to be associated with anthracnose resistance (about 10% of the total phenotypic variance) (Mignouna et al., 2002a), a trait that was reported early as dominant and polygenic (Mignouna et al., 2001a). DNA markers may be useful for aided introgression and selection while dealing with disease-resistance genes in yams (Mignouna et al., 2002d, e).

Outlook

On-going breeding schemes for potato and plantain–banana rely on manipulating ploidy for gene transfer between wild species and cultigens. Other interesting examples of ploidy manipulations are reported in the genetic enhancement of cassava and sweet potato. This breeding approach may be useful in yam, sugar cane, and other berry or fruit crops (Ortiz, 2003b). Ploidy manipulations coupled with incorporation of exotic or wild species germplasm broadens the genetic base of vegetatively propagated crops, thereby enhancing crop adaptation and sustaining genetic gains in respective breeding pools. The success of this breeding approach depends on viable and fertile F_1 hybrids after interspecific crosses, which will allow the integration of chromosome segment(s) or set(s) from a donor species into the cultigen pool.

References

Akano, A.O., A.G.O. Dixon, C. Mba, E. Barrera, and M. Fregene. 2002. Genetic mapping of a dominant gene conferring resistance to cassava mosaic disease. Theor. Appl. Genet. 105:521–525.

Asemota H.N., J. Ramser, C. Lopez-Peralta, K. Weising, and G. Kahl. 1996. Genetic variation and cultivar identification of Jamaica yam germplasm by random amplified polymorphic DNA analysis. Euphytica 92:341–351.

Asiedu, R., N.M. Wanyera, S.Y.C. Ng, and N.Q. Ng. Yams. 1997. Pp. 57–66. In D. Fuccillo, L. Sears and P. Stapleton (eds.) Biodiversity in Trust: Conservation and Use of Plant Genetic Resources in the CGIAR Centres. Cambridge Univ. Press, Cambridge, United Kingdom.

Baker, G.R., S. Fukai, and G.L. Wilson. 1989. The responses of cassava to water deficits at various stages of growth in the subtropics. Australian J. Agric. Res. 40:517–528.

Buteler, M.I., R.L. Jarret, and D.R. Bonte. 1999. Sequence characterization of microsatellites in diploid and polyploid *Ipomoea*. Theor. Appl. Genet. 99:123–132.

Ceballos, H., N. Morante, F. Calle, J.I. Lenis, G. Jaramillo, and J.C. Perez. 2002. Mejoramiento genetico de la yuca. Pp. 295–325. In B. Ospina and H. Ceballos (eds.) La Yuca en elTercer Milenio: Sistemas Modernos de Producción, Procesamiento, Utilización y Comercialización. Publicacion CIAT 327. Centro Internacional de Agricultura Tropical, Cali, Colombia.

CGIAR/TAC (Consultative Group on International Agricultural Research/Technical Advisory Committee). 2001. Contributions made by the CGIAR and its Partners to

Agricultural Development in Sub-Saharan Africa. Document No. SDR/TAC:IAR/01/13. CGIAR Secretariat–The World Bank, Washington DC.

CIP (Centro Internacional de la Papa). 1995 Vines to Roots: Sweetpotato Breeding for Impact. Centro Internacional de la Papa. 46 pp.

Cock, J.H. 1985. Cassava. New Potential for a Neglected Crop. Westview Press, Boulder and London.

Connor, D.J., and J.H. Cock. 1981. Response of cassava to water shortage. II. Canopy dynamics. Field Crops Res. 4:285–296.

Connor, D.J., J.H. Cock, and G.H. Parra 1981. Response ofcassava to water shortage. I. Growth and yield. Field Crops Res. 4:181–200.

Connor, D.J., and J. Palta. 1981. Response of cassava to water shortage. III. Stomatal control of plant water stress. Field Crops Res. 4:297–311.

Craenen, K., and R. Ortiz. 1996. Effect of the black sigatoka resistance gene *bs1* and ploidy level in fruit and bunch traits of plantain-banana hybrids. Euphytica 87:97–101.

Craenen, K., and R. Ortiz. 1997. Effect of the *bs1* gene in plantain-banana hybrids on response to black sigatoka. Theor. Appl. Genet. 95:497–505.

Craenen, K., and R. Ortiz. 2003. Genetic improvement for a sustainable management of the resistance. Pp. 181–198. In Jacome L., P. Lepoivre, D. Marin, R. Ortiz, R. Romero and J.V. Escalant (eds.) *Mycosphaerella* Leaf Spot Diseases of Bananas: Present Status and Outlook. International Network for the Improvement of Banana and Plantain, Montpellier, France.

Craenen, K., R. Ortiz, E. Karamura, and D. Vuylsteke (eds.). 2000. Proceedings of the First International Conference on Banana and Plantain for Africa. Acta Horticulturae 540:589

Crouch, H.K., J.H. Crouch, R.L. Jarret, P.B. Cregan, and R. Ortiz. 1998a. Segregation at microsatellite loci in haploid and diploid gametes of *Musa*. Crop Science 38:211–217.

Crouch, H.K., J.H. Crouch, S. Madsen, D. Vuylsteke, and R. Ortiz. 2000. Comparative analysis of phenotypic and genotypic diversity among plantain landraces (*Musa* spp., AAB group). Theor. Appl. Genet. 101:1056–1065.

Crouch, J.H., H.K. Crouch, H. Constandt, A. Van Gysel, P. Breyne, M. van Montagu, R.L. Jarret, and R. Ortiz. 1999a. Comparison of PCR-based molecular marker analyses of *Musa* breeding populations. Molecular Breeding 5:233–244.

Crouch, H.K., J.H. Crouch, A. Tenkouano, and R. Ortiz. 1999b. VNTR-based analysis of 2*x* and 4*x* full-sib *Musa* hybrids. Electronic J. Biotech. 2(3):99–108 http://www.ejb.org/content/vol2/issue3/full/1/.

Crouch, J.H., D. Vuylsteke, and R. Ortiz. 1998b. Perspectives on the application of biotechnology to assist the genetic enhancement of plantain and banana (*Musa* spp.). Electronic J. Biotech. 1(1):1–12 http://www.ejb.org/content/vol1/issue1/full/2/.

Dansi, A., H.D. Mignouna, M. Pillay, and S. Zok. 2001. Ploidy variation in the cultivated yams (*Dioscorea cayenensis-Dioscorea rotundata* complex) from Cameroon as determined by flow cytometry. Euphytica 119:301–307.

Dansi, A., H.D. Mignouna, J. Zoundjhékpon, A. Sangaré, R. Asiedu, and F.M. Quin. 1999. Morphological diversity, cultivar groups and possible descent in the cultivated yams (*Dioscoroea cayanensis/D. rotundata* complex) in Benin Republic. Genet. Res. Crop. Evol. 46:371–388.

Dansi, A., H.D. Mignouna, J. Zoundjihékpon, A. Sangaré, R. Asiedu, and N. Ahoussou. 2000. Identification of some Benin Republic's Guinea yam (*Dioscorea cayenensis/Dioscorea rotundata* complex) using random amplified polymorphic DNA. Genet. Res. Crop Evol. 47:619–625.

De Schutter, B., P.R. Speijer, C. Dochez, and A. Tenkouano De Waele. 2001. Evaluating host plant reaction of *Musa* germplasm to *Radopholus similis* by inoculation of single primary roots. Nematropica 31:295–299.

Dochez, C., P.R. Speijer, J. Hartman, D. Vuylsteke, and D. De Waele. 2000. Screening *Musa* hybrids for resistance to *Radopholus similis*. InfoMusa 9(2):3–4.

Dumont, R., and P. Vernier. 2000. Domestication of yams (*D. cayenensis D. rotundata*) within the Bariba ethnic group in Benin. Outlook in Agriculture 29:137–142.

Ehlenfenldt, M.K., and R. Ortiz. 1995. On the origins of endosperm dosage requirements in *Solanum* and other angiosperma genera. Sexual Plant Reproduction 8:189–196.

Fregene M., F. Angel, R. Gómez, F. Rodríguez, P. Chavarriaga, W. Roca, J. Tohme, and M. Bonierbale. 1997. A molecular Genetic map of cassava (*Manihot esculenta* Crantz). Theor. Appl. Genet. 95:431–441.

Fregene, M., and J. Puonti-Kaerlas. 2002. Cassava biotechnology. In Pp. 179–207. In R.J. Hillocks, J.M. Thresh and A.C. Bellotti (eds.). Cassava: Biology, Production and Utilization. CAB International, Wallingford, United Kingdom.

Freyre, R., G. Orjeda, and M. Iwanaga. 1991. Use of *Ipomoea trifida* (HBK) G. Don germplasm for sweetpotato improvement. 2. Fertility of synthetic hexaploids and triploids with 2n gametes of *I. trifida* and their interspecific crossability with sweetpotato. Genome 34:209–214.

Gianesi, L., and J. Carpenter. 1999. Agricultural Biotechnology: Insect Control Benefits. National Center for Food and Agricultural Policy, Washington DC.

Golmirzaie, A.M., and R. Ortiz. 2002a. Inbreeding and true seed in tetrasomic potato. III. Early selection on open-pollinated populations. Theor. Appl. Genet. 104:157–160

Golmirzaie, A.M., and R. Ortiz. 2002b. Inbreeding and true seed in tetrasomic potato. IV. Synthetic cultivars. Theor. Appl. Genet. 104:161–164.

Hahn, S.K. 1995. Yams: *Dioscorea* spp. (Dioscoreaceae). In Pp. 112–120. In J. Smartt and N.W. Simmonds (eds.). Evolution of Crop Plants. Longman Scientific & Technical, Harlow, England.

Hahn, S.K., K.V. Bai, and R. Asiedu. 1990. Tetraploids, triploids and 2n pollen from diploid interspecific crosses with cassava. Theor. Appl. Genet. 79:433–439.

Hahn, S.K., D.S.O. Osiru, M.O. Akoroda, and J. Otoo. 1985. Yam production and its future prospects. Outlook Agric. 16:105–110.

Hahn, S.K., E.R. Terry, K. Leuschner, I.O. Akobundu, C. Okali, and R. Lal. 1979. Cassava improvement in Africa. Field Crops Res. 2:193–226.

Hahn, S.K., E.R. Terry, and K. Leuschner. 1980a. Breeding cassava for resistance to cassava mosaic disease. Euphytica 29:673–683.

Hahn, S.K., A.K. Howland, and E.R. Terry. 1980b. Correlated resistance of cassava to mosaic and bacterial blight diseases. Euphytica 29:305–311.

Hamon P., and B. Touré. 1990. Characterization of traditional yam varieties belonging to the *Dioscorea cayenensis-rotundata* complex by their isozymic patterns. Euphytica 46:101–107.

Herren, H.R., and P. Neuenschwander. 1991. Biological control of cassava pests in Africa. Ann. Rev. Entom. 36:257–283.

Hershey, C.H., and D.L. Jennings. 1992. Progress in breeding cassava for adaptation to stress. Plant Breed. Abst. 62:823–831.

Horton, D. 1988. Potatoes: truly a world crop. SPAN 30:116–118.

Huang, J.C., and M. Sun. 2000. Genetic diversity and relationships of sweetpotato and its wild relatives in *Ipomoea* series *Batatas* (Convolvulaceae) as revealed by inter-simple sequence repeats (ISSR) and restriction analysis of chloroplast DNA Theor. Appl. Genet. 100:1050–1060.

Iwanaga, M., R. Freyre, and G. Orjeda. 1991. Use of *Ipomoea trifida* (HBK) G. Don germplasm for sweetpotato improvement. 1. Development of synthetic hexaploids of *I. trifida* by ploidy level manipulations. Genome 34:200–208.

Jarret, R.L., and D.F. Austin. 1994. Genetic diversity and systematic relationships in sweetpotato (*Ipomoea batatas* (L.) Lam.) and related species as revealed by RAPD analysis. Genet. Res. Crop Evol. 41:163–173.

Jarret, R.L., and N. Bowen. 1994. Simple sequence repeats (SSR's) for sweetpotato germplasm characterization. Plant Genetic Resources Newsletter 100:9–11.

Jennings, D.L. 1957. Further studies in breeding cassava for virus resistance. East African Agric. J. 22:213–219.

Jennings, D.L. 1994. Breeding for resistance to African cassava mosaic geminivirus in East Africa. Tropical Sci. 34:110–122.

Jennings, D.L., and C. Iglesias. 2002. Breeding for crop improvement. In Pp. 149–166. In R.J. Hillocks, J.M. Thresh and A.C. Bellotti (eds.). Cassava: Biology, Production and Utilization. CAB International, Wallingford, United Kingdom.

Jones, A. 1967. Theoretical segregation ratios of qualitatively inherited characters of hexaploid sweetpotato (*Ipomoea batatas* L.). USDA. Tech. Bull. 1368

Jones, A. 1986. Sweet potato heritability estimates and their use in breeding. Hort. Science 21:14–17.

Jones, A., P.D. Dukes, and F. Cuthbert Jr. 1976. Mass selection in sweet potato: breeding for resistance to insects and diseases and for horticultural characteristics. J. Amer. Soc. Hort. Sci. 12:701–704.

Kriegner, A., J.C. Cervantes, K. Burg, R.O. Mwanga, and D.P. Zhang. 2003. A genetic linkage map of sweetpotato (*Ipomoea batatas* (L.) Lam.) based on AFLP markers. Molecular Breeding 11:169–185.

Kumagai T., Y. Umemura, T. Baba, and M. Iwanaga. 1990. The inheritance of beta-amylase null in storage roots of sweetpotato, *Ipomoea batatas* (L.) Lam. Theor. Appl. Genet. 79:369–376.

Legg, J.P. 1999. Emergence, spread, and strategies for controlling the pandemic of cassava mosaic virus diseases in east and central Africa. Crop Protection 18:627–637.

Manyong, V.M., A.G.O. Dixon, K.O. Makinde, M. Bokanga, and J. Whyte. 1999. The Contribution of IITA-Improved Cassava to Food Security in Sub-Saharan Africa. IITA Impact Series. International Institute of Tropical Agriculture, Ibadan, Nigeria.

Marra, M.C. 2001. Agriculture biotechnology: a critical review of the impact evidence to date. In: Pp. 155–184. In P.G. Pardey (ed.), The Future of Food: Biotechnology Markets and Policies in an International Setting. International Food Policy Research Institute, Washington, DC.

Martin, F.W., and A. Jones. 1986. Breeding sweet potatoes. Plant Breed. Rev. 4:313–345.

May, G., R. Afza, H. Mason, A. Wiecko, F. Novak, and C. Arntzen. 1995. Generation of transgenic banana (*Musa acuminata*) plants via *Agrobacterium*-mediated transformation. Bio/Technology 13:486–492.

Mba, R.E.C., P. Stephenson, K. Edwards, S. Melzer, J. Nkumbira, U. Gullberg, K. Apel, M. Gale, J. Tohme, and M. Fregene. 2000. Simple sequence repeat (SSR) markers survey of the cassava (*Manihot esculenta* Crantz) genome: towards a SSR-based molecular genetic map of cassava. Theor. Appl. Genet. 102:21–31.

Mignouna, H.D., M.M. Abang, K.R. Green, and R. Asiedu. 2001a. Inheritance of resistance in water yam (*Dioscorea alata*) to anthracnose (*Colletotrichum gloeosporioides*). Theor. Appl. Genet. 103:52–55.

Mignouna, H.D., M.M. Abang, A. Onasanya, and R. Asiedu. 2002d. Identification and potential application of RAPD markers for anthracnose resistance in water yam (*Dioscorea alata*). Ann. Appl. Biol. 141:61–66.

Mignouna, H. D., M. M. Abang, A. Onasanya, B. Agindotan, and R. Asiedu. 2002e. Identification and potential use of RAPD markers linked to yam mosaic virus resistance in white yam (*Dioscorea rotundata* Poir.). Annals of Applied Biology 140:163–169.

Mignouna, H.D., A. Dansi, and S. Zok. 2002a. Morphological and isozymic diversity of the cultivated yams (*Dioscoroea cayanensis*/*D. rotundata* complex) of Cameroon. Genet. Res. Crop Evol. 49:21–29.

Mignouna, H.D., T.N.H. Ellis, M. Knox, R. Asiedu, and Q.N. Ng. 1998. Analysis of genetic diversity in Guinea yams. (*Dioscorea* spp.) using AFLP fingerprinting. Trop. Agric. (Trinidad) 75:224–299.

Mignouna, H.D., R.A. Mank, T.H.N. Ellis, N. van den Bosch, R. Asiedu, M.M. Abang, and J. Peleman. 2002b. A genetic linkage map of water yam (*Dioscorea alata* L.) based on AFLP markers and QTL analysis for anthracnose resistance. Theor. Appl. Genet. 105:726–735.

Mignouna, H.D., R.A. Mank, T.H.N. Ellis, N. van den Bosch, R. Asiedu, S.Y.C. Ng, and J. Peleman. 2002c. A genetic linkage map of Guinea yam (*Dioscorea rotundata* Poir.) based on AFLP markers. Theor. Appl. Genet. 105:716–725.

Mignouna, H.D., P. Njukeng, M.M. Abang, and R. Asiedu. 2001b. Inheritance of resistance to *Yam Mosaic Virus*, genus *Potyvirus*, in white yam (*Dioscorea rotundata*). Theor. Appl. Genet. 103:1196–1200.

Nassar. 2000a. Cassava, *Manihot esculenta* Crantz, genetic resources: their collection, evaluation and manipulation. Adv. Agron. 69:179–230.

Nassar, N.M.A. 2000b. Keeping options alive: a case for averting the extinction of wild cassava in South and Central America. Diversity 16(4):21–23.

Nassar, N.M.A. 2001a. Cassava, *Manihot esculenta* Crantz and wild relatives: their relationships and evolution. Genet. Res. Crop Evol. 48:429–436.

Nassar, N.M.A. 2001b. The nature of apomixis in cassava (*Manihot esculenta* Crantz). Hereditas 134:185–187

New Scientist. 2003. 18 January: 26–29.

Nishiyama, I., T. Miyazaki and S. Sakamoto. 1975. Evolutionary autoploidy in the sweet potato (*Ipomoea batatas* (L.) Lam.) and its progenitors. Euphytica 24:197–208.

Nweke, F.I., J.K. Lynam and D.S.C. Spencer. 2002. The Cassava Transformation: Africa's Best-Kept Secret. Michigan State University Press, East Lansing, MI.

Okada, Y., T. Murata, A. Saito, T. Kimura, M. Mori, M. Nishiguchi, K. Handa, J. Sakai, T. Miyasaki, Y. Matsuda, and H. Fukuoka. 2001. Virus resistance in transgenic sweetpotato.

Orjeda, G. 1989. Development of 4*x* interspecific sweetpotato hybrids as storage root initiators of diploid and tetraploid wild species. CIP Circular 17:12–13.

Orjeda, G., R. Freyre, and M. Iwanaga. 1990. Production of 2n pollen in diploid *I. trifida*, a putative wild ancestor of sweet potato. J. Hered. 81:462–467.

Ortiz, R. 1997a. Breeding for potato production from true seed. Plant Breeding Abstr. 67:1355–1360.

Ortiz, R. 1997b. Morphological variation in *Musa* germplasm. Genet. Res. Crop Evol. 44:393–404.

Ortiz, R. 1997c. Secondary polyploids, heterosis and evolutionary crop breeding for further improvement of the plantain and banana genome. Theor. Appl. Genet. 94:1113–1120.

Ortiz, R. 1998a. Potato breeding via ploidy manipulations. Plant Breed. Rev. 16:15–86.

Ortiz, R. 1998b. AMMI and stability analyses of bunch mass in multilocational testing of *Musa* germplasm in sub-Saharan Africa. J. Amer. Soc. Hort. Sci. 123:623–627.

Ortiz, R. 2000. Understanding the *Musa* genome: an update. Acta Horticulturae 540:157–168.
Ortiz, R. 2001. Dedication: Dirk R. Vuylsteke: *Musa* scientist and humanitarian. Plant Breed. Rev. 21:1–25.
Ortiz, R. 2003a. Plant breeding successes in African rural development. Plant Breeding News 137:3.
Ortiz, R. 2003b. Analytical breeding. Acta Horticulturae 622:235–247
Ortiz, R. 2004. Biotechnology with horticultural and agronomic crops in Africa. Acta Horticulturae 642:43–56.
Ortiz. R. 2005. Bananas and plantains. In J. Janick (ed). Enc. Fruit & Nuts. CABI, Wallingford, United Kingdom. In press.
Ortiz, R., P.D. Austin, and D.Vuylsteke. 1997a. IITA High Rainfall Station: 20 years of research for sustainable agriculture in the West African humid forest. Hort.Science 32:969–972.
Ortiz, R. K. Craenen, and D. Vuylsteke. 1997b. Ploidy manipulations and genetic markers as tools for analysis of quantitative trait variation in progeny derived from triploid plantains. Hereditas 126:255–259.
Ortiz, R., and J.H. Crouch 1997. The efficiency of natural and artificial pollinators in plantain (*Musa* spp. AAB group) hybridisation and seed production. Ann. Bot. 80:693–695.
Ortiz, R., and M.K. Ehlenfenldt. 1992. The importance of endosperm balance number in potato breeding and the evolution of tuber bearing Solanums. Euphytica 60:105–113.
Ortiz, R., L. Frusciante, and D. Carputo. 2005. Stanley J. Peloquin: potato geneticist and cytogeneticist. Plant Breed. Rev. 25:1–19
Ortiz, R., and A.M. Golmirzaie. 2002. Hierarchical and factorial mating designs in quantitative genetics of tetrasomic potato. Theor. Appl. Genet. 104:675–679.
Ortiz, R., and Z. Huamán. 2001. Allozyme polymorphism in tetraploid potato gene pools and the effect of human selection. Theor. Appl. Genet. 103:792–796.
Ortiz, R., S. Madsen, and D. Vuylsteke. 1998. Classification of African plantain landraces and banana cultivars using a phenotypic distance index of quantitative descriptors. Theor. Appl. Genet. 96:904–911.
Ortiz, R., and S.J. Peloquin. 1991. A new method of producing inexpensive 4*x* hybrid true potato seed. Euphytica 57:103–108.
Ortiz, R., and D. Vuylsteke. 1995. Effect of the parthenocarpy gene *P1* and ploidy in bunch and fruit traits of plantain and banana hybrids. Heredity 75:460–465.
Ortiz, R., and D. Vuylsteke 1998a. 'BITA-3': a starchy banana with partial resistance to black sigatoka and tolerance to streak virus. HortScience 33:358–359.
Ortiz, R., and D. Vuylsteke 1998b. 'PITA-14': a black sigatoka resistant tetraploid plantain hybrid with virus tolerance. HortScience 33:360–361.
Ortiz, R., and D. Vuylsteke. 1996. Recent advances in *Musa* genetics, breeding and biotechnology. Plant Breeding Abstr. 66:1355–1363.
Ortiz, R., D. Vuylsteke, H.K. Crouch, and J.H. Crouch. 1998. TM3x: triploid black sigatoka resistant *Musa* hybrid germplasm. HortScience 33:362–365.
Ortiz, R., D. Vuylsteke, R.S.B. Ferris, J.U. Okoro, A. N. Guessan, O. B. Hemeng, D.K. Yeboah, K. Afreh-Nuamah, E.K.S. Ahiekpor, E. Foure, B.A. Adelaja, M. Ayodele, O.B. Arene, F.E.O. Ikiediugwu, A.N. Agbor, A.N. Nwogu, E. Okoro, G. Kayode, I.K. Ipinmoye, S. Akele, and A. Lawrence. 1997c. Developing new plantain cultivars for Africa. Plant Var. & Seeds 10:39–57.
Osuji, J.O., G. Harrison, J.H. Crouch, and J.S. Heslop-Harrison. 1997. Identification of the genomic constitution of *Musa* L. genotypes (bananas, plantains and hybrids) using molecular cytogenetics. Ann. Bot. 80:787–793.
Otim-Nape, G.W., T. Alicai, and J.M. Tresh. 2001. Changes in the incidence and severity of cassava mosaic virus disease, varietal diversity and cassava production in Uganda. Ann. Appl. Biol. 138:313–327.
Pillay, M., D.C. Nwakanma, and A. Tenkouano. 2000. Identification of RAPD markers linked to A and B genome sequences in *Musa*. Genome 43:763–767.
Pillay, M., E. Ogundiwin, D.C. Nwakanma, G. Ude, and A. Tenkouano. 2001. Analysis of genetic diversity and relationships in east African banana germplasm. Theor. Appl. Genet. 102:965–970.
Qaim, M. 1998. Transgenic virus resistance potatoes in Mexico: potential social implications of North-South biotechnology transfer. ISAAA Briefing No. 7. International Service for Acquisition of Agri-Biotech Applications, Ithaca, N.Y.
Ramser J., C. Lopez-Peralta, R. Wetzel, K. Weising, and G. Kahl. 1996. Genomic variation and relationships in aerial yam (*Dioscorea bulbifera* L.) detected by random amplified polymorphic DNA. Genome 39:17–25.
Ramser J., K. Weising, C. Lopez-Peralta, W. Terhalle, R. Terauchi and G. Kahl. 1997. Molecular-marker based taxonomy and phylogeny of Guinea yam (*Dioscorea rotundata-D. cayenensis*). Genome 40:903–915.
Sagi, L., B. Panis, S. Remy, S. Schoofs, K. De Smet, R. Swennen, and B. Cammue. 1995. Genetic transformation of banana (*Musa* spp.) via particle bombardment. Bio/Technology 13:481–485.
Sakamoto, S. 1970. Utilization of related species on breeding sweet potato in Japan. Japanese J. Agric. Res. Quarterly 5:1–4.
Sakamoto, S. 1976. Breeding a new sweet potato variety, Minamiyutaka, by the use of wild relatives. Japanese Agric. Res. Quarterly 10:183–186.
Shiotani, I. 1988. Genomic structure and the gene flow in sweetpotato and related species. Pp. 61–73. In Exploration, Maintenance and Utilization of Sweet Potato Genetic Resources. Centro Internacional de la Papa, Lima, Peru.
Takahata, Y. 1995. Varietal differences in storage root quality and physiological factors in sweetpotato. Japan Agric. Res. Quarterly 29:215–221.
Takahata, Y., T. Noda, and T. Nagata. 1993. HPLC determination of beta-carotene content of sweet potato cultivars and its relationships with color values. J. Breed. 43:421–427.
Tanksley, S.D., M.W. Ganal, J.P. Prince, M.C. de Vicente, M.W. Bonierbale, P. Broun, T.M. Fulton, J.J. Giovanonni, S. Grandillo, G.B. Martin, R. Messeguer, J.C. Miller, A.H. Paterson, O. Pineda, M.S. Roder, R.A. Wing, W. Wu, and N.D. Young. 1992. High density molecular linkage maps of the tomato and potato genomes. Genetics 132:1141–1160.
Tenkouano, A., J.H. Crouch, H.K. Crouch, and R. Ortiz. 1998a. Genetic diversity, hybrid performance and combining ability for yield in *Musa* germplasm. Euphytica 102:281–288.
Tenkouano, A., J.H. Crouch, H.K. Crouch, D. Vuylsteke, and R. Ortiz. 1999a. A comparison of DNA marker and pedigree methods for genetic analysis in plantain and banana (*Musa* spp.) clones. I. Estimation of genetic relationships. Theor. Appl. Genet. 98:62–68.
Tenkouano, A., J.H. Crouch, H.K. Crouch, D. Vuylsteke, and R. Ortiz. 1999b. A comparison of DNA marker and pedigree methods for genetic analysis in plantain and banana (*Musa* spp.) clones. II. Predicting hybrid performance. Theor. Appl. Genet. 98:69–75.
Tenkouano, A., R. Ortiz, and D. Vuylsteke. 1998b. Combining ability for yield and plant phenology in plantain-derived populations. Euphytica 104:151–158.
Tenkouano, A., D. Vuylsteke, J. Okoro, D. Makumbi, R. Swennen, and R. Ortiz. 2003. Registration of TMB2x 5105-1 and TMB2x 9128-3: diploid banana hybrids with good com-

bining ability, partial resistance to black sigatoka and resistance to nematodes. Hort. Science 38:468–472.

Terauchi, R., V. Chikaleke, G. Thottappilly, and S.K. Hahn. 1992. Origin and phylogeny of Guinea yams as revealed by RFLP analysis of chloroplast DNA and nuclear ribosomal DNA. Theor. Appl. Genet. 83:743–751.

Ude, G., M. Pillay, D. Nwakanma, and A. Tenkouano 2002a. Analysis of genetic diversity and sectional relationships in *Musa* using AFLP markers. Theor. Appl. Genet. 104: 1239–1245.

Ude, G., M. Pillay, D. Nwakanma, and A. Tenkouano 2002b. Genetic diversity in *Musa acuminata* Colla and *Musa balbisiana* Colla and some of their natural hybrids using AFLP markers. Theor. Appl. Genet. 104:1246–1252.

Ukoskit, K., and P.G. Thompson. 1997. Autopolyploidy versus allopolyploidy and low-density randomly amplified polymorphic DNA linkage maps of sweetpotato. J. Amer. Soc. Hort. Sci. 122:822–828.

Vuylsteke, D.R., and R. Ortiz. 1995. Plantain-derived diploid hybrids (TMP2x) with black sigatoka resistance. HortScience 30:147–149.

Vuylsteke, D.R., R. Ortiz, R.S.B. Ferris, and J.H. Crouch. 1997. Plantain improvement. Plant Breed. Rev. 14:27–320.

Vuylsteke, D.R., R. Ortiz, R.S.B. Ferris, and R.L. Swennen. 1995. 'PITA-9': A black sigatoka resistant hybrid from the 'False Horn' plantain gene pool. HortScience 30:395–297.

Vuylsteke, D.R., R. Swennen, and R. Ortiz. 1993. Registration of 14 improved tropical *Musa* plantain hybrids with black sigatoka resistance. HortScience 28:957–959.

Wambugu, F.M. 2001. Modifying Africa: how biotechnology can benefit the poor and hungry, a case study from Kenya. A Harvest Biotech Foundation International, Nairobi, Kenya. 76 pp.

Watanabe, K.N., R. Ortiz, and J. Watanabe. 2004. Pp. 83–99. Breeding potential and combining ability in 4x-2x crosses. In Razdan, M.K. and A.K. Mattoo (eds.) Genetic Improvement of Solanaceous Crops, Vol. 1: Potato. Science Publishers, USA.

Woolfe, J.A. 1992. Sweet potato: an untapped food resource. Cambridge Univ. Press, UK & Centro Internacional de la Papa, Lima, Peru. 643 pp.

Zhang, D.P., J. Cervantes, Z. Huamán, E. Carey, and M. Ghislain. 2000. Assessing genetic diversity of sweetpotato (*Ipomoea batatas* (L.) Lam.) varieties from tropical America using AFLP. Genet. Res. Crop Evol. 47:659–665.

Zhang, D.P., M. Ghislain, Z. Huamán, A. Golmirzaie, and R.J. Hijmans. 1998. RAPD variation in sweetpotato (*Ipomoea batatas* (L.) Lam.) varieties from South America and Papua New Guinea. Genet. Res. Crop. Evol. 45:271–277.

Zhang, D.P., Z. Huamán, F. Rodríguez, and M. Ghislain. 2001. Identifying duplicates in sweet potato (*Ipomoea batatas* (L.) Lam.) varieties using RAPD. Acta Hort. 346:535–541.

Zoundjihékpon J., S. Hamon, B. Tio-Touré, and P. Hamon. 1994. First controlled progenies checked by isozymic markers in cultivated yams *Dioscorea cayenensis-rotundata*. Theor. Appl. Genet. 88:1011–1016.

19

Origins of Fruit Culture and Fruit Breeding

Jules Janick
Department of Horticulture and Landscape Architecture
Purdue University

Introduction

In the late Neolithic and Bronze Ages between 6000 and 3000 BCE, the ancient Mediterranean fruits (date, olive, grape, fig, sycomore fig, and pomegranate) were domesticated (Zohary and Spiegel-Roy, 1975). Fruits such as citrus, banana, various pome fruits (apple, pear, quince, medlar) and stone fruits (almond, apricot, cherry, peach, and plum) were domesticated in Central and East Asia and reached the West in antiquity. A number of fruits and nuts were domesticated only in the nineteenth and twentieth centuries (blueberry, blackberry, pecan, and kiwifruit). Some well-known fruits, although extensively collected, remain to be domesticated, such as lingonberry, various cacti, such as pitaya, Brazilnut, and durian. This review will consider the various technologies inherent in the origins of some well-known fruits, emphasizing factors that led to domestication and the genetic changes that ensued. The origin and changes associated with domestication of some well-known fruits is shown in Table 19.1.

Childe (1958) proposed that a second Neolithic Revolution coinciding with the Bronze Age occurred between 6000 and 3000 BCE involving the change from villages to urban communities. The evolution of urban centers is associated with the development of a settled agriculture. This coincides with the beginning of fruit culture, which involved a long-term commitment to a unique piece of ground. In the case of the date and olive, a fruit orchard can remain productive for over a century. It is fruit culture that bonds humans to a particular piece of land and may be a link associated with the concept of territoriality, the development of city–states, and eventually nationhood.

Information on the ancient origins of fruit culture comes from archeological remains of fruit and from pictorial and literary evidence. The high culture of Mesopotamia and Egypt produced a rich art in which fruit is a common motif. A trove of paintings and sculpture is found in Egyptian tombs and monuments. The Sumerian discovery of writing in the third millennium BCE, and Egyptian writings somewhat later, inaugurated the literary tradition that survives today as a result of the near indestructibility of the baked clay tablets used for cuneiform script, the wide use of stone carving for hieroglyphics, and the preservation of papyrus in desert tombs.

Zohary and Spiegel-Roy (1975) proposed that fruit culture, in contrast with mere collection, originated 4000–3000 BCE. Although some information before this period is based on archeological remains, much of it is by inference and conjecture. Perhaps the earliest pictorial evidence of fruit growing occurs on a 1-m tall alabaster vessel known as the Uruk vase found in Jemdet Nasr levels at Uruk dating from about 3000 BCE. Uruk (Erech) is on the Euphrates just north of Basra, Iraq. The imagery depicts water at the bottom of the vase, followed by plants (barley and sesame) and domestic animals, and men bearing baskets of fruit with offerings presented to a female, perhaps the Goddess Innana, later known as Istar (Bahrani, 2002). Unfortunately, the fruits cannot be identified, but they tend to be large and of various shapes. Predynastic drawings of fruit trees in Egypt depict the date palm.

Table 19.1 Genetic changes associated with domestication of fruits

Fruit crop	Species (Chromosome no.)	Family	Origin	Reproduction of wild species	Changes associated with domestication
Mediterranean Fruits					
Date palm	*Phoenix dactilifera* $x=18, 2n=36$	Arecaceae	S. Mediterranean basin	Dioecious	Offshoot production, increased fruit size
Fig—common	*Ficus carica* $x=13, 2n=26$	Moraceae	E. Mediterranean basin	Gynodioecious	Parthenocarpy
Fig—sycomore	*Ficus sycomorus*	Moraceae	East	Monoecious	Parthenocarpy
Grape	*Vitis vinifera* $x=19, 2n=38$	Vitaceae	W. Asia	Dioecious	Hermaphroditic, increased berry size, parthenocarpy
Olive	*Olea europea* $x=23, 2n=46$	Oleaceae	Mediterranean basin	Andromonecious	Increased fruit size, high oil
Pomegranate	*Punica granatum*	Punicaceae	W. Asia	Hermaphroditic	Increased fruit size, sweetness
Asian and European Fruits					
Pome fruits					
Apple	*Malus* × *domestica* $x=17, 2n=34, 51$	Rosaceae	Central Asia	Hermaphroditic, self-incompatible	Combination of size aroma, loss of astringency, sweetness, parthenocarpy, triploidy
Pear—European	*Pyrus communis* $x=17, 2n=34, 51$	Rosaceae	Central Asia	Hermaphroditic, self-incompatible	Combination of size, aroma, loss of astringency, sweetness, parthenocarpy, triploidy
Pyrus–Asian	*Pyrus pyrifolia* *P. bretscheiderii* *P. ussuriensies* $x=17, 2n=34, 51$	Rosaceae	E. Asia	Hermaphroditic, self-incompatible	Combination of size, aroma, loss of astringency, sweetness, parthenocarpy, triploidy
Stone fruits					
Almond	*Prunus dulcis* $x=8, 2n=16$	Rosaceae	SW Asia	Hermaphroditic, self-incompatible	"Sweet" seed, increased kernel size, self-fertility
Apricot	*Prusus armeniaca* $x=8, 2n=16$	Rosaceae	Central & E. Asia	Hermaphroditic, self-incompatible	Increased fruit size
Cherry—sweet	*Prunus avium* $x=8, 2n=16$	Rosaceae	Central Europe & W. Asia	Hermaphroditic, self-incompatible	Self-compatibility (recent)
Cherry—tart	*Prunus cerasus* $x=8, 2n=16, 32$	Rosaceae	W. Asia	Hermaphroditic, self-incompatible	Tetraploidy after interspecific hybridization
Plum–European	*Prunus domestica* $x=8, 2n=48$	Rosaceae	Europe	Hermaphroditic, self-incompatible	Hexaploid after interspecific hybridization
Plum—Asia	*Prunus salicina* $x=8, 2n=16, 32$	Rosaceae	China	Hermaphroditic	Increased fruit size
Plum—American	*Prunus Americana* $x=8, 2n=16$	Rosaceae	North America	Hermaphroditic	Increased fruit size
Peach	*Prunus persica* $x=8, 2n=16$	Rosaceae	China	Hermaphroditic	Freestone, low chill, fuzzless (nectarine), increased size
Vine fruit					
Kiwifruit	*Actinidia delicious* $x=29, 2n=174$ *A, sinensis* $x=29, 2n=58$	Actinidiaceae	China	Dioecious	Unchanged
Subtropical and tropical fruits					
Citrus (orange, mandarin, lemon, lime, pumello, grapefruit)	*Citrus* spp. $x=9, 2n=18$	Rutaceae	Southeast Asia, China	Hermaphroditic	Nucellar embryony, interspecific hybridization, parthenocarpy
Mango	*Mangifera indica* $x=20, 2n=40$	Anacardiaceae	E. Asia	Hermaphroditic	Nucellar embryony, loss of fibers in fruit
Persimmon Asian	*Diospyros kaki* $x=15, 2n=90$	Ebenaceae	China	Polygamodioecious	Loss of astringency, parthenocarpy

Table 19.1 *(continued)*

Fruit crop	Species (Chromosome no.)	Family	Origin	Reproduction of wild species	Changes associated with domestication
American Fruits					
Berry fruits					
Strawberry	*Fragaria* × *ananassa* x=7, 2n=56 Other spp. 2n=14, 28	Rosaceae	Americas	Dioecism	Hermaphroditic, interspecific hybridization
Raspberry	*Rubus idaeus* (red) *R. occidentalis* x=7, 2n=14	Rosaceae	Europe, America	Hermaphroditic	Interspecific hybridization, polyploidy
Blackberry	*Rubus* spp. x=7, 2n=28,35, 42,56,84	Rosaceae	N. America	Hermaphroditic	Interspecific hybridization, polyploidy, thornlessness
Blueberry	*Vaccinium* spp.	Ericaceae	N. America	Hermaphroditic	Increased fruit size, interspecific hybridization, polyploidy
Cranberry	*Vaccinium macroacarpon*	Ericaceae	E. United States	Hermaphroditic	Unchanged
Lingonberry	*Vaccinium vitis-idaea*	Ericaceae	Circumboreal	Hermaphroditic	Unchanged
Subtropical and tropical fruits					
Avocado	*Persea americana* x=12, 2n=24	Lauraceae	Tropical America	Hermaphroditic, synchronous protogynous dicogamy	High oil, smaller seed size
Papaya	*Carica papaya* x=9, 2n=18	Euphorbiacea	Tropical America	Dioecious	Polygamodioecious, reduced fruit size
Pineapple	*Ananas comosus* x=25, 2n=50	Bromeliaceae	Tropical America	Hermaphroditic	Parthenocarpy, seedlessness

Fruit culture and the horticultural arts

Fruit growing involves a more complicated technology than the cultivation of herbaceous annuals such as cereals or pulse crops. Tree crop culture requires a long-term series of horticultural "craft secrets" more or less unique for each species. These include selection of unique clones, vegetative propagation (use of offshoots, cuttings, grafting), continuous irrigation in dry climates, pruning and training, pollination, harvesting, storage, and processing. The cycle of fruit growing is often a year-round activity and must involve orchard establishment in anticipation of production, which may only ensue after a number of years. Current additions to the technology of fruit growing include the use of dwarfing rootstocks, growth regulators, disease and pest control, long-term storage, protected cultivation, and biotechnology. Here we examine the Neolithic and Bronze Age origins of technologies essential to fruit growing.

Species selection

The development of fruit culture in the Fertile Crescent evolved at two loci: the Tigris–Euphrates civilization of Mesopotamia and the Nile valley culture of Egypt. The first cultivated fruits must have been indigenous species that had obvious human value. This is clearly seen in Egypt where the indigenous date palm was the earliest species cultivated, followed by a succession of introduced fruits such as the sycomore fig and pomegranate (Table 19.2). The earliest fruit culture in Mesopotamia included the date and olive (4000 BCE), grape, fig, and pomegranate (third millennium BCE). Later fruit introductions, based on literary sources, include the apple, pear, quince, and medlar (Postgate, 1987). The small-fruited *Malus orientalis* and *Pyrus syriaca* are indigenous to the Fertile Crescent, but these were probably not the forerunners of domesticated apple and pear that were introduced from Western Asia probably via Persia. Contacts between East and West date from as early as 1000 BCE as evidenced by silk strands on Egyptian mummies, but intensified with the incursions of Alexander the Great (356–323 BCE). Thus, by Greek and Roman eras, there was an infusion of Central and East Asian fruits, including the citron and a great variety of stone fruits, including almond, apricot, cherry, peach, and plum. By classical times in Greece and Rome, fruit cul-

Table 19.2 Evidence for fruit crops in Egypt

Fruit crop	Binomial	Earliest record (dynasty or period)	Type of evidence
Date palm	*Phoenix dactylifera*	Pre-dynastic	Archeological
Doum palm	*Hyphaene thebaica*	Pre-dynastic	Archeological
Sycomore fig	*Ficus sycomorus*	Pre-dynastic	Archeological
Jujube (Christ's thorn)	*Ziziphus spina-Christi*	I (Old Kingdom)	Archeological
Fig	*Ficus carica*	II (Old Kingdom)	Artistic
Grape	*Vitis vinifera*	II (Old Kingdom)	Archeological
Hegelig	*Balanites aegyptiaca*	III (Old Kingdom)	Archeological
Persea (lebakh)	*Mimusops shimperi*	III (Old Kingdom)	Archeological
Argun palm	*Medemia argun*	V (Old Kingdom)	Archeological
Carob	*Ceratonia siliqua*	XII (Middle Kingdom)	Archeological
Pomegranate	*Punica granatum*	XII (Middle Kingdom)	Archeological
Egyptian plum	*Cordia myxa*	XVIII (New Kingdom)	Archeological
Olive	*Olea europea*	XVIII (New Kingdom)	Archeological
Apple	*Malus* × *domestica*	XVIIII (New Kingdom)	Literary
Peach	*Prunus persica*	Graeco-Roman	Archeological
Pear	*Pyrus communis*	Graeco-Roman	Archeological
Cherry	*Prunus avium; P. cerasus*	5 BCE	Literary
Citron	*Citrus medica*	2nd century CE	Literary

Source: Janick (2002a) adapted from Darby et al. (1976).

ture had achieved a sophisticated level, not exceeded for over a millennium.

In the Age of Exploration in the late fifteenth and sixteenth centuries, a great exchange took place as fruits of the Americas, including the pineapple, cacao, American species of strawberries, and papaya, and the fruit-bearing solanums such as tomato and pepper, reach Europe, Asia, and Africa, while East Asian fruits, such as the banana, mango, and persimmon, reach the Americas. Some Asian fruits, such as kiwifruit, are relatively recent introductions, and a great many tropical fruits of both Asia (durian, mangosteen, salaak) and the Americas (passion fruit, sapote) are still not extensively commercialized.

Vegetative propagation

Although most fruit species can be produced from seed (except, of course, seedless clones), this is usually an inappropriate technique. Most fruit species are highly cross-pollinated (peach is an exception) and therefore highly heterozygous. Thus, open-pollinated seedlings will consist of a highly heterogeneous mixture of fruit types, most of which will be inferior to the selected clone. Furthermore, there is a long juvenile period in trees grown from seed. The Hebrew Bible is full of references to degenerate plants that clearly indicate that the ancients were aware of the hazards of seed-propagated fruit crops. Thus, the basis of most fruit cultivation is vegetative propagation of unique phenotypes (elite clones) with subsequent improvement based on sexual recombination of elites. Vegetative propagation is accomplished from offshoots in date, layers in grape, cuttings in olive and fig, and runners in strawberries. Fruit crops that can be easily propagated vegetatively have been considered preadapted for domestication by Zohary and Spiegel-Roy (1975). However, many temperate fruits (apple and pear) do not propagate easily by layers or cuttings and are currently multiplied by graftage.

Grafting is ancient (Vöchting, 1892; Mendel, 1953), and Childe (1958) has speculated that it was known before 3000 BCE. Although both root and shoot grafting occur naturally, the technology is not obvious and must be considered one of the horticultural craft secrets. We know this because much of the Roman writings on grafting often confuse which species are graft compatible. Pliny describes a number of ludicrous combinations, such as apple on plane tree, suggesting that he was not writing from real experience.

Recently, a cuneiform description of budwood importation for grape has been uncovered from Mari, Mesopotamia (Harris et al., 2002), dated to about 1800 BCE, which confirms Childe's (1958) speculation on the antiquity of graftage. The next written evidence of grafting comes from the school of Hippocrates (Pseudo-Hippocrates, about 450 BCE) that discusses the graft union (Meyer, 1854), but this reference implies that the technique was very much older. There is speculation that grafting was known in China as early as 1560 BCE (Nagy et al., 1977; Hartmann et al., 1997), but the earliest definitive evidence for grafting occurs in the first century BCE for the bottle gourd (*Lagenaria*) in *The Book of Fan Sheng*, while fruit grafting is referred to in *Qi Min Yao Suj* written by Jia Simiao in the sixth century CE (Guangshu Liu, pers. commun.). Grafting is discussed in detail by Theophrastus, and all the Roman agricultural writers, including Cato, Virgil, Columella, and Pliny, describe it in detail. Grafting is accurately pictured in mosaics in the third century CE. Grafting is not specifically mentioned in the Hebrew Bible but is inferred based on Jewish writings [Mishna in Order Zeraim (seeds), tractate Kilaim written circa third century CE] interpreting prohibitions against mixing of seeds in Leviticus 19:19. Grafting of olive is found in the Christian Bible (*Romans* 11:17, 24), first century):

> And if some of the branches be broken off, and you, being a wild olive tree, were grafted in among them, and with them became a partaker of the root and fatness of the olive tree do not boast against the branches. . . . But if you boast, remember that you do not support the root, but the root supports you. . . . For if you were cut out of the olive tree which is wild by nature, and were grafted contrary to nature into a cultivated olive tree, how much more shall these, which are natural branches, be grafted into their own olive tree.

The development of graftage must have influenced the movement of temperate fruits, such as the apple, from Central Asia to Europe. Grafting is then a pivotal technology in the history of temperate fruits. Precisely when and where detached scion grafting, which made possible the domestication of a new range of fruit trees, was invented is not clear. Barrie Juniper (pers. commun.) has suggested that the initiation of grafting was outside the area of Mediterranean horticulture and was probably introduced from the east, perhaps Persia.

Fruit growing has long been associated with clonal propagation of unique wild seedlings, with subsequent evolutionary progress derived from intercrosses of superior clones plus intercrosses with wild races (Zohary and Spiegel-Roy, 1975), leading to very high seedling diversity. In a number of species, almond for example, the high diversity of wild clones includes segregates that are close to domestic types. This would explain why the ecological adaptation of classic Mediterranean fruits has not exceeded the requirements of their wild ancestors. In most cases, present-day fruit cultivars have undergone far fewer sexual cycles than cereal or pulse crops, and some may be only a few generations from wild clones. Thus, many of these crops have not diverged from their progenitors, in contrast to cereals in which selection has operated for thousands of generations. This has been confirmed in apples, where present-day cultivars are not distinct from elite selections obtained from wild stands in Alma Alta, Kazakhstan (Harris et al., 2002; Forsline et al., 2003).

Pollination and fruit set

Practically all fruit species are naturally outcrossing with variability maintained by natural barriers to avoid self-fertilization. These include dioecy (date palm, grape, fig, strawberry, kiwifruit, papaya), self-incompatibility (pome and stone fruits), and dichogamy, the uneven maturation and receptivity of pistils and stamens (avocado, lychee). In some fruits, cross-pollination is based on unique adaptation with insects (fig) or birds (pineapple).

In dioecious species, accommodation for pollination is necessary. It would be immediately obvious that staminate clones would be nonfruiting. Mass plantings of vegetatively propagated elite pistillate clones would bear few fruit and require either a limited number of staminate plants, proximity to wild pollenizers, or artificial pollination.

This limitation to productivity was solved in various ways—some genetic, some cultural—in different fruit crops. In date palms, early farmers discovered artificial pollination, and this is clearly illustrated in Assyrian bas reliefs, with the practice codified in the Laws of Hammurabi, circa 1750 BCE (Roth 2000):

> §64. If a man give his orchard to a gardener to pollinate (the date palms), as long as the gardener is in possession of the orchard, he shall give to the owner of the orchard two thirds of

> the yield of the orchard, and he himself shall take one third.
> §65 If the gardener does not pollinate the (date palms in the) orchard and thus diminishes the yield, the gardener (shall measure and deliver) a yield of the orchard to (the owner of the orchard in accordance with) his neighbor's yield.

In the fig, the presence of the wild monoecious caprifig that harbored the pollinating blastid fig moth (caprification) was understood to be essential for fig production by Theophrastus (371–287 BCE), but later selection for parthenocarpy eliminated this practice. In grape, strawberry, and papaya, domestication exploited mutations from dioecism to hermaphroditism, and, in some figs and grapes, parthenocarpy reduced the requirement for pollination altogether. In the sycomore fig, introduced to Egypt without the pollinating fig wasp, artificial wounding to ripen parthenocarpic fruit was the solution. In apple and pear, which are self-incompatible and insect pollinated, the problem was solved with the introduction of bees to facilitate cross-pollination along with interplanting of pollenizers. In recent times, pollenizers for self-incompatible sweet cherry were eliminated by introducing self-compatible mutations.

Irrigation

The civilizations that developed in the arid climates of the Fertile Crescent are dominated by large rivers—the Nile in Egypt and the Tigris and Euphrates in Mesopotamia. In Egypt the regular inundations of the Nile, rising in July until the middle of October, followed by rapid subsidence, permitted a unique horticulture based on basin irrigation (Janick, 2002a). The system involved a system of dikes to retain the flood and encourage infiltration into the soil. Earthen banks, parallel to the river, together with intersecting banks, created a checkerboard of dike-enclosed areas, between 400 and 1600 ha each. Canals led the water to areas difficult to flood. The flood waters ran through a series of regulated sluices into each basin, flooding the land to a depth of 0.3–1.8 m. The water could be held for a month or more; the surplus was drained to a lower level and then returned to canals that emptied into the Nile. The advantage of basin irrigation was that no further irrigation was needed for a winter crop of grain, and the silt, rich in organic matter and phosphates, made fertilization unnecessary.

With fruit tree culture, permanent ponds were an important innovation, and the ornamental gardens enclosing ponds testify to their widespread use by the wealthy. In addition, shallow wells, 4–35 m in depth, were dug to be replaced later by deeper artesian wells up to 380 m deep. The culture of fruit crops demands constant and controlled irrigation during the spring and summer drought. At first, irrigation was carried out manually with pots dipped in the rivers, carried on the shoulders with yokes, and poured into field channels. By the time of the New Kingdom (1500 to 1100 BCE), the shaduf, a balanced counterpoise, became the irrigating mechanism for gardens. Later, water-lifting techniques included Archimedes' screw, the sakieh or chain of pots, and siphons.

In Mesopotamia, cultivation in the Tigris–Euphrates flood plain is and always has been dependent on irrigation, and the management of this technology may have been the impetus for the development of nation-states (Pollock, 1999). Irrigation started as small-scale projects but eventually increased in complexity and involved centralized control. The creation of state-controlled irrigation led to a strong central authority requiring conscripted service (corvée) for canal maintenance. The Laws of Hammurrabi richly describe a legal system enforced to maintain the integrity of an irrigated agriculture:

> §53 If a man neglects to reinforce the embankment of the irrigation canal of his field and does not reinforce its embankment, and then a breach opens in its embankment and allows the water to carry away the common irrigated area, the man in whose embankment the breach opened shall replace the grain whose loss he caused.
> §54 If he cannot replace the grain, they shall sell him and his property, and the residents of the common irrigated area whose grain crops the water carried away shall divide the proceeds.
> §55 If a man open the branch of the canal of irrigation and negligently allows the water to carry away his neighbor's field, he shall measure and deliver grain in accordance with his neighbor's yield.
> §56 If a man opens an irrigation gate and releases waters and thereby he allows the water

to carry away whatever work has been done in his neighbor's field he shall measure and deliver 3,000 sila of grain per 18 iku of field.

Because of the braiding character of the Euphrates, short canals, about 1 km in length, could be dug from the numerous river channels and managed by local groups (Pollock, 1999). The natural flow of the river and overflow resulted in natural levees, and in the process the riverbed was gradually raised until it flowed above the level of the surrounding land. This made it relatively easy to cut irrigation channels through the natural levee and allow the water to flow by gravity to cultivated fields and gardens. The natural levees with their good drainage were prized for fruit tree cultivation, but irrigation required water-lifting technology. The natural vegetation of the alluvial plain provided pasturage for sheep and goats; it was once home to game animals such as jackals, lions, gazelles, onagers, and hyenas (as illustrated in the hunting scenes in Babylonian bas reliefs), now hunted to extinction. Long-term irrigation, however, led to unintended consequences, and today much of the area is a vast salty waste as a result of salinization.

Pruning and training

The art of fruit growing is associated with physical techniques to control the shape, size, and direction of plant growth. These include the orientation of the plant in space (training) and judicious removal of plant parts (pruning). In date palm culture the removal of senescing leaves, necessary for both pollination and harvest, is probably the basis of the pruning technique. The dead leaves had a wide variety of uses for shade and basketry. In the case of the grape, vine pruning is essential to control both flowering and yield and to increase fruit quality. Early Babylonian bas reliefs show grapes growing on trees, and the use of arbors and pergolas for grape is well illustrated in late Egyptian paintings. In the Middle East, an ancient training method involved severe pruning, whereby the plant was pruned in the fall and mounded with soil over the winter to avoid cold injury.

Processing and storage

Most of our important domesticated fruits are delicious in the mature state, and indeed this is one of the chief virtues of fruit crops. However, this is not the case with all fruits. The olive, in particular, is bitter and inedible, even in the mature state, and some wild species are toxic. Clearly, the key technology must have been derived from the ameliorating practice of soaking the fruits to make them less bitter. Many primitive societies, such as Amerinds with cassava and Australian aborigines with *Pandanas* (Crib and Crib, 1974), independently came to their detoxification techniques (Johns and Kubo, 1988). Extraction of oil by pressing the fruit transformed olives into the most widely grown fruit crop in the ancient world. The oil was widely used in medicine, cooking, and illumination; the flame with olive oil as a fuel has a very high luminosity. Fruit crops grown for their fruit or seed oil also include oilpalm and avocado (used for soap in Brazil).

Most fruits have a short life after harvest so that processing is required to have a year-round supply. The perishability of many fruits is one of the limiting factors of commercialization. In the case of the date, the high sugar content of the dried fruit permits extended storage, and dried grapes (raisins) have long been prized for their concentrated sweetness and long storage. Sun drying of many types of *Prunus*, especially highly sugared plums (prunes) and apricots, was facilitated by slicing fruits. The conservation of fruits as jams and preserves was based on the addition of sucrose, a substance unknown until the technology of sugarcane production and processing was developed in the Middle Ages. Some wild fruits, such as lingonberry in Nordic countries, have long been collected and stored as preserves. The preservation of fruits by heat (canning) was a nineteenth century technology developed by Nicolas Appert (1750–1841) as a response to the British blockade of France during the Napoleonic wars. Quick freezing and later freeze drying are twentieth-century technologies.

In more temperate climates, fruit life could be extended by common storage in caves or basements. Caves in Cappadocia (Turkey) maintain a temperature of 12.8°C and are still used to store lemons. The modern transformation of this technique led to refrigerated storage and low-oxygen (controlled-atmosphere) storage.

The transformation of fruit juice to an alcoholic product (wine), along with bread making, is a Neolithic discovery. Beer making probably predated wine making. This leap into Bronze Age biotechnology was facilitated by the ubiquitous presence of yeast spores. Although wine can be made from various fruits, the choice species is grape, probably because of its combination of sugars, acids, and tannins. At present, the greatest use of grapes is for wine manufacture. In the East, salting and fermentation technologies were developed as a means of whole-fruit storage. A few tropical fruits (e.g., plantain and breadfruit) are staple starch crops and require cooking.

Genetic changes associated with fruit domestication

Despite some well-known exceptions, fruit crops are characterized by a number of common features. Noteworthy is the obvious appeal of taste—often a combination of sweetness and acidity, which many considered delicious because of aromatic constituents. The appealing sweet taste of many fruits is a naturally selected trait associated with seed dispersal mediated by mammals. Most fruit crops are highly cross-pollinated, and tree fruits generally have long juvenility and long life. Some fruit crops have the ability to propagate vegetatively. Subsequent progress in the improvement of fruit crops resulted from continual selection of seedling populations and from intercrosses among elite clones, or with wild or introduced clones that vastly speeded up the process. This process has been very efficient, and despite progress in plant breeding, replacing grower-selected clones has not been easy.

The origins of fruit growing evolved from an interaction of genetic changes and cultivation technology, often unique for each species. Some idea of how this has occurred can best be inferred from the history of two recent domesticates: cranberry and kiwifruit. What occurred in these crops probably occurred in the past with others, although each crop is unique with its own set of problems and prospects, and each has its own story. Both cranberry and kiwifruit were widely appreciated and entered commerce from wild stands long before domestication. The cranberry had been collected in North America since colonial America, but only became cultivated in the nineteenth century. Successful cultivation involved developing a series of practices to grow a plant adapted to aquatic conditions. Cranberry cultivation has recently been adopted in Chile. The kiwifruit, a dioecious vine native to China but never cultivated there, has been appreciated since the eighth century in China and probably much earlier. It was introduced to England and North America in the beginning of the twentieth century, but New Zealand claims the honor of domestication. While the plant was introduced to England and the United States, the plant languished there, emphasizing the key role of champions (see below). A cultivation system worked out by New Zealand nurserymen and growers involved training and pruning on a trellis, with provision for pollination. The preferred pistillate and staminate clones (Hayward and Bruno, respectively) were selected from seed introduced into New Zealand from China. After the germplasm was selected, cultivation techniques established, and markets developed, the technology was quickly transferred, and kiwifruit became a world fruit crop in less than 25 years.

In both cranberry and kiwifruit, the early elite selections of wild plants were of high quality and could be vegetatively propagated—by cuttings in the case of cranberry and grafting in the case of kiwifruit. Selection, combined with the ability to fix unique combinations by vegetative propagation, was the key breeding technique in these two crops, as in all fruit crops. Breeding work has continued, but even after 100 years, the selections made very early still dominate the industry.

In both cranberry and kiwifruit, related species are under consideration as potential new crops. In kiwifruit, the related yellow-fleshed *Actinidia chinensis* has been introduced, and the small-fruited, hardy *A. arguta* (also known as tara fig) is under consideration as a new domesticate and now widely planted in northern home gardens. In the vacciniums, two related crops—blueberry (especially low-bush types in Maine) and lingonberry—were also widely appreciated and harvested from the wild, but with remarkably different outcomes. Blueberry had more promise as a commercial fruit than did cranberry or lingonberry because the fruit could be consumed fresh as well as processed and there was greater diversity in a number of species. While the domesticates of cranberry and kiwifruit have changed very little from their wild forms, the blueberry has undergone remarkable transformation due to interspecific hybridization and ploidy manipulation. The culture of blueberry was dependent on the understanding that the vacciniums are an acid-loving species and required the ammonium form of nitrogen. Intensive selection and breeding with various species of different ploidy levels transformed this crop into a relatively large industry of wide adaptation. Lingonberry, on the other hand, a large Scandinavia export crop from forest collection, never became domesticated, probably because there was no shortage of collectable fruit. This crop is still based on merely managed wild plantings.

Mutation as an agent of domestication

Many fruit crops differ from their wild progenitors by a few characters that have appeared as mutations (Table 19.3). Typically, these mutations are not advantageous to the plant in its natural setting because they reduce fitness, but would clearly have been immediately selected by humans. The changes from bitter to sweet seed in almond and seeded to seedless fruits along with parthenocarpy (banana and plantain, citrus, fig, grape, persimmon, and pineapple) would have negative fitness but very positive selective value. Parthenocarpy has two advantages: it eliminates the need for pollination, and it is one path to seedlessness that has proved important in grape, banana, and citrus. In dioecious fruit crops, mutations inducing hermaphrodism (strawberry, grape, and papaya) are associated with domestication. Other mutations associated with domestication include loss of spines (brambles, pineapple, pome fruits, and citrus), loss of fruit pubescence (peach), and changes in growth habit mutations (pome and stone fruits). In many fruit crops, fruit color mutations (sports) have become increasingly important, especially in apple, pear, and grapefruit. Many of these mutations are not heritable because they do not occur in the appropriate meristematic layer.

Interspecific hybridization and polyploidization

Many of our fruit crops derived from interspecific hybridization, polyploidization, or both (Table 19.3). This is particularly obvious in *Actinidia*, *Citrus*, *Fragaria*, *Musa*, *Prunus*, *Rubus*, and *Vaccinium*. The evolutionary divergence within these genera into different "species" is often associated with allopolyploidy. These changes represent the divergence of interbreeding populations that became isolated, known as nominalistic species (Spooner et al., 2003). Domestication within these groups and subsequent transfer by human migration would facilitate intercrosses. This development has been well worked out with bread wheat (*Tritcum vulgare*), a hexaploid amphidiploid of the genomic constitution AABBDD. Genomic analysis has identified the three species involved: AA from *Triticum urartu*, or einkorn; BB from *Aegilops speltoides*; and DD from *Aegilops tauschi*, or goat grass. The cross of emmer (AABB) with goatgrass (DD) is presumed to have occurred about 6000 BCE. This process occurred in many fruit crops (cherry, banana and plantains, citrus, brambles) as well as in deliberate chromosome manipulation (blueberry). Recently, interspecific hybridization has been used to create new fruits in citrus (tangors from mandarin × orange), *Prunus* (plumcot from plum × apricot), and *Rubus* (tayberry from blackberry × raspberry).

Table 19.3 Genetic changes associated with domestication in fruit crops

Breakdown of dioecy
Fig, grape, papaya, strawberry
(unchanged, date palm, kiwifruit)
Loss of self-incompatibility
Cherry
Parthenocarpy & seedlessness
Apple & pear, banana & plantain, citrus, fig, grape, persimmon, pineapple
Allopolyploidy
Banana & plantain, blackberry & raspberry, blueberry, citrus, tart cherry, European plum, strawberry
Polyploidy
Triploidy: Banana and plantain, apple, pear
Tetraploid : Tart cherry, raspberry, blackberry, blueberry, kiwifruit (Actinidia sinensis)
Hexaploid: European plums, kiwifruit (A. deliciosa)
Octaploid: Strawberry
Loss of toxic substances
"Sweet" seed: Almond
Nonastrigency: Apple & pear, persimmon, pomegranate
Ease of vegetative propagation
Offshoots: Date palm
Rooting: Apple (rootstock)
Nucellar embryony: Citrus, mango
Loss of spines, thorns, or pubescence
Apple, brambles, citrus, peach, pear, pineapple

Hybridization and selection

Origins of the apple

The apple is cross-pollinated, and most cultivars are self-incompatible. Most cuttings of apple and pear trees do not root from cuttings so that vegetative propagation is difficult without grafting. There are 24 primary species of *Malus*, distributed in Europe, Central Asia, and Eastern Asia, and three in North America (Way et al., 1990). Most of the wild apples are small and bitter. However, the large, sweet-smelling domestic apple (*Malus* × *domestica*) clearly originated in Central Asia, specifically Almaty, Kazakstan (Dzhangaliev, 2003), and was introduced to the West via Persia in antiquity. Recent explorations in Kazakhstan (Forsline et al., 2003) support Vavilov's suggestion that *Malus sieversii* is the major progenitor of the cultivated apple, and this has been confirmed by recent work with molecular markers (Harris et al., 2002). The explorations in Kazakhstan also indicate that elite wild material contains all the characters of modern apples, including size, quality, and various colors from red to yellow to green. Barrie Juniper (pers. commun.) has made the intriguing observation that the selection mechanism for large size may not have been human but rather the result of millions of years of selection for large size by bears, endemic to the area, who consume large numbers of apples; their droppings provide a unique germination medium. All of the other species of apple appear to be distributed by birds, providing no selection mechanism for large size.

Fruits of wild relatives of the domesticated apple occur in the Middle East and Europe and were frequently collected by Neolithic and Bronze Age farmers. However, it is difficult to know precisely when the larger, sweeter apples of Central Asia reached the West because reference to apple (hazur) in the early Sumerian literature may, in fact, refer to the indigenous, bitter, small-fruited species, *Malus orientalis*. The earliest archeological evidence (early dynasty III, about 2200–2100 BCE) of dried apples are rings of small fruits (11–18 mm in diameter), possibly threaded on a string, on saucers in Queen Pu-abi's grave at Ur near present-day Basra, Iraq (Postgate, 1987; Renfrew, 1987). Apples lose their bitterness when dried (Barrie Juniper, pers. commun.), and if wild *M. orientalis* were harvested, it might have been consumed in this manner. A Sumerian cuneiform literary manuscript, entitled *Disputation between the Hoe and the Plow* (Vanstiphout, 2000), dated about 1900 BCE, refers to gardens of apples. In Anatolia (Turkey), the Hittites, who rose to dominance in the second millennium BCE, referred to 40 apple trees on one estate. Later biblical references to sweet-smelling apples such as in the *Song of Songs*, written probably no earlier than 1000 BCE, suggest introduction before the first millennium BCE. In any case, by the first millennium BCE apples were a part of Western agriculture. Homer (eighth or ninth century BCE) refers to apples in the gardens of the King Alcinöus, the king of the Phaeacians, a legendary country. Apples, however, were unknown in Egypt until Greco-Roman times.

How did the apple reach the West from Central Asia? Because apples and pear do not propagate easily from cuttings, the most likely explanation is the introduction of seed carried in saddle bags by caravans along the trade routes, passing through the fabled cites of Bukara and Samarkand to Persia, perhaps facilitated by seed germination in horse droppings, but propagation from root suckers is another possibility. Another explanation is that grafting technology, dated as early as 3800 years ago (Harris et al., 2002), may have been involved with Persia as an intermediary stop. Persia is clearly the source of many fruits from Central Asia and China.

Whatever the precise mechanism, the apple clearly passed through Persia and Greece, and, by the time of the Romans, apple technology, including grafting, pruning, storage, and selection of unique adapted clones, was perfected. Even the use of specific rootstocks seems to have been discovered (a low-growing type is described by Theophrastus), and it is no coincidence that these easily rooted dwarfing clones are called Paradise, the Persian word for garden.

The apple cultivated by the Romans was transported throughout the empire and became naturalized throughout Europe, resulting in thousands of unique types. Possibly some intercrosses with the small, astringent *Malus sylvestris* native to northern Europe gave rise to cider apples that are still cultivated.

Genetic changes in apple include the selection of triploids derived from unreduced gametes, parthenocarpy, mutations involving growth habit, particularly spurry and short internode types, and various forms of pest and disease resistance. Fruit changes include shape and various quality factors such as flesh firmness, sweetness, acidity, flavor, and shelf and storage life (Janick et al., 1996). Triploidy seems to confer an advantage to cultivated apples, for about 10% of cultivars are triploid, yet nonreduction occurs in only about one seed in a thousand. The ability of layered shoots to root is advantageous in rootstocks, which are propagated by mounding soil around the bases of shoots (stooling). Differences in tolerance to many pests are observed in any large collection of apples. Immunity to apple scab has been transferred from *Malus floribunda* by backcrossing, but the problem of races has also appeared in some areas. This places the durability of this gene (*Vf*) in question, although it has held up for many years in the United States (Janick, 2002b).

Zohary and Spiegel-Roy (1975) have concluded that spontaneous hybridization between wild races and cultivated clones was critical to the early domestication of fruits. Selection from sexual recombinants is still the dominant force in the domestication process, as well as modern fruit breeding. The isolation of elite selections, combined with mass plantings, created a situation where mass selection and recurrent selection could operate naturally. This has recently been confirmed in the apple, where elite selections from Kazakhstan are very close to cultivated varieties (Harris et al., 2002).

Selection in the apple

Selection from sexual recombination can be clearly followed in North America, now considered a secondary center of origin for the cultivated apple. In colonial America, starting in the 1600s, the apple was imported by immigrants, some as scions but most as seeds from Holland, Germany, France, and the British Isles (Beach, 1905). In 1634, Lord Baltimore instructed the first Maryland settlers to carry "kernalls of pears and apples, especially of Permains and Deesons, for making thereafter of Cider and Perry" (Calhoun, 1995). Pioneers were encouraged to plant apples, and the requirement for settling Ohio (1787–1788) included that the settler must harvest at least 50 apple or pear trees and 20 peach trees (Morgan and Richards, 1993). Apples, once introduced, were carried far into the wilderness by Native Americans, traders, and missionaries and became naturalized. In 1806, Jonathan Chapman (the legendary Johnny Appleseed), born in 1744 in Leominster, Massachusetts, distributed apple seeds from cider mills in western Pennsylvania and founded a nursery in West Virginia, distributing seeds along the way. The apple flourished in the new territories with the greatest use for hard cider, the distilled product called apple jack, and vinegar for preservation of fruits and vegetables. Many of the imported apple clones were unadapted, and the selection of natural seedlings from orchards became the glory of nineteenth century American pomology. In 1905, 698 apple cultivars were described in Beach's (1905) *Apples of New York*. Roueché (1975) estimated that 100,000 clones have been selected from literally hundreds of millions of seedlings and evaluated by millions of fruit growers. In the United States the screening of open-pollinated, chance seedlings resulted in thousands of selections, many of which proved to be outstanding cultivars, including Golden Delicious, Delicious, Jonathan, McIntosh, Rome Beauty, York Imperial, Stayman Winesap, Yellow Newtown, Winesap, Rhode Island Greening, Northern Spy, and Gravenstein. Golden Delicious had a profound influence on apple growing in Europe in the twentieth century and further proved to be a prepotent parent, producing many important new cultivars from breeding efforts (Janick et al., 1996).

Ease of propagation

Among the characters that are influenced by selection is the ease of propagation. The key to domestication of fruit crops is vegetative propagation of elite types, followed by natural intercrosses between elites and wild species. This can clearly be seen in the date palm, in which domesticates, but not wild species, are naturally propagated by offshoots. It is also occurs in the apple; dwarfing rootstocks are clonally propagated by layering (stooling), and this character, at least before the advent of micropropagation by tissue culture, is an essential trait. A number of nut trees, black walnut and pecan in particular, have been difficult to propagate vegetatively and are often still propagated from seeds, resulting in a negative effect on productivity.

Fruit quality

Selection for fruit quality, based on flavor, color, and shelf life, is the goal of modern fruit-breeding programs. Selection is nowhere more important than in pineapple, where the processing industry was long based on a single cultivar. Cayenne and its spineless sport (Smooth Cayenne) were uniquely adapted to producing canned slices. A sweeter, yellow-fleshed seedling produced from hybridization is now being marketed as DelMonte Gold and is transforming the world fresh fruit industry because of better fresh fruit quality and appearance.

Champions

The decisive contribution to domestication made by individuals is unknown in most fruit crops, and these great horticulturists are largely unremembered and unsung. However, in a few cases of recent domestication, key persons have been identified. These champions are essential to the domestication process. In the case of the kiwifruit this includes the great plant hunter E.F. Wilson, who introduced the fruit to Britain; a missionary, Katie Frazier, who imported seed to New Zealand, possibly derived from Wilson; and H.R. (Hayward) Wright, who selected the pistillate clone that bears his name. The technology for orchard development was made in New Zealand from grafted plants, principally by F.J. Walker in the town of Wanganui.

Domestication of the blueberry was initiated by a single researcher, Frederick Vernon Coville (1888–1937), U.S. Department of Agriculture (USDA) botanist, who recognized the potential of this species, and later had a fortunate collaboration with Elizabeth White, a blueberry enthusiast from Whitesbog, New Jersey. Later influential researchers included George Darrow and Arlen Draper, both USDA researchers. In strawberry, Amédee François Frézier introduced *F. chiloensis* to France, resulting in the fortuitous hybridization with *F. virginiana*, and the great French botanist, Antoine Nicolas Duchesne, who explained the dioecious nature of *Fragaria*, interpreted *F. ananassa* as an interspecific hybrid and identified hermaphroditic types. The origin of the grapefruit (once known as forbidden fruit), which was discovered from natural intercrosses between orange and pummelo, called shaddock in Barbados, can be traced to the trader Philip Chaddock, who in the mid-seventeenth century introduced the pummelo to Barbados. Domestication does not just happen but is carried out by acts of real people, and they need to be honored.

Fruit breeding

Fruit breeding as an organized activity is a nineteenth century innovation. Its origins trace to mass selection efforts in strawberry and pear. The modern strawberry is derived from hybrids between two octoploid ($2n = 56$) native American species, both usually dioecious: *F. virginiana* indigenous to the East Coast of the North America, but reaching Europe in the seventeenth century, and *F. chiloensis*, native from Alaska to Chile. Hybrids between these two species were produced naturally in Brest, France, early in the eighteenth century when a pistillate clone of the large-fruited *F. chiloensis*, introduced by an Amédee François Frézier, a French army officer (and spy), whose family name curiously derives from the French word (fraise) for strawberry, was interplanted with staminate plants of *F. virginiana*. The new hybrids, now known as *Fragaria* × *ananassa* or pineapple strawberry after their shape and aromatic flavor, initiated the modern strawberry industry. Selection through the years has resulted in tremendous changes as the plant evolved from a predominantly dioecious species with male and female plants into a hermaphroditic species, in which flowers contained both stamens and pistils.

Systematic breeding of European pear was first carried out by Jean Baptiste Van Mons (1765–1842), a Belgian physician, pharmacist, and physicist, and an early apostle of selection in plants. He systematically collected clones of pear and sequentially planted (open pollinated) seed of the best material, making new selections for eight generations. An early fruit book, *The American Fruit Culturist* (1863) by John J. Thomas (1810–1895) states that the mean time from seed planting to fruiting in the first cycle was 12–15 years, 10–12 in the second cycle, 8–10 by the third, 6–8 by the fourth, and 5 years by the fifth. By the eighth generation several fruit trees fruited *at the age of four years* (emphasis by the author, presumably based on correspondence of Von Mons). This may be the first example of data on long-term selection in plants.

Thomas Andrew Knight (1759–1838) was the first to improve fruits by selection from genetic recombination derived from interpollinations of clones. An early proponent of the development of plant improvement through cross-breeding and selection, he literally initiated the field of fruit breeding. He released a number of improved cultivars of both fruits (apple, cherry, strawberry, red currant, plum, nectarine, and pear) and vegetables (pea, cabbage, and potato). His studies on the effects of pollen in the garden pea on seed characters presaged the work of Gregor Mendel carried out 40 years later. He describes dominance and segregation, although he failed to make the brilliant leap of Mendel in relating phenotypic characters to the factors we now know as genes. Gregor

Mendel, the father of genetics, was also involved in apple- and pear-breeding programs.

In the United States, fruit breeding became a part of research at the state and federal experiment stations, and a number of important breeding programs were initiated throughout the United States. Fruit breeding also became an activity of the private sector. Luther Burbank (1849–1926) was the first to consider fruit breeding as a commercial endeavor, and although he distrusted Mendelism, he was a staunch believer in the evolutional theories of Darwin.

> A young man once wrote me asking which book he should read by Mendel, the great scientist who concerned himself with the problem of heredity and hereditary laws, to inform himself of the scientific facts in that field. I wrote him in reply something like this: "My advice to you is to start Mendel by reading Darwin, and then let Mendel go and read more Darwin." (Burbank and Hall, 1927)

At present private fruit breeders are an important part of *Prunus* (especially peach and nectarine and plum) and recently strawberry and raspberry.

Although fruit breeding has been a major activity since early in the twentieth century, the results have been uneven and vary from ineffectual to extraordinarily successful (Table 19.4). Many of the world fruit industries are still based on grower-selected clones. The reason for the lack of progress is twofold. First, vegetative propagation permits the genetic fixation of naturally occurring variation. Because of the vast populations involved in seedling orchards, the quality of the selected clones over hundreds and even thousands of years of selection is very high. Second, the difficulties and expense inherent in fruit breeding have inhibited long-term breeding programs. Progress from breeding a number of fruit crops, however, has shown significant advances in the second half of the twentieth century, and selections from controlled crosses are increasingly important in many crops. In apple, although chance seedlings such as Delicious and Golden Delicious have long dominated the world market, Fuji, a seedling derived from a Japanese breeding program (Ralls Janet × Delicious), is now the number one cultivar.

Predicting future changes

Knowledge of domestication should be used to predict future changes and to help domesticate new candidates. Thus, one might anticipate that in kiwifruit hermaphroditic mutations would eliminate the need for staminate clones as pollinators and fruit skin mutations or breeding could lead to nonpubescent clones, with more attractive and more edible fruit surface. The use of interspecific hybridization should lead to improvement in banana and plantain (now in jeopardy because of lack of genetic diversity), the creation of new stone fruits, and the creation of new seedless, easy-to-peel citrus. A number of tropical fruits are candidates for commercialization, provided postharvest technology can be improved. One of the likely candidates for domestication is pitaya (species of *Selenicereus*, *Hylocereus*, and *Cereus*), an extremely attractive fruit of columnar cactus, but breeding efforts to improve quality is required, since many selections are somewhat insipid (Mizrahi et al., 2002).

A number of generalizations can be made concerning the origin and future development of fruit crops. The first is that most fruit crops are little removed from wild species, some perhaps by only a few generations, so that continued progress should be possible. The key breeding system has evolved

Table 19.4 Effects of organized fruit breeding on the commercial world industry

Negligible	Slight	Moderate	Major
Banana & plantain	Tart cherry	Almond	Blueberry
Chestnut	Citrus	Apple	Brambles (raspberry & blackberry)
Date palm	Hazelnut	Apricot	Cherry (sweet)
Fig	Papaya	Avocado	Currants
Grape (wine)	Persimmon	Pear Asian	Grape (table)
Lingonberry	Pear European	Pecan	Strawberry
Olive	Pineapple		Peach & nectarine
Pomegranate			Plum

from selection of elite clones followed by fixation by vegetative propagation. The development of fruit culture is based on an interaction between genetic changes and cultural practices. Indeed, in many fruit crops, once desirable clones are discovered, intensive efforts have been made to prop them up by cultural practices. These include artificial pollination, the use of disease-resistant and size-controlling rootstocks, extensive methods of disease control, including complex schedules of pesticide application, the control of fruit size and annual bearing by manual and chemical fruit and flower thinning, the control of fruit abscission with growth regulators, and extensive pruning and training systems. Despite some intensive breeding programs, extremely successful in *Prunus*, strawberry, and blueberry, many of our fruit cultivars are ancient and based on grower-selected seedlings and somatic mutations. Advances in molecular genetics may overcome some of the limitations to conventional fruit breeding based on sexual recombination by increasing selection efficiency using molecular markers and by transgene technology, whereby individual genes from various sources may be inserted without disturbing unique genetic combinations. Progress has already been achieved in papaya (virus resistance) and apple (resistance to fireblight), but fear of consumer resistance is a problem.

References

Bahrani, Z. 2002. Peformativity and the image: Narrative, representation, and the Uruk vase. Pp. 15–22. *In* E. Ehrenberg (ed.), Leaving no stones unturned: Essays on the ancient Near East and Egypt in honor of Donald P. Hansen. Eisenbrauns, Winona Lake, IN.

Beach, S.A. 1905. The apples of New York. Vol. I. J.B. Lyon Co., Albany, NY.

Burbank, L., with W. Hall. 1927. The harvest of the years. Houghton Mifflin Co., Boston. p. 151.

Calhoun, Jr. C.L. 1995. Old southern apples. The McDonald & Woodward Publishing Co., Blacksburg, VA.

Childe, V.G. 1958. The dawn of European civilization. Knopf, New York.

Crib, A.B., and J.W. Crib. 1974. Wild food in Australia. Fontana, Collins, Australia.

Dzhangaliev, A.D. 2003. The wild apple tree of Kazakhstan. Hortic. Rev. 29:63–371.

Forsline, P.L., H.S. Aldwinckle, E.E. Dickson, J.J. Luby, and S.C. Hokanson. 2003. Collection, maintenance, characterization, and utilization of wild apples of central Asia. Hortic. Rev. 29:1–61.

Harris, S.A., J.P. Robinson, and B.E. Juniper. 2002. Genetic clues to the origin of the apple. Trends in Genetics 18:426–430.

Hartmann, H.T., D.E. Kester, F.T. Davies, Jr., and R.L. Geneve. 1997. Plant propagation: Principles and practices, 6th ed. Prentice Hall, Upper Saddle River, NJ.

Janick, J., J.N. Cummins, S.K. Brown, and M. Hemmat. 1996. Apples. Pp. 1–77. *In* J. Janick and J.N. Moore (eds.), Fruit breeding. Vol. 1. Wiley, New York, NY.

Janick, J. 2002a. Ancient Egyptian agriculture and the origins of horticulture. Acta Hortic. 582:23–39.

Janick, J. 2002b. History of the PRI apple breeding program. Acta Hort. 595:55–59.

Johns, T., and I. Kubo. 1988. A survey of traditional methods employed for the detoxification of plant foods. J. Ethnobiol. 8:81–129.

Mendel, K. 1953. The development of our knowledge on transplantations in plants. Actes du Septième Congres International d'Histoire des Sciences, Jerusalem.

Meyer, E.H.F. 1854. Geschichte den Botanik. Verlag der Gebröder Borntrger, Königsberg, Germany.

Mizrahi, Y., A. Nerd, and Y. Sitrit. 2002. New fruits for arid climates. Pp. 378–384. *In* J. Janick and A. Whipkey (eds.), Trends in new crops and new uses. ASHS Press, Alexandria, VA.

Morgan, J., and A. Richards. 1993. The book of apples. Ebury Press, London.

Nagy, S., P.E. Shaw, and M.E. Veldhuis (eds.). 1977. Citrus science and technology. AVI Publ. Co., Westport, CT.

Pollock, S. 1999. Ancient Mesopotamia: The Eden that never was. Cambridge Univ. Press, Cambridge, United Kingdom.

Postgate, J.N. 1987. Notes on fruit in the cuneform sources. Bull. Sumerian Agric. III:115–144.

Renfrew, J.M. 1987. Fruits from ancient Iraq: The paleoethnobotanical evidence. Bul. Sumerian Agric. III:157–161.

Roth, M., 2000. The laws of Hammurabi. Pp. 335–353. *In* W.W. Hallo (ed.), The context of scripture. Vol II. Brill, Leiden, Boston, Köln.

Roueché, B. 1975. One hundred thousand varieties. The New Yorker, August 11.

Spooner, D.M., W.L.A. Hetterscheid, R.G. van den Berg, and W.A. Brandenburg. 2003. Plant nomenclature and taxonomy: An horticultural and agronomic perspective. Hortic. Rev. 28:1–60.

Thomas, J.J. 1863. The American fruit culturist. C.M. Saxton, New York, p. 26.

Vanstiphout, H.L.J. 2000. The disputation between the hoe and the plow. Pp.578–581. *In* W.W. Hallo (ed.), The context of scripture. Vol I. Brill, Leiden, Boston, Köln.

Vöchting. 1892. Über Transplantation am Pflanzenkörper. Verlag der H. Laupp'schen Buchhandlkung, Tübingen, Germany.

Way, R.D., H.S. Alwinckle, R.C. Lamb, A. Rejman, S. Sansavini, T. Shen, R. Watkins, M.N. Westwood, and Y. Yoshida. 1990. Apples (*Malus*). Acta Hortic. 290:3–46.

Zohary, D., and P. Spiegel-Roy. 1975. Beginning of fruit growing in the Old World. Science. 187:319–327.

20

Sugarcane Genomics and Breeding

Kuo-Kao Wu and Ray Ming, Hawaii Agriculture Research Center
Paul H. Moore, USDA-ARS, Pacific Basin Agricultural Research Center
Andrew H. Paterson, Plant Genome Mapping Laboratory, University of Georgia

Sugarcane breeding

Introduction

Sugarcane is a large, perennial, tropical or subtropical grass widely grown in a zone around the world within 30 degrees of the equator. Sugarcane is vegetatively propagated from axillary buds on stem cuttings. The first, "plant" crop, is generally harvested from 12 to 24 months after planting; thereafter, "ratoon" crops may be harvested at shorter to equal time periods. Ratoon crops may be grown from one to several cycles. The large mature stem contains juice that contains 9–18% sucrose. The juice is extracted by crushing the stems with high-pressure rollers in a mill. Sucrose is crystallized from the juice after removing water by boiling to produce a brown-colored raw sugar. White sugar is produced by recrystallization from raw sugar in a refinery.

There are 80 geographic regions of the world where sugarcane is grown (Tew, 2003). Worldwide cane-sugar production in 2002 was about 100 million tons (Licht, 2003). The largest producing countries ($\geq$ 2 million tons) are South Africa, the United States, Cuba, Mexico, Brazil, China, India, Pakistan, Philippines, Thailand, and Australia. The highest sugar yields recorded were 24.2 tons per hectare per year in Leeward Oahu, Hawaii, and 17.4 tons per hectare per year in the Central District, Australia (Heinz, 1987a and 1987b).

While sugarcane farmers throughout the world face constant challenges to sustain profitability and environmental soundness (Glaz, 2003), breeders face not only constant demand from farmers for higher-yielding cultivars but also a growing challenge as the gap between average farm yield and genetic yield potential is narrowed through improved agronomic practices (Cassman, 1999). This review summarizes sugarcane improvement to date through conventional breeding and then discusses the potential applications of current research in plant biotechnology for sugarcane germplasm improvement. The latter discussion will provide plant breeders with broader knowledge about *Saccharum* germplasm, new selection tools, and new sources of genes for solving future challenges.

History

Sugarcane cultivars (*Saccharum* spp. hybrids) are complex interspecific hybrids primarily between *Saccharum officinarum* and *Saccharum spontaneum* but having contributions from other *Saccharum* taxa and related grass genera such as *Narenga* and *Erianthus*. The first sugarcane-breeding program began in Java in 1888, following the observation independently in the same flowering season in Java and Barbados in 1858 that sugarcane produced viable seed (Stevenson 1965). Other countries soon followed Java's lead.

A key event in early cane breeding was the production of POJ2878, a "nobilized" cane, in 1921 in Java. Nobilized cane varieties, developed in Java and India, are present in the early generations of every modern cane pedigree (Simmonds, 1976). Nobilized canes were derived from Noble canes that are the cultivated clones of *S. officinarum* (x = 10, 2n = 70–140) having thick stalks, high sucrose content, and low fiber (Irvine, 1999). The nobiliza-

Table 20.1 Nobilization crossing scheme

Female × male	→	Progeny (chromosomes)	(% Spontaneum)
NN(80) × SS(64)	→	(80 + 32 = 112)	(S% = 29 = 100 × 32/112)
NN(80) × F_1(112)	→	B_1 (80 + 56 = 136)	(S% = 12 = 100 × 16/136)
NN(80) × B_1(136)	→	B_2 (40 + 68 = 108)	(S% = 7)

NN = noble, 2n = 80; SS = Spontaneum, 2n = 64

tion process involved chromosome nonreduction plus introgression of additional genes through a system of crossing the noble clones with wild clones of *S. spontaneum* (x = 8, 2n = 32–128), having thin stalks, low sucrose content, high fiber, and disease resistance and cold tolerance. More than 90% of the accessions classified as *S. officinarum* have 2n = 80 chromosomes, while the most frequent count of chromosome number in *S. spontaneum* is 2n = 64. Based on these two chromosome complements, one can envision nobilization in the simplified crossing scheme shown in Table 20.1.

Progeny of F_1 and B_1 have the nonreduced somatic complement (2n) of the female parents plus the gametic number (n) of the male. Most cultivated nobilized canes (B_2s, B_3s, etc.) have chromosome numbers in the range 100–130, with about 5–10% consisting of the wild *S. spontaneum* contribution. Clones with chromosome numbers out of this range are rarely suited for commercial production (selection was based on phenotype, not chromosome number.) After 1925–1930, nobilization breeding has been seldom used. Crossing among advanced clones is the main breeding method for producing progeny for the selection of commercial cultivars. Most modern cultivars are essentially derivatives of nobilized canes produced in the early 1900s.

Hybridization and selection

Pollination procedures consist of harvesting tasseling stalks from fields as anthesis begins and moving the stalks to a crossing shelter where they are placed in a weak acid solution for preservation to make either bi-parental crosses or poly-crosses. In a bi-parental crossing shelter, male parents are discarded after 14 days, and female parents are retained for an additional 10–20 days until the seed is ripe. In a poly-crossing shelter, seed from female tassels (panicles) is harvested from 21 days to 35 days. It is necessary to change the acid solutions twice a week to keep the flowering stalks alive. Flowering stalks are shaken each morning at about 0800, shortly after pollen is shed to maximize seed set. Seeds are germinated in a greenhouse, and seedlings are transplanted to fields two to three months after germination (James, 1980).

Procedures for selection of commercial cultivars varies among breeding stations, depending primarily on the length and number of ratoon cropping cycles practiced by the local cane growers (Mamet and Domaingue, 1999). For a long cropping cycle (24 months plant and 24 months ratoon), there are three stages (8 months each) based on visual selection followed by two stages (24 months each) based on numerous highly replicated yield trials in various ecological regions. Selection criteria for long cycles include disease and insect resistance, high sucrose, high cane tonnage with second year suckers, and intermediate fiber (12–15%). For short cropping cycles (18 months plant and 12 months ratoon), in addition to visual and yield trials of the plant crop, several ratoon cycles are evaluated. Selection criteria for short cycles are mainly disease and insect resistance, high sucrose, low fiber (10–13%), and erect growth, with only a few suckers. It may take 12 to 15 years to identify a clone for commercial production, while it may take 7–10 years to select clones as breeding parents (Heinz, 1987a and 1987b).

Breeding achievements

Crop yield is a result of genetic improvements and management practices (Cook, 2001). In general, new sugarcane cultivars need 12–15 years to be identified from a breeding and selection program. A significant contribution of the leading Hawaiian cultivars to sugar yield is shown in the Figure 20.1 highlighting changing of cultivars from 1915 through 2002 (Wu and Gamatero, 2003). The decrease in sugar yield over the last 20 years was mainly due to the closure of plantations and intro-

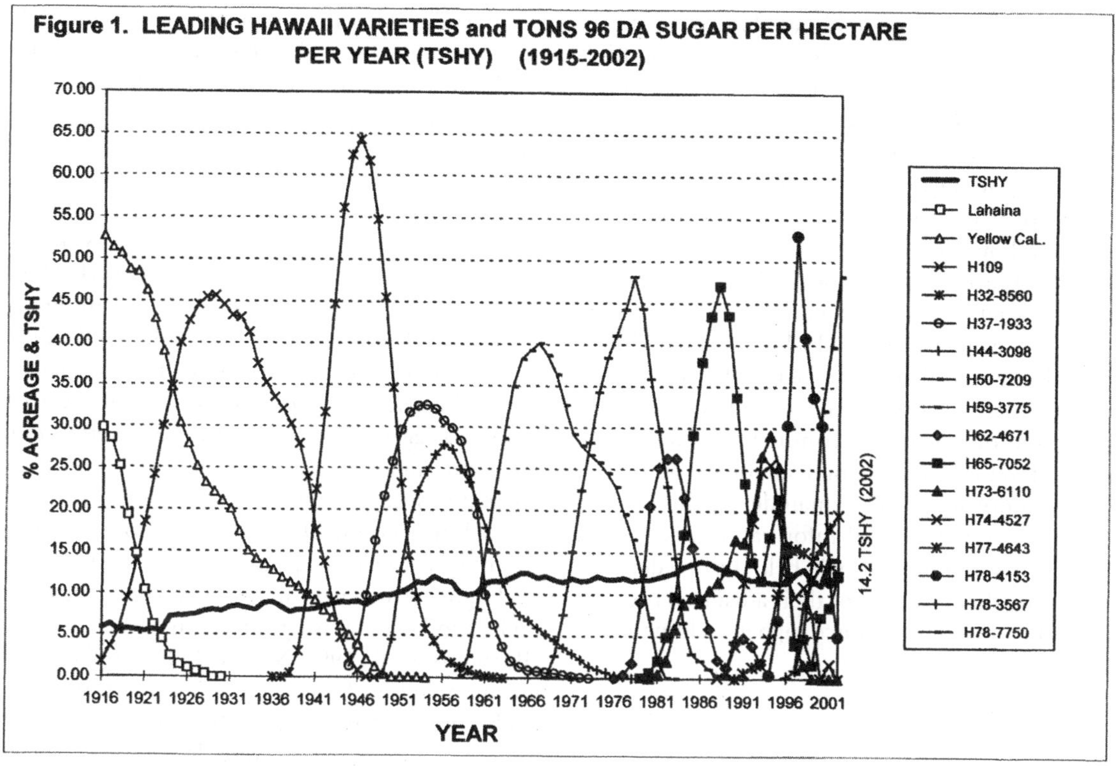

Figure 20.1 Leading Hawaii varieties and 96 tons per day of sugar per hectare per year (TSHY) (1915–2003).

duction of new diseases and pests. Annual production at the end of 2002 was 14.2 tons of sugar per hectare per year.

The rate of increase in yield appears to have slowed over the past 50 years, especially in the 1970s and 1980s. In Australia, sugar yield was static for 20 years from 1970 to 1989. The static trend was likely due to production factors, such as increase in area per grower, industry expansion onto marginal land, and increased losses due to diseases and pests. Sugar yield increased to 12.0 tons per hectare from 1990 to 1995. The increase is believed largely due to increased breeding efforts over the whole industry (Berding et al., 1997). The yield in 2001 was 12.6. (Estimated from Bureau of Sugarcane Experiment Stations (BSES; http://www.bses.org.au), 2001–2002 annual report). In Colombia, sugar yield increased from approximately 5 t sugar/ha/yr at the end of 1950s to 8 t/ha/yr at the end of 1970s and reached 12 t at the end of 1990s (Cook, 2001). Thus, static yields or yield plateaus experienced by several countries may be overcome by intensive breeding for new varieties accompanied by appropriate cultural practices.

Base broadening of advanced breeding population

The initial germplasm base developed in Java and India had no more than 15 genotypes. The genetic base for cane and sugar yield in the advanced breeding populations of today is expected to be much narrower than that in the initial germplasm following more than 100 years of directional selection within the original nobilized germplasm. Attempts to increase this base, called the base-broadening program (BB-program), started in Barbados in 1965 using clones different from those initially used in Java and India. The BB-program has been run parallel to the Barbados contemporary commercial variety-producing program. The BB-program started with nobilization breeding followed with hybridization of nobilized canes. The BB-program has produced many semicommercial-type clones since late 1980s. The gene pool of the semicommercial BB-clones is being incorporated into the gene pool of advanced breeding populations in recent years.

Several other countries have tried BB-programs in the past by crossing wild canes with their commercial clones. However, none of the efforts comprised a long-term and broad-based program such as the BB-program of Barbados. In terms of cultivar production, the BB-programs were generally not successful (Berding et al., 1997).

Perhaps the single greatest hindrance in the base broadening of sugarcane in particular, and in its improvement in general, has been the inability to trace or follow the incorporated germplasm into advanced breeding populations. Large favorable genetic variations among clones of sugarcane species are available (Tai and Miller, 2002). Missing has been tools that can assist breeders in incorporating useful genes from any source into the gene pool of advanced commercial cultivars. Recent developments in biotechnology and genomics, addressed in the remainder of this chapter, are beginning to yield information and technologies that undoubtedly will help the breeders in broadening the gene pool of their advanced breeding populations and produce higher-yielding cultivars in the near future.

Sugarcane genomics

The application of DNA markers to genetic mapping for crop improvement began in the early 1980s. For sugarcane, the groundbreaking theoretical concept that single-dose markers could be used for mapping any polyploid without knowing either the type or level of polyploidy (Wu et al., 1992) opened a new era for sugarcane genetic and genomic research. Now, just a little over a decade after this breakthrough, abundant genomic resources have been established; the basic chromosome numbers of *Saccharum* have been resolved; and knowledge about the genetic diversity and structure of the sugarcane genome is significantly advanced. Here we briefly describe the major events of sugarcane genomics that we have witnessed.

Genetic diversity

Taxonomy of the sugarcane complex, based on morphology, chromosome numbers, and geographical distribution, has been controversial. Recent genomic data for evaluating genetic diversity within the genus are beginning to suggest new relationships among accessions and may ultimately produce a refined classification for the sugarcane species. The first molecular evidence came from restriction fragment patterns of nuclear ribosomal DNA that was used to separate accessions of *S. spontaneum*, which showed the widest within species variation, from accessions of four other taxa often afforded specific status: *S. robustum*, *S. officinarum*, *S. barberi*, and *S. sinense* (Glaszmann et al., 1990). Restriction fragment length polymorphism (RFLP) analyses of the mitochondrial genome showed an identical pattern among 18 *S. officinarum* and 15 of 17 *S. robustum* samples (D'Hont et al., 1993). RFLP patterns were similar among *S. officinarum*, *S. barberi*, *S. sinense*, and *S. edule*, all of which were distinctively different from *S. spontaneum*. Restriction patterns of the chloroplast genome suggested that, except for *S. spontaneum*, the *Saccharum* species all have the same chloroplast restriction sites (Sobral et al., 1994). RFLP analyses of nuclear genomic DNA confirmed observations about the cytoplasmic genomes suggesting distinctively greater diversity in *S. spontaneum* than the high similarity among the four other species (Burnquist et al., 1992; Nair et al., 1999). The most recent analysis, based on genomic *in-situ* hybridization, is compatible with the hypothesis that *S. barberi* and *S. sinense* were derived from interspecific hybridization between *S. officinarum* and *S. spontaneum* (D'Hont et al., 2002). These authors conclude that genetic similarities among *S. barberi* and *S. sinense* accessions do not support the present classification of these being two distinct taxa.

S. robustum has been postulated to be the progenitor of *S. officinarum*. By all lines of evidence, these two species were the most closely related among six *Saccharum* species studied in the multiple independent DNA diversity studies mentioned above. Although it has not been proven, *S. edule* is though to be an intergeneric hybrid between either *S. officinarum* or *S. robustum* with a related genus that may account for its aborted inflorescence (Daniels and Roach, 1987). Because of the phenomena of female restitution, hybrids derived from interspecific or intergeneric crosses involving a female *S. officinarum* conserve the entire genome of *S. officinarum*, which becomes the genetic basis for the high similarity detected among the five species other than *S. spontaneum*. The latest molecular data based on *in situ* hybridization verifies Irvine's (1999) classification of the genus *Saccharum* into two species: *S. spontaneum*, as it is presently recognized, and *S. officinarum*, which includes the other four *Saccharum* species and their interspecific hybrids.

Because of its polyploid nature, interspecific origin, and vegetative propagation, high levels of heterozygosity was detected among modern sugarcane cultivars using RFLP markers (Jannoo et al., 1999a). The major part of this diversity was attributed to the 15–25% chromosome complement that was inherited from *S. spontaneum* by random assortment of half its chromosomes, which have the greatest intraspecific diversity (see above) (D'Hont et al., 1996). Similar patterns of molecular diversity were also detected using amplified fragment length polymorphism (AFLP) markers (Lima et al., 2002). On the other hand, modern sugarcane cultivars, derived from a small germplasm base contributed by only a few genotypes, show strong linkage disequilibrium. Some haplotypes are conserved in segments extending for at least 10 cM (Jannoo et al., 1999b). This is in contrast to the situation in diploid species such as Arabidopsis where linkage disequilibrium decays down to less than 1 cM or 250 kb (Nordborg et al., 2002). The large-scale linkage disequilibrium of modern sugarcane cultivars is likely a consequence of its narrow genetic base and the few generations of hybridization from the original crosses.

Sorting out the basic chromosome numbers

The basic chromosome numbers of *Saccharum* species have long been a topic for speculation among sugarcane scientists. Recent evidence generated from quantitative karyotyping (Ha et al., 1999) and fluorescence *in situ* hybridization (D'Hont et al., 1995, 1996, 1998; Ha et al., 1999) suggests that the *Saccharum* basic chromosome number (x) differs by species (x = 8 for *S. spontaneum* and x = 10 for *S. officinarum* and *S. robustum*). These two sets of basic chromosome numbers correspond to the chromosome numbers in these two horticultural groups. Of the 1,086 *S. spontaneum* samples for which chromosome counts were available, 77% are multiples of eight (2n = 40, 48, 56, 64, 72, 80, 88, 96, 112, 120, and 128). Of the 96 *S. robustum* samples, 72% are multiples of ten (2n = 60, 70, 80, 90, 100, 110, 140, and 170). Moreover, 92% of the 497 *S. officinarum* samples are also multiples of ten (2n = 80) (Irvine 1999). In the Andropogoneae tribe, $x = 10$ is common (Whalen, 1991), but exceptions exist, such as 2n = 6 and 8 for *Iseilema* (Clayton and Renvoize, 1986).

Linkage mapping

Linkage mapping in sugarcane has been conducted on five populations producing seven linkage maps, mostly based on single-dose restriction fragment markers (Wu et al. 1992). The first sugarcane linkage map was constructed from the progeny of a cross between *S. spontaneum* SES208 (2n = 64) and its doubled haploid ADP068 (Da Silva et al., 1993; Al-Janabi et al., 1994). This map consists of 64 linkage groups assembled into eight homologous groups based on 276 restriction fragment length polymorphisms (RFLP) and 208 single dose arbitrarily primed polymerase chain reaction (PCR) loci (Da Silva et al., 1995). The second map was derived from the progeny of a self-pollinated hybrid cultivar R570 (2n = 107 – 115). This map consists of 408 RFLP loci on 96 linkage groups and ten putative homologous groups (Grivet et al. 1996). More recently, AFLP markers were used to construct another linkage map based on the same population but different individuals, resulting in 120 linkage groups. Thirty-four of the linkage groups could be assembled into ten homologous groups (Hoarau et al., 2001). A map of *S. officinarum* Louisiana Purple (2n = 80) is based on a cross with *S. robustum* and consists of 160 randomly amplified polymorphic DNA (RAPD) markers plus one morphological marker assembled in 51 linkage groups (Mudge et al. 1996). Four additional maps were constructed for each of the four parents from two interspecific crosses: *S. officinarum* Green German (GG; 2n = 97 – 117) × *S. spontaneum* IND 81–146 (IND; 2n = 52–56), and *S. spontaneum* PIN 84–1 (PIN; 2n = 96) × *S. officinarum* Muntok Java (MJ; 2n = 140). A total of 72, 69, 72, and 69 linkage groups were assembled from 615, 536, 575, and 418 RFLP markers for GG, IND, MJ, and PIN, respectively (Ming et al., 1998).

Among these seven sugarcane linkage maps, the number of linkage groups in the first two maps is expected to be the same as the number of 2n chromosomes. Because the first map is based on progeny derived from a cross between a doubled haploid and its mother plant and the second map is based on the progeny of a self-pollinated elite cultivar, the full set of 2n chromosomes was transmitted to their progeny. The number of linkage groups in the other five maps, however, is expected to be one-half of the parental 2n chromosome number, since only half the chromosomes were transmitted to the segregating F_1 populations used in mapping. A comparative analysis between the sorghum linkage map and the five sugarcane linkage maps indicates that every one of the sugarcane maps is incomplete (Ming et al., 1998).

Although the basic chromosome numbers of *Saccharum* are known, complete linkage maps reflecting these basic chromosome numbers are still not available. Homology-based RFLP markers would have the advantage of inferring homologous group and chromosome alignment with the more extensively studied close relatives, sorghum and maize, but RFLP maps are labor intensive and slow to generate. Adding PCR-based AFLP markers to existing RFLP maps is a cost-effective approach to increase the genome coverage of sugarcane genetic maps. Genome coverage by any of the sugarcane maps is negatively correlated with chromosome numbers to be mapped. The sugarcane genetic map with the best genome coverage is that of IND 81–146, which has the lowest number of chromosomes (26–28) to be mapped and has about 58% genome coverage (Ming et al., 1998). If sugarcane breeders desire a saturated genetic map of the basic chromosomes, it would be most efficient to work with mapping populations generated from parents having the fewest number of chromosomes. One such population, already partially mapped, derived from a cross between *S. officinarum* Louisiana Purple (2n = 80) and *S. robustum* Molokai 5829 with 40 chromosomes each might be worth additional efforts (Mudge et al., 1996; Ming et al., 1998).

Mapping quantitative trait loci for economic traits

Mapping quantitative trait loci (QTL) in autopolyploids is complicated by the potential for segregation of three or more alleles at a locus and by the lack of preferential pairing. As a consequence, different parental alleles of autopolyploids are not mutually exclusive alternatives. For the subset of polymorphic alleles that show simplex segregation ratios, the effect of an allele substitution can be estimated from the average phenotypic difference between the two possible genotypes (presence versus absence). Large-scale QTL mapping was conducted in two interspecific populations (Ming et al., 2001; 2002a; 2002b) and in a segregating population from a selfed hybrid R570 (Hoarau et al., 2002). Most QTL alleles for sugar content showed phenotypic effects consistent with the parental phenotypes. However, the occasional transgressive QTLs revealed opportunities to purge unfavorable alleles from cultivars or to introgress valuable alleles from exotics (Ming et al., 2001). In many cases, QTLs controlling a given trait were mapped to corresponding genomic locations within the same genotype, across genotypes, and across species. This complex mapping of a given trait suggests that at least some QTLs on the same cluster might be different forms of the same gene or conserved homologous genes (Ming et al., 2001, 2002a, 2002b). Several QTLs mapped for sugar content correspond to approximate genomic locations of previously mapped maize mutants and QTLs for sugar content. This correspondence between loci for sugar content in sugarcane and maize suggests that stem storage and seed storage crops may share a partly overlapping basis of genetic variation for carbohydrate storage (Ming et al., 2001). QTL mapping in the commercial hybrid R570 revealed small effect of individual QTL for sugar yield components that were not conserved between crop cycles (Hoarau et al., 2002).

Multiplex segregation at QTL loci may be partly responsible for phenotypic buffering that is an important factor in the success of many autopolyploid crops. In several sugarcane cases, two or more loci detected by the same DNA probe were each associated with variation in sugar content and plant height, enabling us to investigate the possibility of multiplex phenotypic buffering in sugarcane. "Stacking" of multiple doses of chromosomal segments containing favorable QTLs generally produced diminishing effects on phenotype, especially in cases where high-order duplications could be tested (Ming et al., 2001; 2002a). This is similar to the results reported from stacking unlinked QTLs in the diploid tomato. The tomato results were attributed to epistasis (Eshed and Zamir, 1996). Evaluating epistasis in sugarcane is complicated by the possibility of nonlinear interactions between loci at homologous sites (such as we report) in addition to nonlinear interactions between unrelated loci (Eshed and Zamir, 1996). Detecting this type of phenotypic buffering has potential for variety improvement through marker-assisted selection in autopolyploid crops. Although diagnostic DNA markers enable us to pyramid multiple QTLs in a polyploid, incorporating just one copy of the multiple alleles may be sufficient to achieve most of the desired effect in the breeding population. Nonadditive gene action in multiple dose QTLs may also have contributed to evolutionary opportunities. If a single copy of a gene/QTL is physiologically sufficient, the additional copies are "extra" and thus free to collect mutations, often becoming nonfunctional, but perhaps occasionally resulting in a distinctive new function that improves fitness.

Synteny with other members of the grass family

The conservation of gene repertoire and colinearity of gene order in the genomes of diverse grasses are well established (Freeling, 2001). For sugarcane, the small diploid genome of sorghum has proven an especially facile model—sorghum is the closest relative of sugarcane, and the two grasses diverged from a common ancestor as little as five million years ago. Sorghum and sugarcane genomes share more extensive genomewide colinearity, and fewer chromosomal rearrangements (Dufour et al., 1997; Guimarães et al., 1997; Ming et al., 1998), than either shares with any other known grass. Comparative mapping to establish colinearity between sugarcane and maize is complicated by segmental polyploidy of the maize genome and the resulting mapping of many sugarcane loci to two duplicated loci in maize (Grivet et al., 1996; Dufour et al., 1997). Although it has not been through a genomic duplication event subsequent to its divergence from sugarcane, rice is much more distantly related and numerous chromosomal rearrangements are found when attempting to align their genomes.

Colinearity has been employed to evaluate the correspondence of QTLs affecting related traits in sugarcane and other grasses. Corresponding QTLs controlling plant height and flowering were found in sorghum and sugarcane (Ming et al., 2002a). Several previously mapped maize and rice mutant and QTLs of the sugar metabolic pathway might be candidate genes for controlling sugar content in sugarcane (Ming et al., 2001). Sorghum, rice, and maize linkage maps and physical maps were used to identify potential markers for fine mapping and chromosome walking toward cloning the rust resistance gene in sugarcane (Asnaghi et al., 2000); sorghum RFLP markers played a key role in mapping this gene to a small interval. The close relationship among these grasses, a high degree of colinearity, and cross-hybridization of DNA probes are compelling reasons for using the more abundant information from the small genome of sorghum to guide molecular mapping and positional cloning in sugarcane.

Map-based cloning of the rust resistance gene

The first major gene of sugarcane to be mapped was the gene for resistance to common rust in cultivar R570 (Daugrois et al., 1996). Mapping this gene with sugarcane cDNA probe CDSR29 provided the first opportunity to evaluate the potential for map-based cloning in a complex polyploid plant. With the support from the International Consortium of Sugarcane Biotechnology (ICSB), a bacterial artificial chromosome (BAC) library was constructed with 14 X basic genome or 1.3 X total genome coverage using genomic DNA from R570 (Tomkins et al., 1999). Meanwhile, a fine mapping project began to saturate the region surrounding the rust resistance gene. Using the synteny relationships among sugarcane and sorghum, maize, and rice, and selecting probes in the surrounding regions, this rust resistance gene was mapped to the end of a linkage group corresponding to sorghum linkage group D (Asnaghi et al., 2000). Bulk segregant analysis added eight markers surrounding the rust resistance gene with the two closest flanking markers placed 1.9 and 2.2 cM from the resistance gene (Asnaghi et al., 2004). Flanking markers were narrowed down to 0.3 and 0.6 cM on each side of the target by chromosome walking using sugarcane, sorghum, and rice BAC resources (D'Hont, personal communication). The ability to begin with an unlinked rust resistance gene with a tagged marker 10 cM away to produce a fine-mapped target gene covered with sequenced rice BACs demonstrates the rapid advancement of sugarcane genomics that will ultimately be applied by sugarcane breeders for the benefit of the sugarcane industry.

Sugarcane expressed sequence tags

Current large-scale genome projects on a variety of plants, animals, and microbes are making available vast amounts of information in the form of genomic sequence and expressed sequence tags (ESTs). National and international efforts underway to develop and catalog ESTs for major food crops such as rice, maize, and soybean are paralleled by independent sugarcane EST projects in Australia, Brazil, South Africa, and the United States. Most notable among them, accounting for more than 90% of reported sugarcane ESTs, is the

SUCEST program of the Brazil Organization for Nucleotide Sequencing and Analysis (ONSA) consortium. An initial report about this effort (Arruda, 2001), which was published as a special issue of *Genetics and Molecular Biology*, contains 34 research articles from 74 sequencing and data-mining laboratories relating sugarcane ESTs to factors such as flowering, signal transduction, plant development, aluminum toxicity, pest- and pathogen-defense systems, mitochondria and chloroplast functions, membrane transport and secretion, cell wall metabolism, and cytoplasm metabolism. These analyses were based on the SUCEST database containing 238,000 ESTs from 26 sugarcane cDNA libraries constructed from several tissues—shoot apical meristem, flowers, lateral vegetative buds, unfurled immature leaves, mature leaves, roots, stem (culm) rind, culm internodes, seed, and tissue culture calli—at different developmental stages (Vettore et al., 2001). The ESTs similar in sequence were assembled into 43,000 clusters, of which 38% had no matches in existing public sequence databases. Around 53% of the clusters were formed by ESTs in more than one library, delimiting a group of genes that are coordinately expressed in different tissues, while 47% were formed by ESTs expressed in only one library, delimiting tissue-specific expressed genes.

Because cane sugar is a large, internationally traded commodity, there is often a substantial lag in release of information that might lead to competitive advantages. There are currently only about 10,000 EST sequences in GenBank, although release of many additional sequences is anticipated in the near future. More immediate research needs led Australia, South Africa, and the United States to develop parallel, but reduced-scale, EST programs. The U.S. program, to date, is that of a single laboratory analyzing three cDNA libraries—apex, mature leaf, and mature internode—to develop 9,216 ESTs that were clustered into 3400 nonredundant tags (Ma et al., 2004). About 57% of these ESTs were assigned a putative function based on statistically significant similarity to previously characterized proteins or sequences. Another 28% corresponded to previously identified, but uncharacterized, sequences. Some of the remaining sequences were predicted to be genes that may be new or unique to sugarcane. Comparisons of the sugarcane ESTs to a large sorghum EST database revealed similar compositions of expressed genes between different tissues, suggesting applicability of the more abundant sorghum data. Curiously, a substantial fraction of the U.S. ESTs were absent from SUCEST, suggesting that genotype × environment interactions play an important role in the samples of sequences available to molecular biology.

The published sugarcane EST research has profiled gene expression differences between immature culm internodes that are not storing sucrose versus internodes that are mature and storing sucrose (Carson et al., 2002; Carson and Botha, 2002), or it has focused on internodes that are in the process of maturing and most active in accumulating sucrose (Casu et al., 2003). The underlying hypothesis of both approaches is that knowledge about gene expression associated with high storage of sucrose will be revealed through global analysis and can contribute to a systems approach for increasing yield potential. Approximately one-third of the 400 cDNAs analyzed, 200 from each of the two reciprocal subtractive cDNA libraries, were preferentially expressed in either the immature or mature internodes (Carson and Botha, 2002). ESTs generated from all 132 differentially expressed clones revealed 95 unique transcripts, of which, based on homology, two-thirds were assigned functions such as cell wall metabolism, carbohydrate metabolism, stress responses, and regulatory proteins. ESTs directly associated with sucrose metabolism were found not to be developmentally regulated, suggesting that growth and maturation of the sugarcane culm is associated with the expression of genes for a variety of processes other than sucrose metabolism. Likewise, a sequence survey of 7242 ESTs derived from a sucrose-accumulating, maturing culm revealed that transcripts for carbohydrate metabolism gene sequences (CMGs) were relatively rare in this tissue (Casu et al., 2003). Nevertheless, ESTs of sugar transporter homologues were highly abundant CMG transcripts. The most abundant of the sugar transporter ESTs was associated with phloem companion cells and nearby parenchyma, suggesting a critical role for translocation in sucrose accumulation. The coordinated expression of genes encoding enzymes involved in sucrose synthesis and cleavage as well as glycolysis and the pentose phosphate pathway points toward the need for systems-level approaches toward understanding sucrose accumulation in sugarcane. Recent rapid expansion of sugarcane molecular datasets and the beginning of a systems approach to metabolic modeling of sucrose accumulation point toward future applications for raising potential yields of sugarcane.

Prospects

A decade of progress in sugarcane genomics has provided a wealth of data and formulated at least partial answers to century-old questions that were once considered unanswerable. A growing abundance of genomic information is becoming available to sugarcane breeders and geneticists to assist and advance their research and development programs. Future advancements will provide increasingly practical tools for sugarcane breeders to develop improved sugarcane cultivars and to:

- produce complete sugarcane linkage maps using existing RFLP maps as frame scaffolds upon which to add high-throughput DNA markers such as AFLPs;
- develop the capacity to scan whole sugarcane genomes for QTLs controlling economic traits through use of the completed maps;
- integrate sugarcane and sorghum genetic and physical maps with most or all of the abundant EST on the map to generate the ultimate tools for QTL mapping and gene isolation;
- uncover the mechanism of female restitution in *S. officinarum* to advance our knowledge of basic biology and perhaps find significant applications to crop improvement in other species; and
- solve the last evolutionary puzzle about sugarcane speciation and classification by using genomic *in situ* hybridization to determine the interspecific or intergeneric origin of *S. edule.*

References

Arruda, Paulo. 2001. Sugarcane transcriptome: A landmark in plant genomics in the tropics. Gent. Mol. Biol. 24:0–0.

Asnaghi, C., F. Paulet, C. Kaye, L. Grivet, M. Deu, J.C. Glaszmann, A. D'Hont. 2000. Application of synteny across Poaceae to determine the map location of a sugarcane rust resistance gene. Theor. Appl. Genet. 101:962–969.

Asnaghi, C., D. Roques, S. Ruffel, C. Kaye, J-Y. Hoarau, H. Télismart, J.C. Girard, L. M. Raboin, A. M. Risterucci, L. Grivet, A. D'Hont. 2004. Targeted mapping of a sugarcane rust resistance gene (Bru1) using bulked segregant analysis and AFLP markers. Theor. Appl. Genet. 108:759–764.

Al-Janabi, S.M., R.J. Honeycutt, C. Peterson and B.W.S. Sobral. 1994. Phylogenetic analysis of organellar DNA sequences in the Andropogoneae: *Saccharum*. Theor. Appl. Genet. 88:933–944.

Berding, N., P.H. Moore, and G.R. Smith. 1997. Advances in breeding technology for sugarcane. In Keating, B.A., and Wilson J.R. (eds.), Intensive Sugarcane Production: Meeting the Challenges Beyond 2000. pp. 189–220. CAB International. Wallingford, United Kingdom.

Burnquist, W.L., M.E. Sorrells, and S. Tanksley, 1992. Characterization of genetic variability in *Saccharum* germplasm by means of restriction fragment length polymorphism (RFLP) analysis. Proc. Int. Soc. Sugarcane Technol. 21:355–365.

Carson, D.L., and F.C. Botha. 2002. Genes expressed in sugarcane maturing internodal tissue. Plant Cell Rep. 20:1075–1081.

Carson, D.L., B.I. Huckett, and F.C. Botha. 2002. Sugarcane ESTs differentially expressed in immature and maturing internodal tissue. Plant Science 162:289–300.

Cassman, K.G. 1999. Ecological intensification of cereal production systems: Yield potential, soil quality, and precision agriculture. Proc. Nat. Agron. Soc. USA. 96:5952–5959.

Casu, R.E., C.P.L. Grof, A.L. Rae, C.L. McIntyre, C.M. Dimmock, and J.M. Manners. 2003. Identification of a novel sugar transporter homologue strongly expressed in maturing stem vascular tissues of sugarcane by expressed sequence tag and microarray analysis. Plant Mol. Biol. 52:371–386.

Clayton, W.D., and S.A. Renvoize. 1986. Genera Graminum: Grasses of the world (Kew Bulletin Additional Series XIII). Lubrecht & Cramer, London.

Cook, J.H. 2001. Sugarcane growth and development. Sugar Cane International, p. 5–15.

Da Silva, J., M.E. Sorrells, W.L. Burnquist, and S.D. Tanksley. 1993. RFLP linkage map and genome analysis of *Saccharum spontaneum*. Genome 36:782–791.

Da Silva, J., R.J. Honeycutt, W. Burnquist, S.M. Al-Janabi, M.E. Sorrells, S.D. Tanksley, and B.W.S. Sobral. 1995. *Saccharum spontaneum* L. 'SES 208' genetic linkage map combining RFLP- and PCR-based markers. Mol. Breeding 1:165–179.

Daniels, J., and B.T. Roach. 1987. Taxonomy and evolution. pp. 7–84 in *Sugarcane Improvement Through Breeding*, edited by D.J. Heinz. Elsevier Press, Amsterdam.

Daugrois, J. H., L. Grivet, D. Roques, J. Y. Hoarau, H. Lombard, J.C. Glaszmann, and A. D'Hont 1996. A putative major gene for rust resistance linked with a RFLP markers in sugarcane cultivar 'R570'. Theor. Appl. Genet. 92:1059–1064.

D'Hont, A., Y.H. Lu, P. Feldmann, and J.C. Glaszmann. 1993. Cytoplasmic diversity in sugarcane revealed by heterologous probes. Sugar Cane 1:12–15.

D'Hont, A., F. Paulet, and J.C. Glaszmann. 2002. Oligoclonal interspecific origin of 'North Indian' and 'Chinese' sugarcanes. Chromosome Research. 10:253–262.

D'Hont, A., L. Grivet, P. Feldmann, P.S. Rao, N. Berding, and J.C. Glaszmann. 1995. Identification and characterization of sugarcane intergeneric hybrids, *Saccharum officinarum* x *Erianthus arundinaceus*, with molecular markers and DNA in situ hybridization. Theor. Appl. Genet. 91:320–326.

D'Hont, A., L. Grivet, P. Feldmann, P. S. Rao, N. Berding, and J.C. Glaszmann. 1996. Characterisation of the double genome structure of modern sugarcane cultivars (*Saccharum* spp.) by molecular cytogenetics. Mol. Gen. Genet. 250:405–413.

D'Hont, A., D. Ison, K. Alix, C. Roux, and J.C. Glaszmann. 1998. Determination of basic chromosome numbers in the genus Saccharum by physical mapping of ribosomal RNA genes. Genome. 41:221–225.

Dufour, P., M. Deu, L. Grivet, A. D'Hont, F. Paulet, A. Bouet, C. Lanaud, J.C. Glaszmann, and P. Hamon. 1997. Construction of a composite sorghum genome map and comparison with sugarcane, a related complex polyploid. Theor. Appl. Genet. 94:409–418.

Eshed, Y., and D. Zamir. 1996. Less-than-additive epistatic interactions of quantitative trait loci in tomato. Genetics 143:1807–1817.

Freeling, M. 2001. Grasses as a single genetic system: Reassessment 2001. Plant Physiology 125:1191–1197.

Glaszmann, J.C., Y.H. Lu, and C. Lanaud. 1990. Variation of nuclear ribosomal DNA in sugarcane. J. Genet. Breed. 44:191–198.

Glaz, B. 2003. Integrated crop management for sustainable sugarcane production: recent advances. International Sugar Journal. April, p. 175–186.

Grivet L., A. D'Hont, D. Roques, P. Feldmann, C. Lanaud, and J.C. Glaszmann. 1996. RFLP mapping in cultivated sugarcane (*Saccharum* spp.): Genome organization in a highly polyploid and aneuploid interspecific hybrid. Genetics 142: 987–1000.

Guimarães, C.T., Sills, G.R., and Sobral, B.W.S. 1997. Comparative mapping of Andropogoneae: *Saccharum* L. (sugarcane) in relation to sorghum and maize. Proc. Natl. Acad. Sci. USA 94:14261–14266.

Ha, S., P.H. Moore, D. Heinz, S. Kato, N. Ohmido, and K. Fukui. 1999. Quantitative chromosome map of the polyploid *Saccharum spontaneum* by multicolor fluorescence *in situ* hybridization and imaging methods. Plant Molecular Biology 39:1165–1173.

Heinz, D.J 1987a. Sugarcane Improvement through Breeding. Elsevier, New York, NY.

Heinz, D.J 1987b. Sugarcane improvement: Current productivity and future opportunities. p. 55–70. In Anonymous (ed.), Copersucar International Sugarcane Breeding Workshop. Copersucar, Brazil.

Hoarau, J.-Y., B. Offmann, A. D'Hont, A.-M. Risterucci, D. Roques, J.-C. Glaszmann, and L. Grivet. 2001. Genetic dissection of a modern sugarcane cultivar (*Saccharum* spp.) I. Genome mapping with AFLP markers. Theor. Appl. Genet. 103:84–97.

Hoarau, J.-Y., L. Grivet, B. Offmann, L.-M. Raboin, J.-P. Diorflar, J. Payet, M. Hellmann, A. D'Hont, J.-C. Glaszmann. 2002. Genetic dissection of a modern sugarcane cultivar (*Saccharum* spp.) II. Detection of QTLs for yield components. Theor. Appl. Genet. 105:1027–1037.

Irvine, J. E. 1999. Saccharum species as horticultural classes. Theor. Appl. Genet. 98:186–194.

James, N.I. 1980. Sugarcane. In W.R. Fehr and H.H. Hadley (ed.), Hybridization of Crop Plants. p. 617–629. ASA and CSSA, Madison, WI.

Jannoo, N., L. Grivet, A. Dookun, A. D'Hont, and J.C. Glaszmann. 1999a. Linkage disequilibrium among modern sugarcane cultivars. Theor. Appl. Genet. 99:1053–1060.

Jannoo, N., L. Grivet, M. Seguin, F. Paulet, R. Domaingue, P.S. Rao, A. Dookun, A. D'Hont, and J.C. Glaszmann. 1999b. Molecular investigation of the genetic base of sugarcane cultivars. Theor. Appl. Genet. 99:171–184.

F.O. Licht's International Sugar Report. 2003. Vol. 135, No.7, pp. 108–119.

Lima, M.L.A., A.A.F. Garcia, K.M. Oliveira, S. Matsuoka, H. Arizono, C.L. de Souza Jr., and A.P. de Souza. 2002. Analysis of genetic similarity detected by AFLP and coefficient of parentage among genotypes of sugar cane (Saccharum spp.). Theor. Appl. Genetics. 104:30–38.

Ma, H., S. Schulze, S. Lee, M. Yang, E. Mirkov, J. Irvine, P. Moore, and A. Paterson. 2004. An EST survey of the sugarcane Transcriptome. Theor. Appl. Genet. 108:851–863.

Mamet, L.D., and R. Domaingue. 1999. Shortening the selection process for sugarcane. Experimental Agriculture, 35:391–405.

Ming, R, S.C. Liu, Y.R. Lin, D. Braga, J. da Silva, A. van Deynze, T.F. Wenslaff, K.K. Wu, P.H. Moore, W. Burnquist, M.E. Sorrells, J.E. Irvine, and A.H. Paterson 1998. Detailed Alignment of Saccharum and Sorghum Chromosomes: Comparative Organization of Closely Related Diploid and Polyploid Genomes. Genetics 150:1663–1682.

Ming, R., S.C. Liu, P.H. Moore, J.E. Irvine, and A.H. Paterson. 2001. QTL analysis in a complex autopolyploid: Genetic control of sugar content in sugarcane. Genome Research. 11:2075–2084.

Ming, R., T. Del Monte, E. Hernandez, P.H. Moore, J.E. Irvine, and A.H. Paterson. 2002a. Comparative analysis of QTLs affecting plant height and flowering time among closely-related diploid and polyploid genomes. Genome 45:794–803.

Ming, R., Y.-W. Wang, X. Dryer, P.H. Moore, J.E. Irvine, and A.H. Paterson. 2002b. Molecular dissection of complex traits in autopolyploid: Mapping QTLs affecting sugar yield and related traits in sugarcane. Thoer. Appl. Genet. 105:332–345.

Mudge, J., W.R. Anderson, R.L. Kehrer, and D.J. Fairbanks. 1996. A RAPD genetic map of *Saccharum officinarum*. Crop Science 36: 1362–1366.

Nair, N.V., S. Nair, T.V. Sreenivasan, and M. Mohan. 1999. Analysis of genetic diversity and phylogeny in Saccharum and related genera using RAPD markers. Genet. Res. Crop Evol. 46: 73–79.

Nordborg, M., J.O. Borevitz, J. Bergelson, C.C. Berry, J. Chory, J. Hagenblad, M. Kreitman, J.N. Maloof, T. Noyes, P.J. Oefner, E.A. Stahl, and D. Weigel. 2002. The extent of linkage disequilibrium in Arabidopsis thaliana. Nat. Genet. 30:190–193.

Stevenson, G. C. 1965. Genetics and Breeding of Sugar Cane. London, Longmans, Green and Company.

Simmonds, N.W. 1976. Sugarcane *Saccharum* (Gramineae-Andropogoneae). pp 104–108. In N.W. Simmonds (ed.), Evolution of Crop Plants. Longman, London and New York.

Sobral, B.W.S., D.P.V. Braga, E.S. LaHood, and P. Keim. 1994. Phylogenetic analysis of chloroplast restriction enzyme site mutations in the *Saccharinae* Griseb. Subtribe of the *Andropogoneae* Dumort. Tribe. Theor. Appl. Genet. 87:843–853.

Tai, P.Y.P., and Miller, J.D. 2002. Germplasm diversity among four sugarcane species for sugar composition. Crop Science 42:958–964.

Tew, T.L. 2003. World sugarcane variety census—Year 2000. Sugar Cane International, March/April 2003, pp. 12–18.

Tomkins, J.P., Y. Yu, H. Miller-Smith, D.A. Frisch, S.S. Woo, and R.A. Wing. 1999. A bacterial artificial chromosome library for sugarcane. Theor. Appl. Genet. 99:419–424.

Vettore, A.L., F.R. da Silva, E.L. Kemper, and P. Arruda. 2001. The libraries that made SUCEST. Gen. Mol. Biol. 24(1–4): 1–7.

Whalen, M.D. 1991. Taxonomy of *Saccharum* (Poaceae). Baileya 23:109–125.

Wu, K.K., W. Burnquist, M.E. Sorrells, T.L. Tew, P.H. Moore, and S.D. Tanksley. 1992. The detection and estimation of linkage in polyploids using single-dose restriction fragments. Theor. Appl. Genet. 83:294–300.

Wu, K.K., and E. Gamatero. 2003. 2002/2003 Acreage census of cane varieties. Varieties Report 456, Hawaii Agriculture Research Center.

21

Improving Tolerance to Abiotic Stresses in Staple Crops: A Random or Planned Process?

Gregory Edmeades, Pioneer Hi-Bred International
Marianne Bänziger, CIMMYT, Zimbabwe
Hugo Campos and Jeffrey Schussler, Pioneer Hi-Bred International

Introduction

Three crops, rice (*Oryza sativa* L.), maize (*Zea mays* L.), and wheat (*Triticum aestivum* L.), are vitally important to global food security because they provide almost two-thirds of the calories for human sustenance (Cassman, 1999). By 2020, an additional 20% more people will require food (FAOSTAT, 2003). Global average yields of these three crops (herein referred to as staple crops) continue to rise, though expansion in their planted area is relatively small compared with overall increases in irrigated land and fertilizer consumption (Table 21.1). This suggests that the adoption of improved agronomic practices and varieties is continuing unabated.

Production data at the aggregate level, however, hide significant trends in staple production. One notable example is sub-Saharan Africa, where average crop production environments continue to deteriorate. Although aggregate crop area is slowly rising (Table 21.1), much of the added area is below average quality, and staples are being displaced to lower quality land by urban and industrial development and by higher value crops. In several key production environments, the resource base is deteriorating through compaction, erosion, salinization, or net nutrient export (Cassman, 1999). Irrigated land accounts for 17% of the cropped area but 30% of total production, and water scarcities are entering a critical stage in the Middle East and in much of Asia. Global climate change is now generally considered to be underway (Hillel and Rosenzweig, 2002), providing a long-term trend toward higher temperatures, greater evapotranspiration, and increased incidence of drought in specific regions. Thus, physical environments for staple production are generally becoming harsher, and we cannot depend only on optimum environments to feed the world. Those who live in stress-challenged and risky production environments deserve the added income and security that a more stable food supply brings, and this strategy in turn may spare the use of even more fragile environments for crop production.

Regional and local patterns in abiotic stress

Long-term patterns in key cereal production environments are easily recognized, each with its characteristic pattern of stress. The most common is the summer rainy season, where irrigation is rare, and dry spells of uncertain timing and intensity occur throughout the season. Lowland and upland rice environments are clearly distinguished by water availability, soil compaction, and restricted root growth. Winter wheat grown in Mediterranean environments experiences severe and repeatable terminal drought. Broad areas of acid soils are characteristic of tropical Latin America east of the Andes and in central Africa, and tropical soils are generally more nitrogen deficient than cooler temperate soils. Stress from interplant competition is systematically intensifying in maize, where plant densities are increasing by an average of 700 plants ha^{-1} yr^{-1} in temperate production systems (Edmeades et al., 2003). Since the pattern of genotype × environment interaction in such environments is repeatable over years, breeding programs

Table 21.1 Annual changes for two 11-year periods (1970–1980 and 1990–2000) in area under maize, rice, and wheat, in total irrigated land area, consumption of nitrogenous and phosphatic fertilizer, and in crop yields.

	World	Developed countries	Developing countries
Area of maize, rice, and wheat (m ha yr^{-1})			
1970–1980	5.05*	1.98*	3.06*
1990–2000	0.55 NS	−0.93 NS	1.48*
Irrigated area (m ha yr^{-1})			
1970–1980	4.36*	1.52*	2.84*
1990–2000	2.78*	0.16*	2.62*
Nitrogenous fertilizer use (m ton yr^{-1})			
1970–1980	2.92*	1.30*	1.62*
1990–2000	1.01*	−0.43*	1.45*
Phosphatic fertilizer use (m ton yr^{-1})			
1970–1980	1.04*	0.49*	0.55*
1990–2000	-0.08 NS	−0.72*	0.64*
Grain yields (kg ha^{-1} yr^{-1})			
Maize			
1970–1980	83**	127**	49**
1990–2000	78**	127**	55**
Wheat			
1970–1980	35**	27**	48**
Rice			
1970–1980	40**	−3 NS	41**
1990–2000	45**	59 NS	46**

Source: FAOSTAT, 2003.
Note: Data are linear regression coefficients.
*$P < 0.05$ for each period.

can be designed to provide specific adaptation to these broad environment classes.

Interannual variations in water stress, cold, and heat are much less predictable and are predicted to worsen as global climate change intensifies (Tubiello et al., 2002; Hillel and Rosenzweig, 2002). Significant maize yield losses due to drought occur in about one year in four to five in the western U.S. Corn Belt, and water stress is more frequent and intense in the tropics. Genotype × year interactions can be large and difficult to decompose, and their minimization in otherwise high yielding varieties is a major goal of all plant breeders.

Precision farming research and yield monitors on commercial combines have recently highlighted within-field spatial variability in performance. This reflects variation in soil texture and chemistry, fertility, compaction, plant-available water, and acidity. In general, a single variety sown to such fields must be able to cope with spatial variation in input levels that can result in as much as a 10-fold variation in yield across the field (e.g., Kitchen et al., 1999). Again, a breeding goal will be to minimize yield losses in poor areas of the field without loss of yield in optimal production areas.

Empirical yield gap estimates for maize suggest that more than 50% of potential yield is lost to abiotic constraints. Losses in cereals are mainly due to water and nitrogen, followed by soil acidity, cold, phosphorus, and salinity (Cassman, 2001; Edmeades et al., 2001). Improvements in agronomic practices and in genetic stress tolerance may each reduce this yield gap by 20–30%, but the balance will depend on additional inputs such as water and nitrogen. Genetic improvements are particularly important where agronomic inputs are not available or are unaffordable. For resource-poor farmers, improved seeds are more readily adopted than changed agronomic practices and therefore become an important first step to improving farm family incomes.

Finally, modern plant-breeding systems, starting with the Green Revolution, are increasing our dependence on stable performance of fewer genetic combinations in the field. Many new maize hybrids sold to farmers in the United States, Canada, Argentina, and South Africa are transgenic conversions of a diminishing number of elite genetic combinations, and an undetected level of stress sensitivity in any of these may lead to yield reductions across a significant area.

Choice of selection environments

Target population of environments, multienvironment trials, and their relationship

Environmental classification has focused on the concept of the target population of environments (TPE) that a cultivar might be expected to encounter. Differences among distinct TPEs are defined by long-term patterns of genotype by environment (G × E). However, within each TPE, genotype/year (G × yr) interactions can be expected for stresses such as drought, heat, and cold. Historically, selection for abiotic stress tolerance has relied on performance averaged over multienvironment trials (METs) located in the TPE. If METs provide an inadequate sample of the TPE over space or time (e.g., unusually wet test sites for several years), a cohort of selections may completely escape the stress of interest, and result in stress-sensitive commercial products (Chapman et al., 2003). Many testing schemes are biased toward high-yielding sites because trials are rejected on the basis of their variability or low absolute-yield level. In doing so, valuable performance data are lost from environments where yield stability through stress tolerance could be effectively assayed.

Defining and characterizing the TPE

Experienced breeders have a good feel for the spatial extent of a TPE, but may not appreciate long-term temporal weather patterns. Today, the use of crop simulation with historic weather data, coupled with pattern analysis of model output and weighted by crop distribution, provide tools that describe the TPE by the frequency of occurrence of specific abiotic stresses (Chapman et al., 2003). This approach can "sample" a much greater range of seasons and locations than is feasible through METs. The frequency of principal environment types in a TPE, obtained in this way, can be used to weigh data from any specific MET during selection, and response in the TPE can thus be predicted. Selection of MET sites to sample stresses adequately where G × E and G × yr are major sources of variation is, therefore, a critical step in a successful breeding program. Results from analyses of METs should be monitored and analyzed, and repeatable patterns in the G × E interaction complex accommodated in breeding strategies. TPE/MET approaches have been used successfully to address drought tolerance in wheat and sorghum (*Sorghum vulgare*) in Australia (Chapman et al., 2000).

Repeatable selection environments: MSEs

The stochastic nature of weather-related stresses, and the limited numbers of METs that are possible, have led to the development and use of managed stress environments (MSEs). These are specialized testing sites that allow stringent control of the nature, timing, and intensity of stresses imposed on the target crop. MSEs seek to impose a repeatable stress representative of farmers' fields, yet sufficiently severe that genetic variation for tolerance can clearly be distinguished. With careful management MSEs should decrease environmental variance and increase heritability (or repeatability) for stress-tolerant plant attributes, thereby improving expected genetic gains (e.g., Bänziger et al., 1997). For stresses such as drought, low nitrogen or acidity, MSEs usually depend on rain-free natural environments equipped with irrigation systems, or they comprise spatially uniform, nitrogen-depleted or acidic soil fields. There is an appropriate window of yield reduction targeted in MSEs, and if stress is too intense, genetic variance declines and heritabilities fall (Bolaños and Edmeades, 1996; Bänziger et al., 1997). Traits such as the anthesis–silking interval (ASI) and ears per plant have stable or increasing heritabilities as drought stress at flowering intensifies, even as the heritability for yield declines.

Despite its level of severity, an MSE must represent a significant environment type found in the TPE. Cooper et al. (1997) showed, for instance, that grain yield in low-stress wheat nurseries was a good predictor of optimal conditions in the TPE, but that its value decreased with increasing water stress in farmers' fields. Ceccarelli et al. (1987) noted that the worth of selection in the MSE, when the target environment is the TPE, is given by:

$$CR_{TPE} = (r_A \cdot h_{MSE}/h_{TPE})R_{MSE} \qquad [1]$$

where CR_{TPE} is the correlated response in the TPE, R_{MSE} is the direct response in the MSE, h_{MSE} and h_{TPE} are the square roots of the heritabilities of yield in the MSE and TPE, and r_A is the genetic correlation between grain yields in the two environments. The genetic correlation between yields of the MSE and TPE and heritability in the MSE therefore define the value of the MSE. If r_A is negative, then a crossover interaction between sites has occurred, and, if zero, the MSE has no effect on performance in the TPE. Fortunately there are very few occasions when r_A is < 0.2, and usually it is > 0.5. The intensity of the stress imposed clearly affects r_A (Figure 21.1), and Bänziger and Lafitte (1997) reported values of r_A of zero to 0.71 (mean: 0.38) between 14 pairs of high- and low-nitrogen plots at the same site. The availability of tools such as the genotype + genotype × environment (GGE) biplot (DeLacy et al., 1996) makes the visualization of the correlation among sites and its interpretation fairly straightforward.

Precision phenotyping

The massive increase in capacity to collect genotypic information that has accompanied the genomics revolution generally has not been matched by a concomitant increase in phenotyping capabil-

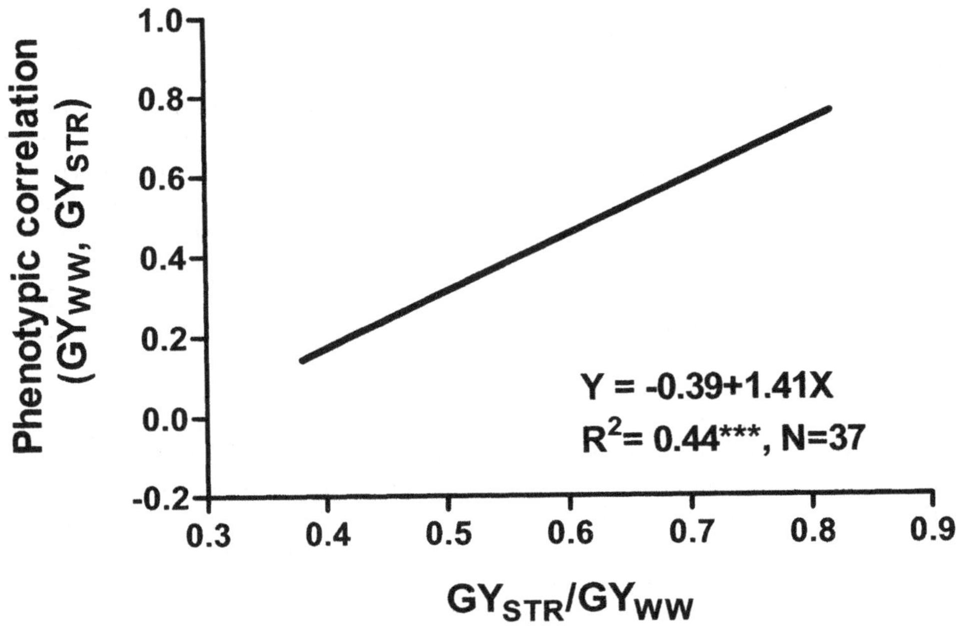

Figure 21.1 Relationship between the phenotypic correlation of yield of temperate maize hybrids under stressed (GY_{STR}) and well-watered (GY_{WW}) conditions vs. the degree to which the stress reduced yield below its well-watered level. Data are from 15 sets of 10–36 modern elite hybrids evaluated under ~3 water levels. Stress was imposed near flowering or early grain filling and in few cases was terminal.

ity. Lack of repeatable gene–phenotype associations may be indicative of biological complexity, but it also reflects inadequacies in phenotyping as well. It is essential to upgrade current phenotyping capabilities in order to identify the most "true to type" phenotype using the least amount of seed, land, and labor, and with a repeatability that is 0.7–0.8 for key traits. This will facilitate the identification of small allelic substitutions without dependence upon a large number of replications and sites, a major consideration for MSEs. Although the same principles apply to laboratory and greenhouse-based tests, our focus here is on field evaluations where the final proof of efficacy must be shown.

Critical elements of precision phenotyping include high throughput, accurate data collection, and improved heritability. Pretrial activities leading to successful studies include

1. selection of fields with a high degree of spatial uniformity in texture and soil depth, cropping history, nutrients, acidity or salinity;
2. choice of appropriate experimental designs, such as incomplete block or augmented designs that reduce within-replicate variation and allow a large number of entries to be tested on small land areas and with limited amounts of seed;
3. uniform maturity among entries to avoid escapes;
4. independent randomization of each environment to eliminate correlated errors; and
5. adequate borders between treatments and among plots, since water and nutrient stress exaggerate edge effects of plots next to alleys.

During the season, repeatability will be increased by

1. uniform application of inputs, such as water, nutrients and herbicides;
2. uniform plant density and spacing within row;
3. knowledge of the yield reduction being targeted and the length of the stress window needed to achieve that yield level at different crop development stages;
4. elimination of unwanted stresses, including insects and disease;
5. maintenance of an unstressed control block to monitor yield potential;
6. use of a harvest system suitable for low-yielding plots;

7. collection of appropriate secondary trait data to enhance selection efficiency and an understanding of critical processes; and
8. use of remote sensing methods to capture data instantaneously so diurnal trends are eliminated.

Postharvest techniques that improve data utility include:

1. generation of adjusted means using augmented or incomplete block analyses;
2. using trelling plots and variograms to visualize residuals and search for outlier points and trends in the data;
3. use of a selection index that weights primary and secondary trait data from several stress levels according to utility; and
4. comparison of MSE trial data with that from the TPE using GGE biplots.

The quality of the information extracted from raw data can be improved using software such as ASREML (Gilmour et al., 1998), which can account for trends existing in the data through spatial modeling and uses the residual maximum likelihood approach to estimate variance components and best linear unbiased predictors of means from unbalanced data sets.

It cannot be stressed enough that statistical techniques are no substitute for careful selection and management of experimental sites. The most powerful statistical analysis can do little to improve data from badly designed or managed trials, and poor-quality data will in turn generate useless or misleading gene–phenotype information.

Understanding abiotic stress tolerance: Putative secondary traits

Attributes associated with high and stable yield under stress are briefly described here so that breeders and molecular geneticists can

- understand traits and their interaction in determining productivity under stress;
- focus on key processes in the selection of candidate genes;
- assess the relationship between stress tolerance and yield potential. Farmers are generally unwilling to sacrifice yield potential for improved stability through tolerance to specific stresses, unless that stress occurs reliably each season; and
- determine if secondary traits could be used with yield during selection.

The physiology and genetics of traits associated with tolerance to specific abiotic stresses have been extensively reviewed (e.g, salinity: Hasegawa et al., 2000; drought: Ludlow and Muchow, 1990; Saini and Westgate, 2000; acidity: De la Fuente-Martinez and Herrera-Estrella, 1999; Hocking, 2001; cold: Thomashow, 1999; heat: Maestri et al., 2002) and will not be considered here in detail. The following is a broad classification of putative traits associated with improved performance under stress.

Yield potential

The ranking of genotypes in stress environments will often be dominated by yield potential until yield declines to around 50% of unstressed levels, when stress-adaptive traits begin to dominate performance (Bänziger et al., 1997). An example is given in Figure 21.1, where variation in yield potential (or unstressed yield) accounts for less than 10% of the variation in a 50% yield reduction environment and the rankings in the two environments are unrelated under a 70% yield reduction. Generally, when stressed yields are < 50–60% of potential, it is more efficient to select for improved yields under the stress itself. Since loss of yield potential is usually an unacceptable price for tolerance to a stress that is stochastic in nature, simultaneous selection under stressed and unstressed conditions serves to combine high yield with stability. Selection in temperate maize based on METs has increased yield potential *and* yield under stress over the past 70 years; Tollenaar and Wu, 1999).

Phenology: Escape versus tolerance

Richards et al. (2002) noted that the single most important factor maximizing yield and adaptation in dry environments is appropriate flowering time, and simulation studies show its importance in drought-prone areas (Chapman et al., 2003). Stable grain production depends on a good match between adequate temperatures, rainfall, nutrient supply, the timing of stress-susceptible developmental phases, and the length of the crop cycle. Earliness per se will reduce yield potential, so combining stress tolerance with full-season maturity for that environment should result in high but stable yield.

Partitioning: Increasing tolerance to stress during kernel set

Around 60% of global crop production is made up of fruit or grains, the products of reproductive growth. Yield of the staples under stress at flowering is closely associated ($r = 0.7$ to 0.9) with kernel number and much less strongly ($r = 0.1$ to 0.5) with kernel weight. In most crops, kernel number per plant is proportional to plant growth rate around flowering, and any stress that lowers photosynthesis per plant at flowering will reduce kernel set. In maize this is associated with ear growth that is inhibited by stress more than tassel growth, leading to an ASI that increases under stress. Grain yield and kernel number per plant show strong associations with ASI ($r = -0.4$ to -0.7) when stresses of several types coincide with flowering (Edmeades et al., 2000b). Carbohydrate flux to the ovaries from current photosynthesis is a major determinant of reproductive success under water stress (Schussler and Westgate, 1991). Inhibition of invertases by water stress and heat is thought to have a role in reducing ovary growth in maize (Zinselmeier et al., 2000; Andersen et al., 2002), and in stress-induced pollen sterility in wheat (Saini and Westgate, 2000). Water deficits and shading both deplete ovary starch reserves, and when spikelet growth falls below some threshold, abortion will occur. This threshold appears to be more genetically variable in maize than in wheat or rice.

Factors other than assimilate flux to the ear can also affect kernel number. An increase in ASI in maize may also indicate that pollen shed is essentially over by the time silks emerge. The coincidence of pollen and emerged silks has been recently used to successfully model kernel set at low pollen concentrations typical of stressed fields (Lizaso et al., 2003). Spikelet sterility has been associated with poor panicle exsertion, delayed flowering, and low panicle water potential in rice (Pantuwan et al., 2002). Kernel abortion can be reduced by enhancing cytokinin levels through upregulation of iso-pentenyl transferase or downregulation of cytokinin oxidase, while ethylene and abscisic acid (ABA) are both thought to increase abortion (Jones and Setter, 2000). Increasing kernel set does not, however, guarantee an increase in yield unless they can be successfully filled.

Selection in tropical maize populations under water stress at flowering has resulted in a reduction in ASI and an increase in kernel set and grain yield under drought. Thus, ASI has proven to be an easily observed trait indicative of improved partitioning to the ear and increased kernel set in maize, and it provides a route to improved general stress tolerance in this species.

Resource capture

Roots

Models of nitrogen and water capture under limiting conditions show significant benefits from fine roots that grow deeper in the profile (King et al., 2003). Root systems have evolved to capture scarce nutrients concentrated in the upper soil strata, and specific placement of nutrients makes this less important. Root length densities of 1 cm cm^{-3} are considered adequate for water capture, and up to 2–5 cm cm^{-3} where nitrogen and phosphorus are limiting. Manipulating the genetic control of root morphology without increasing root biomass is a research priority, though it is well known that root architecture is strongly influenced by nutrient availability (López-Bucio et al., 2003). Roots respond to soil hardness and porosity by reducing their own growth and signaling a reduction in growth of tops as well (Passioura, 2002). Anoxic soils present a contrast: a high root-length density near the surface and aerenchyma formation aid in oxygen exchange in rice and permit tolerance to waterlogging.

Staygreen

Drought and low nitrogen accelerate death of foliage. Staygreen can reflect nitrogen balance in cereals during grain filling (Borrell at al., 2001), but greenness can also be cosmetic rather than functional (Thomas and Howarth, 2000). Green leaf area is associated with improved performance under drought (Borrell et al., 2000) and nitrogen stress (Bänziger and Lafitte, 1997), though its value varies with species.

Input efficiency use

Passioura (1977) first described grain yield as the product of water transpired × water use efficiency (WUE; gram of biomass per gram of water) × harvest index (HI), and the same framework can be applied to inputs such as nitrogen and phosphorus.

WUE

The ratio of stable C isotopes, C^{13}/C^{12} (δ) in the plant reflects the ratio of assimilatory capacity to stomatal conductivity, is negatively associated with WUE in C_3 plants, and is related to yield in a complex manner (Condon et al., 2002). Selection for high WUE using δ has resulted in the recent release of the wheat variety Drysdale in Australia with ~10% yield advantage over sibs with a low WUE under dry conditions (Richards et al., 2002).

Nitrogen use efficiency

Although genetic differences in nitrogen use efficiency (NUE) occur among species, variation within a species is not large (Lafitte and Edmeades, 1994). Improving the recovery of applied nitrogen through improved root exploration is likely to have greater impact than changes in NUE per se in temperate maize where nitrogen is usually plentiful. Under acute nitrogen deficiency, such as in tropical environments, NUE is more important than nitrogen recovery, since little nitrogen remains in the rooting zone at season's end.

Defensive traits

Remobilization of plant reserves

Rice and wheat depend heavily on remobilization as a means of maintaining kernel-filling rates and final kernel weight under stress. This defensive mechanism is much less prevalent in maize, where kernel set (Schussler and Westgate, 1994) and grain filling (Borrás et al., 2004) are much less affected by assimilate reserves.

Osmotic adjustment

Osmotic adjustment (OA), a rather controversial trait, has shown value in wheat (Blum et al., 1999), but modeling approaches suggest it increases yield by an average of only 2% in sorghum under variable rainfall conditions in Queensland (Snell, 2002). Morgan (1999) has described a simple pollen test that identifies the *or* gene in wheat on the basis of absorption of potassium ions from solution and has demonstrated its adaptive value. Serraj and Sinclair (2002) suggest that OA is effective only when the plant is severely stressed and that its principal value lies in maintaining root growth in dry soils.

Antioxidants

Tolerance to drought, cold, heat and salt is associated with an increase in reactive oxygen intermediates (ROIs) and a concomitant rise in antioxidant enzyme activity that protects cell membranes from oxidative damage (Mittler, 2002). Damage to the photosynthetic apparatus in the form of photoinhibition followed by bleaching of chlorophyll is common when stressed plants are exposed to intense radiation. Upright or rolled leaves are less susceptible to this type of damage. Overexpression of the antioxidant superoxide dismutase increases tolerance to chilling in maize and to drought in wheat (Noctor and Foyer, 1998). Other active ROI scavengers include ascorbate peroxidase and catalase, the latter being important under stress.

Protectants

Cellular membrane thermostability, as determined by the conductance of ions leaking from stressed tissue, varies significantly in wheat and was correlated ($r = 0.5^{**}$) with yield under high temperatures (Blum et al., 2001). Heat shock proteins have not been causally related to thermotolerance in cereals, unlike HSP101 in *Arabidopsis* (Maestri et al., 2002). Dehydrin proteins increase sharply in tissues during desiccation or chilling, are upregulated by ABA, and are thought to stabilize plasma membrane structure (Campbell and Close, 1997). The family of *CBF* transcription factors, first identified in *Arabidopsis* (Thomashow, 1999; Seki et al., 2003), has been associated with upregulating many genes associated with tolerance to frost and drought in commercial species, and it shows considerable promise.

Antiporters

Salinity is a common problem in areas where irrigation has been misused or where water is scarce. The gene *AtNHX1* sequesters Na in the vacuole of leaves and, when introduced to tomatoes, has allowed them to grow successfully in a 200-mM NaCl solution (Zhang and Blumwald, 2001).

Abscisic acid

In general, high ABA concentration has been associated with plant survival rather than increased production. Under stress, ABA concentration rises, leading to increased root growth, stomatal closure, leaf shedding, dormancy, and possibly abortion of tip kernels in maize (Jones and Setter, 2000). Mugo et al. (2000) reported a decrease in leaf ABA in maize as drought tolerance increased during recurrent selection, though this could also reflect improved plant water status. Recent reports suggest a role for ABA in limiting ethylene production, thus serving to maintain rather than inhibit growth under stress (Sharp, 2002).

Epigenetic and genetic effects of stress

Abiotic stress appears to have direct effects on the genome. Tsaftaris and Polidoros (2000) reported that high plant density increased the methylation of cytosine bases in maize inbred lines, though not in hybrids. Recently, Verde et al. (2003) reported that the level of meiotic recombination in several populations of maize increases under water stress, suggesting a stress-related response in genetic fitness.

Under what circumstances could secondary traits play a role in selection?

It is highly unlikely that a secondary trait will replace grain yield during selection (see Equation 1 and Falconer and MacKay, 1996, for theory), but adding selected secondary traits to yield in a performance index could improve gain. Bänziger and Lafitte (1997) estimated the heritability of an index comprising grain yield, ASI, ears per plant, and staygreen to be 14% greater than for yield alone for maize selected under low nitrogen. Ideally, a secondary trait should be (1) genetically associated with grain yield under stress; (2) highly heritable; (3) cheap and fast to measure; (4) stable over the measurement period; (5) observed at or before flowering; and (6) not associated with yield loss under unstressed conditions. The need for field validation of stress tolerance mechanisms in terms of improved or stabilized grain yields under farmers' conditions cannot be overstated.

The value of a secondary trait can be estimated by correlation and heritability estimates among progenies of a single population, divergent selection for the trait, or by index theory (Bänziger and Lafitte, 1997). A method showing considerable promise is simulation of breeding programs that include Mendelian control of traits. Chapman et al. (2003) modeled the effects of maturity, staygreen, OA, and transpiration efficiency in sorghum using 108 years of daily weather data from six locations in northern Australia and were able to place a value on each trait for specific crop environments. Tardieu (2003) has proposed combining QTL with crop models, and his group has demonstrated the validity of this approach (Reymond et al., 2003).

Choice of germplasm is perhaps the most important decision taken in plant breeding. In the pursuit of sources of unique secondary traits it is often tempting to move directly to exotic sources. These are often poorly adapted and hence difficult to evaluate, and they carry significant yield drag. Thus, the first step should be to thoroughly evaluate existing elite materials under the assumption that stress-tolerant alleles are present at low frequencies in most elite breeding programs.

Improvements from conventional selection: The baseline for comparison

Conventional selection provides the baseline against which any improved selection method should be compared. Here we consider three examples, two from maize and one from wheat.

Multi-environment testing improves yield under stress in temperate maize

Previous studies have demonstrated improved yield potential and tolerance to high plant density and moderate drought stress through METs and conventional selection (Castleberry et al., 1984; Duvick, 1997; Tollenaar and Wu, 1999).

Density tolerance

Pioneer scientists have expanded the Duvick (1997) Era hybrid set to include the most recent hybrids. This set comprises the leading four to six commercial Pioneer hybrids from each decade from 1930 to present. These were grown under irrigation, high solar radiation, and 300 kg nitrogen ha^{-1} at two plant densities (4.4 and 8.8 plants m^{-2}) in the western United States. Optimum densities for grain yield increased at 760 plants ha^{-1} yr^{-1} from 1920 through 2001 (Edmeades et al., 2003), and predicted yield at that density increased at 109 kg ha^{-1} yr^{-1}. This matches the increase in plant density (690 plants ha^{-1} yr^{-1}) at which farmers are planting maize in the Corn Belt (unpublished data, Pioneer Hi-Bred Int., 2002).

Drought tolerance

A second study of a subset of Era hybrids was conducted in a rain-free location. Eighteen hybrids of similar maturity commercially released between 1953 and present ($n = 3$ hybrids per decade) were chosen and sown at one density (8.5 plants m^{-2}) in a three-replicate incomplete block design in four-row plots. In the second year of this two-year study water was withdrawn for periods of ~512 °Cd (~54 day) in five distinct but overlapping periods 94–132 °Cd apart, starting 258 °Cd after planting. In the latest two intervals leaves senesced before the stress period ended. A well-watered control was included. These six treatments were named control, flower, early-fill, mid-fill, late-fill, and terminal, but each stress period imposed some stress on the previous growth stage as stress intensity increased with time. Flowering occurred 10 days after the end of the first interval, at ~630 °Cd. Leaf rolling and staygreen scores (1 = rolled, dead; 9 = unrolled, green), ASI, yield, and its components were estimated from bordered plot areas. A measure of plant-to-plant variability was obtained from stem volume, estimated from plant height (h) and stalk diameter ($2r$) at the internode below the ear, as $\pi r^2 h$, for 25 consecutive plants in one bordered row per plot.

Results show the relative yields for the six treatments as 100%, 42%, 29%, 30%, 47%, and 73% of the well-watered control. Rates of gain in grain yield, estimated by regression, showed a marked improvement under well-watered conditions (237 kg ha^{-1} yr^{-1}), followed by 102 kg $ha^{-1}yr^{-1}$ for stress coinciding with flowering. Gains then fell off under later stress periods, reaching a minimum in late-fill (Figure 21.2a). Susceptibility, S (Fischer and Maurer, 1978), at flowering was unaffected by selection, but increased under late-fill and terminal stress (Figure 21.2b). The lack of increase in kernel weight and staygreen and low genetic gain under stressed versus unstressed condition during late grainfill indicate that progress in stress tolerance for C assimilation during this growth stage has lagged behind that observed in optimal conditions. This contrasts with clear improvements under flowering stress that were associated with reductions in barrenness (Figure 21.2e) and in ASI of -1.5 °Cd yr^{-1}. Interestingly, variation for stem volume fell with selection (Figure 21.2f), and reduced plant-to-plant variability has been associated with improved stress tolerance (Tollenaar and Wu, 1999). Grain yield and ASI, across water treatments, were significantly correlated ($r = -0.62^{**}$).

Improving tolerance in tropical maize using managed stresses

This has been documented extensively elsewhere (Beck et al., 1996; Edmeades et al., 2000a). In summary, recurrent selection, either as full-sib or S1, was conducted for two to eight cycles in four tropical populations, normally under optimal and two managed-drought stress regimes in rain-free environments in Mexico. Evaluations of cycle bulks and conventionally selected checks took place both in Mexico and other countries (Table 21.2A) and verified that 83% of gains (average 143 kg ha^{-1} $cycle^{-1}$) observed in Mexico in the dry winter season were also observed in normal growing seasons elsewhere. Progress for drought tolerance was significantly greater in populations where selection had taken place in managed stress environments than by the METs of the conventional program.

In a later assessment of progress, C_0 and the most advanced cycles (C_4 to C_6) of three tropical populations were grown within droughted and well-watered Pioneer test sites varying from 19 to 35 °N latitude. As day lengths increased, yields of tropical populations declined, but gains from selection were maintained at 132 kg ha^{-1} $cycle^{-1}$ despite their lack of adaptation. Gains obtained under managed-stress winter conditions, therefore, were approximately 80–90% effective in the normal summer growing environment, even in areas where adaptation was poor for that germplasm.

Selecting under drought also improved performance under moderate nitrogen stress (Table 21.2A; Bänziger et al., 1999) and at high plant densities (Mugo et al., 2003). The existence of a common relationship linking crop growth rate and kernel number under drought and nitrogen stress (Andrade et al., 2002) supports the contention of a single mechanism linking tolerance in maize to any stress that reduces photosynthesis per plant at flowering. When evaluations were conducted under a greater degree of nitrogen stress (71% yield reduction), however, gains declined (Table 21.2B) as tolerance mechanisms specific to nitrogen stress assumed greater importance.

These methods have been applied in a regular maize-breeding program in southern Africa since 1996 with considerable success (Figure 21.3) and have led to the release and use of several maize varieties with significant productivity increases under smallholder farmers' conditions. Progenies are routinely evaluated under optimal conditions,

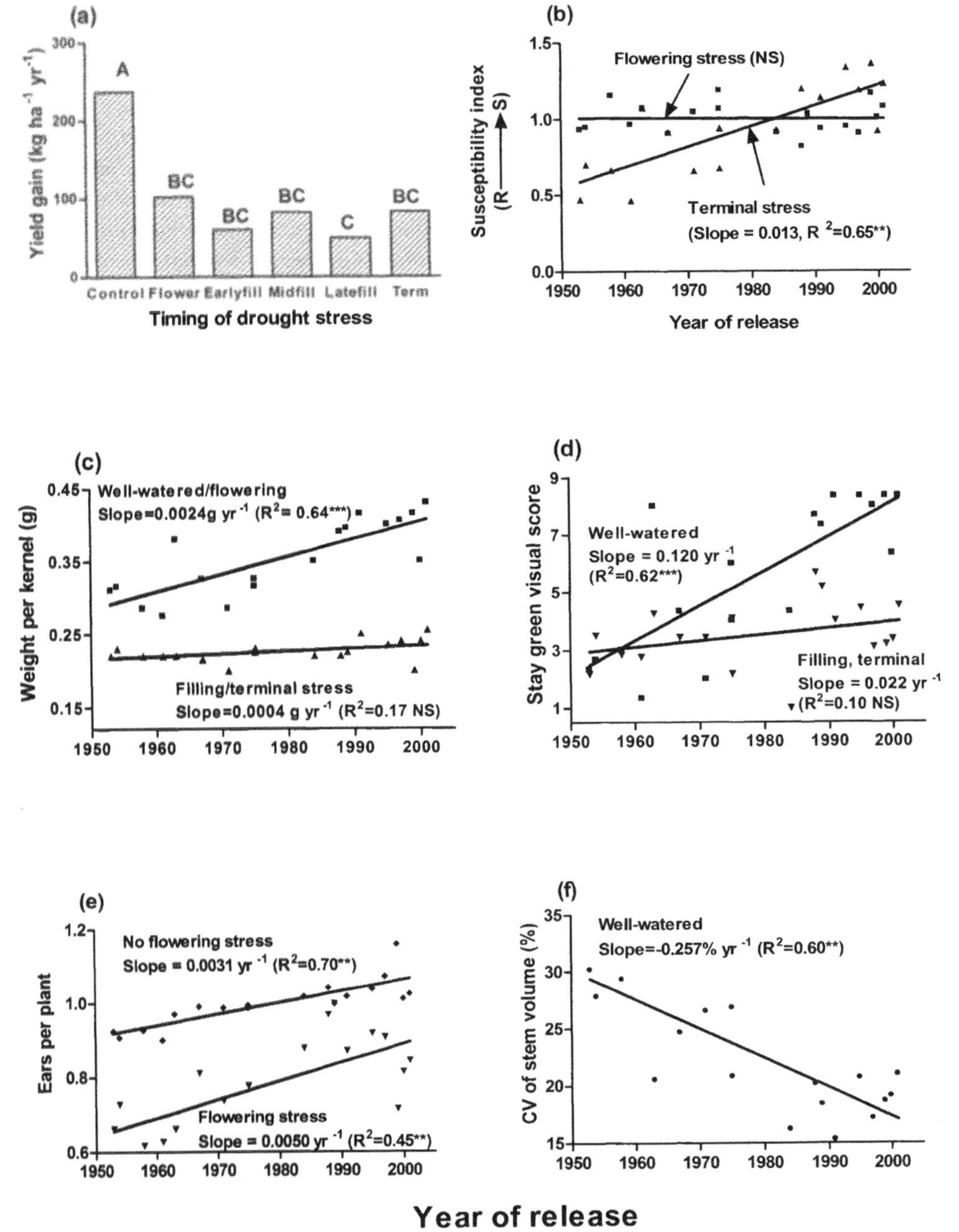

Figure 21.2 Effect of water regime or selection on (a) rate of gain in grain yield; (b) susceptibility index (S; Fischer and Maurer, 1978); (c) weight per kernel; (d) visual staygreen score (9 = green, 1 = senesced); (e) ears per plant; and (f) coefficient of variation of stem volume, in 18 hybrids grown under six water regimes, 2002–2003.

managed drought and nitrogen stress, and performance in the three is equally weighted during selection. Index selection and improved statistical design and analysis techniques, in particular AS-REML, are used to increase gains under stress. Index selection includes yield-stabilizing traits whose heritability is unchanged or increases under stress, such as reduced ASI, barrenness, and leaf senescence (Bolaños and Edmeades, 1996; Betrán et al., 2003). Noteworthy outcomes are (1) uniform gains across stress levels (Figure 21.3), also observed earlier in selection under managed stress

Table 21.2 Effects of selection for drought tolerance on gains per selection cycle in four tropical maize populations when evaluated at 3–6 water-stressed (SS) sites, at 5–8 well-watered (WW) sites, or at two low-nitrogen sites in 1992–94 and evaluations in 1997–98 were conducted at three low-nitrogen and five drought-stressed locations only

Population (cycles completed)	Yield (kg ha^{-1} $cycle^{-1}$) SS	WW	Low N	Anthesis WW (d)	ASI SS (d)	Ears $plant^{-1}$ SS
A. Evaluation 1992/1994						
La Posta Sequía (3)	229**	53 ns	233	−0.52**	−1.18**	0.07**
Pool 26 Sequía (3)	288**	177**	207	−0.93**	−1.50**	0.08**
Tuxpeño Sequía (8)	80**	38**	86	−0.32**	−0.44**	0.02**
Pool 18 Sequía (2)	146**	126**	190		−2.13**	0.05**
Mean gain	186	99	179			
Relative yield	30%	100%	59%			
B. Evaluations 1997/1998						
La Posta Sequía (5)	154**	110 ns	134*	0.33**	−1.01**	0.027
Pool 26 Sequía (3)	163**	142 ns	27 ns	−0.30ns	−1.84**	0.044**
Tuxpeño Sequía (10)	104*	67 ns	21 ns	−0.35ns	−0.58ns	0.018ns
Mean gain	140	106	61			
Relative yield	21%	100%	29%			

Source: (Beck et al., 1996) and (Edmeades et al., 2000a).
*, **, ns, signify significant rate of change per selection cycle at $P < 0.01$, $P < 0.05$, or $P > 0.05$, respectively.

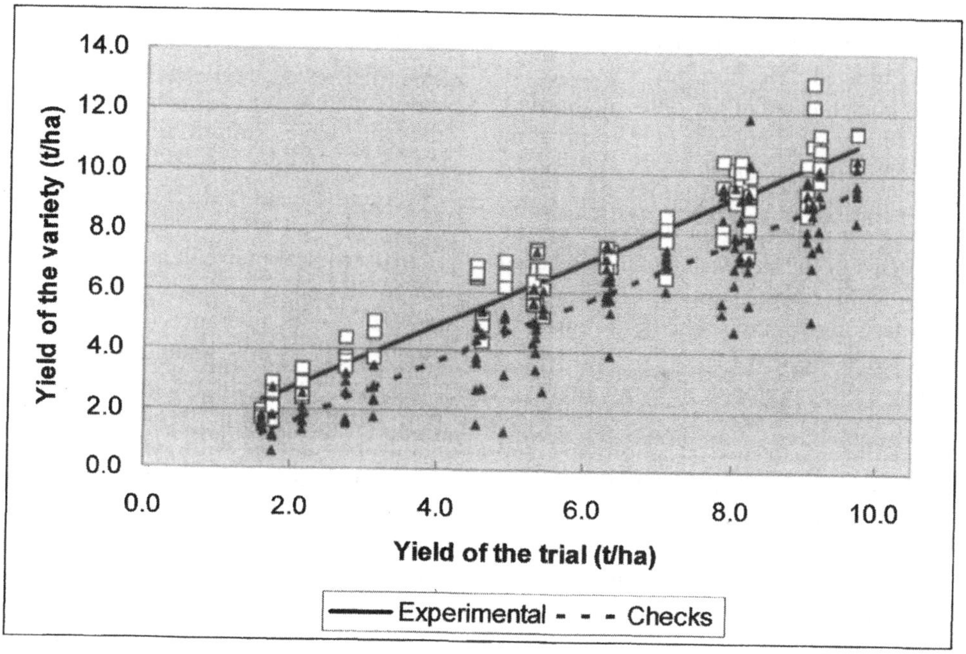

Figure 21.3 Performance of four experimental hybrids (experimental) and five best private company hybrids (checks) of similar maturity across 23 randomly stressed locations in eastern and southern Africa. Experimental hybrids were selected using managed stress environments for drought and low N.

in Mexico (Bolaños and Edmeades, 1993) and (2) adaptation to a wide range of environments in southern and eastern Africa, even though selection is conducted in a few environments and in one country (Zimbabwe).

Improving drought tolerance in spring wheat under managed stress

Beginning in the 1980s, International Maize and Wheat Improvement Center (CIMMYT) wheat breeders, using wheat germplasm that had been bred under optimal conditions, began to screen finished varieties for performance under drought stress. The best of these were entered in a specific MET (SAWYT) targeting semiarid areas. Their performance relative to the trial mean was compared over a seven-year period with those varieties selected strictly under well-watered conditions in CIMMYT's ESWYT MET grown at sites classified as rain-fed and lower yielding. While the best lines from ESWYT and SAWYT both showed a positive increase in productivity with time, the SAWYT lines showed the greatest improvement over the trial mean in low-yielding environments (Trethowan et al., 2002). Directed selection for tolerance to drought was implemented in the 1990s by shuttling segregating materials between two contrasting moisture regimes. In Obregón F_3 lines were exposed to severe terminal stress by sowing on preirrigated beds that were not watered again. The next generation was selected in the well-watered Toluca Valley, and this shuttle was repeated until F_6 (Trethowan et al., 2001). Selections from this scheme are currently being evaluated in the SAWYT.

In summary, these examples illustrate that

- METs have been effective at improving performance under drought in wheat and temperate maize, though less so in tropical maize. Rates of improvement for drought tolerance in the TPE are less than under MSE selection.
- Reduction in heritability for yield under stress has not reduced efficiency of selection substantially, provided drought-adaptive secondary traits with stable heritabilities are also used during selection.
- Improvement in tolerance to one stress (e.g., drought) in maize has resulted in improvements in tolerance to other stresses as well, presumably through a common mechanism linking kernel set with assimilation at flowering.

Manipulating the genome: New ways of identifying, creating and utilizing variability

The integration of genetic and molecular approaches to clarify gene-to-phenotype relationships, leading to improved breeding systems, is often termed molecular breeding. An implicit assumption here is that phenotying will be of the highest quality, since, without this, genetic information will be of limited practical value. Tools associated with quantifying gene-to-phenotype associations are described in detail elsewhere in this book (Cooper et al., Chapter 10, this book; Mackill, Chapter 14, this book) and are considered only briefly here.

QTL analysis (linkage mapping)

This relies upon marker-trait associations, where individual marker genotypes have different expected values for traits influenced by QTLs linked to these markers. Many QTL associated with drought tolerance have been identified in rice, and the colocalization of these for specific traits is beginning to identify key root attributes associated with adaptation to drought (Price et al., 2002). Unfortunately most QTL are population dependent and have limited use as selection tools (Mackill, Chapter 14, this book). Ribaut and coworkers (Ribaut et al., 2003) have recently begun to assemble all QTL associated with stress tolerance in tropical maize in a single consensus map, where "hot-spots" indicate genomic areas that are associated with tolerance in many crosses. Mapping of QTL is now being attempted in complex pedigrees, using pedigree information to circumvent context-specific results. Methods combine within-population with across-population information to better understand epistatic relationships among QTLs, while haplotype analysis helps detect ancestral genomic blocks in parental lines (Jansen et al., 2003).

Detailed QTL analyses also provide entry points to isolate, identify and perhaps clone agronomically important genes. To achieve this, QTLs are Mendelized one by one through recombinant backcross lines and then fine mapped and placed onto physical maps, narrowing their genetic location to intervals under 100 kb. Cloned genes can then be used to explore the genetic control and the physiology of the trait in a controlled manner and modify expression levels with different promoters.

Marker-assisted selection

Marker-assisted selection (MAS) strategies are a means of transferring underutilized QTL alleles associated with stress-tolerant phenotypes into otherwise susceptible genetic backgrounds using flanking molecular markers. QTL × E analysis and an understanding of epistatic interactions affecting QTL expression are needed, however, before attempting QTL transfer. A cost-benefit analysis comparing conventional selection with MAS is a prerequisite before starting a major MAS project (e.g., Morris et al., 2003). Difficulties associated with repeatability of stress levels across seasons and challenging trait assays make the prospects of MAS particularly appealing for abiotic stress tolerance and root-related traits. Alternative approaches include the simultaneous identification and transfer of favorable QTL alleles from unadapted material to elite lines (Tanksley and Nelson, 1996). There are relatively few examples of the successful transfer of a complex trait such as drought tolerance to susceptible target lines (e.g., Ribaut et al., 2003).

Association analysis and linkage disequilibrium

Association analysis is an alternative to linkage mapping and is based on linkage disequilibrium in areas where recombination occurs less frequently than expected (see Flint-Garcia et al., 2003, for description). A correlation therefore exists among neighbor alleles, indicating haplotypes transmitted intact from ancestral genomic regions. Coupled with pedigree information, it can achieve a higher resolution than linkage mapping, since it derives its power from many recombination rounds and scans many alleles at each locus. This is particularly appealing for the analysis of the genetic architecture of abiotic stress in staple crops where the pedigree structure of the population is known. In elite temperate maize, haplotypes have been found that extend up to 100 kb in length (Ching et al., 2002). Association analyses can be conducted on a whole genome basis or can be focused on candidate genes. The latter approach can substantially reduce the amount of genotyping required. MAS becomes cost-effective and efficient using single nucleotide polymorphisms that are usually distributed throughout the haplotype region.

Genomics

The largest long-term impact of genomics may derive from a detailed understanding of the patterns of genetic variation that breeding programs create and on which they depend. In the short term, however, genomics is being used to address two main issues: gathering information from genes through sequencing, expressed sequence tags, and full cDNA-length initiatives, and the use of transcriptional profiling to identify gene networks associated with the expression of traits and physiological processes when the plant or tissue is exposed to a defined stress. Microarrays targeting stress responses are available in rice (e.g., Poroyko et al., 2003) and several other crops. The combined analysis of multiple expression patterns and genomic sequence data can be used to identify genes directly involved in protecting against environmental stresses, as well as those regulating gene expression and signal transduction pathways involved in the stress response. Related approaches such as proteomics and metabolomics can also evaluate posttranslational responses through protein products, pool sizes, and metabolic fluxes associated with specific traits. These tools provide information on structural genes and regulatory sequences that can be used in transgenic approaches to improve stress tolerance.

Transformation

Modern technologies allow us to insert, express and up– or downregulate transgenes, with specificity in expression by tissue and growth stage. Because of its commercial utility and power as a research tool, as transformation in crop plants becomes more predictable it will likely become a dominant methodology. Of special interest are genes that affect the fate of aborting spikelets, the dynamics of grain fill under stress, and their relationship to source activity. When single genes are involved that generate a product such as a hormone that modulates other processes, interaction with genetic background and the environment are likely to be important. Finally, the regulatory and social costs of this approach must be considered carefully during commercial product development.

Delivery of stress-tolerant cultivars in resource-poor farming systems

Adoption of improved crop varieties in the tropics is still well below 50%. Reasons for nonadoption include lack of a viable seed sector and varieties with an advantage under smallholder farmers' conditions. This often requires both abiotic stress tolerance and suitability to consumer preferences (e.g., Smale et al., 1995). Self-pollinated crops retain their genetic integrity for far longer during farmer-to-farmer transfers than do cross-pollinated crops, and, not surprisingly, this reflects in the higher adoption rates of improved wheat (Heisey et al., 2002) versus improved maize varieties (Morris, 2002) in low-income countries.

Public and private breeding programs in less-developed countries are hampered by human and financial resource constraints, low yields, weak infrastructure, lack of attractive markets, an absence of quality-enhancing competition, and inappropriate seed policies (Tripp and Rohrbach, 2001). As well, such programs are often focused on the small proportion of farmers that can afford to purchase seed and other inputs regularly, and the rest are less well served. Donor-driven efforts to support stress breeding tend to be short-term and rarely address delivery of improved seed on a significant scale. Trade liberalization policies have sharply reduced public sector seed production and have resulted in a void. Certainly, where the private sector is willing and able to invest, this is the preferred route because of its experience and proven track record in maintaining supply and seed quality. However, commercial yields in stressed environments often make improved seed uneconomical to adopt (see Pixley, Chapter 17, this book). Innovative approaches to produce and distribute seed of stress-tolerant varieties to those most in need of it are therefore needed.

There is strong evidence that farmer-participatory variety selection may speed up farmer-to-farmer dissemination of self-pollinated crops as reported for rice (WARDA, 2001; Witcombe et al., 2005, Chapter 7 of this book), but will be less effective for cross-pollinated crops. In eastern and southern Africa, systematic institutional collaboration on farmer-participatory variety selection between the public, private, and nongovernmental organization sector has provided an effective model for production and dissemination of improved stress-tolerant open-pollinated maize varieties (Bänziger and De Meyer, 2002). Using a network based on interinstitutional collaboration, new maize varieties that were first evaluated and released in several countries in 1999 occupy an area over 100,000 ha in 2003.

Conclusions

Guiding principles when designing a breeding program to improve abiotic stress tolerance include the following:

- Existence of good evidence that stress tolerance can be improved at no cost to yield potential provided yield potential is monitored during selection.
- Clear definition of the TPE, so optimal choices are made regarding germplasm, a representative set of METs, appropriate mechanisms of stress tolerance, and suitable field screens under managed stress levels.
- Precision phenotyping, where repeatability is high (around 0.5–0.7) and is critical when establishing gene–phenotype relationships. Attention to basic practices of site uniformity, bordering, appropriate design, spatial trend analyses, and use of high-throughput measures of key traits are vital to the detection of genomic regions associated with stress tolerance.
- Research on reproductive success in staple crops as a high priority because of the nutritional importance of grain and the pivotal importance of kernel set under stress.
- Secondary traits and key processes related to stress tolerance that favor productivity rather than survival and that carry little or no yield drag. Mechanisms that provide general stress tolerance, especially in factors affecting kernel set (e.g., ASI in maize; spikelet sterility in rice; role of ABA in kernel abortion) are especially valuable. For all secondary traits and candidate genes and regions, performance in the field remains the definitive proving ground.
- Conventional selection that forms the baseline against which new techniques can be compared. Wide area testing has improved yields under drought and high density in temperate maize. Directed selection for tolerance to drought has accelerated gains in tropical maize and spring wheat. Use of unstressed control plots and of easily observed secondary traits that have stable heritability under stress along with yield under stress in index selection has helped ensure gains of around 80–100 kg ha^{-1} yr^{-1} at yield levels of 20–100% of yield potential in tropical maize populations.
- More progress made in the short term by increasing the frequency of underutilized stress-

tolerant alleles in elite germplasm versus exotics sources. Poor adaptation and yield drag under unstressed conditions in exotics demands a prebreeding approach when using these in commercial product development.

- Promising new techniques in molecular breeding and genomics that will speed the rate at which broadly adapted stable high-yielding varieties are developed. Techniques for identifying context-independent QTLs and haplotypes are being developed, and MAS and transformation will provide the means of testing candidate genes and regions alone or in combination in near-isogenic line comparisons.
- Impact in low-income countries from stress-tolerant germplasm that depends on its effective delivery to resource-poor farmers. This will take significant commitment by the public and private sectors, plus accommodation to each others' constraints. The recent example of maize open-pollinated varieties in southern Africa offers real hope, however, that this can be accomplished.

References

Andersen, M.N., F. Asch, Y. Wu, C.R. Jensen, H. Naested, V.O. Mogensen, and K.E. Koch. 2002. Soluble invertase expression is an early target of drought stress during the critical, abortion-sensitive phase of young ovary development in maize. Plant Physiol. 130:591–604.

Andrade, F.H., L. Echarte, R. Rizalli, A. Della Maggiora, and M. Casanovas. 2002. Kernel number prediction in maize under nitrogen or water stress. Crop Science 42:1173–1179.

Bänziger, M., F.J. Betrán, and H.R. Lafitte. 1997. Efficiency of high-nitrogen selection environments for improving maize for low-nitrogen target environments. Crop Science 37:1103–1109.

Bänziger, M., and J. De Meyer. 2002. Collaborative maize variety development for stress-prone environments in southern Africa. p. 269–296. *In* D. A. Cleveland and D. Soleri (eds), Farmers, Scientists and Plant Breeding: Integrating Knowledge and Practice. CABI, Oxon, United Kingdom.

Bänziger, M., G.O. Edmeades, and H.R. Lafitte. 1999. Selection for drought tolerance increases maize yields over a range of nitrogen levels. Crop Science 39:1035–1040.

Bänziger, M., and H.R. Lafitte. 1997. Efficiency of secondary traits for improving maize for low-nitrogen target environments. Crop Science 37:1110–1117.

Beck, D., J. Betrán, M. Bänziger, G. Edmeades, J.-M. Ribaut, M. Willcox, S.K. Vasal, and A. Ortega C. 1996. Progress in developing drought and low soil nitrogen tolerance in maize. p. 85–111. *In* D. Wilkinson (ed), Proc. 51st Annual Corn and Sorghum Research Conference, Chicago, December 10–11, 1996. ASTA, Washington.

Betrán, F.J., D.L. Beck, M. Bänziger, and G.O. Edmeades. 2003. Secondary traits in parental inbreds and hybrids under stress and non-stress environments in tropical maize. Field Crops Res. 83:51–65.

Blum, A., N. Klueva, and H.T. Nguyen. 2001. Wheat cellular thermotolerance is related to yield under heat stress. Euphytica 117:117–123.

Blum, A., J. Zhang, and H.T. Nguyen. 1999. Consistent differences among wheat cultivars in osmotic adjustment and their relationship to plant production. Field Crops Res. 64:287–291.

Bolaños, J., and G.O. Edmeades. 1993. Eight cycles of selection for drought tolerance in lowland tropical maize. I. Responses in grain yield, biomass, and radiation utilization. Field Crops Res. 31:233–252.

Bolaños, J., and G.O. Edmeades. 1996. The importance of the anthesis-silking interval in breeding for drought tolerance in tropical maize. Field Crops Res. 48:65–80.

Borrás, L, G.A. Slafer, and M.E. Otegui. 2004. Seed dry weight response to source-sink manipulations in wheat, maize and soybean. A quantitative reappraisal. Field Crops Res. (in press).

Borrell, A.K., G.L. Hammer, and R.G. Henzell. 2000. Does maintaining green leaf area in sorghum improve yield under drought? II. Dry matter production and yield. Crop Science 40:1037–1048.

Borrell, A.K., G.L. Hammer, and E. van Oosterom. 2001. Stay-green: a consequence of the balance between supply and demand for nitrogen during grain filling? Ann. Appl. Biol. 138:91–95

Campbell, S.A., and T.J. Close. 1997. Dehydrins: genes, proteins, and associations with phenotypic traits. New Phytol. 137: 61–74.

Cassman, K.G. 1999. Ecological intensification of cereal production systems: yield potential, soil quality, and precision agriculture. Proc. Natl. Acad. Sci. 96:5952–5959.

Cassman, K.G. 2001. Crop science research to secure food security. p. 33–51. *In* J. Nosberger, H.H. Geiger, and P.C. Struik (eds). Crop Science: Progress and Prospects. CABI, Wallingford, United Kingdom.

Castleberry, R.M., C.W. Crum, and C.F. Krull. 1984. Genetic improvements of U.S. maize cultivars under varying fertility and climatic environments. Crop Science 24:33–36.

Ceccarelli, S., M.M. Nachit, G.O. Ferrara, M.S. Mekni, M.Tahir. J. van Leur, and J.P. Srivastava. 1987. Breeding strategies for improving cereal yield and stability under drought. p. 101–114. *In* J.P. Srivastava, E. Porceddu, E. Acevedo, and S. Varma (eds). Drought Tolerance in Winter Cereals. Wiley & Sons, New York, NY.

Chapman, S.C., G.L. Hammer, D.G. Butler, and M. Cooper. 2000. Genotype by environment interactions affecting grain sorghum III. Temporal sequences and spatial patterns in the target population of environments. Aust. J. Agric. Res. 51:223–233.

Chapman, S.C., M. Cooper, D. Podlich, and G. Hammer. 2003. Evaluating plant breeding strategies by simulating gene action and dryland environment effects. Agron. J. 95:99–113.

Ching, A., K.S. Caldwell, M. Jung, M. Dolan, O.S. Smith, S. Tingey, M. Morgante, and J.A. Rafalski. 2002. SNP frequency, haplotype structure and linkage disequilibrium in elite maize inbred lines. Biomed. Central Genet. 3:19–32.

Condon, A.G., R.A. Richards, G.J. Rebetzke, and G.D. Farquhar. 2002. Improving intrinsic water-use efficiency and crop yield. Crop Science 42:122–131.

Cooper, M., R.E. Stucker, I.H. DeLacy, and B.D. Harch. 1997. Wheat breeding nurseries, target environments, and indirect selection for grain yield. Crop Science 37:1168–1176.

Cooper, M., O.S. Smith, R.E. Merrill, L. Arthur, D.W. Podlich, and C.M. Löffler. 2005. Integrating relevant tools and information for more efficient plant breeding. p. xx-xx. In: Lamkey, K.R. and M. Lee (eds). The Arnel R. Hallauer International Symposium on Plant Breeding. Blackwell Publishing, Ames, IA.

DeLacy, I, K.E. Basford, M. Cooper, J.K. Bull, and C.G. McLaren. 1996. Analysis of multi-environment trials—an historical perspective. p. 39–124. *In* M. Cooper and G.L. Hammer (eds). Plant Adaptation and Crop Improvement. CABI, Wallingford and ICRISAT, Hyderabad.

De la Fuente-Martinez, J. M., and L. Herrera-Estrella. 1999. Advances in the understanding of aluminum toxicity and the development of aluminum-tolerant transgenic plants. Adv. Agron. 66:103–120.

Duvick, D.N. 1997. What is yield? p. 332–335. *In* G.O. Edmeades et al. (eds), Developing Drought and Low–N Tolerant Maize. CIMMYT, El Batan, Mexico.

Edmeades, G.O., M. Bänziger, and J.-M. Ribaut. 2000a. Maize improvement for drought-limited environments. p. 75–112. *In* M.E. Otegui and G.A. Slafer (eds), Physiological Bases for Maize Improvement. Haworth Press, New York, NY.

Edmeades, G.O., J. Bolaños, A. Elings, J.-M. Ribaut, M. Bänziger, and M.E. Westgate. 2000b. The role and regulation of the anthesis-silking interval in maize. p. 43–73. *In* M.E. Westgate and K.J. Boote (eds), Physiology and Modeling Kernel Set in Maize. CSSA Special Publication No. 29. CSSA, Madison, WI.

Edmeades, G.O., M. Cooper, H.R. Lafitte, C. Zinselmeier, J.-M. Ribaut, J.E. Habben, C. Löffler, and M. Bänziger. 2001. Abiotic stresses and staple crops. p. 137–154. *In* J. Nosberger, H.H. Geiger, and P.C. Struik (eds). Crop Science: Progress and Prospects. CABI, Wallingford, United Kingdom.

Edmeades, G.O., J. Schussler, H. Campos, C. Zinselmeier, J. Habben, S. Collinson, M. Cooper, M. Hoffbeck, and O. Smith. 2003. Increasing the odds of success in selecting for abiotic stress tolerance in maize. Proc. 5th Australian Maize Conference (in press).

Falconer, D.S., and T.F.C. MacKay. 1996. Introduction to Quantitative Genetics. Prentice Hall, New York, NY.

FAOSTAT. 2003. http://apps.fao.org/

Fischer, R.A., and R. Maurer. 1978. Drought resistance in spring wheat cultivars. I. Grain yield responses. Aust. J. Agric. Res. 29:897–912.

Flint-Garcia, S.A., J.M. Thornsberry, and E.S. Buckler. 2003. Structure of linkage disequilibrium in plants. Annu. Rev. Plant Biol. 54:357–374.

Gilmour, A.R., B.R. Cullis, S.J. Welham, and R. Thompson. 1998. ASREML. NSW Agriculture, Orange NSW, Australia. p. 177.

Hasegawa, P.M., R.A. Bressan, J.–K. Zhu, and H.J. Bohnert. 2000. Plant cellular and molecular responses to high salinity. Annu. Rev. Plant Physiol. Plant Mol. Biol. 51:463–499.

Heisey, P.W., M.A. Lantican, and H.J. Dubin. 2002. Impacts of international wheat breeding research in developing countries, 1966–97. Mexico, D.F.: CIMMYT. p. 74.

Hillel, D., and C. Rosenzweig. 2002. Desertification in relation to climate variability and change. Adv. Agron. 77: 1–38.

Hocking, P.J. 2001. Organic acids exuded from roots in phosphorus uptake and aluminum tolerance of plants in acid soils. Adv. Agron. 74:63–97.

Jansen, R.C., J.L. Jannink, and W.D Beavis. 2003. Mapping quantitative trait loci in plant breeding populations: Use of parental haplotype sharing. Crop Science 43:829–834.

Jones, R.J., and T.L. Setter. 2000. Hormonal regulation of early kernel development. pp. 24–42. *In* M.E. Westgate and K.J. Boote (eds), Physiology and Modeling Kernel Set in Maize. CSSA Special Publication No. 29. CSSA, Madison, WI.

King, J., A. Gay, R. Sylvester-Bradley, I. Bingham, J. Foulkes, P.J. Gregory, and D. Robinson. 2003. Modelling cereal root systems for water and nitrogen capture: Towards an economic optimum. Ann. Bot. 91:383–390.

Kitchen, N.R., K.A. Sudduth, and S.T. Drummond. 1999. Soil electrical conductivity as a crop productivity measure for claypan soils. J. Prod. Agric. 12:607–617.

Lafitte, H.R., and G.O. Edmeades. 1994. Improvement for tolerance to low nitrogen in tropical maize. II. Grain yield, biomass production, and N accumulation. Field Crops Res. 39:15–25.

Lizaso, J.I., M.E. Westgate, W.D. Batchelor, and A. Fonseca. 2003. Predicting potential kernel set in maize from simple flowering characteristics. Crop Science 43:892–903.

López-Bucio, J., A. Cruz-Ramírez, and L. Herrera-Estrella. 2003. The role of nutrient availability in regulating root architecture. Curr. Opinion Plant Biol. 6:280–287.

Ludlow, M.M., and R.C. Muchow. 1990. A critical evaluation of traits for improving crop yields in water-limited environments. Adv. Agron. 43:107–153.

Maestri, E., N. Klueva, C. Perotta, M. Gulli, H.T. Nguyen, and N. Marmiroli. 2002. Molecular genetics of heat tolerance and heat shock proteins in cereals. Plant Molec. Biol. 48:667–681.

Mittler, R. 2002. Oxidative stress, antioxidants and stress tolerance. Trends Plant Sci. 7:405–410.

Morgan, J.M. 1999. Pollen grain expression of a gene controlling differences in osmoregulation in wheat leaves: A simple breeding method. Aust. J. Agric. Res. 50:953–962.

Morris, M.L. 2002. Impacts of international maize breeding research in developing countries, 1966–98. CIMMYT, Mexico, D.F. 54 pp.

Morris, M.L., K. Dreher, J.-M. Ribaut, and M. Khairallah. 2003. Money matters (II): Costs of maize inbred line conversion schemes at CIMMYT using conventional and marker-assisted schemes. Molec. Breeding 11: 235–247.

Mugo, S.N., M. Bänziger, and G.O. Edmeades. 2000. Prospects of using ABA in selection for drought tolerance in cereal crops. p. 73–78. *In*: Ribaut, J.M., and D. Poland (eds). Molecular approaches for the genetic improvement of cereals for stable production in water-limited environments. A strategic planning workshop held at CIMMYT, El Batán, Mexico, June 21–25, 1998.

Mugo, S.N., G.O. Edmeades, and D.T. Kirubi. 2003. Genetic improvement for drought tolerance increases tolerance to high plant density in tropical maize under low input levels. p. 50–51. *In* Abstracts, Arnel R. Hallauer International Symposium on Plant Breeding, CIMMYT, Mexico, 17–20 August, 2003.

Mugo, S.N., M.E. Smith, T.L. Setter, M. Bänziger, G.O. Edmeades, and A. Elings. 1998. Performance of early maturing Katumani and Kito maize composites under drought at the seedling and flowering stages. African Crop Sci. J. 6:329–344.

Noctor, G., and C.H. Foyer. 1998. Ascorbate and glutathione: Keeping active oxygen under control. Annu. Rev. Plant Physiol. Plant Mol. Biol. 49:249–279.

Pantuwan, G., S. Fukai, M. Cooper, R. Rajatasereekul, and J.C. O'Toole. 2002. Yield response of rice (*Oryza sativa* L.) genotypes to drought under rainfed lowland. 3. Plant factors contributing to drought resistance. Field Crops Res. 73:181–200.

Passioura, J.B. 1977. Grain yield, harvest index, and water use of wheat. J. Aust. Inst. Agric. Sci. 43: 117–120.

Passioura, J.B. 2002. Soil conditions and plant growth. Plant,Cell and Env. 25:311–318.

Poroyko, V., M. Fredriksen, S. Shepley, M.S. Pathan, H. Nguyen, R. Bruskiewich, J. Bennett, R. Lafitte, and H.J. Bohnert. 2003. A rice microarray representing stress related transcripts. ASPB Abstracts http://abstracts.aspb.org/pb2003/public/P59/0643.html.

Price, A.H., J.E. Cairns, P. Horton, H.G. Jones, and H. Griffiths. 2002. Linking drought-resistance mechanisms to drought avoidance in upland rice using a QTL approach: Progress and new opportunities to integrate stomatal and mesophyll responses. J. Exp. Bot. 53: 989–1004.

Reymond, M., B. Muller, A. Leonardi, A. Charcosset, and F. Tardieu. 2003. Combining quantitative trait loci analysis and an ecophysiological model to analyze the genetic variability of the responses of maize leaf growth to temperature and water deficit. Plant Physiol. 131:664–675.

Ribaut, J.-M., M. Bänziger, T.L. Setter, G.O. Edmeades, and D. Hoisington. 2003. Genetic dissection of drought tolerance in maize. *In* H. Nguyen, A. Blum (eds), Physiology and Biotechnology Integration for Plant Breeding. Marcel Dekker, NY (in press).

Richards, R.A., G.J. Rebetzke, A.G. Condon, and A.F. van Herwaarden. 2002. Breeding opportunities for increasing the efficiency of water use and crop yield in temperate cereals. Crop Science 42:111–121.

Saini, H.S., and M.E. Westgate. 2000. Reproductive development in grain crops during drought. Adv. Agron. 68:59–96.

Schussler, J.R., and M.E. Westgate. 1991. Maize kernel set at low water potential: II. Sensitivity to reduced assimilates at pollination. Crop Science 31:1196–1203.

Schussler, J.R., and M.E. Westgate. 1994. Increasing assimilate reserves does not prevent kernel abortion at low water potential in maize. Crop Science 34:1569–1576.

Seki. M., A. Kamei, K. Yamaguchi-Shinozaki, and K. Shinozaki. 2003. Molecular responses to drought, salinity and frost: Common and different paths for plant protection. Curr. Opin. Biotech. 14:194–199.

Serraj, R., and T.R. Sinclair. 2002. Osmolyte accumulation: Can it really help increase crop yield under drought conditions? Plant Cell & Env. 25:333–341.

Sharp, R.E. 2002. Interaction with ethylene: Changing views on the role of abscisic acid in root and shoot growth response to water stress. Plant, Cell & Env. 25:211–222.

Smale, M., P.W. Heisey, and H.D. Leathers. 1995. Maize of the ancestors and modern varieties: The microeconomics of HYV adoption in Malawi. Econ. Devel. Cultural Change 43:351–368.

Snell, P.J. 2002. The contribution of osmotic adjustment to grain yield of sorghum in dryland production environments. PhD Dissertation, University of Queensland, Brisbane, Australia.

Tanksley, S.D., and J.C. Nelson. 1996. Advanced backcross QTL analysis: A method for the simultaneous discovery and transfer of valuable QTLs from unadapted germplasm into elite breeding lines. Theor. Appl. Genet. 91:191–203.

Tardieu, F. 2003. Virtual plants: Modelling as a tool for the genomics of tolerance to water deficit. Trends Plant Sci. 8:9–14.

Thomas, H. and C.J. Howarth. 2000. Five ways to stay green. J. Exp. Bot. 51:329–337.

Thomashow, M.F. 1999. Plant cold acclimation: Freezing tolerance genes and regulatory mechanisms. Annu. Rev. Plant Physiol. Plant Mol. Biol. 50:571–599.

Tollenaar, M., and J. Wu. 1999. Yield improvement in temperate maize is attributable to greater stress tolerance. Crop Science 39:1597–1604.

Trethowan, R.M., J. Crossa, M. van Ginkel, and S. Rajaram. 2001. Relationships among bread wheat international yield testing locations in dry areas. Crop Science 41:1461–1469.

Trethowan, R.M., M. van Ginkel, and S. Rajaram. 2002. Progress in breeding wheat for yield and adaptation in global drought affected environments. Crop Science 42: 1441–1446.

Tripp, R., and D. Rohrbach. 2001. Policies for African seed enterprise development. Food Policy 26:147–161.

Tsaftaris, A.S., and A.N. Polidoros. 2000. DNA methylation and plant breeding. Plant Breeding Rev. 18:87–176.

Tubiello, F.N., C. Rosenzweig, R.A. Goldberg, S. Jagtap, and J.W. Jones. 2002. Effect of climate change on US crop production: Simulation results using two different GCM scenarios. Part 1: wheat, potato, maize and citrus. Climate Res. 20:259–270.

Verde, L.A., W. Woodman-Clikeman, M.J. Long, D. Nettleton, and M. Lee. 2003. Increased meiotic recombination in maize genotypes after chronic water-deficit stress. Maize Genet. Conf. Abstr. 45:P67.

WARDA. 2001. NERICA—rice for life. p. 1–8. WARDA, Bouake, Ivory Coast.

Zinselmeier, C., J.E. Habben, M.E. Westgate, and J.S. Boyer. 2000. Carbohydrate metabolism in setting and aborting maize ovaries. p. 1–13. *In* Westgate, M.E., and Boote, K.J. (eds.) Physiology and Modeling Kernel Set in Maize. CSSA Special Publication No. 29. CSSA, Madison, WI.

Zhang, H.-X., and E. Blumwald. 2001. Transgenic salt-tolerant tomato plants accumulate salt in foliage but not in fruit. Nature Biotech. 19:765–768.

22

Breeding for Resistance to Biotic Stresses

R.P. Singh, CIMMYT, Mexico
J. Huerta-Espino, Campo Experimental Valle de Mexico-INIFAP, Mexico
M. William, CIMMYT, Mexico

Abstract

Disease control to achieve stable food production has been one of the challenges to crop scientists for nearly a century. The occurrence of large-scale epidemics, common in the first half of the twentieth century, has decreased thanks to an improved understanding of disease epidemiology and the genetic basis of host–pathogen interactions, to the search for resistance genes, and to the development of cultivars with built-in resistance to important diseases and pests. The race-specific type of resistance based on major genes has been improperly used to control rapidly evolving pathogens (e.g., the rusts of wheat), resulting in boom-and-bust cycles that make it necessary to replace cultivars a short time after their release or to implement chemical control strategies. Durable resistance to important diseases and pests is often based on the interaction of a few minor, additive genes. Breeding strategies to combine such genetic resistance with other, desirable traits must become a primary goal of breeding programs aimed at achieving long-term control of major diseases and pests and ensuring stable food production gains while protecting the environment.

Diseases and pests have posed major threats to stable production of many food crops for centuries. An improved understanding of pathogen epidemiology, combined with the identification and utilization of genetic resistance in crop varieties, has helped prevent or reduce disease epidemics common until 30 or 40 years ago. This success can be attributed to numerous independent research findings that have led to improved genetic control of diseases and pests. Better approaches can be developed from the lessons learned from the successes and failures and from knowledge not available previously.

Historical scientific discoveries that laid the foundations of current resistance breeding

Mendelian inheritance of resistance

Although our ancestors selected for resistance to diseases and pests for centuries, understanding the scientific basis of resistance is only a century old. Determination of the genetic basis of plant disease resistance began soon after the rediscovery of Mendel's laws of segregation and independent assortment in 1901. Rowland Harry Biffen reported the first results in 1905 in Cambridge, UK. He studied the F_1 and F_2 populations from a cross of yellow (stripe) rust (caused by *Puccinia striiformis*) resistant wheat cultivar Rivet with susceptible Red King. The F_1 plants were susceptible, and he observed that 195 plants of the F_2 generation were infected with the disease, while 64 were rust-free. This segregation conformed to a 3:1 ratio, as predicted from Mendelian genetics, and indicated that resistance was recessive and susceptibility dominant. Subsequently, many examples of Mendelian inheritance of disease resistance were documented.

The concept of pathogen races

Following two devastating stem rust epidemics in North America in 1904 and 1916, E.C. Stakman and his coworkers at the University of Minnesota were able to show a biologic form (race) of *Puc-*

cinia graminis (Stakman and Piemeisel, 1917). Soon afterward, wheat cultivars differentiating stem rust and leaf rust populations were identified, and a system of nomenclature based on the resistance and susceptibility response patterns of those cultivars was developed in the United States and adopted in other countries. For 36 years Stakman was extensively engaged not only in determining various aspects of wheat rust epidemiology but also in training numerous scientists who later made major contributions toward achieving effective disease resistance.

The gene-for-gene relationship

H.H. Flor was the first person to initiate simultaneous inheritance studies involving the pathogenicity of flax rust (*Melampsora lini*) and resistance in its host flax (*Linum usitatissimum*). These studies led to the formulation of the "gene-for-gene relationship" concept (Flor, 1956), which states that "for each gene conditioning rust reaction in the host there is a specific corresponding gene conditioning pathogenicity in the parasite." According to this concept, resistance is expressed only when the host cultivar carries resistance alleles and the pathogen race possesses the corresponding avirulence alleles. In the three other alternative situations, the host–pathogen interaction results in susceptibility. Flor's work received more attention when Person (1959) provided a theoretical analysis of the gene-for-gene relationship and concluded that such relationships occur as a general rule in host–parasite systems as a result of selection pressure during evolution. Soon after publication of Person's study, the gene-for-gene concept was demonstrated or hypothesized in a number of host–pathogen systems.

International cooperation in testing for resistance

The International Spring Wheat Rust Nursery Program, initiated in 1950 by B. B. Bayles and R. A. Rodenhiser of USDA-ARS (United States Department of Agriculture–Agricultural Research Services), Beltsville, operated continuously until the mid 1980s. The objectives of the program were (1) to find new genes or combinations of genes in wheat that condition field resistance to rust throughout the world, and (2) to test new varieties and promising selections of wheat developed by plant breeders and pathologists for resistance to rusts. The germplasm and information generated were made available to the global wheat community. This nursery was the foundation of numerous other international nurseries and led to global cooperation to achieve resistance to diseases and pests of several crops.

Concept of utilizing resistance genes in combinations

I.A. Watson and coworkers conducted a very dynamic program to study the evolution of new pathogen races in Australia and used this information in breeding stem rust-resistant cultivars carrying single resistance genes. The deployment of these cultivars contributed to the phenomenon of "boom-and-bust" disease cycles and made it clear that this breeding philosophy would need further refining. The concept of using combinations of resistance genes arose from this experience. Watson was firmly convinced that the development of stable resistance in wheat cultivars requires the use of many distinct resistance genes (McIntosh and Smith-White, 1986). Because not everyone followed the strategy and released cultivars that had those genes singly, resistance breakdowns in a stepwise manner led to a similar problem, as experienced with the use of single genes for resistance.

The multiline approach to deploying resistance genes

Deploying multilines to achieve disease control was proposed in the early 1950s by Jensen (1952) and Borlaug (1953). However, breeders have since realized that although using multilines can control diseases, it is a very conservative, slow approach to achieving yield increases because new varieties may rapidly supersede the recurrent parent. Problems also arose when multiplying each component line and replacing components that became susceptible. For these reasons, the approach did not succeed.

Race-nonspecific and durable resistance

Working with late blight of potato in Mexico, J. S. Niederhauser and colleagues demonstrated the presence of "partial resistance," a term Niederhauser first used in describing the resistance in cultivated varieties and wild species of potato (Niederhauser et al., 1954). In field trials he found that these varieties show a degree of resistance that is exhibited equally toward all races of a pathogen. These varieties remained green longer than varieties without partial resistance.

J.E. Vanderplank (1963) conceptually delineated

resistance into two types, vertical and horizontal. He defined vertical resistance as being effective against some races of the pathogen but ineffective against others; horizontal resistance was defined as being equally directed against all races of the pathogen. In relation to horizontal and vertical resistance, pathogens can have variability for corresponding aggressiveness and virulence, respectively. Vanderplank also pointed out that both types of resistance can, and often do, coexist. This delineation provoked numerous debates from the followers of the gene-for-gene relationship, but now seems to be widely accepted. Several other synonymous terms have been used since then to refer to this type of resistance, namely, race-specific and race-nonspecific resistance.

R.M. Caldwell of Purdue University elucidated the importance of breeding for general resistance to plant diseases (Caldwell, 1968). General resistance is such that no natural variants of a pathogen are able to compensate for the restrictions to their penetration, development, or dispersion that such resistance imposes. This can usually be determined by prolonged testing. Working with leaf rust of wheat, Caldwell (1968) emphasized that slow rusting is a form of general or horizontal resistance. He stated that such resistance to rusts involves mechanisms such as exclusion of the fungus, limitation of the pustule size without hypersensitivity, or, possibly, slow growth and development of the fungus. The joint action of these host characters may drastically slow down a disease epidemic to the point of insignificance. J.E. Parlevliet (1975) used the term "partial resistance" to characterize slow rusting to leaf rust in barley. He maintained that partial resistance is a form of incomplete resistance characterized by a reduced rate of epidemic development despite a high or susceptible infection type. Parlevliet (1976, 1986) also showed that partial resistance to leaf rust involves interaction of minor genes and that the components of resistance may be under pleiotropic genetic control.

R. Johnson of Cambridge, England, described the presence of durable resistance to yellow rust in winter wheat cultivar Cappelle Desprez. This moderate level of adult plant resistance had remained effective for over 20 years in the UK when such resistance was recognized as durable. Durable resistance as defined by Johnson (1978) is that which has remained effective in a cultivar during its widespread cultivation for a long sequence of generations or period of time in an environment favorable to a disease or pest. This term has received wide acceptance and is popularly used in the literature for resistance to several diseases and pests.

Gene identification and utilization

The last quarter of the twentieth century is highlighted by the identification, chromosomal localization, search and transfers from alien sources, and attempts to pyramid race-specific genes in crop improvement. Advances in molecular biology in recent times have resulted in cloning of several resistance genes in many plant species and have shown the striking structural similarity of race-specific resistance genes. Numerous race-specific genes are now known to confer resistance to most diseases and pests. To give an example, 51 resistance genes have been catalogued for resistance to leaf rust of wheat (McIntosh et al., 1998), indicating high genetic diversity for resistance. Except for two genes, all others confer the hypersensitive type of resistance, and virulence for a majority of these hypersensitive resistance genes has been reported. This means useful genetic diversity is much less than would appear from the number of catalogued genes. The *Puccinia triticina* (causal pathogen of leaf rust of wheat) population in Mexico has evolved new virulence within three years, on average, from the time a new semidwarf cultivar is released. A similar phenomenon has been observed in several diseases of wheat and other crops. This experience has prompted some scientists to emphasize the application in crop improvement of race-nonspecific resistance to some major pathogens and pests.

Successful genetic control of diseases and pests in the future

Based on historical scientific knowledge, we can list three factors that dictate the long-term success of breeding for resistance to pathogens and pests.

Variation in pathogen and pest populations

There is significant variation for virulence to specific resistance genes in populations of highly specialized obligate (biotrophic) parasites. The rapid evolution of new virulence through migration, mutation, or recombination and selection of existing virulences is more frequent in some obligate parasites (e.g., wheat and barley rusts, powdery mildew, potato late blight fungi). For this reason, breeding for resistance to these pathogens has always been challenging. Although physiological races are known to occur in most bunts and smuts, evolution and selection of new races are less frequent, which means genetic resistance remains effective longer. Changes in pathogen races are even less common in facultative (necrotrophic) parasites, possibly because there may be no significant advantage to the new races surviving over the old ones on other crops or in crop residues during the off-season. Walker (1966) gave examples of certain diseases where monogenic resistance remained effective for over 50 years against the biotypes that existed in the United States. The genus Fusarium, causal agent of head blight of wheat, although it has several species that cause the same disease, does not appear to have races. Biotypes of insect pests and nematodes are also known to occur, and some of them evolve quite rapidly.

Screening methodology and selection environments

The probability of identifying resistant parents and resistant progenies is increased through the use of a reliable screening methodology and an environment favorable to disease development and pest infestation. Depending on the disease and type of resistance, the methodology may require simple greenhouse tests on seedlings or adult plants, replicated field tests, or even the use of resistance-linked protein and DNA markers. Protocols for screening for resistance to most important diseases and pests are well established. For example, the inclusion of resistant and susceptible check cultivars is important to assess disease pressure and degree of resistance. Choosing appropriate field sites with reliable environmental conditions is crucial when conducting selection in the field. Breeding for resistance to pathogens and pests that are difficult to evaluate using traditional methodologies may require significant investment in developing DNA markers. The advantage of using molecular markers versus traditional resistance screening methodologies will depend on the cost of identifying closely linked markers and the cost of running routine marker assays.

Choosing the type of resistance

Genetic diversity for resistance to most pathogens and pests usually exists either within the crop itself or its related species and genera. Because race-specific genes are relatively easier to identify and select for, they have received the most attention from geneticists, breeders, and, more recently, molecular biologists. Such genes are likely to provide long-term control of pathogens and pests that are slow to evolve, but only short-term control of pathogens that frequently produce new races. Had all diseases been successfully controlled for a long time through single resistance genes, perhaps we would not be discussing the best possible resistance-breeding approach. Combining effective race-specific genes is an attractive strategy for increasing the longevity of this resistance, but, to be successful, its deployment has to be planned at both the regional and global levels. Furthermore, when race-specific resistance is to be used in breeding, a parallel investment is required to monitor the evolution of pathogens and pests and to continually search for new, effective resistance genes.

Diversity of resistance in a particular geographic or epidemiologic region can be maintained by growing cultivars that carry different resistance genes. However, there is a general tendency for farmers to grow only one or a few "choice" cultivars, which, as a result, come to occupy large areas. Growing fewer cultivars that carry race-specific resistance genes can lead to greater genetic uniformity and, consequently, greater disease vulnerability. An effective strategy for avoiding genetic vulnerability is to breed for race-nonspecific resistance that promises to be more durable. Although minor genes that have additive effects mostly confer durable resistance, a single gene, *mlo*, is known to confer a high level of durable resistance to barley powdery mildew fungus.

Basis of durable resistance to important pathogens of major food crops

The wheat rusts

The three wheat rusts (stem, leaf, and stripe) are probably the most-studied diseases due to their historical and continued global importance. Disease epidemics continue to occur despite the fact that numerous race-specific resistance genes have been identified. Improper utilization of race-specific genes has resulted in the appearance of new races even in recent times. Despite these failures, some cultivars have remained resistant for a long time; the basis of their durable resistance is discussed below.

Sr2 and other minor genes for durable resistance to stem rust

Resistance gene *Sr2,* in addition to other unknown minor genes derived from cultivar Hope, provided the foundation for durable resistance to stem rust (caused by *Puccinia graminis*) in germplasm from University of Minnesota in the United States, Sydney University in Australia, and the spring wheat germplasm developed by Dr. N.E. Borlaug as part of a program sponsored by the Mexican Government and the Rockefeller Foundation. Cultivar Yaqui 50, released in Mexico during the 1950s, and other *Sr2*-carrying wheats released since then have stabilized the stem rust situation in Mexico and other countries where modern semi-dwarf wheats were adopted. Changes in stem rust races have not been observed in Mexico in recent years, and natural infections are nonexistent. Released in 1960 in the Indian Subcontinent and subsequently grown on millions of hectares, the cultivar Sonalika has also remained resistant. When present alone, the *Sr2* gene confers slow rusting that is not adequate under heavy disease pressure, but does provide adequate resistance in combination with other minor genes. Unfortunately, not much is known about the other genes in the *Sr2* complex and their interactions. Knott (1988) has shown that adequate levels of multigenic resistance to stem rust can be achieved by accumulating approximately five minor genes. In his studies the genes were different from *Sr2*.

Lr34 and other minor genes for durable resistance to leaf rust

The South American cultivar Frontana is considered one of the best sources of durable resistance to leaf rust, caused by *Puccinia triticina* (Roelfs, 1988). The Mexican-Rockefeller Program first used the variety in the 1950s. Later derivatives such as Penjamo 62, Torim 73, and Kalyan/Bluebird showed slow rusting characteristics possibly derived from Frontana. Genetic analysis of Frontana and several CIMMYT wheats possessing excellent slow-rusting resistance to leaf rust worldwide has indicated that such adult plant resistance is based on the additive interaction of *Lr34* and two or three additional slow-rusting genes (Singh and Rajaram, 1992). Leaf rust severity observed in Mexico on most slow-rusting cultivars can be related to the number of minor genes they carry (Table 22.1). When susceptible cultivars display 100% leaf rust severity, cultivars with only *Lr34* display approximately 40% severity; cultivars with *Lr34* and one or two additional minor genes display 10–15% severity; and cultivars with *Lr34* and two or three additional genes display 1–5% severity. Leaf rust may increase to unacceptable levels on cultivars carrying only *Lr34* or *Lr34* and one or two additional genes. However, cultivars with *Lr34* and two or three additional genes show a stable response in all environments tested so far, with final leaf rust ratings lower than 10%.

Slow rusting resistance to leaf rust is common in spring wheat germplasm. Our studies have shown that at least 10–12 slow rusting genes are involved in the adult plant resistance of CIMMYT wheats.

Table 22.1 Some seedling-susceptible bread wheats that carry good adult plant resistance to leaf rust in Mexico and other countries

Genotype(s)	Usual leaf rust response[a]	Additive genes[b] for resistance
Jupateco 73S	100 S(N)	Highly susceptible
Jupateco 73R	50 MSS	*Lr34*
Nacozari 76	30 MSS	*Lr34* + 1 gene
Sonoita 81, Bacanora 88, Rayon 89	20 MSS	*Lr34* + 1 or 2 genes
Frontana, Parula, Trap, Tonichi 81	10 MSS	*Lr34* + 2 or 3 genes
Chapio, Tukuru, Kukuna, Vivitsi	1 MSS	*Lr34* + 3 or 4 genes
Pavon 76,	40 MSS	*Lr46* + 1 gene
Genaro 81, Attila	40 MSS	2 genes
Amadina	5 MSS	4 genes

[a]Leaf rust response evaluated in Mexico has two components: percentage of severity based on the modified Cobb scale (Peterson et al., 1948) and reaction based on Roelfs et al. (1992). The reactions are: MSS = moderately susceptible to susceptible, that is, medium- to large-sized uredia without chlorosis or necrosis; S = susceptible, that is, large uredia without chlorosis or necrosis; N = necrotic leaves following high leaf rust severity.

[b]Minimum number estimated from genetic analysis.

Table 22.2 Some seedling-susceptible bread wheats that carry good adult plant resistance to stripe rust in field trials in Mexico and other countries

Genotype(s)		Usual yellow rust response[a]	Additive genes[b] for resistance
Jupateco 73S		100 MS	Moderately susceptible
Jupateco 73R		50 M	*Yr18*
Parula, Cook, Trap		15 M	*Yr18* + 2 genes
Tonichi 81, Sonoita 81, Yaco	10 M		*Yr18* + 2 or 3 genes
Chapio, Tukuru, Kukuna, Vivitsi	1 M		*Yr18* + 3 or 4 genes
Amadina		30 M	3 genes
Pavon 76, Attila		20 M	3 genes

[a]Yellow rust response data from Mexico has two components, percentage of severity based on modified Cobb scale (Peterson et al., 1948) and reaction based on Roelfs et al. (1992). The reactions are M = moderately resistant to moderately susceptible, sporulating stripes with necrosis, and chlorosis; S = sporulating stripes without chlorosis or necrosis.

[b]Minimum number estimated from genetic analysis.

We have also identified lines, such as Amadina (Table 22.1), where *Lr34* is absent, but whose level of resistance is high. We therefore believe that durable resistance is feasible even in the absence of *Lr34*. This is the case of variety Pavon 76 (Table 22.1), where we have identified a new gene *Lr46* for slow rusting in chromosome 1BL (Singh et al., 1998).

Yr18 and other minor genes for durable resistance to stripe rust

Singh (1992) and McIntosh (1992) have indicated that the moderate level of durable adult plant resistance to stripe rust (caused by *Puccinia striiformis*) of the CIMMYT-derived U.S. wheat cultivar Anza and winter wheats such as Bezostaja is controlled in part by the *Yr18* gene. This gene is completely linked to the *Lr34* gene. The level of resistance it confers is usually not adequate when present alone. However, combinations of *Yr18* and three to four additional slow-rusting genes result in adequate resistance levels in most environments (Singh and Rajaram, 1994). Cultivars carrying such *Yr18* complexes are listed in Table 22.2. Genes *Lr34* and *Yr18* occur frequently in germplasm developed at CIMMYT and in various countries. Slow rusting gene *Lr46* is completely linked to gene *Yr29*, which confers moderate resistance to stripe rust (William et al., 2003).

Powdery mildew of barley

Although numerous race-specific resistance genes are known to confer resistance against *Blumeria graminis* f. sp. *hordei*, the causal agent of powdery mildew of barley, durable resistance to this pathogen is controlled by recessive alleles at the *Mlo* locus (Jorgensen, 1992). The resistance allele was first induced by mutation in 1942 and also identified in barley landraces collected in 1937–1938. Several alleles, designated as *mlo1* to *mlo11*, are known, and different alleles confer the same kind of resistance phenotype. The resistance is effective against all races and expresses at all growth stages. Only a few compatible mildew colonies are formed on subsidiary cells next to the stomata. Cell defense is not due to a hypersensitivity reaction but through the formation of cell wall appositions without necrosis. The *mlo* alleles are associated with production of necrotic/chlorotic leaf spots even in the absence of disease. This affects grain yield and seed size in some environments and some genetic backgrounds; however, it is possible to breed *mlo*-resistant high-yielding barley varieties through the manipulation of genetic background (Bjornstad and Aastveit, 1990). The *Mlo* gene has been cloned and found to be different from other cloned race-specific resistance genes (Buschges et al., 1997). Such information can now be used to search for additional resistance genes of similar structure and function, which could be accumulated to achieve durable resistance.

Rice blast

Rice blast, caused by *Magnaporthe grisea* (anamorph *Pyricularia grisea*), is the most important and widely distributed disease of rice. Pathogenic variation in the population was first described in 1922 by Sasaki. Several race-specific resistance genes have been identified and used in breeding resistant cultivars. However, such resistance genes used either singly or in combination usually become ineffective within one or two years (Kiyosawa et al., 1984). Race-nonspecific resistance, described as "field resistance" (Ezuka, 1972), "slow blasting" (Ahn and Mukelar, 1986), or "dilatory resistance" (Marchetti, 1983), has been reported to occur. Traditional cultivars, such as Moroberekan, grown in the uplands of West Africa, have remained resistant in farmers' fields over a long period of time (Ahn and Mukelar, 1986). Also, modern, improved cultivars, such as IR 36 in Asia, IR 5 in West Africa, and CICA 7 in South America, show less disease and lower losses than other modern cultivars (Notteghem, 1993). Bonman et al. (1991) showed that in field trials slow disease progress in partially resistant cultivar IR 36 could reduce losses to negligible levels, while susceptible cultivars suffered losses ranging between 20% and 50%. Components of slow disease progress include infection efficiency, latent period, lesion size, and sporulation efficiency, among others (Castano et al., 1989). Cultivars with partial resistance often show polygenic inheritance with low heritability (Notteghem, 1993; Wang et al., 1989). Molecular mapping of durable resistance of the West African cultivar Moroberekan has indicated that nine chromosome regions are associated with reduced lesion number and five quantitative trait loci (QTL) with the strongest effects account for 60% of the variation, with individual contributions of between 8 and 14% (Wang et al., 1993). Breeding for such resistance is therefore slow and further complicated by the presence of a mixture of races in the pathogen population. A well-planned effort, including the use of molecular marker-assisted selection, may be necessary to transfer such resistance to modern high-yielding cultivars.

Late blight of potato

Late blight of potato, caused by *Phytophthora infestans*, is known to cause major yield losses. At least 11 race-specific genes (R genes) are currently known to occur but do not confer resistance to current races of the pathogen. As mentioned in the section on race-nonspecific/durable resistance, J. Niederhauser and colleagues, working in the Toluca Valley of Mexico where the sexual stage of the pathogen was first found, recognized the presence of partial resistance in wild species of potato (Niederhauser et al., 1954). In field trials he found that varieties with partial resistance showed resistance toward all races of the pathogen, remained green longer than varieties without partial resistance, and maintained their resistance over years. Minor genes for partial resistance in some current cultivars are also derived from *S. tubesorum* ssp. *andigena*, *S. demissum*, *S. stoloniferum*, and *S. edinense* (Guzman, 1964; Toxopeus, 1964). Despite such knowledge, cultivation of modern varieties of potato in Toluca Valley, Mexico, is not possible without chemical control. This is also the situation in other countries.

Head scab of wheat

Scab, caused by *Fusarium* spp., is a major production constraint in warm and humid or semihumid wheat-producing areas. *Fusarium graminearum* (perfect stage *Gibberella zeae*) predominates in wheat-growing areas of China and North and South America. Occurrence of races that show host–pathogen interactions is unknown, and at the same time resistance genes having major effects have not been found. The most commonly used sources of scab resistance have originated from China and Japan, Argentina and Brazil, and Eastern Europe. Genetic analyses indicate that a few additive genes confer resistance in Chinese and Brazilian wheats and that genes present in Chinese sources are different from those in Brazilian materials (Singh et al., 1995; Van Ginkel et al., 1996). Recent results from molecular mapping indicate several genomic regions that contribute to resistance (Anderson et al., 2001; Buerstmayr et al., 2002). The resistance from Chinese source Sumai 3 and Brazilian source Frontana is durable. Some synthetic wheats have recently been identified whose moderate resistance must be derived from *Triticum tauschii*, because the *T. turgidum* parents used in generating the synthetics are highly susceptible (Gilchrist et al., 1997). These sources should add new genetic diversity, crucial to enhancing resistance levels present in hexaploid wheats. Because genes for scab resistance are additive, a careful crossing and selection scheme should allow combinations of several genes leading to high levels of resistance and reduced accumulation of fusarium toxins in the grain.

Biotechnology applications

Early biotechnology applications involved the use of tissue culture techniques for plant regeneration aimed at exploiting somaclonal variation. Cytological manipulations coupled with tissue culture techniques were used to make wide hybrids for introgressing important useful traits from wild relatives. The successful transferral of disease-resistance genes through wide hybridization in wheat is well documented (Riley et al., 1968; Banks et al., 1995). Tissue culture–based plant regeneration techniques also resulted in the development of doubled haploids in cereals, which have been effectively exploited in cultivar development in species such as rice, barley, and wheat (Hu and Yang, 1986). Tissue culture-based plant regeneration techniques also formed the basis for genetic engineering procedures aimed at introducing novel resistance genes through transformation techniques (see below).

Molecular genetic analysis

The use of markers to track genes was first suggested by Sax (1923), further advanced by others, and currently known as QTL mapping, which is entirely based on DNA-based markers (Tanksley, 1993). These markers can be used to identify and estimate the genetic effects of major and minor genes based on linkages between the markers and the gene(s) of interest. Although RFLP markers have been used extensively for QTL mapping and gene characterization, currently there are marker systems based on polymerase chain reaction (PCR) such as microsatellites, sequence-tagged sites, or amplified fragment length polymorphisms that are comparatively easy to use and can be optimized for large-scale assays (Hoisington et al., 2002). PCR-based marker systems are ideally suited for use in marker-assisted selection. In self-pollinated crops such as rice, barley, and wheat, populations used for mapping resistance based on minor genes are ideally doubled haploids or recombinant inbred lines derived from crosses between contrasting parents (Tanksley, 1993). Modified procedures such as bulked segregant analysis have also been used successfully (Michelmore et al., 1991). A prerequisite to any successful molecular characterization for any trait is to develop populations and obtain accurate phenotypic data with a high degree of reliability and repeatability.

Advances in molecular approaches for characterizing disease- and pest-resistance genes

Although traditional linkage analysis has been used successfully in mapping and identifying resistance to diseases, advances in molecular genetics and marker systems, supported by statistical software packages, have facilitated the characterization of resistance genes. Especially where resistance is conditioned by major genes, linked markers have been identified for a wide array of biotic stress resistances in many cultivated crop species. For resistances that are quantitative, markers have been used to identify and characterize the QTLs involved (Young, 1996). In a few cases, tightly linked markers have been actively used to introgress and manipulate genes of interest in applied breeding.

In addition to the use of genetic linkage maps in gene characterization, molecular techniques such as transposon tagging involving transposable elements of maize have been used for developing efficient gene-tagging systems in plants (Jones et al., 1989). Transposon-based gene tagging systems have been successfully used in characterizing and cloning resistance genes from species such as maize (Johal and Briggs, 1992) and tomato (Jones et al., 1994). High-density molecular maps of a number of species have been initially developed, especially species with smaller genomes such as Arabidopsis (150 megabases) and tomato (950 megabases). Initial efforts in map-based cloning of disease-resistance genes were successful in Arabidopsis (Bent et al., 1994) and tomato (Martin et al., 1993). Gene *Hm1*, which confers resistance to *Cochliobolus carbonum* in maize, was the first cereal disease-resistance gene cloned (Johal and Briggs, 1992).

Dense genetic maps have been developed along with the physical maps in rice, which has the smallest genome of the cereals (Antonio et al., 1996). Efforts have been made to identify syntenic relationships among major cereals such as rice, maize, barley, sorghum, and oats by cross-hybridizing RFLP probes that have been mapped in related species (Moore et al., 1995). Since the rice genome is well characterized compared with those of other cereals, comparative-mapping tools can be used to define and characterize other cereal genomes in terms of rice linkage blocks and syntenic relationships with rice genes (Gale and Devos, 1998). Currently, given the enormous collections of expressed sequence tags, which are derived from expressed gene sequences, genome relationships can be studied in more detail.

Utilizing an array of techniques such as transposon tagging and map-based cloning using large insert DNA libraries and syntenic relationships, over 30 resistance genes have been cloned in different species (Hulbert et al., 2001). Among cereals, well-known examples of cloned genes include maize: *Hm-1 (Cochliobolus)* and *Rp-1* (*Puccinia*); rice: *Pib* and *Pi-ta* (*Meganaporthe*) and *Xa1* (*Xanthomonas*); barley: *Mla, Mlo* (*Blumeria*), and *Rpg-1* (*Puccinia*). Success in cloning a wide array of resistance genes has enhanced the understanding of the organization (McMullen and Simcox, 1995) and possible mode of action of resistance pathways (Hulbert et al., 2001). Single recessive gene *Mlo* (Jorgensen, 1992) in barley confers mildew resistance that is durable in nature; its structure is different from the other cloned race-specific resistance genes (Buschges et al., 1997).

Although major efforts have been made to clone and characterize genes involved in race-specific resistance, progress on minor genes involved in durable resistance is slow. In wheat, where durable resistance to fungal diseases is important, some efforts have been made to characterize and identify loci associated with durable resistance to leaf and stripe rusts (Messmer et al., 2000; William et al., 2003; Suenaga et al., 2003), powdery mildew (Liu et al., 2001), and fusarium head scab (Anderson et al., 2001). These efforts, combined with advanced novel techniques, should ultimately result in cloning and characterizing these genes as well as finding perfect markers for their use in marker-assisted selection.

For pathogens that evolve frequently, developing molecular tags to facilitate identification and selection of race-specific resistance would not produce a plant variety that would make such efforts worthwhile because such resistance will become ineffective in the short term. Efforts aimed at identifying and cloning genes that confer durable resistance would make breeding and molecular characterization worthwhile.

Genetic engineering approaches

The ability to introduce defined genes in a cultivated species is theoretically the most effective way of modifying the genetic basis of traits of importance. Although transformation techniques are well established for model species such as *Arabidopsis*, and crop species such as tobacco, tomato, and potato, progress in developing efficient transformation procedures for cereals has been more recent. This is mainly due to difficulties encountered in establishing efficient embryogenic cell culture techniques and the lack of efficient DNA delivery methods. Currently, for most cereals such as rice, maize, barley, and wheat, efficient cell culture systems exist from which fertile plants can be regenerated at high frequency. DNA delivery methods have been developed using immature embryo cultures and/or anther or pollen culture techniques based on particle bombardment and, more recently, using *Agrobacterium tumefaciens*. One essential factor determining the success of transformation is the stability of the expression of the genes introduced. Non-Mendelian inheritance and loss of activity of the inserted genes have been observed (Flavell, 1994). Some crops such as cotton, potato, and maize have been transformed with natural or synthetic forms of *cry* genes originally isolated from *Bacillus thuringiensis*, also known as *Bt* genes. Availability of strong promoters such as rice actin or maize ubiquitin has enabled high levels of expression of transgenic products. This has reduced crop losses due to insect damage and helped to avoid adverse environmental and health effects by diminishing the need for insecticides. Several experiments using other gene constructs, such as coat protein (*Cp*) gene of rice stripe virus introduced into rice (Hayakawa et al., 1992) and *thaumatin-like* protein gene introduced into wheat (Chen et al., 1999), have increased the levels of protection from plant pathogens.

Crossing and selection schemes for resistance breeding

There is, in fact, no special crossing and selection scheme specific for resistance breeding. Knowledge of the nature of resistance, sources of resistance, and reliability of resistance screening methodology/selection environment should determine the crossing and selection scheme, including the application of molecular markers. If resistance based on a single major gene is present in a good agronomic background, then selection for resistance is easier, provided the screening methodology is reliable. If the cost of resistance screening is high, selection can be postponed until the advanced generation, because approximately half of the lines would carry the resistance genes in the homozygous state. Single-seed descent or doubled-haploid approaches can be quite beneficial in such circumstances in self-pollinated crops. Backcrossing is recommended when the source of resistance is available in a poor or locally unadapted genetic background. Molecular marker-assisted selection may not be necessary if resistance can be easily and cheaply screened through traditional methodologies. However, if the approach is to generate combinations of major resistance genes, then markers may be essential to distinguish plants that carry gene combinations from those with single genes.

Breeding for resistance based on minor additive genes has been challenging and often slow, for several reasons: (1) a sufficient number of minor genes may not be present in a single source genotype, (2) a source genotype may be poorly adapted, (3) there may be confounding effects from the segregation of both major and minor genes in the population, (4) crossing and selection schemes and population sizes are more suitable for selecting major genes, (5) reliable molecular markers for several minor genes are unavailable, and (6) the cost associated with identifying and using multiple markers, etc., is high. One approach suggested in the literature is to use recurrent selection schemes to accumulate several minor genes in a single genetic background in a self-pollinated crop or in a population in cross-pollinated crops. Such selection schemes have often been of scientific interest more than actually applied in self-pollinated crops. Selection for resistance alone will not generate important popular cultivars, unless it is simultaneously combined with other traits such as high yield and quality. However, such germplasm carrying combinations of minor genes should be very useful in transferring these genes to modern cultivars.

A successful example of breeding for resistance based on minor genes is the case of resistance to leaf and stripe rusts in wheat, which took about 30 years of continuous effort at CIMMYT. In the early 1970s, S. Rajaram, influenced by the concept of slow-rusting resistance to leaf rust in wheat proposed by R. Caldwell and partial resistance to late blight of potato put forth by J. Niederhauser, made a strategic decision to initiate selection for slow-rusting resistance to leaf rust in CIMMYT spring wheat germplasm. In the early phase of breeding he maintained plants and lines in segregating populations that would show 20–30% rust severity with compatible infection type. This strategy led to the release of several successful wheat cultivars, such as Pavon 76 and Nacozari 76, in Mexico and other countries. These slow-rusting lines were used heavily in the crossing program and resulted in a wide distribution of minor genes within CIMMYT spring wheat germplasm.

The genetic basis of such resistance started to become clear in the early 1990s (see section on the basis of resistance to wheat rusts). High-yielding lines that combine four or five additive, minor genes for both leaf and stripe rusts and show near-immune levels of resistance were developed in the 1990s. Three or four lines carrying different minor genes were crossed (three-way and four-way crosses), and plants in large segregating populations were selected under artificially created rust epidemics. Races of pathogens that have virulence for race-specific resistance genes present in the parents were used to create the epidemics (Singh et al., 2000). The resulting highly resistant lines are now being used in a planned manner to transfer these minor resistance genes to well-adapted, "farmers' choice" cultivars that are currently grown over large areas but have become susceptible to pathogen races in Mexico. Based on genetic information on the number of additive, minor genes that must be transferred to achieve the desired level of resistance, the crossing and selection scheme described below was developed and applied. This strategy has allowed simultaneous transfer not only of resistance genes but also other quantitative genes that increase the yield potential or improve the grain quality of an adapted cultivar.

Incorporation of rust resistance based on additive, minor slow-rusting genes into adapted wheat cultivars

To transfer minor gene-based resistance into a susceptible adapted cultivar or any selected genotype, we use the single backcross-selected bulk scheme, where the cultivar/genotype is crossed with a group of about 10 resistance donors (some listed in Tables 22.1 and 22.2); and then, 20 spikes of the F_1 plants from each cross are backcrossed to obtain 400–500 BC_1 seeds. Selection is practiced from the BC_1 generation onward for resistance and other agronomic features under high rust pressure. Because additive genes are partially dominant, BC_1 plants carrying most of the genes show intermediate resistance and can be selected visually. About 1600 plants per cross are space grown in the F_2, whereas population sizes of about 1000 plants are maintained in the F_3–F_5 populations. Plants with desirable agronomic features and low-to-moderate terminal disease severity in early generations (BC_1, F_2, and F_3) and plants with low terminal severity in later generations (F_4 and F_5) are retained. We use a selected-bulk scheme where one spike from each selected plant is harvested as bulk until the F_4 generation, and plants are harvested individually in the F_5 generation. Bulking of selected plants poses no restriction on the number of plants that can be selected in each generation because harvesting and threshing are quick and inexpensive, and the next generation is derived from a sample of the bulked seed. Because high resistance levels require the presence of four to five additive genes, the level of homozygosity from the F_4 generation onward is usually sufficient to identify plants that combine adequate resistance with good agronomic features. Moreover, selecting plants with low terminal disease severity under high disease pressure means that more additive genes may be present in those plants. Selection for seed characteristics is carried out on seeds obtained from individually harvested F_5 plants. Small plots of the F_6 lines are then evaluated for agronomic features, homozygosity of resistance, etc., before conducting yield trials.

Resistant derivatives of several cultivars were recently developed using the above methodology. In each case we could identify derived lines that not only carry high levels of resistance to leaf rust or yellow rust or both, but also show about 5–15% higher yield potential than the original cultivar. We believe this approach to wheat improvement allows us to maintain the characteristics of the original cultivar while improving its yield potential and rust resistance. We are taking the improved versions of the original cultivars back to the area where they are grown to achieve long-term control of rust diseases of wheat. It should be noted that having minor gene-based resistance in several backgrounds should ease future selection for these resistance genes.

Challenges ahead

The challenge in the next 25 years will be to make our crops durably resistant to multiple diseases and pests. Although this will still be based largely on knowledge acquired through conventional plant breeding, qualitative and quantitative genetics, and epidemiology, recent advances in molecular biology should aid in understanding the genetic and functional basis of durable resistance. If carefully applied, they will also aid in selecting and enhancing genetic diversity. It is important to develop molecular markers for numerous durable resistance genes—not just a few—if we are to avoid creating genetic uniformity through marker-assisted selection.

We envision that, as in the past, some widely grown cultivars will be protected against diseases through race-specific, major genes. If so, strong regional collaboration will be necessary to monitor pathogenic virulence and give early warning of potential disease epidemics. This would allow governments enough time to reduce or replace vulnerable cultivars and thus avoid catastrophic losses.

Finally, issues relating to plant genetic resources and intellectual property rights will need to be fully defined to allow the free exchange of germplasm among breeding programs all over the world. We urge caution when defining germplasm that should be freely available to breeding programs and a crop variety suitable for commercial farming to which proprietary rights could be sought. In the latter case, it is essential that the genes contained in the variety continue to be available to all for breeding purposes.

References

Anderson, J.A., R.W. Stack, S. Liu, B. L. Waldron, A.D. Fjeld, C. Coyne, B. Moreno-Sevilla, F.J. Mitchell, Q.J. Song, P.B. Cregan, and R.C. Frohberg. 2001. DNA markers for Fusarium head blight resistance QTLs in two wheat populations. Theor. Appl. Genet. 102:1164–1168.

Ahn, S.W., and A. Mukelar. 1986. Rice blast management under upland conditions. p. 363–373. *In* Progress in upland rice research, Proc. 2nd Int. Upland Rice Conf., Jakarta. IRRI, Los Banos, Philippines.

Antonio, B.A, T. Inoue, H. Kajiya, Y. Nagamura, N. Kurata, Y. Minobe, M. Yano, M. Nakagahra, and T. Sasaki. 1996. Comparison of genetic distance and order of DNA markers in five populations of rice. Genome 39:946–956.

Banks, P.M., P.J. Larkin, H.S. Bariana, E.S. Lagudah, R. Appels, P.M. Waterhouse, R.I.S. Brettell, X. Chen, H.J. Xu, Z.Y. Xin, Y.T. Qian, X.M. Zhou, Z.M.Cheng, and G.H. Zhou. 1995. The use of cell culture for subchromosomal introgressions of barley yellow dwarf virus resistance from *Thinopyrum intermedium* to wheat. Genome 38:395–405.

Biffen, R.H. 1905. Mendel's law of inheritance and wheat breeding. J. Agric. Sci. 1:4–48.

Bent, A.F., B.N. Kunkel, D. Dahlbeck, K.L. Brown, R. Schmidt, J. Giraudat, J. Leung, and B.J. Staskawicz. 1994. *Rps2* of *Arabidopsis thaliana*: a leucine-rich repeat class of plant disease resistance gene. Science 265:1856–1859.

Bjornstad, A., and K. Aastveit. 1990. Pleiotropic effects on the *ml-o* mildew resistance gene in barley in different genetical backgrounds. Euphytica 46:217–226.

Bonman, J.M., B.A. Estrada, C.K. Kim, D.S. Ra, and E.S. Lee. 1991. Assessment of blast disease and valid loss in susceptible and partially rice cultivars in two irrigated lowland environments. Plant Dis. 75:462–466.

Borlaug, N.E. 1953. New approach to the breeding of wheat varieties resistant to *Puccinia graminis tritici*. Phytopathology 43:467 (Abstr.).

Buerstmayr, H., M. Lemmens, L. Hartl, L. Doldi, B. Steiner, M. Stierschneider, and P. Ruckenbauer. 2002. Molecular mapping of QTLs for Fusarium head blight resistance in spring wheat. I. Resistance to fungal spread (type II resistance). Theor. Appl. Genet. 104:84–91.

Buschges, R., K. Hollricher, G. Panstruga, M. Wolter, A. Frijters, R.V. Daelen, T. Lee, P. Diergaarde, J. Groenendijk, S. Topsch, P. Vos, F. Salamini, and P. Schulze-Lefert. 1997. The barley *Mlo* gene: a novel control element of plant pathogen resistance. Cell 88:695–705.

Caldwell, R.M. 1968. Breeding for general and/or specific plant disease resistance. p. 263–272. *In* K.W. Finlay and K.W. Shephard (ed.) Proc. 3rd Int. Wheat Genetics Symp. Australian Academy of Sciences, Canberra, Australia.

Castano, J., D.R. Mackenzie, and R.R. Nelson. 1989. Components analysis of race non-specific resistance to blast disease of rice caused by *Pyricularia oryzae*. J. Phytopathology 127:89–99.

Chen, W.P., P.D. Chen, D.J. Liu, R. Kynast, B. Friebe, R. Velazhahan, and S. Muthukrishnan. 1999. Development of wheat scab symptoms is delayed in transgenic wheat plants that constitutively express a rice thaumatin-like protein gene. Theor. Appl. Genet. 99:755–760.

Ezuka, A. 1972. Field resistance of rice varieties to blast disease. Rev. Plant Protec. Res. 5:1–20.

Flavell, R.B. 1994. Inactivation of gene expression in plants as a consequence of specific sequence duplication. Proc. Natl. Acad. Sci. 91:3490–3496.

Flor, H.H. 1956. The complementary genic systems in flax and flax rust. Adv. Genet. 8:29–54.

Gale, M.D., and K.M. Devos. 1998. Comparative genetics in grasses. Proc. Natl. Acad. Sci. 95:1971–1974.

Gilchrist, L., S. Rajaram, A. Mujeeb-Kazi, M. van Ginkel, H. Vivar, and W. Pfeiffer. 1997. Fusarium scab screening program at CIMMYT. p. 7–12. *In* H.J. Dubin, L. Gilchrist, J. Reeves and A. McNab (ed), Fusarium head scab: global status and future prospects. CIMMYT, Mexico.

Guzman, N.J. 1964. Nature of partial resistance to certain clones of three *Solanum* species to *Phytophthora infestans*. Phytopathology 54:1398–1404.

Hayakawa, T., Y. Zhu, K. Itoh, Y. Kimura, T. Izawa, K. Shimamoto, and S. Toriyama. 1992. Genetically engineered rice resistant to rice stripe virus, an insect-transmitted virus. Proc. Natl. Acad. Sci. 89:9865–9869.

Hoisington, D., N. Bohorova, S. Fennell, M. Khairallah, A. Pellegrineschi, and J-M. Ribaut. 2002. The application of biotechnology to wheat improvement: New tools to improve wheat productivity. p. 175–198. *In* B.C. Curtis, S. Rajaram, and H. Gomez-Macpherson (ed.), Bread wheat—improvement and production. FAO, Rome.

Hu, H., and H.Y. Yang (ed.). 1986. Haploids of higher plants *In Vitro*. China Academic publishers, Beijing.

Hulbert, S.H., C.A. Webb, S.M. Smith, and Q. Sun. 2001. Resistance gene complexes. Annu. Rev. Phytopathol. 39:285–312.

Jensen, N.F. 1952. Intervarietal diversification in oat breeding. Agron. J. 44:30–34.

Johal, G.S., and S.P. Briggs. 1992. Reductase activity encoded by the *Hm1* disease resistance gene in maize. Science 258:985–987.

Johnson, R. 1978. Practical breeding for durable resistance to rust diseases in self pollinating cereals. Euphytica 27:529–540.

Jones, J.D.G., F.M. Carland, P. Maliga, and H.K. Dooner. 1989. Visual detection of transposition of maize element *Activator* (*Ac*) in tobacco seedlings. Science 244:204–207.

Jones, D.A., C.M. Thomas, K.E. Hammond-Kosack, P.J. Balint-Kurti, and D.G. Jones. 1994. Isolation of the tomato *Cf-9* gene for resistance to *Cladosporuium fulvum* by transposon tagging. Science 266:789–792.

Jorgensen, J.H. 1992. Discovery, characterization and exploitation of *Mlo* powdery mildew resistance in barley. Euphytica 63:141–152.

Kiyosawa, S., K. Shimada, and K. Ohta. 1984. Tests for pathogenicity of blast fungus isolates collected from the rice variety, Hama Asahi, in Aomori Prefecture. Japan J. Breed. 34:485–489.

Knott, D.R. 1988. Using polygenic resistance to breed for stem rust resistance in wheat. p. 39–47. *In* N.W. Simmonds and S. Rajaram (ed.), Breeding strategies for resistance to the rusts of wheat. CIMMYT, Mexico.

Liu, S., C.A. Griffey, and M.A. Saghai-Maroof. 2001. Identification of molecular markers associated with adult plant resistance to powdery mildew in common wheat cultivar Messey. Crop Sci. 41:1268–1275.

Marchetti, M.A. 1983. Dilatory resistance to rice blast in U.S.A. rice. Phytopathology 73:645–649.

Martin, G.B., S.H. Brommonschenkel, J. Chunwongse, A. Fray, M.W. Ganal, R. Spivey, T. Wu, E.D. Earle, S.D. Tanksley. 1993. Map-based cloning of a protein kinase gene conferring disease resistance in tomato. Science 262:1432–1436.

McIntosh, R.A. 1992. Close genetic linkage of genes conferring adult-plant resistance to leaf rust and stripe rust in wheat. Plant Pathol. 41:523–527.

McIntosh, R.A., G.E. Hart, K.M. Devos, M.D. Gale, and W.J. Rogers. 1998. Catalogue of gene symbols for wheat. Vol 5: 1–235. *In* A. E. Slinkard (ed.), Proc. 9th Int. Wheat Genetics Symp. Saskatoon, Canada.

McIntosh, R.A., and S. Smith-White. 1986. Irvine Armstrong Watson 1914–1986. Historical Records of Australian Science 7:405–415.

McMullen, M.D., and K.D. Simcox. 1995. Genomic organization of disease and insect resistance genes in maize. Mol. Plant-Microbe Interactions 8:811–815.

Messmer, M.M., R. Seyfarth, M. Keller, G. Schachermayr, M. Winzeler, S. Zanetti, C. Feuillet, and B. Keller. 2000. Genetic analysis of durable leaf rust resistance in winter wheat. Theor. Appl. Genet. 100:419–431.

Michelmore, R.W., I. Paran, and R.V. Kesseli. 1991. Identification of markers linked to disease resistance genes by bulked segregant analysis: A rapid method to detect markers in specific genomic regions by using segregating populations. Proc. Natl. Acad. Sci. 88:9828–9832.

Moore, G., K.M. Devos, Z. Wang, and M.D. Gale. 1995. Grasses, line up and form a circle. Curr. Biol. 5:737–739.

Niederhauser, J.S., J. Cervantes, and L. Servin. 1954. Late blight in Mexico and its implications. Phytopathology 44:406–408.

Notteghem, J.L. 1993. Durable resistance to rice blast disease. p. 125–134. *In* Th. Jacobs and J.E. Parlevliet (ed.), Durability of disease resistance. Kluwer Academic Publishers, The Netherlands.

Parlevliet, J.E. 1975. Partial resistance of barley to leaf rust, *Puccinia hordei*. I. Effect of cultivar and development stage on latent period. Euphytica 24:21–27.

Parlevliet, J.E. 1976. Partial resistance of barley to leaf rust, *Puccinia hordei*. III. The inheritance of the host plant effect on latent period in four cultivars. Euphytica 25:241–248.

Parlevliet, J.E. 1986. Pleiotropic association of infection frequency and latent period of two barley cultivars partially resistant to barley leaf rust. Euphytica 35:267–272.

Person, C. 1959. Gene-for-gene relationship in host:parasite systems. Can. J. Bot. 37:1101–1130.

Peterson, R.F., A.B. Campbell, and A.E. Hannah. 1948. A diagrammatic scale for estimating rust intensity of leaves and stem of cereals. Can. J. Res. Sect. C. 26:496–500.

Riley, R., V. Chapman, and R. Johnson. 1968. The incorporation of yellow rust resistance of *Aegilops comosa* into wheat by genetically induced homoeologous recombination. Nature 217:383–384.

Roelfs, A.P. 1988. Resistance to leaf rust and stem rust in wheat. p. 10–22. *In* N. W. Simmonds and S. Rajaram (ed.), Breeding strategies for Resistance to the Rusts of Wheat. CIMMYT, Mexico.

Roelfs, A.P., R.P. Singh, and E.E. Saari. 1992. Rust diseases of wheat: Concepts and methods of disease management. CIMMYT, Mexico, D.F. 81pp.

Sasaki, R. 1922. Existence of strains in rice blast fungus. I. Journal of Plant Proection, Tokyo 9:631–644.

Sax, K. 1923. The association of size differences with seed-coat pattern and pigmentation in *Phaseolus vulgaris.* Genetics 8:552–560.

Singh, R.P. 1992. Genetic association of leaf rust resistance gene *Lr34* with adult plant resistance to stripe rust in bread wheat. Phytopathology 82:835–838.

Singh, R.P., J. Huerta-Espino, and S. Rajaram. 2000. Achieving near-immunity to leaf and stripe rusts in wheat by combining slow rusting resistance genes. Acta Phytopathlogica Hungarica 35:133–139.

Singh, R.P., A. Kazi-Mujeeb, and J. Huerta-Espino. 1998. *Lr46*: A gene conferring slow rusting resistance to leaf rust in wheat. Phytopathology 88:890–894.

Singh, R.P., H. Ma, and S. Rajaram. 1995. Genetic analysis of resistance to scab in spring wheat cultivar Frontana. Plant Dis. 79:238–240.

Singh, R.P., and S. Rajaram. 1992. Genetics of adult-plant resistance to leaf rust in 'Frontana' and three CIMMYT wheats. Genome, 35:24–31.

Singh, R.P., and S. Rajaram. 1994. Genetics of adult plant resistance to stripe rust in ten spring bread wheats. Euphytica 72:1–7.

Stakman, E.C., and F.J. Piemeisel. 1917. A new strain of *Puccinia graminis.* Phytopathology 7:73 (Abstr.).

Suenaga, K., R.P. Singh, J. Huerta-Espino, and H.M. William. 2003. Microsatellite markers for genes *Lr34/Yr18* and other quantitative trait loci for leaf rust and stripe rust resistance in bread wheat. Phytopathology 93:881–890.

Tanksley, S.D. 1993. Mapping polygenes. Annu. Rev. Genet. 27:205–233.

Toxopeus, H.J. 1964. Treasure-digging for blight resistance in potatoes. Euphytica 13:206–222.

Vanderplank, J.E. 1963. Plant Diseases: Epidemics and Control. Academic Press New York and London.

Van Ginkel, M., W. van der Schaar, and Y. Zhuping. 1996. Inheritance of resistance of scab in two wheat cultivars from Brazil and China. Plant Dis. 80:863–867.

Walker, J.C. 1966. The role of pest resistance in new varieties. p. 219–242. *In* K.J. Frey (ed.), Plant breeding: A symposium held at Iowa State Univ. The Iowa State Univ. Press, Ames, IA.

Wang, G.L., D.J. Mackill, and J.M. Bonman. 1989. Inheritance of partial resistance to blast in Indica rice cultivars. Crop Sci. 29:848–853.

Wang, G.L., D.J. Mackill, J.M. Bonman, S.R. McCouch, and R.J. Nelson. 1993. RFLP mapping of genes conferring complete and partial resistance in a rice cultivar with durable resistance to blast. p. 219–225. *In* Th. Jacobs and J.E. Parlevliet (ed.), Durability of disease resistance. Kluwer Academic Publishers, The Netherlands.

William, H.M., R.P. Singh, J. Huerta-Espino, S. Ortiz-Islas, and D. Hoisington. 2003. Molecular marker mapping of leaf rust resistance gene *Lr46* and its association with stripe rust resistance gene *Yr29* in wheat. Phytopathology 93:153–159.

Young, N.D. 1996. QTL mapping and quantitative disease resistance in plants. Annu. Rev. Phytopathol. 34:479–501.

23

Breeding for Increased Forage Quality

M.D. Casler, Research Geneticist, USDA-ARS, U.S. Dairy Forage Research Center

Introduction

In this chapter, the term *forages* is inclusive of pasture crops and species used to produce conserved feeds, including hay, haylage, and silage, derived from either perennials or annuals. Forage quality can be considered as the ability of a feed to support animal functions such as maintenance, growth, reproduction, and lactation. It has three components: digestibility, intake potential, and energetic efficiency (Raymond, 1969). Digestibility and intake potential can be estimated or predicted from relatively simple laboratory procedures, whereas energetic efficiency is rather difficult to estimate or predict in the laboratory (Hacker, 1982; Van Soest, 1994). Each component can be viewed as a characteristic of a plant, but only as a function of many simpler plant traits, such as its structure, anatomy, morphology, chemistry, and stage of development. In turn, each of these simpler plant traits is under genetic control, that is, they can be manipulated either by natural selection or by human-directed selection. Paradoxically, the three components of forage quality (digestibility, intake potential, and energetic efficiency) cannot be defined in the absence of an animal and thus cannot be strictly defined as plant traits. It is only by arbitrarily classifying a plant as a feed that its genetic traits and their optimal values become defined in terms of animal nutrition concepts.

Forage plants, particularly perennials, coevolved with large mammalian herbivores, evolving numerous traits that serve to limit feed intake by herbivores. Limitations to feed intake are thought to be mechanisms for maintaining fitness of herbaceous plants subjected to grazing pressure. Herbivores, including livestock, can be highly selective grazers, often choosing the most palatable and nutritious plants and plant parts (Casler et al., 1996; Marten, 1989). Lignin, silica, crystalline cellulose, branched hemicelluloses, and phenolic compounds serve, in addition to their physiological functions, as defense mechanisms against large herbivores. These compounds are either indigestible or reduce the digestibility of polysaccharides in consumed feeds, limiting feed intake by herbivores (Buxton and Casler, 1993; Van Soest, 1994). Alkaloids and cyanogenic compounds occur in numerous forage species, causing a wide range of responses in herbivores including nonpreference, diarrhea, weight loss, cancer, and death. Anatomical traits, such as trichomes and silicious dentations, limit feed intake by reducing herbage palatability.

Breeding for increased forage quality begins with a definition of the objective, typically to improve digestibility or intake potential of the forage. In some forage species, intake can be defined in simple terms of nonpreference, often caused by low palatability. Livestock can be used to reliably rank genotypes for palatability or feed preference in populations that possess genetic variability for alkaloid concentration, trichome density, or silicious dentation frequency (Casler et al., 1996; Marten, 1989). These are traits that are immediately obvious to grazing livestock, which can readily sense high-alkaloid concentrations by smell or taste and harsh leaf structures by a touch of the tongue. Soft-leaf tall fescue (*Festuca arundinacea* Schreb.) cultivars have been developed by plant breeders who use their sense of touch to identify plants with the softest leaves.

In many species that lack such obvious antiquality components, the distinction between intake and digestibility is less clear. In ruminant livestock, feed intake is limited by rumen fill, which is, in turn, limited by digestibility of the feed. More digestible feeds will have more rapid particle size breakdown and a faster rate of passage through the rumen, more rapidly eliciting a hunger response to trigger resumed feeding by the ruminant. Fibrous bulk, the plant cell wall, is generally considered to be the factor most limiting to feed intake. Intake of fibrous bulk generally causes rumen fill and satiation before the ruminant has maximized its caloric intake, resulting in a reduced plane of nutrition (Van Soest, 1994). While increased fiber digestibility may partly ameliorate this limitation by increasing rate of particle size breakdown and passage, it does not necessarily contribute to increased intake, because fiber concentration and digestibility are not necessarily correlated.

Digestibility, a measure of energy available to the ruminant, can be measured by one of several methods. In vivo methods are the most reliable and accurate but are impractical for a breeding program due to both animal and feed requirements. Tilley and Terry (1963) developed the first in vitro digestibility criterion, a simple measure of dry matter disappearance during in vitro fermentation in rumen fluid. The Tilley and Terry technique was originally developed for use in forage-breeding programs, as indicated in the first paragraph of their paper. Its first application in a breeding program actually predates publication of the technique itself (Cooper et al., 1962). The Tilley and Terry procedure meets nearly all the necessary characteristics for a reasonable selection criterion in a forage-breeding program: rapid, repeatable, amenable to a relatively small sample size, heritable, and directly correlated with animal performance. Variations of the Tilley and Terry technique include in situ estimation of digestibility by dry matter disappearance in nylon bags immersed in the rumen of a fistulated cow, enzymatic degradation by prepared cellulase enzymes in an in vitro system, or the measurement of gasses produced during in vitro fermentation (Casler, 2001).

The remainder of this review will focus on genetic variability for plant traits related to digestibility and intake potential, including their effects on animal performance and agronomic fitness. Numerous reviews have been written on this topic in recent years, and this review relies heavily on previously published reviews (Buxton and Casler, 1993; Carter et al., 1991; Casler, 2001; Casler and Vogel, 1999; Clark and Wilson, 1993; Coors and Lauer, 2001; Hacker, 1982; Hanna, 1993; Howarth and Goplen, 1983; Hutton, 1971; Marten, 1989; Reed, 1994; Stone, 1994; Van Wijk et al., 1993; Vogel and Sleper, 1994). Direct credit is not always given to the authors of original research due to space limitations and the large number of literature citations that would be required. Citations are given to relevant reviews that contain more detailed information about each topic and contain the proper references to all original research.

Genetic variability and selection progress

Digestibility and related traits

Modified Tilley and Terry procedures represent the most common selection criteria for improving the digestibility of forage crops. As of 1993, genetic variation for in vitro dry matter digestibility (IVDMD) had been documented for 17 species (Buxton and Casler, 1993) and several more species that are described in more recent publications. It is typical to observe a range of variation of 100 g kg^{-1} between the lowest- and highest-ranked individuals, although ranges of up to 377 g kg^{-1} have been observed within populations (Buxton and Casler, 1993). Relatively small differences among clones, families, or cultivars can be detected for IVDMD, with 18 of 32 studies capable of detecting genotypic differences as small as 30% of the range among genotype means with 95% confidence (Buxton and Casler, 1993).

Genetic gains for IVDMD measured by some modification of the Tilley and Terry procedure have been documented in several species, including legumes, cool-season grasses, and warm-season grasses (Casler, 2001). Progress has ranged from 7 to 66 g kg^{-1} $cycle^{-1}$ (1.0–4.7% $year^{-1}$). The nylon bag technique has been used in the longest running program aimed at increasing digestibility, the USDA-ARS bermudagrass [*Cynodon dactylon* (L.) Pers.] breeding program at Tifton, Georgia. Genetic gains in digestibility averaged 2 g kg^{-1} $year^{-1}$ between 1963 and 1993 (Casler, 2001). These studies clearly support the original hypothesis that digestibility can be treated as a repeatable and heritable trait in a plant-breeding program (Tilley and Terry, 1963).

Because digestibility is defined only in terms of

a plant–microbe interaction within an anaerobic environment, genetic regulation of digestibility must be manifested through some plant trait(s) that are directly coded by the plant's genes. Selection studies to date have conclusively demonstrated several mechanisms by which IVDMD can be increased in forage plants: increasing soluble carbohydrate concentration (decreasing structural carbohydrate concentration), decreasing lignin concentration within the cell wall, and reducing the extent of lignin–polysaccharide cross-linking. Genetic changes in maturity, plant anatomy, or plant morphology also may bring about changes in whole-plant IVDMD, but these changes are likely indirect effects resulting from changes in chemical composition of the whole plant.

Digestibility can be increased in forage plants by manipulating the relative amounts of soluble and structural carbohydrates, a topic reviewed by Casler (2001). Soluble carbohydrates are highly and rapidly digestible to ruminants, increasing digestibility and rate of particle passage from the rumen. Soluble carbohydrates are typically estimated by the water-soluble carbohydrate (WSC) criterion, which is heritable, easily amenable to selection, and highly correlated with IVDMD (Humphreys, 1989).

Conversely, digestibility can be increased by selection for decreased concentration of structural carbohydrates, which are often estimated by neutral detergent fiber (NDF) and are closely related to total cell-wall concentration. Structural carbohydrates are, themselves, highly digestible, but their utilization by the ruminant is limited by lignin and phenolic cross-linkages, reducing their availability to enzymatic degradation. Although there are no specific estimates of genetic correlations between structural and soluble carbohydrate, it seems likely that large genetic manipulations of one component will lead to reductions in the other component.

Lignin concentration and composition regulates a considerable amount of genetic variation in IVDMD within forage crops. In smooth bromegrass (*Bromus inermis* Leyss.), a series of experiments showed that each 10-g kg^{-1} increase in IVDMD was apparently caused by a 1.3 g kg^{-1} decrease in acid detergent lignin (ADL) concentration (Casler, 2001). Genetic variation in ADL concentration accounted for over 80% of the genetic variation in IVDMD in each of the above studies. Genetic changes in IVDMD have been associated with cell-wall lignin concentration in other species, as well. Lignin concentration is a heritable trait and can be modified by selection. Most methods of lignin estimation drastically underestimate total lignin, but the Klason method is probably most useful for plant breeding and genetics studies, because it more closely estimates total lignin. Little is known about genetic correlations between the various measures of lignin concentration. Because the lignin estimation methods differ in chemistry, they may result in differential ranking of genotypes that are variable in lignin composition.

Lignin and/or phenolic composition appears to be subject to genetic variability. Although studies to date have been limited to inbred lines or clones and do not include estimates of heritability or selection gains, they suggest a certain range of variability and repeatability of differences among genotypes (Casler, 2001). Ferulic acid appears to be the most important of the phenolics, regulating the degree of cross-linking between cell-wall polysaccharides and lignin. Ferulic acid esterifies to arabinose subunits of arabinoxylan chains. As plants mature, esterified ferulates become etherified to lignin, forming cross-linkages between lignin and cell-wall polysaccharides (Stone, 1994). High IVDMD has been associated with low levels of etherified ferulic acid or high levels of esterified ferulic acid in several forage species (Casler, 2001). Digestibility of the cell wall can be increased by reducing lignin concentration per se, by reducing the degree of ferulate cross-linking, or by a combination of these two factors.

Genetic variability for digestibility may be partially a function of genetic variability for anatomical and morphological structure (Casler, 2001). High-digestibility genotypes may have less slowly and nondegradable tissue types than low digestibility genotypes. Such genetic changes to plant anatomy may also be coincident with changes in lignin concentration or composition, because slowly degradable and nondegradable cells are more highly lignified than readily digestible cells. In species such as alfalfa (*Medicago sativa* L.), for which leaves and stems differ dramatically in digestibility, selection for high digestibility (low lignin concentration) of whole-plant tissue samples results in significant increases in leaf/stem ratio. For grasses that have relatively smaller differences in digestibility between stems and leaves, such changes have not been observed. As such, leafiness per se is not a good indicator of forage quality (Buxton and Casler, 1993; Hutton, 1971).

Changes in plant structure may have a profound effect on growth, development, and fitness of plants. Because slowly and nondegradable cells tend to be involved in plant structure and water relations, reductions in lignin concentration may adversely affect forage yield, seed yield, drought tolerance, carbohydrate transport, and standability. Increasing digestibility by increasing leaf/stem ratio may also have negative consequences, potentially reducing forage and seed-yield potential (Clark and Wilson, 1993). Due to insufficient research on this topic, conclusions about these potential relationships are speculative. Many plant breeders have developed selection criteria and methods to avoid or minimize potential changes in plant structure, such as basing selection on leaf blades or stem segments of a defined age (Buxton and Casler, 1993; Casler, 2001).

Selection for increased digestibility has led to the development and release of numerous cultivars. Several of these cultivars have been evaluated in replicated grazing trials, demonstrating increased liveweight gains compared with parent or check cultivars (Casler and Vogel, 1999). Documentation of improved liveweight gains has been used to successfully promote several of these cultivars in extension, outreach, and marketing programs. Because cultivars with improved animal performance can be delivered to producers without an increase in seed costs, their use represents a highly effective mechanism to increase profitability of livestock production without increasing input costs. It has come as a surprise to many forage breeders and ruminant nutritionists, but relatively small increases in digestibility can be measured as increased animal performance, creating huge improvements in profit potential from improved forage crops (Vogel and Sleper, 1994).

Despite the optimism prompted by these studies, very few cultivars with improved forage quality have been evaluated to provide documentation of improved animal performance. Reed (1994) attributes this to three principal factors: (1) the cost of grazing and/or feeding trials, (2) the relatively poor precision often obtained in grazing and/or feeding trials, and (3) the perception that relatively small differences cannot be detected in these trials. An additional factor may be the lack of interaction and collaboration of breeders with agronomists or nutritionists who have an interest in genetic improvements of forage quality. The most productive breeding programs, with forage quality as a goal, have been team efforts with expertise in breeding/genetics, agronomy and/or plant pathology, and ruminant nutrition.

Computer technology can provide an alternative to grazing or feeding trials for some species. Simulation software and ration-balancing spreadsheets can aid breeders in predicting animal responses of genotypes or populations that differ in digestibility or fiber concentration (Clark and Wilson, 1993; Donnelly et al., 1994; Shenk, 1977; Undersander et al., 1993).

Traits related to intake potential

Genetic improvements in voluntary intake potential of forage crops may be potentially more valuable than genetic improvements in digestibility, given the greater importance of intake to animal performance (Fahey and Hussein, 1999). For most forage diets, intake cannot be maximized due to limitations in feed quality (Van Soest, 1994). Fibrous bulk, the plant cell wall, is generally considered to be the factor most limiting to feed intake. Intake of fibrous bulk generally causes rumen fill and satiation before the ruminant has maximized its caloric intake, resulting in a reduced plane of nutrition (Van Soest, 1994).

Intake potential has received far less attention than digestibility from forage-plant breeders. There has never been a "silver bullet" developed for intake potential, a single selection criterion that is clearly understood and accepted as a satisfactory and reliable predictor of intake potential, such as the Tilley and Terry technique for predicting digestibility. Intake traits are complicated by the diversity in feeding systems for forage crops. For example, intake may be improved by designing a grazed plant to be more accessible to livestock or by engineering a conserved forage to have more favorable chemical or physical characteristics, neither trait having an effect on intake in the alternative management system (Hutton, 1971). Despite the lack of breeding activity in this area, there are several plant traits that show promise for improving intake potential of forage crops.

The concentration of NDF is typically used as the most reliable predictor of intake potential for forage crops, as it is a direct measure of fibrous bulk of the feed. Ranges of variation within populations, realized heritability, and genetic gains, although reported for fewer species, are similar to those for IVDMD. Genetic progress toward reduced NDF concentration has been reported in three grass species, ranging from 5 to 13 g kg^{-1} $year^{-1}$, 0.8 to 2.0% $year^{-1}$ (Casler, 2001).

Alternative strategies to increase intake potential include the measurement of the energy required to grind samples through a defined screen size, the energy required to shear leaf tissue, and a measure of particle-size reduction during ball milling (Casler, 2001; Marten, 1989). Each of these potential selection criteria appear to be regulated by heritable plant traits and amenable to selection. However, they have either not been validated with animal performance results or results to date have been inconclusive (Casler, 2001). Furthermore, little is known about the underlying mechanisms that are responsible for genetic variability of these characteristics of forage plants.

Reed canarygrass (*Phalaris arundinacea* L.) and phalaris (*P. tuberosa* L.) both contain alkaloids that reduce intake potential by reducing palatability of the forage. In extreme cases, ruminants can detect these compounds by olfactory mechanisms and will completely avoid grazing these grasses. Tryptamines and β-carbolines are the most toxic and objectionable of these compounds, causing diarrhea and, in extreme cases, death to animals grazing these grasses. Gramine is a less-toxic alkaloid that seems to inhibit palatability but has no adverse health effects on grazing ruminants. The concentration of tryptamines and β-carbolines has been drastically reduced in the forage of new cultivars of both reed canarygrass and phalaris, increasing the popularity of both grasses for use in permanent pasture. Elimination of tryptamines and β-carbolines and a reduction in gramine concentration in reed canarygrass has been documented to result in increased animal health and liveweight gains (Marten, 1989). These cultivars have excluded all others with unfavorable alkaloid profiles from the reed canarygrass pasture seed market in the United States. Because alkaloids are metabolized in hay of these two grasses, they do not present problems to hay-fed livestock.

Some toxins have a hidden effect on forage quality, reducing intake by negatively impacting animal health and well-being, often causing mortality. Cyanogenic glycosides are produced in numerous grasses and legumes. Genetic variation is present for these compounds, and their levels can be reduced by breeding and selection (Hacker, 1982), creating greater margins of safety for livestock grazing cyanogenic forage crops, such as sudangrass [*Sorghum bicolor* (L.) Moench ssp. *drummondii* (Nees ex Steud.) de Wet & Harlan] and white clover (*Trifolium repens* L.). Nitrates often accumulate to toxic levels in grasses, particularly under high-nitrogen fertilization. Genetic variability exists for nitrate concentration, but this has received little or no attention by breeders because it can be readily controlled by management (Hacker, 1982). In sweetclover (*Melilotus officinalis* L.), a long-term breeding program reduced the concentration of coumarin, which produces the deadly compound dicoumarol (Howarth and Goplen, 1983). Other toxins produced by forage species can have detrimental effects on reproduction and metabolism, without direct effects on intake potential (Hacker, 1982; Hutton, 1971).

Hypomagnesaemia is likely the most serious ruminant disease caused by mineral imbalance of forage crops. It is caused by a deficiency of Mg and/or an excess of K in the forage tissue, either of which results in decreased intake, severe Mg deficiency, and eventual death of grazing livestock. Two breeding programs have attacked this problem by selecting for increased Mg uptake. In Italian ryegrass (*L. multiflorum* Lam.), a 56% increase in Mg concentration of the forage led to a 13% increase in blood serum Mg and a 12% increase in dry matter intake of grazing ewes. Ewe and lamb liveweight gains increased 10%, while the proportion of clinical cases of hypomagnesaemia was reduced 88%, with no fatalities among 120 ewes grazing the high-Mg population. In tall fescue (*Festuca arundinacea* Schreb.), selection for increased Mg and a reduction in the mineral ratio K/(Ca + Mg) led to a reduction in the mineral ratio of 18%. This resulted in a 12% increase in blood serum Mg levels of grazing cattle and an 8% reduction in the blood serum mineral ratio K/(Ca + Mg) (Casler, 2001).

Although their effects are not well documented, leaf diseases of forage crops can have a dramatic adverse effect on intake of forages. Heavily diseased leaf tissue reduces palatability and digestibility (Casler, 2001; Casler and Vogel, 1999), reducing intake potential by grazing livestock (Hacker, 1982; Hanna, 1993). Thus, genetic resistance to disease-causing pathogens is a mechanism of preserving quality of forage crops and is particularly valuable for pasture crops.

Selection methodology

Phenotypic recurrent selection

Genetic improvement of forage quality is typically accomplished using phenotypic recurrent selection. Realized heritability for forage quality traits is typically in the range of 0.2–0.4, obviating the need for complex genotypic selection systems. For perennial forages, most breeders plant large nurseries of plants and preselect on the basis of agronomic traits, such as vigor, persistence, disease resistance, and growth habit. These plants are then sampled for laboratory analysis and selection for some defined forage quality trait.

Selection for forage quality traits is typically conducted on the basis of unreplicated plants. Clonal replication can be accomplished in most perennial forage crops, but seldom results in sufficient improvement in the efficiency of selection to warrant the cost and effort. Only in the case of severe genotype × environment (GE) interactions and when individual-plant heritability is very low will replicated selection be more efficient than unreplicated selection. Furthermore, the use of replicated selection may cause the breeder to reduce selection intensity because of the increase in the number of samples that must be processed.

Many breeders will sample plants for forage quality analyses at multiple stages or harvests to improve their phenotypic assessment of the selection units. Such a practice is a form of replication in time, but does not take the place of replication in space. Nevertheless, replication in time can provide a more accurate assessment of phenotype by averaging across multiple harvests that occur across the range of environmental conditions encountered during the growing season. As with replicated selection, the breeder should be careful to avoid reducing the selection intensity simply because of the large number of samples generated by this practice.

Genotype × environment interactions

The growing body of literature suggests that GE interactions are relatively unimportant in determining phenotype for forage-quality traits (Carter et al., 1991; Casler, 2001). Genotypes or populations that differ in forage quality traits are generally consistent in ranking across different environments or managements. This relationship can often break down when genetic variability for maturity is confounded with genetic variability for forage quality traits. For maize lines evaluated in a short-season environment, differential maturity can cause differential grain fill, resulting in a disruption or masking of genetic variability for forage quality traits of the stover (Carter et al., 1991). In perennial forage crops, extreme variability in maturity can cause differential regrowth rates, leading to severe genotype × harvest interactions (Coors et al., 1986).

Stronger evidence for this assertion comes from selection experiments themselves. Documentation of genetic gains from recurrent selection is a two-stage process. In stage I, the breeder conducts one or more cycles of selection, preserving seed of the original population and of the intercrossed population representing each cycle of selection. In stage II, each population is planted in replicated and randomized field experiments, typically using multiple locations and years. For field-based selection, it is impossible to duplicate the environmental conditions of the original selection nursery. Soil type, moisture conditions, weather patterns, and many other environmental factors may all vary between the selection and evaluation environments. Thus, the measurement of significant gains from selection in the evaluation environment is evidence, in itself, that genotypic effects of selection are more important than GE interaction effects. Furthermore, because selection is typically conducted under spaced-planted conditions, evaluations that are conducted as sward plots represent particularly robust tests of GE interaction.

Despite the above generalizations, there are some examples of GE interactions for forage quality traits (Buxton and Casler, 1993; Casler, 2001; Casler and Vogel, 1999). In some cases, these interactions seem to be particularly severe, causing considerable difficulty in accomplishing effective selection and/or the measurement of genetic gains. Many early genetic variability studies used either the acid-detergent or permanganate methods to estimate lignin concentration. There has been recent interest in use of the Klason lignin method as a more accurate representation of total lignin in plant cell walls. Initial studies suggest that Klason lignin measurements seem particularly sensitive to GE interactions, although more research is required to verify this tentative conclusion.

Near-infrared reflectance spectroscopy

The application of near-infrared reflectance spectroscopy (NIRS) to forage breeding ranks with the Tilley and Terry digestibility procedure as one of the most important events in the history of this discipline. As with the Tilley and Terry procedure, forage breeding played a key role in the development of NIRS technology to predict forage quality of feeds. Dr. J.S. Shenk, a forage grass breeder at The Pennsylvania State University, and several colleagues were largely responsible for application of the technology, software and statistical development, hardware improvement, and training of forage researchers around the world. The improvements to NIRS technology have resulted in an ability to predict animal response variables directly from forage spectra with greater accuracy and precision than from standard laboratory reference methods.

The use of NIRS improves the efficiency of forage-breeding programs by three mechanisms (Casler, 2001). First, it reduces the cost associated with laboratory analysis 70–90%, assuming a 10-year life span of the NIRS equipment and nearly constant use. Second, NIRS is much faster than wet-laboratory analysis, reducing the time required to generate data for selection by as much as 90%, depending on the size of each experiment. Third, the first two factors combine to allow the breeder to analyze many more genotypes or families, increasing selection intensity and/or the number of populations that can be cycled through the selection process. These factors also combine to create additional flexibility in a breeding program, allowing the breeder to develop imaginative methods to increase the efficiency of the breeding program, such as sampling several weeks prior to anthesis to facilitate selection on both male and female gametes without lengthening the selection cycle. Such flexibility is impossible with traditional wet-laboratory methods.

Finally, the reduction in laboratory expense associated with NIRS is further enhanced if the forage nutritional value selection program is based on multiple traits, such as the Cornell University alfalfa-breeding program (crude protein and acid detergent fiber) or the University of Wisconsin maize silage-breeding program (silica, lignin, and NDF). Because forage sample spectra can be used to predict a wide array of plant traits, the only additional expense associated with multiple-trait selection would be the expense of wet-laboratory analysis of the additional traits on the calibration subset of forage samples. The NIRS procedure is sufficiently versatile that it can efficiently predict a wide array of chemical traits, including digestibility, fiber or cell-wall constituents, protein, and mineral elements.

Maturity and plant parts

Nearly all measures of forage quality are influenced by reproductive maturity (or growth stage) and by the relative plant-part composition of the plant at the time of sampling. Positive measures of forage quality (such as protein and digestibility) decline and negative measures of forage quality (such as fiber and lignin) increase as plants mature and cell-wall development progresses. Furthermore, in many forage crops, leaves have considerably higher quality than stems or stalks. Depending on the methods used in the selection program, genetic variation for maturity and/or plant-part composition in the population may be one of the more important causal factors in bringing about changes in forage quality. Such changes do not necessarily reflect genetic variability for forage quality traits, particularly if genetic variability for maturity or plant-part composition is large. Numerous examples can be found in the literature, demonstrating apparent differences in forage quality among cultivars, but due only (or largely) to differences in maturity or plant structure (Van Wijk et al., 1993). Such differences may have unforeseen consequences, such as changes in management, harvest frequency, or growth cycles. While some of this genetic variation may be of value to agronomists and livestock producers, it does not necessarily reflect a genetic improvement in forage quality per se.

In populations that contain genetic variability for timing of reproductive maturity, selection for increased forage quality based on sampling at a fixed harvest date generally results in later heading. Some breeders have attempted to correct this problem by sampling all plants at a fixed growth or maturity stage, a process that tends to shift the population in the opposite direction, toward earlier heading. Such a method results in a tendency to select plants harvested on the earliest dates, suggesting that their cell-wall development is delayed relative to reproductive maturity. In smooth bromegrass, there is no genetic variation for maturity, but a large amount of genetic variation for cell-wall development, suggesting that reproductive maturity and cell-wall development can be decoupled by directed selection. For many species and populations, selection for increased forage quality without a shift in maturity will require some form of maturity rating and selection for increased forage quality, either statistically adjusted to a constant maturity rating or within maturity classes.

Similarly, in populations that contain genetic variability for plant-part composition, selection for forage quality on the basis of whole-plant samples may result in significant changes in plant-part composition. For example, selection for reduced NDF of smooth bromegrass whole-plant samples at the heading growth stage resulted in large increases in leaf/stem ratio, because stems have much high NDF than leaves (Casler, 2001). Changes in NDF of individual plant parts were relatively minor in comparison. Similar responses have been observed following divergent selection for lignin concentration in alfalfa. While such changes may have minor agronomic consequences in the short term, they may have potentially large consequences if allowed to accumulate over many cycles of selection. To solve this potential problem, most breeders use specifically defined stages of maturity and plant parts as sampling units. Examples include leaf blades in pasture grasses, lower stem segments in alfalfa, and defined internode segments in maize. Selection for low NDF on the basis of leaf blades alone resulted in no changes in plant-part composition when three cycles of selection were evaluated at the heading growth stage (Casler, 2001).

Breeding maize for increased silage quality is a relatively recent phenomenon. Most maize hybrids and populations have been developed by selection for increased grain yield. Increased grain yields have been associated with minimal or no changes in stover yield, but increased harvest index (grain-to-stover ratio) (Coors and Lauer, 2001). There has been relatively little effort to select and breed maize hybrids or populations for increased forage yield or quality. The lack of selection pressure for forage quality traits may be partly responsible for the large amount of genetic variability observed in maize populations and breeding programs. The variability observed among maize inbreds is similar in magnitude to that observed in most perennial forage populations (Coors and Lauer, 2001).

Agronomic consequences

Although the Tilley and Terry procedure was developed in the early 1960s, very few breeders actively sought to select and breed forage crops for improved forage quality until the late 1970s and early 1980s. One reason for this lag was that many forage breeders and agronomists were fearful of negative agronomic consequences, such as forage yield drag, seed yield reductions, increases in lodging, and loss of natural pest resistance. In retrospect, some of these fears were justified, but we know this only because of the efforts of several forage breeders during the past 40 years.

The USDA-ARS bermudagrass (*Cynodon dactylon* L.) breeding program at Tifton, Georgia, under the leadership of G.W. Burton, and later W.W. Hanna, is the longest-running breeding program designed to improve quality of a forage crop. Early cultivar releases from this program were characterized by large increases in digestibility, but reduced forage yield compared with Coastal, the most prominent cultivar at the time (Figure 23.1). Fortunately for bermudagrass growers in the southeastern United States and elsewhere in the world, the team of scientists at Tifton recognized the value of these cultivars, despite a small reduction in forage yield potential. Hacker (1982) correctly predicted that such negative correlations between forage yield and quality could be broken by the persistent efforts of breeders. While the losses in digestibility of the last two cultivars suggest that the negative genetic correlation between forage yield and quality still exists in this bermudagrass gene pool, these two cultivars validate Hacker's prediction that these negative correlations need not be an impediment to long-term breeding progress. Combined selection for forage yield and quality in alfalfa and reed canarygrass has also resulted in populations with improved forage quality without sacrifices in forage yield (Casler, 2001).

Conversely, in smooth bromegrass, selection for increased intake potential by reduction in NDF concentration has resulted in consistent decreases

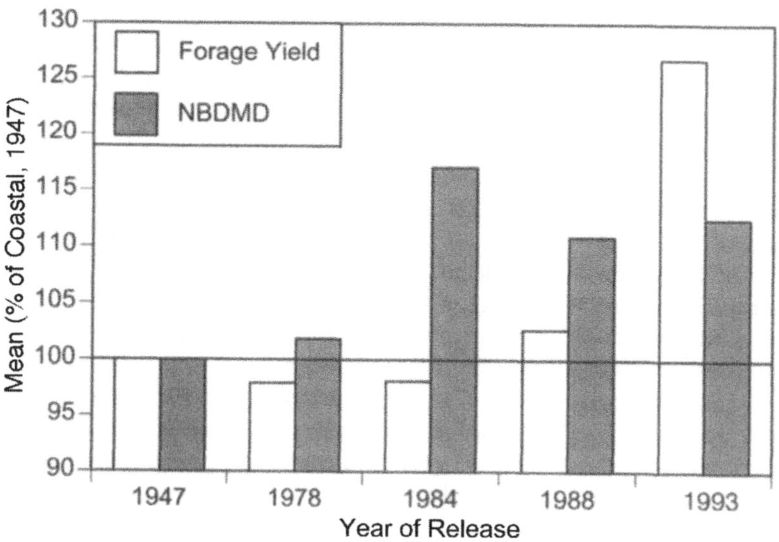

Figure 23.1 Progress in breeding bermudagrass for nylon bag dry matter digestibility (NBDMD) and forage yield as a function of year of cultivar registration. Cultivars were Coastal (1947), Tifton 44 (1978), Tifton 68 (1984), Tifton 78 (1988), and Tifton 85 (1993). Adapted from Hill et al. (1993).

in forage yield. Results are consistent across diverse germplasms, suggesting a possible pleiotropic effect (Figure 23.2). The plant cell wall represents a physical frame upon which numerous plant functions and processes are built. Cell walls are responsible for the retention of upright growth as tillers grow taller, larger, and heavier. Cell walls also function in the transport of nutrients, photosynthate, and water through the vascular system of a tiller. Reduced cell-wall concentration probably represents a direct limitation to a plant's ability to accumulate dry matter as it matures.

Three cycles of recurrent phenotypic selection for IVDMD in switchgrass (*Panicum virgatum* L.) resulted in a gradual increase in plant mortality, attributed to a loss in cold tolerance (Casler et al., 2002). Similar results were observed for one cycle of divergent selection for IVDMD in orchardgrass (*Dactylis glomerata* L.). These results are similar to changes observed in plant mortality of alfalfa populations divergently selected for lignin concentration (Buxton and Casler, 1993). However, when survivors of the high-lignin and low-lignin populations were intercrossed and evaluated for mor-

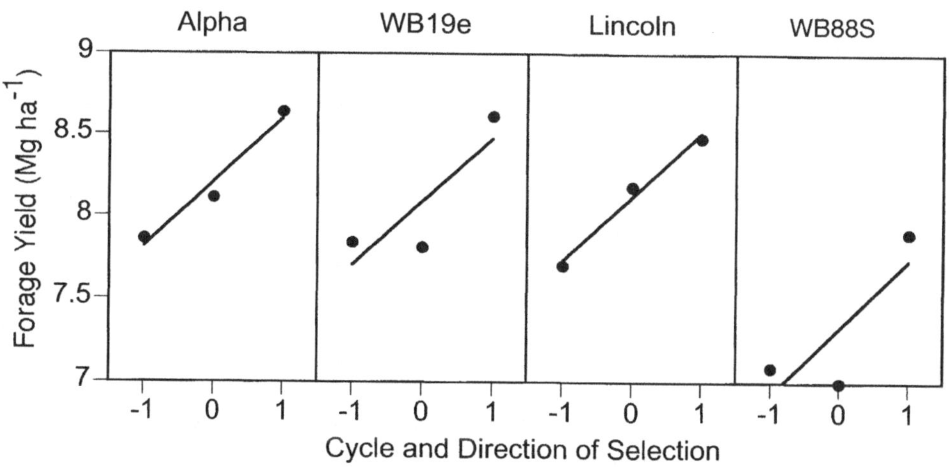

Figure 23.2 Effect of one cycle of divergent selection for high (+1) or low (−1) NDF (neutral detergent fiber) concentration on forage yield of four smooth bromegrass populations (Casler, 2001, unpublished data). Selection responses were: b = 0.39 ($P < 0.01$) for Alpha, b = 0.39 ($P < 0.01$) for WB19e, b = 0.20 ($P = 0.05$) for Lincoln, and b = 0.40 ($P < 0.01$) for WB88S.

tality, there were no differences between the high-lignin and low-lignin progeny populations, despite a large level of plant mortality across the study (Casler et al., 2002). These results suggest that the association of cold tolerance and IVDMD in these species is likely due to linkage.

In maize, there are two major mechanisms to improve the quality of maize silage: (1) increase the harvest index, the proportion of highly digestible grain, or (2) increase stover quality per se. Stover traits, such as fiber, lignin, silica, and digestibility, can be improved by selection and breeding (Coors and Lauer, 2001). Maize breeders face a challenge of improving stover quality without sacrificing grain yield so that improved populations and hybrids can serve a dual purpose of silage or grain production. A general lack of correlation between stover quality traits and grain yield suggests that these traits can be simultaneously improved.

The effects of selection for increased forage quality on lodging potential have not been well documented, probably because little lodging has been observed in agronomic field studies designed to evaluate selection responses and agronomic studies have not been designed to measure lodging per se. However, stalk strength and lodging resistance can be improved in maize without sacrificing stover quality (Coors and Lauer, 2001). Stalk-lodging resistance appears to be more a function of overall plant health, rather than on chemical or physical traits that may be considered to be anti-quality traits. Similarly, studies in small grains have suggested that low-lignin lines may have greater stem-lodging resistance due to greater stem flexibility and elasticity (Casler, 2001).

Conversely, selection for increased resistance to second-generation European corn borer (ECB; *Ostrinia nubilalis* Hübner) in maize resulted in increased fiber, lignin, and silica of the stover (Casler, 2001; Coors and Lauer, 2001). Increased leaf toughness, conferred by increased fiber, lignin, and silica, appears to be a mechanism for increased resistance to second-generation ECB in maize.

While disease resistance can protect forage plants from reduced forage quality associated with disease infection, there is no overwhelming evidence that disease resistance is compromised by genetic increases in forage quality. Although increased lignification has been suggested as a mechanism of fungal disease resistance by numerous authors, there is little evidence that reduced lignification or increased digestibility has affected disease resistance in several forage-crop pathosystems (Casler, 2001). The one possible exception to this is crown rust (*Puccinia coronata* Corda) reaction, which has been associated with high digestibility, high WSC, or low lignification in several perennial grasses (Casler, 2001). However, most forage grass breeders consider these relationships to be due to linkage, because they can be broken up by selection and recombination, although these observations are not well documented.

Finally, Clark and Wilson (1993) argued that agronomic problems associated with genetic increases in forage quality were probably more perceptual than real. For example, a cultivar with improved digestibility may result in a forage yield reduction or an increase in disease symptoms when evaluated as a hay crop, but the yield reduction or disease incidence may be small and unimportant if the cultivar is used in short-rotation pastures where forage is distributed among several growth cycles. Breeders must carefully define both the target population of environment and management and sample from these populations for all evaluation experiments to ensure that observed genetic gains and correlated responses are real and meaningful.

Gene discovery

Very few genes have been discovered to have a direct and measurable effect on forage quality traits. The most notable of these are the brown-midrib (*bmr*) genes, which act as Mendelian recessives and have arisen either naturally or through mutagenesis (Cherney et al., 1991). Although the agronomic effects of the various *bmr* genes differ, they all have a large negative effect on agronomic performance, reducing forage yield, grain yield, and stalk or stem strength (Cherney et al., 1991; Coors and Lauer, 2001). Despite these effects, brown-midrib hybrids have been released and marketed in the United States. Unfortunately, economic analyses are insufficient in power and scope to act as a reliable decision-making tool to informatively choose between brown-midrib and normal forages.

There has been relatively little effort to identify quantitative trait loci (QTL) affecting forage quality traits (Casler, 2001). Most of the efforts in this regard have led to the general conclusion that QTL-marker associations are highly population specific and cannot be generalized across populations. Thus, the utility of any particular QTL for marker selection or marker-assisted selection is limited to the population in which it is discovered. Conversely, in perennial ryegrass (*L. perenne* L.), numerous studies have shown an association between the *b* allele of the *Pgi-2* locus and high WSC concentration (Casler, 2001). Because the product of this gene, phosphoglucose isomerase (PGI), catalyzes the reversible isomerization of glucose-6-phosphate and fructose-6-phosphate, an essential step preceding carbohydrate metabolism in plants, the PGI marker may be the WSC-QTL itself.

Transgenic technology has tremendous potential to create novel and useful genetic variability for forage quality traits (Boudet et al., 1995; Boudet and Grima-Pettenati, 1996; Campbell and Sederoff, 1996). Most efforts have focused on downregulation of enzymes in the lignin biosynthetic pathway to reduce or modify lignin, or on upregulation of novel proteins to improve protein quality. Most of the enzymes in the lignin biosynthetic pathway have been cloned and used to create antisense constructs for transformation. Many of the early studies were based on a very small number of plants that were frequently characterized by stunted and/or abnormal growth patterns (Casler, 2001). As activity increased in this area during the late 1990s, reports of normal growth patterns for novel lignin transgenics increased in frequency. Most likely, as transformation systems became more efficient, more transgenics were generated, allowing researchers greater opportunity to select plants with more normal phenotypic appearance, combined with the desired novel lignin transgenic phenotype.

Most research on downregulation of lignin biosynthetic enzymes has been based on model plant systems that have little direct applicability to forage crops, including maize. The usefulness of plant transformation to improve forage quality will depend on the development of efficient transformation systems that can generate large numbers of transformed plants. This will allow selection of plants that have favorable expression of the transgene(s) combined with favorable overall phenotype. Evaluations should be made in field environments to provide an accurate and realistic assessment of phenotype. Greenhouse and growth chamber evaluations do not provide a reliable assessment of phenotype of transgenic plants under field conditions (Baucher et al., 1999).

Major genes, marker selection for specific QTL, and plant transformation are all appealing technologies for improving forage quality, but all suffer from an often ignored characteristic. These approaches are expensive and time-consuming, each in its own way. Marker selection and transformation require sophisticated instrumentation and software that is not available to all plant breeders. This is particularly true for many forage-breeding programs in which profit margins are insufficient to fund such activities. Major genes and transformation both require years of backcrossing to develop lines or synthetics of the proper genetic construction, followed by years of agronomic testing to fully characterize these lines. Breeding programs incur opportunity costs associated with these activities, although these costs are difficult or impossible to quantify. These approaches are not short-cuts to the development of cultivars with improved forage quality.

By far, most of the gains made in developing new cultivars with improved forage quality have been the result of quantitative genetic variation, likely associated with a large number of loci. While quantitative trait-breeding systems can be cumbersome and time-consuming, they offer the opportunity to simultaneously select for forage quality and agronomic traits, improving the probability of obtaining agronomically favorable progeny. Most cultivars are developed as routine products of recurrent selection systems, showing small gains associated with selection for increased forage quality. Nevertheless, small increases in forage quality can provide huge increases in animal performance, often without the negative consequences of reduced agronomic performance and without the opportunity cost of more sophisticated approaches. While it is scientifically interesting to know the genes that are the target of our selection protocols aimed at increasing forage quality, this information is not essential to make breeding gains in forage quality.

References

Baucher, M., M.A. Bernard Vailhé, B. Chabbert, J.M. Besle, C. Opsomer, M. Van Montagu, and J. Botterman. 1999. Down-regulation of cinnamyl alcohol dehydrogenase in transgenic alfalfa (*Medicago sativa* L.) and the effect on lignin composition and digestibility. Plant Mol. Biol. 39:437–447.

Boudet, A.M., and J. Grima-Pettenati. 1996. Lignin genetic engineering. Mol. Breed. 2:25–39.

Boudet, A.M., C. Lapierre, and J. Grima-Pettenati. 1995. Biochemistry and molecular biology of lignification. New Pytol. 129:203–236.

Buxton, D.R., and M.D. Casler. 1993. Environmental and genetic factors affecting cell wall composition and digestibility. p. 685–714. *In* H.G. Jung et al. (eds.), Forage cell wall structure and digestibility. Amer. Soc. of Agron., Madison, WI.

Campbell, M.M., and R.R. Sederoff. 1996. Variation in lignin content and composition. Mechanisms of control and implications for the genetic improvement of plants. Plant Physiol. 110:3–13.

Carter, P.R., J.G. Coors, D.J. Undersander, K.A. Albrecht, and R.D. Shaver. 1991. Corn hybrids for silage: An update. p. 141–164. *In* 46th Ann. Corn Sorghum Res. Conf. Amer. Seed Trade Assoc., Washington, DC.

Casler, M.D. 2001. Breeding forage crops for increased nutritional value. Advan. Agron. 71:51–107.

Casler, M.D., D.R. Buxton, and K.P. Vogel. 2002. Genetic modification of lignin concentration affects fitness of perennial herbaceous plants. Theor. Appl. Genet. 104:127–131.

Casler, M.D., J.F. Pedersen, G.C. Eizenga, and S.D. Stratton. 1996. Germplasm and cultivar development. p. 413–469. *In* L.E. Moser et al. (eds.), Cool-season forage grasses. ASA Madison, WI.

Casler, M.D., and K.P. Vogel. 1999. Accomplishments and impact from breeding for increased forage nutritional value. Crop Science 39:12–20.

Cherney, J.H., D.J.R. Cherney, D.E. Akin, and J.D. Axtell. 1991. Potential of brown-midrib, low-lignin mutants for improving forage quality. Advan. Agron. 46:157–198.

Clark, D.A. and J.R. Wilson. 1993. Implications of improvements in nutritive value on plant performance and grassland management. p. 165–171. *In* M.J. Baker (ed.), Grasslands for our world. SIR Publ. Wellington, New Zealand.

Cooper, J.P., J.M.A. Tilley, W.F. Raymond, and R.A. Terry. 1962. Selection for digestibility in herbage grasses. Nature 195:1276–1277.

Coors, J.G., and J.G. Lauer. 2001. Silage corn. p. 347–392. *In* A.R. Hallauer (ed.), Specialty corns. 2nd ed. CRC Press, Boca Raton, FL.

Coors, J.G., C.C. Lowe, and R.P. Murphy. 1986. Selection for improved nutritional quality of alfalfa forage. Crop Science 26:843–848.

Donnelly, J.R., M. Freer, and A.D. Moore. 1994. Evaluating pasture breeding objectives using computer models. N.Z. J. Agric. Res. 37:269–275.

Fahey, G.C., Jr., and H.S. Hussein. 1999. Forty years of forage quality research: Accomplishments and impact from an animal nutrition perspective. Crop Science 39:4–12.

Hacker, J.B. 1982. Selecting and breeding better quality grasses. p. 305–326. *In* J.B. Hacker (ed.) Nutritional limits to animal production from pastures. Comm. Agric. Bureau, Farnham Royal, Great Britain.

Hanna, W.W. 1993. Improving forage quality by breeding. p. 671–675. *In* D.R. Buxton et al. (eds.), International Crop Science I. Crop Sci. Soc. Amer., Madison, WI.

Hill, G.M., R.N. Gates, and G.W. Burton. 1993. Forage quality and grazing steer performance from Tifton 85 and Tifton 78 bermudagrass pastures. J. Anim. Sci. 71:3219–3225.

Howarth, R.E., and B.P. Goplen. 1983. Improvement of forage quality through production management and plant breeding. Can. J. Plant Science 63:895–902.

Hutton, E.M. 1971. Plant improvement for increased animal production. J. Aust. Inst. Agric. Sci. 37:212–225.

Humphreys, M.O. 1989. Water-soluble carbohydrates in perennial ryegrass breeding. III. Relationships with herbage production, digestibility and crude protein content. Grass Forage Sci. 44:423–430.

Marten, G.C. 1989. Breeding forage grasses to maximize animal performance. p. 71–104. *In* D.A. Sleper et al. (eds.), Contributions from breeding forage and turf grasses. Crop Sci. Soc. Amer. Spec. Publ. 15. CSSA, Madison, WI.

Raymond, W.F. 1969. The nutritive value of forage crops. Advan. Agron. 21:1–108.

Reed, K.F.M. 1994. Improved grass cultivars increase milk and meat production—a review. N.Z. J. Agric. Res. 37:277–286.

Shenk, J.S. 1977. The role of plant breeding in improving the nutritive value of forages. J. Dairy Sci. 60:300–305.

Stone, B.A. 1994. Prospects for improving the nutritive value of temperate, perennial pasture grasses. N.Z. J. Agric. Res. 37:349–363.

Tilley, J.M.A. and R.A. Terry. 1963. A two-stage technique for in vitro digestion of forage crops. J. Br. Grassl. Soc. 18:104–111.

Undersander, D.J., W.T. Howard, and R.D. Shaver. 1993. Milk per acre spreadsheet for combining yield and quality into a single term. J. Prod. Agric. 6:231–235.

Van Soest, P.J. 1994. Nutritional ecology of the ruminant. 2nd ed. Cornell Univ. Press, Ithaca, NY.

Van Wijk, A.J.P., J.G. Boonman, and W. Rumball. 1993. Achievements and perspectives in the breeding of forage grasses and legumes. p. 116–120. *In* M.J. Baker (ed.), Grasslands for our world. SIR Publishing, Wellington, New Zealand.

Vogel, K.P., and D.A. Sleper. 1994. Alteration of plants via genetics and plant breeding. p. 891–921. *In* G.C. Fahey et al. (eds.), Forage quality, evaluation, and utilization. Amer. Soc. Agron., Madison, WI.

24
Breeding for Grain Amino Acid Composition in Maize

Audrey Darrigues, Department of Agronomy, Iowa State University
Kendall R. Lamkey, Department of Agronomy, Iowa State University
M. Paul Scott, USDA-ARS, Corn Insects and Crop Genetics Research Unit

Improving the amino acid balance of grain has been a long-standing objective of plant-breeding research. In this chapter, we review the history of maize breeding for improved amino acid balance. Following this, we present results of our experiments involving divergent selection for the levels of the amino acids tryptophan and methionine in random-mated populations.

The majority of maize produced worldwide is used for food and feed, so one of the best ways to improve the value of this grain is to improve its nutritional quality. The main nutritional limitation of maize is that it is not a good source of protein. While maize grain typically contains 4–10% protein, this protein has less dietary value than protein from animal sources. This is because plant proteins tend to be digested less efficiently and are more likely to cause antigenic responses than animal proteins commonly used in diets. More importantly, plant proteins are deficient in certain amino acids, while other amino acids are in excess relative to the needs of animals (Figure 24.1). This imbalance in amino acid content decreases the nutritional value of plant proteins in animal diets.

Essential amino acids

Monogastric animals (including humans) require specific dietary essential amino acids. By definition, amino acids are deemed essential if an organism does not synthesize them and they must therefore be supplied in the diet. If one of the essential amino acids is limiting, the deficiency results in a negative nitrogen balance (Berg, 2002). Nonessential amino acids are not required per se in the diet but are required for protein synthesis (Cheeke, 1999) and therefore must be either supplied in the diet or synthesized from dietary components. Therefore, a nonessential amino acid may be the limiting factor in growth if its level in the diet is insufficient and if the essential amino acids from which it is made are present in marginal amounts (Wiseman, 1987).

In animal diets, the efficiency of protein utilization is dependent upon two types of factors: external factors, which relate to rearing conditions, and internal factors, relating directly with the protein itself (Berg, 2002). The nutritional value associated with a protein may be estimated by comparing the ingested nitrogen to that which is actually retained for protein synthesis. Requirements for a specific monogastric animal depend on its metabolic peculiarities and are dependent on the genotype, performance, method of feeding, and the environment in which the animal is reared (Wiseman, 1987). Also, the protein quality requirement is different for growth than for maintenance and is affected by sex and by species (Cheeke, 1999). Signs of protein deficiency include anorexia, reduced growth rate, reduced or negative nitrogen balance and reduced efficiency of feed utilization. Specific lesions may appear with deficiencies for certain amino acids: tryptophan deficiency produces eye cataracts; methionine deficiency produces fatty liver (Pond, 1995). In humans, diets with imbalanced amino acid levels contribute to the malnutrition conditions of Kwashiorkor and Marasmus.

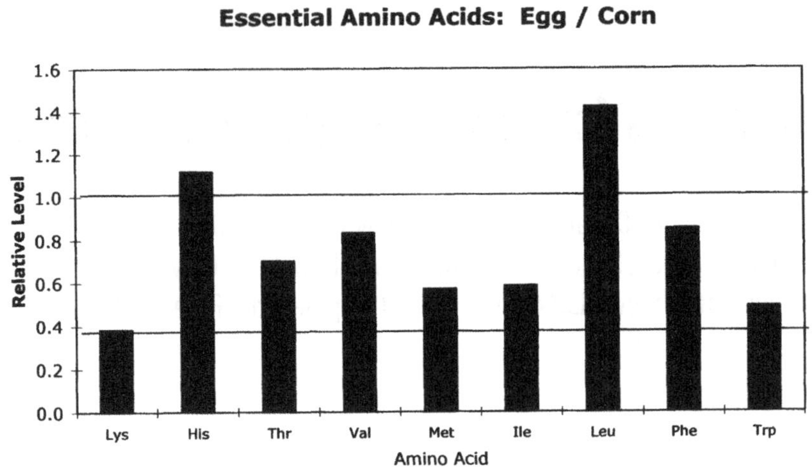

Figure 24.1 For each amino acid, the ratio of the level in egg to the level in corn is presented. Egg is considered a nearly balanced source of protein, so this illustrates the deficiencies and surpluses of corn protein.

Amino acid levels in maize

What determines amino acid levels

Osborne (1924) classified seed proteins into four classes based on their solubilities. These classes are the albumins (water soluble), the globulins (salt soluble), the prolamins (aqueous alcohol soluble), and the glutelins (not soluble in water, saline solutions, or aqueous alcohol). This nomenclature is still used today. In most cereals the most abundant seed storage proteins are prolamins, while in dicots the most abundant seed storage proteins are usually globulins. In each genus, the major seed storage protein is named on the basis of the genus name, thus the major seed storage proteins in maize are called the zeins for the genus *Zea* and belong to the prolamin class of proteins. Similarly, the major seed storage proteins of soybean are globulins called glycinins after the genus *Glycine*.

The zeins can account for 40–60% of the total protein in the maize endosperm, and, because of their abundance, they are the primary determinants of the amino acid composition in maize kernels (Larkins et al., 1993). Osborne and Clapp (1908) characterized the amino acid composition of the zein proteins and reported that they lack two essential amino acids, lysine and tryptophan. This deficiency is reflected in the amino acid balance of maize, as illustrated in Figure 24.1. Therefore, increasing the levels of lysine and tryptophan are important goals for plant-breeding efforts directed to improving grain amino acid balance.

Maize-based feed is often supplemented with oil-seed by-products such as soy protein. These by-products complement the amino acid balance of maize protein somewhat, resulting in a more balanced diet. However, because the globulin storage proteins of dicots are deficient in the sulfur amino acids cysteine and methionine, increasing methionine levels is another important goal of plant-breeding programs.

Genes involved in determining amino acid balance

Several mutant genes affect the amino acid balance in maize. Generally, these mutants alter the accumulation of zeins. An example is *opaque-2*, first described in 1935. Homozygous kernels carrying this mutation have a low density and do not transmit light because of their floury nature. Kernels with this phenotype have elevated levels of lysine (Mertz et al., 1964). This change is accompanied by a reduction in the levels of alpha zeins (Mertz et al., 1964), which are low in lysine content. The gene product of *Opaque-2* is a transcription factor involved in regulation of zein synthesis (Hartings et al, 1989; Schmidt, et al, 1990).

The *floury-2* mutant (Mumm, 1935) has a phenotype similar to *opaque-2* and has an altered amino acid balance with high concentrations of lysine and methionine (Nelson et al., 1965). In this mutation, a modified zein that cannot be processed properly interferes with accumulation of proteins dependent on the secretory system, including the zeins (Coleman and Larkins, 1995).

Thus, *opaque-2* and *floury-2* both achieve similar phenotypes with floury kernels and improved lysine content by similar processes. Both mutations result in reduced zein deposition, although *opaque-2* contains a transcriptional defect and *floury-2* contains a translational defect. Because the zeins that are reduced in these mutations have very little lysine, the overall effect is an increase in lysine concentration.

A mutation that results in increased levels of kernel methionine has been characterized. This mutation results in overproduction of 10-kDa delta zein that is rich in methionine (Phillips and McClure, 1985). The overproduction of this zein is attributed to a regulatory gene *Zpr10/22* (Benner et al., 1989) that regulates delta zein accumulation posttranscriptionally (Cruz Alvarez et al., 1991). The *Zpr10/22* locus was later renamed *Dzr1* (Chaudhuri and Messing, 1994).

Studies on quality traits in maize

Traditional breeding for quality traits

Selection for protein quantity

A number of maize breeders have used traditional breeding methods to increase the level of protein in maize kernels. In 1896 Hopkins initiated a selection program for protein and oil content (Hopkins, 1899) that developed into the Illinois long-term selection experiment. This experiment consisted of repeated cycles of divergent selection for oil and protein and was highly successful in changing kernel composition in maize. After 70 cycles of selection, the protein content of the high-protein (IHP) and low-protein (ILP) strains was 26.6% and 4.4%, respectively (Dudley, 1974).

The major agronomic difference between the IHP and the ILP populations is grain yield. IHP plants yield substantially less than ILP plants. The chemical kernel composition of strains derived from the high and low selection for protein and oil has been studied extensively for physiological and biochemical modifications. Differences have been associated with the nitrogen (N) and carbon (C) metabolisms in the plant, which is influenced by the uptake, assimilation, translocation, and utilization of N and C.

IHP strains are more efficient at absorbing and translocating N (Lorenzoni et al., 1978; Wyss et al., 1991) and have a higher capability to assimilate nitrate in the roots (Lohaus et al., 1998), a higher capacity for amino acid transport to the grain (Reggiani et al., 1985; Lohaus et al., 1998), elevated asparagine levels (Lohaus et al., 1998), higher N-metabolism enzyme activity (Lohaus et al., 1998), and limited remobilization of leaf N (Wyss et al., 1991). In addition, the IHP strains have a higher level of seed phytic acid than ILP (Raboy et al., 1989), higher levels of amino acids and lower levels of sugars than ILP (Reggiani et al., 1985), higher enzyme activity (Reggiani et al., 1985), higher enzyme activity associated with starch accumulation (Lorenzoni et al., 1978), greater endoreplication, and higher ploidy level (Cavallini et al., 1995). ILP have lower levels of zein (Lorenzoni et al., 1978).

Frey et al. (1949) suggested that selecting for an increase in total protein in the maize kernel results in an increase in the zein fraction in the endosperm proteins. Similarly, Reggiani et al. (1985) concluded that in maize, long-term selection for diverging levels of protein in the grain has resulted in diverging levels of storage proteins in the endosperm. Dr. Fred Below of the University of Illinois suggests that it is primarily the accumulation of 19- and 22-kDa alpha zeins that have been altered by selection in the Illinois long-term selection experiment (personal communication). Given that the zein fraction has poor nutritional quality due to its lack of tryptophan and lysine, it seems likely that selection for protein content will result in lower protein quality. In 1951, Frey concluded from his study on the interrelationships of proteins and amino acids in corn that the protein in selections for low protein was more nutritionally balanced than the protein in selections for high protein. Thus, in order to improve protein quality by selecting for protein quantity, it would be best to select for low protein (Frey, 1951).

Selection for protein quality

Zuber and Helm (1972) studied the improvement of protein quality, defined as an increase in the lysine content of open-pollinated varieties, without the use of endosperm mutants. They used a recurrent selection method as a means of improving the amino acid balance. They were able to increase the level of lysine using two cycles of selection, though the mean protein values remained essentially the same. They also suggest that different environmental conditions could have caused such changes because the two cycles were grown in different seasons (Zuber and Helm, 1972).

Mutation breeding and QPM

The findings of Mertz, Bates, and Nelson in 1964 that the *opaque-2* mutant has a lower content of the zein proteins in the endosperm and also provides 69% more lysine than wild-type maize kernels (Mertz et al., 1964) changed the emphasis of plant breeding for amino acid balance from recurrent selection to mutation breeding. Unfortunately, pleiotropic effects of the *opaque-2* phenotype complicated breeding efforts. These effects are reduced grain yield; soft and chalky kernel phenotype; greater vulnerability to ear rot; greater moisture content, which conflicted with the dry-down of the seed; and lower rate of germination (Vasal, 2001). Efforts to improve the *opaque-2* phenotype were initiated at CIMMYT in the mid-1970s. The combination of two mutants, *sugary-2* and *opaque-2*, was found to be slightly better than *opaque-2* maize in terms of kernel hardness and ear rot tolerance, but grain yield and germination rate was not improved. However, the protein quality of the double-mutant combination was sometimes better than that of the *opaque-2* maize (Mertz, 1992).

In the 1980s, CIMMYT engaged in developing quality protein maize (QPM) by combining the *opaque-2* gene with genetic modifiers that improved the hardness of the maize kernel. Eventually, the scientists at CIMMYT were able to develop QPM material that yielded as well as their normal counterparts and contained the improved amino acid balance conditioned by the *opaque-2* mutation (Vasal, 2001).

Biotechnology approaches to improving protein quality

With the advent of genetic engineering, a number of studies have proven the feasibility of improving the methionine content in a variety of crops. Lai and Messing (2002) constructed a transgene based on a chimeric *Dzs10* gene by replacing the 3′ UTR with a transcript of the cauliflower mosaic virus that would enhance the level of expression in maize endosperm cells. The level of methionine was increased as a result of the accumulation of the Dzs10 protein, a high-methionine zein. Because milk protein has the potential to provide good nutritional enhancement with its excellent amino acid profile, Yang et al. (2002) sought to synthesize a porcine α-lactalbumin gene construct. Expression of this synthetic gene in maize kernels resulted in a 20% increase in lysine levels (Bicar et al., unpublished). Transformation of narrow-leafed lupin (*Lupinus angustifolius* L.) seeds expressing the sunflower seed albumin (SSA) gene resulted in a 94% increase in methionine when compared with the wild-type (Molvig et al., 1997). Molvig et al. reported that not only was the protein quality improved in the transgenic seeds, but also the true protein digestibility, the biological value, and the net protein utilization. In tobacco, a group of researchers created and transformed a chimeric gene encoding a Brazil nut methionine-rich seed protein (Altenbach et al., 1989). The accumulation of the protein in the tobacco seeds resulted in a 30% increase in the levels of methionine.

Biotechnological approaches in improving the nutritional quality of crops may be promising, though both advantages and disadvantages have to be elucidated. Benefits of such technology are that genes expressing protein from a different organism than the target host can have beneficial nutritional attributes. In the study conducted by Yang et al. (2002), a porcine milk protein with good digestibility, bioavailability, and amino acid balance was introduced into maize. Disadvantages of using biotechnological tools are the difficulties associated with plant transformation and expression of foreign proteins and the potential introduction of allergenic properties. Several studies have reported such difficulties. Molvig et al. (1997) reported that molecular approaches in improving the amino acid balance were hindered by the difficult regeneration of grain legumes and by the unstable expression of the modified protein in the target host. The methionine-rich Brazil nut protein expressed in soybean may have had unfavorable allergenic properties (Nordlee et al., 1996).

Methods for quantifying amino acids in maize kernels

In any plant-breeding program aimed at improving the content of amino acids, it is critical to have a method for quantifying the amino acids of interest accurately and inexpensively. Recent advances in automation, especially liquid-handling technology, greatly facilitate these measurements. Small-scale, replicated assays can be conducted efficiently in 96-well-plate format for a fraction of the cost of older analytical methods.

Amino acid analysis consists of two parts, hydrolysis of the protein to amino acids and the quantitation of the level of the amino acid in the hydrolysate. Standard methods generally use a chemical hydrolysis procedure; however, these methods tend to be expensive and time consuming, requiring strong acid or base solutions, high temperatures, and reaction conditions that are not suited to high-throughput analyses. Enzymatic hydrolyses are much more amenable to high-throughput procedures. Maize is problematic because the zeins are not soluble in the conditions under which most proteases function optimally. To alleviate this problem, we hydrolyze maize protein at pH 2, a condition that solubilizes maize proteins efficiently. We use the digestive enzyme pepsin, which functions well in these conditions.

Three types of methods are normally used to quantify target amino acids: bioassay, chemical assay, or chromatography. The American Organization of Analytical Chemists recognizes ion-exchange chromatographic methods for the determination of amino acids; however, the relatively low throughput and high cost of this method make it poorly suited to primary screening in plant-breeding programs. Chromatographic methods are well suited to verifying the results of more high-throughput, lower-cost methods because of their accuracy and acceptance by the scientific community.

Bioassays are low-cost high-throughput methods that are well suited to analysis in plant-breeding programs. Shankman et al. (1943) used strains of *Lactobacillus arabinosus* that are auxotrophic for specific amino acids to determine the concentrations of eight amino acids. The content of the amino acids was determined based on the amount of lactic acid produced by the bacteria. In 1995, Wright and Orman proposed another microbiological method for the analysis of methionine in maize and soybean seeds. They used the bacteria *Pediococcus cerevisiae,* which is auxotrophic for methionine, and measured the turbidity, a representation of bacterial growth, as an indication of the methionine content in the sample. Although this method may not provide the best analytical accuracy, it provides the high throughput required by plant breeders (Wright and Orman, 1995).

Hernandez and Bates (1969) determined that microbial assays and chromatographic techniques used in the determination of tryptophan were expensive, tedious, and time consuming. They developed a chemical method using iron chloride to characterize papain-hydrolyzed protein in terms of tryptophan content. This method was used extensively in the maize-breeding program at CIMMYT. In 1985, Sastry and Tummuru proposed a different method for analyzing the protein hydrolysates for tryptophan. After alkali hydrolysis of the sample, this method takes advantage of the colored product of the reaction between tryptophan, thioglycolic acid, and sucrose under acid conditions to measure tryptophan levels spectrophotometrically. This method is highly sensitive, rapid, and simple (Sastry and Tummuru, 1985).

Divergent selection for tryptophan and methionine in two maize populations

Materials and methods

Populations used in this study

Two different maize populations were used in this study. One population was derived from BS11, a population originally designated as Pioneer Two-Ear Composite. It was developed by crossing southern prolific material and Corn Belt lines (Hallauer, 1967). The second population was derived from BS31, another random-mated synthetic population derived from FS8A(T)C4 (Lamkey, 2002). The FS8A population was initially developed at the Florida Agricultural Experiment Stations and released in 1988. Germplasm from southeastern United States, Corn Belt, and tropical sources, respectively, account for approximately 30%, 22%, and 48% of FS8A(T). The initial development of this population consisted of intermating a wide range of accessions with resistance to southern corn leaf blight (Horner, 1990). The BS11 and BS31 material used in this study has been under selection for agronomic performance for several cycles of recurrent selection.

Breeding strategy

One hundred and two hundred half-sib ears from the populations BS11 and BS31, respectively, were produced in the summer of 2000 at the Iowa State University Agronomy Farm, analyzed, and categorized based on their methionine and tryptophan content. The five ears with the highest value of each amino acid and the five ears with the lowest value for each amino acid were selected from each population, giving eight categories, each containing five selected ears. These categories were called BS11HT, BS11LT, BS11HM, BS11LM, BS31HT, BS31LT, BS31HM, and BS31LM. Thus, the BS11HT category represents the ears from the BS11 population with the highest content of tryptophan (HT), whereas BS31LM represents the ears from BS31 with the lowest content of methionine, and so on.

In the summer of 2002, a balanced bulk was made from each of the five ears selected in each category in 2000. Each of these eight bulks was planted in five adjacent rows with 25 kernels per row. The plants in each bulk were randomly intermated so that each plant that was used as a male was also used as a female, giving about 40 half-sib ears in each category. The resulting ears were harvested individually and the tryptophan and methionine content was analyzed as described below. Five selections from the approximately 40 ears in each category were chosen on the basis of their amino acid content as before. Taken together, these five selected ears constitute the cycle 1 population for each category.

Preparation of samples for analysis of methionine and tryptophan levels

Each ear of maize was shelled and packaged individually. From each ear, five randomly selected whole kernels were ground to a fine powder using a Wiley Mill with a 40-mesh screen. This powder was stored in Eppendorf tubes. With the flap of the tube open, the samples were then dried for four hours at 65°F, after which the tubes were closed and stored in ambient conditions. Samples were analyzed in 96-well plates using a Randomized Complete Block Design including two checks (B101 and B45o2) and six standards consisting of known concentrations of commercially prepared amino acids. The B101 inbred was chosen as a check for its exceptionally high levels of methionine (Hallauer and Wright, 1995). The B45o2 inbred, an *opaque-2* mutant, was used as a check for high tryptophan. The standard concentrations were 5, 20, 35, 60, 75, 100 μM for methionine and 0, 100, 240, 300, 480, 600 μM for tryptophan. The experiment was replicated on three plates (i.e., three blocks). The checks were replicated twice within a plate and the standards three times within a plate. Ten milligrams of each ground sample and checks were weighed into the well of a V-bottom, 96-well microtiter plate.

Protein hydrolysis

Each sample was subjected to enzymatic hydrolysis using pepsin. To each well, 200 μL of 0.2 mg/mL pepsin solution in a KCl-HCl pH 2 buffer was added. The plate was then sealed, covered with a lid, and placed in a 37°C shaking incubator for approximately 15 hours. After the incubation period, the plate was centrifuged at 3000 rpm for 20 minutes, after which the supernatant was removed for analysis.

Assay for tryptophan

The method for the determination of tryptophan in maize kernels is a modified version of the one originally described by Sastry and Tummuru (1985). Twenty microliters of hydrolysate or standard was transferred directly into the wells of a flat-bottom, 96-well assay plate. The plates were sealed between operations to prevent evaporation. For each plate, the assay solution consisted of 9.5 mL of concentrated HCl, 250 μL 2.5% thioglycolic acid, and 250 μL 10% sucrose. This solution was prepared and warmed to 42°C for 23 minutes to allow the solution to turn yellow. Eighty microliters of this assay solution was added to the hydrolysate in the assay plate. The plate was then shaken for three minutes, after which the optical density at 510 nm was immediately determined with a microplate reader.

Assay for methionine

The microbiological method for the determination of methionine in maize kernels is similar to that described by Wright and Orman (1995). An auxotrophic strain of *Escherichia coli*, P4x, was used in this assay. The inoculum was prepared in M9 media (Maniatis et al., 1982) supplemented with 10 μL of 1 mg/mL methionine solution per 5 mL of M9 media and grown to late log phase. Ten microliters of hydrolysate or a standard was transferred directly into a flat-bottom, 96-well assay plate. The plates were sealed between operations to prevent evaporation. To each well, 100 μL of M9 media and 2 μL of the inoculum were added. The plate was then sealed, covered with a lid, and placed in a 37°C shaking incubator for seven hours. After the incubation period, the plates were placed on a plate shaker for three minutes, and the 595 nm light scattered by the sample was determined using a microplate reader.

Statistical analysis

In our amino acid assays, the greatest source of error is plate-to-plate variation between each replication of a sample. To correct for this, the mean value of the samples on each plate and the grand mean of the samples on all three plates in each experiment were calculated. Each value from a given plate was then normalized by multiplication by the value required to make the mean of that plate equal to the grand mean of the experiment.

Results

Tryptophan and methionine concentrations in starting populations

To determine the feasibility of direct selection for tryptophan and methionine content, populations derived from BS11 and BS31 that are under investigation for their agronomic traits were chosen on the basis of their protein content and their variability. One hundred and 200 individuals from the respective populations were analyzed for their tryptophan and methionine content. Figures 24.2

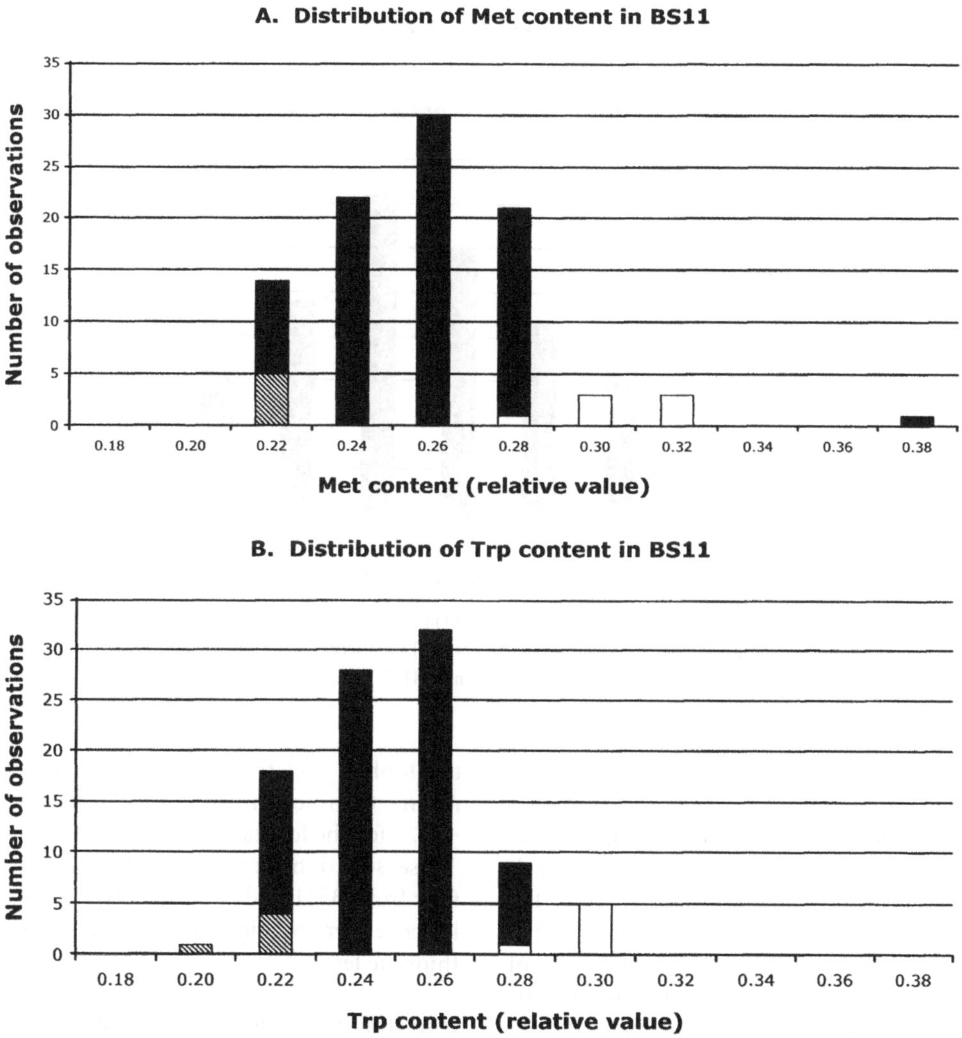

Figure 24.2 Distribution of methionine content (a) and tryptophan content (b) in the initial BS11 population. The levels of methionine and tryptophan represent the optical density measurement corrected for the mass of the sample. The selections made for cycle 1 of the recurrent selection program for the low categories (BS11LM and BS11LT) are cross-hatched. The selections for the high categories (BS11HM and BS11HT) are in white. The overall population mean methionine content was 0.25 and the mean tryptophan content was 0.24.

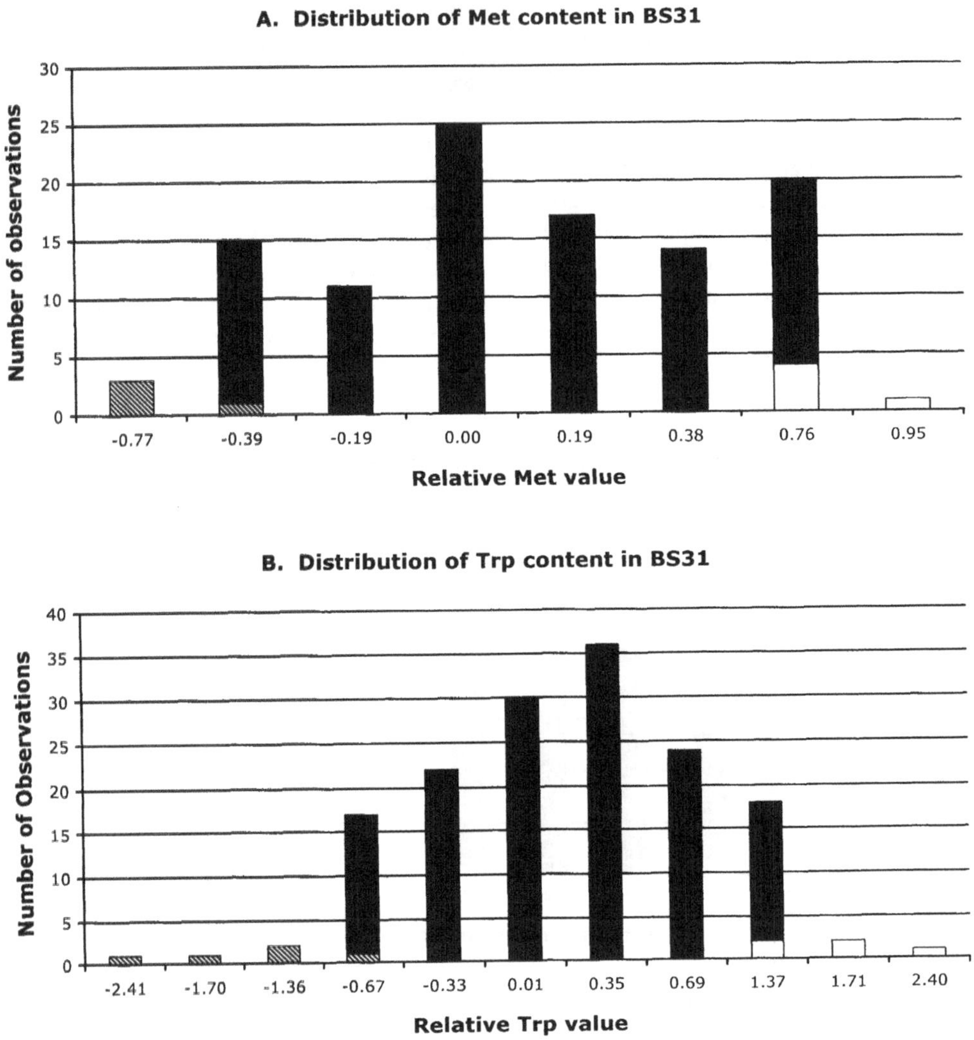

Figure 24.3 Distribution of methionine content (a) and tryptophan content (b) in the initial BS31 population. The methionine and tryptophan values are relative to the overall mean value for the population. The selections made for cycle 1 of the recurrent selection program for the low categories (BS31LM and BS31LT) are crosshatched. The selections for the high categories (BS31HM and BS31HT) are in white.

and 24.3 show the distributions of tryptophan and methionine content in these individuals. Selections were made within these two populations to generate high and low subpopulations from each starting population. In the BS11 population, the mean tryptophan content of the selections in the high category was 28% higher than the mean of the selections in the low category. Similarly for the methionine content, the mean of the high category was 29% higher than the mean of the low selections. The distribution of the starting BS31 population is given relative to the mean of the population for each trait. There was more variation for tryptophan (relative values of ±2.40 from the mean) than for methionine (relative values of −0.77 for the low tail and +0.95 for the high tail). These selections formed eight new populations, four from BS11 and four from BS31, that were selected either for high or low tryptophan or methionine levels.

Effect of selection on random-mated populations

We completed one full cycle of selection by intermating among the selections within each population in the summer of 2002. This allowed us to evaluate the potential of recurrent selection for

Table 24.1 Mean Met and Trp content for each category of the cycle 1 populations derived from BS11 and BS31 and their statistical analysis for mean comparisons of the categories for each trait in each experiment

A. Mean Met and Trp values

		Population[a]	
Trait	Category	BS11	BS31
Met	HM	0.1643	0.1757
	LM	0.1533	0.1736
Trp	HT	0.3030	0.2985
	LT	0.2849	0.2916

B. Results of F tests for mean comparisons

	Population	
Contrast	BS11	BS31
HM vs. LM	b	n.s.
HT vs. LT	b	n.s.

Note: All statistically significant differences show that the mean value of the high category is higher than the mean value of the low category.

[a]Relative values for the trait represented by the optical density measurement corrected for the mass of the sample.

[b]significant at alpha = 0.05

n.s., not significant.

changing amino acid levels. Approximately 40 ears resulting from intermating among the selections within each population were analyzed for their tryptophan and methionine content. The mean tryptophan and methionine values for each category are reported in Table 24.1. Statistically significant differences in methionine and tryptophan levels were observed between the high and low populations derived from BS11. For both tryptophan and methionine levels, the high category was found significantly higher than the low category. The mean methionine content of BS11HM was 7.15% higher than the mean of BS11LM. Similarly, the mean of BS11HT was 6.37% higher than the mean of BS11LT. The tryptophan and methionine populations derived from BS31 did not have statistically significant differences in methionine and tryptophan levels. However, the mean tryptophan content of BS31HT was 2.39% higher than the mean of BS31LT, and the mean methionine content of BS31HM was 1.23% higher than the mean of BS31LM.

Discussion

Effect of selection

To investigate the effect of selection for tryptophan and methionine content in maize populations, we completed one cycle of divergent selection for tryptophan and methionine concentrations in two populations. The data suggest that a divergence in tryptophan and methionine levels is possible.

The response to selection for methionine and tryptophan was greater in the BS11-derived populations than in those derived from the BS31 populations. Several possibilities may explain the different responses to selection in these two populations. There may be greater genetic variability in BS11 than BS31. A second possible explanation could be that we are observing genetic drift because of our small population sizes (Keeratinijakal and Lamkey, 1993). These issues will be clarified upon completion of more cycles of selection.

These data illustrate the feasibility of direct selection for tryptophan and methionine in two maize populations. The differences in gain from selection in the two populations underscore the importance of identifying an appropriate population for this type of experiment. Upon completion of more cycles of selection, these populations will be valuable tools for the development of inbred lines with altered amino acid balance. They will also serve as a tool for studies of genetic factors controlling tryptophan and methionine levels in grain.

References

Altenbach, S.B., K.W. Pearson, G. Meeker, L.C. Staraci, and S.S.M. Sun. 1989. Enhancement of the methionine content of seed proteins by the expression of a chimeric gene encoding a methionine-rich protein in transgenic plants. Plant Mol Biol Int J Mol Biol Biochem Genet Eng 13:513–522.

Benner, M.S., R.L. Phillips, J.A. Kirihara, and J.W. Messing. 1989. Genetic analysis of methionine-rich storage protein accumulation in maize. Theor Appl Genet 78:761–767.

Berg, J.M., J.L. Tymoczko, and L. Stryer. 2002. Biochemistry. Fifth Edition ed. W.H. Freeman & Co.

Bicar, E.H., M. Lee, and M.P. Scott. Unpublished. Molecular and genetic characterization of a synthetic porcine a-La lactalbumin transgene in maize. PhD, Iowa State University, Ames, IA.

Cavallini, A., L. Natali, C. Balconi, E. Rizzi, M. Motto, G. Cionini, and F. D'Amato. 1995. Chromosome endoreplication in endosperm cells of two maize genotypes and their progenies. Protoplasma 189:156–162.

Chaudhuri, S., and J. Messing. 1994. Allele-specific parental imprinting of dzr1, a posttranscriptional regulator of zein accumulation. Proc Natl Acad Sci USA 91:4867–4871.

Cheeke, P.R. 1999. Applied Animal Nutrition: Feeds and Feeding. Second Edition ed. Prentice Hall.

Coleman, C.E., and B.A. Larkins. 1995. MNL 69:124–125

Cruz Alvarez, M., J.A. Kirihara, and J. Messing. 1991. Post-transcriptional regulation of methionine content in maize kernels. Mol Gen Genet 225:331–339.

Dudley, JW. 1974.Crop Sci. Soc. Am., Madison, Wisconsin

Frey, K.J. 1951. The Interrelationships of Proteins and Amino Acids in Corn. Cereal Chem. 28:123–132.

Frey, K.J., B. Brimhall, and G.F. Sprague. 1949. The Effects of Selection Upon Protein Quality in the Corn Kernel. Agronomy Journal 41:399–403.

Hallauer, A.R. 1967. Development of Single-Cross Hybrids from Two-Eared Maize Populations. Crop Science 7:192–195.

Hallauer, A.R., and A.D. Wright. 1995. Registration of B101 maize germplasm. Crop Science 35:1238–1239.

Hartings, H., M. Maddaloni, N. Lazzaroni, N. Di Fonzo, M. Motto, F. Salamini, and R. Thompson. 1989. The O2 gene which regulates zein deposition in maize endosperm encodes a protein with structural homologies to transcriptional activators. EMBO J. 8:2795–2801.

Hernandez, H., and L.S. Bates. 1969. A Modified Method for Rapid Tryptophan Analysis of Maize, p. 1–7 International Maize And Wheat Improvement Center Research Bulletin.

Hopkins, C.G. 1899. Improvement in the chemical composition of the corn kernel. Illinois Agricultural Experimentation Station Bulletin 55:205–240.

Horner, E.S. 1990. Registration of maize germplasms FS8A(S), FS8A(T), FS8B(S), and FS8B(T). Crop Science 30:964.

Keeratinijakal, V., and K.R. Lamkey. 1993. Responses to reciprocal recurrent selection in BSSS and BSCB1 maize populations. Crop Science 33:73–77.

Lai, J., and J. Messing. 2002. Increasing Maize Seed Methionine by mRNA stability. The Plant Journal 30:395–402.

Lamkey, K.R. 2002. Minutes of NCR-167 Meeting [Online] http://www.agron.iastate.edu/corn/ncr167/Minutes/2002_NCR167_Official_Minutes.pdf.

Larkins, B.A., C.R. Lending, and J.C. Wallace. 1993. Modification of maize-seed-protein quality. Am. J. Clin. Nutr. 58:264S–269S.

Lohaus, G., M. Buker, M. Hubmann, C. Soave, and H.W. Heldt. 1998. Transport of amino acids with special emphasis on the synthesis and transport of asparagine in the Illinois Low Protein and Illinois High Protein strains of maize. Planta 205:181–188.

Lorenzoni, C., P. Alberti, A. Viotti, C. Soave, N. Di Fonzo, E. Gentinetta, T. Maggiore, and F. Salamini. 1978. Biochemical characterisation of high and low protein strains of maize, p. 173–197, *In* B. J. Miflin and M. Zoschke, eds. Carbohydrate and protein synthesis. Commission of the Eur. Communities, Brussels.

Maniatis, T., E.F. Fritsch, and J. Sambrook. 1982. Molecular cloning: A laboratory manual Cold Spring Harbor Laboratory, New York, NY.

Mertz, E., L. Bates, and O.E. Nelson, Jr. 1964. Mutant Gene that Changes Protein Composition and Increases Lysine Content of Maize Endosperm. Science 16:279–280.

Mertz, E.T., (ed.) 1992. Quality Protein Maize, pp. 1–294. The American Association of Cereal Chemists.

Molvig, L., L.M. Tabe, B.O. Eggum, A.E. Moore, S. Craig, D. Spencer, and T.J.V. Higgins. 1997. Enhanced methionine levels and increased nutritive value of seeds of transgenic lupins (Lupinus angustifolius L.) expressing a sunflower seed albumin gene. Proc. Natl. Acad. Sci. USA 94:8393–8398.

Mumm. 1935. p. 1–83, *In* R. Emerson, et al., eds. Cornell University Agricultural Experiment Station Memoir, Vol. 180.

Nelson, O.E., E.T. Mertz, and L.S. Bates. 1965. Second Mutant Gene Affecting the Amino Acid Pattern of Maize Endosperm Proteins. Science 150:1469–1470.

Nordlee, J.A., S.L. Taylor, J.A. Townsend, L.A. Thomas, and R.K. Bush. 1996. Identification of a Brazil-Nut Allergen in Transgenic Soybeans. The New England Journal of Medicine 334:688–692.

Osborne, T.B. 1924. The Vegetable Proteins Longmans, Green, London.

Osborne, T.B., and S.H. Clapp. 1908. Hydrolysis of the Proteins of Maize, *Zea mays*. Am. J. Physiol. 20:477–493.

Phillips, R.L., and B.A. McClure. 1985. Elevated protein-bound methionine in seeds of a maize line resistant to lysine plus threonine. Cereal Chem. 62:213–218.

Pond, W.G., D.C. Church, and K.R. Pond. 1995. Basic Animal Nutrition and Feeding. Fourth Edition ed. John Wiley & Sons.

Raboy, V., F.E. Below, and D.B. Dickenson. 1989. Alteration of maize kernel phytic acid levels by recurrent selection for protein and oil. Journal of Heredity 80:311–315.

Reggiani, R., C. Soave, N. Di Fonzo, E. Gentinetta, and F. Salamini. 1985. Factors affecting starch and protein content in developing endosperms of high and low protein strains of maize. Genet. Agrar. 39:221–232.

Sastry, C.S.P., and M.K. Tummuru. 1985. Spectrophotometric determination of tryptophan in proteins. J. Food Sci. and Tech 22:146–147.

Schmidt, R.J., F.A. Burr, M.J. Aukerman, and B. Burr. 1990. Maize regulatory gene opaque-2 encodes a protein with a "leucine-zipper" motif that binds to zein DNA. Proc. Natl. Acad. Sci. 87:46–50.

Shankman, S., M.S. Dunn, and L.B. Rubin. 1943. The Analysis of Eight Amino Acids by a Microbiological Method 477–478.

Smith, O.S. 1983. Evaluation of recurrent selection in BSSS, BSCB1, and BS13 maize populations. Crop Science 23:35–40.

Vasal, S.K. 2001. High Quality Protein Corn, p. 85–129, *In* A.R. Hallauer, ed. Specialty Corns, Second Edition ed. CRC Press LLC.

Wiseman, J., (ed.) 1987. Feeding of Non-Ruminant Livestock, p. 1–214. Butterworth & Co.

Wright, A., and B. Orman. 1995. Rapid screening procedure for methionine levels in maize and soybean. Crop Science 35:584–586.

Wyss, C.S., J.R. Czyzewicz, and F.E. Below. 1991. Source-sink control of grain composition in maize strains divergently selected for protein concentration. Crop Science 31:761–766.

Yang, S.H., D.L. Moran, H.W. Jia, E.H. Bicar, M. Lee, and M.P. Scott. 2002. Expression of a synthetic porcine alpha-lactalbumin gene in the kernels of transgenic maize. Transgenic Res. 11:11–20.

Zuber, M.S., and J.L. Helm. 1972. Approaches to Improving Protein Quality in Maize Without the Use of Specific Mutants, p. 241–252. High-quality protein maize. Halsted Press.

25

Derivation of Open-Pollinated Inbred Lines and Their Relation to Z-Lines for Cyclic Hybridization

Fidel Márquez-Sánchez, professor/researcher, Centro Regional Universitario de Occidente, Universidad Autónoma Chapingo at Guadalajara

Abstract

A new kind of inbred line can be obtained from several plants from one family under open pollination, including selfing. The pollen of the plants is collected, mixed, and applied to their shoots; this results in ears with seed from selfing and crossing. The inbreeding value depends on the number of plants involved; in this study two, three, and four plants are included. Cyclic hybridization, where hybrids and their lines are obtained simultaneously, requires a pair of two-eared plants in each of two heterotic populations. In each plant one shoot is used for selfing and the other shoot is used for crossing to make the hybrid. When there are only one-eared plants, two pairs of sister plants of an initial family of two heterotic populations are needed. For cyclic hybridization, Z-lines are proposed. The pollen of the two plants of the family is mixed and applied to one plant, while the shoot of the other plant is used for crossing. Z-lines can also be obtained with three and four plants using a similar methodology: pollen of the plants is collected, mixed and applied to one plant while the rest of the plants are used for hybridization and vice versa. With any number of plants it is demonstrated that OP lines and Z-lines have the same inbreeding, higher than for conventional full-sib and half-sib lines, but lower than for selfed lines.

Introduction

Selfing is the most common type of pollination used to obtain inbred lines. Lines are early tested and used to make single-cross hybrids. When double-cross hybrids are required, prediction of their yields is made by means of single-cross hybrids. Other methods of pollination, such as full-sibing and half-sibing, may be used to develop inbred lines. Hallauer and Miranda Fo. (1981) found that there was little indication that full-sibbing would give more vigorous lines than selfing.

In this chapter a new method of obtaining inbred lines with two or more plants from a family under open pollination (OP), including selfing, is presented. The inbreeding coefficient of the OP lines was compared with the inbreeding coefficient of Z-lines proposed for cyclic hybridization. Cyclic hybridization was proposed and worked out by Hallauer (1967) to obtain inbred lines and their hybrids simultaneously; however, it requires pairs of two-eared plants. With one-eared plants, Márquez-Sánchez (1982) defined a Z-line as one in which pollen of a pair of plants of an initial family is mixed and applied to one of the plants (the shoot of the other plant being used for crossing to make the hybrid). It is the purpose of this chapter to compare the inbreeding values of OP lines with the inbreeding values of Z-lines, using two, three, and four plants per line.

Materials and methods

OP lines

OP lines are obtained by hand pollination as follows. The pollen of a small number of plants of a family is collected and mixed. The mixture is then applied to the shoots of the same plants. If there are m plants, m selfings (s) take place, and P_2^m (P being the number of permutations in the combinatorial analysis) plant-to-plant crosses (b), the total number of pollinations being m^2.

Inbreeding of an OP line is obtained from the inbreeding formula for mass selection of an OP variety (including selfing), made with n families and m plants per family (Márquez-Sánchez, 1998).

$$F(MS)_t = \left(\frac{1}{2nm}\right)\left(1+2m(n-1)F_{t-1}+2(m-1)F_{t-2}+F_{t-3}\right)$$

To obtain OP lines, the second term within the brackets of this equation refers to plant-to-plant crosses (b), and the fourth term to self-fertilization (s). With $n = 1$, since we are trying to obtain inbreeding from a family, we have:

$$F(MS)_t = \left(\frac{1}{2m}\right)\left(1+2(m-1)F_{t-1,b}+F_{t-2,,s}\right) \quad (1)$$

and with the values of $m = 2, 3$, and 4,

$$F(OP-2)_t = \frac{1}{4}(1+2F_{t-1,b}+F_{t-2,s}) \quad (2)$$

$$F(OP-3)_t = \frac{1}{6}(1+4F_{t-1,b}+F_{t-2,s}) \quad (3)$$

$$F(OP-4)_t = \frac{1}{8}(1+6F_{t-1,b}+F_{t-2,s}) \quad (4)$$

If in Eq. (1) we examine inbreeding in generation 1, we have

$$F(MS)_1 = \frac{1}{2m}(1+2(m-1r_{OW}+F_0)$$

in which r_{OW} is the coancestry within b families. But the inbreeding coming from r_{OW} is present until generation 2; thus Eqs. (2), (3), and (4) become

$$F(OP-2)_t = \frac{1}{4}(1+2F_{t,b}+F_{t-1,s}) \quad (5)$$

$$F(OP-3)_t = \frac{1}{6}(1+4F_{t,b}+F_{t-1,s}) \quad (6)$$

$$F(OP-4)_t = \frac{1}{8}(1+6F_{t,b}+F_{t-1,s}) \quad (7)$$

As can be appreciated, Eqs. (5) and (7) are not the formulae for regular systems of inbreeding of conventional full-sib and half-sib families.

Z-lines

Cyclic hybridization was invented by Hallauer (1967) using two-eared plants of two populations, A (plant 1) and B (plant 2). In population A, the first ear of plant 1 is selfed, while its second ear is crossed by plant 2. The same procedure is followed in population B (Figure 25.1). Therefore, selfed lines and hybrids (A × B) are obtained simultaneously. Only in the best hybrids are their lines preserved for the continuation of selfing and hybridization.

Márquez-Sánchez (1982) proposed a modification of cyclic hybridization using one-eared plants. He used two related lines in each of the populations A (plants 1 and 2), and B (plants 1 and 2). Pollen of plant 1 and plant 2 of population A was collected, mixed, and applied to plant 1; the same was made with plant 1 and plant 2 of population B (Figure 25.2). For population A (and so for population B), for any advanced generation (t) its inbreeding is

$$F(Z-2)_t = \frac{1}{2}\left(\frac{1}{2}(1+F_{t-1,s})+\frac{1}{4}(1+2F_{t-1,b}+F_{t-2,b})\right)$$
$$= \frac{1}{4}\left(\frac{3}{2}+F_{t-1,b}+\frac{1}{2}F_{t-2,b}+F_{t-1,s}\right) \quad (8)$$

With three and four plants per family, the field procedure is similar.

With three plants per family (plants 1, 2, and 3) of family A, pollen of the three plants is collected, mixed, and applied to plant 1 of family A; plants 2 and 3 are used as females crossed with a mixture of pollen of plants 1, 2, and 3 of family B; the same procedure is used in family B (Figure 25.2); therefore, the calculation of inbreeding of a Z-line in each of the two populations is the average of three events: one selfing and two crosses.

For $m = 3$ in each of families A and B

$$F(Z-3)_1 = \frac{1}{3}\left(\frac{1}{2}(1+F_{0.s})+2\frac{1}{4}(1+2F_{0,b}+F_{-1,b})\right)$$
$$= \frac{1}{3}\left(1+F_{0.b}+\frac{1}{2}F_{-1,b}+\frac{1}{2}F_{0,s}\right)$$

and for advanced generations

$$F(Z-3)_t = \frac{1}{3}\left(1+F_{t-1,b}+\frac{1}{2}F_{t-2,b}+\frac{1}{2}F_{t-1,s}\right) \quad (9)$$

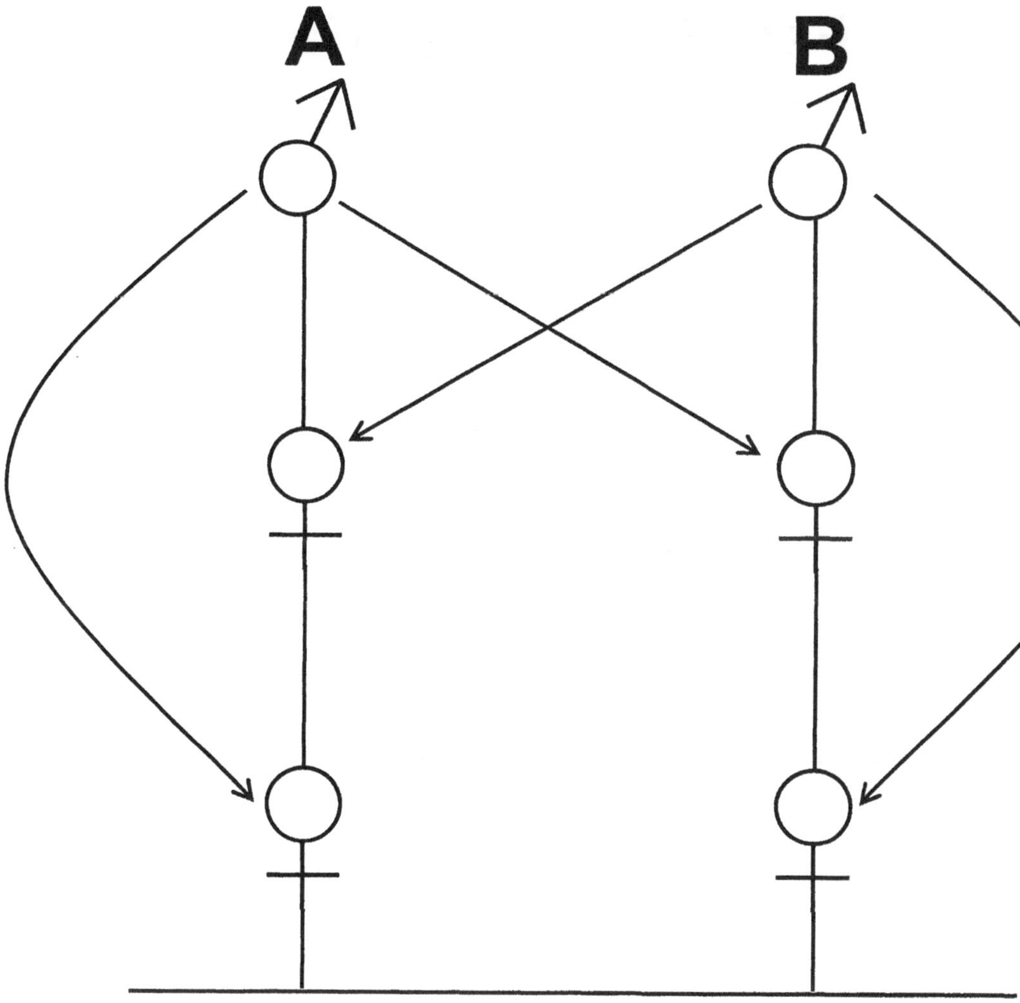

Figure 25.1 A tassel pollinates A ear and crosses to B ear, producing hybrid A × B. B tassel pollinates B ear and crosses to A ear, producing hybrid B × A.

With four plants (plants 1, 2, 3, and 4) of family A, pollen is collected, mixed, and applied to plant 1 of family A; plants 2, 3, and 4 are used as females crossed with a mixture of pollen of plants 1, 2, 3, and 4, of family B. The same procedure is used in family B (Figure 25.3); therefore, the calculation of inbreeding of a Z-line in each of the two populations is the average of four events: one selfing and three crosses. Because inbreeding of these Z-lines has been not been previously calculated, we will do so as follows.

$$F(Z-4)_1 = \frac{1}{4}\left(\frac{1}{2}(1+F_{0,s}) + 3\frac{1}{4}(1+2F_{0,b}+F_{-1,b})\right)$$
$$= \frac{1}{4}\left(\frac{5}{4}+\frac{3}{2}F_{0,b}+\frac{3}{4}F_{-1,b}+\frac{1}{2}F_{0,s}\right)$$

and for advanced generations

$$F(Z-4)_1 = \frac{1}{4}\left(\frac{5}{4}+\frac{3}{2}F_{t-1,b}+\frac{3}{4}F_{t-2,b}+\frac{1}{2}F_{t-1,s}\right) \qquad (10)$$

Comparison of OP lines with Z-lines

For $n = 2$, Eq.[8] must be equal to Eq. [5].

Eq. [8]:

$$F(Z-2)_t = \frac{1}{4}\left(\frac{3}{2}+F_{t-1,b}+\frac{1}{2}F_{t-2,b}+F_{t-1,s}\right)$$

but from $F_{t,b} = \frac{1}{4}\left(1+2F_{t-1,b}+F_{t-2,b}\right)$, we have

$$F_{t-1,b} = \frac{1}{2}\left(4F_{t,b}-F_{t-2,b}-1\right)$$, then

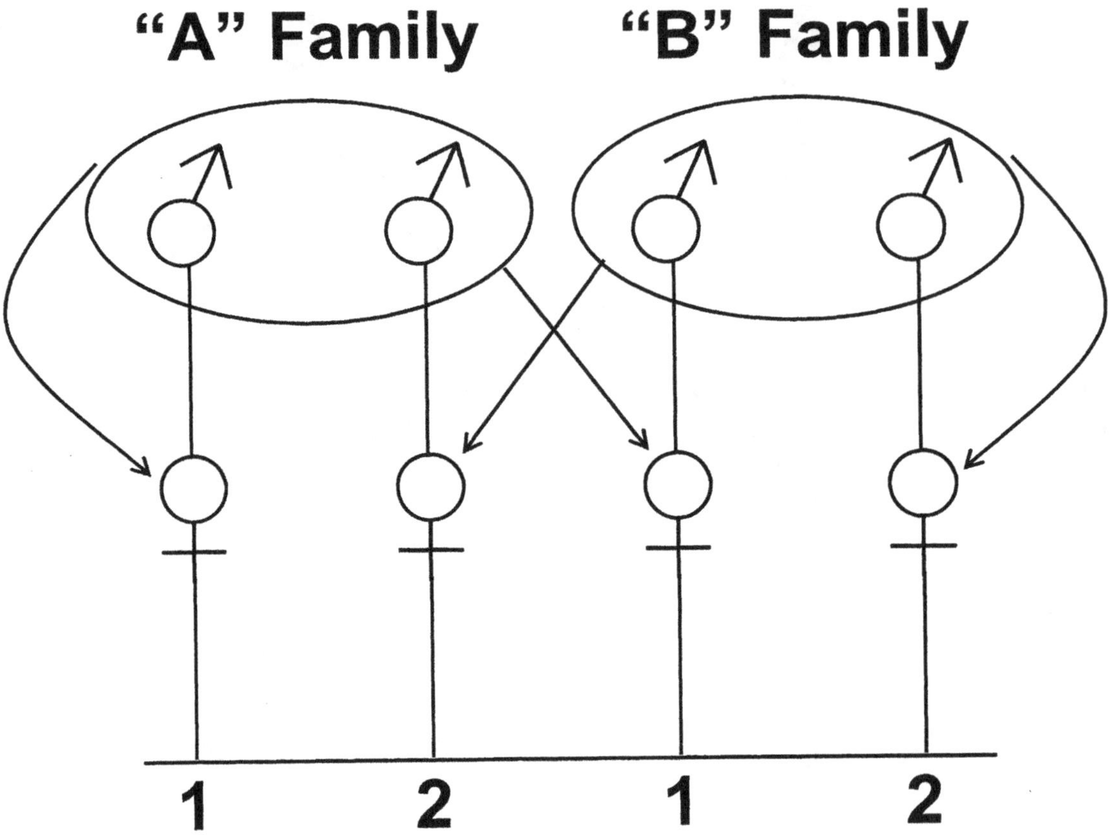

Figure 25.2 Plants (A1 + A2) pollinate plant A1 and cross to plant B1, producing hybrid B × A. Plants (B1 + B2) pollinate plant B2 and cross to plant A2, producing hybrid A × B.

$$F(Z-2)_t = \frac{1}{4}\left(\frac{3}{2}+\frac{1}{2}(4F_{t,b}-F_{t-2,b}-1)+\frac{1}{2}F_{t-2,b}+F_{t-1,s}\right)$$

$$=\frac{1}{4}\left(1+2F_{t,b}+F_{t-1,s}\right)$$

Eq. (5) = $F(\text{OP-2})_t$].
For $n = 3$, Eq. [9] must be equal to Eq. [6].
Eq. [9]:

$$F(Z-3)_t = \frac{1}{3}\left(1+F_{t-1,b}+\frac{1}{2}F_{t-2,b}+\frac{1}{2}F_{t-1,s}\right)$$

but from $F_{t,b} = \frac{1}{4}\left(1+2F_{t-1,b}+F_{t-2,b}\right)$, we have

$$F_{t-1,b} = \frac{1}{2}\left(4F_{t,b}-F_{t-2,b}-1\right),$$

$$F(Z-3)_t = \frac{1}{3}\left(1+\frac{1}{2}(4F_{t,b}-F_{t-2,b}-1)+\frac{1}{2}F_{t-2,b}+\frac{1}{2}F_{t-1,s}\right)$$

$$=\frac{1}{6}\left(1+4F_{t,b}+F_{t-1,s}\right)$$

and then
and Eq. [6] = $F(\text{OP-3})_t$.
For $n = 4$, Eq.[10] must be equal to Eq. [7].
Eq. [10]:

$$F(Z-4)_1 = \frac{1}{4}\left(\frac{5}{4}+\frac{3}{2}F_{t-1,b}+\frac{3}{4}F_{t-2,b}+\frac{1}{2}F_{t-1,s}\right)$$

but from $F_{t,b} = \frac{1}{4}\left(1+2F_{t-1,b}+F_{t-2,b}\right)$,

$$F(Z-4)_t = \frac{1}{4}\left(\frac{5}{4}+\frac{3}{2}\frac{1}{2}(4F_{t,b}-F_{t-2,b}-1)+\frac{3}{4}F_{t-1,b}+\frac{1}{2}F_{t-1,s}\right)$$

$$=\frac{1}{8}\left(1+6F_{t.b}+F_{t-1,s}\right)$$

$F_{t-1,b} = \frac{1}{2}\left(4F_{t,b}-F_{t-2,b}-1\right)$, then
and Eq. [7] = $F(\text{OP-4})t$.

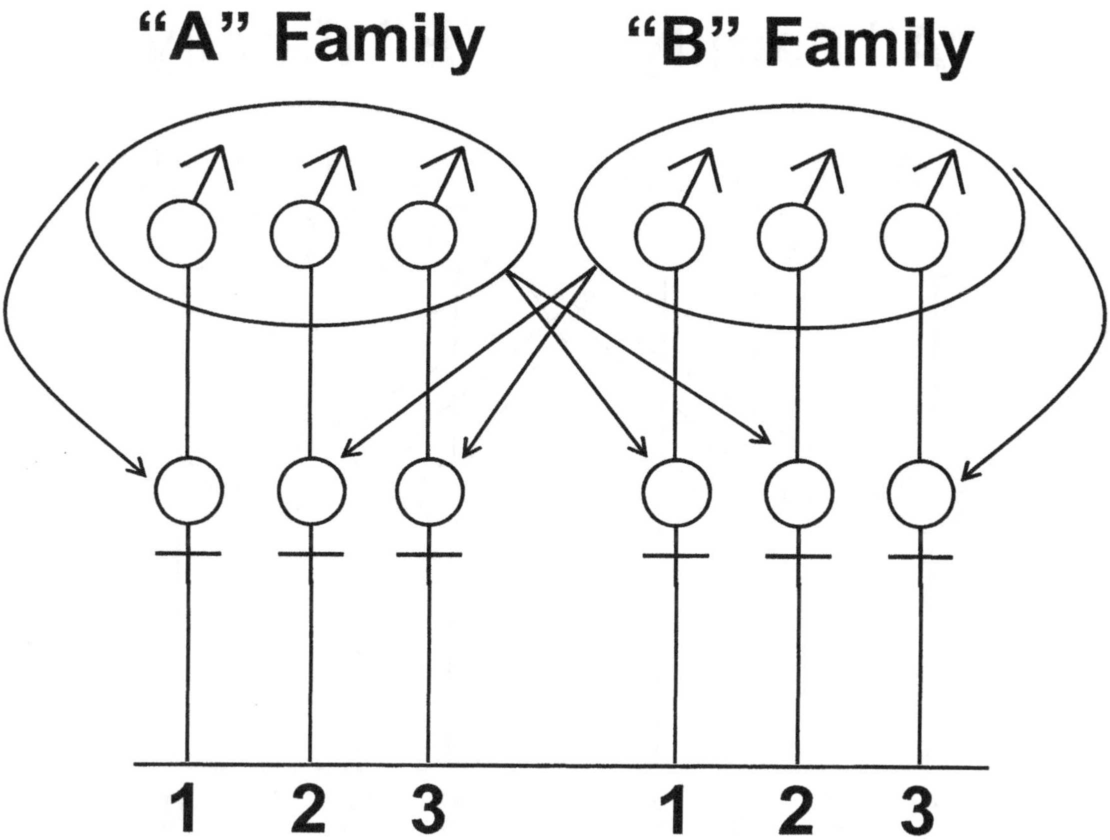

Figure 25.3 Pollen of plants 1, 2, and 3 of family A is collected, mixed, and applied to silks of plant 1 of family A and to silks of plants 1 and 2 of family B. The same procedure is used for family B.

Discussion

In each of the three cases studied, with two different equations we arrive at the same result. All this can be explained easier if we have in mind how random mating by hand-pollination is made. With OP-2 lines, we have the average of four events: two selfings and two crosses that is, $\frac{1}{4}(2S+2C)=\frac{1}{2}(S+C)$, which is the same for Z-2 lines. With OP-3 lines, we have the average of nine events: three selfings and six crosses, that is, $\frac{1}{9}(3S+6C)=\frac{1}{3}(S+2C)$, which is the same for Z-3 lines. Finally, with OP-4 lines, we have the average of 16 events: 4 selfings and 12 crosses, that is, $\frac{1}{16}(4S+12C)=\frac{1}{4}(S+3C)$, which is the same for Z-4 lines. Equations [5], [6], and [7] look somewhat similar to those for inbreeding of full-sibs, of a line with no name (NN; that is, an imaginary line since there is no way of creating it in the field), whose inbreeding formula is $F(NN)_t=\frac{1}{6}\left(1+4F_{t-1}+F_{t-2}\right)$, and of half-sibs. In Figure 25.4, inbreeding of OP-lines is compared with inbreeding of the regular lines (FS, NN, and HS'); it can be appreciated how the former have higher inbreeding than the latter due to self-fertilization.

Finally, the derivation of both OP lines and Z-lines is correct according to the algebraic equality and pollination similarity. However, in practice it would be necessary that full-sib plants and selfed plants be sown separately. In practice this can not be done since these two lines are planted as a mixture. Thus, for the equations already obtained for $n = 2$, 3, and 4 initial plants, to become real recurrence equations, plants used for pollination must be randomly chosen.

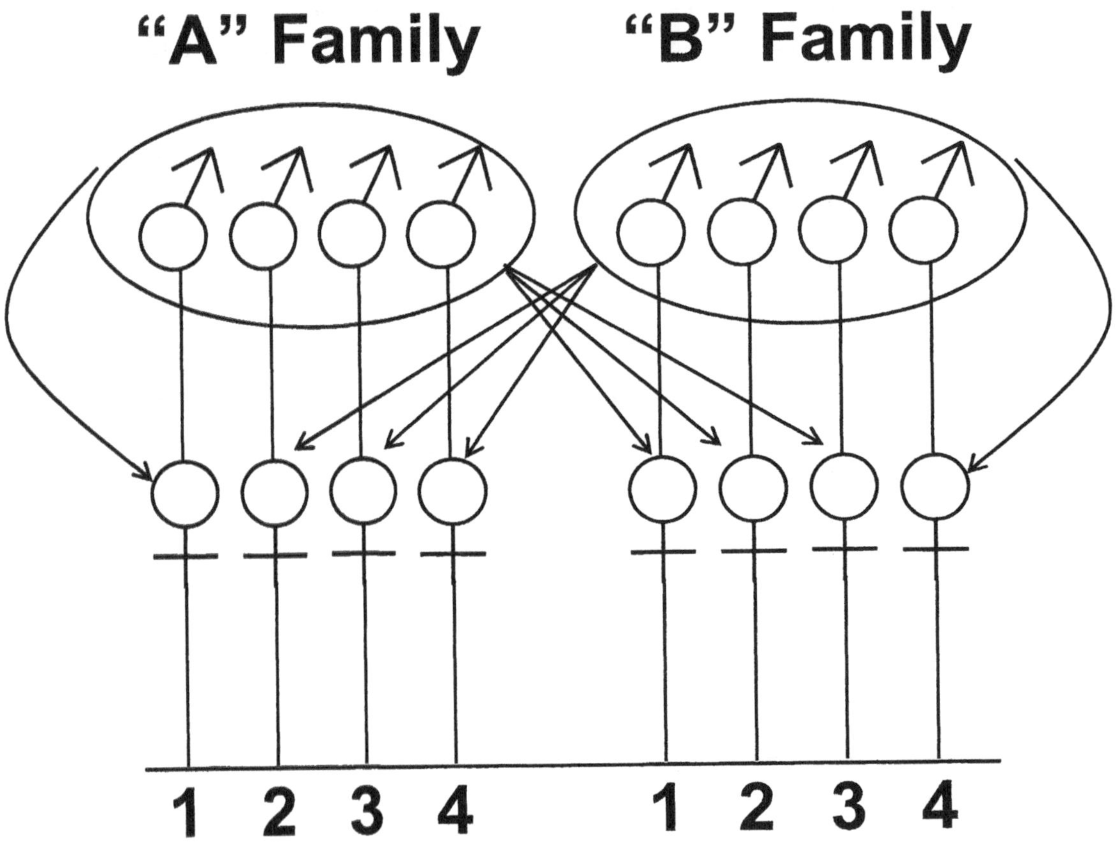

Figure 25.4 Pollen from the four plants of family A is collected, mixed, and applied to silks of plant 1 of family A and to silks of plants 1, 2, and 3 of family B. Pollen from the four plants of family B is then collected, mixed, and applied to plant 4 of family B and to plants 2, 3, and 4 of family A.

We must follow Falconer's (1961) equation under random mating in one line, whose m individuals undergo selfing and crossing, to calculate the inbreeding in each generation. The Falconer (1961) equation is

$$F_t = \frac{1}{2m} + \left(1 + \frac{1}{2m}\right)F_{t-1} = \frac{1}{2m}\left(1 + (2m-1)F_{t-1}\right) \quad (11)$$

and the respective Eqs. [5], [6], and [7] will be according to Eq. [11].

$$F(OP-2)_t = \frac{1}{4}\left(1 + 3F_{t-1}\right) \quad (12)$$

$$F(OP-3)_t = \frac{1}{6}\left(1 + 5F_{t-1}\right) \quad (13)$$

$$F(OP-4)_t = \frac{1}{8}\left(1 + 7F_{t-1}\right) \quad (14)$$

In Table 25.1, the inbreeding values of the conventional lines and of the OP lines are presented. It can be appreciated that the difference between conventional lines and OP lines is smaller as the number of individuals in a line increases.

Table 25.1 Inbreeding coefficients of conventional lines and OP-lines

FS	OP-2	NN	OP-3	HS′	OP-4
0.0000	0.0000	0.0000	0.0000	0.0000	0.0000
0.2500	0.2500	0.1666	0.1666	0.1250	0.1250
0.3750	0.4375	0.2777	0.3055	0.2187	0.2343
0.5000	0.5781	0.3795	0.4212	0.3046	0.3300
0.5937	0.6835	0.4660	0.5177	0.3808	0.4138

Final note

The calculation of inbreeding of Z-2 lines has a mistake in the original article (Márquez-Sánchez, 1982); here the corrected formula is presented. Graphs were also drawn as if plant-to-plant crosses and selfings were planted separately.

References

Falconer, D.S. 1961. Introduction to Quantitative Genetics. The Ronald Press Company, New York.

Hallauer, A.R. 1967. Development of single-cross hybrids by selfing and full-sib mating. Crop Sci. 7:192–195.

Hallauer, A.R., and J.B. Miranda Fo. 1981. Quantitative Genetics in Maize Breeding. Iowa State University Press. Ames, Iowa.

Márquez-Sánchez, F. 1982. Modifications to cyclic hybridization in maize with single eared plants. Crop Sci. 22:314–319.

Márquez-Sánchez, F. 1998. Expected inbreeding with recurrent selection in maize: I. Mass selection and modified ear-to-row selection. Crop Sci. 38:1432–1436.

26

Breeding Maize Exotic Germplasm

F.J. Betrán, K. Mayfield, T. Isakeit, and M. Menz
Texas A&M University

Introduction

Maize (*Zea mays* L.) exotic germplasm can be defined as genetic material (inbreds, hybrids, populations, bank accessions, etc.) that has not been previously evaluated and/or selected in the target breeding area. What is local in one target area is considered exotic in another. For example, temperate U.S. Corn Belt germplasm is exotic for a breeding program in Zimbabwe, and tropical germplasm is exotic for a breeding program in Iowa. Here, we will focus on U.S. maize-growing areas and consider exotic maize germplasm material with tropical and subtropical origin.

Breeding programs in the United States exploit a small fraction of the genetic diversity available worldwide in maize. Maize breeders, predominantly in the private sector, have focused increasingly on short-term breeding objectives, creating a relative narrow genetic base for commercial maize hybrids. The genetic diversity of maize in the United States is limited to mainly one race (i.e., Corn Belt Dent) of the approximately 300 races that exists worldwide. In other words, U.S. maize breeders use less than 1% of the total germplasm available. Only a small number of popular open-pollinated varieties (OPVs) constitute the genetic background and foundation for most of parental inbred lines used in the United States (Troyer, 1999). These OPVs were used to develop the first generation of inbreds. After this initial sampling of OPVs, the emphasis of inbred development shifted to the use of breeding populations created by crossing complementary inbreds and the selection of progenies possessing desirable traits from both parents (Hallauer, 1990). In this line recycling crosses "good × good" among elite lines from the same heterotic group are emphasized. Selection within F_2 and backcross populations using pedigree breeding became and remains the most important breeding method to develop maize inbreds. Breeding in these populations became cyclical and second, third, fourth, etc. generation recycled inbreds were developed. A backcross or multiple backcrosses to the best parent are used commonly. As a consequence of this recycling of elite lines, the pedigrees of most hybrids grown in the United States trace back to derivatives of just six to eight inbred lines (e.g., B14, B37, B73, Oh43, Mo17, and C103)(Goodman, 1992). Genetic diversity studies with DNA molecular markers support this observation (Smith et al., 1999). With such a restricted base, maize in the United States may be vulnerable to new diseases and pests and may not contain all the desirable and favorable alleles to maintain selection progress. Lack of alleles for adequate resistance could be devastating, as demonstrated by the 1970 epiphytotic of southern maize leaf blight (*Bipolaris maydis* race T).

Although genetic gains for maize apparently have not diminished for several decades, the consequence of continuous, long-term, and intensive selection in the same germplasm base over almost one century may be a narrower genetic variance for grain yield in U.S. hybrid germplasm. Breeding efforts to incorporate and combine exotic sources into the existing germplasm base could reduce the impact of future, unforeseen threats to production, as well as enhance current or alternate uses of maize. Exotic germplasm constitutes a reservoir of alleles and allele combinations that, once identified, can be incorporated and combined with elite local material. Some seed companies have germplasm-enhancement programs that emphasize the use of elite proprietary lines from stations outside the United States. Exotic germplasm is a genetic resource that may contribute alleles for grain yield, disease and pest resistance, tolerance to abiotic stresses, and value-adding traits to temperate maize. The use of exotic germplasm in temperate areas has been suggested for many years (Wellhausen, 1965), but has received limited attention. Exotic germplasm has not been used extensively because of poor agronomic performance, less-intensive breeding history, and lack of adaptation. The amount of exotic germplasm used in commercial breeding programs in the United States remains small, but it is increasing: about 1% in 1984 to almost 3% in 1996 (Goodman, 1985, 1999). Tropical sources have been used mainly as sources for disease and insect resistance.

Sources of exotic germplasm

The three main sources of exotic germplasm for temperate areas to date are from other temperate areas (mainly Argentina, Europe, and South Africa), lowland tropics (landraces such as Cuban Flint, Suwan, and Tuxpeño), and highland tropics (landraces such as Cuzco, Chalqueño) (Goodman, 1999). Temperate germplasm comprises most of the exotic germplasm used (e.g., Maíz Amargo from Argentina, French lines F2 and F7).

There is a wide range of improved exotic material available from international (e.g., International Maize and Wheat Improvement Center [CIMMYT], International Institute of Tropical Agriculture [IITA]), national (e.g., The National Institute for Forestry, Agriculture and Livestock Research [INIFAP] in Mexico, Brazilian Agricultural Research Corporation [EMBRAPA] in Brazil, etc.,) and seed industry (Asgrow, Dekalb, Pioneer, etc.) breeding programs. For example, CIMMYT and other institutions have open-pollinated cultivars, inbred lines, hybrids, and narrow-based synthetics that possess tolerance to abiotic stresses (drought, low nitrogen and phosphorus, and aluminum toxicity), resistance to a wide range of pests (e.g., armyworms, maize borers, storage pests, and parasitic weeds) and diseases (e.g., downy mildews, leaf rusts, leaf blights, Phaeosphaeria and gray leaf spot, stalk rots, ear rots, bacterial and viral diseases). They also have nutritional attributes such as high lysine and tryptophan in QPM (quality protein maize) and elevated iron, zinc, and beta-carotene levels. In addition, there are unimproved germplasm accessions and related species (teosinte and tripsacum) characterized for several plant descriptors and agronomic traits (Taba et al., 2004). Parra and Hallauer (1997) describe sources of tropical and temperate-adapted exotic germplasm.

The type of exotic germplasm available includes open-pollinated varieties, synthetics, hybrids, inbred lines, and genetic stocks (Figure 26.1). There has been an evolution in the type of exotic germplasm used in incorporation or introgression efforts. Exotic broad-base populations have been a common source, but with more exotic inbreds characterized and available the trend is to use more inbred-based breeding populations. Biotechnology can facilitate a more precise introgression of desirable alleles at specific genomic regions or genes. It is likely that the discovery of genes or genomic regions affecting target traits will bring more attention to allele mining. The use of exotic germplasm with a history of inbreeding either as breeding lines or hybrids is preferred, because they have been selected for relevant traits and reduced inbreeding depression during their development (Goodman et al., 2000). Over 400 CIMMYT maize lines (CMLs) are available with information about their performance, attributes, and hybrid performance (CIMMYT, 1999).

Choosing among sources of exotic germplasm

The increasing emphasis of hybrid development in tropical and subtropical areas, the effort in characterizing germplasm accessions, and data management have increased the amount of information available to select exotic material based on different criteria. Information about characteristics, such as broad or general adaptation (temperate, tropical, subtropical, highland), grain characteristics (color, texture), selection history, photoperiod sensitivity, geographic origin, performance per se, combining ability and heterotic response, degree of tolerance to abiotic stresses, and resistance to diseases and pests, is useful for identifying promising materials.

After making the first selection, based on information available, exotic germplasm can be evalu-

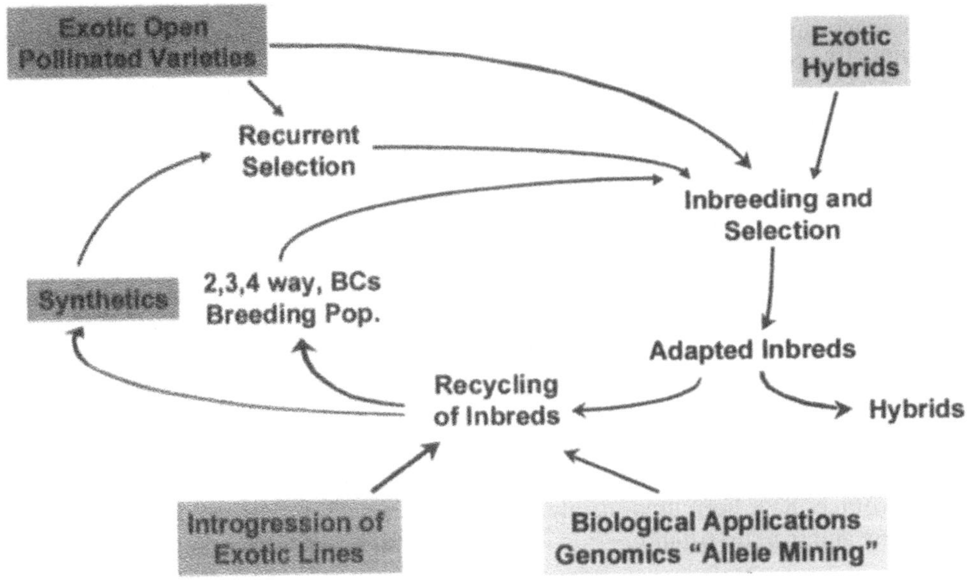

Figure 26.1 Development of first cycles and recycled inbred lines in maize with different types of exotic germplasm.

ated in target environments. Evaluations made initially can include: (1) populations or inbreds, (2) cultivar or line diallel crosses, and (3) population or line × tester testcrosses (Geadelmann, 1984). Evaluations of selfed progenies can identify favorable recessive alleles for some qualitative traits (e.g., resistance to diseases) and eliminate unfavorable alleles otherwise masked by dominant favorable alleles. Inbred line evaluation can serve as preliminary assessment of the degree of adaptation and help in the characterization of highly heritable traits such as maturity. Cultivar or inbred cross diallels have been used to estimate combining ability of populations and lines with different levels of exotic germplasm (Crossa et al., 1987; Michelini and Hallauer, 1993; Goodman et al., 2000). Testcrosses of exotic populations or inbreds with elite testers of known heterotic groups are a way to estimate yield potential and to assign exotic material into established heterotic groups.

Estimates of general (GCA) and specific (SCA) combining ability effects are useful to identify desirable exotic inbreds with good performance in hybrids. The "transfer of alleles" methods for identifying populations and inbred lines to improve parents of elite single crosses have been described by Dudley (1982, 1987a, 1987b), Gerloff and Smith (1988), Bernardo (1990), Metz (1994), and Hohls et al. (1995). The exotic germplasm should have favorable alleles not present in current elite germplasm. These methods describe different statistics to estimate which exotic source has the greatest number and frequency of favorable alleles to improve the hybrid. Several practical studies have been conducted following these theoretical methods (Dudley et al., 1996; Kraja and Dudley, 2000).

Goodman et al. (2000) suggested the evaluation of testcrosses in fall/winter nurseries (short days) as an efficient method to choose which tropical hybrids or accessions are best for temperate breeding. Proper comparisons can be made with the best domestic commercial hybrids and exotics that do not have desirable agronomic performance based on yield, maturity, height, lodging, heterotic response, or other characteristics.

Introgression, incorporation, and combination of exotic germplasm

The amount of exotic germplasm in the development of adapted maize lines can be variable. Selected lines can be developed directly from exotic germplasm without any hybridization with temperate adapted material or from populations with high percentage of exotic germplasm (>75%). This approach is called *incorporation* (Simmonds 1993;

Goodman et al., 2000). Inbreds NC300, NC298, Tx772, B114, and populations BS28 and BS29 are examples of incorporation. Alternatively, exotics can be hybridized with temperate material in different proportions. If one or more backcrosses with temperate material as the recurrent parent are used, the approach is called *introgression* (<25% exotic germplasm) (e.g., NC250, NC272, NC278). If the proportion of exotic germplasm is between 25% and 75%, this approach is called *combination* (e.g., Mo47, Tx770). The productivity and adaptation developed from the cross between northern flints (early maturity, short plant, tillers, and hard-texture kernels) and southern dents (late, tall, deep kernels, and many-kernel rows) to generate the Corn Belt Dent race represent a good successful example of combination of germplasm with different backgrounds.

Theoretical studies examining the introgression of exotic into adapted germplasm indicated that at least one generation of backcrossing to the adapted parent (25% exotic) would be appropriate, with additional backcrossing possibly justified when diversity between the parental populations is large (Geadelmann, 1984). Dudley (1982, 1984) also found that there are advantages of backcrossing if one parent has more loci with favorable alleles than the other, the parents are diverse, or the level of dominance is high. Useful alleles either absent in local germplasm or probably at greater frequencies in the exotic germplasm are more likely to be lost as the percentage of exotic material in the germplasm cross declines. On the other hand, desirable alleles present in local germplasm can be lost when the proportion of exotic germplasm increases. Crossa (1989) calculated the probability of fixation for one favorable gene (one locus) and for two independent genes (two loci) from different base populations (F2, BC1, BC2) at different effective population sizes (*Ne*), gene frequencies in local and exotic germplasm, and intensities of selection (i). When *Ne* is large enough to reduce genetic drift and local and exotic populations have similar performances, the F2 population should be used as a base breeding population. If populations differ strongly in performance, one backcross to the superior germplasm is recommended.

Another relevant factor in the proportion of exotic germplasm to use is the geographic area or target environment. In general, in the southern United States, populations with >50% exotic germplasm can be successfully used in breeding, while areas in the Midwest may require proportions <50%. However, empirical results demonstrate that the use of 100% exotic germplasm is also possible. Several tropical populations have been successfully adapted to temperate areas after several cycles of mass selection for early flowering (Hallauer, 1978 and 1994), and other exotic populations have been used directly as a source of inbred lines (Goodman et al., 2000).

There are beneficial effects of several generations of intermating with mild phenotypic selection following the introgression or combination process before intense inbreeding and selection is initiated (e.g., separation of useful and undesirable linked alleles) (Eagles et al., 1989). High linkage disequilibrium, along with intense selection, could result in the loss of favorable alleles. More recombination has the obvious disadvantage of requiring more time and resources, although this could be partially compensated by initiating mild selection for adaptation. Goodman (1999) raised the question of which generation of selfing from exotic × local breeding cross should be used for topcross testing. F1 and S1 generations have a greater average genetic variation from the accession. However, the genetic variation in subsequent generations of inbreeding (>S2) is dominated by new genetic variance arising from the recombination between the accession and the local inbred. Therefore, Goodman (1999) suggests testing individual F1 plants or S1 families if efficient selection within the exotic accessions is pursued. In general, average grain yield has been lower for inbred progenies (S1, S2) developed from populations with some degree of exotic germplasm than for lines from adapted populations. However, introgression has increased genetic variation resulting in higher predicted gains (Crossa and Gardner, 1987; Eagles and Hardacre, 1990).

Projects using exotic germplasm

The collection and classification of maize landraces from Mexico, the Caribbean, and Central and South America were conducted during the 1950s. These accessions were classified and stored in germplasm banks. The Latin America Maize Regeneration Project (LAMRP) was established to rescue and regenerate maize germplasm accessions held in Colombia, Peru, and Mexico. The project successfully regenerated approximately 40% of a total of 18,298 accessions, which were stored at the North Central Plant Introduction Station, Ames, Iowa (Goodman, 1988). The LAMRP served to emphasize the importance of maintaining viable accessions of maize germplasm.

The Latin American Maize Project (LAMP) was a cooperative project involving 12 countries to evaluate and characterize maize germplasm accessions included in the gene banks of Latin America and the United States for yield and agronomic traits (Salhuana, 1995). Collection and preservation of maize genetic resources is important, but their use in breeding programs depends on the information about the traits they carry, their performance, and agronomic characteristics. LAMP evaluated over 12,000 accessions in locations divided into five homologous areas covering latitudes from 34° S to 41° N and altitudes from 29–3300 meters above sea level (Salhuana and Sevilla, 1995; Salhuana et al., 1997). LAMP was the first international project dealing with the evaluation of genetic resources in maize.

As LAMP was nearing its conclusion in 1992, an organized effort, Germplasm Enhancement of Maize (GEM), was put into place in 1994 to enhance the best LAMP accessions and incorporate them into breeding programs. GEM is a cooperative effort of the USDA-ARS, land-grant universities, and industry. Almost all large U.S. companies and public maize-breeding programs participate in GEM. The objective of this program is to provide the maize industry with enhanced materials developed from useful exotic germplasm. The ultimate goal is to improve and broaden the germplasm foundation of maize hybrids grown by U.S. farmers. The basic breeding protocol of GEM is a modified pedigree breeding system to develop S_3 lines or synthetics of recombined S_2 lines. One private cooperating company crosses an exotic material by a proprietary inbred line to make a 50% exotic breeding cross that is crossed by another company with a different proprietary line of the same heterotic group to make a 25% exotic breeding cross. Breeding crosses are evaluated and selected to develop inbred lines by cooperators. GEM-enhanced lines and synthetics are or will be freely available through National Genetic Resources Program after their release. Private-sector collaborators have contributed in-kind support for the breeding effort (winter and summer nursery rows, yield trial plots, and disease observation rows) and made the crosses of exotic accessions to their proprietary inbred lines. Agronomic productivity, disease and insect resistance, and value-added traits are considered. Diseases and insects with high priority include maize root worm, second generation of European corn borer (ECB), gray leaf spot, anthracnose stalk rot, aflatoxin, and fusarium ear mold. Grain quality traits such as composition (percent oil, protein, and starch), starch quality, oil quality (fatty acid composition), and protein quality (amino acid composition) are also traits under consideration (http://www.public.iastate.edu/~usda-gem/). In 2001, the first GEM-derived release GEMS-0001, germplasm resistant to ECB, was announced. Since then, other GEM lines have been made available for cooperators and have been publicly released. GEM is an example of collaboration among public institutions and private seed companies to effectively utilize diversity.

The North Carolina State University maize program, led by M. Goodman, has developed temperate adapted inbred lines that are of exotic origin. This program demonstrated that derivatives of tropical hybrids and intercrosses among them can compete with domestic lines (Uhr and Goodman, 1995; Holland et al., 1996; Goodman, 1999; Tallury and Goodman, 1999). Visual selection emphasized early flowering, good synchrony between silking and pollen shedding, standability, prolificacy, low ear placement, and ear quality. Inbreds such as NC296, NC296A, NC298, and NC300, all with tropical origin, are today used as source of new alleles and allele combinations for grain yield. Tropical germplasm generally introduced more lodging, taller plants, and higher ear placement, higher moisture at harvest, and susceptibility to smut. On the other hand, better yields and increased gray leaf spot and southern rust resistance were identified in the exotic germplasm (Goodman, 1999). Goodman (1999) identified poor germination and seedling vigor under adverse spring growing conditions and the lack of high-yielding, early maturing tropical inbreds as the main restrictions to make additional progress with tropical derivatives.

The cooperative federal–state maize-breeding program at Iowa State University (ISU) has emphasized introgression of broad-base exotic populations. BS16, BS27, BS28, and BS29 are temperate adapted populations, insensitive to photoperiod, derived from tropical populations ETO, Antigua, Tuxpeño, and Suwan-1, respectively, by mass selection for earlier flowering applied directly in the tropical accessions (Hallauer, 1994). The selection intensity was high: 1.5–2.5% (300–500 plants selected among approximately 20,000 plants evaluated). On average, a response of three days per cycle was observed. Correlated responses were also in the desired direction with reduced plant and ear heights, reduced lodging, and tassel size. Average yields for these adapted populations have been similar to temperate populations and have shown good combining ability (heterosis of 34.4%) with temperate populations (Echandi and Hallauer, 1996). Another example of recent introgression efforts at ISU, in collaboration with CIMMYT, is the Maize Germplasm Conversion Program, a project that has involved crosses and backcrosses between elite tropical inbreds and temperate populations, as well as between elite temperate lines and tropical populations (Hallauer, personal communication). ISU has also developed temperate-adapted inbreds from exotic germplasm such as B103, B107, B108, and B114 derived from CIMMYT Pool 41.

Breeding exotic germplasm at Texas A&M University

Breeding approaches

At Texas A&M University's maize-breeding program at College Station, Texas, we are using tropical and subtropical white and yellow inbred lines. Texas is located between the U.S. Corn Belt and subtropical–tropical maize-growing areas extending from latitudes 26° 30′ to 36° 30′ north. This offers a wide range of different environments, with a transition from subtropical environments in the south (Rio Grande Valley) to temperate environments in the northwest (High Plains), and a unique opportunity to screen exotic germplasm. Our breeding goal is to develop maize inbreds adapted to the southern United States from exotic material. The target environments in the southern United States are noted for long growing seasons, mycotoxin occurrence (particularly aflatoxin), ear-feeding insects, and post-flowering drought and heat stress.

Before initiating breeding activities, exotic germplasm (mainly in the form of inbreds) goes through a preliminary phase of characterization and screening in our nurseries. Crosses among selected exotic inbreds, as well as between exotic and temperate inbreds, are used to develop breeding populations that will serve as sources for new inbreds carrying desirable alleles from exotic germplasm. Once the segregating population is created, we define three stages for inbred development:

1. Selection and advancement of early generation lines
2. Testcross evaluation of selected lines
3. Advanced hybrid evaluation and release

In stage 1, lines are advanced and selected from S_0 to S_4. A combination of pedigree and bulk methods is used (Figure 26.2). Seed from selected plants are harvested and kept separate. In the next generation, families are created by bulking a few (two to three) previous generation plants. Each bulk is planted in one row and identified by a pedigree symbol (B1, B2, B3, etc.). This method maintains the simplicity of the bulk method, requiring less space and time than the pedigree method, while preserving a balanced representation of selected individuals and genetic variability. With the same resources, it is possible to handle more breeding populations than with a pure-pedigree breeding scheme. Breeding methods such as the use of short plots with 20–30% higher densities; pollination of end or desirable plants as suggested by Troyer (2001); the use of breeding populations derived from multiparent crosses, for example, (B104 × CML285) × (Tx601y × NC300); and the utilization of direct pollinations facilitate the incorporation of a high number of exotic lines. Nevertheless, probably the most important factor during stage 1 is the ability to conduct fall-winter nurseries in neutral short day length in South Texas. Our summer nursery at College Station, Texas, is planted early March and harvested sometime in July. The fall–winter nursery at Weslaco, Texas, is planted mid-August and harvested late December. The early planting in our summer nursery and the neutral day-length environment at Weslaco permit an adequate screening of exotic germplasm, which is less affected by photoperiod that commonly masks their real genetic value in

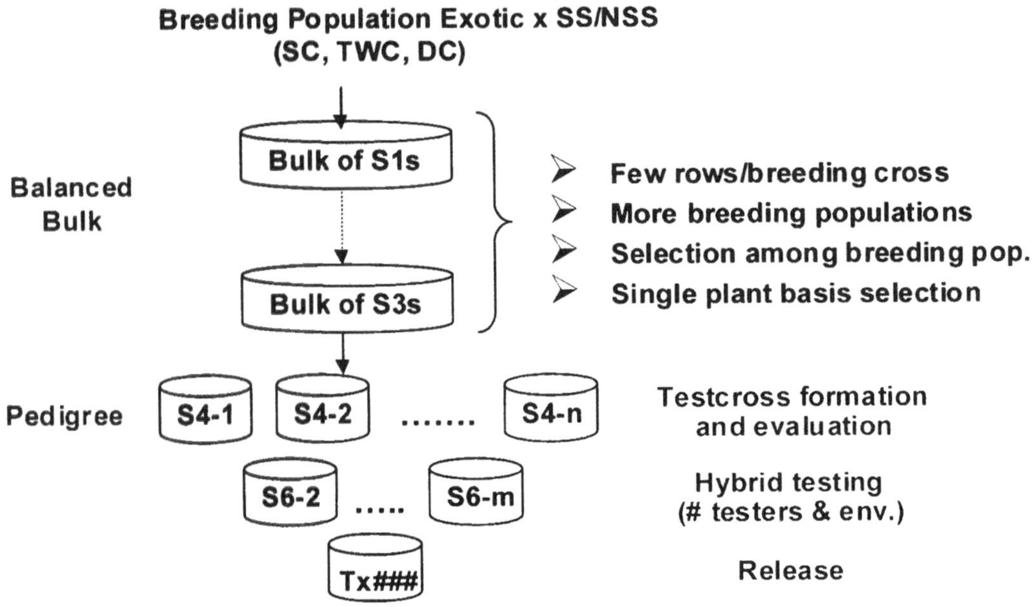

Figure 26.2 Breeding scheme combining bulk and pedigree breeding methodologies used to develop inbred lines from exotic germplasm.

more temperate long-day environments. Selection for maturity, kernel characteristics (hard endosperm), plant morphology (low ear placement), standability, kernel integrity and early vigor is imposed at the family level.

In stage 2, performance of selected exotic lines in hybrids and their adaptation to major growing areas in Texas is estimated in testcross evaluations in a range of three to five subtropical and temperate environments. Advanced S4 lines are used to make testcrosses with representative testers for the heterotic groups commonly used in the United States, Stiff Stalk Synthetic (SSS) and non-Stiff Stalk Synthetic (NSSS). Heterotic response of exotic inbreds is examined by comparing grain yields with the two testers (Figure 26.3). These testers are elite lines commonly used in commercial hybrid production (e.g., currently we are using LH195 and LH210 as the SSS and NSSS testers, respectively). In general, exotic lines have shown better combining ability with SSS testers than with NSSS testers. Different overall performance of the two testers in the tested environments can affect the initial heterotic classification, and further characterization with additional crosses (stage 3) is required (Figure 26.3). The testcross data, taken together with the inbred per se performance, are considered to select the most promising inbreds (S6 now) that are advanced to stage 3.

In stage 3, more testers and environments are used to evaluate the lines in hybrid combinations. Representative locations including farmers' fields, involvement of the seed companies in the evaluation, and proper environmental characterization are important factors in this stage. In addition to grain yield, the hybrids are evaluated for aflatoxin accumulation, response to drought stress, grain and processing characteristics, and regional adaptation. Factorial designs exotic × temperate and diallel analysis are commonly used. Biplot representation with testers represented by arrows and testing lines as dots are used to determine the relationships between lines and classify them into potential heterotic groups (Figure 26.4). This classification is relevant to identify potential superior hybrids and to decide best breeding crosses for line recycling. In southern U.S. environments, we have observed that hybrids between temperate by exotic inbreds, both in yellow and white-grained maize, are competitive with commercial hybrids (Table 26.1 and Figure 26.5). These positive heterotic responses between exotic and temperate germplasm have been reported in populations (Michelini and Hallauer, 1993) and hybrids (Goodman, 1999).

Inbreds surviving after stage 3 are proposed for release and made available to private and public breeding programs. For example, Tx772 is an Argentine line recently released that has shown low afla-

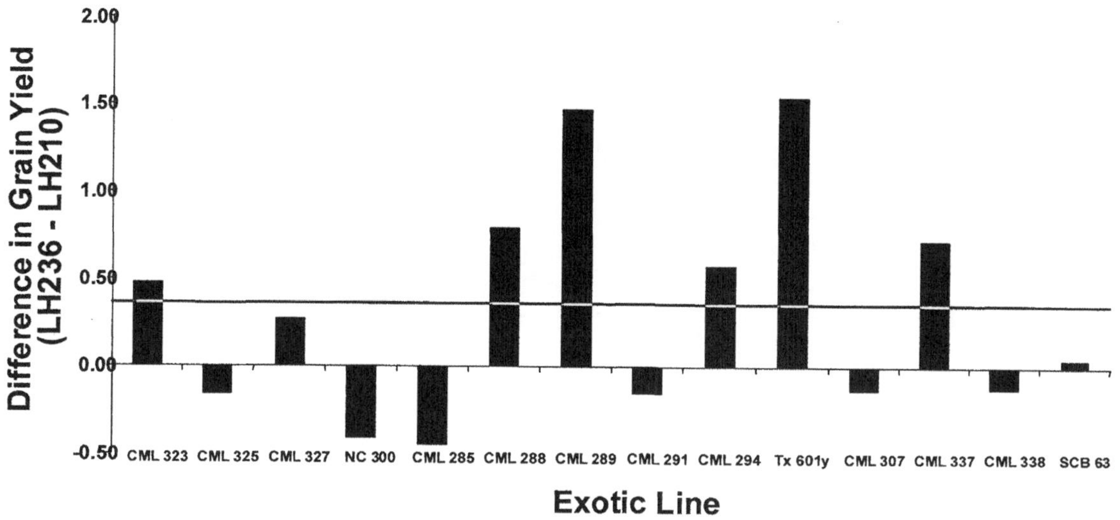

Figure 26.3 Differences in grain yield (t ha^{-1}) response between testcrosses of exotic yellow lines with temperate testers LH236 and LH210 across six environments in Texas. Green line indicates the difference in grain yield across hybrids between LH236 over LH210.

Table 26.1 Performance of commercial and white hybrids with exotic parents in the state food corn performance tests at College Station, Texas, in 2002 and at Weslaco, Texas, in 2000

COLLEGE STATION, TEXAS 2002			
Hybrid	Silking Date (d)	Grain Moisture (%)	Yidld (t/ha)
Tx114 × CML78	82	15.7	11.8
2259W	86	14.6	11.6
Zimmerman 1851W	83	11.9	11.5
Zimmerman WX 7812	83	11.9	11.3
Zimmerman WX 8272	84	12.2	11.2
DynaGro 5518RR	83	11.0	11.2
Fill	81	12.6	11.1
2250W	82	13.6	10.8
Rx 953W	84	14.0	10.7
Fill	81	12.5	10.7
2254W	81	14.6	10.5
Asgrow Rx 949W	83	14.9	10.4
Pioneer 32Y52	80	12.3	10.2
2247W	82	13.0	10.0
Fill	81	12.0	9.8
Pioneer 32A85	81	12.3	9.7
Fill	81	12.7	9.6
Fill	81	12.5	9.2
DynaGro 5505RR	81	11.4	8.9
DynaGro 5475RR	80	11.3	7.9
Mean:	81.80	12.83	10.4
C.V.:	1.24	2.84	8.26
L.S.D.:	1.43	0.52	1.2

WESLACO, TEXAS 2000			
Hybrid	Silking Date (d)	Grain Moisture (%)	Yidld (t/ha)
Tx114 × CML379	68	17.5	8.8
CML343 x Tx114	71	16.9	8.7
Pioneer 30G40	72	18.6	8.5
2K109W	69	16.5	8.3
Check	64	14.4	8.2
Check	71	16.3	8.1
Pioneer 30G54	72	17.1	8.0
Pioneer 30R39	72	20.5	7.3
2K107WQ	72	15.2	7.2
Zimmerman 1851W	70	13.7	7.2
2K201W	68	13.9	7.1
Asgrow RX 949W	70	15.8	6.6
Zimmerman N71-T7	65	13.5	6.4
TRX 6821W	67	13.2	6.2
Zimmerman Z75W	70	13.1	6.1
Garst 8277W	66	13.9	5.5
Garst 8122W	66	13.8	5.2
Zimmerman Z62W	70	13.2	3.2
Mean:	69	15.4	7.0
L.S.D.:	1.70	0.71	1.05
C.V.:	1.74	3.25	11.2

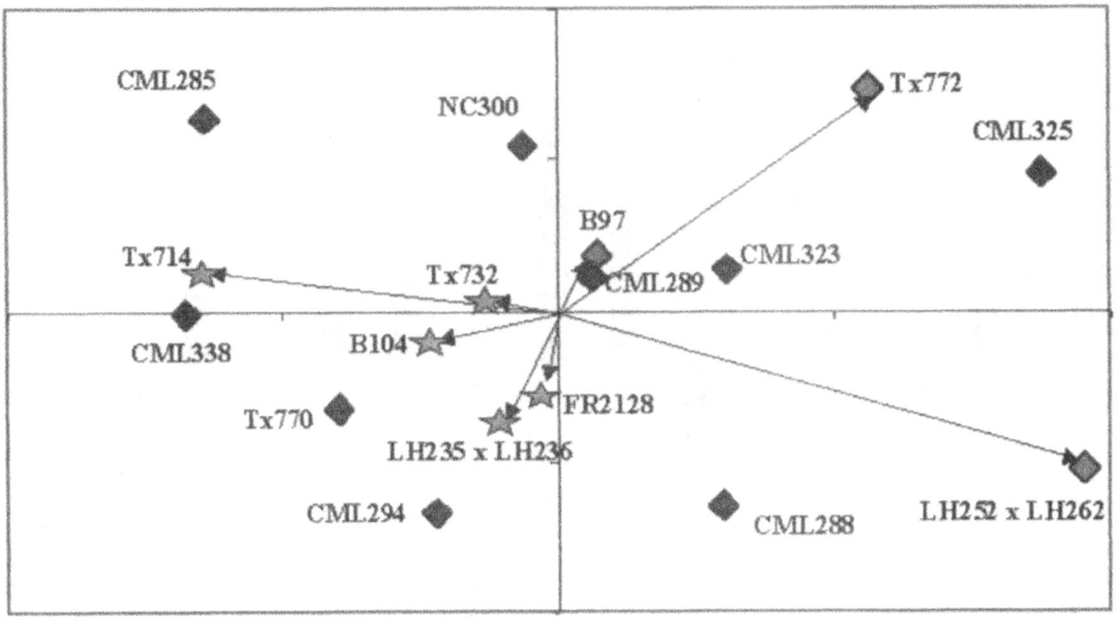

Figure 26.4 Biplot representation of factorial crosses between exotic lines (blue diamonds) and temperate SSS (stars) and NSSS (green diamonds) testers for average grain.

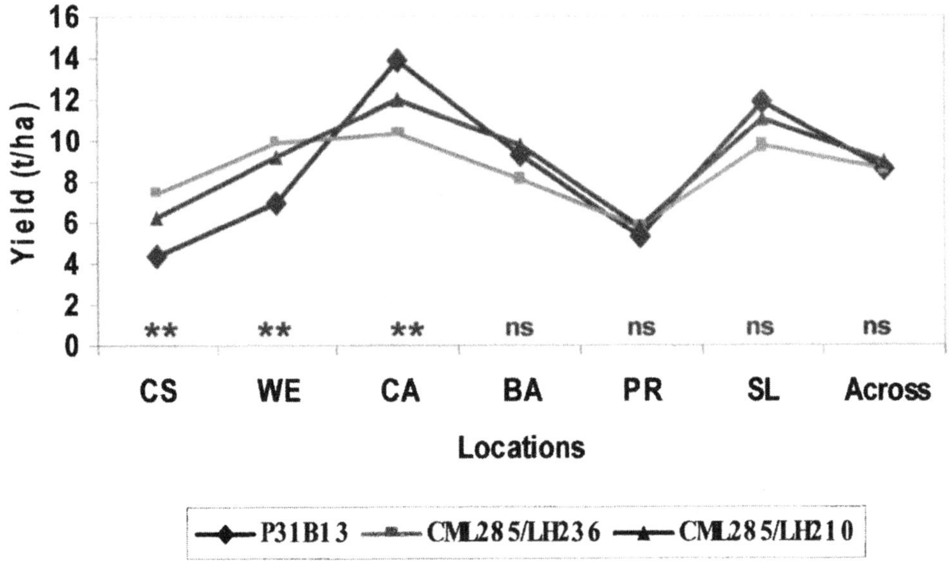

Figure 26.5 Grain yield for commercial hybrid Pioneer Brand P31B13 and exotic × temperate hybrids CML285/LH236 and CML285/LH210 at six locations in Texas (CS, College Station; WE, Weslaco; CA, Castroville; BA, Bardwell; PR, Prosper; SL, Springlake) and across locations.

Table 26.2 Aflatoxin accumulation in parts per billion (ppb) under inoculation with Aspergillus flavus in white hybrids at College Station (CS), Corpus Christi (CC), and Weslaco (WE), Texas, in years 2001 and 2002, in yellow hybrids at WE and CC in year 2002, and at WE and CS in year 2003

White Hybrids

Hybrid	CS2001 Aflatoxin ppb	(rank)	CC2001 Aflatoxin ppb	(rank)	WE2001 Aflatoxin ppb	(rank)
CML269/CML176	35.0	(6)	133.3	(1)	73.0	(1)
CML269/Tx807	73.5	(10)	333.3	(7)	320.5	(10)
CML269/CML384	45.3	(3)	156.7	(3)	399.5	(17)
CML384/CML176	156.3	(25)	210.0	(4)	205.0	(3)
Tx807/Mp313E	106.5	(15)	136.7	(2)	750.0	(26)
CML322/Tex6	23.0	(1)	1326.7	(25)	290.0	(9)
P32H39 Pioneer	74.5	(12)	440.0	(12)	380.0	(14)
RX901W Asgrow	82.3	(11)	363.3	(8)	580.0	(25)
RX921W Asgrow	117.3	(20)	4800.0	(30)	383.5	(15)
Mean (30 entries)	126.9		872.7		487.9	

Hybrid	WE2002 Aflatoxin ppb	(Rank)	CC2002 Aflatoxin ppb	(Rank)
CML269/CML176	0.0	(1)	25.6	(2)
CML78/CML269	10.9	(2)	72.2	(7)
CML311/CML176	39.4	(7)	47.3	(3)
CML78/CML176	19.9	(4)	93.2	(6)
Tx114/CML176	104.9	(14)	78.6	(5)
Tx807/CML176	22.1	(5)	130.3	(11)
P30G54 Pioneer	817.2	(30)	422.0	(21)
1851W Wilson	163.2	(19)	225.0	(17)
RX949W Asgrow	81.6	(10)	611.2	(26)
RX951W Asgrow	16.5	(3)	438.1	(22)
1910W	165.2	(20)	93.4	(9)
Mean (30 entries)	192.2		319.6	

Yellow Hybrids

Weslaco 2002	AF(ppb)	Corpus Christi 2002	AF(ppb)	Weslaco 2003	AF(ppb)	College Station 2003	AF(ppb)
CML323 × NC300	9.8	FR2128 × NC300	112.26	FR2128 × NC300	4.66	CML338 × Tx772	9.50
CML323 × CML288	11.5	NC300 × CML288	181.83	CML338 × NC300	33.66	B104 × Tx772	11.91
FR2128 × NC300	16.5	CML288 × CML285	212.71	FR2128 × CML288	37.56	LH195 × Tx772	17.57
(LH235 × 236) × CML288	33.4	Tx770 × CML288	255.94	Tx770 × Tx745	41.92	Tx772 × CML288	20.50
(LH235 × 236) × CML285	34.3	(235x236) × CML288	332.72	CML323 × CML288	43.33	NC300 × Tx745	23.57
P31B13	1200.0	P31B13	2569.99	P31B13	338.00	P31B13	506.71
P32R25	413.7	P32R25	724.65	P32R25	168.00	P32R25	576.71
RX897	100.7	RX897	628.26	RX897	50.55	RX897	156.91
DK687	634.4	DK687	212.77	DK668	90.60	DK668	154.00
Mean (30 entries)	220.1		925.00	Mean (45 entries)	217.49		186.80
Max.	1200.0		2852.62	Max.	1373.28		896.71

toxin accumulation, good combining ability with SSS lines, high-protein content, low ear placement, and good husk coverage (Llorente et al., 2004).

Target traits

There is potential for exotic germplasm to provide desirable alleles for input and output traits, in addition to grain yield. We are emphasizing specific traits that have desirable expression in exotic germplasm and which are relatively lacking in current commercial hybrids in the Southern United States. Some of these traits are aflatoxin resistance, husk coverage, kernel integrity, endosperm hardness and test weight, long maturity, cob and grain color, nutritional value, and drought and heat tolerance.

We have evaluated aflatoxin accumulation of exotic and temperate inbreds and hybrids following artificial inoculation with *Aspergillus flavus* in three Texas locations (Weslaco, Corpus Christi, and College Station). The inbreds with less susceptibility to aflatoxin in hybrid combinations were of exotic origin. Yellow inbreds CML323, CML326, CML288, CML289, CML338, Tx772, and experimental QPM lines TxX69s, and white lines CML269, CML176, CML78, Tx130, and Tx807, are some of most promising maize inbreds for reducing the risk of aflatoxin under Texas growing conditions (Betrán et al., 2002a). Most of these inbreds have subtropical or tropical origin and hard endosperm. Their hybrids were less susceptible to aflatoxin than commercial hybrids currently grown in Texas (Table 26.2).

Husk coverage is a trait that influences grain moisture content, kernel integrity, damage produced by insects, and fungi colonization. For example, we found a significant positive correlation between poor husk coverage and aflatoxin content

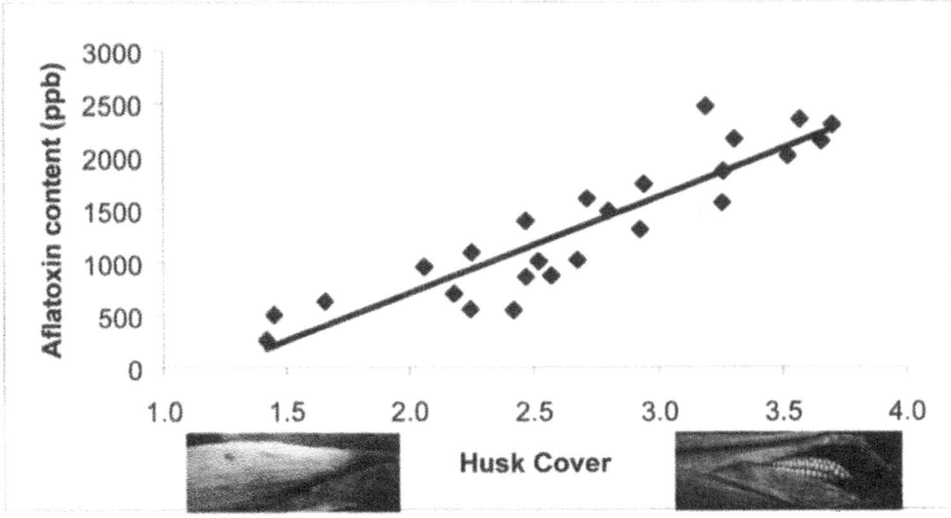

Figure 26.6 Relationship between aflatoxin concentration (ng/g) and husk coverage rating (1 = well-covered ear; 5 = poorly covered ear) of full-, intermediate-, and early-season commercial hybrids inoculated with A. flavus in College Station and Weslaco, Texas.

($r = 0.77$), evaluating commercial hybrids with different maturities (Figure 26.6) (Betrán and Isakeit, 2004). Early-maturing hybrids with loose, poorly covered husks had more aflatoxin than full-season hybrids that had better husk coverage. This trait is selected in opposing directions between the midwestern and the southeastern United States. In the Midwest, loose husks are preferred to allow faster drying in the field before harvest, while in the southeastern United States, tight and long husks are preferred to reduce or prevent insect and fungal damage. Maturity and husk coverage are closely correlated and their effects are difficult to separate because later hybrids have more plant and husk leaves than early hybrids.

The capability of kernels to maintain physical integrity is an important trait in grain quality and food safety. Kernel injuries in the pericarp and endosperm caused by insects [e.g., corn earworm, *Heliothis zea* (Boddie)] and abiotic stresses (stress cracks) reduce processing quality and predispose them to fungal infection and mycotoxin contamination. Stress-cracked, broken and damaged kernels take up moisture more rapidly and disintegrate during processing. Kernel integrity is associated with good husk coverage, endosperm hardness, and insect resistance. Hybrids with loose husks are more vulnerable to loss of kernel integrity (Odvody et al., 1997). Endosperm hardness is defined as the proportion of hard versus soft endosperm. Flinty hard endosperm is very dense, whereas floury soft endosperm is less dense. There are genetic differences for horny/floury ratios, pericarp thickness, and cell structure among temperate and exotic germplasm. Hardness can be estimated visually in the field or with a light box. It can also be quantified using a tangential abrasive dehulling device that uniformly removes pericarp and endosperm. The amount of material removed is related to the relative proportions of hard to soft endosperm, kernel size and shape, and type of denting. Test weight, a measured of bulk density obtained by weighing a specific volume of grain (kg/hl), is an important indicator of grain quality and endosperm hardness. A low test weight is an indication that the grain has a high proportion of soft endosperm. Hard endosperm maize has higher test weights and less stress cracks than commodity maize (U.S. Grains Council, 2001). Visual ratings for kernel integrity and endosperm hardness were associated with aflatoxin contamination in inbreds and their testcrosses (Figure 26.7) (Betrán et al., 2003). These traits can be used in indirect selection to reduce aflatoxin, because they are easy to screen in big populations and have high repeatabilities. Hybrids with softer endosperm normally have worse kernel integrity than hybrids with flinty endosperms. In general, we have observed better kernel integrity and harder endosperm in exotic inbreds and their hybrids than in some popular commercial hybrids in Texas. There are good sources of endosperm hardness,

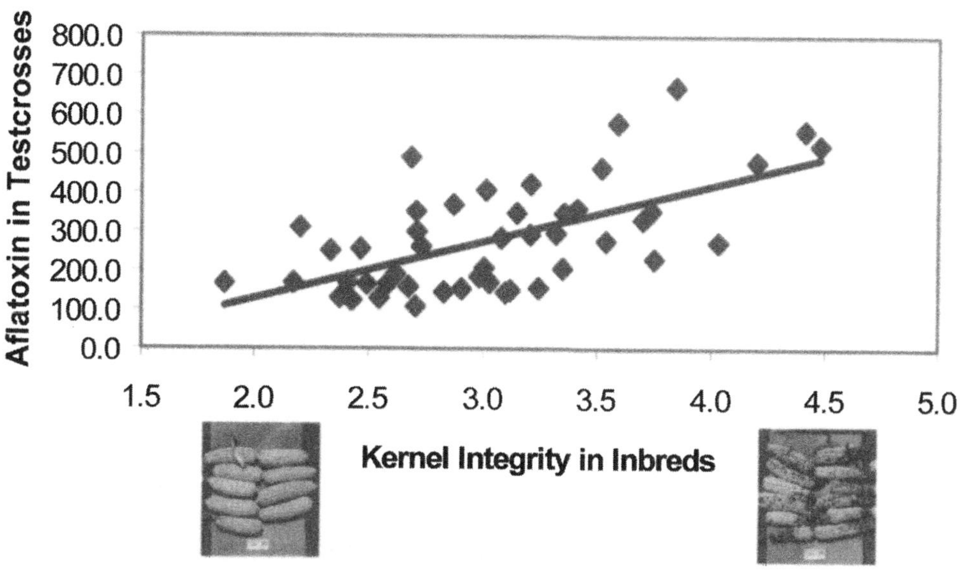

Figure 26.7 Relationship between average aflatoxin concentration (ng/g) for testcross hybrids under inoculation with Aspergillus flavus at three locations in Texas and kernel integrity ratings (1 = good; 5 = poor) for their parental lines at Weslaco, Texas.

high test weights, and good kernel integrity among tropical and subtropical (Caribbean flints, Cateto, Suwan) and temperate exotic (e.g., red-flinty germplasm from Argentina) germplasm.

Hybrids in the southern United States are full-season hybrids with longer maturities (DRM > 120 days) as compared with hybrids in the U.S. Midwest (DRM < 110 days) (Troyer, 2001). We have found a positive correlation between grain yield and maturity in our evaluations in south-central Texas. Full-season hybrids have yielded more than intermediate and early hybrids. Because the growing season is not a limiting factor (>300 days of frost-free period) the use of longer-maturity tropical and subtropical exotic germplasm is a feasible approach to increase yield potential. In addition, a negative correlation between maturity and aflatoxin accumulation has been observed with full-season hybrids being less susceptible than early and intermediate hybrids (Betrán and Isakeit, 2004).

Grain color is an important attribute, particularly in white maize, because it affects the color of the finished product. For white maize, the color should be a clear white. The grayish, dingy whites or yellowish colors are undesirable. For yellow maize the color should be a good, clean, bright yellow. Some final users, such as the poultry industry, prefer orange or dark yellow with high contents of carotenoids. White color is preferred in many tropical and subtropical areas over yellow. In the United States, most of the production is yellow maize. Hence, conversion of white exotic to yellows is a possibility to broaden the exotic germplasm base. Cob color, determine by the *P* gene, affects quality for alkaline cooking. A white cob (P-*ww*) is desirable for food purposes. Pink and red cobs are common in the U.S. Corn Belt yellow hybrids, while white cobs are predominant in exotic inbreds.

Quality protein maize (QPM) is a high-lysine maize homozygous o_2o_2 with vitreous endosperm that increases the nutritional value of food and feed maize products (Vasal, 2001). Successful development of competitive QPM hybrids using conventional breeding procedures by CIMMYT, as well as institutions in Brazil, South Africa, and China, among others, indicates the potential of current exotic QPM germplasm (Vasal, 2001). We are using several approaches to develop temperate-adapted QPM hybrids: (1) selection within tropical and subtropical QPM germplasm to reduce ear placement, maturity, photoperiod-sensitivity, and tassel size, and to improve standability, grain yield, and quality; (2) conversion of elite U.S. inbreds to QPM using opaque temperate and QPM exotic inbreds as donors for the *o2* allele and the endosperm modifiers; and (3) selection within populations obtained by hybridizing soft endosperm temperate germplasm with exotic QPM with hard endosperm. Our goal here is to combine desirable

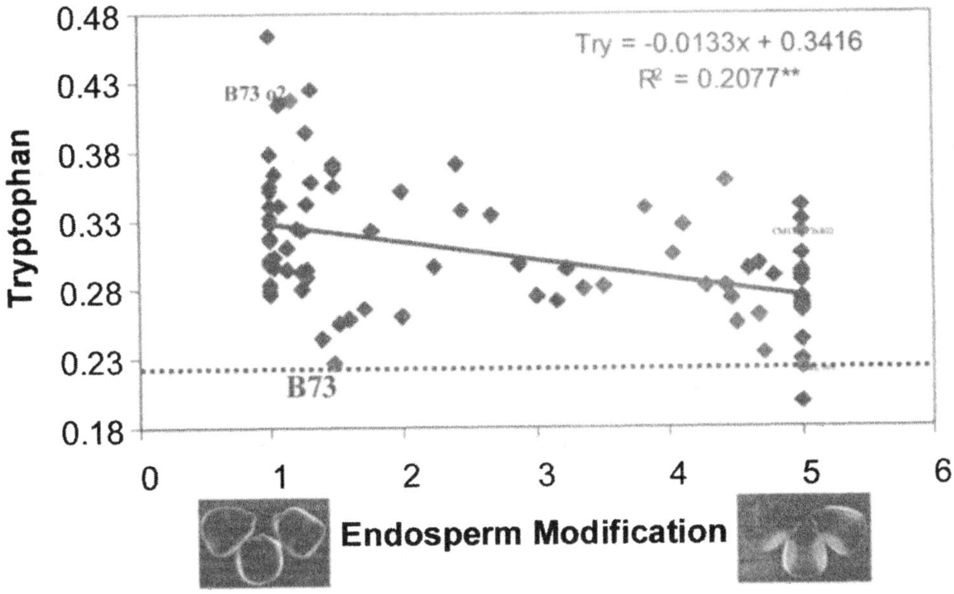

Figure 26.8 Relationship between tryptophan relative content and endosperm modification (1 = opaque, 5 = hard) for 86 high lysine experimental inbreds.

traits from temperate opaque (early maturity, good stalks and roots, heterotic grouping, plant morphology) and QPM exotic (endosperm modification, aflatoxin resistance, kernel integrity and husk coverage). Endosperm modification and tryptophan and lysine contents have been negatively correlated (Figure 26.8). Tryptophan relative contents are used to select lines with high degrees of endosperm modification but maintaining high levels of essential amino acids. In addition to its enhanced nutritional value, QPM is a promising source to reduce aflatoxin (Betrán et al., 2002).

Drought is one of the main environmental limitations on yield in rain fed areas in the southern United States. Drought and high temperatures at flowering and grain-filling periods are common in Texas. Commercial temperate hybrids have been improved for drought and high-density tolerance for more than five decades. There are also significant improvements in tropical maize for drought tolerance at flowering (Beck et el., 1996; Edmeades et al., 1999). Drought-tolerant germplasm in the form of populations (e.g., TS6, La Posta Sequia), synthetics (ZM521), inbreds (CML339, CML444), and hybrids have been developed in exotic maize. Furthermore, significant quantitative trait loci (QTLs) consistent across genetic backgrounds and environments identified for different yield components, morphological traits, and physiological parameters have been identified (Ribaut et al., 2002). This germplasm and QTL from these exotic sources can be used to introduce new alleles for drought tolerance in temperate germplasm. In our program, we are using drought-tolerant exotic inbreds in combination with temperate lines to enhance drought tolerance.

Genomics and introgression/incorporation of exotic germplasm

Genomics have the potential to increase efficiency of the introgression/incorporation process in various ways:

- *Provide additional information about the relationships among exotic and temperate germplasm and the amount of genetic diversity present.* This information can be used to establish potential heterotic groups based on genetic distances or similarities, to assign inbreds to established heterotic groups, and to choose parents for breeding (Melchinger, 1999). Also, molecular markers can be useful in creating core collections of germplasm accessions and populations.
- *Allele characterization of specific genomic regions ("haplotyping") and genes.* Haplotyping can be used to select exotic germplasm that generates

maximum diversity and minimum duplication with local germplasm at specific regions (Menz et al., 2003). Furthermore, if phenotypic information is available for target traits, it is feasible to conduct association genetics to identify genomic regions involved in their expression (Cardon and Bell, 2001). In the case of known genes, it becomes a search for the best alleles in what it is known as "allele mining."

- *Mapping of QTL or genes by hybridization between exotic and temperate germplasm.* Different background of exotic and temperate materials increases the chances of different expression and genetic composition for the target traits and the creation of sufficient degree of linkage disequilibrium suitable to identify QTLs (Edwards, 1992). Goodman (1999) suggested the use of temperate-adapted exotic inbreds to conduct QTL mapping because they can exploit and detect more exotic genetic contributions than unadapted sources that require additional backcrosses prior to phenotypic evaluation.
- Marker-assisted selection to introgress specific genomic regions, previously identified in exotic populations, into temperate germplasm. If QTLs responsible for the target trait(s) in the exotic germplasm are known, markers can be directly used to recover progenies carrying the desired genomic regions in a temperate background (Lee, 1995). Molecular markers can also help to increase the probability of recovering transgressive segregants, or lines with high degree of exotic genome while maintaining adaptation, to increase intensity of selection while maintaining variability, to increase parental control by selecting before pollination, and to reduce number of seasons by selecting outside the selection environments.
- *Change of diversity and selection in breeding materials over time either in recurrent selection or pedigree breeding.* Molecular marker pedigree analysis can estimate parental contribution and potential genomic regions selected over time ("retrospective mapping"). The study of the genomic composition of ancestors of inbreds permits tracing the inheritance of genomic segments and genes in pedigrees (haplotyping) through multiple generations allowing the analysis of genetic contribution of ancestors and to examine the effects of selection and genetic drift (Melchinger, 1999).

The combination of phenotypic evaluation and molecular technologies is making the introgression/incorporation of exotic germplasm more efficient in speed and precision. However, there is a debate about how much effort to devote to biotechnology applications in maize breeding and, in particular, to the introgression of germplasm (Goodman and Carson, 2002). The relative resource allocation between genomics and phenotypic conventional approaches depends on several factors: the stage of the breeding project (early stages can be dedicated to screening germplasm and advance stages to QTL mapping once material is well characterized), the nature of the target trait(s) (genetics, efficiency of phenotypic selection), amount of genetic variation available, short- versus long-term objectives, type of biotechnology application (e.g., mapping requires more time and effort than DNA fingerprinting or associative genetics), and amount of resources and personnel.

Conclusion

The principles of breeding exotic germplasm are similar to those of breeding local temperate germplasm. Good data from environments that allow discrimination among breeding values of testing genotypes are extremely important to making improvement (Cooper et al., 2003). Especially relevant is the selection of environments that allows full expression of the genetic potential harbored in exotic germplasm. One particular advantage of breeding exotic germplasm is that for most traits the amount of genetic variation observed is generally greater than in "good × good" breeding populations commonly used in temperate maize. Consequently, realized genetic progress, which is proportional to the genetic variance available, is relatively easy. On the other hand, genetic progress can be affected by a large genotype × environment variance component if contrasting environments (e.g., tropical and temperate) are used in the evaluations. In addition to the implementation of breeding principles and biotechnology tools, we would like to emphasize the importance of becoming acquainted with breeding material by careful, continuous, and patient observation. This attachment with the germplasm facilitates breeding decisions, "germplasm engineering," and development of superior progenies.

We are searching for new alleles in exotic germplasm for resistance to mycotoxins, tolerance to drought and heat, high test weight, high proportion of hard flinty endosperm, kernel integrity with kernels free of fissures or stress cracks, white cobs, and nutritional value for food or feed. Transitional areas between tropical and temperate areas, such as Texas, represent excellent opportunities to incorporate, combine and introgress exotic germplasm in temperate material, and vice versa. By developing inbreds adapted to southern United States, we expect to facilitate the connection and genetic flow between these gene pools.

Finally, breeding exotic germplasm increases networking and cooperation among scientists and breeders around the world. This collaboration brings exchange of information and germplasm (despite increasing limitation due to Intellectual Property Rights) and creates a good support for plant breeding education.

Acknowledgments

We would like to congratulate Dr. Hallauer for a successful career full of contributions. Our deep gratitude for your impressive legacy and for educating plant breeders and scientists around the world. Dr. Hallauer (2002, personal communication) said: "Life is like a bike, you need to keep pedaling to not fall down. I just kept pedaling as hard as I could." On behalf of all the generations that will follow your trace, thank you.

References

Beck, D., F.J. Betrán, M. Bänziger, G.O. Edmeades, J.M. Ribaut, M. Willcox, S.K. Vasal, and A.C. Ortega. 1996. Progress in developing drought and low soil nitrogen tolerance in maize. Proceedings of 51st Ann. Corn and Sorghum Ind. Res. Conf. 1996. ASTA, Washington.

Bernardo, R. 1990. Identifying populations useful for improving parents of a single cross based on net transfer of alleles. Theor. Appl. Genet. 80:349–352.

Betrán, F.J., T. Isakeit, and G. Odvody. 2002a. Aflatoxin accumulation of white and yellow inbreds in diallel crosses. Crop Science. 42:1894–1901.

Betrán, F.J., T. Isakeit, G. Odvody, and K. Mayfield. 2002b. Maize germplasm evaluation for aflatoxin resistance in Texas. Mycopathologia. 155:81.

Betrán, F.J., and T. Isakeit. 2004. Aflatoxin Accumulation in Maize Hybrids of Different Maturities. Agronomy J. 96:565–570.

Betrán, F.J., T. Isakeit, G. Odvody, S. Bhatnagar, and K. Mayfield. 2003. Aflatoxin accumulation and associated traits in maize inbreds and their testcrosses. Aflatoxin/Fumonisin Workshop 2003, October 13–15, 2003, Savannah, GE.

CIMMYT. 1999. A compilation of 424 CIMMYT maize lines (CMLs). CIMMYT internal publication.

Cardon, L.R., and J.I. Bell. 2001. Association study designs for complex diseases. Nature Reviews Genetics 2:91–99.

Crossa, J., C.O. Gardner, and R.F. Mumm. 1987. Heterosis among populations of maize (*Zea mays* L.) with different levels of exotic germplasm. Theor. Appl. Genet. 73:445–450.

Crossa, J. 1989. Theoretical considerations for the introgression of exotic germplasm into adapted maize populations. Maydica 34:53–62.

Crossa, J., and C.O. Gardner. 1987. Introgression of an exotic germplasm for improving an adapted maize population. Crop Science 27:187–190.

Dudley, J.W. 1982. Theory for transfer of alleles. Crop Science 22:631–637.

Dudley, J.W. 1984. Theory for identification and use of exotic germplasm in maize breeding programs. Maydica XXIX:391–407.

Dudley, J.W. 1987a. Modification of methods for identifying inbred lines useful for improving parents of an elite single crosses. Crop Science 27:944–947.

Dudley, J.W. 1987b. Modification of methods for identifying populations to be used for improving parents of an elite single crosses. Crop Science 27:940–943.

Dudley, J.W., K.R. Lamkey, and J.L. Geadelmann. 1996. Evaluation of populations for their potential to improve three maize hybrids. Crop Science 36:1553–1559.

Eagles, H.A., A.K. Hardacre, and R.K. Bansal. 1989. Testcross performance of maize lines from backcross populations containing highland mexican or highland Peruvian germplasm. Euphytica 41:263–272.

Eagles, H.A., and A.K. Hardacre. 1990. Genetic changes from introgression of highland Mexican germplasm into a corn belt dent population of maize.

Echandi, C.R., and A.R. Hallauer. 1996. Evaluation of U.S. Corn Belt and adpated tropical maize cultivars and their diallel crosses. Maydica 41:317–324.

Edmeades, G.O., J. Bolaños, S.C. Chapman, H.R. Lafitte and M. Bänziger. 1999. Selection improves tolerance to mid/late season drought in tropical maize populations. I. Gains in biomass, grain yield and harvest index. Crop Science 39:1306–1315.

Edwards, M. 1992. Use of molecular markers in the evaluation and introgression of genetic diversity for quantitative traits. Field Crops Research 29:241–260.

Geadelmann, J. 1984. Using exotic germplasm to improve northern corn. 39th Annual Corn & Sorghum Research Conference. ASTA. Washington, DC.

Gerloff, J.E., and O.S. Smith. 1988. Choice of method for identifying germplasm with superior alleles. Theor. Appl. Genet. 76:209–216.

Goodman, M.M. 1992. Choosing and using tropical corn germplasm. Annu. Corn Sorghum Res. Conf. Proc. 47:47–64.

Goodman, M.M. 1985. Exotic maize germplasm: status, prospects and remedies. Iowa State J. Res. 59:497–527.

Goodman, M.M. 1988. Final report of USDA-NCSU specific coop. Agreement no. 58-7b30-573 of the Latin American Maize Regeneration Project. Dept. of Crop Sciences, North Carolina State Univ., Raleigh, NC.

Goodman, M.M. 1999. Broadening the genetic diversity in breeding by use of exotic germplasm. *In* J.G. Coors and S. Pandey (eds.). The genetics and exploitation of heterosis in crops. ASA, WI.

Goodman, M.M., J. Moreno, F. Castillo, R.N. Holley, and M.L. Carson. 2000. Using tropical maize germplasm for temperate breeding. Maydica 45:221–234.

Goodman, M., and M.L. Carson. 2002. Reality vs. myth: Corn breeding, exotics, and genetic engineering. Proceedings of 57st Ann. Corn and Sorghum Ind. Res. Conf. 2002. ASTA, WA.

Hallauer, A.R. 1978. Potential of exotic germplasm for maize improvement. In International Maize Symposium. W.L. Walden, ed., p. 229–247. John Wiley & Sons, New York, NY.

Hallauer, A.R. 1990. Methods used in developing maize inbreds. Maydica 35:1–16.

Hallauer, A.R. 1994. Registration of BS28 and BS29 maize germplasm. Crop Science 34:544–545.

Hohls, T., G.P.Y. Clarke, P.E. Shanahan, and H.O. Gevers. 1995. An unbiased estimator for identifying lines useful for the improvement of elite single crosses, based on a combining ability model. Heredity 75:155–163.

Holland, J.B., M.M. Goodman, and F. Castillo-Gonzalez. 1996. Identification of agronomically superior Latin American maize accessions via multi-stage evaluations. Crop Science 36:778–784.

Kraja, A.T., and J.W. Dudley. 2000. Identification of tropical and temperate maize populations having favorable alleles for yield and other phenotypic traits. Crop Science 40:941–947.

Lee, M. 1995. DNA markers and plant breeding programs. Adv. Agron. 55:265–344.

Llorente, C.F., F.J. Betrán, A. Bockholt, and F. Fojt III. 2004. Registration of Tx772. Crop Science 44:1036–1037.

Melchinger, A.E. 1999. Genetic diversity and heterosis. p. 99–118. *In* J.G. Coors and S. Pandey (eds.). The genetics and exploitation of heterosis in crops. ASA, WI.

Menz, M.A., R.R. Klein, N.C. Unruh, W.L. Rooney, P.E. Klein, and J.E. Mullet. 2003. Genetic Diversity of Public Inbreds of Sorghum Using Mapped AFLP and SSR markers. Crop Sci. 2004 44:1236–1244

Metz, G. 1994. Probability of net gain of favorable alleles for improving an elite single cross. Crop Science 34:668–672.

Michelini, L.A., and A.R. Hallauer. 1993. Evaluation of exotic and adapted maize (*Zea mays* L.) germplasm crosses. Maydica 38:275–282.

Odvody, G.N., N. Spencer, and J. Remmers. 1997. A description of silk cut, a stress-related loss of kernel integrity in preharvest maize. Plant Dis. 81:439–444.

Parra, J.R., and A.R. Hallauer. 1997. Utilization of exotic maize germplasm. Plant Breeding Reviews 14:165–187.

Ribaut J-M, M. Bänziger, F.J. Betrán, C. Jiang, G.O. Edmeades, K. Dreher, and D. Hoisington. 2002. Use of molecular markers in plant breeding: Drought tolerance improvement in tropical maize in "Quantitative Genetics, Genomics, and Plant Breeding," Manjit S. Kang (ed.), chapter 7, CABI Publishing, Wallingford, United Kingdom.

Salhuana, W. 1995. Latin American maize project leaves untapped legacy of agricultural riches. Diversity 11(4):6.

Salhuana, W., and R. Sevilla. 1995. Latin American Maize Project (LAMP) Stage 4 Results from homologous areas 1 and 5. Natl. Seed Storage Lab., Fort Collins, CO.

Salhuana, W.R., R. Sevilla, and S.A. Eberhart. 1997. Final reports, Latin American Maize Project. Pioneer Hi-Bred Int., Johnston, IA.

Simmonds, N.W. 1993. Introgression and incorporation. Strategies for the use of crop genetic resources. Biol. Rev. 68:539–562.

Smith, J.S.C., D.N. Duvick, O.S. Smith, A. Grunst, and S.J. Wall. 1999. Effect of hybrid breeding on genetic diversity in maize. p. 119–126. *In* J.G. Coors and S. Pandey (eds.). The genetics and exploitation of heterosis in crops. ASA, WI.

Taba, S., S.A. Eberhart, and L. Pollak. 2004. Germplasm resources. Chapter 1.3. *In* W. Smith, F.J. Betrán and E. Runge (eds.). Corn: Origin, history, technology, and production. John Wiley & Sons, New York, NY.

Tallury, S.P., and M.M. Goodman. 1999. Experimental evaluation of the potential of tropical germplasm for temperate maize improvement. Theor. Appl. Genet. 98:54–61.

Troyer, A.F. 1999. Background of U.S. hybrid corn. Crop Science 39:601–626.

Troyer, A.F. 2001. Temperate corn—background, behavior, and breeding. p. 393–466. *In* A.R. Hallauer (ed.), Specialty corn. CRC Press, New York, NY.

Uhr, D.V., and M.M. Goodman. 1995. Temperate maize inbreds derived from tropical germplasm. I. Testcross yield trials. Crop Science 35:779–784.

U.S. Grains Council, 2001. Value-enhanced grains quality report. Washington, DC.

Vasal, S.K. 2001. High quality protein corn. p. 85–129. *In* A.R. Hallauer (eds.). Specialty Corns. 2nd ed. CRC Press, Boca Raton, FL.

Wellhausen, E.J. 1965. Exotic germplasm for improvement of Corn Belt maize. Annu. Corn Sorghum Res. Conf. Proc. 20:31–45.

27

Development of a Heterotic Pattern in Orange Flint Maize

Guillermo H. Eyhérabide, INTA—Instituto Nacional de Tecnología Agropecuaria,
Graciela Nestares, Universidad Nacional de Rosario
María José Hourquescos

Introduction

Maize production in Argentina

Argentina is one of the main corn-producing countries of the world, and the volume of its export ranks second only to the United States. The average annual corn production estimate for Argentina is 14 million tons, of which around 50% is exported, and the remainder is used for food, feed, seed, and industrial purposes. Nontillage agriculture has reached a significant percentage of the cropped land in Argentina. Thirty-eight percent of the corn crop area was under nontillage husbandry in the 1999/2000 season (Lorenzatti, 2000). Productivity per unit of land (6.05 Mg ha^{-1} in 2001/2002) has increased at a large enough magnitude to more than compensate for the decrease in the area covered by the crop (Macagno et al., 1993). A recent study holds that a genetic gain of 2.9% per year (0.17 Mg $ha^{-1}year^{-1}$) has occurred in the Argentine Corn Belt across the past 20 years, but the rate of gain has not been uniform. The rate of genetic gain during the 1990s was larger than that during the 1980s (Eyhérabide and Damilano, 2001). The replacement of double and three-way crosses by single crosses that occurred at the end of the century seems to be strongly associated with the improvement of the rate of genetic gain, as it happened in the United States of America and other countries (Duvick, 1996; Hallauer, 1999). Farmers have a wide array of cultivars available, including transgenic hybrids resistant to both sugarcane borer (Diatraea sacharallis Fab.) and herbicides. Transgenic Bt-hybrids have broadened the alternatives for including maize in crop rotations. Late planting of Bt-hybrids of maize (late December) as a second crop after wheat withstands attacks of a third generation of D. saccharalis.

A common strategy of breeding programs in Argentina is to take advantage of the strength of the Argentine Orange Flint versus U.S. Yellow Dent germplasm heterotic pattern (Maunder, 1992). Crosses between lines of both heterotic groups often become very highly productive cultivars. Both the U.S. heterotic groups, Reid Yellow Dent (RYD) and Lancaster Sure Crop (LSC), and the heterotic group composed of native germplasm of the Cristalino Colorado (Cateto) race will be referred to as the Flint versus Dent heterotic pattern. Dent hybrids either developed in or introduced to Argentina following the RYD versus LSC heterotic pattern exhibit quite good performance, especially in favorable environments. They are appreciated for their fast dry-down of grain moisture, especially when grown in the southeast of the Province of Buenos Aires. These cultivars are more suitable to the wet milling industry. Flint hybrids are appreciated in the market not only because of the hardness of the endosperm, which makes the kernel less susceptible for breakage and more suitable for dry milling industry (Eyhérabide and Gonzalez, 1997), but also for their greater resistance to drought stress, MRCV (Maize Río Cuarto Virus), and ear rots. In general, chemical composition provides flint kernels larger biological value when compared with dent kernels (Schang et al., 1993). Despite breeding efforts, flint hybrids still show average grain yields behind those of flint × dent and dent hybrids across a range of environmental conditions; breeding efforts should be aimed at the enhancement of the performance of crosses between local flint inbred lines.

Heterotic groups

Appropriate knowledge about the performance of maize genotypes in crosses is required to organize germplasm in applied breeding programs. The observed positive association between grain yield and genetic divergence of the parents of a cross within certain range (Moll et al., 1965) and interest in reducing genetic vulnerability to stress conditions justify the exploitation and characterization of new sources of germplasm to broaden the crop genetic base. In species that exhibit heterosis, information about combining ability with genetically divergent testers is useful when classifying the germplasm in heterotic groups. Germplasm classification based on genetic distances is important because crosses between lines extracted from more divergent sources will probably exhibit greater levels of heterosis (Ordás, 1991). Relative performance in testcrosses of lines to a group of divergent testers can be used as an estimation of genetic distance. Schon et al. (1994) indicated that the genetic correlation among testcross performance of a group of inbred lines to different testers depends on the number of loci with similar effects that the testers have in common. Hallauer et al. (1988) reported that heterotic pattern among populations could be assessed on the basis of testcross evaluation, and Comstock and Moll (1963), cited by Pandey and Gardner (1992), pointed out that testcross evaluation in several locations in a single season would provide the information needed to identify the most promising germplasm. Thus, a collection of 79 Argentine landraces was testcrossed to two pairs of divergent testers (Eyhérabide and Gonzalez, 1997). Testcrosses were evaluated in several locations in northern Buenos Aires, and differences for landraces, testers, and landrace × tester interactions were found to be significant for grain yield and other agronomic traits. Yield data were further analyzed following additive main effects (landrace and tester) and multiplicative interaction (landrace × tester) models (Crossa et al., 1990). The landrace × tester interaction sum of square was partitioned into principal component axes. The first two principal components were significant and accounted for 74% of the interaction sum of square. The first axis was consistent with the Flint versus Dent major heterotic pattern. The second axis exhibited a contrast between two local flint heterotic groups. The pattern of landrace × tester interaction was considered in order to identify five landraces to form each one of two composites.

Reciprocal recurrent selection

It has been determined that the heterotic response in a maize hybrid requires the existence of differences in gene frequency between the parents for alleles having dominance effects. Reciprocal recurrent selection (HSRRS) (Comstock et al., 1949) is a breeding procedure designed to improve the heterosis between two populations by maximizing the genetic divergence between them for loci with dominance effect. The method uses half-sib interpopulation progenies as selection units. Each base population is used as the tester of its counterpart. Hallauer (1967) suggested a method for developing single crosses using full-sib interpopulation progenies from two, two-eared populations. Alternate rows of both populations are planted and individuals from one population are crossed to individuals from the other population. Both plants involved in the crosses are also selfed. Full-sib interpopulation progenies are evaluated in replicated trials in different environments. S_1 seed from the parental plants that performed better in crosses is planted in paired rows to initiate a new cycle of inbreeding and full-sib crossing. A cycle of reciprocal population improvement also can easily be accomplished if S_1 progenies derived from the selected parents in each population are intermated separately (Hallauer and Eberhart, 1970). By providing some modifications, the reciprocal full-sib recurrent selection method can be also applied to one-eared base populations. It would require the development of S_1 progenies before making any cross for yield testing. In the second year, S_1 progenies are planted ear-to-row in pairs, each pair including one S_1 progeny from each one of the base populations. Individual plants from one row are selfed and used to pollinate several plants from the opposite row and vice versa. This step provides an opportunity for selection for agronomic traits between and within S_1 progenies.

The reciprocal full-sib recurrent selection (FSRRS) allows an appropriate integration of long- and short-term breeding objectives, since enhancement of breeding populations can be accomplished with the development of single-cross hybrids (Hallauer and Miranda Filho, 1988). The uniqueness of reciprocal full-sib selection is that it selects pairs rather than individuals, in such a way that both parents of the crosses are identified. The superiority of these combinations probably relies on nonadditive effects. Although general combining ability effects are exposed to selection, those effects are emphasized by the reciprocal full-sib recurrent selection procedure. The premise for achieving an efficient recurrent selection program is the availability of good estimates of the breeding values of the individuals sampled from each population. For this aspect, the estimation of the breeding value of an individual is affected by the breeding value of its mate, and thus, full-sib families would estimate breeding values with less accuracy than half-sib families (Jones et al., 1971). Nevertheless, full-sib families are better at detecting nonadditive effects than are half-sib families. If performance of desirable crosses relies on favorable interallelic interactions, FSRRS will select both parents of the cross, while HSRRS will select just one of them. Although recombination of the selected parent in each population will break those favorable gene combinations, they will have a higher probability of occurrence with FSRRS (Jones et al., 1971). The importance of nonadditive effects in hybrid breeding makes FSRRS a suitable procedure since it increases probabilities of identifying inbred lines from both base populations with better performance in their reciprocal crosses (Hallauer, 1999).

Empirical data indicate that the improvement of cross performance can be achieved by applying reciprocal recurrent selection (Hallauer, 1999). Efficiency of both methods of recurrent selection has been reported by Eyhérabide and Hallauer (1991) (FSRRS) and Keeratinikajal and Lamkey (1993) (HSRRS).

Direct response measured in the population cross as well as indirect responses measured as changes in the performance of the parent populations per se have been also reported (Hallauer, 1999). Improvement in the performance of the interpopulation cross comes along with improvement in the mid-parent heterosis. This last improvement over cycles of selection is expected to be associated with a further increase in the heterosis expressed by hybrids between selected lines extracted from the parent populations. Improvement of the mid-parent heterosis has been attributed to the increase of the heterozygosity in the interpopulation cross-over cycles of selection. Apparently, the FSRRS selects different isoalleles or different sets of loci exhibiting dominance in each population. Full advantage of FSRRS is provided when there is an initial level of heterosis in the cross of both base populations. According to Lópes de Souza and Miranda Filho (1985), positive changes in mid-parent heterosis are expected over cycles of selection assuming complete dominance when frequencies for favorable alleles are different in both populations, unless the corresponding loci are fixed in one base population. Assuming overdominance, differences in gene frequency between base populations would cause even greater improvement of mid-parent heterosis.

Development of heterotic flint populations

A cooperative maize-breeding program between INTA and a consortium of seed companies was established in 1992. Besides other objectives, long-term breeding efforts are aimed at the development of new flint strains. Since performance of orange flint hybrids is lower than that of dent and flint × dent hybrids, a project was initiated in order to develop a Flint versus Flint heterotic pattern from the available genetic material. Several stages were fulfilled accordingly: (a) to characterize a collection of inbred lines, most of them orange flint lines developed by INTA, for grain yield combining ability; (b) to classify the lines according to their relative performance in crosses with genetically divergent testers; (c) to develop new orange flint synthetics, based upon their pattern of specific combining ability effects; (d) to evaluate the performance and heterosis of crosses between these synthetics and another synthetic representative of the U.S. Dent heterotic group; and (e) to initiate a long-term FSRRS program using the new synthetics as base populations. In the following sections, main results of each stage of this research project will be presented and discussed.

Table 27.1 Inbred lines testcrossed to four divergent testers and germplasm source from which they were extracted

Line	Source population	Id.[a]	Line	Source population	Id.
ZN6	Local population	LocPop	LP109	Selección Masal	SelM
P1338	Argentine × Exotic	ArgxExot	LP110	Selección Masal	SelM
LP1	Composite 3:3A	3:3A	LP113	Argentino Caribe	CAC
LP2	Composite 3:3B	3:3B	LP117	Argentino Caribe	CAC
LP13	Synthetic Colorada Dura	SCD	LP122	Argentino Caribe	CAC
LP19	Synthetic Colorada Dura	SCD	LP123	Argentino Caribe	CAC
LP22	Synthetic Colorada Dura	SCD	LP125	Castaño × Klein	CxK
LP25	Synthetic Colorada Dura	SCD	LP128	F.M.P. 1712	FMP
LP32	Synthetic Colorada Dura	SCD	LP131	F.M.P. 1714	FMP
LP33	Synthetic Colorada Dura	SCD	LP134	Canario-í	C-í
LP34	Synthetic Colorada Dura	SCD	LP136	Colección exótico	Exot
LP38	Poblaciones Coloradas Argentinas	PCA	LP138	Colección exótico	Exot
LP41	Poblaciones Coloradas Argentinas	PCA	LP140	Resistente Paraná	CRP
LP44	Poblaciones Coloradas Argentinas	PCA	LP146	Resistente Paraná	CRP
LP45	Poblaciones Coloradas Argentinas	PCA	LP147	Resistente Paraná	CRP
LP56	Synthetic A	SA	LP152	Cross of P578	P578
LP62	Synthetic A	SA	LP153	Cross A1 × LP70	LP70
LP68	Synthetic A	SA	LP199	Composite II	CII
LP70	Synthetic A	SA	LP521	Synthetic Colorada Dura	SCD
LP86	Synthetic A	SA	LP662	Single cross Ax252	Ax252
LP87	Synthetic A	SA	L196	Composite II	CII
LP98	Poblaciones Coloradas Argentinas II	PCAII	L687	Single cross Ax252	Ax252
LP103	Selección Masal	SelM	B73	BSSS(C8)	BSSS
LP108	Selección Masal	SelM	B87	BS22	BS22

[a]Abbreviation of population names.

Characterization of the collection of inbred lines

A collection of 48 inbred lines was considered for this project. It included two U.S. dent lines and 46 inbred lines developed by INTA from 20 different sources. Source germplasm covered a wide array of synthetics, composites, landraces, planned crosses, and a commercial hybrid (Table 27.1). Two local flint testers (synthetics HP3 and P5L2) and two U.S. dent testers (synthetics SB73 and SMo17) were crossed as a male parent to the 48 inbred lines in isolation plots. Flint synthetics derived from a short-term reciprocal recurrent selection program applied to the advanced generation of flint single crosses P578 × L256 (P5L2) and H38 × P465 (HP3). Dent synthetics resulted from intermating inbreds related to B73 or Mo17. Seed samples were kindly supplied by Arnel Hallauer.

A diallel cross among the four testers was evaluated in four environments in the north of the Province of Buenos Aires (Pergamino I, Pergamino II, Colón, and Arrecifes) during the 1991/1992 season and data analyzed following the Griffing IV fixed model.

Significant differences for grain yield across environments ($P < 0.05$) among testers were found for general combining ability effects. Specific combining ability was declared not significant in the combined analysis, although it was significant ($P < 0.05$) in the environment with the highest mean yield (Pergamino I). The two highest-yielding crosses in Pergamino and across environments involved Argentine flint (HP3) versus U.S. Dent combinations (SB73 and SMo17) (Table 27.2).

Table 27.2 Grain yield performance of diallel cross among testers at Pergamino (above diagonal) and across environments (below diagonal)

Tester	SMo17	SB73	P5L2	HP3
	Mg . ha^{-1}			
HP3	11.17	9.82	9.27	—
P5L2	9.32	9.42	—	7.07
SB73	9.64	—	6.56	7.50
SMo17	—	7.36	7.28	7.99

Pergamino: Mean: 9.50 Mg.ha^{-1}; CV(%): 7.5, Lsd (.05): 1.05 Mg.ha^{-1}.
Across environments: Mean: 7.32 Mg.ha^{-1}; CV(%): 13.6; Lsd (.05): 1.62 Mg.ha^{-1}.

Table 27.3 Significance of mean squares for different traits of testcrosses of 48 inbred lines to four testers in Pergamino and across environments

	Mean squares							
Source of variation	Grain yield[b]	Ear height[b]	Time to anthesis[b]	Grain yield[a]	Kernel weight[a]	Kernel rows[a]	Ear length[a]	Test weight[a]
Environments Sets								
Env. × sets								
Reps(env.x set)								
Lines (set 1)	ns	**	**	ns	**	**	**	**
Lines (set 2)	**	ns	**	ns	**	**	**	**
Lines (set 3)	ns	**	**	*	**	**	**	**
Lines (set 4)	ns	**	**	ns	**	**	**	*
Lines (set 5)	ns	**	ns	ns	**	**	**	ns
Env.x Line(set)	ns	ns	ns					
Error (a)								
Testers	*	**	**	**	ns	**	**	**
Testers × Set	ns	**	**					
Line × tester(set)	**	**	**	**	*	**	**	**
Line × tester(set1)	**	ns	*	**	*	**	**	ns
Line × tester (set2)	**	ns	*	**	*	**	ns	ns
Line × tester (set3)	ns	*	ns	*	ns	ns	ns	**
Line × tester (set4)	ns	**	ns	ns	ns	*	ns	**
Line × tester (set5)	**	**	**	**	**	**	**	**
Environment × Tester	**	**	ns					
Env. × Set × Tester	ns	ns	ns					
Env. × Line × Tester(set)	ns	ns	ns					
Error (b)								

*,** Significant at the 0.05 and 0.01 probability levels, respectively.
ns, Not significant at $P < 0.05$.
[a]Pergamino I.
[b]Across environments.

With a few exceptions due to the amount of seed available, the line testcrosses were evaluated in the same four environments as the diallel cross mentioned above. The combinations of inbreds and testers were randomly assigned to five sets, each set including testcrosses tracing to 10 inbred lines. A two-replication within-set design with a split-plot arrangement of entries was used in the experiments. Main plots corresponded to inbred lines and consisted of four-row experimental units, each row planted with a different testcross of the same inbred line. To make the mean of testcrosses assigned to different sets comparable, a covariance adjustment was used, taking the mean of each set and trial as a covariate.

Differences in general combining ability effects for testers were found for grain yield, as revealed by the significance of testers in Pergamino and across environments. Grain yield differences among lines for grain yield were detected in set 2 across environments and in set 3 in Pergamino I (Table 27.3). Significant line × tester interactions for grain yield and other traits found in Pergamino I and across environments suggest the presence of differences in specific combining ability due to nonadditive effects. The combined analysis of variance for ear height and time to 50% pollen shedding detected significant differences for lines, testers, and line × tester interactions. Vasal et al. (1992a, 1992b) reported similar results for the same traits in testcrosses of tropical and subtropical inbred lines. A similar pattern of significance was detected for yield components, although the contribution of line × tester interaction was less important than for grain yield.

Line × tester interaction was the source of variation that contributed the most to the testcrosses' sum of squares for grain yield in each environment (47–56%) (Table 27.4). Local flint lines testcrossed to SB73 had the greatest mean grain yield, followed by testcrosses to SMo17 (data not shown). Godshalk and Kaufmann (1995) found a similar response of Argentine flint lines in crosses with testers representative of the Reid Yellow Dent and Lancaster Sure Crop heterotic groups.

Table 27.4 Contribution of line, tester, and line × tester interaction to the sum squares of testcrosses for different traits by environment and across environments

Environment	Trait	Line	Tester	Line × tester
			%	
Across Env.	Grain yield	33.20	12.90	53.20
Pergamino I		33.70	10.70	55.60
Pergamino II		39.70	13.00	47.30
Colón		37.30	0.70	62.00
Arrecifes		39.00	6.70	54.30
Across Env.	Ear height	46.75	17.57	35.68
Pergamino I		47.20	12.80	40.00
Pergamino II		47.70	14.90	37.40
Across Env.	Days to pollen shedding	62.00	17.00	21.00
Pergamino I		67.60	11.50	20.90
Pergamino II		49.00	14.70	36.30
Pergamino I	Kernel weight	75.00	0.56	24.44
	Row Number	51.20	37.30	11.50
	Ear length	55.40	14.60	30.00
	Ear diameter	53.85	30.40	15.75
	Test weight	42.70	21.60	35.70

Low or negligible Spearman correlation estimates were found among mean grain yield of testcrosses to different testers, except between both U.S. dent testers in Pergamino and also HP3 and SB73 across environments. Even in these cases, the estimates of correlation coefficient were not high enough (around 0.4) to have predictive value (Table 27.5). These results and the significance of line × tester interaction across environments suggest the presence of differences in frequency of alleles with nonadditive effects that could be used to generate heterotic populations.

Classification of inbred lines

Specific combining ability effects for each line and tester combination were submitted to a principal component analysis. The first principal component accounted for 40% of the total variance and was interpreted as a contrast between the performance of Argentine flint lines testcrossed to U.S. dent testers versus the performance of the same lines testcrossed to Argentine flint testers (Table 27.6). The second and third principal components accounted for 36% and 24% of the total variation, respectively. Second axis was interpreted as a contrast between the performance of lines crossed to HP3 and SMo17 versus the performance of the same lines crossed to P5L2 or SB73 testers and the third axis as a contrast between performance of testcrosses to HP3 and SB73 versus testcrosses to P5L2 and SMo17. A nonhierarchical procedure for clustering was used (k-means), and an optimum number of four clusters was determined by cubic criteria.

Table 27.5 Spearman correlation coefficients between grain yield of testcrosses of a collection of 48 inbred lines testcrossed to four testers at Pergamino I (above diagonal) and across environments (below diagonal)

Tester	SMo17	SB73	P5L2	HP3
HP3	0.19^{ns}	−0.00^{ns}	0.15^{ns}	—
P5L2	0.13^{ns}	0.18^{ns}	—	0.20^{ns}
SB73	0.39**	—	0.07^{ns}	0.42**
SMo17	—	0.00^{ns}	0.06^{ns}	0.14^{ns}

*, **Significant at the 0.05 and 0.01 probability levels, respectively.
ns, Not significant at $P < 0.05$.
Source: Adapted from Nestares et al., 1999.

Table 27.6 Eigenvectors for principal component analysis based on the relative performance of 48 inbred lines testcrossed to testers HP3 (Argentine flint), P5L2 (Argentine flint), SB73 (U.S. dent), and SMo17 (U.S. dent)

	Principal components		
Tester	cp 1	cp 2	cp 3
HP3	−0.64	−0.35	-0.41
P5L2	−0.28	+0.66	+0.51
SB73	+0.61	+0.30	-0.53
SM017	+0.37	−0.59	+0.54
Eigenvalue	1.63	1.42	0.94
Variance (%)	40	36	24

Lines derived from the same hybrid cultivar (L687, LP662), or lines more closely related according to pedigree data (not shown), such as LP109 with LP110, LP122 with LP123, LP146 with LP147, LP70 with LP153, or LP199 with L196, classified into the same group (Table 27.7). The level of genetic relationship between some inbred lines and the testers affected their classification, accordingly. Lines extracted from broad genetic base populations did not show a clear pattern of classification as those derived from narrow base genetic populations, such as planned crosses or hybrid cultivars.

The four clusters of inbred lines were relatively homogeneous. Three of them had the same number of elements (lines). Groups named II, III, and IV had low dispersion and deviations. Instead, group I was the most scattered, since it presented the largest root-mean-squared distance between observations in the cluster. In addition, the largest

Table 27.7 Classification of 48 inbred lines in four clusters based upon their relative performance in testcrosses to four divergent testers

Group I		Group II		Group III		Group IV	
Line	Source[a]	Line	Source	Line	Source	Line	Source
LP2	3:3B	P1338	ARGXEX	ZN6	LocPop	LP38	PCA
LP32	SCD	LP22	SCD	LP1	3:3A	LP56	SA
LP33	SCD	LP34	SCD	LP13	SCD	LP62	SA
LP117	CAC	LP44	PCA	LP19	SCD	LP86	SA
LP521	SCD	LP45	PCA	LP25	SCD	LP103	SelM
B73	BSSS	LP70	SA	LP41	PCA	LP109	SelM
		LP98	PCAII	LP68	SA	LP110	SelM
		LP113	CAC	LP87	SA	LP125	CsxK
		LP122	CAC	LP108	SelM	LP131	FMP
		LP123	CAC	LP128	FMP	LP138	Exót
		LP136	EXOT	LP134	Cn-í	LP140	CRP
		LP153	A1XLP70	LP146	CRP	LP152	P578
		LP662	HSC	LP147	CRP	B87	BS22
		L687	HSC	LP199	CII		
				L196	CII		

[a]Source of germplasm from which the inbred lines were derived.

Table 27.8 Number of elements, average root-mean-square (RMS) standard deviation, centroid distance between pairs of nearest clusters, and nearest cluster for the groups as determined by relative performance of 48 inbred lines testcrossed to four divergent testers

Group	Frequency	RMS Std deviation	Centroid distances	Nearest group
I	6	1.10	2.10	IV
II	14	0.87	1.85	IV
III	15	0.86	1.93	IV
IV	13	0.73	1.85	II

distance between centroids of pairs of nearest cluster occurred with group I (Table 27.8, Figure 27.1).

The clusters presented a distinctive pattern of specific combining ability of their lines with the testers (Figure 27.2).

As a part of a preliminary research (Morales and Ornella, personal comm.), 32 out of the 48 inbred lines included in this study were fingerprinted using 21 single-sequence repeats evenly distributed in the maize genome. An average of six alleles per locus was obtained. The average locus polymorphic information index was 0.65, with a range from 0 to 0.90. Jaccard's similarity index was calculated (Jaccard, 1908) between pairs of inbred lines. Cluster analysis of lines using unweighted pair group arithmetic average applied to the similarity matrix resulted in several groups with different elements compared with the groups obtained using principal components based on specific combining ability effects between lines and testers.

Development and evaluation of synthetics

Four synthetics (named SPF1, SPF2, SPF3, and SPF4) were developed by intermating the inbred lines assigned to each of the four different groups whenever their general combining ability effects were positive. The number of lines included in each synthetic were two (SFP1), six (SPF2 and SPF3), and ten (SPF4). Except for SPF1, which was generated using controlled pollinations, intermating was done in isolation plots. Parental lines were detasseled and then pollinated by a bulk of all lines included in each synthetic. Seed samples from all crosses were planted again in isolation plots for two more recombination cycles. Each cross was planted as identified rows, detasseled before silking, and pollinated by a balanced bulk of all crosses from each group.

Seed samples of synthetics SPF1 through SPF4 along with BS13P were included in a diallel cross. BS13P derives from BS13 (synthetic developed by the USDA-ARS and Iowa State University), after one cycle of half-sib recurrent selection for grain yield conducted by INTA using flint line LP662 as tester. The synthetics and their crosses were evaluated in three environments in northern Buenos Aires (Pergamino, Junín, and Ferré) in 1997/1998 using incomplete block designs with two replications. Grain yield of parent populations across en-

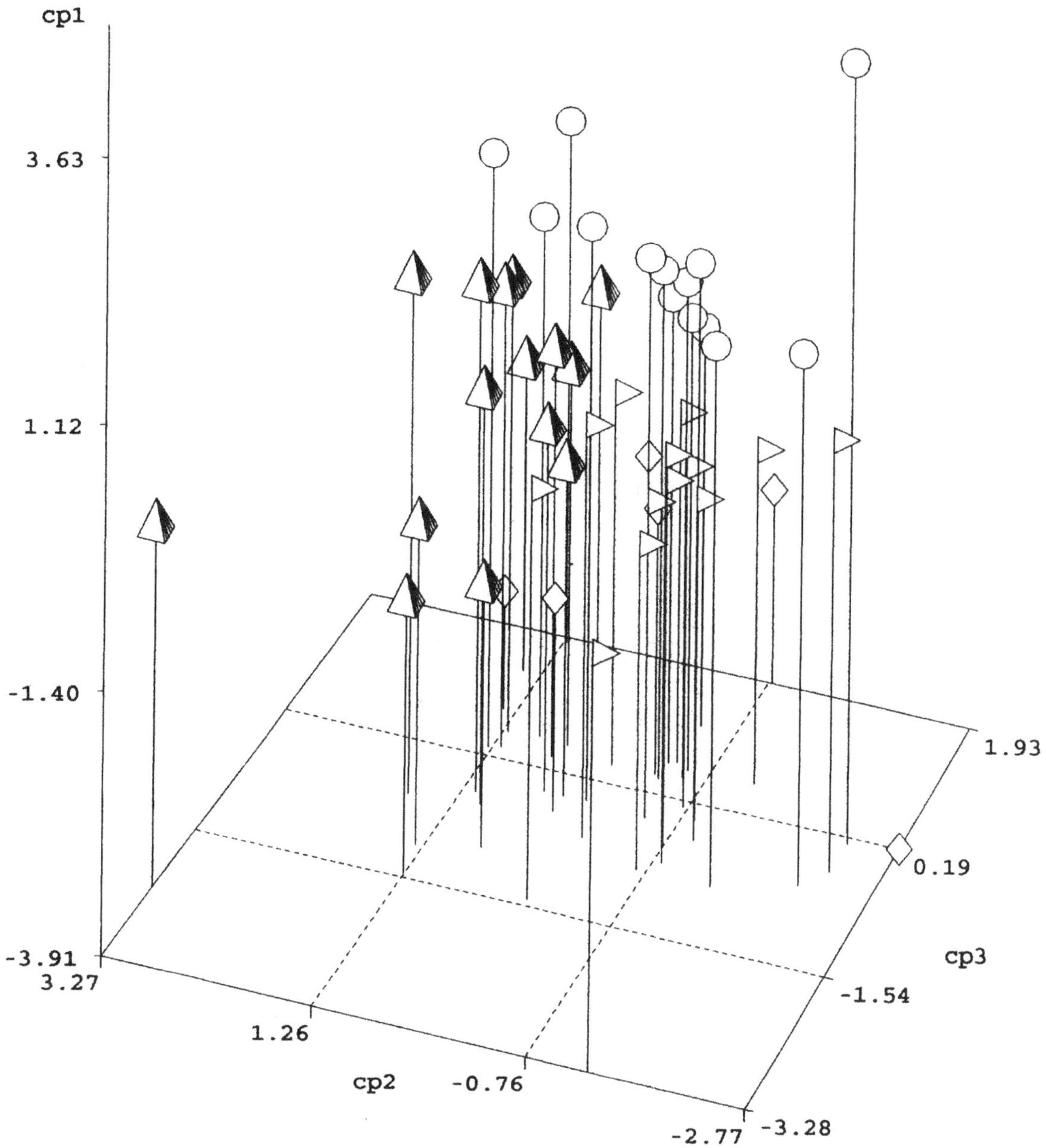

Figure 27.1 Spatial arrangement of four clusters of lines as determined by principal component based on specific combining ability effects.

vironments ranged 4.5 Mg ha^{-1} (SPF2) to 6.4 Mg ha^{-1} (SPF4), and yield of interpopulation crosses ranged 6.5 (SPF2 × SPF4) to 8.8 Mg ha^{-1} (SPF1 × BS13P) (Table 27.9). Per se grain yield of synthetics and their interpopulation crosses were approximately 40% and 10–15% lower than grain yield of check hybrids (Table 27.9).

Partitioning of the source of variation due to heterosis was done following Gardner and Eberhart (1966). Significant differences across environments were found for both populations and heterosis ($P < 0.01$) (Table 27.10). Populations and heterosis were significant for grain yield components ($P < 0.05$ or $P < 0.01$). Average heterosis was

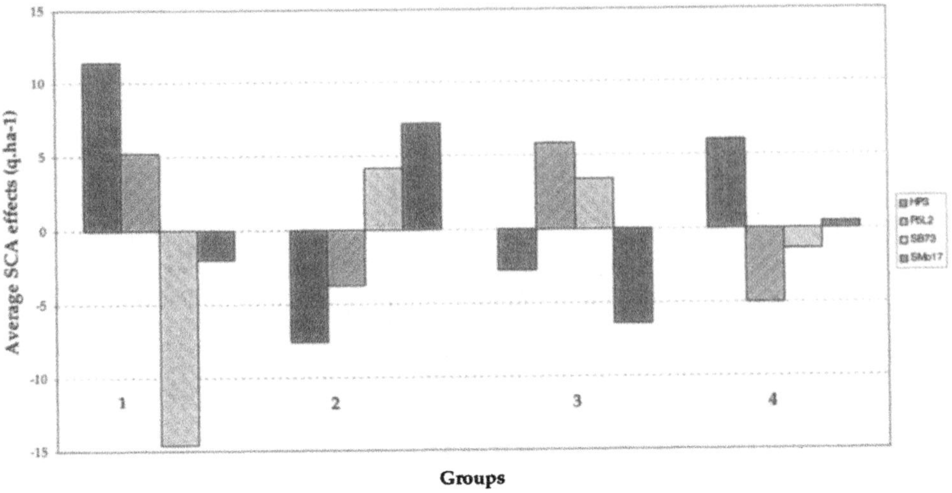

Figure 27.2 Average specific combining ability effects with flint (HP3, P5L2) and dent (SB73, SMo17) testers of lines from groups 1 through 4, as defined by cluster analysis.

Table 27.9 Means of parent populations and their crosses across environments for days from planting to 50% anthesis, ear height, test weight, grain yield and yield components

Population	Time to anthesis	Ear height	Ear length	Ear diameter	Kernel row	Kernel weight	Grain yield	Test weight
	d	cm	cm	cm	no.	g	Mg. ha^{-1}	kg hl^{-1}
SPF1	97	90.8	16.1	4.2	13.0	85.9	5.17	77.3
SPF2	97	98.6	14.1	4.4	13.1	82.2	4.51	78.7
SPF3	95	117.3	16.3	4.4	14.3	79.8	5.58	77.7
SPF4	92	114.3	17.4	4.3	13.2	87.4	6.39	78.3
BS13P	97	94.2	16.8	4.6	14.9	86.0	5.45	76.6
SPF1 × SPF2	92	112.0	16.3	4.4	13.5	85.3	7.20	78.2
SPF1 × SPF3	92	116.6	17.5	4.5	14.1	83.7	7.29	79.0
SPF1 × SPF4	92	111.0	17.7	4.3	13.1	90.0	7.73	78.4
SPF1 × BS13P	92	117.7	17.1	4.7	14.4	94.8	8.81	76.5
SPF2 × SPF3	92	112.9	16.2	4.6	13.9	85.7	7.16	78.5
SPF2 × SPF4	92	117.4	16.6	4.3	12.6	90.7	6.52	78.6
SPF2 × BS13P	95	126.9	16.1	4.7	14.5	91.7	7.90	77.8
SPF3 × SPF4	92	119.0	16.8	4.3	13.3	87.1	7.25	78.1
SPF3 × BS13P	92	120.8	17.5	4.6	14.4	91.3	8.43	78.6
SPF4 × BS13P	93	116.1	17.5	4.6	14.7	92.0	8.37	78.2
DK752	92.5	94.1	17.7	5.1	18.1	83.2	10.71	76.9
ACA923	96.0	121.0	18.5	4.6	14.2	95.5	9.62	78.4
Mean	94.4	119.2	16.9	4.6	13.8	92.3	7.54	77.7
CV%	1.2	18.1	4.8	2.5	4.2	6.0	8.7	1.9
LSD (0.05)	2.2	42.8	0.9	0.1	0.7	6.3	0.75	1.7

significant ($P < 0.01$) only for grain yield and accounted for an important percentage of the heterosis sum of squares. It reflects the superiority of interpopulation crosses over the mean of the parental populations. Average heterosis was significant for kernel weight ($P < 0.01$). Significant varietal heterosis for grain yield ($P < 0.01$) and kernel row number ($P < 0.05$) reflect differences of heterotic patterns of at least one synthetic crossed to the others (Hallauer and Miranda Filho, 1988). Specific heterosis was not significant for grain yield, kernel weight, and ear length and significant for ear diameter, kernel row number, and test weight ($P < 0.05$).

Table 27.10 Analysis of variance for a diallel cross among five populations (SPF1, SPF2, SPF3, SPF4, and BS13P) for test weight, grain yield, and yield components across environments

Source of variation	d.f.	Mean squares					
		Ear length	Ear diameter	Kernel rows	Kernel weight	Test weight	Grain yield
		——— cm ———		no.	— g —	Kg hl^{-1}	q.ha^{-1}
Treatments	14	6.64**	0.27*	5.06**	158.27**	5.21^{ns}	1337.16**
Populations	4	16.65**	0.69*	14.00**	276.78**	8.54^{ns}	983.71**
Heterosis	10	2.63*	0.11*	1.48**	110.87**	3.88^{ns}	1478.54**
Average heterosis	1	15.94^{ns}	0.50^{ns}	0.62^{ns}	609.68**	5.90^{ns}	12110.23**
Varietal heterosis	4	0.95^{ns}	0.08^{ns}	1.05*	59.01^{ns}	1.46^{ns}	573.15**
Specific heterosis	5	1.33^{ns}	0.05*	2.00*	52.60^{ns}	5.42*	76.50^{ns}
Treatments × Env.	28	0.80^{ns}	0.03*	0.48^{ns}	42.07^{ns}	3.94**	67.06*
Popn's × Env.	8	0.67^{ns}	0.00^{ns}	0.49^{ns}	75.70**	6.00**	92.43*
Het × Env.	20	0.85^{ns}	0.03*	0.47^{ns}	28.62^{ns}	3.12^{ns}	5.69^{ns}
Avg. het. × Env	2	2.15*	0.13*	0.17^{ns}	22.00^{ns}	0.17^{ns}	45.44^{ns}
Var het. × Env	8	0.65^{ns}	0.02^{ns}	0.44^{ns}	20.80^{ns}	3.52^{ns}	53.99^{ns}
Spec. het × Env.	10	0.74^{ns}	0.01^{ns}	0.56^{ns}	36.20^{ns}	3.35^{ns}	61.55^{ns}
Pooled Error	197	0.66	0.01	0.34	30.80	2.16	43.38

**, *Mean squares significant at the 0.01 and 0.05 probability level, respectively.
ns: Not significant.

Mid-parent heterosis for grain yield was highly significant (P < 0.01) for all crosses. Range of mid-parent heterosis expressed in actual units was 1.07–3.5 Mg ha^{-1} (Table 27.11) or 19.6–66 expressed as a percentage (Table 27.12). For crosses between flint synthetics, that range was 1.07 Mg ha^{-1} to 2.36 Mg ha^{-1} (19.6% to 48.7%). The greater level of mid-parent heterosis corresponded to SPF1 × BS13P (3.5 Mg ha^{-1} or 66%). BS13P was the population with the highest average mid-parent heterosis (2.95 Mg ha^{-1}). The range of mid-parent heterosis of crosses involving BS13p was 2.55 Mg ha^{-1} to 3.5 Mg ha^{-1}.

Observed values for high parent heterosis were also highly significant (P < 0.01), except for the SPF2× SPF4 cross. Range of high parent heterosis was 0.86 Mg ha^{-1}–3.36 Mg ha^{-1} (13.5–61.6%). High parent heterosis for crosses between flint synthetics varied from 0.86 Mg ha^{-1} to 2.03 Mg ha^{-1} (13.5–39.3%). BS13P exhibited the greatest high parent heterosis when crossed to SPF1 (3.36 Mg ha^{-1} or 61.6%) and an average of high-parent heterosis of 2.66 Mg ha^{-1}1.

Predicted grain yield of crosses between all possible composites using the four flint populations (Hallauer and Miranda Filho, 1988) indicated that composites with best performance would be those derived from recombination of SPF1 with SPF4 and SPF2 with SPF3. Estimates obtained suggest that both composites will exhibit heterosis in crosses with dent synthetic BS13P.

Final considerations

Two populations, named SPF14 and SPF23, developed from the recombination of SPF1 with SPF4 and SPF2 with SPF3, respectively, are being submitted to a reciprocal full-sib recurrent selection program in order to enhance the performance of the interpopulation cross and thus to develop a new Flint-versus-Flint heterotic pattern. Since frequency of two-eared plants in both populations is very low, the modified approach (Hallauer, 1967) of FSRRS had to be adopted. Approximately 500 S_0 plants were selfed in each population and date of pollination recorded. Phenotypic selection for high-heritability traits was practiced among selfed plants. Approximately 350 S_1 ears were saved at harvest from each population. During the 2002/2003 season, S_1 progenies from SPF14 and SPF23 were planted ear-to-row in alternate rows in order to make $S_1 \times S_1$ interpopulation crosses. About 290 crosses will be evaluated for grain yield, maturity, and standability during 2003/2004 as a step of the first cycle of FSRRS.

Expectations are that by applying FSRRS the heterosis expressed by the interpopulation cross

Table 27.11 Mid-parent and high-parent heterosis exhibited by a diallel among five populations across environments, expressed in actual units for test weight, time to 50% pollen shedding, grain yield, and yield components

	Grain yield Mg.ha^{-1}	[1]Test weight Kg.hl^{-1}	Kernel weight g	Ear length cm	Ear diameter cm	Kernel rows no.	Ear height cm	Time to anthesis d
				Mid-parent heterosis				
SPF1 × SPF2	2.36**	0.25ns	1.30ns	1.20**	0.15**	0.50*	17.35ns	−4.5**
SPF1 × SPF3	1.91**	1.55*	0.90ns	1.30**	0.20**	0.45ns	12.50ns	−3.5**
SPF1 × SPF4	1.96**	0.65ns	3.42ns	1.00**	0.10ns	0.00ns	8.50ns	−2.5**
SPF1 × BS13P	3.50**	−0.40ns	8.90**	0.70*	0.30**	0.45ns	25.20ns	−5.0**
SPF2 × SPF3	2.11**	0.30ns	4.75ns	1.00**	0.20**	0.20ns	4.95ns	−4.0**
SPF2 × SPF4	1.07**	0.10ns	5.95*	0.90**	0.00ns	−0.55*	11.00ns	−2.5**
SPF2 × BS13P	2.91**	0.20ns	7.60**	0.70*	0.25**	0.55*	30.55*	−1.5ns
SPF3 × SPF4	1.27**	0.15ns	3.55ns	0.00ns	0.00ns	−0.45ns	3.25ns	−1.5ns
SPF3 × BS13P	2.91**	1.45*	8.45**	0.95**	0.15**	−0.20ns	15.10ns	−3.5**
SPF4 × BS13P	2.45**	0.75ns	5.30*	0.40ns	0.20**	0.65**	12.05ns	−1.5ns
LSD (0.05)	0.53	1.44	5.03	0.65	0.11	0.47	30.25	1.54
LSD (0.01)	0.69	1.89	6.62	0.85	0.15	0.61	40.05	2.03
				High-parent heterosis				
SPF1 × SPF2	2.03**	−0.45ns	−0.55ns	0.20ns	0.05ns	0.45ns	13.64ns	−4.5**
SPF1 × SPF3	1.71**	1.35ns	−2.15ns	1.20**	0.10ns	−0.20ns	−0.75ns	−4.5**
SPF1 × SPF4	1.34**	0.15ns	2.65ns	0.35ns	0.05ns	−0.10ns	−3.25ns	−5.0**
SPF1 × BS13P	3.36**	−0.75ns	8.85**	0.35ns	0.10ns	−0.50ns	23.5ns	−5.0**
SPF2 × SPF3	1.58**	−0.20ns	3.55ns	−0.10ns	0.20**	−0.40ns	−4.40ns	−5.0**
SPF2 × SPF4	0.13ns	−0.10ns	3.37ns	−0.75ns	−0.05ns	−0.60ns	3.15ns	−5.0**
SPF2 × BS13P	2.44**	−0.85ns	5.70ns	−0.65ns	0.15*	−0.35ns	28.35ns	−1.5ns
SPF3 × SPF4	0.86**	−0.15ns	−0.25ns	−0.55ns	−0.05ns	−1.00**	1.75ns	−3.0**
SPF3 × BS13P	2.85**	0.90ns	5.35ns	0.70ns	0.05ns	−0.50ns	3.55ns	−4.5**
SPF4 × BS13P	1.99**	−0.10ns	4.60ns	0.10ns	0.05ns	−0.20ns	1.85ns	−4.0**
LSD (0.05)	0.64	1.77	6.17	0.80	0.14	0.57	37.0	1.88
LSD (0.01)	0.85	2.32	8.11	1.04	0.18	0.75	49.0	2.49

[1]Test weight is a weight/volume ratio. Weight is measured in kilograms (kg). Volume is measured in hectoliters (hl, 1 hl = 100 liters).
*, **Significant at the 0.05 and 0.01 probability levels, respectively.
ns, Not significant at $P < 0.05$.
_Mid-parent heterosis estimated as $F_1 - (P_1 + P_2)/2$; high parent heterosis estimated as $F_1 - P_i$, where P_i is the parent of the cross with the highest mean for the trait.

will improve over cycles of selection, and, consequently, the probabilities of identifying parent lines of promising flint × flint hybrids will also increase. We also expect that the high combining ability of these new flint materials with BS13P, which is currently being improved by intrapopulation recurrent selection, will allow the isolation of new inbred lines to take advantage of the Flint-versus-Dent heterotic pattern as well.

References

Comstock, R.E., H.F. Robinson, and P.H. Harvey. 1949. A breeding procedure designed to make maximum use of both general and specific combining ability. J. Am. Soc. Agron. 41:360–367.

Comstock, R.E., and R.H. Moll. 1963. Genotype-environment interactions. In: W.D. Hanson, H.F. Robinson /eds. Statistical genetics and plant breeding. Pub. 982. NAS-NRC, USA.

Crossa, J., H.G. Gauch Jr, and R.W. Zobel. 1990. Additive main effects and multiplicative interaction analysis of two international maize cultivar trials. Crop Science 30:493–500.

Duvick, D.N. 1996. Plant breeding, an evolutionary concept. Crop Science 36:539–548.

Eyhérabide, G.H., and A. S. Gonzalez. 1997. Interaction between landraces and testers. Maydica 42:29–38

Eyhérabide G., and A.L. Damilano. 2001. Comparison of genetic gain for grain yield of maize between the 1980s and 1990s in Argentina. Maydica 46:277–281.

Eyhérabide, G., and A.R. Hallauer. 1991. Reciprocal full-sib recurrent selection in maize. I. Direct and indirect responses. Crop Science 31:952–959.

Gardner, C.O., and S.A. Eberhart. 1966. Analysis and interpretation of the variety cross diallel and related populations. Biometrics 22:439–452.

Godshalk, E.B., and K.D. Kaufmann. 1995. Performance of exotic x temperate single-cross maize hybrids. Crop Science 35:1042–1045.

Hallauer, A.R. 1967. Development of single-cross hybrids from two-eared maize populations. Crop Science 7:192–195.

Hallauer, A.R. 1999. Temperate maize and heterosis. Pp 353–361. In: J. Coors and S. Pandey (Eds). Genetics: an exploitation of heterosis in crops. ASA-CSSA-SSSA, Madison, WI.

Table 27.12 Mid-parent and high-parent heterosis exhibited by a diallel among five populations across environments, expressed as percentage for test weight, time to 50% pollen shedding, grain yield, and yield components

Interpopulation Crosses	Grain yield	Test weight	Kernel weight	Ear length	Ear diameter	Kernel rows	Ear height	Time to anthesis
				Mid-parent heterosis				
				%				
SPF1 × SPF2	48.7	0.32	1.55	7.95	3.49	3.83	18.32	−4.64
SPF1 × SPF3	35.6	2.00	1.09	8.02	4.65	3.30	11.90	−3.65
SPF1 × SPF4	33.9	0.84	3.92	5.97	2.35	0.00	8.29	−2.65
SPF1 × BS13P	66.0	−0.52	10.35	4.26	6.82	3.23	27.24	−5.15
SPF2 × SPF3	41.9	0.38	5.86	6.58	4.55	1.46	4.59	−4.17
SPF2 × SPF4	19.6	0.13	7.02	5.71	0.00	−4.18	14.53	−2.65
SPF2 × BS13P	58.5	0.26	9.04	4.53	5.56	3.93	31.69	−1.55
SPF3 × SPF4	21.2	0.19	4.25	0.00	0.00	−3.27	2.81	−1.60
SPF3 × BS13P	52.9	1.88	10.19	5.74	3.33	−1.37	14.28	−3.65
SPF4 × BS13P	41.5	0.97	6.11	2.34	4.49	4.63	5.72	−1.59
				High-parent heterosis				
				%				
SPF1 × SPF2	39.3	−0.57	−0.64	1.24	1.14	3.44	13.64	−4.64
SPF1 × SPF3	30.6	1.74	−2.50	7.36	2.27	−1.40	−0.00	−4.64
SPF1 × SPF4	21.1	0.19	3.03	2.01	1.16	−0.76	−2.84	−5.15
SPF1 × BS13P	61.6	−0.97	10.29	2.08	2.17	−3.36	24.95	−5.15
SPF2 × SPF3	28.3	−0.25	4.32	−0.61	4.55	−2.80	−3.75	−5.15
SPF2 × SPF4	2.0	−0.13	3.83	−4.31	−1.14	−4.50	2.76	−5.15
SPF2 × BS13P	44.8	−1.08	6.63	−3.87	3.26	−2.35	28.75	−1.55
SPF3 × SPF4	13.5	−0.19	−0.29	−3.16	−1.14	−6.99	1.49	−3.16
SPF3 × BS13P	51.2	1.16	4.52	4.17	1.09	−3.36	3.03	−4.64
SPF4 × BS13P	31.1	−0.13	5.26	0.57	1.09	−1.34	0.02	−4.12

Hallauer, A.R., and S.A. Eberhart. 1970. Reciprocal full-sibselection. Crop Science 10:315–316.

Hallauer, A.R., J.B. Miranda Filho. 1988. Quantitative genetics in maize breeding. Iowa State Univ. Press, Ames, IA. 1981. p. 468.

Hallauer, A.R., W.A. Russell, and K.R. Lamkey. 1988. Corn Breeding. Pp 463–564. In: G.F. Sprague and J.W. Dudley (Eds). Corn and corn improvement. ASA-CSSA-SSSA, Madison, WI.

Jaccard, P. 1908. Nouvelles recherches sur la distributionflorale. Bull. Soc. Vaud Sci. Nat. 44:223–270.

Jones, L.P., W.A. Compton, and C.O. Gardner. 1971. Comparison of full and hal-sib reciprocal recurrent selection. Theor. Appl. Genet. 41:36–39.

Keeratinikajal, V., and K.R. Lamkey. 1993. Responses to reciprocal recurrent selection in BSSS and BSCB1 maize populations. Crop Science 33:73–77.

Lópes de Souza, C., and J.B. Miranda Filho. 1985. Alteracoes na heterose via selecao intra e interpopulacional. Pesq. Agropec. Bras. 20(10):1197–1201.

Lorenzatti, S. 2000. El cultivo de maíz en siembra directa. In: Maíz en siembra directa. Jornadas de intercambio técnico 2000. Asociación Argentina de Productores en Siembra Directa (AAPRESID), Argentina.

Macagno L., J. Pizarro, G. Eyhérabide, and G. Fernández. 1993. Retornos a la investigación en maíz. Evaluación económica de la ganancia genética en la Argentina. p. 147–185. In: F. Cirio and A. Castronovo (eds). Actas del Simposio Internacional "La investigación agrícola en la República Argentina: Impactos y necesidades de inversión". INTA-IICA-University of Minnesota, Buenos Aires, República Argentina.

Maunder, A.B. 1992. Identification of useful germplasm for practical plant breeding programs. p. 147–169. In: H.T. Stalker, J.P. Murphy (eds). Plant breeding in the 1990s. Proceedings of the Symposium on Plant Breeding in the 1990s. CAB International, United Kingdom.

Moll, R.H., J.H. Lonnquist, J.V. Fortuna, and E.C. Johnson. 1965. The relation of heterosis and genetic divergence in maize. Genetics 52:139–144.

Nestares, G., E. Frutos, and G. Eyhérabide. 1999. Evaluación de líneas de maíz flint Colorado por aptitud combinatorial. Pesq. Agropec. Bras. 34(8):1399–1406.

Ordás, A. 1991. Heterosis in crosses between American and Spanish populations of maize. Crop Science 31:931–935.

Pandey, S., and C.O. Gardner. 1992. Recurrent selection for population, variety, and hybrid improvement in tropical maize. Adv. Agron. 48:1–87.

Schang, M.J., J.O. Azcona, D. Cardona, P. Gutovnik, A. Irazusta, and E.M. Pierson. 1993. Nutritional value of red, semident and dent corn. In: Poultry Science 72 (Supplement 1). p. 216.

Schon, C.C., A.E. Melchinger, L. Boppenmaier, E. Brunklaus-Jung, R.G. Herrmann, and J. F. Seitzer. 1994. RFLP mapping in maize: quantitative trait loci affecting testcross performance of elite European flint lines. Crop Science 34:378–389.

Vasal, S.K, G. Srinivasan, G.C. Han, and F. Gonzalez. 1992a. Heterotic patterns of eighty-eight white subtropical CIMMYT maize lines. Maydica 37:319–327.

Vasal, S.K., G. Srinivasan, S. Pandey, H.S. Cordova, G.C. Han, and F. Gonzalez. 1992b. Heterotic patterns of ninety-two white tropical CIMMYT maize lines. Maydica 37:259–270.

Printed and bound by CPI Group (UK) Ltd, Croydon, CR0 4YY
16/09/2026
14963449-0003